72nd World Foundry Congress (WFC 2016)

Nagoya, Japan
21-25 May 2016

ISBN: 978-1-5108-3312-8

Printed from e-media with permission by:

Curran Associates, Inc.
57 Morehouse Lane
Red Hook, NY 12571

Some format issues inherent in the e-media version may also appear in this print version.

Printed by Curran Associates, Inc. (2016)

For permission requests, please contact the World Foundry Organization Ltd
at the address below.

World Foundry Organization Ltd
Winton House, Lyonshall
Kington HR5 3JP, United Kingdom

Phone: 0044 121 601 6976
Fax: 0044 1544 340332

www.thewfo.com

Additional copies of this publication are available from:

Curran Associates, Inc.
57 Morehouse Lane
Red Hook, NY 12571 USA
Phone: 845-758-0400
Fax: 845-758-2633
Email: curran@proceedings.com
Web: www.proceedings.com

TABLE OF CONTENTS

SL-1

■ In the modern world, cast products are indispensable. They are used as automobile and machine parts, in construction machinery as well as commodities. I think the casting industry is one of the key industries for manufacturing, and it greatly contributes to the development of the world economy.

■ Currently, its major trading partner is the automobile industry, which is characterized by a constant drive for lighter, more safe, more energy efficient vehicles. This direction in the automobile industry means that the needs for cast products continue to change. To expand the market, the casting industry needs to develop new applications.

■ And, the casting industry historically and currently consumes a large amount of energy. This results in risks and instability, given the rapidly fluctuating energy price changes. In addition to this, discussions surrounding the sustainable management of carbon dioxide emissions on a global scale has developed. The casting industry should contribute to this discussion for the betterment of both the industry itself and the environment more broadly.

■Also, the production process has changed greatly in the casting industry because of the development of IT. To produce high quality products, of course, it must utilize new analytical tools in planning, bring further automation to manufacturing, and accumulate and analyze data on both the products and the production process. In addition, additive manufacturing is ready for practical use, not only with resin but with various other materials as well. Utilizing this technology is expected to contribute to the development of value-added products.

■On the other hand, many developed countries are aging, their birthrates are falling, and young people are increasingly moving away from manufacturing. In the future, it is possible that this move will become a global trend.

■Taking Japan as an example, this lecture looks at the current situation and issues surrounding the casting industry.

Proceedings of the 72nd World Foundry Congress
21-25th May 2016, Nagoya, Japan

Interfacial Phenomena in Casting Process

Youngjig Kim
School of Advanced Materials Engineering, Sungkyunkwan University, Suwon 16419, Korea

The control of the interface phenomena and reactions might lead to a wide range of structures, properties and performances of the castings. This presentation is concerned with the research on the practical utilization of the interface phenomena, occurring between liquid metal and atmosphere, in commercial Al, Mg and Ti cast alloys. Alpha-case formation in Ti castings, which results in surface-related embrittlement, shows interface phenomena between liquid Ti and solid mold. The range of alpha-case layer could be controlled by thorough understanding of formation mechanism and interface behavior. As an example of practical utilization of interface phenomena between liquid metal and gas atmosphere, Eco-Mg development is presented. The application of ultrasound for grain refinement and degassing of Al and Mg alloys may be expected the use of interface phenomena among liquid metal, solid substrate and gas bubble.

Keywords: *Interface phenomena, Ti alloy, Eco-Mg, Al alloy, High intensity ultrasound treatment*

1. Introduction

There are many cases of interface phenomena involved in casting process. However, most of our knowledge regarding the control of interface phenomena and morphology has developed in recent years, as casting practices have transformed from the mysteries of art to an engineering science.

The control of the interface phenomena and/or reactions might lead to a wide range of structures, properties and performances of the castings.

The formation of interface in casting process may occur in a number of ways, including: A) surface oxidation, melts contamination from crucible, and degassing treatment during melting and holding, as well as B) heterogeneous nucleation, grain growth, and melts-mold reaction during solidification.

The morphology of interface depends upon numerous factors involved in one or more of the following aspects:

1. Melts composition (including segregation)
2. Presence of surface active elements
3. Melt treatment (including effective nucleants)
4. Time-temperature history (including diffusion)
5. Growth rate
6. Mold treatment

This presentation is concerned with the research on the practical utilization of the interface phenomena in commercial Al, Mg and Ti cast alloys.

2. Melting and casting of Ti

An example of interface phenomena between liquid metal and solid atmosphere is exhibited in melting and casting of Ti.

Despite the superior properties of Ti and Ti alloys, applications have been limited by several factors, including the cost-affordability, low production yield, and limited shape in isothermal forging, etc. Therefore, investment casting (lost-wax casting) has been used in order to cost-effectively satisfy complex shape requirements. However, what distinguishes Ti casting from all other metal casting is that Ti and Ti alloys exhibit extremely high reactivity in a molten state with atmosphere and form an alpha-case layer on the surface of the Ti castings (Fig.1), which can lead to crack initiation and propagation [1, 2].

In order to alleviate or mitigate the alpha-case reaction, one must have a thorough understanding of the mechanism by which liquid Ti reacts to the mold material. At the same time, structural factors such as surface-related embrittlement that occurs on the degradation of mechanical properties should be identified.

This study suggests a newly developed alpha-case controlled mold material for Ti investment casting.

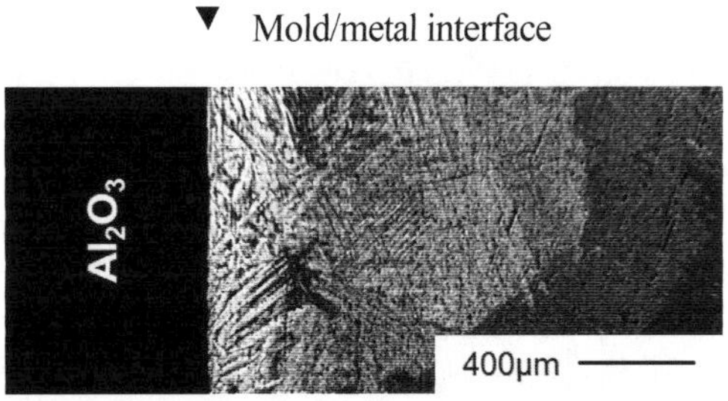

Fig. 1 Formation of alpha-case layer in Ti castings

3. Eco-Mg casting

It is well recognized that environmental benefits provided by lightweight, unlimited, and recyclable Mg alloy have the potential to grow significantly in the future. It could be realized practically by eliminating global warming SF_6 or other protective gases such as SO_2 and Be addition for environment and health issues as well as by ensuring safety during manufacturing and application. It should be done, however, without sacrificing process ability and mechanical property and increasing the life-cycle cost of Mg alloy [3-4].

As an example of practical utilization of interface phenomena between liquid metal and gas atmosphere, Eco -Mg development is presented in this section.

The concept of Eco-Mg (Environment conscious magnesium) approach is to incorporate simply CaO into conventional Mg alloys and generate the reaction of CaO on the surface of molten Mg alloys. The oxide film on the surface of molten Mg alloy changes from porous to dense and complex films. Therefore, the Eco-Mg approach exhibits the improved oxidation and ignition resistances, safety of parts, castability (i.e. increase of fluidity and reduction of hot tearing), and mechanical properties, mainly due to the improvement of melt cleanliness.

In this presentation, the background and recent research and development on Eco-Mg is introduced.

Fig. 2 shows one example of Eco-Mg castings.

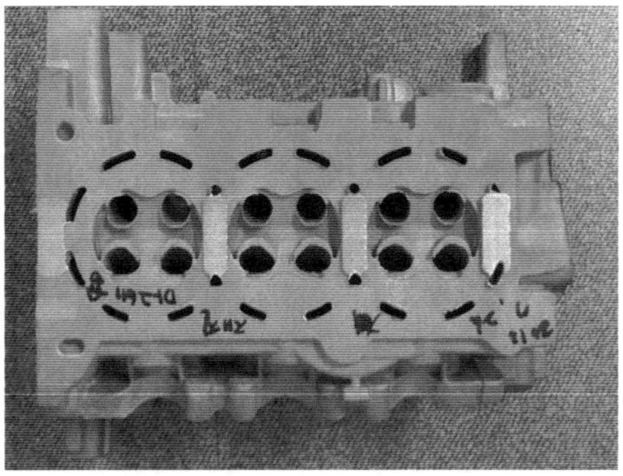

Fig.2 Sand cast cylinder head of AZ-type Eco-Mg by using general sand mold

4. Ultrasound treated Al casting

Generally, grain refinement is one of the effective method achieving improvements in structural uniformity, strength and ductility, reducing the size of defects and increasing the resistance to hot tearing in castings.

The application of ultrasound for grain refinement and degassing of Al and Mg alloys is concerned with interface phenomena among liquid metal, solid substrate and gas bubble.

When high intensity ultrasonic waves are introduced prior to solidification and propagated through molten metal, it has been found that they are effective in grain refinement, uniform dispersion of composition, degassing resulting in reduced porosity and in increasing fluidity.

The results may be explained by the cavitation effects and the influence of the fluid-flow phenomena (acoustic streaming).

In this study, nucleation enhanced ultrasonic melt treatment (NEUMT) using ultrasonic vibration prior to solidification is suggested for grain refinement and degassing in Al casting.

Fig. 3 illustrates grain refinement of Al alloy by ultrasonic melt treatment.

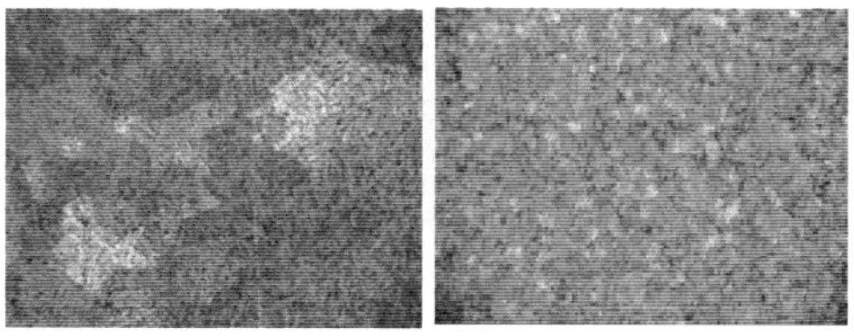

Fig.3 Grain refinement by ultrasonic treatment (left; without UST, right; with UST)

References

[1] S. Y. Sung and Y. J. Kim, Mater. Sci. Eng. A 405 (2005) 173-177.

[2] B. J. Choi and Y. J. Kim, Met. Mater. Int., 19 (2013) 439-444.

[3] S. K. Kim, T. W. Hong and Y. J. Kim, Mater. Trans. 42 (2001) 539-542

[4] S. K. Kim, Magnesium Alloys – Design, Processing and Properties, Ed. by Frank Czerwinski, (InTech, 2011) pp. 431-468

Proceedings of the 72nd World Foundry Congress
21-25th May 2016, Nagoya, Japan

Superalloys for Jet Engine and Gas Turbine Applications: Evolution and Revolution for the Future

Hiroshi Harada
Senior Scientist - High Temperature Materials
National Institute for Materials Science (NIMS)
1-2-1 Sengen, Tsukuba, Ibaraki 305-0047,
Japan

Conventionally cast (CC), directionally solidified (DS) and single crystal (SC) Ni-base superalloys are used for high temperature components, typically turbine blades and vanes that determine the thrust and thermal efficiency of jet engines and power generation gas turbines. The SC superalloys have the highest temperature capabilities; a 6th generation SC superalloy TMS-238, containing Re and Ru, exhibits world highest 1120°C temperature capability. New superalloys beyond Ni-base superalloys, such as Ir-base refractory superalloys, are also being developed as possible alternative materials in the future.

Keywords: *Ni-base Superalloy, Single Crystal, Turbine Blade, Jet Engine, Gas Turbine, Refractory Superalloy*

1. Introduction

Turbine blade and vane materials with higher temperature capabilities are crucial to improving thermal efficiency in jet engines and gas turbines for reducing fuel consumptions and CO_2 emissions. The turbine blades and vanes are made of Ni-base superalloys and produced by vacuum investment casting, with ceramic moulds holding complex shaped cores, to hollow components for air-cooling inside. The cooling air flow, however, must be minimised not to lose the efficiency, and for this reason, the temperature capability of superalloys as well as casting technologies must be improved.

2. Ni-base Superalloy Development

The superalloy was first developed and used in jet engines during the World War 2. The first Ni-base superalloy, Nimonic 80, was a wrought alloy and had temperature capability as high as 730°C (Fig.1). In 1950s, vacuum induction melting and casting technology was established for conventionally cast (CC) superalloys, and then for directionally solidified (DS) and single crystal (SC) superalloys. In NRIM

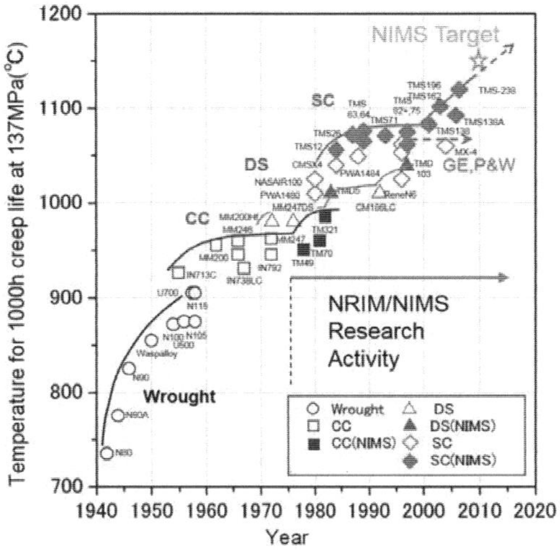

Fig.1 Improvement in temperature capability of Ni-base superalloys, with NIMS project target for new Ni-base SC superalloys.

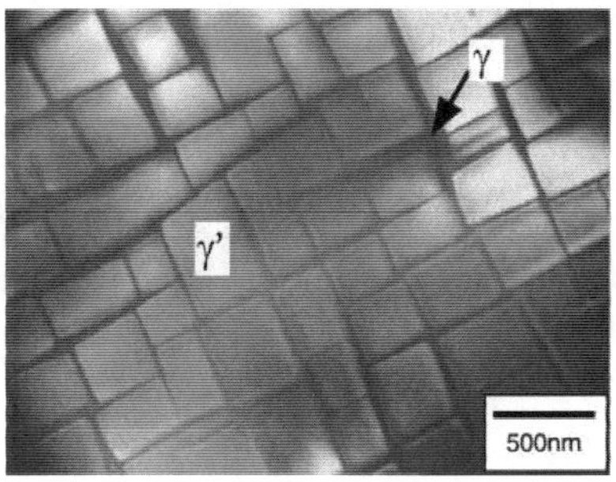

Fig.2 Typical microstructure of the Ni-base SC superalloy with a γ/γ' coherent structure.

(National Research Institute for Metals, Tokyo) we started our alloy design and development work for Ni-base superalloys in 1975. We developed an alloy design computer program (ADP) and, using the program, developed CC, DS, and SC superalloys [1]. A typical microstructure of the SC superalloy is shown in Fig.2. The NRIM work has been succeeded in NIMS. Newer generation SC superalloys, up to 6th generation SC superalloy TMS-238 (Ni-6.5Co-4.6Cr-1.1Mo-4W-5.9Al-7.6Ta-0.1Hf-6.4Re-5Ru,wt%) with world highest 1120°C temperature capability, have been developed [2] and further improvements are expected.

Some of the superalloys we developed so far are now being brought into actual practical applications. Typically, an SC superalloy (TMS alloy, the alloy number is not disclosed) has been used in Rolls-Royce Trent1000 engines powering more than 100 Boeing 787 airplanes operated by All Nippon Airways (Fig.3), British Airways and many others. A high temperature high pressure turbine blade with thermal barrier coating is shown in Fig.4.

Fig.3 First application of NIMS advanced SC superalloy was made in Rolls-Royce Trent1000 engine powering Boeing 787.

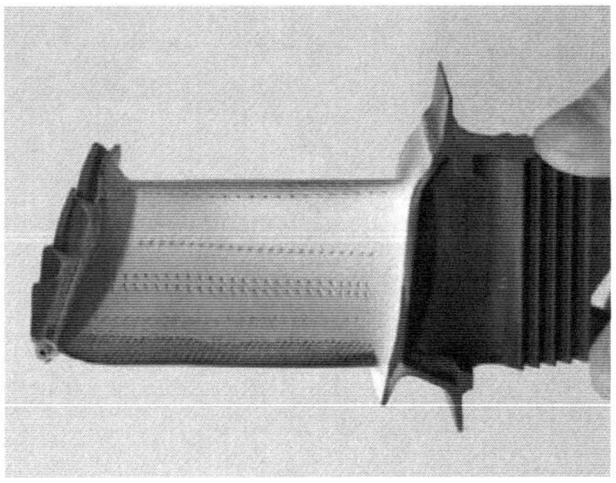

Fig.4 An example of Rolls-Royce hollow turbine blade with thermal barrier coating and holes for film cooling.

A high cost-performance SC superalloy TMS-1700 /MGA1700 (Ni-9Cr-0.6Mo-7.6W-5.4Al -10Ta, wt%) has been developed under collaboration with Mitsubishi Heavy Industries, Ltd. [3], and is going to be applied into a 1700°C gas turbine for combined cycle power generation system of 57% (HHV) thermal efficiency in the year 2020. A casting trial to SC blade of 300mm long and 10kgs weight has been successfully made. Casting defects such as freckles, stray grains and re-crystallisation have been eliminated by selecting right alloy chemical composition and casting condition.

For the cost reduction of the alloys containing many rare metals, a new recycling technology, that is, direct recycling process, has been developed in our group. By remelting in CaO crucible, scrapped turbine blades are refined to a quality as good as the genuine superalloy in laboratory scale [4]. Scale up trials targeting finally 2 tons are planned with ingot makers.

3. Beyond Ni-base superalloys

Much effort has also been paid to develop new alloys beyond Ni-base superalloys. They include refractory metals and alloys such as Mo or Nb silicide base alloys, Co-base γ/γ' alloys and some others.

In NIMS the author and his group developed "refractory superalloys" composed of the γ/γ' structure based on platinum group metals (PGMs). For instance, alloys using Ir (iridium, 2447°C melting point) as a base metal exhibit temperature capability as high as 1750°C [5]. We are trying to lower the alloy density (~22) and also reduce cost by using such techniques as mixing it with Ni-base superalloys. PGMs base refractory superalloys may become a new field of high temperature materials in the future.

4. Conclusions

Ni-base superalloys, especially single crystal (SC) superalloys, and their casting technologies have been the key technologies in jet engine and power generation gas turbine systems. For higher thermal efficiency to reduce CO_2 emission, these technologies are becoming even more important. Alloys beyond Ni-base superalloys are also expected in future, though more research and development work will be needed.

References

[1] H.Harada, et.al: Journal of The Gas Turbine Society of Japan, Vol.43, No.5 (2015), 349-356.

[2] K.Kawagishi, et.al: Superalloys 2012, 189-195.

[3] H.Oguma, et al: Proc. Intern. Gas Turbine Congress 2015 Tokyo, 173-178.

[4] S.Utada, et.al: Superalloys 2016, to be published.

[5] H.Harada: Kinzoku, Vol.75 (2005), 619-625.

Proceedings of the 72nd World Foundry Congress
21-25th May 2016, Nagoya, Japan

Trace elements and graphite shape degeneracy in nodular graphite cast irons

Jacques Lacaze[1]
[1] CIRIMAT, Université de Toulouse, France

Graphite degeneracy in spheroidal graphite cast iron is just a common feature faced by foundries. It is generally associated with the presence of so-called poisoning elements, and may in some cases be suppressed by addition of other elements. Mastering these additions is all except simple in practice since industrial alloys do generally contain many elements that can affect graphite shape even when present at low or trace level. In this work, trace and low level elements are considered in relation with the three successive steps of microstructure formation: i) nucleation of graphite; ii) growth of graphite; and iii) solid-state transformations.

Keywords: *nodular cast iron, graphite degeneracy, trace elements.*

1. Introduction

The role of low level elements on solidification of cast irons has been the subject of many reviews, e.g. by Lux [1]. One obvious problem is quantifying low level elements as they may be effective below the detection limits of standard or even advanced analytical means. Further, chemical analyses are global and do not give any hint on where and how trace elements have affected microstructure formation. Emphasis has thus often been put on characterizing compounds in which they are tied and to make some assumptions on the mechanism of their action. This work will first discuss this aspect when dealing with nuclei formation, then the discussion will be extended to graphite growth during solidification. Finally, some comments will be made on redistribution of elements during solid-state transformation.

2. Graphite nucleation

The inoculation mechanism in spheroidal graphite cast irons has been studied by Skaland et al. [2] who proposed a three-step model where a core of sulfide gets surrounded by a shell of oxide on top of which a complex XO.MgO.SiO_2 oxide precipitates, where X may be Al, Ca or Ba. The understanding of the role of oxi-sulfides in the building up of appropriate nuclei led to the definition of new inoculants doped with S and O and containing elements reacting with them such as Al, Ba, Ca, Ce, Mn, Sr, Zr [3]. A somehow simpler mechanism involving a core made of MgO encapsulated in MgS has been proposed [4]. In this work, particles of (Mg,Si,Al)N nitride were also found at the periphery of the nuclei, which could eventually turn to become nuclei in low-S melts [5, 6].

Figure 1 shows the nucleus of a graphite spheroid found in an industrial nodular cast iron [7]. The same phases as mentioned above could be identified as well as RE-bearing phases already reported [6]. This nucleus consists mainly of (Ca,Mg)S, but is amorphous. Further, no epitaxy could be found between any of the dark crystalline spots and the graphite around [7]. It thus appears that graphite could certainly nucleate on any heterogeneous substrate, and that the most important step in the spheroidizing process is graphite growth.

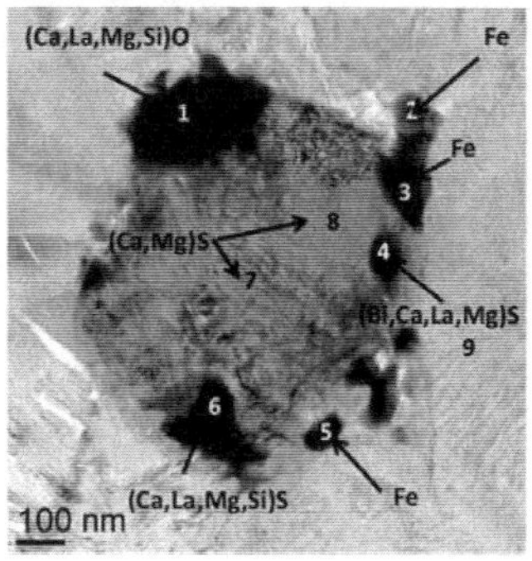

Fig. 1 Graphite nucleus
with an amorphous sulfide core.

3. Graphite growth

Provided spheroidizing elements (Mg and RE) are present, graphite will easily surround these foreign particles due to its recognized high capability to bend. A consensus has been raised that this is due to the fact spheroidizers are prone to pick up oxygen and sulfur out of the melt. It may then be easily accepted that half-treatment of the melt would lead to a mix of lamellar and spheroidal graphite (the so-called compacted graphite), but it should be stressed that overtreatment leads to degenerate graphite [1]. In other words, magnesium and RE not only ties O and S as stable compounds precipitating in the melt, they also affect graphite growth.

As reviewed previously [8], many attempts have been made since long not only to quantify the amount of trace elements in graphite precipitates but also to locate them at a smaller scale. The conclusions are that any element present in the melt is prone to enter to some level in graphite, and that there is no significant accumulation at the graphite melt interface.

To get further insight in the growth mechanism of graphite, Fe-C-X samples have been prepared with excessive X addition [9]. With Ce, it is possible to get graphite growing as isolated sectors or exploded nodules, see Fig. 2. It could thus be concluded that spheroidizing elements adsorb on graphite and slow down the growth along the prismatic directions. From such results, growth of spheroids may be described by continuous nucleation of new blocks at the circumference of the spheroids that then grow along the prismatic planes [10].

Fig. 2 Thin foil of an exploded nodule.

4. Solid-state transformations

An increase in concentration of trace or alloying elements at the graphite/matrix interface has sometimes been reported to explain graphite degeneracy or pearlite promoter effect. If such a film had formed during solidification, it should have quickly stopped transfer of carbon to graphite spheroids leading to metastable solidification. In other words, this enrichment if it exists must have appeared in solid-state, which seems difficult to imagine in the case of substitutional solutes such as Cu, Sn and Sb. This appears still controversial but other explanations have been offered to the pearlite promoter effect of these elements [11]. However, no explanation has been suggested up to now for the effect of many trace elements as reported long ago by Thielemann [12].

5. Conclusion

Plotting the critical level of various elements for graphite degeneracy in nodular irons versus the corresponding atomic mass shows a clear correlation: the heavier are the atoms the lower is their critical limits (Fig. 3). As the atomic weight relates to the size of the atoms and to the number of their outer electrons, such a relationship can be easily associated with adsorption of these elements on the graphite surface. This confirms that the main role of so-called "active" elements is modifying graphite growth by simple adsorption and absorption.

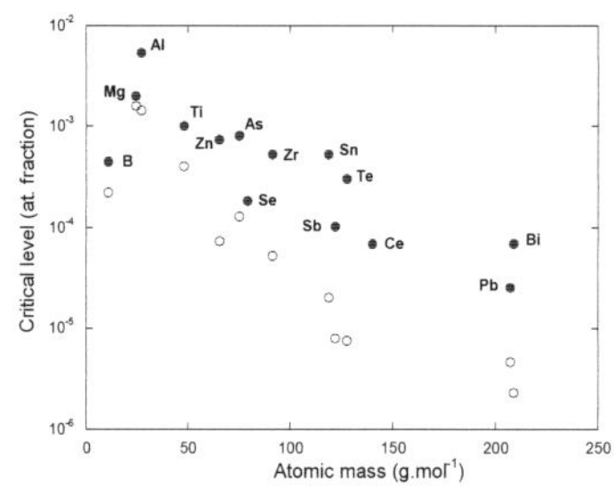

Fig. 3 Relation between atomic mass of elements poisoning nodular irons and their reported critical levels [12]. Open symbols relate to minimum values seen to affect graphite shape, solid symbols to maximum admissible values.

References

[1] B. Lux: AFS Cast Metals Research J. 8 (1972) 25-39.
[2] T. Skaland, O. Grong and T. Grong: Metal. Trans. 24A (1993) 2321-2345.
[3] T. Skaland: AFS Trans., 2001, paper 078
[4] Y. Igarashi and S. Okada: Int. J. Cast Metals Res. 11 (1998) 83-88.
[5] J.K. Solberg and M.I. Onsoien: Mater. Sci. Tech. 17 (2001) 1238-1242.
[6] H. Nakae and Y. Igarashi: Mater. Trans. 43 (2002) 2826-2831.
[7] K. Theuwissen et al.: to appear in Int. J. Cast Metals Res.
[8] J. Lacaze et al.: Adv. Mater. Sci. Eng. (2013) ID 638451
[9] K. Theuwissen, J. Lacaze and L. Laffont: Carbon 96 (2016) 1120-1128.
[10] R. Ghergu, L. Magnusson Åberg and J. Lacaze: Materials Science Forum 790-791 (2014) 435-440.
[11] J. Lacaze and J. Sertucha: to appear in Int. J. Cast Metals Res.
[12] T. Thielemann: Giessereitechnik 16 (1970)16-24.

Research on ELI Grade Casting Titanium Alloy and Its Precision Forming Technology

Lou Yanchun[1], Liu Shibing[1], Xie Huasheng[1], Zhao Jun[1], and Jin Lei[1]
[1] Shenyang Research Institute of Foundry, Shenyang 110022, Liaoning, China

Extra low interstitial (ELI) grade titanium alloy castings are very important in the manufacture of high-end equipment, especially in the areas of aviation and aerospace. In this research, two types of ELI grade casting titanium alloy Ti-5Al-2.5Sn ELI and Ti-6Al-4V ELI are investigated on their chemical compositions, microstructure, mechanical properties and the process of precision forming. Result shows that the optimal chemical composition for Ti-5Al-2.5Sn ELI is Al 4.5-5.5 wt.%, Sn 2.0-3.0 wt.% and for Ti-6Al-4V ELI is Al 5.5-6.5 wt.%, V3.5-4.5 wt.%. The testing results, tensile strength R_m ≥ 780 MPa, elongation A ≥ 15%, impact value KU_2 ≥ 67 J for Ti-5Al-2.5Sn ELI and R_m ≥ 910 MPa, A ≥ 10 %, KU_2 ≥ 62 J for Ti-6Al-4V ELI, in room temperature indicate favorable mechanical properties. Also, these two ELI alloys show a high fracture toughness and damage resistance. In addition, two methods, ceramic mould investment process and machining graphite mould combined with ceramic core process, are developed for the precision forming process of ELI alloys. With these methods, castings with minimum surface roughness R_a 3.2 μm, highest dimensional tolerance IT14 and surface contamination layer ≥ 6 μm can be achieved.

Keywords: *Extra Low Interstitial (ELI); Casting Titanium Alloy; Ti-5Al-2.5Sn ELI; Ti-6Al-4V ELI; Precision Forming*

1. Introduction

Extra low interstitial (ELI) grade titanium alloy, which contains low-level amount of interstitial elements (C, N, H, O) and impurity elements (Fe, Si etc.) is designed for the requirement of high damage tolerance of parts. Since 1960s, many countries have developed their own designations and standards of ELI grade titanium alloy[1],such as standards AMS4907,AMS4931,ASTM367, ASTM F136 (medical implants) for Ti-6-4 ELI alloy(U.S.), AMS4909, AMS4924 for Ti-5-2.5 ELI alloy(U. S.); IMI318 ELI(Britain), TA6V ELI(France); TAF 6400E(Japan) and TC4-DT (China).Due to the prospective applications of ELI grade titanium alloys in aviation and aerospace, a lot of research have been carried out, such as effects of oxygen content on the thermal deformation of Ti-6Al-4V ELI, effects of surface roughness and notch on the low temperature fatigue property of Ti-5Al-2.5Sn ELI, effects of cooling rate on the microstructure and properties of TA15 ELI[2-5], etc. Most work focused on the aspects of ingredients, structures and properties of ELI grade titanium alloys shaped by forging or powder metallurgy. However, in recent years, as the rapid development in aviation and aerospace, larger amount and higher properties of ELI grade titanium alloys are required for casting applications in rockets, airplanes and missiles. However, related study has rarely been done so far. In this paper, the factors which influence the performance of ELI grade cast titanium alloy and precision forming technology have been studied to provide important technical support for the development and application of casting technology of ELI grade titanium alloys.

2. Experimental procedure

In this paper, samples were melted and poured in vacuum consumable electrode arc skull furnace, with 40-50 V melting voltage, 5000-8000 A electric current and 15 Pa vacuum degree. The mother alloy is cast ingots of Φ200mm Ti-6Al-4V ELI and Ti-5Al-2.5Sn ELI alloy, which are melted twice in vacuum consumable electrode arc furnace. The cast mold is made of graphite block by machining, and used for pouring after vacuum degassed at the temperature of 800 ℃ . Samples were poured by centrifugal casting with a speed of 150 r/min. After HIP treatment (920±10 ℃ , 2 h, 130 MPa), the chemical component, mechanical properties and metallographic structure of samples are analyzed. For element testing, the sample with a size of Φ3×30 mm is tested by meter TCH-600 for

N, H, O and direct-reading spectrometer QSN750 for other elements. Also, universal material testing machine CSS-1120, impact testing machine JB-300W and electro-hydraulic servo fatigue testing machine Instron8801 are used for the testing of tensile property, impact property, fracture toughness and fatigue crack growth rate at room temperature. The phase analysis was conducted on X-ray diffraction analysis meter D/max 2500pc. For optical testing, the metallographic specimen is etched in a solution with a component ratio of HF:HNO_3:H_2O=1:3:7 for 2 min and then observed under optical microscope OLYCIAm3 and transmission electron microscope JEM200CX.

3. Results and discussion

3.1 Chemical composition of cast ELI grade titanium alloy

Recently, the standards of ELI grade titanium alloy on element content are different in different countries (Table 1). The current standard of the content range is too wide for ELI grade titanium, which makes it difficult to ensure a steady quality of castings because of the large dispersion of mechanical properties. Basically, Al and Sn can increase the strength of Ti alloy and decrease the plasticity. In Fig. 1, we can see that the content ratio of Al is lower than 7 wt.% for most Ti alloys, except for TiAl, because a brittle-phase Ti_3Al can be easily formed and seriously weaken the plasticity when the weight ratio is higher than this value. This phenomenon also occurs when the content of Sn increases in Ti alloy. In this paper, optimal component ratios Al = (4.5-5.5) wt.%, Sn= (2.0-3.0) wt.% for Ti-5Al-2.5Sn ELI alloy and Al = (5.5-6.5) wt.%, V= (3.5-4.5) wt.% for Ti-6Al-4V ELI casting are proposed for the balance of strength and plasticity. However, the interstitial impurity elements in alloy don't satisfy the requirement of damage tolerance design because it needs strict control of the upper limit (Table 2).

Table 2 Composition of cast ELI grade titanium alloys proposed and their castings (wt.%)

Nominal code	Forming method	Al	Sn	V	C	N	H	O
Ti-5Al-2.5Sn ELI	cast	4.5-5.5	2.0-3.0	-	≤ 0.05	≤ 0.03	≤ 0.0120	≤ 0.12
Ti-6Al-4V ELI	cast	5.5-6.5	-	3.5-4.5	≤ 0.05	≤ 0.03	≤ 0.0120	≤ 0.12
Ti-5Al-2.5Sn ELI	cast	5.1	2.7	-	0.02	0.01	0.009	0.05
Ti-6Al-4V ELI	cast	6.2	-	4.1	0.01	0.02	0.0010	0.09

Table 1 Typical ELI grade titanium alloys and their composition (wt.%)

Country	Shaping method	Standard and grade	Nominal code	Al	Sn	V	C	N	H	O
America	Deformation	ASTM348 (Grade 23)	Ti-6Al-4V ELI	5.5-6.5	-	3.5-4.5	≤ 0.08	≤ 0.03	≤ 0.0125	≤ 0.13
France	Deformation	NF L14 (TA6V ELI)	Ti-6Al-4V ELI	5.5-6.75	-	3.5-4.5	≤ 0.08	≤ 0.05	≤ 0.01	≤ 0.13
England	Deformation	IMI318 ELI	Ti-6Al-4V ELI	5.5-6.75	-	3.5-4.5	≤ 0.08	≤ 0.03	≤ 0.0125	≤ 0.13
Japan	Deformation	JIS H4657 (TAF 6400E)	Ti-6Al-4V ELI	5.5-6.5	-	3.5-4.5	≤ 0.08	≤ 0.03	≤ 0.0125	≤ 0.13
China	Deformation	GB/T 3620.1 (TC4 ELI)	Ti-6Al-4V ELI	5.5-6.5	-	3.5-4.5	≤ 0.08	≤ 0.03	≤ 0.0120	≤ 0.13
China	Deformation	(TC4-DT)	Ti-6Al-4V ELI	5.6-6.3	-	3.6-4.4	≤ 0.05	≤ 0.03	≤ 0.0120	≤ 0.12
America	Deformation	AMS4909 (Ti-5-2.5 ELI)	Ti-5Al-2.5Sn ELI	4.5-5.75	2.0-3.0	-	≤ 0.05	≤ 0.035	≤ 0.0125	≤ 0.12
China	Deformation	GB/T 3620.1 (TA7 ELI)	Ti-5Al-2.5Sn ELI	4.5-5.75	2.0-3.0	-	≤ 0.05	≤ 0.035	≤ 0.0125	≤ 0.12

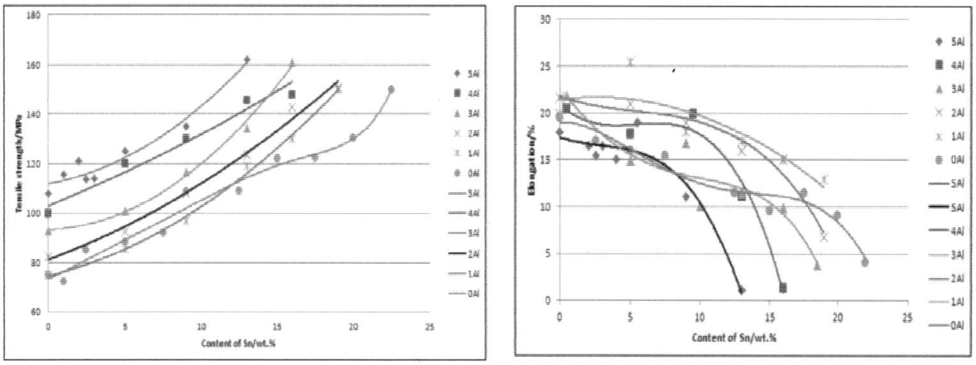

(a) influence of Sn on strength (b) influence of Sn on plasticity

Fig. 1 Influence of content of Sn on mechanical properties of titanium alloys[6]

The final contents of the poured ELI grade titanium alloy samples are shown in Table 2. It shows that the contents of impurity elements, especially C, N, H, and O, are controlled in a low level.

3.2 Structure of cast ELI grade titanium alloy

3.2.1 Microstructure of cast Ti-5Al-2.5Sn ELI alloy

Microstructure of Ti-5Al-2.5Sn ELI alloy in different state are shown in Fig. 2. It shows that some original β phase is kept in the microstructure of Ti-5Al-2.5Sn, which is a typical Widmanstatten structure. Inside the grain, there are lath-shaped colonies arraying in a certain orientation, among which the boundaries are visible clearly. There is also continuous or discontinuous coarse grain boundary phases distributed on the boundaries of the original β phase, which are irregular curves. XRD analysis results of these two kinds of microstructure are shown in Fig. 3, in which, the microstructure of Ti-5Al-2.5Sn ELI alloy is mainly composed of α phase (colonies in Fig. 2a). The grain boundary phase is also shown to be α phase in Fig. 4 TEM analysis. In HIP state, the microstructure of Ti-5Al-2.5Sn is still composed of lath-shaped colonies and grain boundary of original β phase (Fig. 2b), and the major phase is still platelet α phase (Fig. 3b). However, comparing with as-cast structure, the colonies are less but larger in HIP state structure, accompanied with thinner and more straight grain boundaries.

3.2.2 Microstructure of cast Ti-6Al-4V ELI alloy

Fig.5 shows the microstructure of cast Ti-6Al-4V ELI alloy, which is mainly composed of equiaxed grains with lamellar phase (Fig. 5a). In β phase there are very clear grain boundaries around with either continuous or discontinuous intergranular phases. The β grain, which also

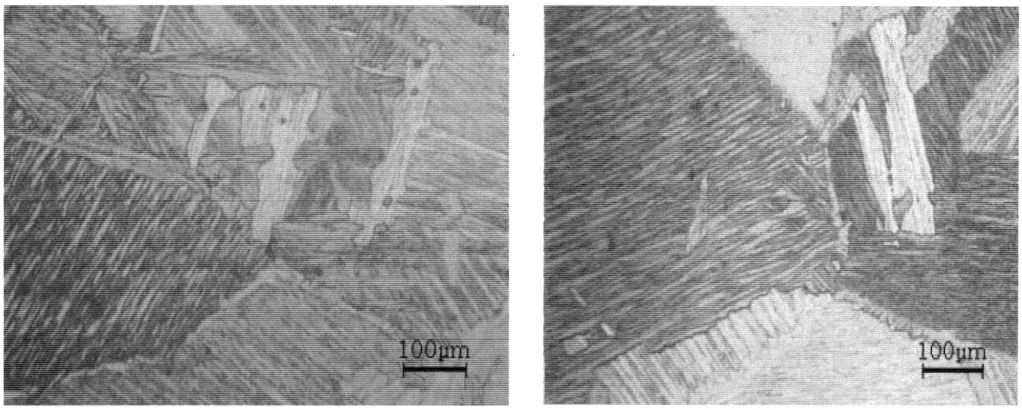

(a) as-cast (b) HIP treated

Fig. 2 Microstructure of cast Ti-5Al-2.5Sn ELI alloy

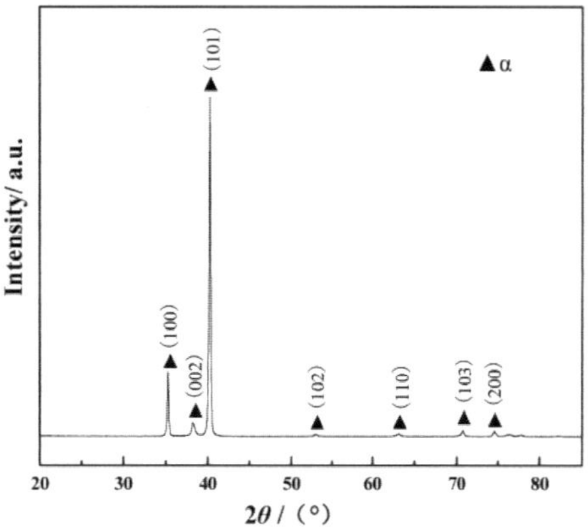

(a) as-cast

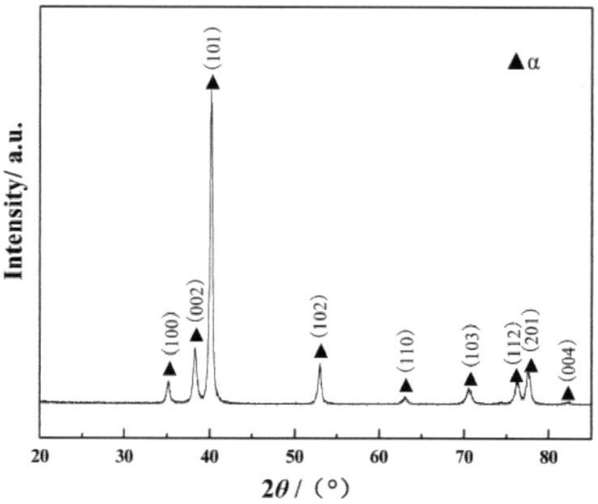

(b) HIP treated

Fig. 3 XRD patterns of cast Ti-5Al-2.5Sn ELI alloy

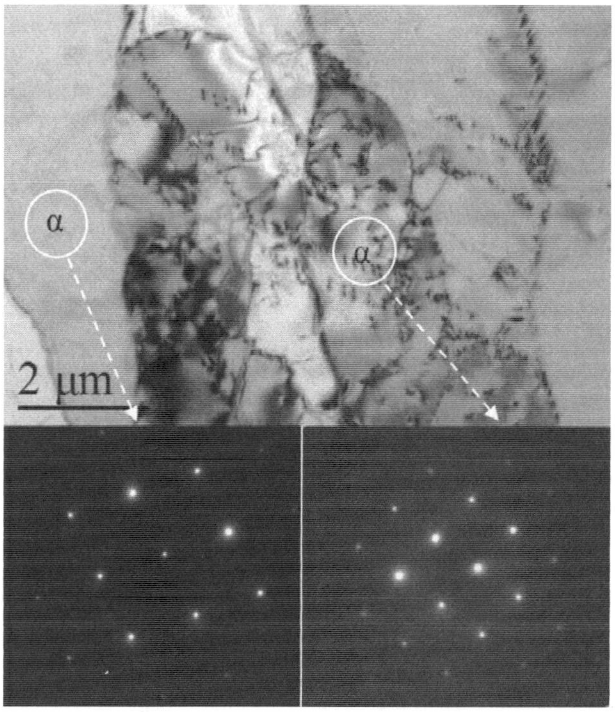

Fig. 4 TEM analysis of Ti-5Al-2.5Sn ELI alloy

contains some subgrains inside, is relatively large with a size of 200 μm. The XRD analysis (Fig. 6) illustrates that the microstructure is composed of both α and β phase. And the TEM analysis in Fig. 7 shows that the white lamellar phase in Fig. 5a is α phase, while the black phase, which distributes among α phase, is β phase, and the phase on the original β grain boundary is intergranular α phase. Fig. 5b shows the microstructure of Ti-6Al-4V ELI alloy after HIP treatment. It is clear that the lamellar phase after HIP treatment is coarser and thicker, compared with that of casting alloy.

Furthermore, the biggest change is the appearance of basketweave structure within the grain, which is quite different from platelet α phase in lath colonies. The results of XRD and TEM analysis show that the lamellar phases whether in lath or in basketweave structure are α phase. Also, the phases on the grain boundary are α phases, while β phases are actually distributed among α phases. This phenomenon is caused by the change of hole defects. For example, in the HIP treatment, shrinkage of cavity and porosity accompany with creep of alloys occur under high

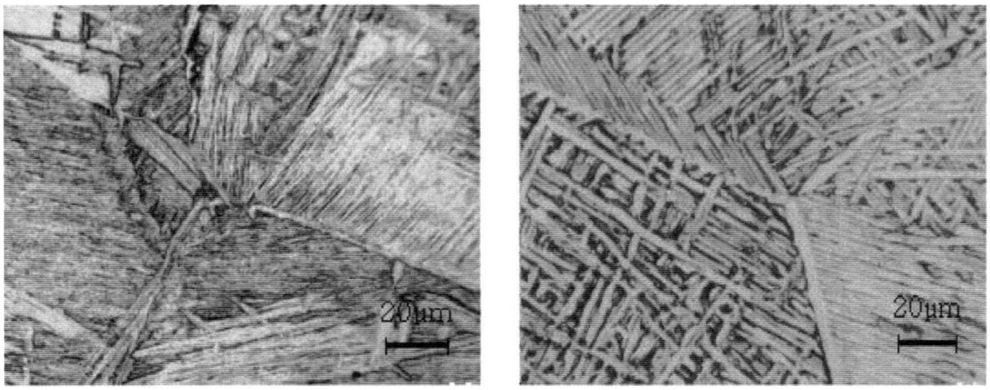

(a) as-cast (b) HIP treated

Fig. 5 Microstructure of cast Ti-6Al-4V ELI alloy

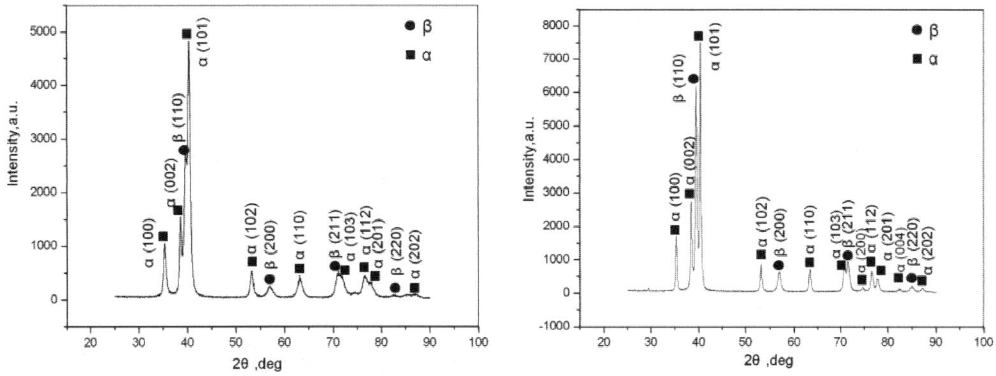

(a) as-cast (b) HIP treated

Fig. 6 XRD patterns of cast Ti-6Al-4V ELI alloy

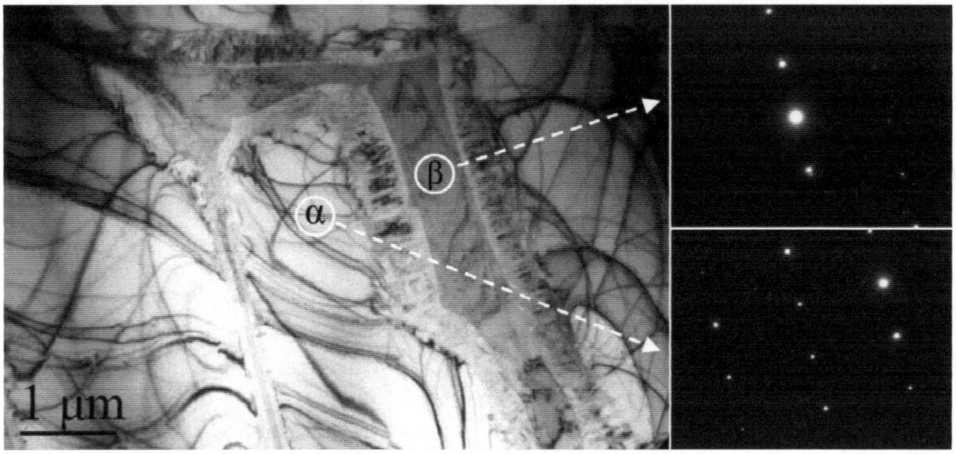

Fig. 7 TEM image of cast Ti-6Al-4V ELI alloy

temperature and high pressure. In these areas, where voids close, recrystallization occurs because of the creep deformation[8-9], and allows part of the lamellar structure to be equiaxed. The lamellar α phases grow under the effect of the second diffusion of element, which can release some stress concentrated on the edge of lamellar α grain and benefit improving the plasticity of alloys.

3.3 Mechanical properties of cast ELI grade titanium alloy

3.3.1 Tensile properties

Tensile properties of cast Ti-5Al-2.5Sn ELI alloy are shown in Table 3. It can be seen that as-cast Ti-5Al-2.5Sn ELI alloy has high tensile strength and good plasticity. After treatment of HIP, its yield strength and elastic modulus have no obvious change while the elongation increases in a certain degree. This change is ascribed to the changes of the microstructure. On one hand, the macro- and micro-void defects such as gas holes, shrinkage and cavity are closed and disappear after HIP treatment, and the crack source caused by defects during the process of tensile test decreases significantly. On the other hand, in the area of defects closed, equiaxed grains cross each other, the lap joints of them are firm, and the crack is difficult to propagate, as illustrated in Reference 10. The tensile fracture morphologies of cast Ti-5Al-2.5Sn ELI alloy in the condition of casting and HIP treatment are shown in Fig. 8. It can be seen that in the condition of casting, most of the dimples are small in size and shallow in depth, the deep one is scare, but after HIP treatment, there are many large and deep dimples, and there are many small dimples in the large dimples. In general, the size, amount, and depth of dimple are directly related with the energy absorbed during fracturing; many large and deep dimples can absorb more energy, which is not easy to fracture. These indicate again that the plasticity of Ti5Al2.5Sn ELI alloy has been improved obviously after HIP treated.

Table 3 Tensile properties of cast Ti-5Al-2.5Sn ELI alloy at room temperature

Condition	Sample	R_m/MPa	$R_{p0.2}$/MPa	A/%	Z/%	E/GPa
Casting	#1-1	785	745	12.0	24.5	121
	#1-2	790	750	11.5	22.0	123
HIP	#2-1	795	760	17.5	29.0	123
	#2-2	790	755	15.5	28.0	122

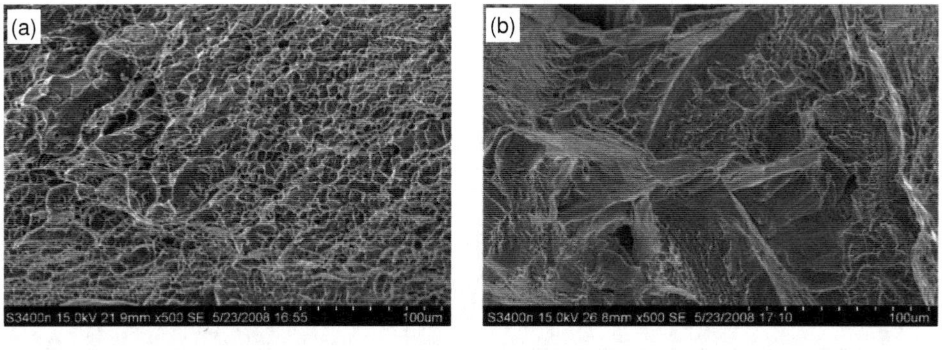

Fig. 8 Tensile fracture morphology of Ti-5Al-2.5Sn ELI alloy after (a) casting and (b) HIP treatmen

Table 4 shows the tensile properties of cast Ti-6Al-4V ELI alloy after casting or HIP treated. It can be seen that cast Ti-6Al-4V ELI alloy has high strength and good plasticity. Both in the conditions of casting and HIP treatment, the tensile strengths of cast Ti-6Al-4V ELI alloy are higher than that of cast Ti-5Al-2.5Sn ELI alloy by about 100 MPa. Both the strength and the elongation of cast Ti-6Al-4V ELI alloy increase after HIP treatment, which is also ascribed to the changes of chemical composition and microstructure. As mentioned above, the defects and the casting stress are both eliminated, and the composition segregation is reduced during HIP treatment. So the homogeneity of composition and microstructure are improved significantly, and the tensile properties are improved.

Table 4 Tensile properties of cast Ti-6Al-4V ELI alloy at room temperature

Condition	Sample	R_m/MPa	$R_{p0.2}$/MPa	A/%	Z/%	E/GPa
Casting	#1-1	890	845	10.5	18.0	110
	#1-2	895	850	12.0	17.0	109
HIP	#2-1	910	855	15.5	22.0	115
	#2-2	920	860	16.5	24.0	118

3.3.2 Impact properties

Table 5 shows the comparison among impact values of cast Ti-6Al-4V alloy, cast Ti-5Al-2.5Sn alloy and their ELI grade alloys. The data shows that the impact property of ELI grade alloy is higher than that of the same alloy in non-ELI grade. For different kinds of alloys with the same level of impurity, the impact value of Ti-5Al-2.5Sn alloy is higher than that of Ti-6Al-4V alloy. Also, for alloys with the same chemical composition and same impurity level, the impact value of HIP-treated one is higher than that of as cast. Generally, the impact value of alloy is related to both the type of microstructures and the contents of interstitial impurities.

Table 5 Impact properties of ELI and non-ELI cast titanium alloys

Alloy	Condition	KU_2/J	Alloy	Condition	KU_2/J
Ti-5Al-2.5Sn ELI	cast	61	Ti-6Al-4V ELI	cast	58
Ti-5Al-2.5Sn ELI	HIP	67	Ti-6Al-4V ELI	HIP	62
Ti-5Al-2.5Sn	cast	54	Ti-6Al-4V	cast	51
Ti-5Al-2.5Sn	HIP	58	Ti-6Al-4V	HIP	54

Previous research[6] shows that impurity elements such as C, N, H, O exist in the lattice gaps of titanium alloys. These impurity elements reduce the amount of slip system and also lead to crack sources caused by the stress concentration. Therefore, the ductility and toughness of ELI grade cast titanium alloys are higher than those of non-ELI grade alloys generally. In addition, the influence of interstitial impurities is different for different types of microstructures of alloys. For α titanium alloys, the crystal lattice is close-packed hexagonal structure, and interstitial elements exist in its tetrahedral interstice and octahedral interstice. But for α+β titanium alloys, the interstitial impurity elements exist in both tetrahedral α phase and octahedral interstice of β phase simultaneously. These two types of crystal have same type and number of lattice interstice, but the largest interstice radius of close-packed hexagonal lattice, which is the radius of octahedral interstice, 0.146a (a is lattice constant), is larger than that of body-centered cubic lattice, which is the radius of tetrahedral interstice, 0.126a. So in the same condition, the number of interstitial impurity element atoms contained in body-centered cubic lattice is smaller than that of close-packed hexagonal lattice. For titanium alloys, comparing with close-packed hexagonal lattice α phase, the distortion of β phase with body-centered cubic lattice is more serious when the content of interstitial elements is the same. And the interstitial elements that can not be accommodated by the structure are easier to precipitate as compound, which leads to a decrease in plasticity and toughness.

3.3.3 Fatigue crack growth rate

Fig. 9 shows how fatigue crack growth rates da/dN of above four kinds of cast alloys, which are HIP treated, change with the stress intensity factor ΔK. The curves indicate that within the range of ΔK value, the growth rates of ELI grade alloys are lower than those of non-ELI grade alloys with the same chemical composition. And Ti-5Al-2.5Sn (ELI) alloy has lower growth rate than Ti-6Al-4V (ELI) alloy at the same level of interstitial impurity. The Ti-6Al-4V alloy is shown to have the highest growth rate, while the Ti-5Al-2.5Sn ELI alloy has the lowest. The growth rate Da/dN is one of the most important indexes to measure the reliability of materials. Generally, the lower the growth rate, the better the anti-fatigue crack growth ability of material could be. Therefore, in terms of the anti-fatigue crack growth ability, ELI grade titanium alloy is better than non-ELI grade alloy, and Ti-5Al-2.5Sn (ELI) is better than Ti-6Al-4V (ELI) alloy.

Actually, there are many factors that influence the da/dN value of titanium alloy, such as load ratio, microstructure, environment parameter, etc. This paper mainly focuses on the factor of the content of interstitial element C, N, H, O in alloys. As we know, elements C, N, H and O exist in the tetrahedral interstice and octahedral interstice of titanium alloy and form into solid solution with the matrix. They can increase the lattice constant a and c, even c/a for α phase, and lead to lattice distortion, which prevents the dislocation and reduces the slid system of alloy, leading to an enhancement of tensile strength. Furthermore, the lattice distortion decreases the plastic area of

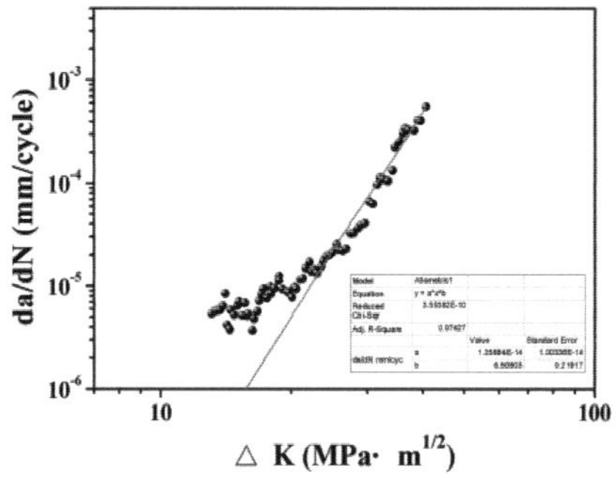

(a) Ti-5Al-2.5Sn ELI

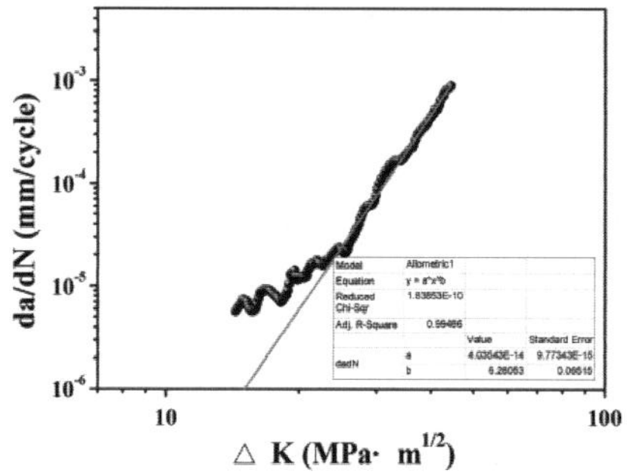

(b) Ti-5Al-2.5Sn

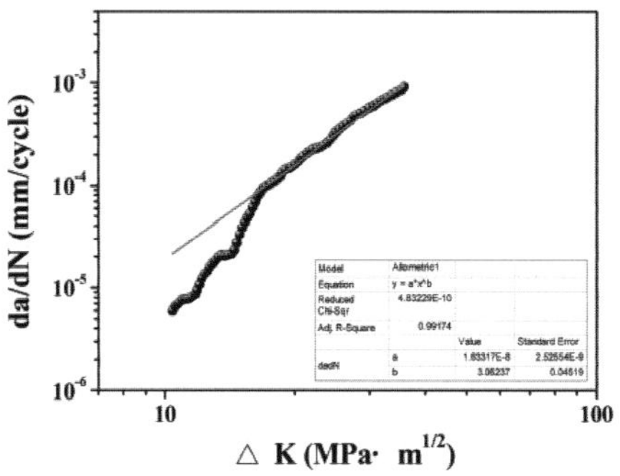

(c) Ti-6Al-4V ELI

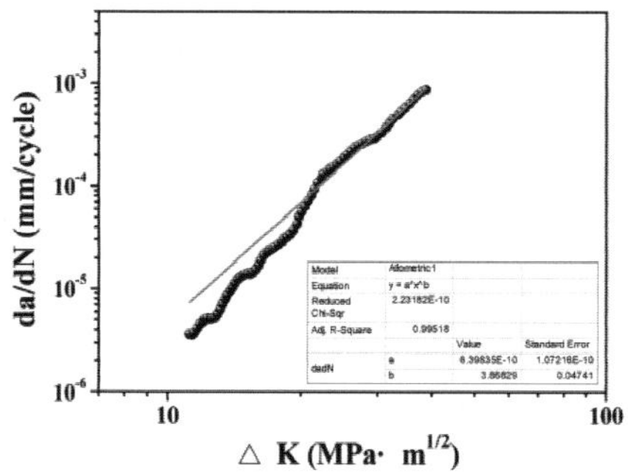

(d) Ti-6Al-4V

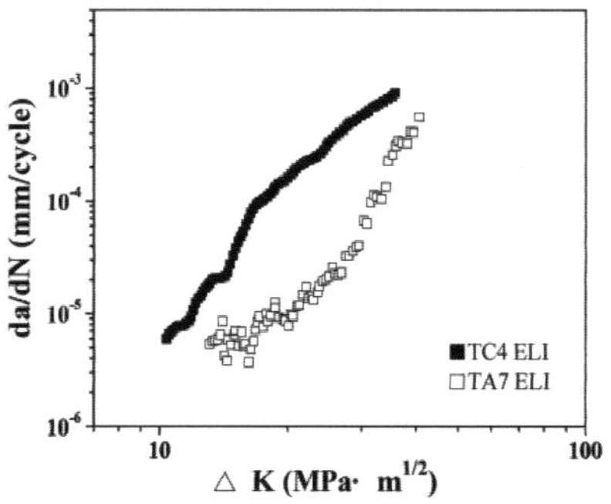

(e) ELI grade alloys

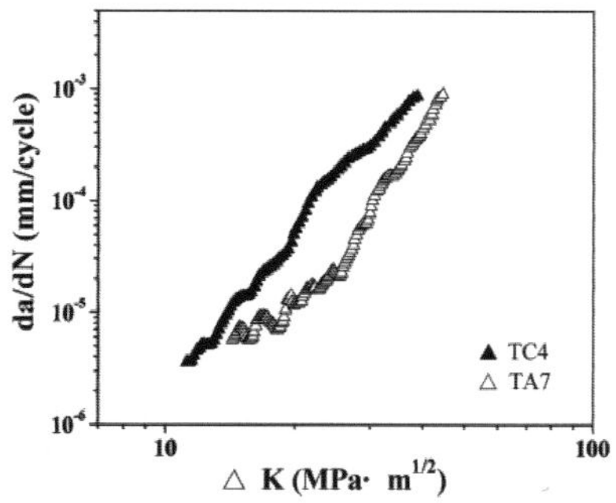

(f) non-ELI grade alloys

Fig. 9 Fatigue crack growth rate of titanium after HIP treated

crack tip, which results in a stress concentration. It provides a driving force for the propagation of crack on the interface of α/β phase and lamellar α phase, where lattice defects exist. Besides, the HIP treated Ti-5Al-2.5Sn alloy and Ti-6Al-4V alloy are both in lamellar structure, no matter whether they are ELI grade or not. The previous research also shows that for single-phase α titanium alloy, fatigue crack propagates along the interface of α phase lamellas, and for α+β titanium alloy, fatigue crack propagates along the α/β interface of the lamellar structure, but the propagation can either be parallel or perpendicular to the lamellas. Compared with single phase titanium alloy, two phase titanium alloy has larger atomic misfit on the α/β interface, where the fatigue crack is easier to propagate because of the weak binding force. That's why two-phase alloy Ti-6Al-4V (ELI) has higher fatigue crack growth rate than single-phase alloy Ti-5Al-2.5Sn (ELI).

3.3.4 Fracture toughness

Test data of the four alloys' fracture toughness are shown in Table 6. In the field of engineering, K_{Ic}, the plane strain fracture toughness of material, is usually used as the basis for damage tolerance design. In this study, K_q, the conditional value of K_{Ic}, is obtained under the test condition of K_{Ic}. As a valid value of K_{Ic}, K_q needs to meet two requirements of plane strain fracture toughness. First, the specimen sizes should be B, a_0 and $(W-a_0) \geq 2.5(K_q/R_{p0.2})^2$. Second, the load ratio should be $P_{max}/P_q \leq 1.10$ (P_{max} is the maximum load, Pq is the conditional load, a0 is the length of crack). Values of P_{max}/P_q in Table 6 are all greater than 1.1, and B, a_0, $(W-a_0) <2.5(K_q/R_{p0.2})^2$, so the data doesn't meet the requirement and is not able to calculate the criterion of valid K_{IC}. Therefore, K_q is used as the criterion of fracture toughness in this paper. As shown in Table 6, cast Ti-6Al-4V ELI alloy has the maximum fracture toughness. The second and third are cast Ti-6Al-4V alloy and cast Ti-5Al-2.5Sn ELI alloy, respectively. And cast Ti-5Al-2.5Sn alloy has the minimum fracture toughness. In terms of microstructure of alloys, dual-phase Ti-6Al-4V (ELI) has higher fracture toughness than that of single-phase Ti-5Al-2.5Sn (ELI) alloy. Previous research[12] indicates that fracture toughness of material is related to the size of plane strain plastic zone, r_0 ($r_0 = K_q/R_{p0.2})^2/(4\sqrt{2}\pi)$, and yield strength, $R_{p0.2}$. In general, fracture toughness increases with the raise of r0 and the decline of $R_{p0.2}$. Because as the yield strength decreases, the area of bearing failure load as well as the maximum bearing load of materials will increase. Also, with the expansion of plastic zone, more energy from external work can be absorbed. Therefore, the material can achieve a higher load bearing ability and higher fracture toughness. Table 3 and 4 show that trend of how the fracture toughness K_q of the four types of alloys change is consistent with that of r_0, while it is opposite to that of $R_{p0.2}$ for alloys with same microstructures. This result also consists with the existing research results.

Table 6 Fracture toughness of ELI grade and non-ELI grade cast alloys

Alloy	NO.	B /mm	W /mm	a_0 /mm	$(W-a_0)$ /mm	P_{max} /kN	P_q /kN	P_{max}/P_q	$2.5(K_q/R_{p0.2})^2$/mm	K_q /(MPa.m$^{1/2}$)	r_0 /mm
Ti-5Al-2.5Sn ELI	1	25.47	50.13	27.05	23.08	89.5	50.44	1.77	39.26	97.12	0.88
Ti-5Al-2.5Sn ELI	2	25.45	50.21	26.88	23.33	89.52	51.39	1.74	39.42	97.32	0.89
Ti-5Al-2.5Sn	3	25.3	50.73	24.90	25.83	97.13	56.64	1.71	32.10	87.83	0.72
Ti-5Al-2.5Sn	4	25.45	50.22	26.24	23.98	88.9	47.25	1.88	30.61	85.75	0.69
Ti-6Al-4V ELI	5	25.19	50.89	25.34	25.55	76.77	64.1	1.20	50.71	108.25	1.14
Ti-6Al-4V ELI	6	25.17	50.94	25.62	25.32	75.93	64.39	1.18	52.85	110.50	1.19
Ti-6Al-4V	7	25.3	50.76	25.32	25.44	77.82	61.54	1.26	46.73	103.90	1.05
Ti-6Al-4V	8	25.32	50.85	25.23	25.62	75.2	61.56	1.22	45.83	102.90	1.03

4. Precision forming technology of cast ELI grade titanium alloys

4.1 Melting process

At present, available melting processes for titanium alloys are vacuum consumable electrode arc skull melting, vacuum induction melting (ISM) and electron beam melting, etc[1], among which the first two are usually used for casting. Due to the content of many low melting point elements, narrow range of composition, and low content of impurity, there is high requirement for the melting process of ELI grade titanium alloy in casting. In terms of the element evaporation of Ti-5Al-2.5Sn and Ti-6Al-4V titanium alloy when melting in vacuum, the influence of melting temperature and element evaporation rate Nm has been studied by Guo Jingjie et al[13~15]. And they point out that for the influence on *Nm* , there is an obvious critical point of vacuum chamber pressure, above which the *Nm* drops quickly. Also, the higher the melt temperature is, the higher the critical point could be. Since evaporation is actually a thermal dynamic phenomenon, so for metals of the same element, higher melting point means higher evaporation rate, and for metals of different elements, lower melt point facilitates the evaporation. For example, the evaporation rate of Al is tens or hundreds times higher than that of Ti and V in Ti-6Al-4V alloy. But due to the complex mechanism in real evaporation, sometimes the evaporation rate doesn't really decrease with the increase of melt point in the same melting condition. It is shown in Reference 13 that evaporation rate of element Al is greater than that of element Sn at the same temperature.

According to above discussion, the vacuum

consumable electrode arc skull melting is considered to be more suitable for melting ELI grade titanium alloy in this paper. On one hand, the arc melting process is faster than ISM process, which shortens the evaporation time; on the other hand, the superheat, which is usually 50-150 ℃ higher than the melting point, of melt alloy is lower, and the alloy has a lower critical point (*Nm*), so the evaporation rate can be reduced. In addition, to satisfy the requirement of ELI grade alloy on the control of gas element, the vacuum condition is necessary in the melting process. Since the evaporation of element Al, Sn and V is inevitable, to control the composition accurately, it is important to improve the original content of mother alloy or to add elements in the process of melting. In this study, to control the composition of ELI grade titanium alloy efficiently, a melting process with low voltage, high current, low pressure and inert ambient is applied. (Specific melting process parameters are shown in Section 2 and the final chemical composition is shown in Table 2.)

4.2 Molding process

Because of the high requirement of impurity controlling for ELI grade titanium, the cavity of mold needs good surface quality and high temperature stability. Currently, there are three kinds of casting molds available for the forming of titanium alloys by casting: block graphite mold, investment shell and metal mold. Among them, the graphite mold made of high-purity graphite block has high strength, high temperature stability but poor withdrawing capability, which results in a poor forming quality. And for the investment shell, sintered refractory oxide is applied as surface layer because it has good stability and forming quality in high temperature and also easy to be made in complex structure. But the process of the preparation is too complicate and too sensitive to the environment parameters. The method of metal mold uses cheap metals such as cast iron and is usually for castings with simple parting surface. The forming quality of this method is between those of graphite mold and investment shell.

In this paper, a new casting process, which applies coated graphite mold and oxide ceramic core, is developed. It can cover the shortage of above molds and realize the casting forming and chemical composition accurate control of ELI grade titanium alloy. In this process, graphite mold is used as the outer shell and refractory oxide coating is sprayed on the surface of cavity to reduce the cooling rate of the graphite mold and prevent carbonaceous dust and particles from getting involved in the melt. Also, oxide ceramic blocks are used as cores of complex cavity to reduce the cooling rate of thin-walled structure, thereby improving its formability and surface quality. This process combines advantages of graphite mold and investment shell, resulting in an ELI grade titanium alloy casting with pure chemical composition, smooth surface and even no cold shut, flow, hole or other defects. Fig. 10 shows graphite mold, oxide ceramic core and their combination. Fig. 11 shows the ELI grade titanium alloy casting made in this process.

Fig. 10 Pattern of mold figure

Fig. 11 ELI grade titanium alloy casting

4.3 Mold filling process

Accounting for the limitation of vacuum consumable electrode arc furnace melting, the superheat of ELI grade titanium alloy melt is low and the fluidity of melt is poor just like conventional titanium alloys. To complete the forming of ELI grade titanium alloy complex casting, its liquidity needs to be increased. At present, there are two technologies to improve titanium alloy liquidity which are centrifugal casting and pressure casting, and each one has its respective advantages and disadvantages. In this study, to improve the liquidity of melt further, vacuum infrared heating centrifugal casting is developed based on the thought of decreasing cooling rate of melt, which is heating mold by infrared heating device inside the vacuum chamber and pouring in centrifuge. Fig. 12 shows the principle diagram of this method. The advantage is that the mold can be heated to 400-600 °C real-timely, and heat loss of the melt is little in the process of filling, meanwhile, the melt can fill the mold rapidly under the action of centrifugal force. By applying this technology, the titanium alloy fluidity and forming ability are enhanced obviously. Fig. 13 shows the titanium alloy casting with complex curved surfaces and thin walls formed by the technology, and the minimum thickness of wall is 1 mm.

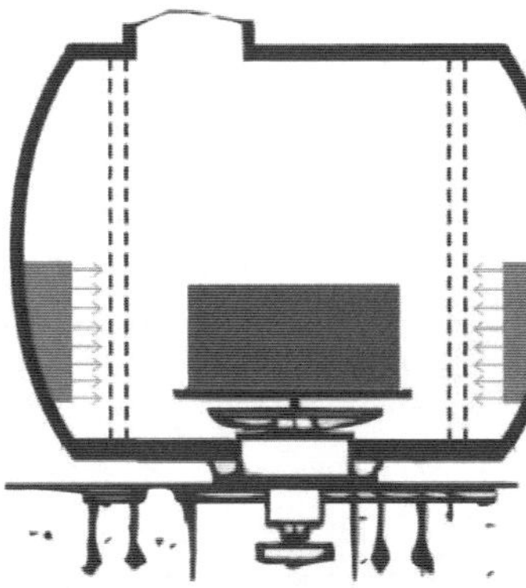

Fig. 12 Principle diagram of mold by infrared heating in vacuum

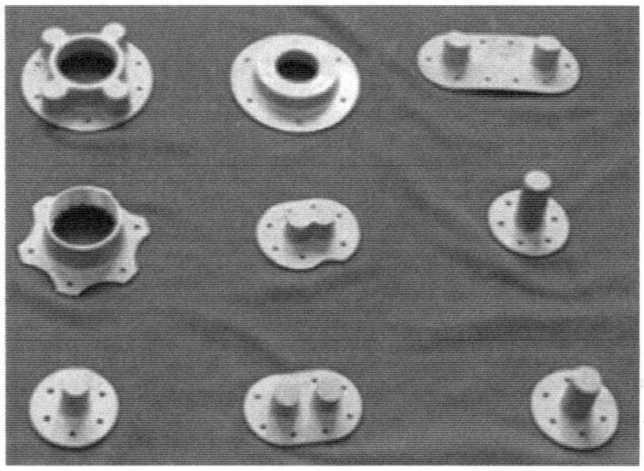

Fig. 13 Thin-wall titanium alloy castings poured

4.4 Quality of forming

Surface defects, roughness, contamination layer, etc. of the ELI grade titanium castings manufactured with above technology have been analyzed. Fig. 14 shows the surface quality of castings with mold coating or without mold coating, and the data in Table 7 is about surface roughness of the casting. The result indicates that most of the cold shuts, flow marks and blowholes are eliminated; meanwhile the surface roughness of casting, which is reduced significantly, can reach a level of R_a 3.2 μm. Fig. 15a shows that the thickness of surface contamination of casting is about 6 μm,

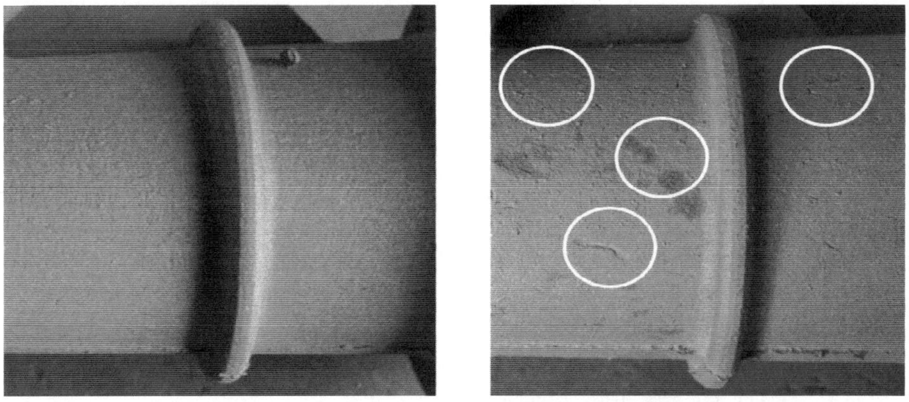

(a) coating (b) no coating

Fig. 14 Surface of ELI grade titanium alloy castings by different technologies

Table 7 Roughness of ELI grade titanium alloy casting by coating mold

Mold	Roughness, Ra /μm				
	1	2	3	4	5
Coating	3.05	3.20	3.28	3.32	3.25
	3.11	3.19	3.35	3.18	3.22
No coating	12.22	12.25	12.10	12.62	11.10
	12.35	11.90	12.40	12.48	11.56

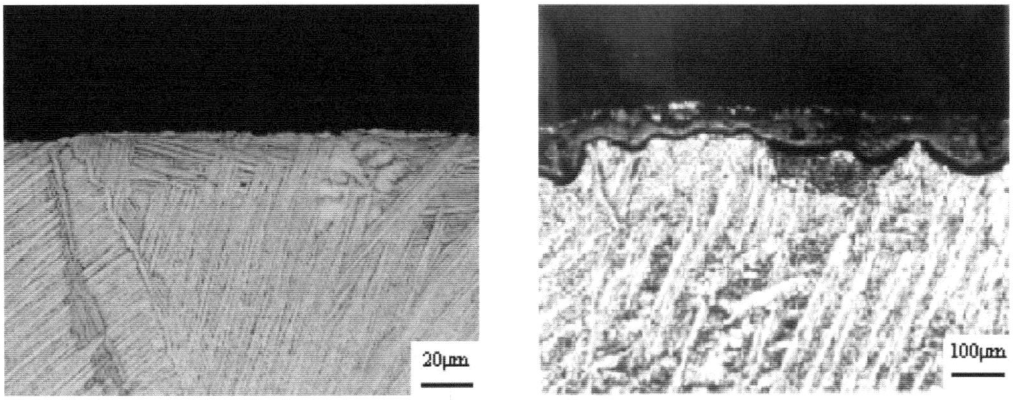

(a) coating (b) no coating

Fig. 15 Surface contamination layer of titanium alloy castings by different technologies

which is also decreased greatly compared with that of general graphite mold casting, 100 μm (Fig. 15b). Moreover, the dimension tolerance of casting is proved to be in level IT14 with the measurement of different wall thickness. As a conclusion, the technology developed in the paper provides an effective way to form high quality cast ELI grade titanium alloys precisely.

5 Conclusions

(1) The optimal composition of cast ELI titanium alloys is Al (4.5-5.5) wt.%, Sn (2.0-3.0) wt.% for cast Ti-5Al-2.5Sn ELI alloy and Al (5.5-6.5) wt.%, V (3.5-4.5) wt.% for cast Ti-6Al-4V ELI alloy. The content of interstitial impurity is C $\leq$ 0.05 wt.%, N $\leq$ 0.03 wt.%, H $\leq$ 0.0120 wt.%, O $\leq$ 0.12 wt.%.

(2) For the Ti-5Al-2.5Sn ELI alloy, microstructures of both as-cast and HIP treated cast actually consist of single α phase. Compared with the as-cast, the structure of the HIP-treated one has coarser α phase, smaller amount of lath colonies, equiaxed parts and in which, the original grain boundary disappears or narrows. However, for Ti-6Al-4V ELI alloy, the as-cast and the HIP treated cast consist of α and β phases. In as-cast condition, α phase is arranged as platelet westergren; while in the condition of HIP treatment, α phase is arranged as basketweave besides platelet westergren.

(3) Tensile strength of cast Ti-6Al-4V alloy is higher than that of cast Ti-5Al-2.5Sn alloy, no matter it is ELI grade or not. ELI grade alloy has higher impact toughness than non-ELI grade alloy, while the fatigue crack propagation rate of ELI alloy is lower than that of non-ELI grade alloy. For alloy with same major composition, fracture toughness of ELI grade alloy is higher than that of non-ELI grade alloy. Therefore, the damage resistance of ELI grade alloy is higher than that of non-ELI grade alloy.

(4) Precision forming of complex thin wall casting of ELI grade titanium alloy can be realized by using following techniques: the process of melting in consumable electrode arc skull furnace with low pressure, molding with combination of graphite shell and ceramic cores, centrifugal casting by infrared heating. With these technologies, an ELI grade titanium alloy casting with a minimum thickness of 1 mm, roughness of Level R_a 3.2 μm, contamination layer thickness of 6 μm and dimension tolerance of Level IT14 can be achieved with good forming quality.

Acknowledgments

This study is supported by the National Natural Science Foundation of China (51375318), Natural Science Fund Project of Liaoning Province of China (2013010439-401), and Science and Technology Plan Project of Shenyang City of China (F14-178-1-00).

References

[1] Zhu Zhishou. Research and development of new-brand titanium alloys of high performance for aeronautical applications[M]. Beijing: Aviation Industry Press, 2013.

[2] Y.V.R.K Prasad, T Seshacharyulu, S.C Medeiros, W.G Frazier. Influence of oxygen content on the forging response of equiaxed (α+β) preform of Ti–6Al–4V: commercial vs. ELI grade[J]. Journal of Materials Processing Technology, 2001, vol.108(3), pp.320-327.

[3] T. Yuri, Y. Ono, T. Ogata. Effects of surface roughness and notch on fatigue properties for Ti–5Al–2.5Sn ELI alloy at cryogenic temperatures[J]. Science and Technology of Advanced Materials, 2003, vol.4(4), pp.291-299.

[4] Li Shikai, Hui Songxiao, Ye Wenjun, Yu Yang, Xiong Baiqing. Effect of cooling rate on the microstructure and properties of TA15 ELI alloy[J]. Rare Metal Materials And Engineering, 2007, Vol.36(5), pp.786-789.

[5] Wu Jun, Xu Lei, Lei Jiafeng, Liu Yuyin. Preparation of powder metallurgy TA7 ELI alloy and its interface reaction layer[J]. The Chinese Journal of Nonferrous Metals, 2010, vol. 20(S1), pp.s299-s302.

[6] C.R.Brooks. Heat treatment, structure and properties of nonferrous alloys[M]. American: American Society for Metals, 1984.

[7] Liu Wei, Du Yu, Yu Zhentao, Cai Xuezhang, Yang Guanjun, Sun Qiaoyan. Analysis of relationship between interstitial content and impact properties of a type of near α titanium alloy[J]. Journal of Xian Institute of Technology, 2005, vol.25 (4), pp.370-373.

[8] Wang Xiaolin, Zhao Zhilong, Zhang Yansheng, Zhao Fangli, Mao Qi, Tian Weihua. Discussion on closing mechanism of TC10 alloy casting defects under HIP[J]. Founrdy, 2011, vol.60(8), pp.731-734.

[9] Wang Xiaolin, Wang Lanfang, Zhu Zhongren, Zhang Duhuai, Ding Wuzhou, Dou Yongqing. Effects of HIP(heat isostatic pressing) treatment on as-cast microstructure and properties of TC10 alloy[J]. Special casting & nonferrous alloys, 2011, vol.31(8), pp.760-761.

[10] Cui Zhongqi. Physical metallurgy and hat treatment[M]. Beijing: China Machine Press, 2002.

[11] Ma Yingjie, Li Jinwei, Lei Jiafeng, Tang Zhenyun, Liu Yuying, Yang Rui. Influences of microstructure on fatigue crack propagating path and crack growth rates in TC4 ELI alloy[J]. Acta Metallurgica Sinica, 2010, vol.46 (9), pp.1086-1092.

[12] Ma Yingjie, Wang Dingchun, Wang Hongwu, Lei Jiafeng, Wang Jian, Liu Yuying, Gao Qi, Yang Rui. Factors influencing fracture toughness of TC4 ELI alloys[J]. The Chinese Journal of Nonferrous Metals, 2010, vol.20 (S1), pp.s414-s418.

[13] Guo Jingjie, Liu Yuan, Su Yanqing. Evaporation of Al and Sn during ISM Process of Ti-5Al-2.5Sn Alloy[J]. Material Engineering, 2000, (2), pp.6-9.

[14] Guo Jingjie, Liu Yuan, Jia Jun. Behavior of volatilizing loss of Al element in Ti-6Al-4V melt during ISM melting[J]. Foundry, 1999, (2), pp.14-18.

[15] Liu Guizhong, Su Yanqing, Guo Jingjie, Ding Hongsheng, Jia Jun, Fu Hengzhi. Evaporation tendency of components in Ti-6Al-4V alloy melt[J]. Materials Science & Technology, 2001, vol.9(2), pp.175-177.

[16] Xie Chengmu. Titanium and titanium alloy casting[M]. Beijing: China Machine Press, 2005.

Proceedings of the 72nd World Foundry Congress
21-25th may 2016, Nagoya, Japan

Further Study on Application of the Book of Changes in Metal Casting

Jiang Yi Xiang
Xihua University, Sichuan Chengdu, China

Abstract

With mathematical mode, the existence, uniqueness and limitless of metal casting have been revealed. The problem " where is metal casting from? where it go?" has been answered. The research result indicated that metal casting is not only technology,but also science, as well as civilization. Specially, afresh evaluation on metal casting, finally, metal casting has become fully deserved an ancient first great invention in China. It had never been really understanding before.

Keywords: origin of casting, a large scale system of metal casting, mathematical mode of casting

1. Introduction

In China there though was full evident 5000 years foundry history, there were a few of foundry technical literature in history records. Many traditional technology have almost been lost. We should like to ask, where has been the metal casting from? where will it go? What is essence of metal casting? Does metal casting be technology or science and civilization?

In 69thWFC the author has written an article about "Yi Jing and Casting"(2),in which a new theory and new concept has been put forward.

2. The New Theory and It's Mathematical Mode

Using Yi Jing's concept viewing the situation as a whole,we regard all castings of Chinese 5000 years as a large scale system of castings.From this time on,we may able to study metal casting from the overall, other than one by one,therefore to explore their changes and develop.

According to the view of yin and yang in the Book of Changes,we have the Dao of metal casting.

The Dao of metal casting is Law of yin and yang,
Yang is the pattern (or melted metal),
Yin is the mold cavity.
The yang from among yin is core,
The yin from among yang is cave.
Yin cavity and yang pattern,
Like a seamless heavenly robe.
Through their interactions,
Till a casting was produced.
Thus a cycle finished.
All castings correspond theirs-self endless cycles.

In addition,the Dao of metal casting has been expressed as following Three Laws in mathematical mode.

The Three Laws of Metal Casting

First Law. (existence)
The Castability Really Exist and Follows the Dao of Casting.
The necessary conditions of existence of metal casting are three factors: heaven,earth and people. Not a single one of three conditions can be dispensed with. This is concrete application of holism of human beings and universe of the Book of Changes.

Second Law.(uniqueness)
The Metal Casting is an Unique 3D Duplicate Technique in Metal Manufacture.
Please see part 3 about essence of metal casting.

Third Law.(limitless)
All Shapes of All Castings May Be Arbitrary. All Weights of All Castings May Be Arbitrary,but Bounded.
Here limitless means arbitrary.

3. Further Understanding Origin of Metal Casting

This section we will answer question: Where metal casting came from? And where it go?

According to the Book of Changes,one is called as the supreme ultimate. Two is yin and yang. We do not possible and not necessary to know which day metal casting star at? However, we can affirm that origin of casting has stared at one period or at initial period. About 6000 years ago, we supposed that someday a worker accidentally discovered a pit,which can shape melted copper. The initial shape was fuzzy. But it carried within itself the seeds of casting. Many years had passed. Once birth of casting technology, it has been divided in yin and yang, two opposite sides: yin cavity and yang pattern. This was one produced two. It is mentioned that the process of one to two experienced a long evolution in the remote antiquity. Archaeological result tell us that from the earliest excavated fuzzy brass plate in Banpo Shanxi to a shape bronze weapon in the village of Erliton Henan (started at the place Chinese casting), the period have just proved our true supposition: the evolution of one to two was about 2000 years and more.

In ancient China the metal casting was richly endowed by nature and possessed vast reserves of manpower. It was necessary condition that metal casting have created a brilliant 5000 years foundry history. We may suppose next 5000 years or 10000 years, If the human beings exhaust all mineral resources, or do not need any more castings, then metal casting will not be sustainable and lose itself.

4. Further Understanding the Essence of Metal Casting

Our investigation indicated that the metal casting in China is not only a technology,but also science, as well as civilization.

4.1 Metal Casting is a Duplicate Technology

Please be specially aware that metal casting is 3D-duplicate technology. This is different from any other manufacturing method. Here, 3D means three-dimensional space.

Mathematically the duplicate method surpasses

processing ability of all different machine-manufacturing. Drilling machine only can make a hole. Planning machine can make only a plan. Turning lathe can only make a circular bar. No matter what levels of machine, locus of cutter's motion will not be able to change. However, the duplicated method can make products of any shapes. This method even surpasses finite element method in mathematical principle. It can make some most exquisite and most complicated castings, and can not be replaced by any other methods.

4.2 Metal Casting also is a Science

In Chinese foundry dictionary metal casting has been regarded as a science long ago.

4.3 Metal Casting is Civilization as Well

4.3.1 Metal Casting Created Chinese Civilization

The Chinese civilization has began with the Xia Dynasty(approx 21st century BC to 16th century BC), when the Bronze Age of China started. The Xia was a slave society. All bronze utensils have been owned by aristocrats.The slaves have not right to possess any bronze utensils. In that time the bronze have become a symbol of property and right. Therefor the appearance of classes and form of nation both were inseparable from the bronzes. China has a splendid Bronze Age of 2000 years which laid the foundation of the Chinese civilization. Just in the Bronze Age of great prosperity, the Book of Change was produced.

So we have every reason to think that metal casting created the Chinese civilization, both material and spiritual.

4.3.2 Metal Casting Promoted Development of the Chinese Civilization

There are many examples. We selected typical one. Following the Bronze Age was the Iron Age of about 2000 years, in which iron casting was mainly used to make tools and weapons. Among them the ploughshare was a very special and very amazing farm implement . Up to present the ploughshare still is applied in some rural areas of China. The tool made output of food to increase by 2-3 times or more. As a big agricultural country with a history of thousands years,ploughshare, a iron castings,played a key great pushing effect on agricultural production in China.

4.3.3 Metal Casting Witnessed the Chinese Civilization

Sichuan Chengdu has formerly been recognized a city of 2000 years history. However in 2001 year Jinsha site and thousand bronze tools has been discovered.The great important discover made the history of Chengdu as the same step as the Chinese civilization. The history of Chengdu has been pushed back for 1000 years.

4.3.4 Metal Casting Disseminated Chinese Civilization

Ding has special meaningful casting article. In the Chinese history Ding was deeply respected. According to records of the Grand Historian of China,long long ago Fuxi has made a value ding.

After hundreds years yellow emperor also has made a value ding. Up to Dayu of the Xia Dynasty the 9 Dings has been made as a symbol to conger and rule the country.

Later the 9 Dings have been disseminated from generation to generation, but to the First Emperor Qin the 9 Dings have lost. In China casting dings were continued for thousands years.

It is mentioned that the dings made by the honor of top leader have almost lost over 5000 years.

Only the Century Valuable Ding and the United Valuable Ding exist now. The former sets in the United Nations which is the symbol of the world peace, the later is in memory of the Chinese nation union. Two significant articles have become the best examples of casting disseminating the Chinese civilization.

Therefore, in China metal casting has been recognized as an important part of Chinese civilization

5. Afresh Evaluation of Metal Casting in the View of New Theory and New Concept

Now we can get following several "the most" easily.

5.1 Metal casting was the oldest technology and invention in China. Its long history of 5000 years was unparalleled.

5.2 Metal casting have the most long life.

5.3 Metal casting is the most primitive and the most modern.

5.4 Metal casting has the most wide-spread application in Chinese nation economy.

5.5 Metal casting created the most many miracles: historical, real and living.

5.6 The number of casting products were and are the most great in China industry.

5.7 The shapes of casting products were and are the most complicated. The changes of casting shapes were and are the most.

5.8 It is the most Chinese. Among four big civilization ancient countries around the world, the Chinese civilization was the most successive. Because China had a period of the Bronze Age of 2000 years, which made no division of Chinese history into periods.

In a word, in Chinese ancient science and technology metal casting was incomparable in space, in time, and in social effects. It is has fully deserved an ancient Chinese first great invention. It had never really understood before. We have to say, in some instance metal casting also was a lost civilization.

6. Conclusions

6.1 Using Yi Jing's concept viewing the situation as a whole, we have designed a large scale system of all metal castings of Chinese 5000 years history, in which three laws have been induced.

6.2 Using the holism of human beings and universe of the Book of Changes, we have analyzed a close relationships between metal casting and Chinese civilization, and metal casting have made great influence for civilization.

6.3 We have afresh evaluated a great effect of metal casting. Finally, the metal casting has become fully deserved an ancient Chinese first great invention. It is really a new great discover.

Acknowledgments

Very special thinks to :

Sichuan Xiang Ye Mechanical Casting co., Ltd. Chengdu China

Chongqing Changjiang River Molding Material (Group) Co.,Ltd. Chongqing China

Suzhou Xingye Material Technology Co.,Ltd. Suzhou China

Chengdu Foundry Association

References

[1] Jiang Yixiang: Green Sand System Design and Analysis for Improved Iron Casting Quality. AFS Transactions. (1999)813-818.

[2] Jiang Yixiang: An Approach to the Application of the Book of Changes in Metal Casting. The 69th World Foundry Congress.Hangzhou China.(2010)0805-0809.

O-2

SUBMERSIBLES OF CASTED MARAGING STEEL

Olavi Piha[1], M.Sc. (Tech.), Tanja Saarelainen[1], Lic.Sc. (Tech.) and Juhani Orkas[1], Professor, D.Sc. (Tech.).
[1] Aalto University School of Engineering, Department of Engineering Design and Production, Foundry Engineering. PO Box 14100, FI-00076 AALTO, FINLAND.
Corresponding author, E-mail: olavi.piha@aalto.fi

In 1982, the Soviet Union Academy of Sciences enquired whether a Finnish company Rauma-Repola would be interested in manufacturing two research vehicles capable of diving to a depth of 6 kilometres. The contract was signed on May 1985 and the vehicles signed off into service in December 1987. The most critical challenge was to determine how the personnel's spheres could be formed as they were required to protect personnel from pressures reaching 600 bar. Maraging steel was finally selected as the material for their construction. Large parts had never before been poured in this material. The foundry developed ultra-high strength MVD steel for the manned sphere. This material contained low level of impurities and underwent a specific heat treatment to attain the required properties: Yield strength R_e 1650 N/mm^2, Tensile strength R_m 1730 N/mm^2, Elongation A_5 6 %. The chemical analysis was C < 0.02 %, Ni 17-19 %, Mo 4-6 %, Co 10-13 %, Ti < 0.35 %. Those values can only be reached using vacuum technology.

Keywords: *steel foundry, ultra-high strength maraging steel, vacuum technology, centerline shrinkage, Tevo Lokomo Ltd.*

1. Introduction

Rauma-Repola shipyards had after World War II delivered around a thousand different types of vessels to the Soviet Union. By 1972, the firm had started the production of oilrigs. The Lokomo works, which was part of the machinery sector of Rauma-Repola, had delivered casted parts for the oilrigs. Rauma-Repola considered the Lokomo works to be the most suitable designer and producer of the research vehicles. Lokomo's potential was also well known to the Soviet Union. The Soviet Union Academy of Sciences had previously ordered two vessels with a diving capacity of 2 kilometres from a Canadian company. The company however, could not offer vehicles of 6 kilometres due to insolvency. France, Sweden and Switzerland had not been willing to offer them at the time. It was told, that, these types of vessels were required to research and collect nodules from the bottom of sea. It is uncertain how productive this exercise had been, as no results have been disclosed to the public. A rumour even stated that the vessels were needed due to a Soviet navy nuclear submarine having sunk to a depth of 6 kilometres in the Atlantic Ocean. [1]

2. Quality Control

Before precipitation glowing of the spheres, the company used penetration liquid due to the non-magnetic austenite structure. After precipitation glowing, it was possible to use magnaflux control, as the structure became martensitic. Ultrasound control was verified by random checks using radioisotope imaging (iridium 192). When asked about the required level of the NDT control, Igor Mikhaltsev – a professor of the Soviet Union Academy of Sciences and supervisor of production, stated: "No visible defects are found with modern control methods". [2]

Figure 1: The heat treatment of the sphere halve at the Lokomo steel foundry. [3]

The production of the pressure-resistant personnel sphere was a real engineering triumph for Rauma-Repola. It succeeded thanks to top metallurgy and the diligent work of the whole project team. The company bravely signed the contract before there was absolute certainty that the method would work. Therefore the company took a considerable albeit well controlled risk both in technical and commercial respects. Due to the high purchase price, there was room for error thus reducing pressure on succeeding on the first attempt.

The material used was maraging steel, of which Americans had forged small and demanding pieces. It had exceeded the required yield strength (min 1650 N/mm^2). Large pieces had never before been cast out of this material. The metallurgists of Lokomo saw an opportunity to succeed, above all thanks to the VODC vacuum converter purchased in 1982, and the modern testing equipment (spectrometers, electronic microscopes, highly developed pressure testing equipment) manned by highly skilled and professional technicians.

3. Vacuum Oxygen Decarburization Converter

To fulfil the high quality standards of the cast and forged steel parts of the oilrigs made by Rauma-Repola, the Lokomo steel works purchased a VODC converter (Vacuum Oxygen Decarburization Converter) in 1982. [4]

The benefits of the VODC converter:

- Low gas contents, eg. hydrogen < 2 ppm
- Low carbon contents, C < 0.02 %
- Low sulphur contents, S < 0.007%
- Low other impurities
- Low emissions due to a closed process

4. The production of the personnel spheres

The spheres were cast in two halves, and these halves were joined together at the assembly stage using bolts. After pouring, the walls of the halves were 200 mm thick. The final wall thickness was 40 mm, so there was a large amount of additional material. Most of this additional material was removed (machined) from the inside, so that the potential areas of centreline shrinkage and impurities associated with thick sections during solidification could be removed. The result was a fine granular and clean convex-like outer surface. [5]

Heat treatment of the sphere's halves:

- Homogenising treatment 1200 °C
- Soaking treatment 850 °C
- Precipitation treatment 450 °C

With the help of the first two heat treatment phases, the chemical analysis and microstructure was homogenised. This treatment also allowed homogenisation of the remaining impurities, and the precipitation treatment gave the sphere the required hardness or toughness. This final microstructure was a low carbon slat martensitic steel. The hardness increased by 500-800 HV. [6]

Figure 2: The sphere's halves joined together with bolt joints. The remaining hemisphere is exhibited at the Lokomo steel foundry which is nowadays called TEVO Lokomo Ltd. [7]

Due to the high price of the vessels, it was also possible to produce large pieces out of this steel of similar quality as in small pieces. Because of the purity requirements of this steel, the converters and ladles had to be relined. Some further modifications of the normal casting methods were also required. On the other hand, the steel could easily be cast, which made testing significantly easier. The successful heat treatment also required separately cast test pieces. Thanks to the equipment available and knowhow of the Lokomo steel foundry these tests could be undertaken on site alongside the production.

Acknowledgements

The authors thank TEVO Lokomo Ltd., who participated to the study. In addition authors thank Association of Finnish Foundry Product Industries and Finnish Foundry Technical Association. Special thanks are directed to Mr. Paavo Tennilä, former Director of Lokomo steel works for his excellent comments.

References

[1] The Finnish Foundry Journal Valimoviesti 1(2007)
[2] Mr. Tennilä, Paavo; M.Sc. (Mech.), interviews 2015
[3] TEVO Lokomo Ltd, archives
[4] Mr. Norvasto, Timo; M.Sc. (Mech.) TEVO Lokomo
[5] The Technical Magazine Metallitekniikka 10 (2003)
[6] Mr. Tennilä, Paavo; M.Sc. (Mech.), interviews 2015
[7] Writers own archives

Proceedings of the 72nd World Foundry Congress
21-25th May 2016, Nagoya, Japan

Entire Foundry Production "Management" to achieve "Quality Castings"

Hiromu Yamaguchi[1], Koichi Kato[1] and Nobuyuki Kawai [1]
[1] Sintokogio, Ltd, Toyokawa, Aichi, Japan

Foundries constantly have been improving to achieve "high quality", "high productivity", "saving energy", "near net shape casting", and "manpower saving". However, there has been little unitary data management system for entire foundry production. The authors have developed the entire foundry production "management" system to support the total operation.

The developed management system enables to grasp casting conditions by monitoring production data. For this system, various measuring devices are equipped in each process, and the production data is digitized, monitored and stored as integrated database. Then, the database is utilized unitary. In order to realize "quality castings", it is very important to grasp and feedback the operation condition data such as pouring, molding and green sand properties.

Moreover, this system is also able to cover many other purposes; for example to save consumption of electric power, water, and compressed air. The digitization of the energy consumption contributes to reduce total energy consumption. In addition, it also enables immediate response towards abnormal energy consumption.

The developed foundry production management system enable to build strong management foundation and produce high quality castings.

Keywords: *production management, digitization, quality casting, high productivity, saving energy and manpower*

1. Introduction

In recent years, foundry companies have required the advanced production management system with the decline in the labor force population and the aging of skilled engineer .

This presentation explains a developmed production management system to contribute the future foundry management.

2. The Contents of development

This development consists of two fields.

1) Management Support Software

- Production Management System

-Traceability Device / Monitoring System

2) Utility Total Monitor

- Electricity, water, compressed air

"Management Support Software" monitors information gotten individually such as cooling condition, molding line and sand properties and utilizes them as a database of production trace data in addition to the operation rate or production order.

"Utility Total Monitor" is the system to manage energy such as electric, water and compressed air. It contributes to reduce the total energy with digitizing the consumption at the points to be measured. Moreover, this system monitors the abnormality such as sand properties causing the casting defect. From the above, it realizes the preventive maintenance towards the abnormal energy consumption or the casting defect.

3. Utilization in foundry

3.1 Quality of casting

The efficient production of quality casting is one of the most important task for the foundry.

Production Management System, one of management support software, is able to support making high quality casting. Using Production Management System, various information of each mold is monitored and digitized such as pouring, cooling, molding condition, sand properties and so on.

An example is shown in Fig. 1.

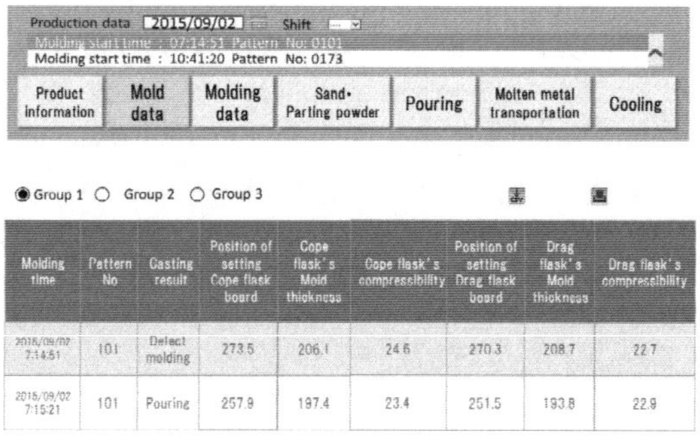

Fig. 1 Production Management System

It is integrated and utilized unitary as a database. The monitored data is used in various applications, for example, linking with data to eliminate casting defect. In addition, this system alerts the inspection department to the casting defect produced by the abnormal property sand, etc.

As mentioned above, this system provides the production system with fewer casting defects.

3.2 High productivity

The productivity is also important factor for foundry as much as the quality.

Production Management System also contributes to improve the productivity.

Figure 2 shows molding machine operation rate.

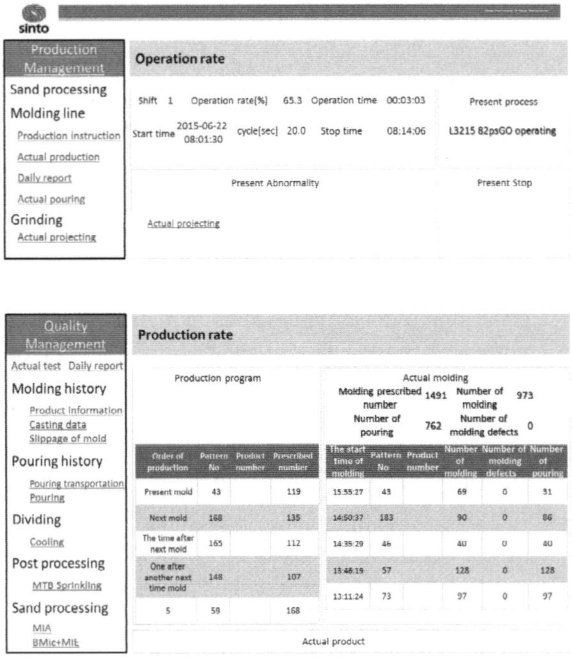

Fig. 2 Molding machine operation rate

Operator can grasp the current production condition with checking molding machine operation rate.

In addition, "Remote Monitor" is utilized for the management of entire foundry production system.

An example is shown in Fig. 3.

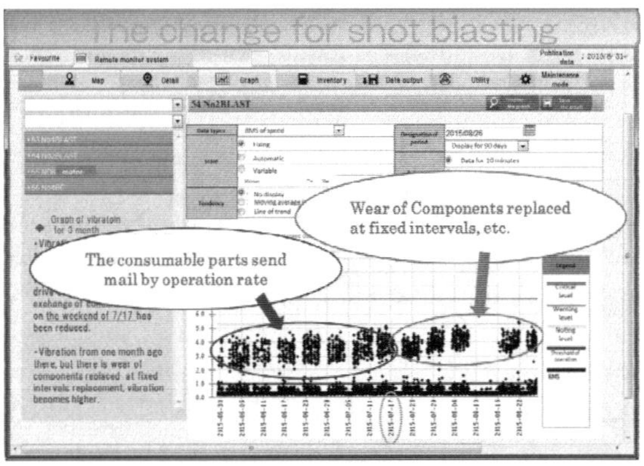

Fig. 3 Remote Monitor

This figure shows the monitoring of vibration at shot blasting. When the value of vibration exceeds the certain value, this monitor system shows a warning.

Therefore, foundry can continue stable operation with Remote Monitor which continuously monitors the operation condition to react immediately for the abnormality.

3.3 Energy saving

In foundry, large quantity of energy has been consumed. Digitization of energy consumption contributes to save total energy consumption.

Figure 4. shows digitized electric power consumption in foundry.

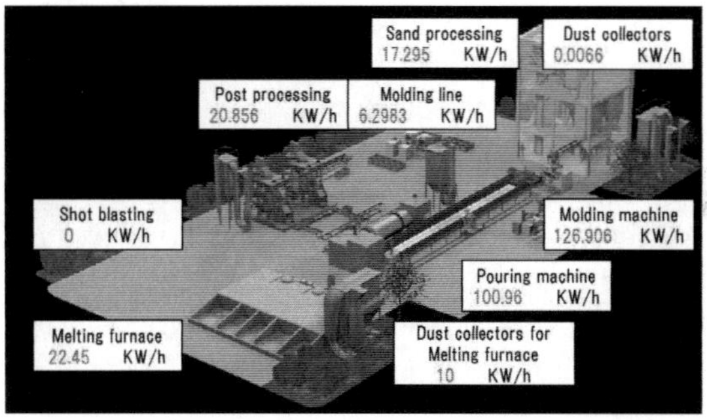

Fig. 4 Electric power consumption

In addition, it also enables immediate response towards abnormal energy consumption.

4. Conclusion

The developed foundry total management system is first step for "Industry 4.0". The developed system would contribute to reduce a casting defect dramatically and produce high quality castings with low energy.

Proceedings of the 72nd World Foundry Congress
21-25th May 2016, Nagoya, Japan

Fabrication of Fe-Based Self-Lubrication Composite Containing Graphite Particles by Centrifugal Mixed-Powder Method

Hisashi Sato[1], Kentaro Tanaka[1] and Yoshimi Watanabe[1]
[1] Graduate School of Engineering, Nagoya Institute of Technology,
Gokiso-cho, Showa-ku, Nagoya 466-8555, JAPAN

As self-lubrication composites, there are metal-based composite containing graphite particles. However, many of the self-lubrication composites have pure Cu or Cu alloy matrix, and investigation of Fe alloy-based self-lubrication composite is few. This is because that Fe alloys can form carbide easily with graphite. In this study, SUS430-based composites containing graphite particles are fabricated by centrifugal mixed-powder method. By using this method, the SUS430-based composites containing graphite particles are successfully obtained. Although very thin carbides are formed at interface between a graphite particle and SUS430 matrix, graphite particles are remained in the SUS430 matrix. Comparing with wear resistance of as-cast SUS430 without graphite particles, the SUS430-based composites have better wear resistance. In addition, frictional coefficient of the SUS430-based composites is much lower than that of the as-cast SUS430. Therefore, the centrifugal mixed-powder method is the effective processing for the fabrication of the SUS430-based composites containing graphite particles.

Keywords: *Self-lubrication composite, Stainless steel, Centrifugal casting, Centrifugal mixed-powder method, Wear.*

1. Introduction

As one of metal-based composites with high wear properties, self-lubrication composite has been proposed [1-3]. The self-lubrication composite contains solid lubrication particles in metallic matrix. Kováčika *et al.* have made Cu-based composite containing graphite particles by sintering process [2]. According to their study, dispersion of the graphite particles in pure Cu matrix decreases frictional coefficient and improves wear resistance. Also, it has been reported that composites fabricated by sintering of graphite particles and Cu alloys matrix powder have high wear resistance [1]. However, strength of these self-lubrication composites is relatively low, because graphite particles with low strength are homogeneously distributed in the matrix. To overcome this problem, parts of authors have recently developed the Cu-based composite containing graphite particles by using centrifugal mixed-powder method [3]. Since the centrifugal mixed-powder method can distribute graphite particles only around the sliding part, this composite has high strength comparing with the Cu-based composite fabricated by sintering process with homogeneously distribution of graphite particles. Hence, if the self-lubrication composite would be applied for bearing, the self-lubrication composite by the centrifugal mixed-powder method is suitable.

Here, considering about materials used for bearing, many of bearings are made from Fe alloy. However, matrix metal of the self-lubrication composites developed in the previous studies is not Fe alloys but Cu or Cu alloys [1-3]. This would be because that Fe alloys easily form carbide with graphite particles. If the composite can be cooled down before formation of carbide, it is expected that Fe alloy-based composites containing graphite particles can be obtained.

In this study, Fe alloy-based composite containing graphite particles is developed by the centrifugal-mixed powder method. For matrix of this composite, SUS430 ferritic stainless steel is used because this stainless steel is stable under high temperature. Using the fabricated Fe-based composites, its wear properties are evaluated.

2. Experimental Procedure

Mixed-powders of SUS430 powder (50-150 μm) and graphite powder (50 μm) were prepared. The prepared mixed-powders were SUS430-35vol.% graphite and SUS430-50vol.%graphite. Using these mixed-powders, SUS430-based composites are fabricated by the centrifugal mixed-powder method. The centrifugal mixed-powder method is the centrifugal casting combined with powder metallurgy [4]. At first, the SUS430-graphirte mixed-powder is inserted into the mold. Then, the mold with the mixed-powder is set into the vacuum centrifugal casting machine and subsequently the SUS430 ingot was melted by melting furnace. After that, the molten SUS430 is poured into the mold by inducing centrifugal force, and the SUS430 powder is melted by heat of the molten SUS430. Finally, SUS430-based composites containing graphite particles are obtained. Details of the casting condition are shown in Table 1. Using fabricated composites, wear tests were carried out under reciprocal movement using ball-on-disc type wear machine. The counter ball was bearing steel (SUJ2) with 5 mm in diameter. Load and sliding distance were 4.9 N and 288 m, respectively. After the wear tests, wear resistance and frictional coefficient were measured. The wear resistances were evaluated by measuring cross-sectional area of worn groove formed on the composites. Also, microstructural observations were performed by using scanning electron microscopy (SEM) and energy dispersive X-ray spectroscopy (EDX).

Table 1 Casting conditions of composites.

Name of composites	Weight		Centrifugal force	Melting temperature
	Mixed -powder	SUS430 ingot		
SUS430-35vol.%Gr	9.5 g	95 g	80 *G*	1600 ℃
SUS430-50vol.%Gr	7.5 g			

3. Results and discussion

Figures 1 (a) and (b) are SEM backscatter compositional images of SUS430-35vol.%Gr and SUS430-50vol.%Gr composites, respectively. As can be seen in Fig. 1, graphite particles are successfully distributed in SUS430 matrix for the both composites. Although very thin carbide is formed at interface between a graphite particle and SUS430 matrix, most part of the graphite particle is un-reacted. Frictional coefficient and cross-sectional area of worn groove for the both composites are presented in Fig. 2. As reference, wear properties of as-cast SUS430 are also shown in this figure. The cross-sectional area of worn groove formed on composite is used to evaluate wear resistance. Larger worn groove means lower wear resistance. Frictional coefficient is decreased by dispersion of graphite particles. In addition, wear resistances of the composites are drastically improved by the addition of graphite particles. From these

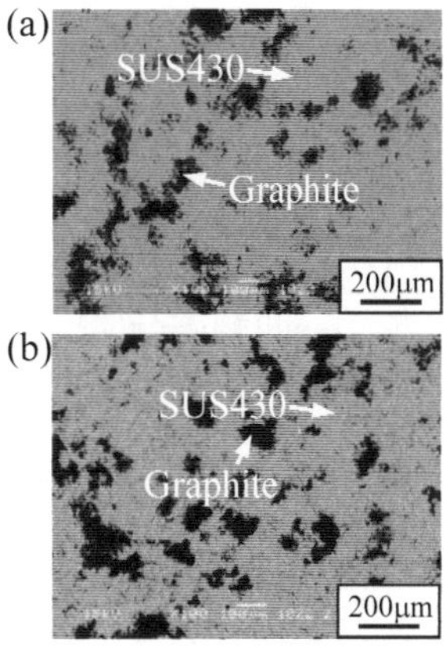

Fig. 1 SEM backscatter compositional images of (a)SUS430-35vol.%Gr and (b)SUS430-50vol.%Gr.

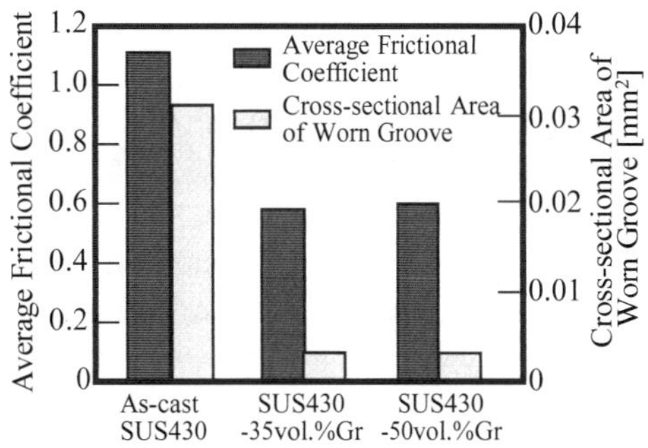

Fig. 2 Wear properties of SUS430-based composites.

results, it is found that the centrifugal mixed-powder method is effective processing of Fe alloy-based composites containing graphite particles.

Acknowledgements

This work was supported from "Adaptable and Seamless Technology Transfer Program Through Target-driven R&D (AS262Z00286K)" by Japan Science and Technology Agency. This financial support was gratefully acknowledged.

References

[1] H. Kato, M. Takama, K. Washida, Y. Sasaki and S. Miyashita: J. Jpn. Soc. Powder Powder Metallurgy 50 (2003) 968-972.

[2] J. Kováčik, Š. Emmer, J. Bielek and L'ubomír Kelešic: Wear 265 (2008) 417-421.

[3] H. Sato, W. Wei, K. Oguri, M. Yamada and Y. Watanabe: Mater. Sci. Forum 783-786 (2014) 1579-1584.

[4] Y. Watanabe, Y. Inaguma , H. Sato and E. M.-Fujiwara: Materials 2 (2009) 2510-2525.

Metallic Fuel Slug prepared by a Modified Injection Casting Method for Reducing Volatilization

Hoon Song[1], Jong-Hwan Kim[2], Young-Mo Ko[3], Yoon-Myung Woo[3], Ki-Hwan Kim[3], and Chan-Bock Lee[3]
[1] Korea Atomic Energy Research Institute, 1045 Daedeokdaero, Yuseong, Daejeon, Korea
[2] Korea Atomic Energy Research Institute, 1045 Daedeokdaero, Yuseong, Daejeon, Korea
[3] Korea Atomic Energy Research Institute, 1045 Daedeokdaero, Yuseong, Daejeon, Korea

Metallic fuels, such as the U-Pu-Zr alloys, have been considered as a nuclear fuel for a sodium-cooled fast reactor (SFR) related to the closed fuel cycle for managing minor actinides and reducing the amount of highly radioactive spent nuclear fuels since the 1980s. Management of minor actinides (MA) became an important issue because direct disposal of the long-lived MA can be a long-term burden for a tentative repository up to several hundreds of thousand years. In order to prevent evaporation of volatile elements such as Am, one of the main MAs, and improve the quality of fuel slugs, U-Zr alloy fuel slugs were fabricated through a modified injection casting method instead of a vacuum injection casting method, which is a method of fabricating a metallic fuel slug conventionally by modifying the casting process conditions.

Keywords: *Metallic fuel, sodium-cooled fast reactor, injection casting, U-Zr alloys.*

1. Introduction

The fabrication technology of metal fuel for an SFR has been under development in Korea as a national nuclear R&D program since 2007 [1]. Although injection casting has been a well-established fabrication method for metal fuel for decades, Am addition to the metal fuel hampers conventional fuel fabrication processes because of the high vapor pressure of Am at the melting temperature of the uranium alloys. In this study, a fuel slug fabrication method has been introduced to develop an innovative fabrication process of metal fuel of an SFR for preventing the evaporation of volatile elements such as Am. Metal fuel slugs were fabricated using an improved injection casting method in KAERI. Volatile species can be retained through the use of a cover gas with over pressure, covered crucibles, and short cycle times under this fuel fabrication method, and melted under an inert atmosphere [2].

2. Experimental procedure

The feasibility of an improved injection casting method including melting under an inert atmosphere was evaluated in a small-size induction-melting furnace. For a preliminary test, pure copper was selected as a surrogate material, which has a melting temperature similar to uranium. After a surrogate fuel slug was generally soundly cast by the improved injection casting method under an inert atmosphere, fuel slugs in U-Zr alloy by improved casting method have been successfully fabricated in KAERI for the prevention of evaporation of volatile elements such as Am.

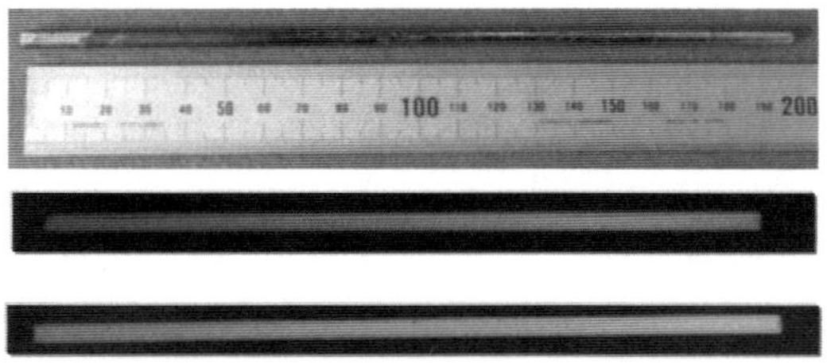

Fig. 1 Typical U-10Zr-5Mn fuel slug (upper), gamma-ray radiography at 0 degrees (middle) and gamma-ray radiography at 90 degrees (bottom) (#S13-06).

3. Experimental results and discussion

Through the experience of the surrogate fuel slug, U-10wt.%Zr-5wt.%Mn fuel slugs containing a volatile surrogate element, Mn, shown in Fig. 1, were soundly cast by improved injection casting for the prevention of evaporation of volatile elements such as Am under various atmospheres such as a vacuum state, reduced atmosphere, and an inert atmosphere. The gamma-ray radiography of the as-cast surrogate slug was performed to detect internal defects such as cracks and pores. The general appearance of the slug was smooth, and the diameter and length were 5.4 mm and about 200-250 mm, respectively. The density variations according to the location of the fuel slugs are shown in Fig. 2.

The melting and casting parameters of the fuel slug casting process such as the melting batch size, coating method, melting and casting temperature, and mold pre-heating temperature were investigated to obtain the sound fuel slugs.

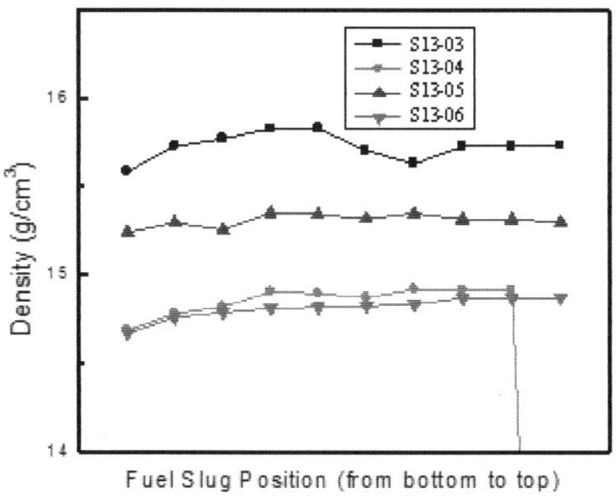

Fig. 2 Alloy density of U-10Zr fuel pins prepared by injection casting method.

The density of the fuel pins prepared using an injection casting method shows a different way in accordance with the casting process conditions. The U-10Zr (#S13-03) fuel pins in which the surrogate Mn for volatile radioactive constituents was not added showed the highest density because of the lighter density of Mn elements, but the others such as U-10Zr-Mn (#S13-04, #S13-05, #S13-06) by adding an equal amount of Mn elements such as 5wt,% showed a difference owing to the conditions of the casting process. The density of the fuel pins in an inert atmosphere (#S13-06) showed the lowest value, which means the lowest evaporation because of the consumption of light elements. However, the U-10Zr-Mn fuel pins (#S13-04) for a reduced atmosphere in which a vacuum was applied for a short time showed a higher density even when the volatile amount of Mn was small compared to the other casting method. The U-10Zr-Mn fuel pins (#S13-05) in a vacuum state showed the highest density because of a high evaporation.

The alloy compositions of U-10 wt.%Zr-5 wt.%Mn fuel slugs are shown representatively in Table 1. From the variations of Mn compositions, the same trend was observed as the density results. Based on these results there is a high level of confidence that Am losses will also be effectively controlled by application of a modest amount of overpressure.

These limited experiment results show that the Mn was not volatilized and conserved in the inert gas conditions compared to the vacuum condition. The volatility of Mn can be controlled by changing the casting process, and minimal Mn (and Am) loss is possible. An improved casting method under an inert atmosphere is more effective in the prevention of vaporization than casting under a vacuum and reduced atmosphere. In addition, improved casting under a reduced atmosphere shows a considerable effect on the prevention of vaporization.

Table 1 Alloy compositions of U-10Zr fuel slugs (#S13-06).

		U (wt.%)	Zr (wt.%)	Mn (wt.%)	C (ppm)	O (ppm)	N (ppm)
Upper	1	85.2	10.7	5.10	70	480	10
	2	84.9	10.7	4.99	70	500	10
(g/cm^3)	3	84.7	10.6	5.11	60	510	10
Middle	1	86.0	10.5	4.96	50	610	10
(g/cm^3)	2	84.0	11.0	5.08	70	340	10
	3	85.5	10.5	5.02	60	770	10
Bottom	1	84.4	10.5	5.04	60	1190	10
(g/cm^3)	2	84.1	10.8	5.02	90	820	10

4. Conclusion

In order to prevent the evaporation of volatile elements such as Am, a fabrication method of metal fuel slugs was examined using an improved injection casting method. U-10wt.%Zr and U-10wt.%Zr-5wt.%Mn fuel slugs were soundly fabricated for the retention of the volatile surrogate element. Based on these results, there is a high level of confidence that Am losses will also be effectively controlled by the application of a modest amount of overpressure. The Mn element was almost recovered with the prevention of evaporation. It was seen that the losses of these volatile elements such as Am can be effectively controlled to below the detectable levels using modest argon pressure.

References

[1] Lee, C.T., Oh, S.J., Ryu, H.J., Kim, K.H., Lee, Y.S., Kim, S.K., Jang, S.J., Woo, Y.M., Ko, Y.M., Lee, C.B., "Casting Technology Development for SFR Metallic Fuel", Global-2009, Paris, (September, 2009).

[2] C.L. Trybus, "Injection Casting of U-Zr-Mn, Suurogate Alloy for U-Pu-Zr-Am-Np", J. Nucl. Mater., 305(224), 1995.

Proceedings of the 72nd World Foundry Congress
21-25th May 2016, Nagoya, Japan

Evaluation of casting defect and analysis on mechanical properties for die cast by using 3D computed tomography

In-Sung Cho* and Si-Young Kwak
e-Design R&D Center, Korea Institute of Industrial Technology, Korea
156 Getpearl-ro, Yeonsu-gu, Incheon, 21999, Korea

In the present study, statistical evaluation of casting defects in die cast and effect of the defect on physical properties such as yield strength, tensile strength, and elongation is investigated. Die Casting parts are widely used in automotive and electronics industry. Defects in the die cast is inevitable, and influence the mechanical properties of the parts. Nowadays, computed tomography (CT) was widely used to investigate the defect of the die cast, but there are still no standardization for CT analysis. In the present study, test specimen with various casting defects were manufactured by using die casting process, and all specimens were CT scanned and mechanically tested. Statistical analysis was carried out for the relation analysis between the characteristics of casting defects in the die cast and mechanical properties.

Keywords: *computed tomography, casting defects, die casting, physical properties*

1. Introduction

Die Casting parts are widely used in automotive and electronics industry. Defects in the diecast is inevitable, and influence the mechanical properties of the parts. Usually the metal products have the defects during manufacturing and are inevitable. In the NDT system, ability for evaluate the casting defects such as cracks and pores is also very important. Nowadays, computed tomography (CT) was widely used to investigate the defect of the die cast, but there are still no standardization for CT analysis. – The increasing use of computed tomography (CT) as a diagnostic tool creates the need for an efficient means of evaluating the performance of the CT systems now in use.

2. Experiment

2.1 Diecasting Experiment and preparation of specimen

In order to make specimens for the die cast defects, control of defects in the die cast is necessary. Generally defects in the die cast is affected by the existense of vacuum channel and thickness of the die cast. A mold die was prepared. The mold equipped vacuum block at the top side, and 6 specimen cavities with various thickness from 4mm to 14mm. Diecast specimens were produced by using the mold, as shown in Fig. 1. The materials used were Silafont 55 and ALDC12.

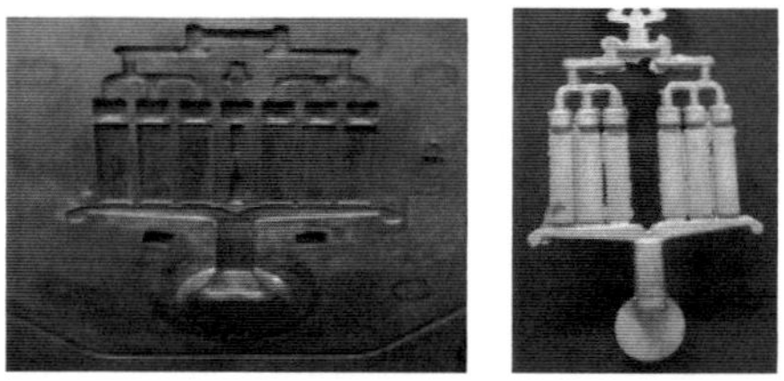

Fig. 1 Diecast mold for specimen preparation.

The die cast specimens were inspected by selected and machined by 3mm thickness test specimen with regards to the ASTM standard[1]. The CT scanned data were analysed by using VX3D software. VX3D has a function of labelling the object (Fig. 2). It is very useful when multiple specimen are scanned for one time. VX3D also has a fast analysis module for porosity detection by using GPU in the NVIDIA graphics card.

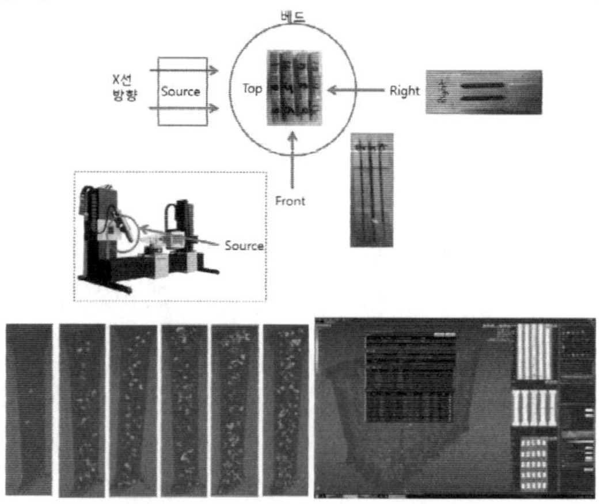

Fig. 2 CT scan for test specimen.

3. Results and Discussion

3.1 Silafont 55

Fig. 3 shows the mechanical properties with repect to the defect grades. 48 test specimens were prepared. The defect grades is categorized by number and size of porosity defects. Determination of defect grades was refenenced and modified by using the standard[2]. Grade 1 has small size of the defects (<0.5 mm) and Grade 4 has large size (>1mm) of porosity. With increase the grade number, decrase of yield strnegth, tensile trength, elongation were found.

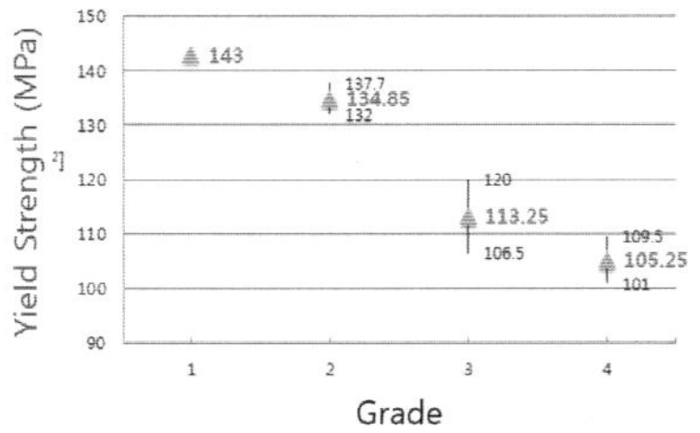

(a) Yield strength

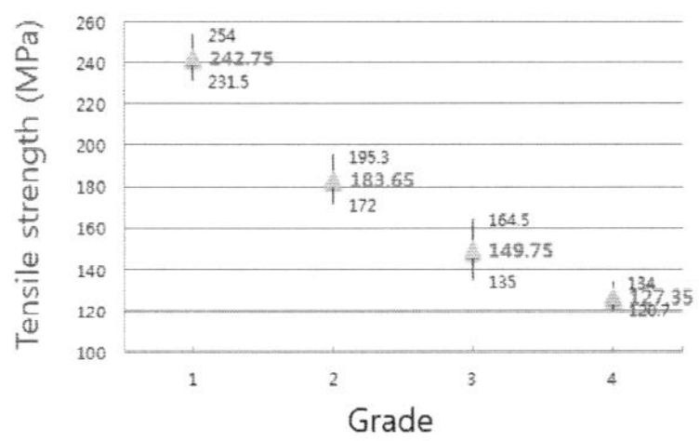

(b) Tensile strength

Fig. 3 Mechnical properties with Grade number of Silafont 55.

3.2 ALDC 12

Fig. 4 shows the mechanical properties with defect grades of ALDC12. 216 test specimens were used and analysed. With increase of the defect grade number, decrease of mechanical properties such as tensile strength, yield strength and elongation decrease. However, elasticity has no dependency with the defect grades. Compared to the case of Silafont 55, wider range of mechnical properties of each grades were found. It is considered that other defects than porosity may affect the mechanical properties.

Fig. 5 shows the in the specimen. CT scan Figure 10 shows the relation of mechanical properties with total porosity volume. Both analyses show same result. Therefore, CT volume analysis is considered to be used and replaced by density measurement experiment.

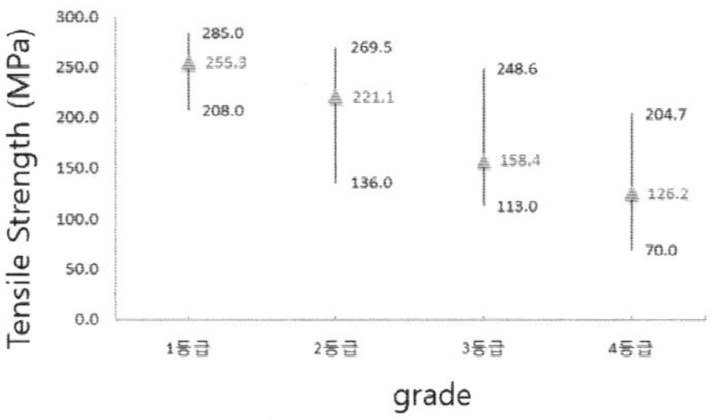

(a) Tensile strength

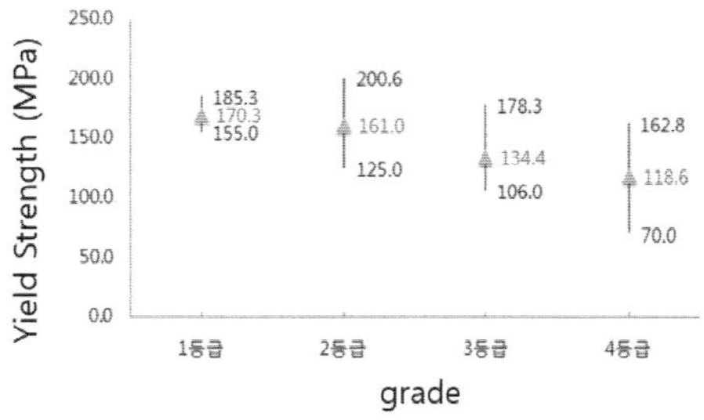

(c) Yield strength

Fig. 4 Mechanical properties with defect grades of ALDC12 test specimens.

	(a) Density measurement	(b) CT volume analysis
Tensile Strength	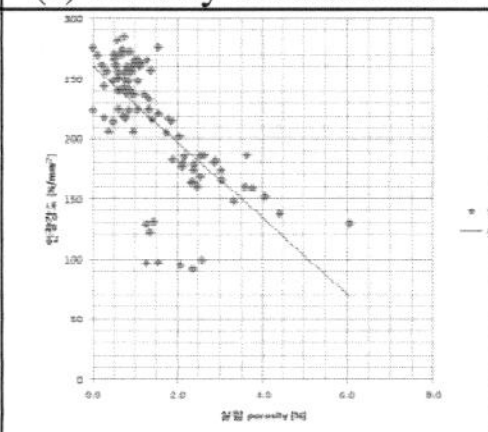	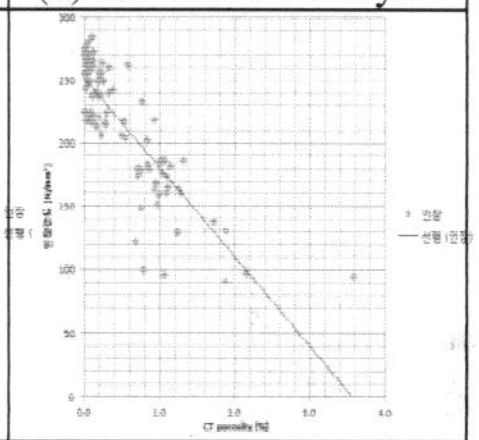
Yield Strngth	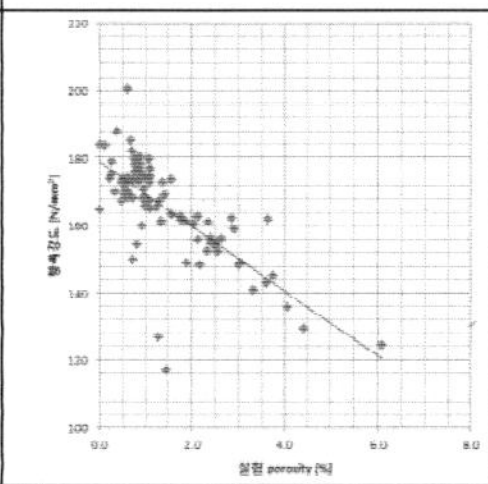	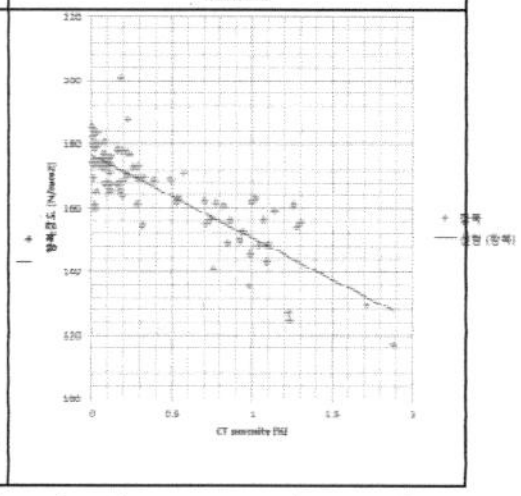

Fig. 5: Relationship of mechanical properties with total porosity volume by density measurement and CT volume analysis.

REFERENCES

[1] ASTM B557M, p. 2.
[2] ASTM E 93, p.3.
[3] http://www.3dii.kr

Proceedings of the 72nd World Foundry Congress
21-25th May 2016, Nagoya, Japan

Effects of nitriding and nitro-carburizing on the fatigue properties of ductile cast iron

Tohru NOBUKI[1], Yusuke KAWASAKI[2], Minoru HATATE[1], Akihiko IKUTA[1] and Naoji HAMASAKA[3]
[1] Faculty of Engineering, KINKI University, 1 Takaya-Umenobe Higashi-Hiroshima, Hiroshima, JAPAN
[2] Graduate School Student, KINKI University, 1 Takaya-Umenobe Higashi-Hiroshima, Hiroshima, JAPAN
[3] Manufacturing Engineering Department Center, KOMATSU Ltd., 3-3-1 Ueno, Hirakata, Osaka, JAPAN

The effects of nitriding or nitro-carburizing treatments on rotating-bending fatigue properties were investigated on five kinds of pearlitic ductile cast iron samples. We produced containing of V, Al-Cr and Al-V alloyed ductile iron samples. Tensile and hardness characteristics of the nitro-carburizes samples compared to those of the nitrided ones and as-cast ductile irons without alloying element. The iron nitride formed on the surfaces of the nitride samples and the iron complex nitrides formed on the surfaces of the nitro-carburized ones. As a result, the fatigue lives existing in the higher stress range from 500 to 650 MPa was found to be longer in the order of as-cast, nitro-carburized and nitrided samples. However, the fatigue limit of the lower stress zone existing at 410 to 450 MPa and became no longer significant differences on endure limit by differences of nitiriding method.

Keywords: nitriding, nitro-carburizing, fatigue properties, ductile cast iron

1. Introduction

Nitrided ductile cast iron is used for precisely controlled components by oil pressure with complicated shapes for construction machineries because of not only its mechanical characteristics but also its superior machinability and castability. Recently, to changing green foundry, the pollution-free liquid nitro-carburization method (called Multinite) has been developed[1]

The nitriding process and the nitro-carburizing process are an industrial surface treatment process employed for the improvement of wear/friction and fatigue resistance of ductile cast irons, since it is associated with high production accuracy required in high precision controlled alloying elements. However, the fatigue properties of nitrided and nitro-carburized ductile cast iron with some alloying elements to control surface hardened layer are still under investigation. In this study, samples of pearlitic ductile cast iron (FCD700) with alloying elements such as Al, V and Cr were produced by nitriding and nitro-carburizing. These surface-improved ductile cast iron samples were observed by optical and EPMA elemental analysis and also done the rotating-bending fatigue test, to comparing and to optimize the chemical compositions for fatigue resistance.

2. Experimental Procedures

2.1 Materials

Ductile cast iron sample (FCD700) without alloying element is produced from high purity pig iron, mild steel, pure cupper, Fe-Si alloy and Fe-Mn alloy using a high frequency induction furnace. For the production of Cr 0.1%, V 0.1% and Al 0.1% samples, Fe-62%Cr alloy, Fe-83%V alloy and the pure aluminum are added individually at 1723 K in the same manner. The chemical composition of samples is shown in Table 1.

The gas nitriding processing was done in the industrial furnace used on the production site[2]. The nitrocarburizing processing was done in salt bath nitrocarburizing by the pollution-free liquid carburization furnace of 633K for holding time of 1.8ks and 833K for holding time of 18 ks.

2.2 Hardness test and tensile test

To investigate the hardness distribution from the surface of test piece, micro-Vickers hardness test was done at the load of 0.49 N. Tensile test was using Instron type universal testing machine, testing condition was crosshead speed of 8.3×10^{-3} mm/s and testing temperature was room temperature (R.T.).

2.3 Fatigue test

The fatigue testing was carried out using the test piece with 8mm diameter and the rotating-bending

Table 1 Chemical composition of samples (mass%).

	Sample name	C	Si	Mn	Cu	Cr	V	Al
A	0.1%Al, 0.1%V	3.50	2.59	0.38	0.60	-	0.07	0.11
B	0.1%Al, 0.1%Cr	3.55	2.59	0.38	0.61	0.12	-	0.11
C	0.1%Al	3.53	2.48	0.37	0.60	-	-	0.10
D	0.1%V	3.57	2.63	0.39	0.60	-	0.08	-
	FCD	3.39	2.68	0.37	0.58	-	-	-

fatigue testing machine, that testing condition the rotating speed of 3400 rpm and performed at R. T.

3. Results and discussion

Figure 1 shows the microstructure on near surfaces of nitride and nitro-carburizing samples. The matrix microstructure of all specimens are consists of pearlite matrix and ferrite structure around the spheroidal graphite. The white about 10μm thickness layer of specimen surface have observed which occurred from nitriding or nitro-carburizing heat treatment.

Table 2 shows the tensile strength and elongation of samples. These obtained values of surface-improved samples show lower value comparing to FCD 700 as cast condition. This reason considering that the heat affect to the sample surface by nitriding heat treatment, but occurred the hardened thin layer by nitriding or nitro-carburizing heat treatment.

Figure 2 shows the mapping of N distribution near surfaces of nitride and nitro-carburized samples. The nitrogen element was strongly distributed near the 10 μm from the specimen surfaces, considering forms complex nitrides. From the XRD analysis, the nitride layer consists of mainly nitrides of γ'-Fe_4N for nitriding samples, γ'-Fe_4N and ε-$Fe_{2-3}N$ for nitro-carburizing samples. The comparing to the thickness of the nitride layer and nitrogen diffusion depth of nitrided samples, the nitride layer of the nitriding sample seems more deeply nitrogen distribution comparing with nitro-carburizing one.

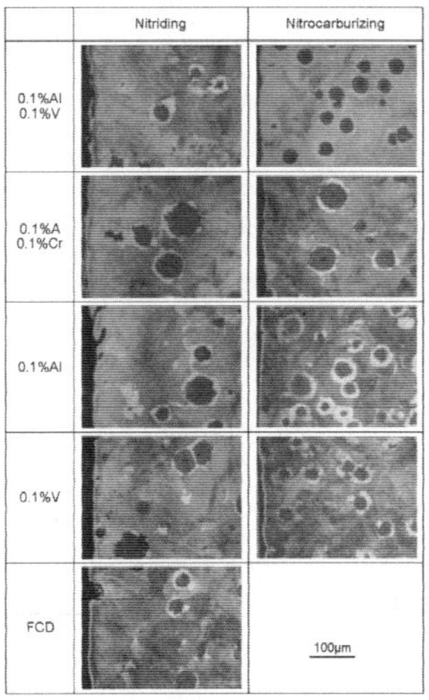

Fig. 1 Microstructure on near surfaces of samples.

Table 2 Tensile strength and elongation of samples.

	Sample name	Nitriding		Nitrocarburizing	
		Tensile strength [MPa]	Elongation [%]	Tensile strength [MPa]	Elongation [%]
A	0.1%Al, 0.1%V	620	1.1	656	2.3
B	0.1%Al, 0.1%Cr	650	1.2	692	2.4
C	0.1%Al	585	1.0	644	2.4
D	0.1%V	610	1.4	632	2.0

Figure 3 shows the S-N curves of as-cast, nitride and nitro-curburizied samples. The fatigue properties of nitride samples as indicated in outline marker shifted up to high stress side comparing to as-cast FCD700 sample as indicated double circle marker. The fatigue strength and endure limit of nitro-carburizied sample exhibits the intermediate region between nitrided sample and as cast sample.

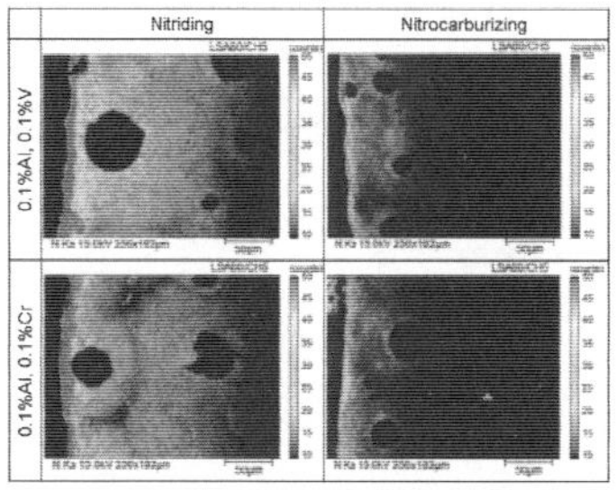

Fig. 2 Mapping of nitrogen distribution near surfaces of nitride and nitro-carburized samples.

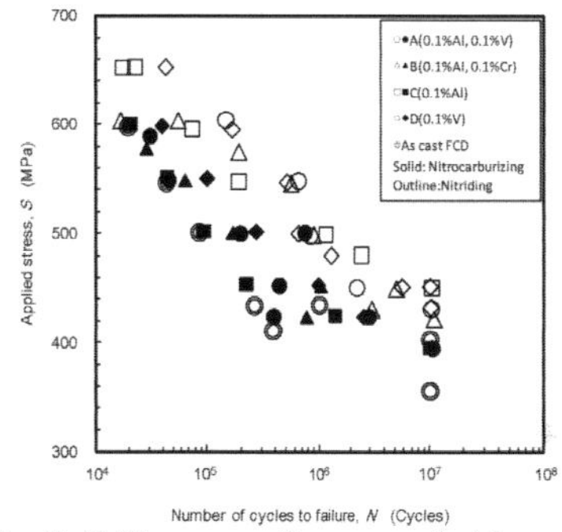

Fig. 3 S-N curves of as-cast, nitride and nitro-carburized samples.

4. Conclusions

1. The gas nitriding treatment is given to the ductile cast iron, the layer of γ'-Fe_4N is formed on the surface. And the nitro-carburizing treatment one is γ'-Fe_4N and ε-$Fe_{2-3}N$ phase formed.
2. The nitrogen element is diffused to matrix more deeply by using nitriding than nitro-carburizing heat treatment.
3. Fatigue property of nitride samples improves more effective than not only as-cast but also nitro-carburizing, because of owing larger width of hardened nitrides layer.

References

[1] M. Hatate, T. Nobuki, A. Ikuta, N. Hamasaka and Y. Kawasaki: J.JFS, Vol.87(2015) pp.382-387.

[2] Y. Kawasaki et al: Proc. 71st WFC, paper no. P-11 (2014, Bilbao)

Proceedings of the 72nd World Foundry Congress
21-25th May 2016, Nagoya, Japan

Optimal design of pure lanthanum based MgFeSi alloy for high performance ductile iron ladle treatment

Gunnar Maltsev Hansen[1], Chris Ecob[2], Joe Zhou[3]
[1] Elkem Foundry R&D, Kristiansand, Norway
[2] Elkem Foundry Technical Customer Service, Birmingham, UK
[3] Elkem Foundry Technical Customer Service, Shanghai, China

Lanthanum containing MgFeSi has been used in ductile iron production for 40 years. First solely for "in the mould" processes, and then in the last 12 years it has seen increased use as a ladle alloy. There are several reasons why these alloys have gained popularity, such as shrinkage reduction, improved microstructure properties, and a lower level of La in the alloy compared to the total rare earth (RE) level from mischmetal based alloys. During the rare earth price crisis in 2010-2012, this turned into an appreciable cost advantage for the lanthanum based alloys. This article will explore some of the benefits of using La as the sole source of rare earth in the MgFeSi.

Keywords: *Ductile iron, MgFeSi-alloys, Lanthanum.*

1. Introduction

In 2004, a paper by Dr. Torbjørn Skaland was presented to the WFC, winning the best paper award. The title of the work was "A new method for chill and shrinkage control in ladle treated ductile iron".

The paper was a study of the effects of La and Ce based MgFeSi in ladle treatment [1] and it clearly demonstrated that La based alloys had a far wider application than in mould.

In the 12 years that have passed since WFC 2004, there has been a profound shift, where La based MgFeSi have gone from a niche product to becoming the basis of over 1.5 million tons of ductile iron production based on current market figures.

2. Iron and Rare Earth (RE) elements

After RE lost to Mg as the main route for nodularizing ductile irons in the 1940's, many benefits of RE have since been discovered, and these elements are now an essential part of ductile iron production. Controlled levels were shown to reduce edge carbides and increase nodule count [2,3,4]. Trace element control is deemed crucial for making good ductile iron, especially in present times as steel scrap quality deteriorates and RE has been found to neutralize the effect of deleterious trace elements, such as Pb, Sb, Bi, As [5].

The first use of La as the sole source of RE in MgFeSi dates back to the early days of in the mould process, where reduced shrinkage tendency was demonstrated by the introduction of pure La MgFeSi [6]. Since their introduction as ladle alloys 12 years ago they have grown in market shares based upon industry success with reduced shrinkage tendency, improved mechanical properties and improved machining surface finish.

So what is special about La? It is part of the Rare Earth family, of similar molecular size and reactivity as Ce. As other RE's, it is a strong deoxidant and desulphuriser, helping Mg in tying up S and O. It gives an improved inoculation effect, higher nodule counts and reduced carbide formation. It shares the particular ability of other RE's to tie up subversive elements [5].

However, it has shown a tendency to give a characteristic nodule distribution, linked to its efficiency at reducing and eliminating shrinkage and chill [1]. The use of lanthanum in the nodulariser has been seen to give a skewed nodule distribution with small late forming nodules adding a late expansion effect to reduce shrinkage porosity.

3. Optimal La level in alloys

The original development work was carried out at a MgFeSi addition rate of 1.5% and the optimum chill reduction, nodule count and thermal analysis properties were found at an addition of 0.5% La to the nodulariser. Since that time, there has been a significant industry shift towards lower MgFeSi addition rates and this has been driven by improvements in ladle design, increases in Mg recovery, the desire to reduce costs and the RE crisis previously mentioned. The average MgFeSi addition rate in ladle practice is now estimated to be around 1.2%. Thus, in many cases the total ppm of La has been reduced and it is pertinent to determine if this La level in the MgFeSi is still valid.

Below is the result of three studies of La concentration in MgFeSi and the effect on nodule count nodule count [1,10,11].

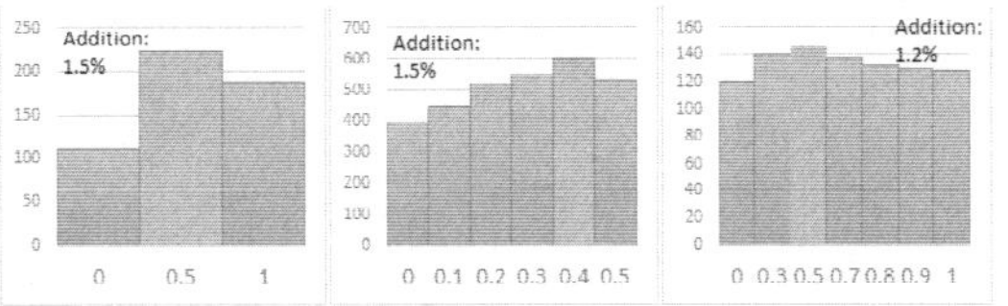

Fig. 1: Optimum. La in MgFeSi.
(%La in MgFeSi on x-axis, nodule count on y-axis No/mm²)

Adjusted for addition rate, studies 2[10] and 3[11] show the exact same peak level, corresponding to a total addition of *0.006% La to the iron via the alloy*. Also noted during the experimental studies were significant changes measured using thermal analysis. The most nucleated irons were found at the 0.5% La in the MgFeSi, lower nucleation being seen at both higher and lower La levels.

There are, however, reasons to be slightly cautious about using the same level for any process. Given the importance of La in balancing detrimental elements, the quality of the base iron should always be taken into consideration. At a minimum it seems reasonable that a stoichiometric balance should be obtained, which can be estimated as *RE/Subversive Elements=0.5*. (Sub.Elem. = Pb, As, Bi and Sb) [12]. A total concentration of Pb, As, Bi, Sb of 0.008 would require a La addition of 0.004%.

4. Case study: Shrinkage in steering knuckle

An automotive foundry had been having shrinkage issues in a steering knuckle. The foundry was using a 1.3% addition of MgFeSi (5.5%Mg, 1.5%RE). After treatment, the iron goes into a pouring ladle, with a 0.4% addition of a CaBa inoculant used.

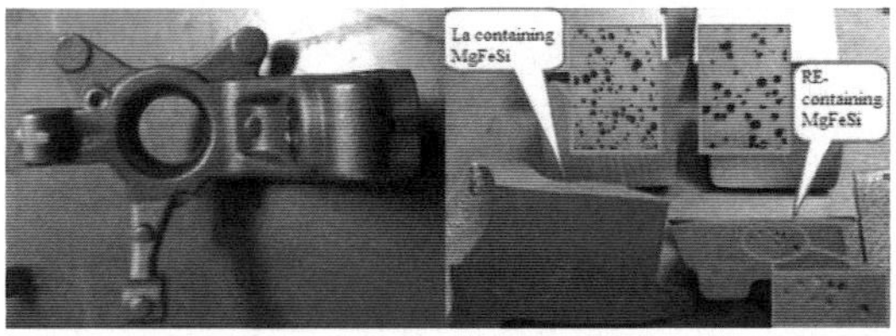

Fig. 2: Steering knuckle (L), result comparison (R)

The foundry then changed the MgFeSi to one containing 5.9%Mg and 0.5%La, and reduced the addition rate to 1.2% to aim for the same residual Mg. The result was better microstructure and a strong reduction in shrinkage tendency as shown in fig 2.

5. Conclusion

Much has changed in the decade that has passed since the original presentation and foundries have improved their processes. However, the original observations about the ideal levels and the advantages of lanthanum based MgFeSi still holds true.

Foundrymen still need to be aware that decreasing addition rates, less pure raw materials and higher amount of trace elements, both will increase the need for La needed in the alloy. As always flexibility in alloy design is important.

When changing from a RE based MgFeSi it is better not to overcompensate the reduction in RE when changing to La based grades. By having La levels in the alloy above optimum, many of the observed benefits regarding shrinkage and structure will be reduced, in addition to causing a higher alloy cost.

Overall, when applied correctly, there are strong benefits from La-based MgFeSi, therefore there is no reason why these grades should not continue to increase in popularity in the years to come.

References:

[1] T. Skaland, *A New Method for Chill and Shrinkage Control in Ladle Treated Ductile Iron*, WFC, (2004)
[2] M.J Lalich, *Effective Use of Rare Earths..., AFS Transac, 74-111*
[3] A.S Amin, et.al, "Cerium and Rare Earths in Ductile Cast Irons", *AFS Transactions 78-154*
[4] W.A.Henning, *Commercial experience with 5% magnesium-ferrosilicon alloys...*, AFS Transactions 83-01
[5] H. Morrogh, *Influence of some residual elements and their neutralization...* AFS Transactions 52-60
[6] Dunks, C.M., *In-the-mold Worldwide – Today and Tomorrow*, AFS Transaction 82-83
[7] D.M. Stefanescu, et.al, *Effect of lanthanum and cerium on the structure of eutectic cast iron*, AFS transactions, #81
[8] E. Døli, *Ce-La fading*, Elkem internal report, 2015
[9] M.Onsøien, T. Skaland, *Effects of cerium and lanthanum on the graphite formation...*, DIS, 1998.
[10] D.S.White, *Studies of Nuclei in Ductile Iron*, presentation, DIS conference, Vancouver, (2010)
[11] Internal report, Elkem (2015)
[12] G.M.Hansen, et.al, *Evolving foundry strategies and alloy design to counter the rare earth price...*, AFS, (2012)

Proceedings of the 72nd World Foundry Congress
21-25th May 2016, Nagoya, Japan

Effect of Si Content on the Ferrite-to-Austenite Transformation during Austenitizing in Spheroidal Graphite Cast Iron

Tatsuya Tokunaga[1], Tatsuki Furukawa[2], Nozomu Oda[3], Hidenori Era[1]
[1] Department of Materials Science and Engineering, Kyushu Institute of Technology, Kitakyushu 804-8550, Japan
[2] Graduate School of Engineering, Kyushu Institute of Technology, Now at Saveway Japan Co., Ltd., Kitakyushu 805-0012, Japan
[3] Graduate School of Engineering, Kyushu Institute of Technology, Kitakyushu 805-8550, Japan

The effect of Si content on the ferrite-to-austenite transformation during austenitization in spheroidal graphite cast iron with a ferrite matrix was investigated. In this study, ferritic ductile cast irons with two different Si contents, 2.3 mass%Si (JIS-FCD400) and 3.8 mass%Si (SSFDI), were used. Microstructural observation revealed that the rate of the ferrite-to-austenite transformation for SSFDI was slower than for FCD400. In addition to the experiments, a simulation of the austenitization was performed for FCD and SSFDI using the DICTRA software package employing kinetic and thermodynamic databases. The simulations reproduced the experimental results on the austenitization behavior for both FCD and SSFDI well.

Keywords: *spheroidal graphite cast iron, transformation, silicon, segregation, diffusion.*

1. Introduction

Spheroidal graphite cast iron is widely used for various structural components because of its high strength and ductility compared to flake graphite cast iron. Recently, solution-strengthened ferritic ductile cast iron (SSFDI) has been developed by increasing the Si content in the matrix [1]. The austempering process is one of the strengthening methods for spheroidal graphite cast iron, and the rate of austenitization is controlled by carbon diffusion from the graphite nodules [2]. However, microsegregation of constituent elements occurs in the matrix, and this may affect the austenitization of the cast iron. In this study, the microsegregation in spheroidal graphite cast irons with varying Si content was evaluated. In addition, the effect of Si content on the ferrite-to-austenite transformation during austenitization in spheroidal graphite cast iron with a ferrite matrix was investigated.

2. Methods

2.1 Experimental procedures

The materials used in this study were two types of spheroidal graphite cast iron with different Si content, namely 2.3 mass% (JIS-FCD400) and 3.8 mass% (EN-GJS-500-14), hereafter referred as to FCD and SSFDI, respectively. The chemical composition and number of graphite nodules of the two cast irons are listed in Table 1. The samples were immersed in molten Al alloy baths heated at temperatures between 850 °C and 900 °C and held at these temperatures for various periods, after which the samples were quenched in water to interrupt the ferrite-to-austenite transformation. Microstructural observation was carried out using optical microscopy and the fraction of martensite obtained after quenching was evaluated as the fraction transformed to austenite during austenitization. The microsegregation of constituent elements was investigated by means of electron probe micro analysis (EPMA).

Table 1 Chemical composition (mass%) and number of graphite nodules (mm^{-2}) of specimens.

Samples	C	Si	Mn	P	S	Cu	Mg	Graphite nodules
FCD	3.6	2.3	0.2	0.02	0.01	0.02	0.04	397
SSFDI	3.2	3.8	0.3	0.02	0.01	0.02	0.04	372

2.2 Calculation procedure

To evaluate the diffusion behavior of carbon and other alloying elements, the simulations were performed using the DICTRA software package [3] employing the mobility database MOB2 and the thermodynamic database SGTE SSOL4. The geometrical model used in the DICTRA simulations was a closed system with one spherical cell with a graphite nodule at the center surrounded by the ferrite matrix. Austenite was assumed to form between the graphite nodule and the ferrite matrix. The size of the graphite nodule and ferrite matrix and the composition of the ferrite were taken from the present experimental data. The simulation was performed for the austenitizing temperatures between 850 °C and 900 °C and was stopped at the time step of 3600 s (1 h).

3. Results and discussion

EPMA showed that both the as-cast FCD and SSFDI samples exhibited microsegregation of Si and Mn near the graphite nodules and at the eutectic cell boundary, respectively. The progress of the isothermal austenitic transformation at 850 °C for the FCD and SSFDI specimens based on the microstructural observation is plotted with symbols in Fig. 1, together with fitted curves with the Johnson–Mehl–Avrami equation [4,5]. The obtained result indicated that the rate of ferrite-to-austenite transformation for SSFDI was slower than for FCD. It is well known that the addition of Si in steels raises the austenitizing temperature. In addition, the carbon concentration in austenite at both the ferrite/austenite and austenite/graphite interfaces, which is one of the factors controlling the austenitizing rate, is affected by the increase in the transformation temperature. Thus, the slow-down of the rate of the ferrite-to-austenite transformation for SSFDI compared with FCD may be caused by the differences in the carbon concentration in austenite at the interfaces.

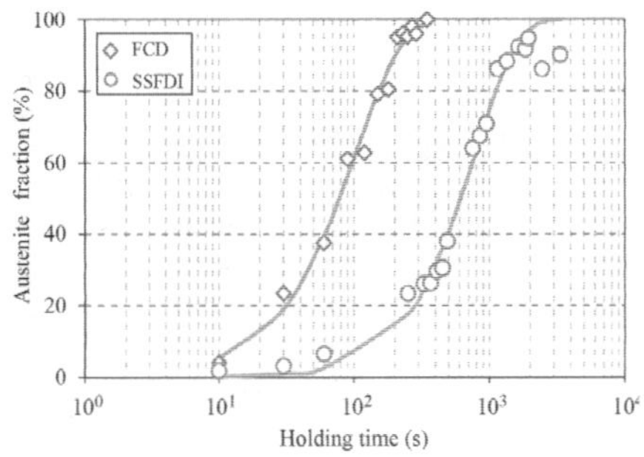

Fig. 1 Isothermal austenitic transformation at 850 °C for the FCD and SSFDI specimens.

Figure 2 shows the calculated carbon concentration profiles in SSFDI after holding at 850 °C for 10, 100, and 1000 s. The data are plotted as a function of distance from the center of the graphite nodule. The discontinuity in the profiles around the distance of 20–30 μm corresponds to the interface between ferrite and austenite, and it was found that the austenite region became larger with an increase in the time step. The calculated progress of the isothermal austenite transformation for FCD and SSFDI at various temperatures was found to reproduce the observed curves well.

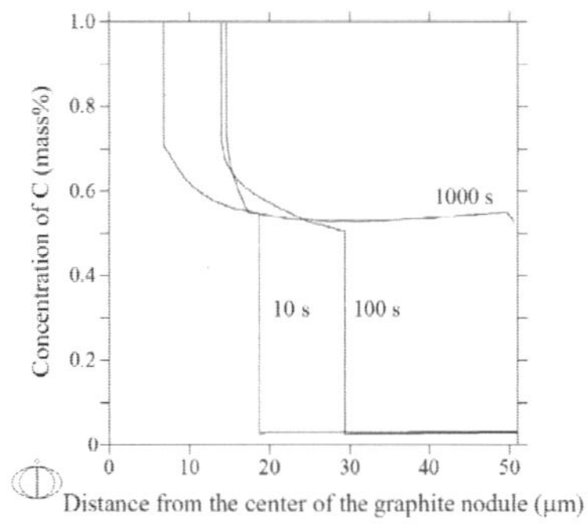

Fig. 2 Calculated carbon concentration profiles in SSFDI after holding at 850 °C for 10, 100, and 1000 s.

4. Summary

The effect of Si content on the ferrite-to-austenite transformation in ferritic ductile cast iron with a ferrite matrix was investigated. The results obtained are summarized as follows.

(1) The progress of isothermal austenite transformation for SSFDI was retarded compared with the conventional FCD.

(2) The DICTRA simulations reproduced the experimental austenite transformation for both FCD and SSFDI.

Acknowledgements

The authors are grateful to Hinode, Ltd. for providing the materials used in this study.

References

[1] R. Larker: China Foundry 6 (2009) 343–351.

[2] T. Oowadano and Y. Miyazaki: J. JFS. 46 (1974) 592–595, (in Japanese).

[3] J.-O. Andersson, L. Höglund, B. Jönsson and J. Ågren: *Fundamentals and Applications of Ternary Diffusion*, Ed. by G. R. Purdy (Pergamon Press, New York, 1990) pp. 153–163.

[4] W.A. Johnson and R.F. Mehl: Trans. AIME. 135 (1939) 416–442.

[5] M. Avrami: J. Chem. Phys. 7 (1939) 1103–1112.

Proceedings of the 72nd World Foundry Congress
21-25th May 2016, Nagoya, Japan

A new metallurgical process for the ductile iron foundry

Emmanuel Berthelet[1], Shinya Sano[2], Toshihiko Hatano[3]
[1] Vesvius UK Limited, Foseco Foundry Division,
Drayton Manor Business Park Tamworth Staffordshire B78 3TL, UK
[2] Foseco Japan Limited, Marketing Division,
Orix Kobe Sannomiya Bldg., 6-1-10, Gokodori, Chuo-ku, Kobe 651-0087, Japan
[3]Tomita Co.,Ltd, Foundry Division,
Kanen 4-31, Yatomi city , Aichi 498-0043, Japan

Abstract: There are limitations in the sandwich or tundish methods of producing ductile iron. The new INITEK process, which includes a unique converter system and INODEX initializer alloy that produces exceptional metal quality with significant reduction in magnesium use. INITEK gives 2 major options for the ductile iron foundry. Firstly cost savings in charge metal, treatment and refractory costs. Secondly to improve casting quality with specific casting defect reductions such as shrinkage cavities, inclusions, misruns, or abnormal graphite.

Keywords: *INITEK, converter, initializing, ductile*

1. Introduction

FOSECO's recently developed nodularizing and inoculating treatment process "INITEK" consist of

1. A specially designed converter ladle which has a high thermal efficiency, reducing temperature losses and increasing the recovery of magnesium. (Fig. 1)
2. Initializing treatment of the iron with INODEX initializer which has the effect of neutralizing the Oxygen and Sulphur activities, and nucleating the iron before the addition of magnesium. (Fig. 2-B)
3. Treatment with magnesium. (Fig. 2-C)
4. Controlled timing of all the process steps to ensure that reactions are complete and independent.
5. Late inoculation for a final adjustment of the metallurgy, if needed.
6. Process control using thermal analysis.

The efficiency of the process results in magnesium recovery of around 70%, meaning that lower Mg additions are needed and the inoculation step is reduced. The economics of the process are greatly improved.

The problems resulting from too high magnesium additions, i.e. carbide tendency, risk of shrinkage, non-metallic inclusions and high cost are reduced and the metal produced by this process has exceptional mechanical properties, with a greater tendency to ferrite and high elongation values. Foundries are able to use raw materials which are less pure and cheaper (higher Mn%) and still get the specified properties. They can work with lower amounts of pig iron and cheaper grades of steel scrap.

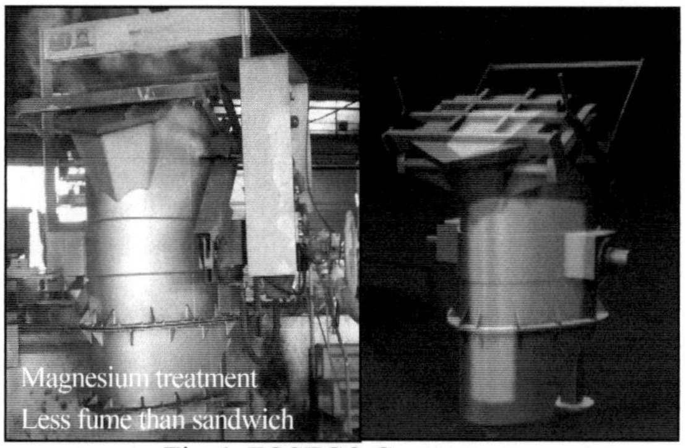

Fig. 1 FOSECO Converter

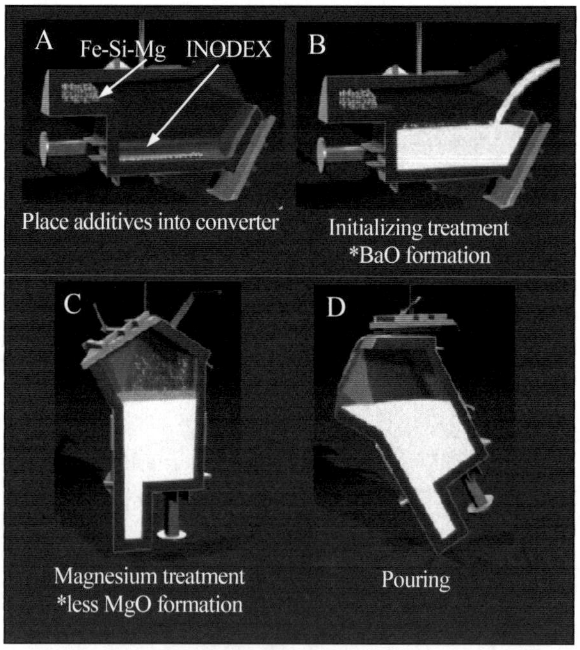

Fig. 2 INITEK process flow

2. Own studies

2.1 Comparison test

INITEK process was compared to Sandwich process using conditions outlined in Table 1.

Table 1 Comparative test conditions (mass%)

	Sandwich	INITEK
Fe-Si-Mg addition rate	1.5% (4 % Mg/2% RE)	0.9% (6 % Mg/1% RE)
1st Inoculation	0.2% (Ca-Al-Ba) *with Fe-Si-Mg	0.4% (INODEX) *before Mg treatment
2nd Inoculation	0.05% (Ca-Al-Ba)	0.05% (Ca-Al-Ba)
Cover material	0.3%	nil
Total additives	2.05%	1.35%
Treatment temp	1773K	1773K
C	3.65%	3.70%
Si	2.35%	2.35%
Residual Mg	0.040%	0.035%

No cover material for INITEK and lower addition rate with Fe-Si-6%Mg against Fe-Si-4%Mg.

Little higher C%, lower residual Mg% are targeted.

2.2 Test result

Comparative test results are shown in Table 2 and Fig. 3, 4.

Table 2 Physical properties*

Parameter	Sandwich	INITEK
Nodule count (n/mm^2)	164	240
Nodularity	92.5%	94%
Elongation	17.1%	19.4%
Tensile Strength (MPa)	520	491
HB	179	171
Residual Mg (mass%)	0.039	0.037
Mg recovery	65%	69%

*Average value of 9 'Y' block samples

INITEK increased nodule count, nodularity and elongation. In addition, INITEK gave better Mg recovery without cover material than Sandwich.

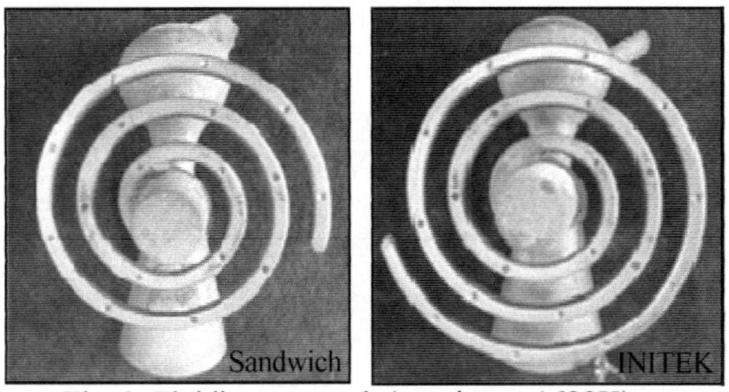

Fig. 3 Fluidity test result (pouring at 1623K)

Fluid distance of Sandwich was 440mm, INITEK was 560mm. INITEK improved fluid distance by 27%.

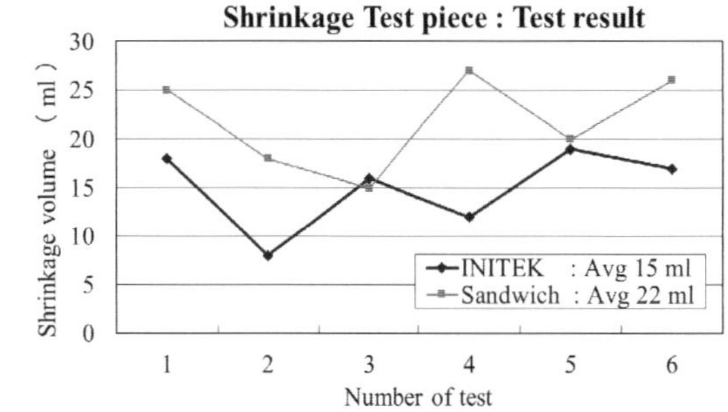

Fig. 4 Shrinkage cavity volume comparative result

The averaged shrinkage cavity volume of Sandwich was 22ml, INITEK was 15ml. INITEK reduced 32% of shrinkage volume.

3. Practical benefits

As Fig.5 shows, shrinkage cavity defect were eliminated for 19 castings manufactured in DCI 450.

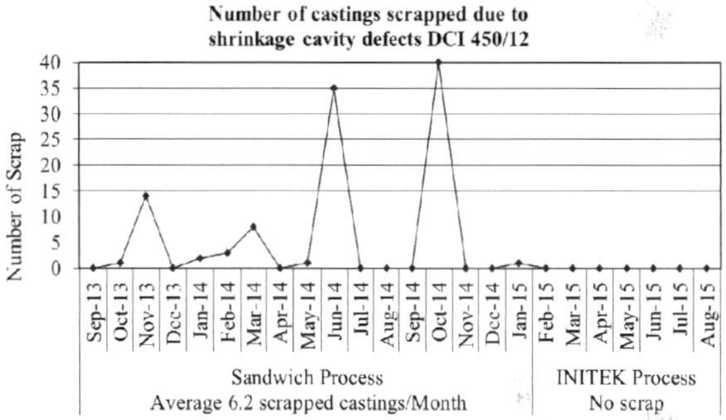

Fig.5 Shrinkage cavity defect occurrences over 2 years

4. Summary

INITEK process gave several benefits

1) Reduced additives which contributes less slag formation and better fluidity of the molten metal
2) Increased nodule count and elongation.
3) Better magnesium recovery
4) Reduced shrinkage cavity volume

References

[1] E.Barthelet, patent publication n° WO2010/128273A1,Treatment ladle (Nov,11,2010)

[2] S.N.Lekakh, D.G.C.Robertson, C.R.Loper Jr, "Thermochemistry of nodulization & inoculation of irons" AFS (Sept, 2005)

[3] F.Mampaey, D.Habets, F.Seutens, J.Plessers, "The use of oxygen activity measurement to determine optimum properties of ductile iron during production" Heraus Electro (2007)

[4] R.Siclari "Preconditioning for pure Mg treated ductile iron" 14th DISA/GF international licensee conference(2000)

Influence of pouring temperature on the microstructure in ductile cast iron

Jonas Åberg, Gunnar Maltsev Hansen, Lena Magnusson Åberg
Elkem AS Foundry R&D, Fiskaav. 100, 4621 Kristiansand, Norway

It is advised to keep the melt temperature of cast iron as low as possible to avoid problems during casting. Most casting properties except mould filling is improved by keeping the temperature down. Nodule density, nodularity and ferrite increase with lower melt temperature, but this has not been quantified earlier. This paper attempts to fill this void by describing a suitable evaluation method, and using it on data from controlled experiments and foundry trials to demonstrate that nodule density increase with 0.3 - 1% per ℃ pouring temperature decrease, while ferrite increased 0.2% per ℃ decrease. Nodularity increased 0.2% per ℃ decrease in temperature in one trial.

Keywords: *Melt temperature, nodule density, nodule count, nodularity*

1. Introduction

Foundrymen know the importance of keeping the melt temperature as low as possible, and to reduce the time from melting to solidified cast. The melt temperature should be low to control oxidation [1,2], and the treatment temperature should be kept low to keep a high and consistent magnesium yield, and by that reduce variation in residual magnesium during spheroidization treatment [3]. A high pouring temperature increase the dissolution of various elements such as hydrogen and nitrogen. These elements precipitates when the temperature drops during the cooling of the melt before and during solidification, which may cause gas porosities [4]. A high pouring temperature can also cause metal penetration into the mould and cores [5].

What is less known is the quantitative effect of the melt temperature on nodule density (ND), nodularity and matrix structure.

The ND is important since it reduce both the primary carbide formation as well as inverse chill [6,7,8]. High ND is also reported to reduce porosity [**Error! Bookmark not defined.**,9] and to improve toughness [10], although this is disputed in [11]. The nodularity is interesting since there is a direct link to mechanical properties such as Young's modulus and fracture elongation. Nodularity also affects the thermal conductivity and damping characteristics of the material [12].

2. Measurement and Method

2.1 Lab scale melting and casting process

In the experimental set up, raw material was melted in an induction furnace, an MgFeSi containing 4.6% Mg added to the pocket in a sandwich ladle and covered with steel chips. The melt was treated at 1500 ℃, followed by slag removal. The treatment ladle melt temperature was measured several times before and after it was poured into 12-14 pouring ladles with 30 s interval, each holding 33 kg of melt with 0.2% inoculant added. The pouring ladles were held 60 s each before pouring component moulds. The temperatures were measured using dip thermocouples (t.c.) or interpolated between measurements. The difference in temperature drop from inoculation to mould pouring between the pouring ladles have been computer simulated, as well as being verified by measurements in a separate trial and found to be negligible. All temperature readings were made using t.c.'s and time stamped. All pouring times were time stamped for accurate temperature interpolations.

2.2 Microstructure evaluation

The microstructure was evaluated in the centre of 10 mm thick plates from step blocks and in rod samples of 25 and 40 mm diameter. The variation in quantified microstructure within a plate was evaluated to estimate the uncertainty due to variations in the position of the cut, since it varies from the hot to the cold part of a step (1.5 nodules/mm change in position). To minimize the uncertainty of the microstructure analyses themselves, each sample was examined using 25 images totalling 14.2 mm^2 each in the carefully selected identical position within each sample. The observed range of ND, nodularity and ferrite levels were 105-212 nodules/mm^2, 70-90%, and 46-69%. The uncertainty in the measurements, expressed as 95% confidences, were 5-12 nodules/mm^2, 2-6% and 2%.

2.3 Normalization

To be able to study the effect of the temperature variation only in the trials, the results were normalized. As an example, the ND for one sample was divided by the average result of the two repetitions made per melt to get a normalized ND (ND_n). This method makes it possible to compare the relative influence of only the temperature variation on the output between different melts, inoculants, etc.

3. Results

The chemical composition changed slightly between the two trial melts, C 3.6 - 3.9 %, Si 2.2 - 2.3%, P 0.015 - 0.024%, Mn 0.33 - 0.3% and Mg 0.047 - 0.040%.

The left plot in Fig 1 shows that the ND_n increase with 0.3%/°C drop in inoculation/pouring temperature. The plot in the middle show that the ferrite level increase with 0.2%/°C. The colours represent different melts and sample geometries, and show that the behaviour is consistent.

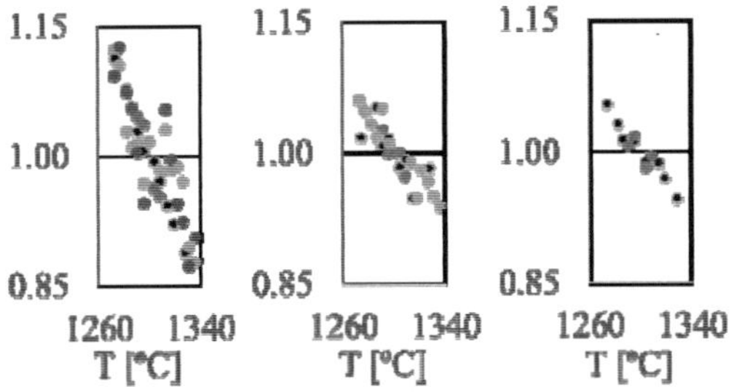

Fig. 1 Normalised ND (left), ferrite (middle) and nodularity (right) vs temperature in a lab trial.

The literature e.g. [13] describes that increased ND creates nucleation sites for austenite, which transforms to ferrite. This was confirmed in this study by the correlation between ND_n and normalized ferrite, Fig 1.

The nodularity dependence of temperature was evaluated from two trials with different results. In one trial, there was a near linear 0.2% increase in normalized nodularity per degree drop in temperature, Fig 1, while in the other there were no correlation observed.

3.1 Results from a foundry trial verification

The behaviour of the ND_n was confirmed in a foundry trial, Fig 2, that show that the behaviour of ND_n vs temperature is similar as in the lab scale trial.

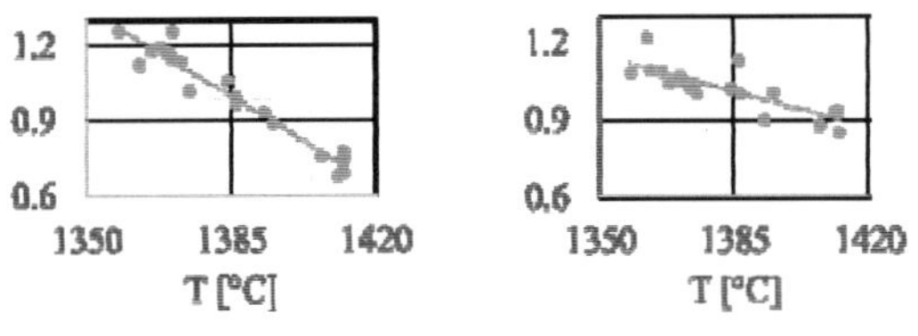

Fig. 2 ND_n, 10 mm (left) and >50 mm section (right).

The left plot is from sections of components with about 10 mm thickness, the right plot is for a thicker section of the same components equivalent to >50 mm thick, positioned near a feeder.

In the foundry trial, the variation in ND_n with temperature was more than twice as large in 10 mm sections compared to >50 mm sections. The slope was 1%/°C and 0.4%/°C respectively.

4. Analysis and discussion

The literature indicates that the pouring temperature [14] and section thickness (solidification rate) [15] influence ND, nodularity and matrix. This effect has not been quantified earlier. By carefully measuring the temperature during experiments and normalizing the results, it is possible to quantify this effect, and use it to compensate for the variation in temperature during the trial. This needs to be done for each individual trial for best accuracy. Adjusting for the effect of temperature can considerably improve the accuracy of the results in comparative trials. The observations made with different geometries, melts and trial location support each other despite the overlaid stochastic variations in the results.

5. Conclusion

ND, nodularity and matrix vary with the pouring temperature. This was verified for ND_n in a foundry trial. For best accuracy this effect should be evaluated, and if needed, adjusted for per trial.

References

[1] R.W.Heine et al, AFS Trans. (1966) No 71
[2] S. Katz, AFS Trans. (2004) No 132
[3] *The Sorelmetal Book of Ductile Iron*, (Rio Tinto & Titanium Inc., 2004) 65
[4] H. Fredriksson and U. Åkerlind, *Materials Processing During Casting*, (Wiley, Chichester, 2006) 258-261
[5] *Casting Design and performance, H.W. Stoll: Casting Design Issues and Practises,* (ASM International, Ohio, 2009) 2
[6] C.R.Loper Jr., Foundry (Sept 1966) 75-81
[7] K.M. Pedersen and N.S. Tiedje, WFC (2008) 167-171
[8] A. Javaid et al, AFS Trans. (1974) No 99
[9] N. Rizzo, AFS and DIS: Keith Millis Symposium on Ductile Cast Iron (2008) 148-156
[10] C. Labrecque, DIS Conference (Oct. 27-29, 2010)
[11] G.M. Goodrich, Modern Casting, (May 1990) 44
[12] J. Åberg, PhD Thesis, (KTH Stockholm, 2006) 19
[13] K.M. Pedersen, N.S. Tiedje, Mater. Charact. 59 (2008) 1111-1121
[14] G.M. Goodrich, AFS Trans. (1993), No 224
[15] R.W. Heine, AFS Trans. (1993), No 84

Proceedings of the 72nd World Foundry Congress
21-25th May 2016, Nagoya, Japan

Advanced Properties of Ausferritic Ductile Iron Obtained in As-cast Conditions

Susana Méndez[1], Urko de la Torre[1], R. González-Martínez[1], and Ramón Súarez[1,2]
[1] Engineering, R&D and Metallurgical Processes, IK4-AZTERLAN, Aliendalde Auzunea 6 E-48200 Durango (Bizkaia), Spain
[2] Veigalan Estudio 2010, E-48200 Durango (Bizkaia), Spain

Abstract

In previous studies the process parameters to obtain ausferritic ductile iron in as-cast conditions by means of engineered cooling were defined, that is without an austempering heat treatment. It was demonstrated that obtaining fully ausferritic microstructures by means of engineered cooling was feasible and that the properties met the requirements of the conventionally produced austempered ductile iron (ADI).

Keywords: *As-cast ausferrite, Fatigue, Corrosion, Mechanical properties, Impact resistance, Three-point bending test.*

1. Introduction

Austempered ductile iron (ADI) has an excellent combination of high resistance, good ductility and high fatigue strength [1, 2]. With the aim of improving production sustainability, a new process based on "engineered cooling" was developed to reduce the energy consumption during the production cycle [3, 4].

The aim of this work is to carry out an advanced characterization of the materials produced according to this alternative methodology and compare them with the properties cited in the literature or standards for conventional ADI processes in terms of low temperature, dynamic mechanical properties and corrosion behavior.

2. Experimental Details

2.1 Material

Three ductile iron melts were prepared in a medium frequency induction furnace to obtain standard keel blocks Y2 as per EN1563. The preparation method is described in details elsewhere [5].

The shake out temperature was 800ºC and three different isothermal transformation temperatures, 400, 350 and 300ºC (AUS-40, AUS-35, AUS-30) were chosen to compare the properties achieved with upper and lower ausferritic microstructures.

2.2 Tests

The static mechanical parameters measured were the ultimate tensile strength (U.T.S.), yield strength (Y.S.), elongation (E.), Brinell hardness (HB) and impact energy at 20ºC and -20ºC. Three-point bending experiments were conducted at the same temperature conditions, and fatigue tests were performed to build a Wohler diagram with a ratio S_{min}/S_{max}=0.1 and 20Hz in frequency. Corrosion behavior of the specimens was evaluated using electrochemical and weight loss methods in 0.03M NaCl solution.

3. Results and Discussion

The chemical composition of the three batches is shown in Table 1. Mo and Ni contents were added to avoid the pearlitic nose.

Table 1 Chemical composition (wt.-%)

Test	C	Si	Mn	Mg	Mo	Ni	Cu
AUS-40	3.61	2.10	0.15	0.044	0.21	2.80	0.04
AUS-35	3.61	2.02	0.16	0.054	0.33	2.84	0.03
AUS-30	3.52	2.10	0.17	0.045	0.22	2.87	0.06

The microstructures consist of ferrite needles (α) in a stabilised high carbon austenite (γ_{HC}) matrix (ausferritic microstructure) with nodular graphite: broad ferrite needles of upper ausferritic microstructure for isothermal transformation temperature of 400 ºC and very fine needles of ferrite for isothermal transformation temperature of 300 ºC Samples treated at 350 ºC present a mixture of both structures.

The values of the static mechanical properties are shown in Table 2 and fulfill the standards of the conventional ADI materials (UNE EN 1564).

Table 2 Static mechanical properties of the engineered cooled samples.

Property	Average values		
	AUS-40	AUS-35	AUS-30
Y.S.(MPa)	573	666	676
U.T.S.(MPa)	830	869	974
E.(%)	10.6	6.2	6.4
HB	273	289	312
Charpy at 20 °C	12/10/11	12/13/13	9/9/12
Charpy at -20 °C	7/8/8	8/8/8	9/9/8

Regarding the fatigue tests results, Fig. 1 shows a Wohler diagram that was plotted comparing the results from the AUS-40 and AUS-30 materials.

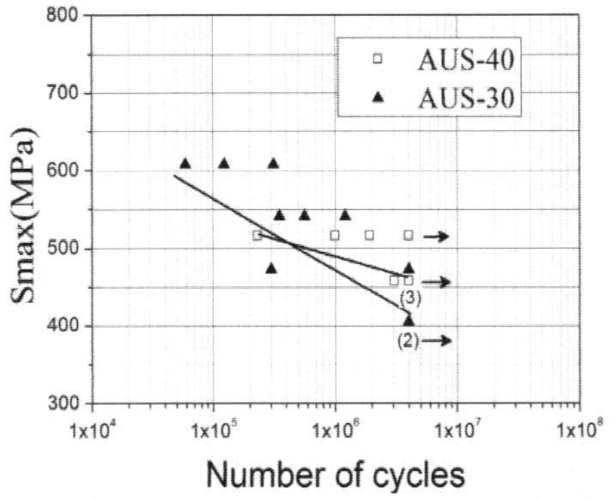

Fig. 1. Fatigue response of AUS-40 and AUS-30.

The results of the bending tests are shown in Fig. 2. All specimens present the same behavior: the load reaches a maximum and then the sample breaks suddenly which indicates that the fracture initiation is of brittle nature.

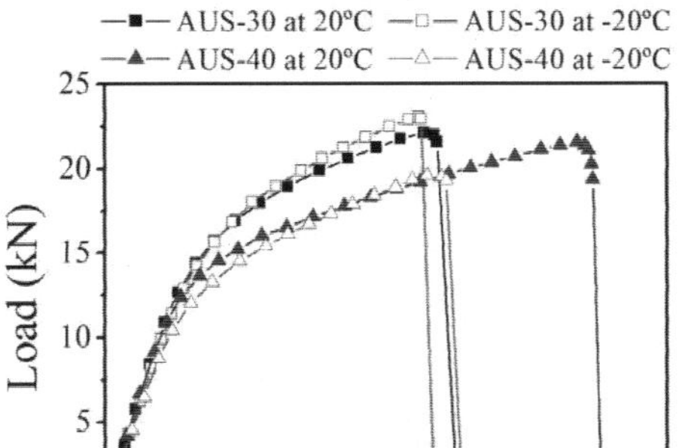

Fig. 2. Load-displacement curves of the bending tests.

The corrosion behavior of samples tested for the 300 and 400 °C isothermal transformation temperatures is similar in terms of E-t answer but it can be seen that higher isothermal transformation temperatures shift the corrosion potential to the anodic direction. Besides, AUS-30 corrodes quicker than AUS-40 due to its higher I_{corr} value.

The behavior of the weight loss test agrees with the results of the electrochemical results tendency.

Conclusions

Based on the results obtained in this work, the following conclusions have been drawn:

- The material obtained by means of engineered cooling met the requirements of conventional ADI.
- The material AUS-40 presents a higher fatigue limit than the AUS-30 material.
- The bending tests show that the AUS-40 presents a higher displacement for the same applied load, but the maximum load is lower than in the case of the material tempered at 300 °C.
- The effect of lowering the temperature in terms of impact and bending resistance is significant for the samples tempered at 400 °C, while the ones tempered at 300 °C are slightly affected.
- Upper ausferrite microstructure is shown less active than lower ausferrite in terms of corrosion behavior in NaCl.

References

[1] J.R. Keough, K.L. Hayrynen and G.L. Pioszak: AFS, 10-129 (2010) 1-5.
[2] Y. Tanaka and H. Kage: Mat. Trans. JIM 33 (1992) 543-557.
[3] U. De la Torre, D. M. Stefanescu, D. Hartmann and R. Suárez, AFS, (2013) 233-234.
[4] S. Méndez, U. de la Torre, R. Suarez, P. Larrañaga, and D. M. Stefanescu: AFS 15-010, (2015) 1-7.
[5] J. Sertucha, J. Lacaze, J. Serrallach, R. Suárez and F. Osuna: Mater. Sci. Technol. 28 (2012) 184–191.

Kinetics of Nucleation of Graphite at different Stages of Solidification for Spheroidal Graphite Iron

G. Alonso[1], P. Larrañaga[1], D.M. Stefanescu[2], E. De la Fuente[1], A. Natxiondo[3] and R. Suarez[1,3]
1. Ingeniería, I+D y Procesos Metalúrgicos, IK4-AZTERLAN, Aliendalde Auzunea 6, E-48200 Durango, Spain
2. University of Alabama and Ohio State University, Columbus, OH
3. Veigalan Estudio 2010 SLU, Aliendalde Auzunea 6, E-48200 Durango (Bizkaia), Spain

Keywords: *spheroidal graphite iron, nucleation, solidification, inclusion.*

The importance of the nucleation and growth phenomena that control the solidification of castings on the mechanical properties and soundness of cast iron cannot be overemphasized. The graphite nucleation mechanism is directly related to the carbon content of the iron and the inoculation treatment. Over the years, a multitude of theories have been developed with the objective to explain the nucleation process of spheroidal graphite (SG) iron. The most important include the gas bubble theory [1], the graphite or carbon rich cluster theory [2,3], the salt-like carbide theory [4], the silicon carbide theory [5,6], the sulphide/oxide theory [7,8,9,10,11,12], the silicate theory [13,14,15,16].

The heterogeneous nucleation theory is focused on the non-metallic inclusions present in all commercial cast irons. These particles must satisfy some specific conditions to act as possible nucleation sites, including: good crystallographic compatibility, low lattice disregistry or mismatch (no more than 3%), fine dispersion in the melt (1-3 μm) and high stability at elevated temperatures [14,16].

In this work, interrupted solidification experiments were conducted on spheroidal graphite irons (SG) at three different levels of carbon equivalent, with and without addition of a commercial inoculant. After superheating to 1500ºC, the induction-melted iron was treated into the pouring ladle with 1.1 mass % of a FeSiMg alloy (47.2% Si, 6% Mg, 1.15 Ca, 0.24% Al, 0.3% Mn, and 0.88% RE) by the sandwich method. Standard thermal analysis cups were poured. The solidification of the iron in the cup was interrupted by quenching in brine at increasing times (immediately after pouring, after 10 seconds and after 60 seconds). Inoculation was made directly in the cups through the addition of 0.2% of a commercial inoculant (62.6% Si, 0.22% Mg, 1.01% Al, 1.79% Ca, 5.96% Mn, 0.13% Ti, 6.77% Zr, 0.65% Ba and less than 0.07% lanthanides). The chemical analysis of the experimental irons is reported in Table 1. In addition to the elements listed in the table, the alloys contained 0.017% P, 0.04% Cr, 0.01% Mo, 0.07% Cu and less than 0.01/% Al and 0.005% Sn.

Table 1. Chemical composition (mass%) of cast irons

heat	CE	C	Si	Mn	Mg	Ti	S
QI	4.38	3.78	1.93	0.22	0.038	0.030	0.009
QII	4.20	3.56	2.03	0.19	0.039	0.021	0.011
QIII	3.93	3.30	2.00	0.19	0.041	0.020	0.010

Scanning electron microscopy (spectrums, mappings, line scan and different detectors) was carried out to analyze and quantify the possible nucleation sites at different solid fractions, as well as the influence of the inoculant in their formation. An example of the use of different detectors to identify nuclei is given in Fig. 1.

Non-metallic inclusions of varying composition have been observed in the matrix and at the centers of graphite nodules of the different irons (Fig. 2). They include silicates, oxides, carbo-nitrides. The complex oxides include Mg, Ce, La (Fig. 2-d). Numerous Ce and lanthanides inclusions were found at the grain boundary without any contact with the graphite.

Over 70% of the inclusions assumed to play a role in graphite nucleation were identified through X-ray composition maps and spectra as (Mg,Ca)S neighboring TiC (Fig. 2-b). This tendency was confirmed in un-inoculated and inoculated irons, independent of iron melt chemistry and of solid fraction. Calcium appears in a great number of sulphides, but less frequent than magnesium. Barium was never found in any nucleation site. A significant number of nuclei included Ti carbo-nitrides that contained Zr in many cases (Fig. 2-c). Most of the time the sulphides appear to act as nucleation sites for the carbides and at the same time seem to nucleate the graphite.

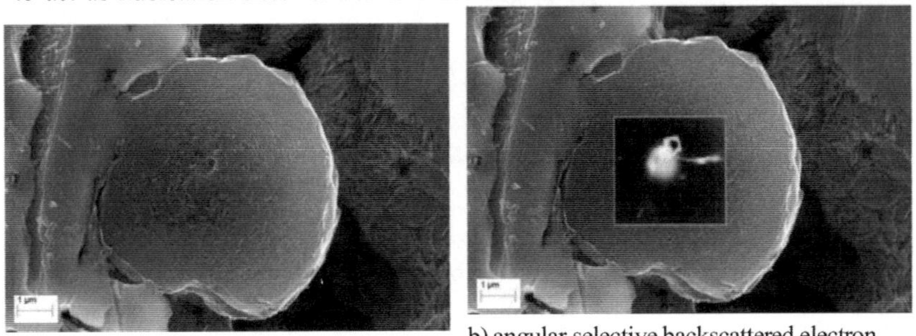

a) in-lens detector

b) angular selective backscattered electron detector

Fig. 1. SEM images of an iron from heat QI under different detectors: sulphide and carbide inclusions were detected

Based on the frequency of occurrence of elements in the 2456 graphite spheroids investigated, and on the thermodynamics free energy of formation calculated with Factsage, it appears that the majority of the nuclei are complex oxy-sulfides of Mg, Ca and sometimes Al (O was found in 76% of the inclusions and S in 17%). While 13.7% inclusions contained MgO, and 11.7% included MgS, 10.3% where Mg oxy-sulfides. A large amount of Ti carbides and carbo-nitrides were also identified (Ti was found in 9.3% of the inclusions). As TiC has a relatively high energy of formation, the Ti compounds are probably more complex carbo-nitrides containing Zr, Al, Si and Mg. Indeed, 6.4% of the inclusions were found to be TiN, out of which 1.8% were (Ti,Zr)N. The number of silicates was very low as only 1% of the inclusions included both Si and O.

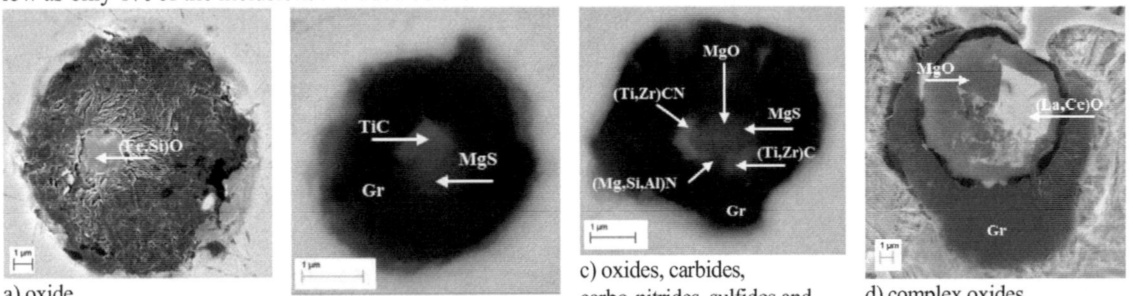

a) oxide

b) carbides and sulfides

c) oxides, carbides, carbo-nitrides, sulfides and nitrides

d) complex oxides

Fig. 2 SEM observations of graphite with different inclusions in the core

Some of the findings in this work are not explained by theories inferring that graphite spheroids nucleate on inclusions that contain an MgS core surrounded by an oxide shell, or with an outer shell of complex Mg silicates, as in many instances the nucleus was made of two or three different compounds, and all of them were in contact with the graphite. MgS and TiC compounds were the major sites for SG nucleation in this work.

References

1 S.I. Karsay, "Ductile Iron, The State of the Art". Uitgave QIT, 1976 and 1980

2 A. Boyles, "The Structure of Cast Iron", Ohio: Metals Park, 1947

3 D.M. Stefanescu, G. Alonso, P. Larrañaga, R. Suarez, Acta mater. 103 (2016) 103-114

4 B. Lux, Modern casting, 45 (1964) 222-232

5 C.H. Wang and H. Fredriksson, in 48th International Foundry Congress, Varna, Bulgaria, 1981

6 H. Fredriksson, Mat. Sci. and Eng. 65 (1984) 137-144

7 R.L. Naro, J.F. Wallace, AFS Transactions, 78(1970) 229-238

8 M.H. Jacobs, T.J. Law, D.A. Melford, M.J. Stowell, Metals Technology, 1 (1974) 490-500

9 M.J. Lalich, J.R. Hitchings, AFS Trans., 84 (1976) 653-664

10 G.X. Sun, C.R. Loper, AFS Trans. 91(1983) 639-646

11 R.J. Warrick: AFS Trans. 74 (1966) 722-733.

12 Y. Igarashi and S. Okada, Int. J. Cast Metals Res., 11(1998) 83-88

13 M.H. Jacobs, T.J. Law, D.A. Melford, M.J. Stowell, Metals Technology, 3(1976) 98-108

14 T. Skaland, T. Grong, T. Grong.. Metall. Trans. 24A(10) (1993) 2321–2345

15 Y. Igarashi and H. Nakae, J. Jap. Foundry Eng. Soc. 74 (2002) 30-35

16 M. Chisamera, I. Riposan, S. Stan and T. Skaland. In: Proceedings of the 64th WFC, Paris, France, 2000 (Paper no. 62)

Proceedings of the 72nd World Foundry Congress
21-25th May 2016, Nagoya, Japan

Possibility on Estimation of Fatigue Limit Using Non-Distractive Testing Method in Spheroidal Graphite Cast Iron

Naoto Shiraki[1], Ryosuke Fujimoto[2] ,Masami Tomizawa[3] ,Takumi Hara[3] ,Tsubasa Hirose[4]

[1]Department of Mechanical Engineering, Tokyo City University, Setagaya-ku, Tokyo, Japan
[2]Toshiba Machine Co.,LTD., Numazu-shi, shizuoka, japan
[3]Toshiba IT & Control Systems Corporation, Shinjuku-ku, Tokyo, Japan
[4]Graduate student, Tokyo City University, Setagaya-ku, Tokyo, Japan

In this study, the axial fatigue test specimen was machined from large size ingot of spheroidal graphite cast iron. Axial fatigue test was also carried out using an electric-hydraulic fatigue tester (made by Shimadzu Corp. Ltd.) with a capacity of ±19.6 kN. The test was conducted using a sinusoidal load control system with a load repetition frequency of 17Hz, and stress ratio, R=-1. The maximum number of repetitions for the fatigue test was 1×10^7 cycles. The test was carried out at room temperature. The approximation of S-N diagrams obtained by the axial fatigue test and the fatigue limits were calculated in conformity with JSMS-SD-6-02. The fracture origins and their peripheral areas of all specimens fractured after the fatigue test were observed using a scanning electron microscope.

The defects in specimen were detected using X-rays CT and estimated fatigue limit by equation of Murakami. And we compare fatigue limit which was estimated from fracture analysis using an equation of Murakami with the fatigue limit of experimental results. The possibility on estimation of the fatigue limit was investigated by the nondestructive technique.

Keywords: *Spheroidal graphite cast iron, Fatigue limit, High-resolution X-ray CT, Fracture origin, Defect size*

1. Introduction

It is desired to establish a safe and simple method using for strength design because experimental determination of fatigue limit spends enormous time and effort [1]. Murakami et al. proposed the 4-parameters method to estimate the fatigue limit in consideration of the hardness of the matrix and defect size in a material, and showed that the fatigue limit can be evaluated with a good accuracy for a material with defects whose sizes are less than 1,000 μm [2].

Hardness of the matrix can be obtained conveniently, but for defect size, many fatigue specimens are required and also the statistical analyzing of defect size from fracture surface observation is necessary.

In this study, authors estimate the fatigue limit by using the defect size of the specimens observed by X-ray CT and actual value of fatigue limit obtained by carrying out axial fatigue test in the observed specimens. We will compare measured fatigue limit with estimated value and estimation of fatigue limit using Non-distractive testing method in spheroidal graphite cast iron carried out.

2. Sample materials

Specimens targeted chemical components shown in Table 1 and Rare-earth element in spheroidizing agent was performed casting by adding so as to be 200 ppm in the melt. Figure 1 shows the shape of the axial load fatigue test specimen followed by JIS Z2273 [3]. The specimens were taken from h500×w500×t500 mm ingot.

Table 1 Chemical composition of material used
(mass%)

C	Si	Mn	P	Mg	S
3.4~3.5	2.5	0.9	≤0.03	0.05	≤0.01

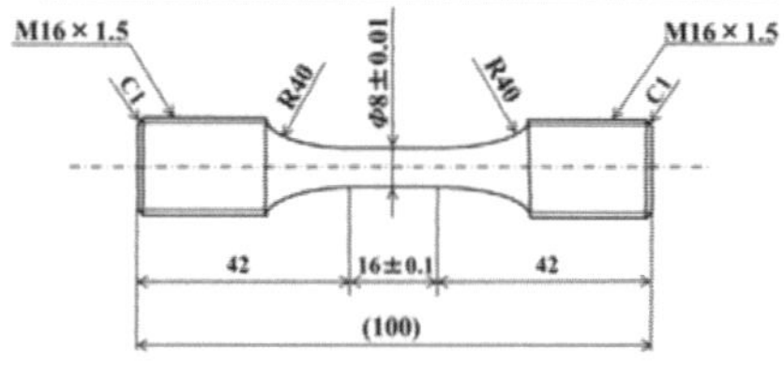

Figure 1 Shape and dimension of specimen in axial fatigue test

3. Experimental methods

Two in Axial fatigue test specimens (We call as FCD-1 and FCD-2) before fatigue test were observed using high-resolution X-ray CT, TOSCANER-32300μFD manufactured by Toshiba IT & control systems Co.,LTD.

In the axial fatigue test, electro-hydraulic servo pulsar was used with a loading frequency of 17 Hz and stress ratio R=-1 at room temperature in the atmosphere. The number of cycle to discontinue test was set to 1×10^7 cycles. The fracture origins were observed by scanning electron microscope (SEM).

4. Experimental Results and Discussion

Figure 2 shows the maximum defects in two test specimens observed by high-resolution X-ray CT and secondary electron image of fractured surfaces observed by SEM.

In the measurement of the defect size of the maximum defect observed by X-ray CT and fracture origins observed by SEM, they were approximated by a square and the square root of the area assumed to be the defect size $\sqrt{area}$. As a result of observation, the kinds of maximum defects in two specimens observed by X-ray CT were confirmed micro-shrinkages. The fracture origin in FCD-1 was confirmed aggregate graphite and the fatigue of FCD-2 was confirmed micro-shrinkage that was the same as maximum defect observed by X-ray CT. As the reason in FCD-1 that fracture origin was different from a maximum observed defect, the maximum defect may not become the fracture origin to lead to destruction that the cracks that occurred from plural defects transmit. Fatigue limit was obtained by experimental axial fatigue test was 145 MPa.

It is known that the fatigue limit of a material with a defect (crack) can be estimated using the 4-parameter model [4].

Region II is a region mainly ruled by the resistance against the propagation of a small defect (crack), the fatigue limit could be accurately estimated using Eq. (1) for the material with defects of the size smaller than 1,000μm[2].

$$\sigma_w = \frac{\alpha(HV+120)}{\sqrt{area}^{1/6}} \tag{1}$$

HV is the Vickers hardness of matrix, α is a coefficient, which depends on the position of a defect. Table 2 shows the fatigue limit, the defect size $\sqrt{area}$,Vickers hardness, the fatigue limit

	FCD-1 (X-ray CT)	FCD-1 (Fracture surface by SEM)	FCD-2 (X-ray CT)	FCD-2 (Fracture surface By SEM)
$\sqrt{area}$	918μm	636μm	900μm	1028μm

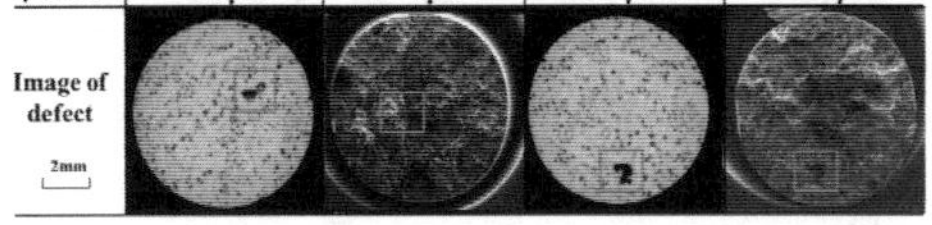

Figure 2 Maximum defects observed by X-ray CT and fracture surfaces observed by SEM

Table 2 Comparison of experimental fatigue limit, defect size, fatigue limit estimated by equation (1) and ratio of σ_W/σ_{WR}

Material		Fatigue limit determined by S-N curve σ_w,MPa	$\sqrt{area}$ μm	Vickers hardness $HV_{1.96}$	Fatigue limit estimated by eq.(1) $\sigma_{wR}(1)$,MPa	σ_w/σ_{wR} (1)
FCD-1	X-ray CT	145	918	184	139	1.06
	Fracture Surface		591		150	0.98
FCD-2	X-ray CT		900		140	1.05
	Fracture Surface		1028		137	1.07

estimated using Eq.(1) and the estimate accuracy of the fatigue limit. The evaluation is rated as precise as the ratio comes close to 1. The Fatigue limit estimated using Eq.(1) (Region II) in the 4-parameter method obtained good result.

5. Conclusions

A axial load fatigue testing was carried out and specimen observed by high-resolution X-ray CT. Obtained conclusions as follows.

(1) X-ray CT is possible to observe internal defects of spheroidal graphite cast iron.

(2) Fatigue limit estimated by Eq.(1) using size of the defects observed by X-ray CT and hardness of the matrix was a good result.

Acknowledgements

We are deeply grateful to Katsuhiko Taki, Nihon Visual Science Inc. offered analysis data.

References

[1] K.Ohmichi : Strength of fracture Materials, (Ohm-sha, 1985)p.194.

[2] Y.Murakami : *Metal Fatigue, Effects of Small Defects and Nonmetallic Inclusions*, (Yokendo, 1993)p.43.

[3] JIS Z 2273 : General Rules for Fatigue Testing of Metals, (1978)

[4] Y.Sugiyama, K,Asami and S.Matsuoka : Transactions of JSME(A)(in Japanese)588(1992)2287-2292.

Fatigue characterization and optimization of the production process of heavy-section ductile iron castings

Danilo Lusuardi[1], Eugenio Foglio[2], Stefano Goffelli[1], Annalisa Pola[2] and Marcello Gelfi[2]
[1] Fonderie Ariotti S.p.A. via E.Fermi 5, Adro (Italy)
[2]Università degli Studi di Brescia, Department of Mechanical and Industrial Engineering via Branze 38, Brescia (Italy)

Proper design of heavy-section ductile iron castings requires a database nowadays absent or very poor. In this paper, it will be presented the mechanical characterization, carried out at Fonderie Ariotti S.p.A., of ductile irons with solidification time up to 20 hours.

The characteristics of the various types of cast iron can widely change varying the solidification time. It was observed that the ferritic cast iron is more suitable for the production of large thickness castings subjected to fatigue loads, because decrement in ductility and its effect on other properties is less. For these reasons, the study of high silicon ferritic cast irons will be deepened.

It has been verified that data dispersion (especially significant in heavy castings and rarely found in the literature) of ferritic cast irons is on average lower than that of the pearlitic ones, probably because of the higher ductility of the matrix.

Keywords: *ductile cast iron, heavy-section, long solidification times, mechanical properties.*

1. Introduction

Ductile cast irons are widely used because they represent low cost ferrous alloys with high fluidity and very low shrinkage [1].

Increasing the castings size, decreases the cooling rate and consequently the internal microstructure can heavily change, with possible formation of defects and anomalous structures and decrease of the mechanical properties of the material [2]. The fatigue behavior of a ductile cast iron primarily depends on the matrix microstructure, on the presence of surface defects such as oxides, inclusions and porosity [3,4] and on the graphite nodules count and morphology [5]. In heavy-section castings, the desired spherical morphology of graphite often degenerates into different shapes, such as spiky, vermicular, exploded and chunky [6].

The substantial lack of technical data for heavy-section castings design forces the designers to oversize the dimensions to reduce any risk.

This work is aimed at obtaining a redefinition of the rules for heavy-section structural elements fatigue design in presence of defects.

2. Experimental procedure

2.1 Materials

The first tests results shown in this work are included in a larger project that involves a full characterization in terms of microstructural and mechanical properties of different types of cast irons for the production of heavy-section castings in order to establish the application areas of each of them.

The studied materials are standard ferritic ductile cast irons compared to ferritic cast irons with high silicon content and pearlitic matrix ductile irons.

The chemical composition of the first tested samples is reported in table 1.

Table 1 Chemical composition of the samples (mass%, Fe the rest).

Sample	C	Si	Mn	Cu	Mg	Cr	P
A	3.59	2.54	0.25	0.22	0.06	0.05	0.03
B	3.58	2.62	0.22	0.14	0.05	0.07	0.04
C	3.58	2.48	0.21	0.16	0.06	0.04	0.03
D	3.39	3.35	0.29	0.07	0.05	0.05	0.03

2.2 Techniques and methods

A direct comparison between two identical castings of the weight of 9t, one in high silicon ferritic iron with 3.4% Si and one in pearlitic cast iron with 1.0% Cu, showed nearly identical values of fatigue limit, but with a data dispersion 60% higher for the pearlitic one.

Based on these results, in order to deeply investigate the behavior of different materials with low cooling rates, an innovative production line has been then devised to reproduce the characteristics of castings of tens of tons, with different solidification times up to 20 hours, on a small scale sample in an electric furnace. The whole cooling cycle of the samples has been monitored using a thermocouple inserted in the casting .

2.3 Mechanical characterization

The first analysis performed was an ultrasonic control to detect any possible internal defects. Tensile and fatigue tests bars were then obtained from the castings.

Tensile tests were performed at room temperature, according to the UNI EN ISO 6892-1: 2009 and Brinell hardness was measured by using a durometer B3000J with a sphere diameter of 10 mm and an applied load of 29.4 KN.

The fatigue properties were studied by rotating bending fatigue tests, using the stair-case method according to UNI 3964/85 standard to calculate the fatigue limit with a survival probability of 50%. All tests were carried out at room temperature with a frequency of 100 Hz. The fatigue tests were stopped at sample failure or after $5 \cdot 10^6$ cycles.

Metallographic analysis were carried out on samples taken from various mechanical test bars for all the studied castings and, in some cases, also SEM (Scanning Electron Microscope) analysis of the fracture surfaces were performed to define the different fracture mechanisms.

3. Result and discussion

Table 2 Cooling curves and metallographic analysis results.

Sample	Matrix	Solid. time (h)	Eutectic T(°C)	Eutectic t(h)	Nodule count (n/mm^2)
A	Ferritic	3	1153	2.5	8
B	Ferritic	6	1160	5	9
C	Ferritic	20	1153	18	5
D	Ferritic High Si	12	1164	10	9

Table 3 Mechanical tests results.

Sample	R_m (MPa)	$R_{p0.2\%}$ (MPa)	A%	HB	$\sigma_{f,50\%}$ (MPa)
A	329±17	286±3	3.1±1.4	153±4	145±6
B	353±8	298±2	4.2±0.8	165±8	167±18
C	249±29	/	0	162±8	130±17
D	347±9	342±6	0.9±0.9	165±8	157±10

From the results of the tests described above, it was found that cooling curves, microstructures and mechanical properties of the samples are similar to those of the heavy-section production castings already tested. The validity of the process was confirmed.

According to these results, many trials were performed to check some changes in the production process. For so long solidification times, modifications in the preconditioning and inoculation treatments (quantity and type) did not led to improvements.

4. Conclusions

- In heavy-section castings, the fatigue behavior of a fully ferritic matrix cast iron is better compared to an even partially pearlitic one, because of the higher ductility and the lower data dispersion.

- The first presented results confirm the similarity between the samples obtained according to the developed system and some production castings with weights of tens of tons, in terms of microstructures and mechanical properties.

- For long solidification times, all the different preconditioning and inoculation treatments tested loose any effect.

- Solidification time is critical to the mechanical properties in the centre of the castings. Microstructure and mechanical properties have been greatly improved by active cooling systems, applied and checked on castings up to 30t of weight, and actually extended at much higher weight, up to 60t.

References

[1] J. R. Davis: *Fatigue and fracture properties,* (ASM Specialty Handbook – Cast Irons, Metals Park, Ohio, 1996) p. 450.

[2] P. Ferro, P. Lazzarin and F. Berto: Mater. Sci. Eng. A. 554 (2012) 122-128.

[3] A. Mourujärvi, K. Widell, T. Saukkonen and H. Hänninen: Fatigue Fract. Engng. Mater. Struct. 32 (2009) 379–390.

[4] B. Atzori, F. Bonollo and G. Meneghetti: Key Eng. Mater. 457 (2011) 181–186.

[5] M. Shirani and G. Härkegård: Eng. Fail. Anal. 18 (2011) 1496–1510.

[6] H. W. Jr. Hoover: Trans. Am. Foundryman's Soc. 94 (1986) 601–608.

Proceedings of the 72nd World Foundry Congress
21-25th May 2016, Nagoya, Japan

Giga-cycle Fatigue Properties for Spheroidal Graphite Cast Iron

Hide-aki Nishikawa[1], Etsuo Takeuchi[1], Yoshiyuki Furuya[1], Liu Zhimin[2] and Takuo Handa[2]
[1] National Institute for Materials Science, Materials Reliability Unit, 1-2-1 Sengen, Tsukuba-shi, Ibaraki, 305-0047 Japan
[2] Japan Casting co., LTD., Products Development Section, 2-1 Shiraishi-cho, Kawasaki-ku, Kawasaki, 210-9597 Japan

In this study, Giga-cycle fatigue properties were investigated for spheroidal graphite cast irons of FCD400 and FCD800. Fatigue tests were carried out by ultrasonic fatigue testing at 20 kHz and conventional servo-hydraulic fatigue testing at 30 Hz using specimen with 3 and 6 mm diameters under stress ratio of R = -1. The results of the ultrasonic fatigue testing showed negligible difference from those of conventional fatigue testing. Accordingly, ultrasonic fatigue testing applicable to spheroidal graphite cast iron. Decrease of fatigue strength in the giga-cycle region was in range from 10 to 20%, namely, fatigue limit was not confirmed even up to 1×10^{10} cycles.

Keywords: *Spheroidal Graphite Cast Iron, Giga-cycle fatigue, Micro shrinkage*

1. Introduction

Fatigue properties for spheroidal graphite cast iron were mainly investigated up to 10^7 cycles conventionally. On the other hand, it is known that fatigue limit becomes disappear and internal fracture from inclusions appears in more than 10^7 cycles for high strength materials such as spring steel. This is so-called Giga-cycle fatigue. Although graphite and micro shrinkage of spheroidal graphite cast iron leads similar internal fracture possibly, their giga-cycle fatigue properties were not investigated enough.

In this study, giga-cycle fatigue properties for graphite cast irons of FCD400 and FCD800 were investigated with ultrasonic fatigue testing.

2. Material and experimental procedure

Materials used in this study were spheroidal graphite cast irons FCD400 (3.6C -2.2Si- 0.09Mn -0.031Mg) and FCD800 (3.2C -3.09Si -0.35Mn -0.37Cu -0.036Mg). Tensile strength were 401 MPa and 867 MPa respectively. Microstructure of these materials were shown in figure 1. Base metal were composed by ferrite for FCD400 and perlite for FCD800. FCD800 were normalized and stress relief heat treated before machining fatigue specimen.

To evaluate size effect, two types of fatigue specimen with different risk volume (region subjected to a stress amplitude above the 90 % of the maximum stress) were prepared. One was hourglass type 3 mm diameter and the other was straight type 6 mm diameter specimen. Specimen risk volume were 33 mm^3 and 504 mm^3 respectively.

Fatigue tests were carried out by ultrasonic fatigue testing at 20 kHz and conventional servo-hydraulic fatigue testing at 30 Hz. The specimen were air-cooled to suppress any temperature increase during ultrasonic fatigue testing.

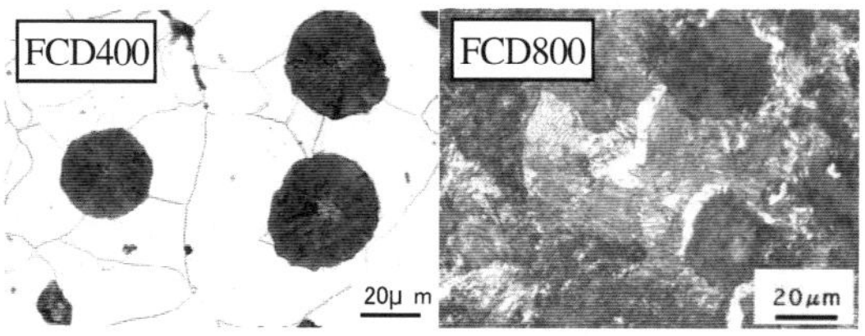

Fig. 1 Microstructures.

3. Results and discussion

Figure 2 shows S-N diagram. The results of the ultrasonic fatigue testing showed negligible difference from those of conventional fatigue testing for FCD400 and FCD800. However specimen were broken even up to 10^8 cycles and obvious fatigue limit was not confirmed. In addition fatigue strength of ϕ6 mm specimen were lower than that of ϕ3 mm specimen related to existence probability of larger defect as described below.

Figure 3 shows typical fracture surface image for crack initiation site. Fracture origins were mainly micro shrinkage at specimen surface, while several internal origin and graphite origin fracture were appeared in FCD800.

Authors pointed out that ultrasonic fatigue testing is applicable for materials which fractured from small defects (e.g. [1]). This report clarified that ultrasonic fatigue testing is applicable for graphite cast iron which also broken from small defect.

Fig. 4 shows Fatigue limit (10^{10} cycle fatigue strength) versus tensile strength. Although fatigue strength of FCD800 was decreased by micro shrinkage, FCD400 shows fatigue strength as excellent as hot rolled materials which has no defects.

Table 1 shows predicted maximum $\sqrt{area}$ size of micro shrinkage and fatigue limit. Fatigue limit were calculated with $\sqrt{area}$ parameter model proposed by Murakami [2]. Maximum $\sqrt{area}$ size were calculated by Gumbel probability plot evaluation for each risk volume [2] based on micro shrinkage size observed at crack initiation site of fracture surface. Predicted fatigue limit is reasonable for FCD800, while it is too much lower than test results for FCD400. This results indicate that FCD400 has excellent fatigue resistance for small defects.

In summary, applicability of ultrasonic fatigue testing for spheroidal graphite cast iron and its giga-cycle fatigue properties which has no fatigue limits were clarified. Furthermore, excellence of fatigue strength of FCD400 were indicated by fatigue limit prediction based on $\sqrt{area}$ parameter model and Gumbel probability plot evaluation.

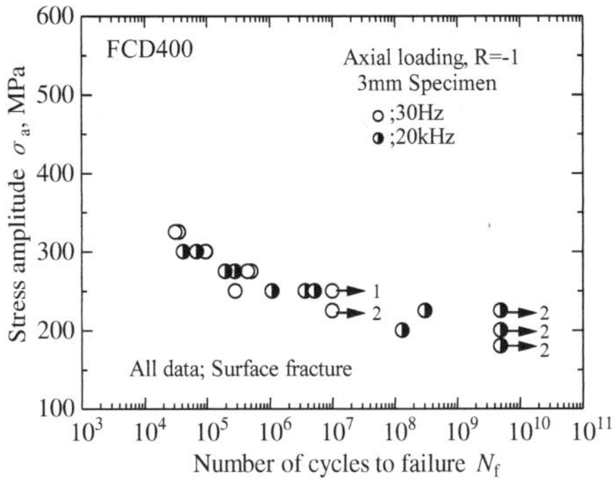

(a) FCD400, 3mm specimen

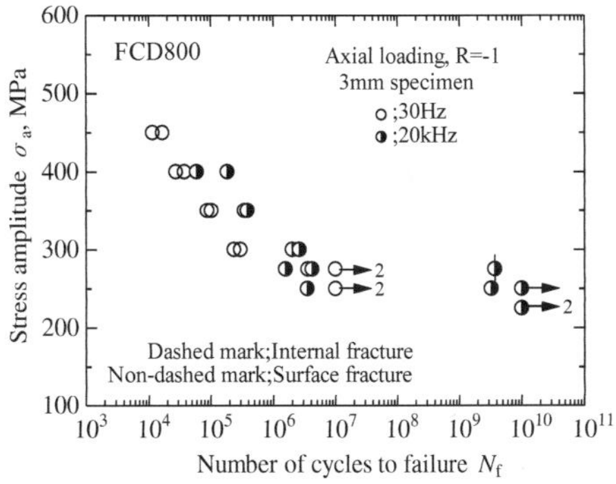

(b) FCD800, 3mm specimen

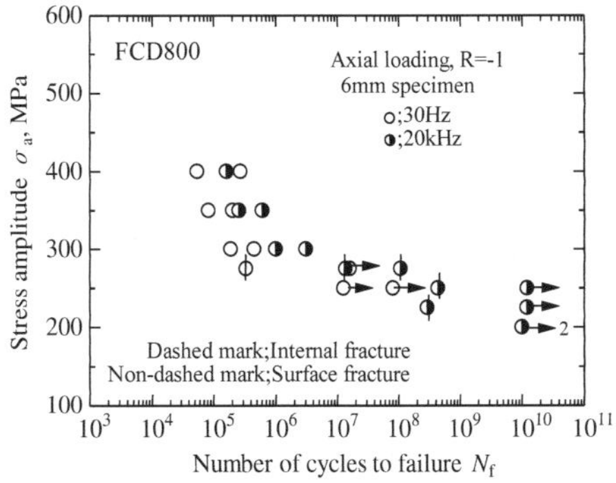

(c) FCD800, 6mm specimen

Fig. 2 S-N diagram.

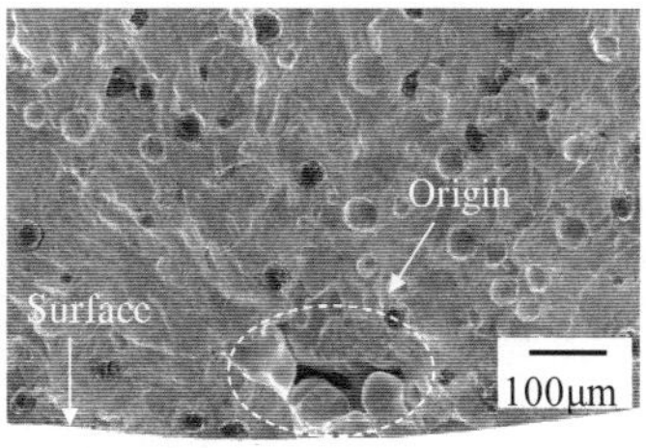

Fig. 3 Typical fracture surface image (FCD800, N_f=1.6×10^5).

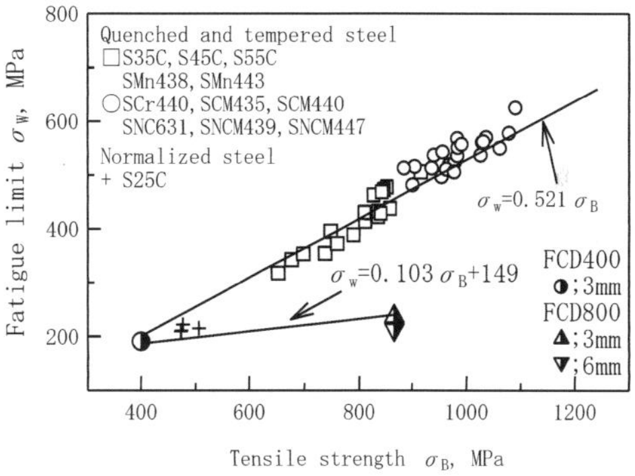

Fig. 4 Fatigue limit versus tensil strength.

Table 1 Predicted maximum $\sqrt{area}$ size of micro shrinkage and fatigue limit.

	Specimen type	Risk volume (mm^3)	Predicted $\sqrt{area}_{max}$ (μm)	Predicted σ_w (MPa)	[4]Experi-mental σ_w (MPa)
FCD400	φ3mm	[1]825	214	[2]155	190
FCD800	φ3mm	[1]825	289	[2]217	238
	φ6mm	[1]12600	434	[3]225	213

[1] Equivarent to total risk volume for 25 specimen
[2] Calucrated from surface origin model equation
[3] Calucrated from internal origin model equation
[4] 10^{10} cycle fatigue strength

References

[1] Y. Furuya et. al., Scripta. Mater., 46 (2002), 157.
[2] Y. Murakami, METAL FATIGUE, Elsevier, (2002).

O-18

Solubility of Hydrogen and Nitrogen in liquid cast iron during melting and mold filling

Attila Diószegi[1], Jessica Elfsberg[2] and Zoltán Diószegi[3]
[1] Materials and Manufacturing - Foundry Technology, Jönköping University, SE-551 11, Jönköping, Sweden
[2] Scania CV AB, SE-151 87 Södertälje, Sweden
[3]Volvo Group Trucks Operation, SE-541 87 Skövde, Sweden

Defect formation like gas- and shrinkage porosity at cast iron component production is related to the content of gaseous elements in the liquid metal. The present work investigate the solubility of hydrogen and nitrogen in liquid iron aimed for production of lamellar and compacted graphite cast iron. The used methods and instruments are a combination of commercial measuring devices and novel experimental assemblies for measuring solubility of hydrogen and nitrogen during melting and mold filling of a complex shaped cast component. The obtained results reveal the role of the charge material and the mold filling on the solubility of the investigated elements.

Keywords: *gas porosity, shrinkage porosity, lamellar cast iron, compacted graphite cast iron, hydrogen, nitrogen.*

1. Introduction

Hydrogen and Nitrogen are two important elements solved in liquid cast iron influencing the formation of gas- and shrinkage porosity formation [1]. Previous investigations indicate that the solubility of hydrogen and nitrogen is temperature dependent with a remarkable lowering of solubility at solidification and transformation from liquid to solid state [2]. Investigations in different iron foundries, performed under various production conditions, indicate that the content of hydrogen and nitrogen in the liquid iron is well below the solubility limit of hydrogen respective the nitrogen in solid state [3]. Liquid iron prepared for compacted iron production have a lower level of solved H and N than liquid iron aimed for production of lamellar graphite iron. Mold filling experiments of artificial cast samples in combination with measurement of gas content were indicating increase of the hydrogen and nitrogen content during the mold filling as a function of the average velocity of the liquid metal in the gating system [4]. The research question of the present work is how the mold filling influence the hydrogen and nitrogen content of the liquid iron at casting a complex shaped automotive cast component.

2. Experimental procedure

2.1 Experimental setup of a complex shaped cast component

An experimental setup according to Fig. 1 and Fig. 2 was created to measure the hydrogen and nitrogen concentration in the liquid iron after filling a complex shaped automotive cast component. In the top of the mold cup the yellow surface indicate where the liquid metal is poured into the mold cavity while the purple colored cylindrical volume indicate where the liquid iron was collected for measurement after passing the mold cavity.

Fig.1 Molding box

Fig.2 Mold cavity

2.2 Experimental description

The mold box presented in Fig.1 was molded on a Künkel-Wagner molding line in green sand and the quartz sand made internal cores were hardened by epoxy harts. The cylindrical cavity aimed for the direct measurement of hydrogen and the collection of samples for nitrogen measurement was drilled in the cope. Liquid metal melted in an induction furnace were transferred to a pouring ladle before casting. At casting the compacted graphite iron the induction melted base iron was transferred to a special ladle where a standard Mg treatment was performed prior casting. Melt temperature measurement, direct hydrogen measurement and collection of samples for nitrogen measurements were performed in the pouring ladle before casting and in the cylindrical cavity above the casting after the mold filling. Experiments casting lamellar graphite iron was repeated five times while compacted graphite iron were cast four times.

2.3 Experimental instruments used for measuring the concentration of the solved gaseous elements in liquid iron.

Hydrogen concentration were measured using the measurement system HYDRIS from Heraeus Electro-Nite based on immersing a measuring unit into the liquid metal. The detailed measuring principle is described in the Multi-Lab Hydris Operational and Instruction Manual [5]. Nitrogen concentration was measured by an OE spectrometer ARL 3460 on specimens collected by SaF400P6 samplers from Heraeus Electro-Nite.

3. Experimental results

The measured hydrogen content is presented in Tab. 1 which indicate an average low level of hydrogen in the ladle, below the hydrogen solubility in austenite (4 ppm) [3] for both LGI and CGI production. The hydrogen increase after pouring varies from 22% to 116% at casting LGI and from 26% to 153% at casting CGI, approaching the solubility limit in solid state. The main source of the increased hydrogen is the humidity in the molding material and the absorption mechanisms has, in an earlier work, been demonstrated to be the extended interface between the liquid iron and the gases filling the mold cavity [4].

Table 1. Measured hydrogen levels (ppm).

<table>
<tr><th rowspan="2">Sample</th><th colspan="2">LGI</th><th colspan="2">CGI</th><th rowspan="3">Hydrogen solubility in austenite [3]</th></tr>
<tr><th>Ladle</th><th>Mold</th><th>Ladle</th><th>Mold</th></tr>
<tr><td>1.</td><td>1,8</td><td>3,65</td><td>1,5</td><td>1,9</td></tr>
<tr><td>2.</td><td>2</td><td>2,45</td><td>1,3</td><td>3,1</td><td rowspan="4">4</td></tr>
<tr><td>3.</td><td>1,8</td><td>2,45</td><td>1,2</td><td>3,2</td></tr>
<tr><td>4.</td><td>1,7</td><td>2,55</td><td>1,3</td><td>3,3</td></tr>
<tr><td>5.</td><td>1,8</td><td>3,9</td><td></td><td></td></tr>
</table>

The measured nitrogen content is presented in Tab. 2 indicating an average high level of nitrogen already in the ladle at LGI production. After pouring the nitrogen content exceeds the nitrogen solubility limit in austenite (90 ppm) [3]. At production of CGI the nitrogen content remains at about 65% of the solubility limit and no significant change in the nitrogen content can be observed after mold filling.

Table 2. Measured nitrogen levels (ppm).

<table>
<tr><th colspan="6">Nitrogen measurement</th></tr>
<tr><th rowspan="2">Sample</th><th colspan="2">LGI</th><th colspan="2">CGI</th><th rowspan="3">Nitrogen solubility in austenite [3]</th></tr>
<tr><th>Ladle</th><th>Mold</th><th>Ladle</th><th>Mold</th></tr>
<tr><td>1.</td><td>84</td><td>92</td><td>61</td><td>59</td></tr>
<tr><td>2.</td><td>86</td><td>96</td><td>62</td><td>58</td><td rowspan="3">90</td></tr>
<tr><td>3.</td><td></td><td></td><td>62</td><td>66</td></tr>
<tr><td>4.</td><td></td><td></td><td>71</td><td>67</td></tr>
</table>

Comparing the results, lower level of both hydrogen and nitrogen at melting iron aimed to CGI production is attributed to the cleaner charge content including returned CGI and SGI chips and pig iron. The increase of hydrogen is significant during the mold filling independent on the type of liquid iron. Local segregation of hydrogen and nitrogen can easily exceed the solubility limit of those elements in the austenite fulfilling one of the criteria for porosity formation at solidification.

Acknowledgements

The present work is a part of the SPOFIC project, financed by the Swedish Governmental Agency for Innovation Systems (VINNOVA). Cooperating parties are Jönköping University, Swerea SWECAST AB, Scania CV AB and Volvo Group Truck Operation. Participating persons from these institutions/companies are acknowledged

References

[1] I. L. Svensson, H. Fredriksson: SolidificationTechnology in the Foundry and Cast House; Coventry; England; 15-17 Sept. 1980, pp 376-380.

[2] I.L. Svensson: Solidification Technology in the Foundry and Cast House, Coventry, England, 15-17 Sept. 1980, pp 253-262.

[3] J. Orlenius, L. Elmquist, A. Diószegi: Transactions of the American Foundry Society, Vol. 115, pp. 617-623. 2007.

[4] A. Diószegi, T. Björklind, Z. Diószegi: Key Engineering Materials Vol. 457. pp 422-427. Trans Tech Publications, Switzerland, 2011.

[5] Multi-Lab Hydris Operational and Instruction Manual, Heraeus Electro-Nite AB

O-19

Evaluation of Anti-sticking Effect of Microstructure Components in Cast Iron by Friction Thermal Shock Test

Nozomu Oda[1], Yasunori Nozaki[1], Hiroshi Yamamoto[2], Tatsuya Tokunaga[2] and Hidenori Era[2]
[1] Hitachi Metals Wakamatsu Ltd., 1-9-1 Kitahama, Wakamatsu-ku, Kitakyusyu, 808-8558 Japan
[2] Kyushu Institute of Technology, 1-1 Sensui-cho, Tobata-ku, Kitakyusyu, 804-8550 Japan

Anti-sticking resistance is an important property to avoid deep cracks for hot strip mill work roll materials, especially in the later stands where mill incidents like cobble are likely to occur.

Effect of microstructure components on anti-sticking property was investigated with cast steel and iron samples using friction thermal shock test. Graphite was found to show higher effect than carbides, although they also show preferable effect, depending on its carbide type.

Keywords: *casting roll, cobble incident sticking property, microstructure*

1. Introduction

HSS rolls were started to be applied to HSM work roll in the end of 1980's. They have much higher wear resistance than high chrome steel, high chrome iron and ICDP based rolls. In the past, many trials have been made to use HSS rolls in later stands, however they were not successful due to their inferior sticking resistance which lead to rapid crack propagation. In the late 1990's, enhanced ICDP rolls which contain MC carbides (vanadium carbide, niobium carbide), have been introduced in HSM finishing trains. Although they show higher wear resistance than conventional ICDP rolls, still much less compared with HSS rolls[1]. Lower wear resistance roll leads to reducing HSM production efficiency by frequent roll changes. Accordingly there is still high demand for new roll materials having both high wear resistance and anti-sticking property.

HSS contains several types of carbide such as MC, M_2C, M_7C_3 (M: Metal) and so on, but no graphite. Enhanced ICDP contains higher amount of carbide, mostly M_3C with smaller content of MC and graphite[1]. The research about the relationship between microstructure of rolls and sticking property is expected to be utilized for the development of new roll materials for HSM later stands.

2. Experimental Procedure

2.1 Sample preparation

Sample ingots containing three series of microstructure component variation were prepared. White pig iron, steel scrap, electrolytic nickel and ferrous-based compounds of alloys were used for raw materials. They were melted in high frequency induction furnace and cast into molds composed with iron chiller at the bottom and heat-insulated sleeve at side part in order to obtain directionally solidified 4kg ingots. Ingots were quenched and then tempered at 623 to 823 K to adjust to hardness of 50-55HRC, however some samples could not be controlled within this hardness range.

Table 1 shows basic chemical composition of each series. C-series (Cs) has 0.8 to 3.0mass% range of carbon aiming to obtain various contents of cementite (M_3C). C-Si-series (CSis) has 1.5 to 3.4 mass% range of carbon and 1.7 to 2.0mass% of silicon range to obtain various contents of M_3C and graphite. Cr-series (CCrs) has 9.2 to 23.4 mass% chromium range and 1.2 –to 3.7 mass% carbon range to obtain various contents of chromium carbide.

The microstructure observation by optical microscopy and sticking property evaluation were made at the distance of 30mm from the chilled surface of each ingot. The microstructures were quantitatively measured as area ratio of each structural component using image processing software.

Table1 Basic chemical composition (mass%).

C	Si	Mn	Ni	Cr	Mo	V
*[1]	0.75*[2]	0.75	1.0	0.9*[3]	*[4]	0.5

*[1]:0.8 – 3.4, *[2]:Cs and CCrs, *[3]:Cs and CSis, *[4]: Cs and CS-s <0.15, CCrs 1.0

2.2 Sticking property evaluation

The sticking property of samples were evaluated using friction thermal shock test (FTS test) [2] which simulates cobble incident phenomenon[3]. Mild steel rod is forced into a narrow gap between the specimen and a steel cylinder connected with pinion gear by dropping of a weight. The steel rod is pressed strongly on specimen surface and friction heat is generated. The specimen surface is heated up by the friction and the steel rod stick to it. Sticking property was evaluated as "sticking surface ratio", the area ratio of the stuck area against the outline of steel rod contact on the specimen .

3. Results

Figure1 shows the relationship between carbide content "carbide area ratio" and sticking property "Sticking surface ratio". Sticking surface ratio decrease as carbide area ratio increase in both Cs and CCrs. Especially Cs show larger decrease rate and reaching to same level as CSis at 20% of carbide area ratio. In case of CSis, no clear correlation between the sticking surface and carbide area ratio is found and sticking surface ratio shows stably low level around 20%.

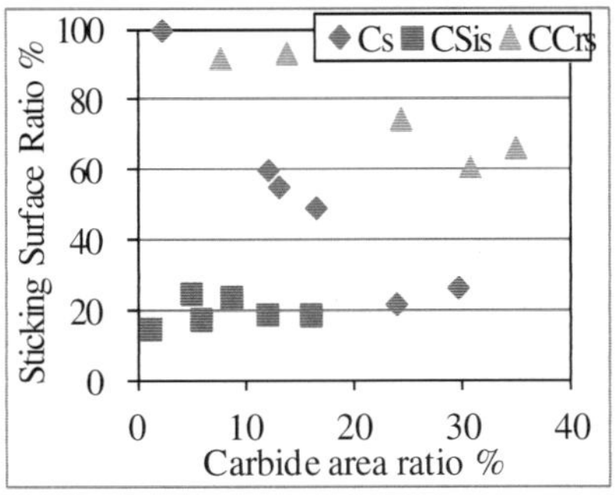

Fig.1 Relationship between carbide amount and sticking property.

Graphite amounts and distributions of all CSis group samples were observed with non-etched samples by optical microscopy. There were various shapes of graphite found in the samples like grain type or flake type. The range of graphite area of CSis was measured as from 0.5 to 6.7%. Their hardness were within 50-53HRC, except the sample containing 6.7% graphite area ratio with 35HRC.

In both of Cs and CCrs samples, no obvious graphite has been recognized by microscopy, only 10 - 20 μm - sized very small dark points were observed with non-etched polished state, which could be graphite but also inclusions or shrinkages. The area ratio of this dark point part in Cs samples were less than 0.3%. Electron micro scope analysis has been conducted to check the dark points and found that those in the samples showing lower sticking surface ratio were around 10μm - sized small graphite.

4. Discussion

It is well known that graphite is a excellent solid lubricant. In this study, it was confirmed that graphite is more effective to prevent steel from sticking to cast iron than carbides under FTS test condition. And 0.5% area ratio of graphite can be concluded to be enough to get the anti-sticking effect, considering the experiment results of no further reduction of sticking surface ratio above this amount.

In case of two of Cs samples, which show the same level of sticking surface ratio as CSis samples in spite of containing less than 0.3% area ratio of graphite, rather high carbide area ratio may have some effect for anti-sticking resistance.

Carbides in Cs and CSis samples are mainly M_3C and those in CCrs are M_7C_3. CCrs samples show higher sticking surface ratio than Cs, compared at same carbide areas. From this result, we can conclude that M_3C is more effective for anti-sticking component than M_7C_3. Further study is necessary to locate the reason of this difference effect of them.

5. Conclusions

Effect of anti-sticking property of each component in cast steel and iron samples was investigated and the following results were obtained ;

(1) Graphite is the most effective component for anti-sticking property even very small amount like 0.5% of area ratio.

(2) Carbide is also effective component for anti-sticking property, and its effectiveness depends on carbide type.

References

[1] N. Oda: *ISIJ Nishiyama Kinen Kouza No221&222*, (2015) pp. 47-66.

[2] S. Ohhashi,T. Ishiguro and H. Gotoh: *Tetsu - to - Hagane* 77 (1991) 652-659.

[3] A. Noda, E. Matsunaga, T. Hattori and Y. Sano: Hitachi Metals Technical Review 13 (1997) 89-92.

Proceedings of the 72nd World Foundry Congress
21-25th May 2016, Nagoya, Japan

Effect of vanadium addition and cryogenic treatment on the development of carbidic austempered ductile iron

Subham, Parth Shukla and Indrajit Chakrabarty
Department of Metallurgical Engineering, Indian Institute of Technology
(Banaras Hindu University), Varanasi-221005, India.
e-mail: ichakraborty.met@iitbhu.ac.in

The present study has been carried out with an objective to evaluate the effects of additions of vanadium to develop CADI structure. Further improvement in wear properties may be achieved with a cryogenic treatment. In view of this, the iron was subjected to cryogenic quenching in dry ice bath at -50^{0}C subsequent to austempering. The effects are characterized by optical and scanning electron microscopy, hardness and impact toughness measurement and the study of dry sliding wear behavior.

Keywords: *Carbidic austempered ductile iron, cryogenic treatment, impact toughness, dry sliding wear.*

1. Introduction

Hard carbides in a tough ausferritic matrix gives a superior quality wear resistant material which is known as carbidic austempered ductile iron (CADI). f CADI can be developed by alloying with Cr,Mo,V,B or Ti either singly or in combination so that approximately half of the carbon is bound as eutectic carbides [1,2]. Control over the microstructure is exercised in order to obtain the optimum balance between wear resistance and toughness. Deep cryogenic treatment at liquid nitrogen temperature has been reported to improve the wear properties of certain materials [3]. But use in ADI has been rather limited except a few [4,5]. Deep cryo-treatment has been found to cause a complete conversion of the retained austenite into martensite.The aim of the present paper is to study the effect of cryotreatment on the microstructure and mechanical properties of vanadium containing Carbidic Austempered Ductile Iron.

2. Experimental

The as-cast irons were subjected to austenitization at 900^{0}C for 1 h and subsequently austempered at 350^{0}C and 420^{0}C in a salt bath (KNO_3+ $NaNO_3$ in 1:1 ratio) for 15,30, and 45 minutes. Subsequently samples were either air cooled or quenched in dry ice in alcohol bath of temperature -50^{0}C for 6h.

Table 1 Chemical Compositions of the alloy irons.

Alloy Designation	%C	%Si	%V
Alloy 1	3.6	2.0	0.1
Alloy 2	3.5	2.2	0.3

3. Results and Discussions

The austempered structures of alloy 1 at 350^{0}C with varying austempering times followed by cryotreatment are shown in Fig.1. The structures show acicular ausferritic and considerable amount of retained austenite. With increase in austempering time acicularity is gradually decreasing and the retained austenite has increased with 30 min; further increase in causes increase in fine precipitate and decrease in retained austenite. On increasing austempering temperature from 350^{0}C to 420^{0}C, the morphology of ausferrite no longer remaining acicular in nature and fine secondary alloy carbides are precipitated in the matrix. When ferrite forms within the austenite during the austempered process of nodular or ductile cast iron, the carbon is rejected from these regions and goes into solution in the surrounding austenite. As more and more ferrite forms, the carbon content of austenite (γ_{HC}) increases. However, if sufficient austempering time is provided this γ_{HC} initiate rejection of alloy content and Ms temperature is raised. During subsequent cooling it will transform to martensite. Since the cryotreatment is shallow type the M_f is not reached and a considerable retained austenite is observed.

Although the amount of retained austenite differs, similar trend of transformation has been observed with and without cryo-treatment.

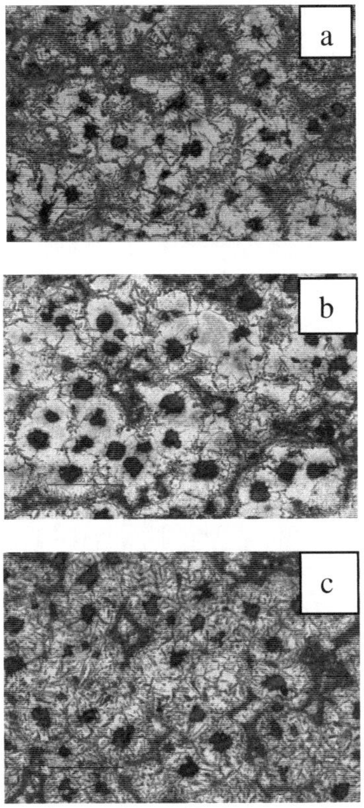

Fig. 1 Microstructures of austempered CADI at 350°C for a)15, b)30 and c)45 minutes followed by cryotreatment.

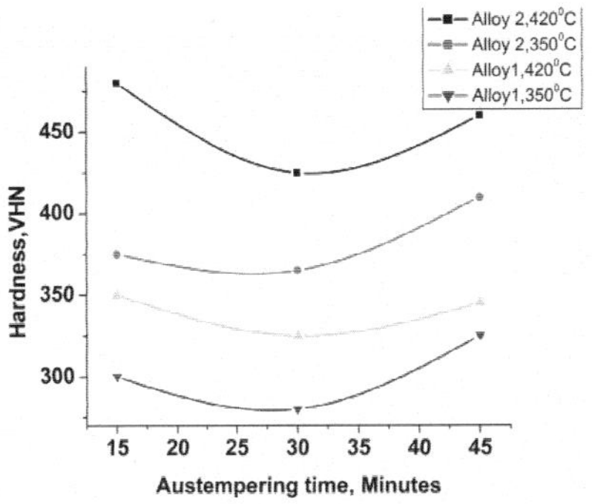

Fig. 2 Hardness vs. austempering time plot with cryotreatment.

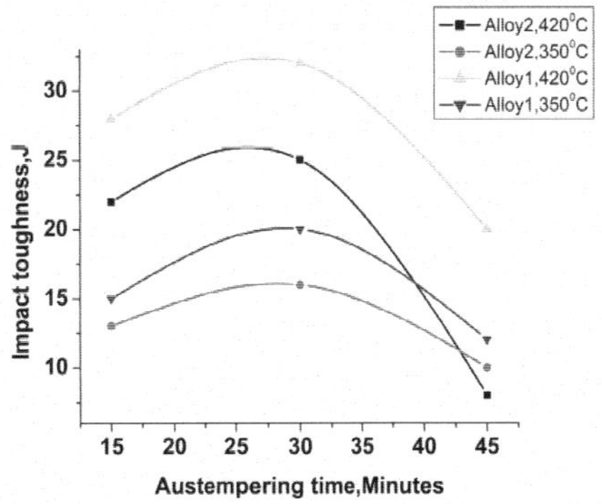

Fig. 3 Charpy impact toughness vs. Austempering time plot with cryotreatment.

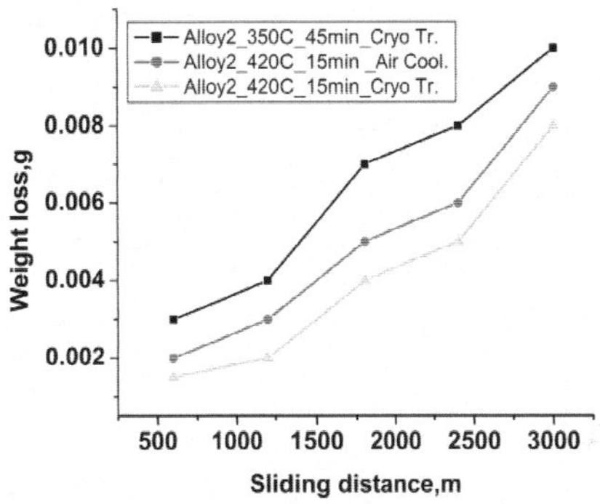

Fig. 4 Dry sliding wear loss vs. distance plot (velocity:1m/s; load:30N).

References

1. K.L. Hayrynen, K.R. Brandenberg: AFS Transactions,111 Paper No.03-088, (2003) 845-850.
2. S. Laino, J.A. Sikora and R.C. Dommarco: Wear 265 (2008) 1–7.
3. D.N. Collins, J. Dormer, Heat Treat.Met.24(1997) 71–74.
4. Susil Putatunda, C. Martis, R. Papp and F. Diekman, in: B.L.Ferguson, R. Jones, D.S. MacKenzie, D.Weires (Eds.), Proceedings of the 26th ASM Heat Treating Society Conference (2011) 44–49.
5. Saranya Panneerselvam, Codrick J. Martis, Susil K. Putatunda and James M. Boileau, Mater.Sci.Eng.A 626(2015)237–246.

Proceedings of the 72nd World Foundry Congress
21-25th May 2016, Nagoya, Japan

The Influence of Strain Induced Martensitic Transformation on Machinability of (ADI)

Adel Nofal[1], Mostafa Ahmed[1], Abdel Hameed Hussein[2], and Mohamed Waly[1]

[1]Central Metallurgical Research & Development Institute (CMRDI)
[2]Faculty of Engineering, Cairo University

Abstract

This work discusses the machinability properties of different ADI grades measured in terms of the cutting force experienced during machining under different machining conditions of cutting depth and cutting speeds.

Four different grades of ADI with different microstructures and hardness levels were subjected to machinability tests by measuring the turning cutting force using three-dimensional force dynamometer with different machining parameters.

The influence of machining parameters on the cutting force was investigated and correlated to the strain induced martensitic transformation from retained austenite during machining.

Keywords: *austempered ductile iron (ADI), TRIP Phenomenon, machinability, cutting force.*

1. Introduction

The recently found wide spectrum of applications is related to ADI's unique combination of mechanical and physical properties [1-2]. However, the serious problems experienced during machining are still limiting its volume production.

The hard grades of ADI would represent a real challenge for high volume machining operations, whereas the soft grades does not represent better machining options due to high retained austenite content, which would transform to martensite by TRIP mechanisms as a result of excessive strains associated with machining [3-5].

In this study, the machinability properties of four grades of ADI in terms of cutting force are evaluated. The interaction between different machining parameters like depth of cut and cutting speed is discussed and correlated to the TRIP phenomenon.

2. Experimental Work

Ductile iron was melted in medium frequency 100 Kg capacity induction furnace. The chemical composition of the D.I bars is shown in table 1.

Table 1. Chemical Composition of D.I bars

C	Si	Mn	P	S	Mo	Cu	Fe
3.68	2.52	0.28	0.0128	0.0139	0.3	0.6	Bal

The base ductile iron (D.I) was subjected to four austempering cycles:-

1. Two conventionally austempered ductile irons *(ADI-275 & ADI-375)* austenitizaed at 900°C and then austempered at 275°C and 375°C for 90 min with fine and coarse ausferritc microstructures, representing hard and soft grades consequently.
2. Two – step austempered ductile iron (ADI-2step), austenitized at 900°C and austemepered at 275°C for 10 min. then at 375°C for 90 min with mixed fine/coarse ausferritic structures.
3. Intercritically austenitized ADI (IADI) at 820°C (where $\gamma + \alpha +$ graphite coexist). The presence of proeutectoid ferrite in the microstructure would contribute to better machinability.

The cutting force during turning was measured by a three-dimensional force dynamometer type 9255B (Kistler, Coherent). The machining parameters varied between 0.5 – 2.0 mm depth of cut and 55 – 180 m/min cutting speed with 0.187 mm/rev feed rate. The martensite content in the microstructure of machined surface was determined using XRD (Rietveld method-MAUD software), and heat tinting techniques

3. Results and Discussion

The retained austenite, martensite content and the hardness of the four investigated ADI grades before machining are shown in Fig. 1.

The maximum cutting force experienced by some ADI alloys at cutting depth of 1.5 mm (Fig. 2) with further decrease of the force with increasing the cutting depth to 2.0 mm suggests that another mechanism plays an important role in formulating the correlation between the machining parameters. With increasing the cutting depth over 1.5 mm, the cutting tool seems to avoid the strain hardened layer and the cutting action proceeds through the deeper unhardened material with a drop in the cutting force.

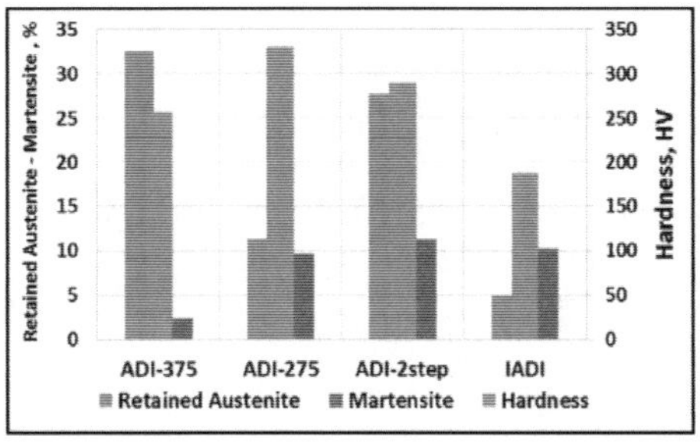

Fig. 1 XRD analysis and hardness of the machinability bars

At higher cutting speeds, the cutting force continues to increase with cutting depth, indicating that the hardened layer is thicker than 2.0 mm due to higher rate of material deformation during machining. With all cutting parameters, the cutting force of IADI was significantly lower than that needed for the other alloys, apparently due to the existence of proeutectoid ferrite in the structure of this grade.

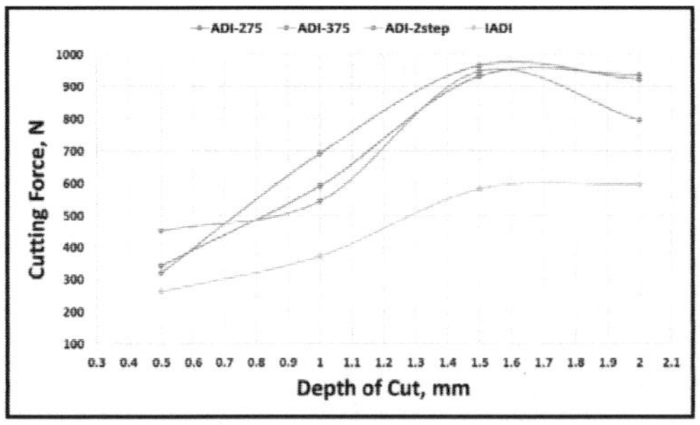

Fig 2 The change in cutting force with depth of cut for different ADI-grades at 55 m/min cutting speed

The increase in cutting force was always associated with a corresponding increase in the martensite content for all investigated alloys. It is interesting to observe that though the ADI-375 and ADI-275 had martensite contents 2.46% and 9.71 % respectively before machining, the two alloys, being subjected to a cutting depth of 2.0 mm had a very similar contents of total martensite (~25%), which is the sum of original content before machining + martensite formed by deformation induced transformation from the retained austenite.

The retained austenite was entirely transformed in ADI-275 and only partially transformed in ADI-375, which indicates that retained austenite formed at higher austempering temperature is more mechanically stable than that formed at lower austempering temperature [6].

The fact that the two alloys reached the same cutting force at 2.0 mm depth of cut confirms that the martensitic transformation due to induced strains is closely related to the machinability behaviour of ADI-alloys. Such correlation is more evident with the ADI-2step alloy (Fig.3), where at 1.5 mm cutting depth, the martensitic content and the force of cutting reached almost the same levels observed in ADI-275 and ADI-375. However, with increasing the cutting depth to 2.0 mm, both martensite content and the cutting force decreased in very similar manner, obviously 2.0 mm depth of cut is deeper than the hardened layer and so, the cutting tool avoids that layer and penetrates deeper in the base metal.

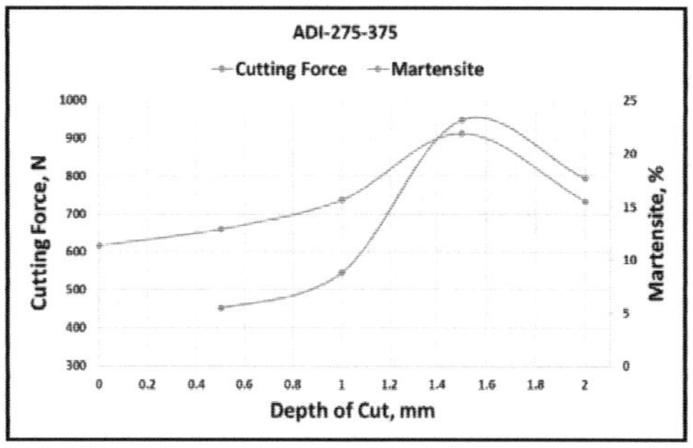

Fig 3 Variation of both martensite content and cutting force with the cutting depth for ADI – 275-375

4. Conclusions

- Retained austenite transformation to martensite through the TRIP effect was found to play a profound role in determining the cutting force and machinability behaviour of ADI-grades.
- The thickness of the hardened layer by martensite formation is a function of different factors, among them are the cutting depth and cutting speed.
- It was found that optimum machinability of different ADI-grades requires different combinations of cutting speed and depth of cutting.

5. References

[1] A. Nofal, 10th International Symposium on the Science and Processing of Cast Iron–SPCI10- Argentina (2014).

[2] Harding RA,Asia-Europe Environment Forum Conf, Jakarta, Indonesia (2005).

[3] M. Carvalho, D. Montenegro, J.Gomes, J. of Materials Processing Technology 213 (2013) 560– 573.

[4] A. Meena & M. El Mansori, International Journal of Advanced Manufacturing Technology 69, (2013) 2833–2841.

[5] K.M. Kumar, P. Hariharan, J. Procedia Engineering 64 (2013), 1495 – 1504.

[6] Saranya P, Codrick J., Susil K. Putatunda, J. of Materials Science and Engineering: A 626 (2015), 237–246.

Proceedings of the 72nd World Foundry Congress
21-25th May 2016, Nagoya, Japan

A contribution to new material standards for Ductile Irons and Austempered Ductile Irons

Franco Zanardi[1], Franco Bonollo[2], Nicola Bonora[3], Andriew Ruggiero[3], Giuliano Angella[4]
[1] Zanardi Fonderie SpA, via Nazionale 3 - 37046 Minerbe (VR), Italy
[2] Dpt. of Management and Engineering, University of Padova, Str. San Nicola, 3 – 36100 Vicenza, Italy
[3] Dpt. of Civil and Mechanical Engineering, University of Cassino and Southern Lazio, Via G. Di Biasio 43 – 03043 Cassino, Italy
[4] CNR IENI, U.O.S. of Milano, via R.Cozzi 53 – 20125 Milano, Italy

Some results of material characterization activities, dedicated to classical and fracture mechanics fatigue and elasto-plastic properties, have already been published. Others have not yet been published.
The possible use of all of these results in new standards is here discussed.

Keywords: *Ductile Iron, ADI, IDI, Notch Mechanics, Material Standard.*

1. Introduction

The structural design of castings is mainly based on normative properties measured with the conventional tensile test and other informative properties (the most important is fatigue strength).

The conventional tensile test is representative of the material properties when applied to test probes taken from separately cast test bars, supposed free from defects.

Defects in castings are principally influencing the local fatigue strength. The plastic deformation of the casting as a whole is usually affected at a minor extent by small defects. For this, the tensile test done on test probes taken from the casting not always represents the material design properties.

This paper consider the ductile iron material family, collectively referred as XDIs, including the new "Silicon Solution Strengthened Ferritic Ductile Iron" grades and the " Isothermed Ductile Iron IDI", a patented pearlitic-ferritic grade obtained by heat treating (austenitization in the intercritical temperature range and quenching into a salt bath) a unalloyed ferritic ductile iron casting.

All material properties are referred to test probes taken from Lynchburg 25 mm dia and/or Y 25 mm cast test samples.

2. Fatigue design in presence of defects

The Kitagawa diagram and the Murakami approach for the determination of the equivalent defect size offer a practical approach for the estimation of the fatigue strength in presence of small defects. For this, supported by the Dpt. of Mechanical Engineering, University of Padova, Italy, we calculated some "threshold value for the stress intensity factor" ΔK_{th} [MPa $m^{0.5}$] [2,3,4].

3. The uniaxial tensile curve

When comparing XDIs with steels, little consideration is generally given to the fact that the plastic deformation pattern of XDIs and steels are different. Generally speaking, the plastic deformation design should consider the engineering tensile test up to the UTS R_m and not in the necking plastic instability range.

Using the Voce approach for different grades of XDIs and for different grades of commercial Q&T steels, co-author[4] developed indicators able to discriminate the strain hardening shape of ADIs from those of ferritic-pearlitic DIs and/or steels.

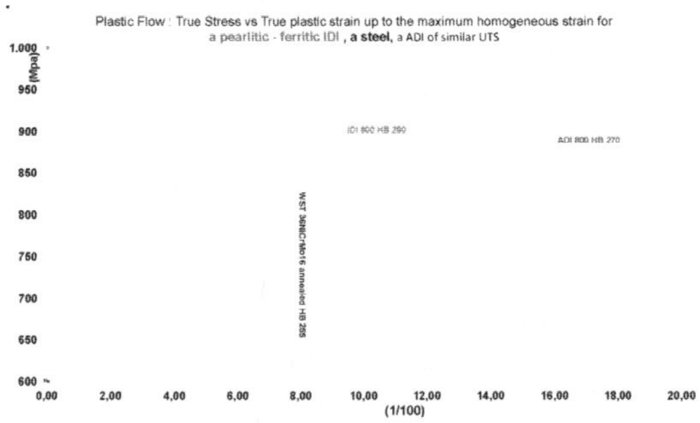

Fig. 1 Plastic flow of an IDI, a steel and a ADI

4. The low temperature and high strain behavior

Co-authors[3] performed uniaxial tensile characterization for the grade ADI UTS min 1050 MPa over the temperature range comprised between

-60°C and +70°C, and for a strain rate ranging from 0.001 and 1000 s^{-1}). Results showed 15% increase of the flow stress at 1000 s^{-1}.

5. Proposals for an improved approach to the standards

A new XDIs material standard is proposed, based on the following dimensions:

1. GRADE: minimum R_m and/or HB
2. UNIFORMITY: range $HB_{max} - HB_{min}$
3. QUALITY

In most ferritic-pearlitic DIs existing material standards, the designation table communicates to designers a brittle or quasi-brittle behavior for all grades, except for the fully ferritic ones. The designation values are so poor because they consider a wide range of hardness. Since the beginning (90s) of his activity in the standardization committees, the author claims [1] (unsuccessfully) that the existing designation approach is not appropriate, proposing the use of a " R_mA_5 Material Quality Index MQI".

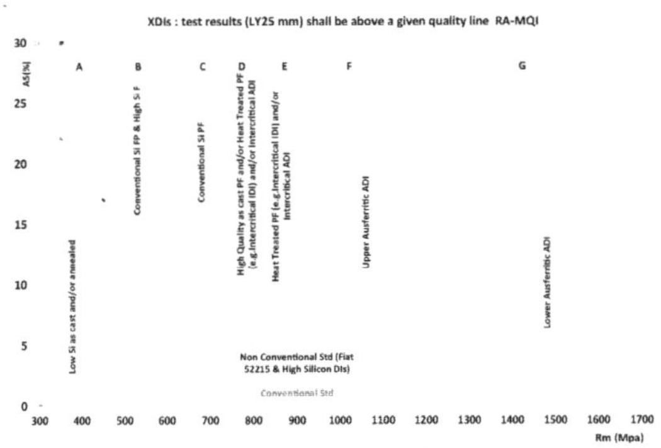

Fig. 2 XDIs plane R_m A_5 and RA-MQI index

The inappropriate definition of grades in most standards is one of the reasons for the predictable success of the "Silicon Solution Strengthened Ferritic Ductile Irons SiSSFDIs". The main difference between SiSSFDIs and conventional Ferritic Pearlitic DIs lies in the different ratio between UTS (Rm) and Yield Strength (Rp0.2).

Because of the importance in safety factors definition, it should be recommended also a specified range for the Rm/Rp0.2 ratio.

The highest SiSSFDI grade Rm min = 600 MPa Rp0.2 min = 470 MPa A5 min = 10% has no alternative in the conventional standards. In fact, to get a similar Rp0.2 min, it is necessary to adopt a fully pearlitic grade, which cannot achieve a satisfactory elongation. This is anyway possible with the adoption of the patented IDI grade $R_{m\,min}$ = 800 MPa $R_{p0.2\,min}$ = 480 MPa $A_{5\,min}$ = 6%, offering a lower elongation but a higher UTS.

When material properties (including possible defects influence) are required from test probes taken from the casting, a finite life minimum number of cycles at a given stress level should be required, instead of minimum properties to be achieved on a tensile test.

Conclusions

A new normative approach, based on material grade and material quality, has recently been submitted to the attention of CEN TC 190 WG 07 working on prEn1563.

A fatigue strength calculation based on the Kitagawa diagram, generalized to blunt notches by mean of the University of Padova Atzori-Lazzarin approach, is included in the proposal.

Acknowledgements

The author is grateful to the Dpt. of Mechanical Engineering, University of Padova, particularly to prof. Giovanni Meneghetti who in the last 15 years strongly contributed in the field of fatigue design in presence of defects.

Thank you to dr. PhD Kathy Hayrinnen for her kind help in reviewing the text.

References

[1] ISO/TC 25/SC 2/WG 1 NYYY 2001-12-10 R.Gast John Deer USA, F.Zanardi Zanardi Fonderie Italy : Procedure for the adoption of a Material Quality Index

[2] Elsevier Sc.Dir.Proc.Eng. 00 (2009) 000-000 – Fatigue 2010 – Stefano Masaggia, Zanardi Fonderie : The development of ADI and IDI in Italy

[3] Key Eng.Mat. © (2011) Trans Tech Pubbl.Switzerl.doi: 10.4028 www.scientific.net/KEM.547.181, Sc. and Proc.of Cast Iron IX B.Atzori, F.Bonollo, G.Meneghetti: Notch Fatigue and Fracture Mechanics of Austempered Ductile Irons"

[4]© 2012 World Foundry Congress Monterrey Mexico G. Meneghetti University of Padova, S. Masaggia Zanardi Fonderie Spa : Estimation of the fatigue limit of components made of Austempered Ductile Iron weakened V-shaped notches

Proceedings of the 72nd World Foundry Congress
21-25th May 2016, Nagoya, Japan

Chemistry and Process Control in Production of High-Alloy Graphitic Irons

Delin Li[1] and Clayton Sloss[2]
[1] CanmetMATERIALS, Natural Resources Canada (NRCan), Hamilton, ON, Canada
[2] Wescast Industries Inc., Brantford, ON, Canada

High-alloy graphitic irons include ferritic high-silicon molybdenum irons (referred to as SiMo) and austenitic high-nickel irons (referred to as Ni-resist). The high alloy contents greatly influence the eutectic carbon content and the constituents of these cast irons. This paper highlights some examples of the latest developments in SiMo and Ni-resist irons produced for automotive exhaust components with section thickness < 20 mm (3/4"). The graphite shapes of SiMo typically consist of spheroidal graphite (SG), compacted graphite (CG), and the recently-developed mixed graphite (MG) with medium graphite nodularity. MG iron constitutes a new class of cast iron that can improve manufacturability and material properties as compared to CG and SG irons.

Keywords: *SiMo, mixed graphite, carbon equivalent, Ni-resist, chunky graphite.*

1. Introduction

There are four groups of cast ferrous alloys produced for elevated-temperature service: Group 1 - ferritic high-silicon irons, e.g., SiMo; Group 2 - austenitic ductile irons, e.g., D5S; Group 3 - ferritic stainless steels, e.g., CB30 per ASTM A743; and Group 4 – austenitic stainless steels, e.g., HK per ASTM A297.

In the 1990s, the SAE Iron and Steel Castings Committee generally documented the four groups of alloys in regard to chemistry, properties, machinability, and temperature limits [1]. Recently, more comprehensive investigations of exhaust manifold materials were performed [2]. High-alloy graphitic irons from Groups 1 and 2 continue to be the material of choice for automotive exhaust components where castability, machinability and oxidation resistance become the important selection criteria.

Previous publications mainly focused on lab-scale studies. The primary objective of the present paper is to facilitate the transfer of development results from research lab to production floor for SiMo and Ni-resist iron castings.

2. SiMo irons

2.1 Mixed graphite (MG)

The graphite nodularity is typically specified as ≥80% for SG iron, ≤ 30% for CG iron, and in a range of 40% to 70% for MG iron. Historically, MG iron was treated as scrap iron (it was neither CG nor SG iron), but recently it has been intentionally developed for elevated temperature applications [3]. As both SG and CG SiMo irons are currently used for automotive exhaust components, the present work proved that MG iron can integrate the advantages of both SG and CG structures. Extensive process development and material evaluations have been performed for MG SiMo iron. Progress to date has shown that MG SiMo provides multiple benefits such as improving manufacturability and thermal durability in comparison with CG and SG SiMo. A new ASTM standard A1095, "Specification for High-Silicon Molybdenum Ferritic Iron Castings" with SG, MG, and CG structures has been recently created.

2.2 Brittleness at medium temperature (BMT)

SG SiMo irons can be susceptible to BMT where the minimum ductility occurs at temperatures between 310 °C and 490 °C. BMT is greatly influenced by the residuals of magnesium, sulfur, and phosphorus. The tensile testing results for elongation at room temperature (RT) and 425 °C are presented in Fig. 1. As the P content increased from 0.016% to 0.047% (wt. % used in this paper), the elongation at RT was in a range of 10% to 16% which is typical for SG SiMo. However, the elongation at 425 °C responded significantly as a function of P content. The baseline samples with 0.016% P exhibited pronounced BMT. When exceeding 0.025% P, BMT is suppressed. Specifications previously did not control the minimum P content and have allowed huge variations in BMT of SG iron. Therefore, the new standard ASTM A1095 specifies the P concentration from 0.02% (min.) to 0.05% (max.) in order to eliminate BMT of SG iron. MG iron was immune to BMT because of moderate magnesium content in the range studied.

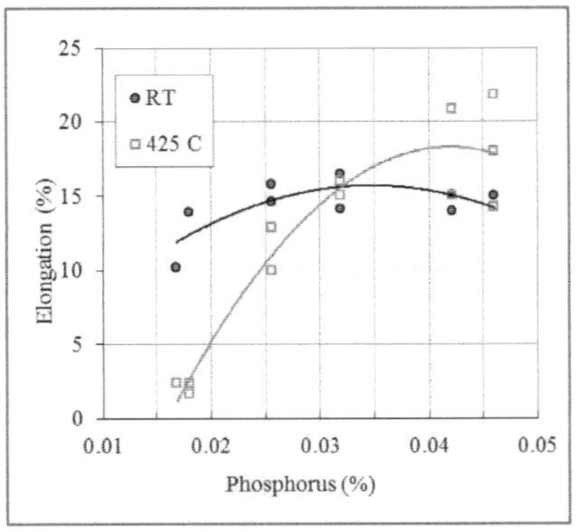

Fig. 1 Results of tensile testing at RT and 425 °C of SG SiMo samples containing 0.010% S [3].

2.3 Carbon equivalent (CE)

It has been proven that CE (C+1/3Si) can alter the microstructure and quality of castings. Controlling CE with C and Si contents would be a great advantage for foundries; they could optimize microstructure, mechanicals, and yield of their castings. Figure 2 is an example of a nomogram to provide guidance on an optimum CE from 4.50 to 4.75 for SG SiMo. Slightly lower CE is recommended for CG and MG irons.

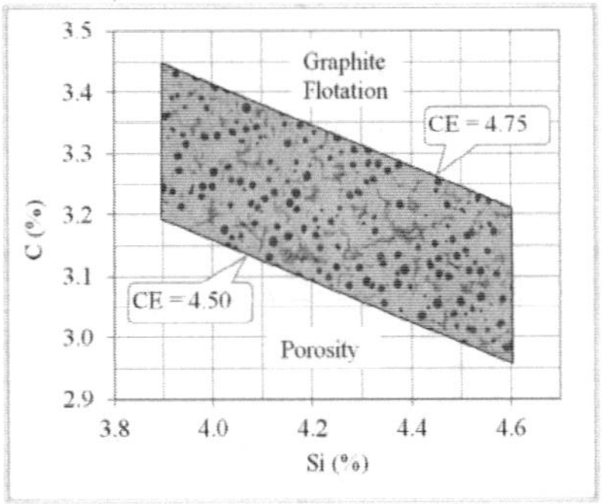

Fig. 2 Carbon, silicon, and optimum CE (micrograph zone) ranges for SG SiMo irons.

3. Ni-resist irons

3.1 Chemistry control

Three grades of Ni-resist are commonly produced for use at high temperatures: D2S (25% Ni), D4A (30% Ni) and D5S (35% Ni) per ASTM A439. Ni-resist chemistry control ranges, especially C and CE must be tighter than ferritic SiMo irons. The CE of Ni-resist iron is calculated by Eq. 1 [4]. The optimum CE was found between 4.20 and 4.40 to achieve uniform microstructure (Fig. 3a). Castings are prone to shrinkage or porosity if CE < 4.20. Degenerated graphite structures occur if CE > 4.40.

$$CE = C + 0.33Si + 0.047Ni - 0.0055Si \times Ni \quad (1)$$

3.2 Chunky Graphite

Lower nickel D2S iron alleviated susceptibility to chunky graphite as compared to D5S. A very small amount of rare earth metals increased the graphite nucleation undercooling and thus caused massive chunky graphite in D5S iron as shown in Fig. 3b. Over inoculation, e.g., double inoculation in both ladle and mold, augmented the propensity to chunky graphite, especially when the melt is treated with Fe-Si-Mg instead of Ni-Mg nodularizer. Chunky graphite could happen in thick sections (20 mm), as well as thin sections (4 mm) of castings. Formation of chunky graphite in Ni-resist irons can reduce the tensile testing results, especially the elongation.

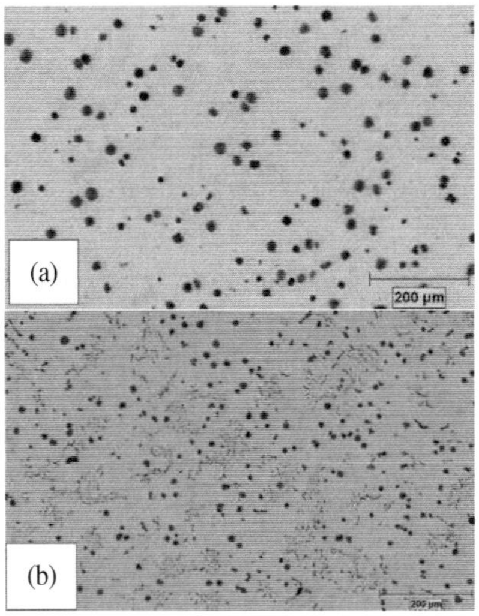

Fig. 3 Micrographs of D5S samples: (a) normal, and (b) chunky graphite observed when 80 ppm La added.

Acknowledgements

The authors acknowledge the support from Canada PERD program. NRCan (Crown Copyright) and Wescast permission for publishing this work is gratefully acknowledged.

References

[1] SAE ISTC, "High Temperature Materials for Exhaust Manifolds," SAE J2515, 1999.

[2] M. Ekstrom, "Oxidation and Corrosion Fatigue Aspects of Cast Exhaust Manifolds," Ph.D. Thesis, Sweden, 2015.

[3] D. Li and C. Sloss, "Brittleness at Medium Temperature of High-Silicon Molybdenum Cast Irons," AFS Proceedings 2015, Paper 15-018.

[4] G. Cox, "Spheroidal Graphite, Austenitic Irons for Use at High Temperatures," The British Foundryman, Jan. 1970.

Proceedings of the 72nd World Foundry Congress
21-25th May 2016, Nagoya, Japan

Microstructure and Mechanical properties of Y_2O_3-Strengthened Fe-Cr Alloy Castings for High Temperature Applications

Shimaa El-Hadad[1], Adel Nofal [1], Mervat Ibrahim [1], Mohamed Mourad[1]
[1] Central Metallurgical Research and Development Institute, Box. 87 Helwan-Egypt

The oxide dispersion strengthenened alloy (ODS), MA956, is an ideal candidate in sulfur containing high temperature environments. This alloy is an iron-based alloy prepared by adding Y_2O_3 via powder metallurgy. The challenge is to produce an alloy close to MA956 by casting technology. This is because Y_2O_3 which has lower density than steel floats on the surface of the melt and accumulate in the cast riser. In the current work, a 20% Cr-Fe alloy strengthened by 0.5% Y_2O_3 was fabricated through casting process. The microstructure and mechanical properties of this alloy after casting and heat treatment were investigated. Energy dispersive spectrometry (EDS) showed some randomly distributed particles containing Y. The tensile strength was doubled after forging and heat treatment of the cast alloy. These results confirmed that Y_2O_3 can be added to the steel matrix during casting, which is a more economical and practical method for ODS alloy fabrication.

Keywords: Oxide strengthened alloys, casting, microstructure characterization, mechanical properties.

1. Introduction

High operating temperature, combined with sulfur environment is likely to be encountered in service. This creates a severe operating condition for materials used in high temperatures. In some Egyptian boiler power plants, a sulfur containing fuel was recently used as an alternative in case of shortage in natural gas. In the burner area, where the S containing fuel circulates, the diffusers fail after 6 months of usage as shown in Fig. 1. These diffusers were fabricated of Ni50:Cr50 alloy. The failure analysis showed that formation of Ni-sulfides was the main reason of failure. It was reported that the melting point of Ni-sulfide eutectic ($Ni-Ni_3S_2$) is as low as 635°C, compared to eutectics of cobalt and iron which are 880°C and 985°C respectively [1,2]. This low melting point sulfide cause voluminous scales and lead to damage. Therefore, an iron–based high

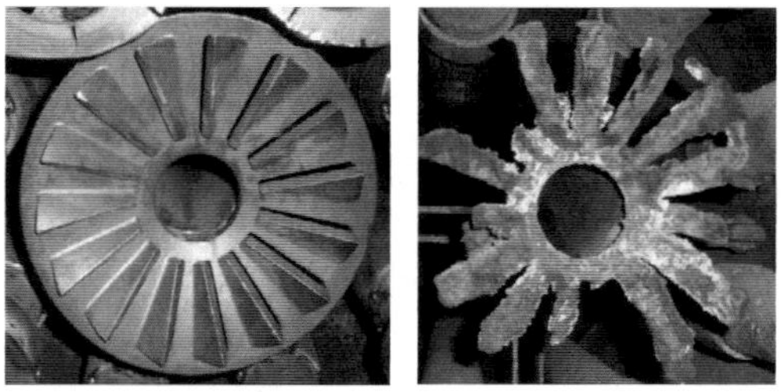

Fig. 1 Diffuser before and after failure.

temperature alloy which resists S containing gases was selected. Alloy MA956, is known for its high-temperature creep resistance in S containing environments [3,4]. However, this alloy is prepared by mechanical alloying followed by hot pressing which is a complicated process. In the current study, a cast alloy close in its chemical composition to MA956 was prepared and its microstructure and mechanical properties were investigated.

2. Experimental work

The charge materials were added as follows; steel scrap, low carbon ferro chromium followed by addition of pure Al metal, and finally Y_2O_3 powder. Melting was done in an induction furnace at 1650 °C. The frequency was controlled to allow the melt stirring in order to prevent the oxide particles from floating on the surface. The chemical composition of the Fe alloy is shown in Table 1. Some cast bars were then forged to distribute the oxide particles and refine the cast microstructure. After forging, the samples were heat treated by normalizing at 1100 °C for 15 minutes. Microstructure investigation and mechanical properties evaluation of the cast, heat treated and forged samples were performed.

Table 1 Chemical composition of the cast alloy.

C	Si	Mn	P	S	Cr	Mo	Ni	Al
0.059	0.51	0.587	0.0187	0.011	18.22	0.0538	0.221	>3.6
Co	Cu	Nb	Ti	V	W	Pb	As	B
0.316	0.074	0.014	0.22	0.0735	0.0149	<0.003	0.005	<0.0001

3. Results and discussion

3.1 Microstructure investigation

Microstructure of the samples was investigated using SEM and EDS. Figure 2 shows the microstructure of the cast and forged specimens before and after heat treatment (HT). The grain refinement of the ferritic matrix due to forging is shown in Fig. 2 (c). Since the elemental yttrium couldn't be analyzed (in Table 1) using the available spectrometer, EDS was used to analyze some particles which were randomly distributed in the matrix as shown in Fig. 3. It is clear that Y_2O_3 could successfully be introduced in the matrix. The amount of added Y_2O_3 was planned to be 0.5%, however the low amount of Y observed by EDS is due to floating of such fine particles during melting.

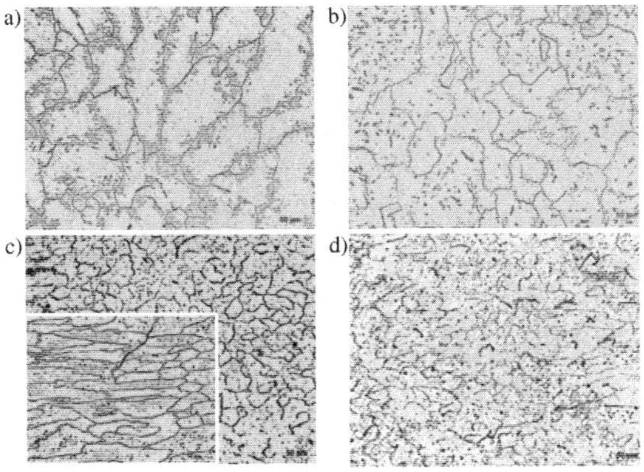

Fig. 2 Microstructure of cast samples (a, b) and forged specimens (c,d) before and after heat treatment respectively.

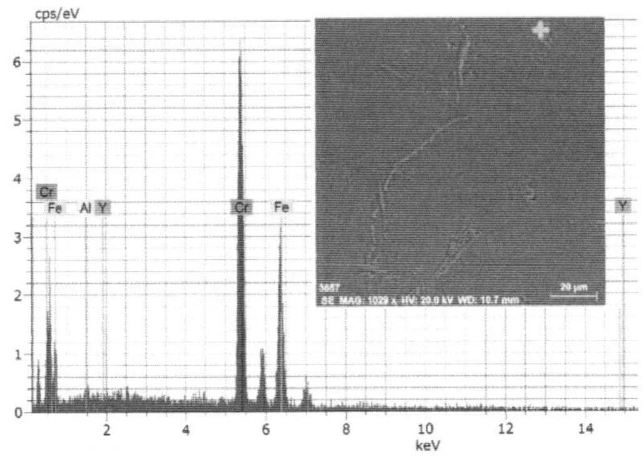

El	AN	Series	unn. C [wt.%]	norm. C [wt.%]	Atom. C [at.%]	Error [wt.%]
Cr	24	K-series	36.78	54.84	56.07	1.2
Fe	26	K-series	29.57	44.09	41.96	1.1
Al	13	K-series	0.65	0.97	1.91	0.1
Y	39	L-series	0.07	0.10	0.06	0.1
		Total:	67.06	100.00	100.00	

Fig. 3 EDX analysis of some Y containing particles.

3.2 Mechanical properties

In order to evaluate the properties of this Y_2O_3 added cast alloy, hardness measurements and tensile test were performed, as shown in Table 2. It is observed that the cast samples have low hardness and low ultimate tensile strength as well. After forging and heat treatment, UTS was relatively increased. It was reported that [5], the ferritic steels show lower strength at temperatures exceeding 600°C, but are more resistant to thermal shocks than high temperature austenitic stainless steels. With the thermal conductivity higher and the thermal expansion lower than the respective values for austenitic steels, equal thermal shocks will result in lower thermal stresses in the ferritic material. In these terms, ferrites allow greater tolerances for design and operation . Therefore, the designed alloy in this work is believed to work properly at high temperature and S containing atmosphere.

Table 2 Hardness and tensile strength measurements.

Condition	Hardness, *Hv*	UTS, N/mm^2
As-cast	104	251
Cast-HT	133	310
As-forged	168	316
Forged-HT	251	502

4. Conclusions

The current research showed that Y_2O_3 could be added to the steel matrix during casting process. Subsequent deformation and heat treatment could significantly enhance the mechanical properties of the castings. Further research works are needed to increase the percentage of the oxide dispersions in the steel matrix using casting technology.

Acknowledgements

The authors would like to thank the fund granted from the "Egyptian Academy of Scientific Research".

References

[1] Elliott, P., "Practical Guide to High-Temperature Alloys," Materials Performance, 29 (4), p. 57 (Apr. 1989).

[2] Lai, G. A., "High-Temperature Corrosion of Engineering Alloys," ASM International, Metals Park, OH (1990).

[3] A technical report from URL: www.specialmetals.com.

[4] Nickle Development Institute ,NiDi pub, No. 266.

[5] Technical report from http://www.outokumpu.com.

Proceedings of the 72nd World Foundry Congress
21-25th May 2016, Nagoya, Japan

Development of Cast Steel Brake Disc with Heat Shock Resistance

Naoki Harada[1], Yukio Tanida[1], Tsuyoshi Fukuda [1], Masanori Takuma[2],
Masato Tsujikawa[3], Kenji Higashi[3]
1KURIMOTO Ltd., Shibatani 8-45 Shibatani 2-chome, Suminoe-ku, Osaka-shi 559-0021, JAPAN
2Kansai University Yamatecho 3-35, Suita-shi 564-8680, JAPAN
3Osaka Prefecture University Gakuen-cho 1-1, Naka-ku, Sakai-shi 599-8531, JAPAN

Recently, the maximum running speed of railway cars and braking load continues to increase leading to the need to improve the heat shock resistance of railway cars discs. In this study, the heat shock resistance of two materials with different hardenability was compared in the braking tests. In discs with good hardenabilty, hardness near the friction surface was increased and many cracks were observed. The white layer was clarified to be martensite. On the other hand, in discs with poor hardenabilty, hardness hardly changed and the heat shock resistance was good. The white layer was clarified to be tempered martensite.

Keywords: *cast steel, brake disc, heat shock resistance, heat crack, hardenability.*

1. Introduction

Recently, the running speed of railway car has become fast and the load to the brake disc has increased. It is said that the transformed phase so called white layer has something relations to crack initiation not only in the brake disc but also in a wheel and rail[1-3], however the relation between the microstructure and the crack resistant has not been clarified yet.

In this study, discs with two materials which have different hardenabilities (G: good hardenability and P: poor hardenability) were subjected to braking tests and the heat cracks and the white layer on the friction surface were observed by OM and TEM to investigate the relation between crack resistance and microstructure change.

2. Experimental Procedure

2.1 Manufacturing the discs

Brake discs whose chemical compositions were shown in Table 1. were melted in RF furnace and cast into the sand mold. Thereafter the discs were heat-treated (1223K×3h Annealing, 1233K×3h Water Quenching, and 903K×3h Temperig) and were machine worked.

Table 1 Chemical compositions of discs with different hardenabilities (mass %).

Material	C	Si	Mn	Ni	Cr	Mo	Al
G	0.23	0.80	0.80	1.70	0.80	0.50	0.03
P	0.23	0.80	0.50	0.50	0.80	0.80	0.03

2.2Braking test and microstructure observation

Brake discs were subjected to braking-test assuming Shinkansen equivalent conditions: maximum running speed is 300km/h)

Following the braking test, the discs were cut and microstructures near friction surfaces were observed by OM, and hardness changes were measured by Vickers hardness tester (Hv0.10).

White layers confirmed by OM were observed by TEM in particular.

3 Results and Discussion

3.1 OM observation and hardness distribution

Fig.1 shows the microstructures observed by OM and Fig.2 shows the hardness distribution near friction surface.

OM observations clarified that both discs have white layer and heat cracks. However hardness near the friction surface of good hardenability disc has increased than that of poor hardenability.

3.2 TEM observation

Fig.3 shows the micrograph observed by TEM.

White layer with good hardenability disc was fresh martensite with high density dislocation. On the other hand, acicular cementite precipitation has confirmed in white layer with poor hardenability disc. The material of poor hardenability has less Mn content than the material of good hardenability, and it is considered that it is easily tempered in the material of poor hardenability.

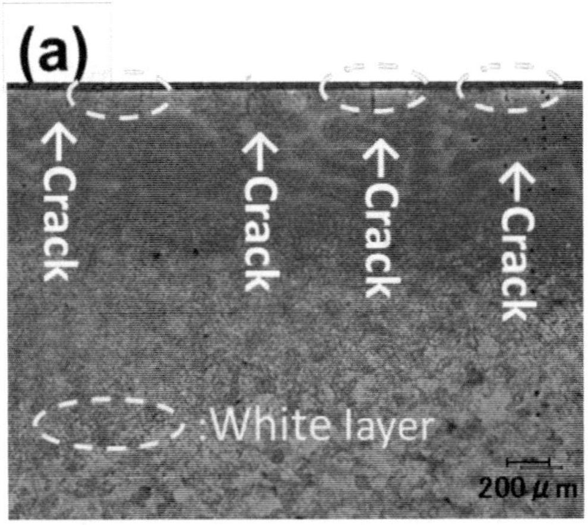

(a) Good hardenability disc

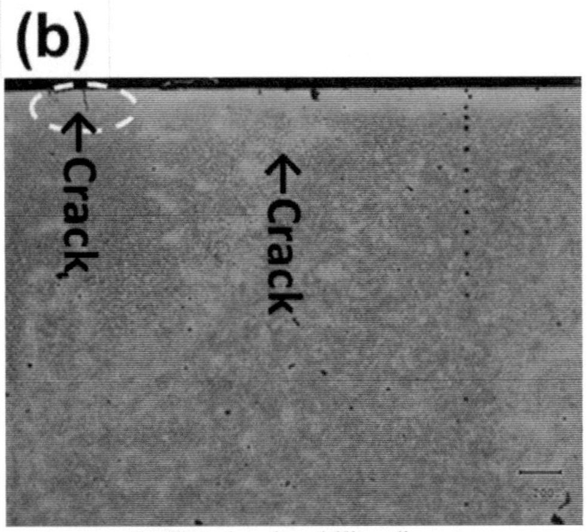

(b) Poor hardenability disc

Fig.1 Microstructures observed by OM.

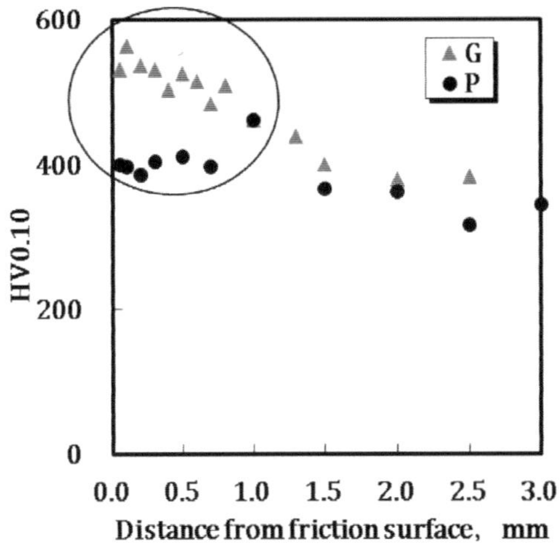

Fig.2 Hardness distribution near friction surface.

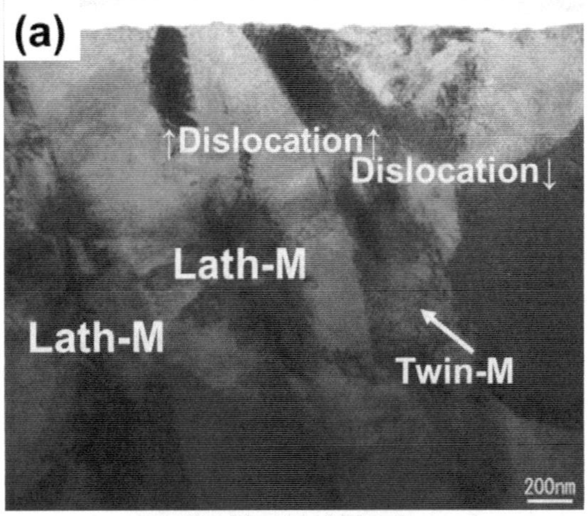

(a) Good hardenability disc

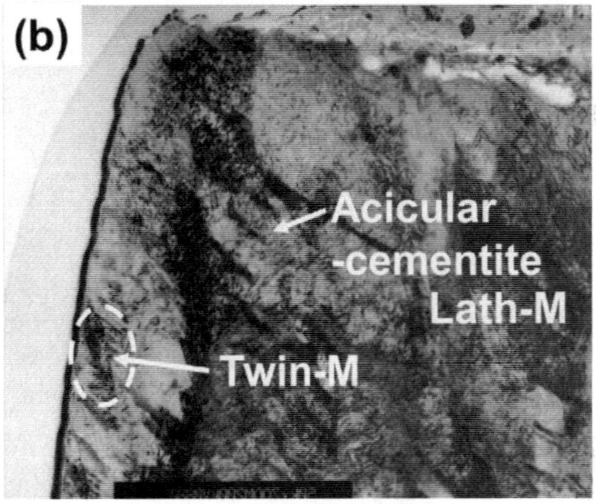

(b) Poor hardenability disc

Fig.3 TEM images of white layer observed by TEM.

4. Conclusions

(1) Heat shock resistance of disc has correlate strongly with its hardenability.

(2) White layers were observed in both discs regardless of their hardenabilities.

(3) White layers of disc with good hardenability were lath-martensite including high-density dislocations, the others were lath-martensite with acicular cementite because of less content of Mn.

References

[1]M.Tsujie and H. Mori Materia Japan p.445-463,49(2010)10

[2]H. Mori and M Matsui Tetsu to Hagane p.64-66,91(2005)12

[3]K. Handa, Ph. D. Thesis, p.64-p.77, Tokyo Institute of Technology, Tokyo (2010).

Proceedings of the 72nd World Foundry Congress
21-25th May 2016, Nagoya, Japan

Hot Ductility Behavior of Continuous Casting Slab of Copper and Nickel Alloyed Steel

Seokhwan Lee[1], Sunryong Yoon[1], Jiyeon Jeong[1], Kwang Suk Son[1] and Donggyu Kims[1]
[1] Dept. of Materials Science and Engineering, Dong-A University, Busan 49315, Korea

Copper is deliberately added to low carbon steels to improve corrosion characteristics and increase strength. However, copper as a tramp elements has been reported to encourage transverse cracking by hot shortness of slab surface upon straightening operation during continuous casting. Nickel can be added to compensate for this loss by both increasing Cu extent of solid solution in austenite and decreasing Cu enriched phase in oxide scale. In a continuous casting steelmaking operation, the surface of a slab is under a condition that can be characterized as high temperature, low-cycle fatigue in which the tensile and compressive stress is repeatedly developed. For this reason, for the evaluation of the hot ductility of a slab, considering the fatigue deformation is more feasible before a tensile or compressive test. And also the oscillation mark developed on the surface of slab due to the oscillation of copper mold during continuous casting would results in reduced hot ductility. In this study, the effects of low cycle fatigue and oscillation mark on the hot ductility of copper and nickel containing steels are investigated at various temperatures.

Keywords: *continuous casting, slab, surface crack, hot ductility, fatigue test*

1. Introduction

During continuous casting processes, there are many problems related to the hot ductility of the cast slab. Surface defects such as transverse cracks and edge cracks are regarded to potentially deteriorate the properties of the slab and lead to waste slab during the removal of defects [1,2].

The low carbon steels containing Cu and Ni have been used in ocean structure. Copper has been reported to encourage surface cracks during continuous casting, since iron is preferentially oxidized on the surface of the slab resulting in the formation of copper enriched phase between scale and matrix. At high temperature, copper enriched phase start to melt and penetrate into the matrix along the grain boundary of austenite, thereby hot ductility is reduced. On the other hand, nickel has been added to copper alloyed steel to neutralize the detrimental effect of copper in steel. The strain accumulated during continuous casting and slab surface morphology including oscillation mark should be considered to get more precise hot ductility data for the application in the field.

2. Experimental Procedure

2.1 Specimen

The composition of the experimental specimen is given in Table 1. Two ingot samples were prepared using a vacuum induction melting furnace. Fatigue specimens were machined from the surface of solidified ingot so as to obtain material with a cast microstructure and oscillation mark.

Table 1 Chemical composition of specimens (wt.%).

	C	Si	Mn	S	Cu	Ni
Steel A	0.08	0.22	1.43	0.002	0.23	0.25
Steel B	0.07	0.20	1.56	0.001	0.28	0.86

2.2 Hot ductility test

A cast simulator was used to measure the hot ductility via a simple tensile test in addition to a tensile test after fatigue deformation at a high temperature under an Ar atmosphere.

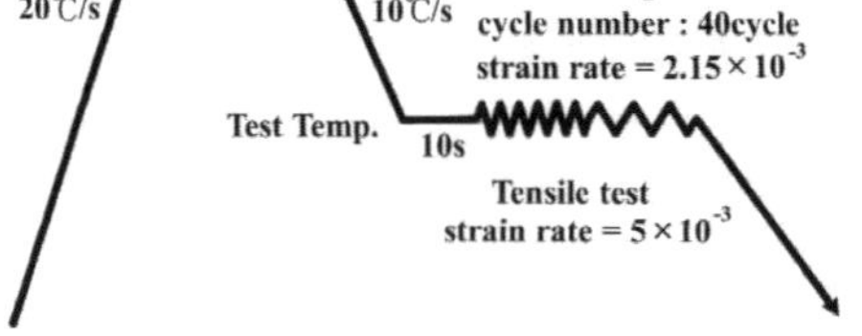

Fig. 1 Thermal history of hot ductility test.

Fig. 1 illustrates the thermal profile that the specimen passes through during a mechanical test as well as the fatigue and tensile tests. The specimens were heated to 1300 °C and held at this temperature for 5 min for solutionizing. They were then cooled to the testing temperature at the rate of 10 °C /s, which is similar to the cooling rate during thin slab casting. At the testing temperature, the specimen was held for 10 s in order to reduce the temperature gradient before tensile or tensile after fatigue test.

The high temperature tensile test was conducted in a temperature range of 600 to 1000 °C, with 50 °C intervals. The strain rates were 5×10^{-3}/s for those in both continuous casting and thin slab casting. A number of specimens were quenched for microstructural analysis in various conditions.

Fatigue test was performed under conditions in which 10 % fatigue deformation by 40 cycles, and a deformation rate of 7.16×10^{-3} were loaded before the tensile test at the testing temperature. The number of fatigue cycles, 40, was chosen as the number of rolls in the secondary cooling zone is approximately 40.

3. Results and Discussion

Fig. 2 show the hot ductility curve of Steel A. Ductility trough is developed at 800 °C and ductility reversion is arisen at high and low temperature region in both test conditions. There exists no significant difference of hot ductility between simple tensile test and fatigue test conditions.

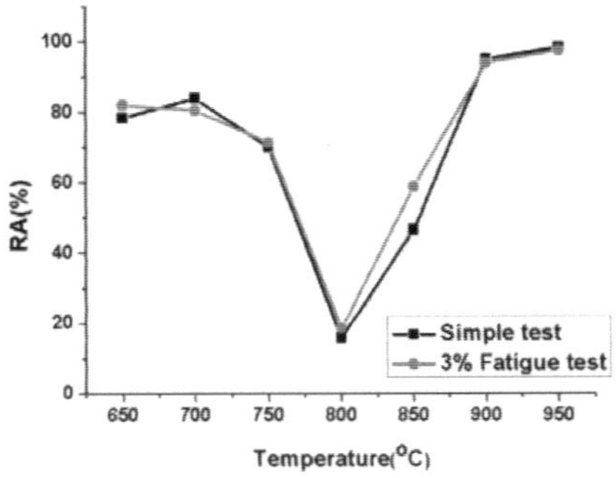

Fig. 2 Hot ductility curves of Steel A.

Microstructures of Steel A quenched at various temperature after fracture is shown in Fig. 3. Ferrite thin film is developed along the prior austenite grain boundary at 800 °C at which ductility shows lowest value. Increased amount of ferrite is formed at 750 °C, thereby hot ductility reversion arises at this temperature.

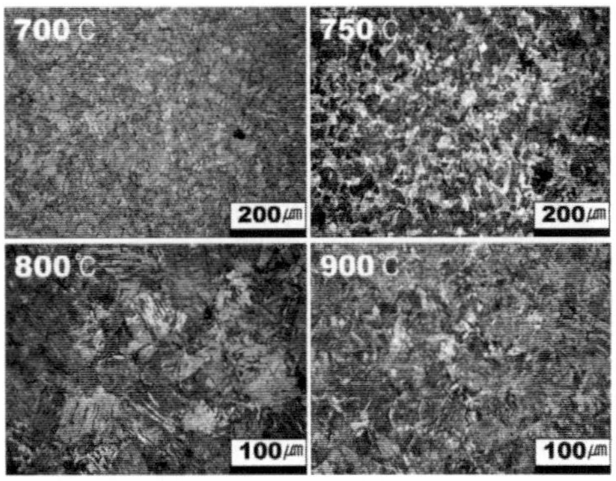

Fig. 3 Microstructures of Steel A quenched at various test temperature after fracture.

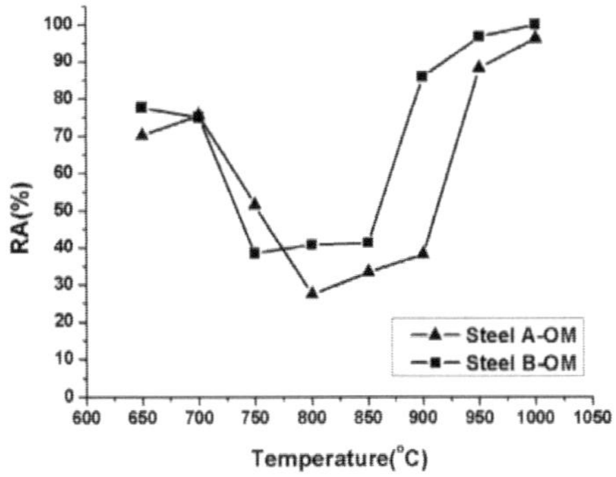

Fig. 4 Hot ductility curves of Steel A and B.

Fig. 4 shows the hot ductility curves of Steel A and B, from which Steel B has increased hot ductility at 900 °C while Steel A represent lower ductility value. The difference in hot ductility value at this temperature can be ascribed to the lower A_{r3} temperature due to the increased amount of nickel in Steel B.

4. Summary

With increasing nickel to copper ratio, hot ductility of copper containing steel was improved. No meaningful difference of hot ductility between simple tensile test and fatigue test conditions can be ascribed to the result of an increased amount of ferrite due to the low carbon content.

References

[1] . B. Mintz, S. Yue, and J. J. Jonas: Int. Mater. Rev. 36 (1991) 187.

[2] B. Mintz: ISIJ Int. 39 (1999) 833.

Development of New Casting Process for Heat-Resistant Cast Steel

Ph.D. Zhong-Zhi ZHANG, Eiji YAMAMOTO, Takahiro SATO, Naruo MATSUMOTO
AISIN TAKAOKA CO.LTD.
1 Tennoh, Takaoka shin-machi, Toyota, Aichi, 473-8501 Japan

Regarding the casting of small steel casting, such as the turbine housing (referred to as T/H), molten metal is poured manually in most foundries, compared to cast iron which is poured automatically. So the productivity of steel iron is lower than that of cast iron. The casting technology of cast steel is far behind that of cast iron.

Pressure casting technology is effective in ensuring the fluidity of the cast steel, but the machine required is expensive and difficult to maintain, so it is not suitable for production. Therefore, we developed a new automatic pouring process which pours at a constant temperature and speed to produce a high quality heat-resistant cast steel product at low cost.

Keywords : *heat-resistant cast steel, turbine housing, casting process, automatic pouring machine, constant pouring temperature*

1. Introduction

The number of turbocharger equipped vehicles is increasing in order to reduce vehicle emissions and to increase fuel efficiency. As a result the temperature of exhaust gas is increasing. Therefore the T/H needs to be increasingly heat-resistant, and a higher level of heat-resistant cast steel is required for the T/H.
In general cast steel technology is not as advanced as that of cast iron. So we have developed a new casting process for good quality heat resistant cast steel.

2. Experiment

We made molds of 2mm thickness T/H with the green sand molding process, the shell molding process and the cold box process. We examined the resulting casting quality for fluidity and sand burning of the mold, casting method (gravity casting, low pressure casting) and the casting conditions (pouring temperature and speed). As the result of these studies, a new casting process to obtain low cost high quality heat resistant cast steel product was established.

3. Results and Discussion

Generally, there are two casting methods to ensure the fluidity of cast steel: the vacuum casting method and the pressure casting method. Pressure casting was considered in this paper, as we developed a new low pressure casting machine, which could maintain stable fluidity. In a typical low pressure casting machine as shown in Figure 1, the life of the stalk is too short which is the main disadvantage of this method. Therefore, we devised a new low pressure machine for producing cast steel in which the stalk is made of refractory material (see Figure 1). With this machine we produced 2mm T/H test pieces. We concluded that pressure produced good fluidity results. However, the fluidity is too good, contains the into 0.5mm or less of the small gap between the combination molds, and made a lot of casting fins when mold accuracy is bad even a little. The cost of the low pressure casting machine is high and it is difficult to maintenance. For these reason, it is judged as not suitable for mass production, we addressed to develop the gravity casting process.

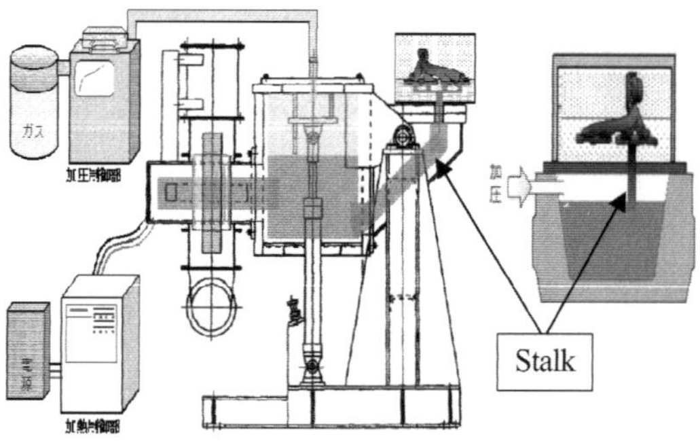

Fig.1 New low pressure casting machine (left), Conventional low pressure casting machine (right)

When casting cast steel by gravity casting manually with foundry ladle, tend to hold high pouring temperature at first mold pouring in order to ensure the molten metal fluidity at last mold since the molten metal temperature drop during pouring.
The high pouring temperature cause sand burning. Furthermore, it causes large variation of pouring temperature and pouring speed. So this makes the quality of cast steel vary and unstable.
Therefore developed the automatic holding pouring furnace with minimal variability in pouring temperature and speed, and then succeed to produce more quality stable cast steel product. Figure 2 shows the results of the pouring temperature and speed of continuous casting the T/H in JIS SCH5 material (equivalent GX40CrSi17) of 1 charge.
The pouring temperature vary 1,524-1,535C, the pouring speed vary 3.1-3.5kg/sec. Both properties are very stable. In addition, it is not necessary to increase the pouring temperature at the first mold, such as when cast manually with foundry ladle, and also it led to prevent the sand burning.

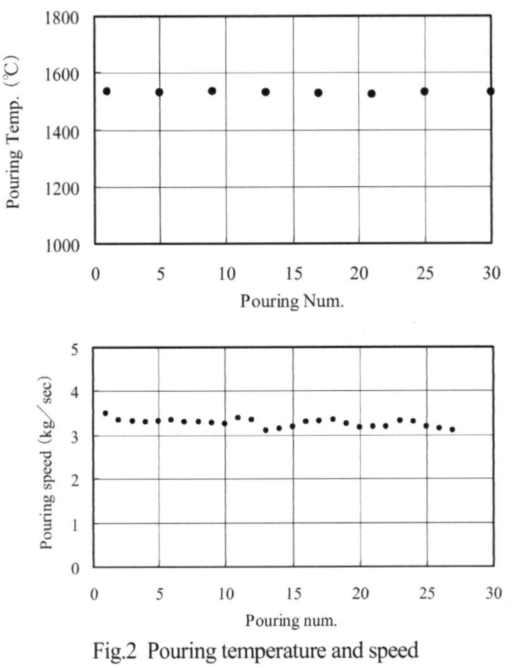

Fig.2 Pouring temperature and speed in the continuous casting

Green sand molds, shell molds and cold box molds were investigated. More burning occurred with the green sand molds than with the other molds.
It was difficult to reduce the mold weight using the cold box molds. So we finally selected the shell molding process which is suitable for producing lightweight molds. As shown in Figure 3, the shell molding process enabled to produce a mold nearer the net shape of shape of the T/H, so this provide 20-30% lightweight mold compare to conventional mold and reduce mold cost.

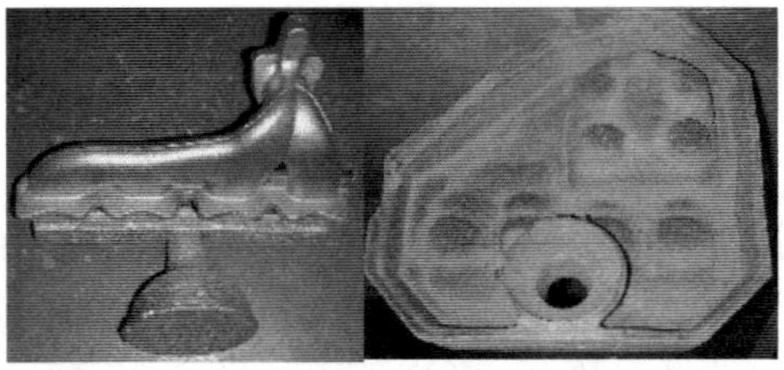
Fig.3 Steel castings (left) produced using the lightweight shell mold (right)

4. Summary

The developed concept is shown in Figure 4.
We developed an automatic holding pouring furnace with minimal variability in pouring temperature and speed. By utilizing this developed process with the lightweight shell molds, success to produce high quality thin-walled cast steel products at low cost.

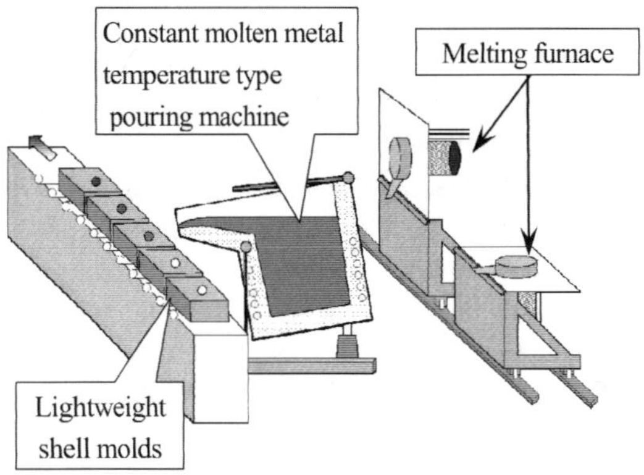

Fig.4 The concept of the newly developed process

O-28

Effect of heat treatment condition on microstructure and mechanical properties of 2.5% Ni cast steels

Takehito Hagisawa, Makoto Shikano, Hideki Narita and Fumio Takahashi
Muroran Research Laboratory, The Japan Steel Works, LTD., Muroran 051-8505, Japan

The effect of the cooling rate (WQ:Water-Quenching, AC: Air-Cooling and FC: Furnace-Cooling) after austenization on the microstructure and mechanical properties of 2.5%Ni cast steels was investigated. The WQ sample exhibited a fully bainite microstructure, while the AC and the FC samples exhibited a ferrite-bainite microstructure and ferrite-pearlite microstructure, respectively. The mechanical properties were significantly improved with increasing cooling rate after austenization. SEM fractography and EBSD measurement results showed that this improvement in mechanical properties was caused by grain refinement of the bainite microstructure.

Keywords: *austenization; cooling rate; bainite; grain refining; fracture appearance transition temperature (FATT);*

1. Introduction

JSW supplies large cast-steel products with heavy-sectioned parts for structural elements and power generation plants. Since the giant earthquakes in Kobe and the east coast of Japan, the expectations for higher strength and higher toughness of these products have increased. For example, ASTM A148 Grade 210-180 (JIS G5102, SCW620) is a high-strength cast steel with high toughness and high weldability. Although the microstructure and heat treatment condition must be optimized to ensure the mechanical properties of heavy sectioned casting parts made from ASTM A148 Grade 210-180, they have not previously been reported on. In the present study, the effect of the cooling rate after austenization on the microstructure and mechanical properties of 2.5%Ni cast steel with a similar composition to that of ASTM A 148 Grade 210-180 were therefore investigated.

2. Experimental procedures

Table 1 shows the chemical composition of the alloy ingot. The alloy ingot was melted by induction melting and cast in a sand mold in a vacuum. After stripping the mold, the following heat treatment was carried out.

- Annealing : 1323 K × 18 hours⇒FC
- Austenization : 1163 K × 22 hours
- Tempering : 888 K × 18 hours ⇒FC
- Stress Relief annealing : 878 K × 24 hours⇒FC

Three patterns (WQ, AC and FC) of cooling rate after austenization were tested. After the heat treatment, microstructural observation, a tensile test at room temperature, and the Charpy impact test were carried out. The microstructure was observed using an optical microscope and a field-emission scanning electron microscope (FE-SEM) with an electron back scatter diffraction (EBSD) analyzer.

Table 1 Chemical composition of alloy ingot (mass%)

C	Si	Mn	Ni	Cr	Mo	V	N
0.18	0.28	0.86	2.45	0.48	0.18	0.04	0.009

3. Results and Discussion

3.1. Microstructure

Figure 1 shows the optical micrographs of the samples after heat treatment. The WQ sample exhibits a fully bainite microstructure, while the AC and FC samples exhibit the ferrite-bainite microstructure and the ferrite-pearlite microstructure, respectively.

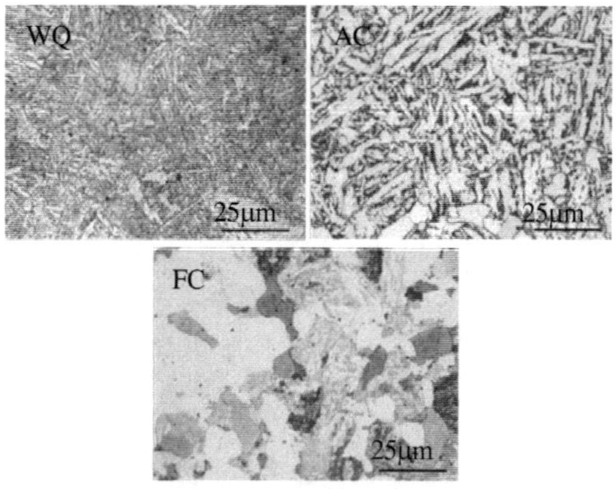

Fig.1 Optical micrographs of the samples after heat treatment.

3.2. Mechanical properties

Table 2 shows the tensile properties of the samples. The WQ sample showed the highest strength in all samples, and the strength decreased with decreasing cooling rate after austenization.

Figure 2 shows the temperature dependence of percent ductile fracture. The fracture appearance transition temperature (FATT) of the WQ sample shifted to a lower temperature than the AC and FC samples.

Table 2 Mechanical properties of the samples

	Tensile test result				Charpy test
	0.2%Y.S. /MPa	T.S. /MPa	EL/%	RA/%	FATT/K
WQ	613	718	22.9	63.5	213
AC	540	662	22.9	55.9	285
FC	436	583	30.3	65.3	287

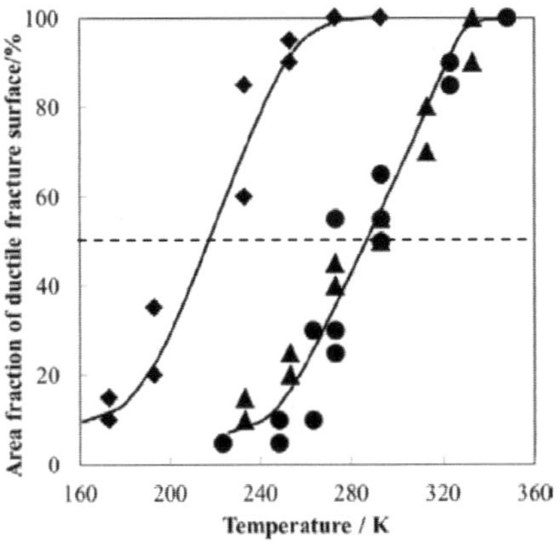

Fig. 2 Temperature dependence of percent ductile fracture.

3.3. SEM fractography and EBSD result

Figure 3 shows the SEM fractography of the samples. Although all samples exhibited a typical cleavage-like fracture surface, the size of fracture unit was significantly different among the samples. The fracture unit of the WQ sample was the smallest of all the samples.

Figure 4 shows the EBSD microstructure (inverse pole figure map) of the WQ and FC samples, where the black line shows the high angle grain boundary with misorientation larger than 15 degrees. It was found that the grain size of the bainite microstructure in the WQ sample was significantly smaller than that of the AC sample. It has been reported by Izumiya et al. [1] that the fracture unit in the Charpy impact test was almost the same as the average size of grain covered by the high angle boundary. The present result showed a tendency similar to their result [1]. Kadoya et al. reported that the bainite microstructure becomes smaller with decreasing Bs temperature [2]. In the present study, it was also confirmed that the transformation behavior and the distribution of precipitates change with increasing cooling rate after austenization. Therefore, it was concluded that refining the bainite microstructure by increasing the cooling rate after austenization is an effective means of improving mechanical properties of 2.5%Ni cast steels.

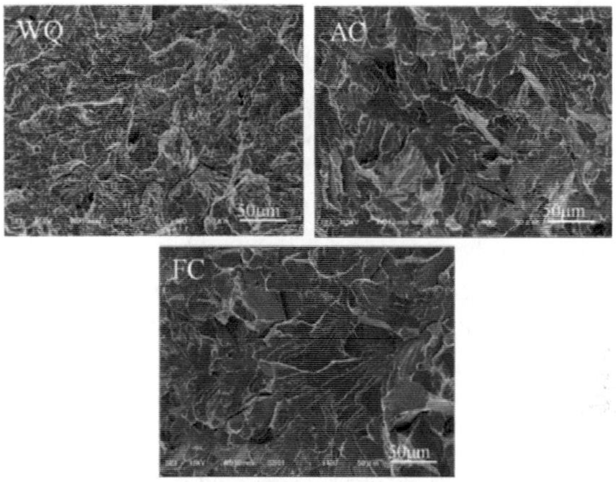

Fig. 3 Typical SEM fractgraphy of the samples

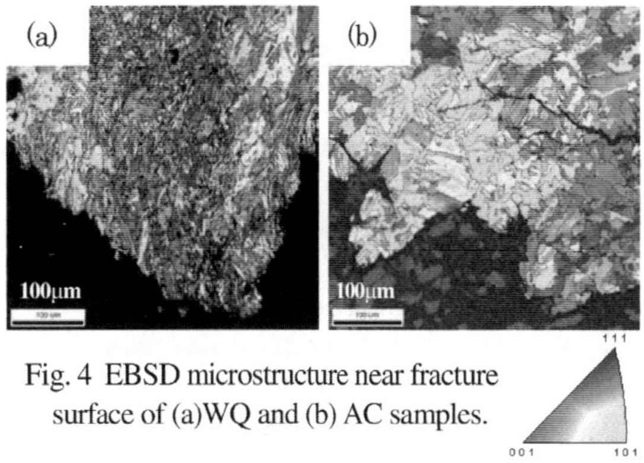

Fig. 4 EBSD microstructure near fracture surface of (a)WQ and (b) AC samples.

4. Conclusion

・The WQ sample exhibits the bainite microstructure. The AC sample exhibits the ferrite-bainite microstructure, and the FC sample exhibits the ferrite-pearlite microstructure.

・The mechanical properties of 2.5%Ni cast steel were improved by the refinement of the bainite microstructure with increased cooling rate after austenization.

References

[1] Y. Izumiyama et al. : Tetsu to Hagane, 100 (2014), 704-712.

[2] Y. Kadoya et al. : Tetsu to Hagane, 79 (1993), 980-987.

Proceedings of the 72nd World Foundry Congress
21-25th May 2016, Nagoya, Japan

The mold-less casting technique for production of the wrought aluminum alloy components

Jun Yaokawa[1], Yasushi Iwata[1], Yoshio Sugiyama[1], Naoaki Sugiura[2] and Yuichi Furukawa[3]
[1] Material & Processing Dept. I, Toyota Central R&D Labs., Inc., Nagakute 480-1192 Japan
[2] Foundry Technology Development Dept., Toyota Motor Corporation, Toyota 471-8571, Japan
[3]Kamigo Plant Engine Component Casting Div., Toyota Motor Corporation, Toyota 471-8571, Japan

This paper describes design freedom, dimensional accuracy, solidification structures and mechanical properties of hollow products of A6063 aluminum alloy fabricated by the mold-less casting technique. Kinds of geometries, such as a pipe with inner ribs, were fabricated by this process. High thickness accuracy was achieved when a thermally stable condition was maintained. Ultimate tensile strength, proof strength and fracture strain of T6 heat treated A6063 alloy fabricated by this technique were as high as those of extrusions.

Keywords: *Wrought aluminum alloys, Complex geometry, Capillary shaping, Directional solidification, Strength*

1. Introduction

In recent years, there have been increasing demands for manufacturing automotive frame components of wrought aluminum alloys (from 2XXX to 7XXX) with complex geometries, in order to improve ride comfort and reduce weight. However, it is difficult to fabricate these frame components by conventional manufacturing techniques.

Mold-less casting is one of the capillary shaping techniques by which products with a hollow and complex geometry are fabricated directly from a melt [1-3]. In this process, a melt column is formed above the melt surface using a starting device and shaping devices (Fig. 1). Then the starting device is withdrawn upward, cooling the position higher than the melt column, so that a top of the melt column continuously solidifies. Product geometries are controlled by a pulling path and a movement of the shaping device. The directional solidification in this process permits shaping of the wrought aluminum alloys with high strength and elongation, despite their high hot-cracking sensitivity.

In this study, design freedom and the dimensional accuracy of A6063 aluminum alloy were investigated. An effect of Ti and B additions into A6063 alloy melt on the solidification structure and mechanical properties were determined.

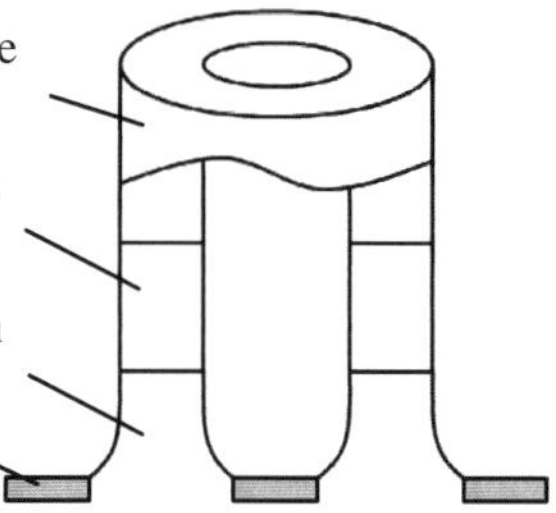

Fig. 1 A schematic of the mold-less casting to fabricate a hollow product.

2. Experimental procedure

Melts of A6063 alloy and grain refined A6063 alloy by inoculation of Al-5Ti-1B master alloy (A6063-Ti) listed in Table 1 were used in this study. Fig. 2 shows an apparatus. Shaping device was placed on the melt, and aluminum alloy melt was withdrawn through an opening of the shaping device using a starting device. Coolant air was blown onto the solidified part by inner and outer cooling nozzles. By changing a pulling path and an opening dimension of the shaping device, specimens with kinds of geometries were fabricated. During the pulling process, melt position was monitored by a melt level sensor in order to control the pulling rate and positioning of the shaping device. In some conditions, the vertical temperature distribution from 140 mm under the melt level to the solidified part was recorded by thermocouples fixed to the starting device.

Solidification structure of the hollow specimens was observed by polarized microscope after anodization treatment (Barker's etching). The hollow specimens of A6063 alloy and A6063-Ti alloy were machined to the tensile test plate specimens with size of 11 mm in the gauge length, 4 mm in width and 2 mm in thick after the T6 heat treatment. Tensile tests were carried out at the 0.005 mm/s in the crosshead speed.

Table 1 Chemical composition of alloys (mass%).

Alloy	Cu	Si	Mg	Fe	Ti	Al
A6063	0.01	0.58	0.74	0.08	0	bal.
A6063-Ti	0.01	0.56	0.74	0.08	0.06	bal.

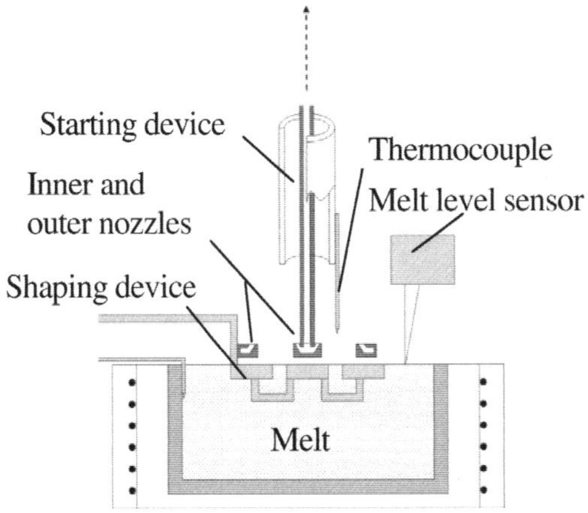

Fig. 2 A schematic of the mold less casting.

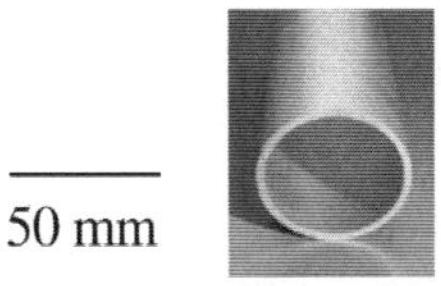

Fig. 4 A specimen with high thickness accuracy.

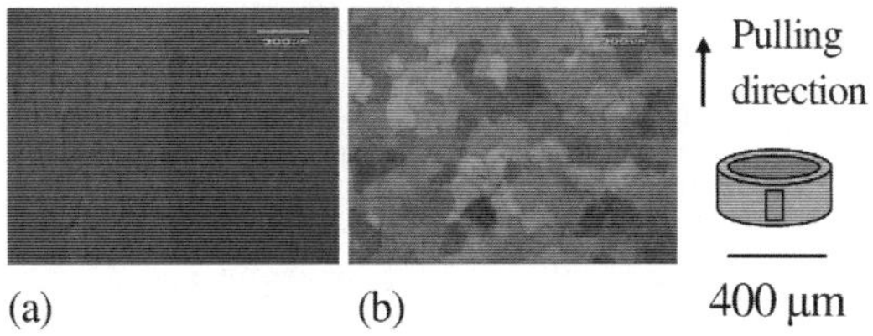

Fig. 5 Microstructures of (a) A6063 alloy and (b) A6063-Ti alloy specimens.

3. Results and discussion

Fig. 3 shows examples of A6063 alloy specimens produced by this process. Hollow specimens with inner ribs (Fig. 3a) and bent geometries (Fig. 3b, 3c) were fabricated by changing the opening dimension of the shaping device and the pulling path.

High thickness accuracy was achieved when the pulling process was conducted under thermally stable condition [3]. A pipe of A6063 alloy with size of 50.6 ± 0.15 mm in outer diameter and 2.6 ± 0.1 in thickness was fabricated at a pulling rate of 1.4 mm/s (Fig. 4).

Fig. 5 shows microstructures of the A6063 and A6063-Ti alloys. The primary dendrite trunks aligned almost antiparallel to the pulling direction for the A6063 alloys specimen, whereas equiaxed grain structure was obtained for the A6063-Ti alloy specimen. Since the directional solidification was macroscopically maintained during the pulling process, no shrinkage porosity was found in the both specimens.

The reported value of tensile properties for the T6 heat treated A6063 extrusions were 240MPa in the ultimate tensile strength, 215MPa in the 0.2% proof strength and 12% in fracture strain [4]. The tensile properties of the T6 heat treated A6063 alloy and A6063-Ti alloy specimens fabricated in this study were as high as those of the extrusions (Fig. 6).

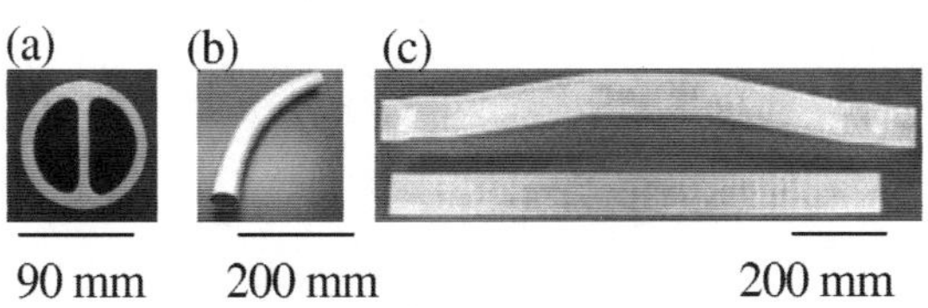

Fig. 3 Examples of (a) a pipe with an inner rib, (b,c) hollow specimens with bent geometries.

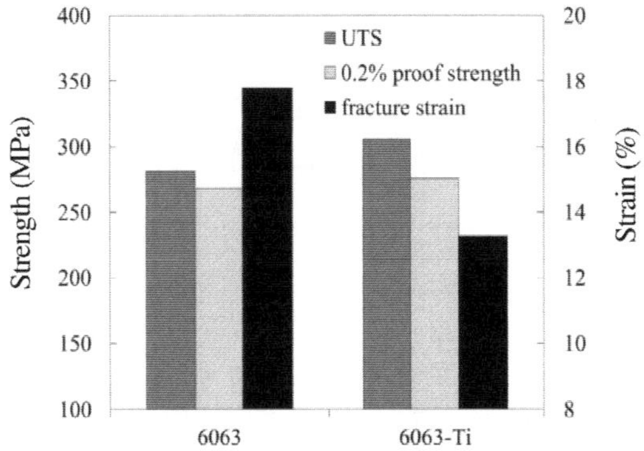

Fig. 6 Tensile properties of the T6 heat treated A6063 alloy and the A6063-Ti alloy specimens.

4. Conclusion

The mold-less casting technique were investigated in order to develop new manufacturing process for the automotive frame components of wrought aluminum alloys with high geometrical complexity. Hollow products with inner ribs and bent geometries of A6063 alloy were fabricated by this technique and high thickness accuracy was achieved. Also high tensile properties were obtained for the T6 heat treated A6063 specimens due to the absence of the shrinkage porosity.

References

[1] A.V. Stepanov, Zh. Tech. Fiz. 29 (1959) 381-383

[2] V. V. Peller, S. P. Nikanorov: *Handbook of Aluminum vol. 1: Physical Metallurgy and Processes*, Ed. by G. E. Totten, D. Scott, (CRC Press, 2003) pp. 695-736

[3] J. Yaokawa, et.al: Acta Materialia, accepted

[4] *Structure and Properties of Aluminum*, (The Japan Institute of Light Metals,1991) pp. 480-481

O-30

Proceedings of the 72nd World Foundry Congress
21-25th May 2016, Nagoya, Japan

The mold-less casting technique for applying to auto mobile parts

Naoaki Sugiura[1], Mitsuhiro Kobayashi[1], Yuichi Furukawa[2], Jun Yaokawa[3] and Yasushi Iwata[3]
[1] Foundry Technology Development Dept., Toyota Motor Corporation, Toyota Aichi 471-8571, Japan
[2] Kamigo Plant Engine Component Casting Div., Toyota Motor Corporation, Toyota Aichi 470-1217, Japan
[3]Material & Processing Dept. I, Toyota Central R&D Labs., Inc., Nagakute Aichi 480-1192, Japan

This paper deals with a newly-developed "mold-less casting technique" which permits high design freedom of aluminum alloy components. Some automotive parts e.g. a crash absorption component and a side frame, which are typically made of wrought aluminum alloys were fabricated by the mold-less casting technique. The crash absorption component with a wavy design fabricated by this process showed high absorption energy per unit weight.

Keywords: *Wrought aluminum alloys, Complex geometry, Capillary shaping, , Strength, Automotive components*

1. Introduction

The mold-less casting is one of the capillary shaping technique by which complex shaped aluminum products can be fabricated from an aluminum alloy melt. In this technique, a starting device is immersed into the melt through an opening of a shaper located on the melt surface. Then, the shaper is withdrawn upward, forming a melt column with a pre-determined shape, cooling a part above the solid-liquid interface by air blowing, allowing the wrought aluminum alloys to be cast without the hot-cracking. In general, it is difficult to fabricate wrought aluminum alloy components with high quality by the conventional casting processes. By this technique, however, kinds of geometries, such as a hollow, variable thickness and bent products can be fabricated and high tensile strength and elongation are obtained.

In this study, we investigated the mold-less casting technique to manufacture newly designed automotive components such as a crash absorption component and a side frame. In the case of the crash absorption component, a wavy design was introduced to a hollow structure with inner ribs (Fig. 1). The wavy design increased a number of buckling deformations and thus increased the energy absorption per unit weight. In order to fabricate the wavy design, factors affecting shape stability and defect formation were investigated using experimental and numerical techniques. We also investigated to fabricate aluminum frame components with inner ribs and a bent geometry in order to increase structural stiffness, enhance design freedom and reduce weight.

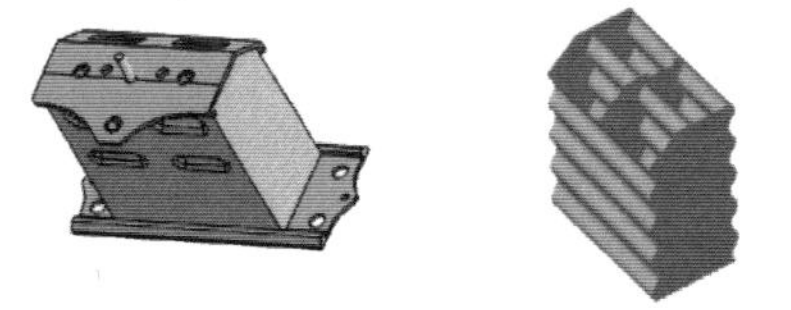

(a) Conventional design (b) New design

Fig. 1 Illustrations of crash absorption components.

2. Experimental procedure

A6xxx series aluminum alloys were used. The crash absorption components of the A6063 alloy were fabricated using the shaping device with an opening as shown in Fig. 2. The melt column was formed on this opening. Cooling air was blown to the outer shells from outer nozzles, while the inner ribs were cooled by air blown from the inner ribs. The wavy design was formed by control of a pulling path (Fig. 3). The specimens were T6 heat treated. Drop-weight tests were conducted to determine crash absorption energy (Fig.4).

A frame component of A6005C with inner ribs and a bent geometry (Fig.5) was fabricated using vertical and rotary pulling paths combined with a horizontal movement of the shaper. Because the pulling rate at an outer position was faster than that at an inner one during the rotary pulling, the flow rate of coolant air from outer and inner nozzles were varied to keep the pulling condition within the thermally stable conditions [1].

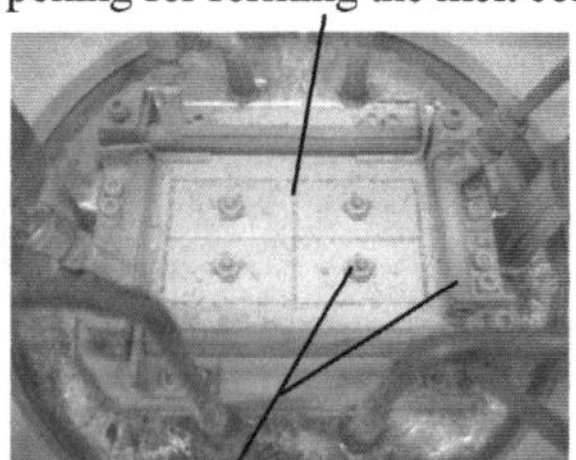

Fig. 2 A shaping device.

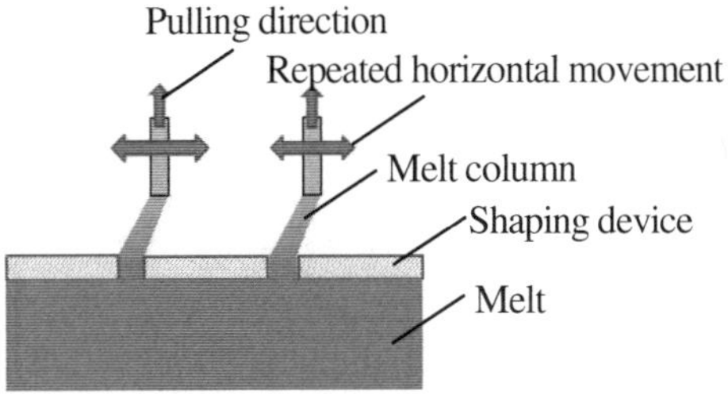

Fig. 3 Illustration of pulling of the wavy design.

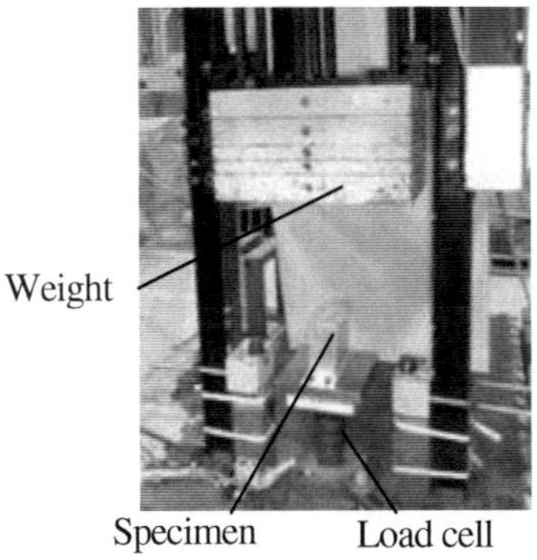

Fig. 4 Illustration of the drop-weight test.

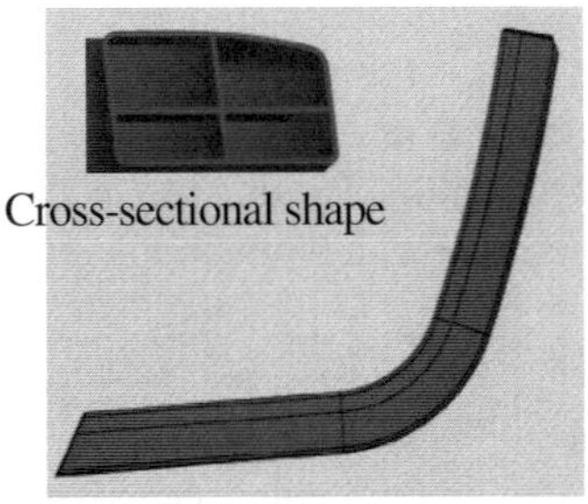

Fig. 5 A frame component.

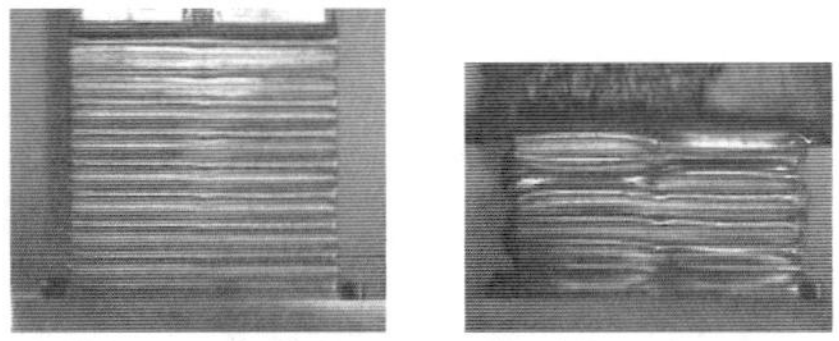

(a) Wavy specimen (b) After the test

Fig. 6 Photos of the crash absorption component with the wavy design before and after the drop-weight test.

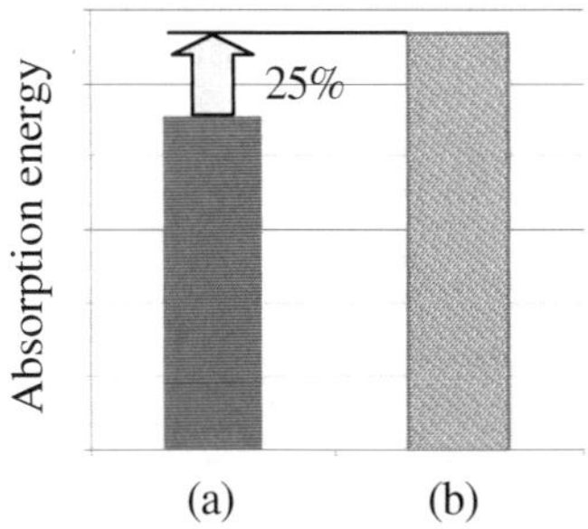

Fig. 7 Crash absorption energy

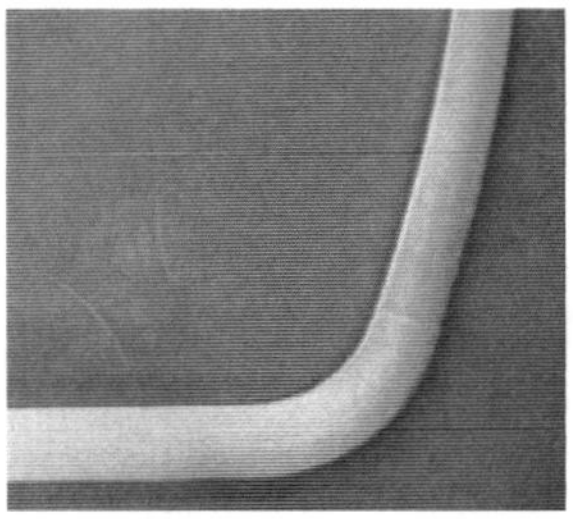

Fig. 8 A photo of the frame component with inner ribs and a bent geometry.

3. Results and discussion

Fig. 6 shows photos of the crash absorption components before and after the drop-weight tests taken by a high-speed camera. Buckling deformations appeared at antinodes of the wavy design. The crash absorption energy per unit weight of the wavy specimens increased approximately 25 pct. comparing to that of the specimens with a conventional design as shown in Fig. 7.

The frame components with inner ribs and a bent geometry could be fabricated by this technique (Fig.8).

4. Conclusion

In this study, the mold-less casting technique to manufacture newly designed automotive components was investigated. The wavy shaped crash absorption components with inner ribs could be fabricated by this technique. The absorption energy per unit weight of the wavy specimens increased approximately 25 pct. comparing to that of the specimens with a conventional design. A frame component with inner ribs and a bent geometry was also fabricated by this technique.

References

[1] J. Yaokawa, et.al: Acta Materialia, accepted.

Proceedings of the 72nd World Foundry Congress
21-25th May 2016, Nagoya, Japan

Effect of Si content and Melt superheat on the liquid aluminum viscosity

Young Ki LEE[1], Jeong IL YOUN[1], Seok Min HWANG[1], Young Jig KIM[1*],
Kyu Chul HWANG[2], Hwi Jun KIM[2]
[1] School of advanced materials science and engineering, Sungkyunkwan University,
2066 Seobu-ro, Jangan-gu, Suwon, Gyeonggi-do 16419, Korea.
[2] Region Advanced Manufacturing Technology Agency, Korea Institute of Industrial Technology,
156 Getpearl-ro, Yeonsu-gu, Incheon 21999, Korea

In the casting processes, fluidity is defined as the maximum length which liquid metal will flow before it is stopped by solidification. The main factor determining the fluidity of liquid metal is the viscosity because which affect in many rheological properties such as surface tension, thermal conductivity and diffusivity. Viscosity is affected mainly on alloy composition and temperature, especially in Al-Si alloy casting alloy, evaluation of viscosity for Si content and pouring temperature is very important because which are affected directly on fluidity. However the researches on the effect of Si content and temperature on viscosity are not sufficient, because methods for measurement of viscosity have limitations such as liquid metal's low viscosity, high temperature and error of measurement. So in this study, the effect of Si contents and melt superheat on the liquid aluminum viscosity is studied.

Keywords: *Aluminum, Silicon, Viscosity, Fluidity, Solidificiation*

1. Introduction

In the casting processes, fluidity is defined as the maximum length which liquid metal will flow before it is stopped by solidification. Fluidity of liquid metals is significantly important property in producing sound castings. The fluidity is affected by many factors that can be divided into metallurgical variables, such as composition, superheat, latent heat, viscosity, surface tension and mould/castings variables, such as part configuration, cooling rate, degree of super heat, mould material and its surface characteristcs [1]. Especially viscosity is an important rheological parameter for understanding the hydrodynamics and kinetics of reactions in metal casting [2]. Viscosity directly influence many metallurgical variables such as surface tension, thermal conductivity and diffusivity, and moreover indirectly influence the mould/castings variables such as mould material and its surface characteristics. Viscosity is affected mainly on alloy composition and temperature, and especially in Al-Si alloy casting alloy, evaluation of viscosity for Si content and pouring temperature is very important because which are affected directly on fluidity. Nevertheless the viscosity is indispensable parameter to predict important transport phenomena of Al-Si casting alloy, only a few researches have paid attention to the viscosity of Al-Si alloys because methods suitable for liquid metals are limited by metal's low viscosities (of the order of 1 mPa·s), their chemical reactivity and generally high melting points [3]. So in the present study, we tried to evaluate the effect of Si contents and melt superheat on the liquid aluminum viscosity.

2. Experimental Procedure

2.1 Viscometer set up

Generally Capillary, oscillating vessel, rotational bob or crucible, oscillating plate have been developed or modified for this purpose [4]. These methods generate the thixotrophic microstructure during measurement of viscosity because these are torque type viscometer. So in this study, we proposed non-torque viscometer which was based on conventional falling ball viscometer and Stokes' law. In this method, a body (normally a sphere) was either allowed to drop due to gravity or drag upwards through the liquid [5].

2.2 Viscosity Measurement

The melt was prepared in a cylindrical crucible with the inner diameter of 60 mm and length of 350 mm. To evaluate of the Si content on viscosity of aluminum, different content of Si (0, 3, 7, 12, 17 wt%) were added in aluminum melts. And to evaluate of the superheat of melts on viscosity of aluminum, degree of superheat are established 45, 80 °C at liquidus temperature of Al-Si alloys. Measurement of velocity of the falling object was performed to infer viscosity of the liquid.

3. Result and Discussion

Fig. 1 shows a schematic view of the falling ball viscometer. The falling ball made of steel with parting agent because of suppress of chemical react with melts. And to minimize the fraction of device, block and rail are used with pulley.

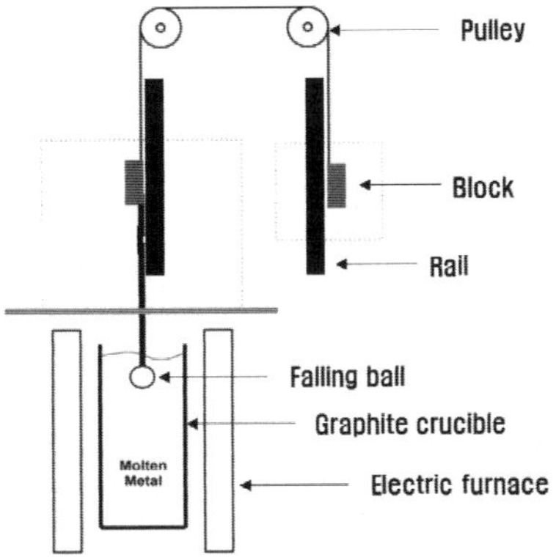

Fig. 1 Schematic view of the falling ball viscometer

The experimental values for the viscosity of the Al–Si alloy at different temperature are shown as Fig. 2. Si plays an important role in lowering the viscosity of Al melt. The viscosity of Al-Si alloy decreases as the Si concentration increases [6], and especially in Al-12Si, viscosity change irregularly with temperature varies. The work by Persion reports complex behaviour especially near the eutectic composition, which they associate with changes in the structure of the liquid [7]. And in the case of Al-7Si, temperature dependence of viscosity is larger than other Al-Si alloys.

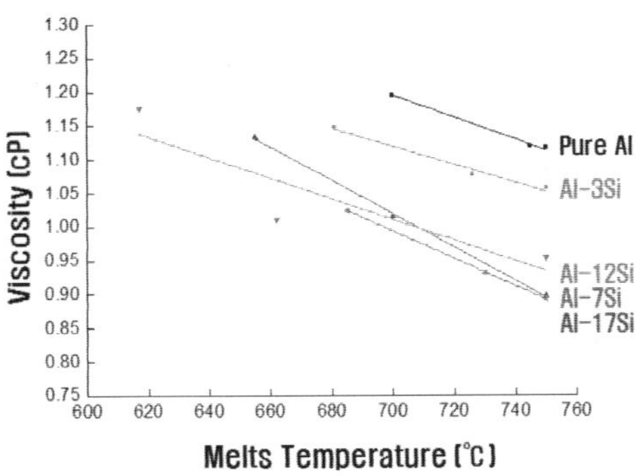

Fig. 2 Temperature dependence of the experimental viscosity data of Al–Si alloy with different Si content.

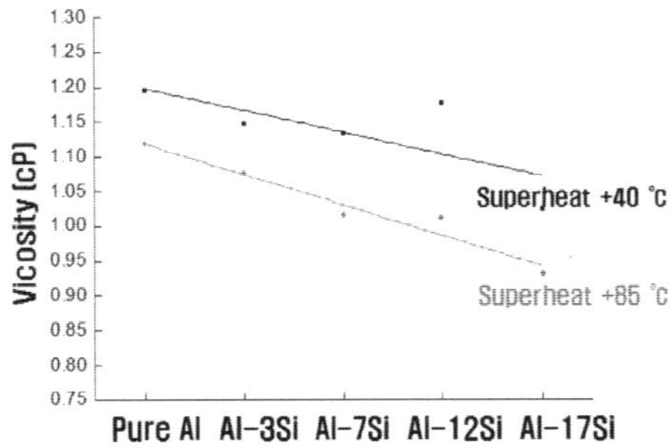

Fig. 3 The experimental viscosity data of Al–Si alloy with different superheat at liquidus temperature.

Fig. 3 shows the experimental viscosity data of Al-Si alloy with different superheat at liquidus temperature. Generally the viscosity decreases as the temperature of liquid metal increases, so when superheat of liquid metal increases, the viscosity of Al-Si alloy decreases. Especially, the degree of superheat increases, the rate of change of viscosity increases.

4. Conclusion

In this study, we evaluate the effect of Si contents and melt superheat on the liquid aluminum viscosity. The viscosity of Al-Si alloy decreases as the Si concentration increases generally. And when superheat of liquid metal increases, the viscosity of Al-Si alloy decreases.

References

[1] K. R. Ravi, R. M. Pillai, K. R. Amaranatha, B. C. Pai and M. Chakraborty: J. Alloy. Compd. 456 (2008) 201-210

[2] S. I. Bakhtiyarov and R. A. Overfelt: Acta Mater. Vol. 47, No. 17 (1999) 4311-4319

[3] A. T. Dinsdale and P. N. Quested: J. Mater. Sci 39 (2004) 7221-7228

[4] R. F. Brooks, A. T. Dinsadale and P. N. Quested: Meas. Sci. Technol. 16 (2005) 354

[5] S. Hossein Elahi, Hadi Abdi and H. R. Shahverdi: Rev. Sci. Instrum. 85, 015115 (2014) 1-4

[6] J. M. Vignau, P. Azou and P. Bastein, Compte Rendu Acad.Sci. 264 (1967) 174

[7] A. T. Dinsdale, P. N. Quested: J. Mater. Sci. 39 (2004) 7221-7228

Description of Al-Si10-Mg1 Alloys By Advanced Thermal Analysis Based on Their Known Chemical Compositions

I.Vicario[1], M.B. Djurdjevic[2], G. Huber[2], E. Villanueva[2], A. Meléndez[2]
[1] Tecnalia Research & Innovation, Derio, Spain
[2] Nemak Linz, Linz, Austria

Thermal analysis is a quality control tool in aluminium casting plants. The cooling curve and its derivatives analysis offer parameters that can be used in simulation software programs to improve simulation accuracy and to control the metallurgical quality of cast parts. This paper shows the calculation of thermo-physical, solidification and precipitated phases characteristics of AlSi10Mg1 cast aluminium alloy. Minor alloying elements are included, comparing the results with the results obtained by using other methods as the Silicon Equivalent (SiEQ), showing a very good correlation with the different developed formula.

Keywords: *Thermal analysis, Aluminium-Silicon alloys, Silicon Equivalent.*

1. Introduction

In the last years new alloys have been developed in order to obtain aluminium structural parts by casting, especially High Pressure Die Casting (HPDC). Between the different alloys, the AlSi10Mg1 alloy has been one of the most studied, in order to obtain parts with improved ductility and crash performance.

Thermal analysis is a quality control tool in aluminium casting plants. Some methods have been developed to determinate solidification main temperatures. The results obtained by using the Silicon Equivalent (Si_{EQ}) method [1] are very accurate for the liquidus temperature (R^2=0.98) and more than other previous methods [2-4], but smaller for the Al-Si eutectic (R^2=0.88) and Al-Si-Cu eutectic (R^2=0.77).

This work demonstrates how the statistical analysis of the results obtained for a wide range of alloy chemical compositions shows a better correlation for the different eutectic temperatures.

2. Experimental conditions

The approach used in the present work has been based on the identification of the effect of 12 main alloying elements in solidification parameters through the Taguchi methodology. Two orthogonal matrices have been used, a L16 matrix and a modified L8 matrix. The former employs two levels that are related to the maximum and minimum amounts of the alloying element. Modified matrix incorporates intermediate values.

The base alloy for the developments has been chosen from the most common used alloys for HPDC. The selected alloy is the AlSi10Mg according to standard EN AC-43.400, included in the EN 1706:2010 standard. The obtained compositions are resumed in Table 1.

Table 1 Composition of base alloy (mass %).

Ref.	Si	Fe	Cu	Mn	Mg	Cr	Ni	Zn	Ti	Pb	Sn	Sr	Al
1	9,00	0,38	0,03	0,34	0,30	0,01	0,00	0,01	0,02	0,00	0,00	0,02	89,89
2	8,02	0,29	0,02	0,21	0,19	0,01	0,00	0,00	0,01	0,00	0,00	0,00	91,24
3	8,66	0,30	0,02	0,21	0,14	0,01	0,00	0,29	0,20	0,27	0,04	0,01	89,85
4	10,01	0,34	0,02	0,67	0,69	0,15	0,23	0,01	0,02	0,00	0,00	0,06	87,79
5	9,75	0,34	0,02	0,72	0,68	0,15	0,23	0,35	0,12	0,14	0,06	0,06	87,39
6	8,77	0,85	0,19	0,21	0,15	0,16	0,21	0,16	0,12	0,21	0,07	0,01	88,89
7	8,43	0,91	0,19	0,18	0,11	0,14	0,19	0,18	0,19	0,19	0,07	0,05	89,18
8	9,02	1,05	0,29	0,81	0,38	0,07	0,21	0,06	0,17	0,21	0,02	0,05	87,67
9	9,26	0,73	0,09	0,53	0,56	0,07	0,00	0,21	0,02	0,01	0,00	0,01	88,50
10	11,65	0,34	0,20	0,30	0,58	0,02	0,20	0,03	0,24	0,07	0,03	0,02	86,33
11	10,54	0,34	0,16	0,31	0,52	0,02	0,15	0,23	0,17	0,26	0,03	0,05	87,22
12	11,49	0,91	0,42	0,67	0,40	0,14	0,00	0,15	0,23	0,18	0,04	0,05	85,33
13	11,60	0,83	0,18	0,74	0,46	0,18	0,00	0,19	0,02	0,23	0,00	0,01	85,55
14	11,64	0,96	0,08	0,08	0,53	0,16	0,08	0,13	0,27	0,08	0,03	0,01	85,96
15	11,82	0,96	0,11	0,11	0,52	0,14	0,11	0,18	0,11	0,11	0,05	0,02	85,77
16	11,41	0,95	0,27	0,69	0,35	0,09	0,30	0,09	0,25	0,25	0,03	0,04	85,29
17	12,07	0,83	0,13	0,49	0,28	0,03	0,17	0,02	0,08	0,16	0,06	0,03	85,64
18	10,21	0,43	0,05	0,33	0,28	0,07	0,00	0,08	0,02	0,00	0,00	0,01	88,51
19	10,37	0,50	0,11	0,44	0,28	0,14	0,00	0,01	0,02	0,00	0,00	0,01	88,12
20	10,64	0,41	0,05	0,33	0,63	0,07	0,00	0,10	0,02	0,00	0,00	0,01	87,74
21	10,31	0,54	0,09	0,35	0,29	0,11	0,00	0,01	0,01	0,00	0,00	0,01	88,28
22	10,80	0,48	0,05	0,33	0,52	0,06	0,00	0,10	0,03	0,00	0,00	0,01	87,61
23	10,90	0,51	0,10	0,47	0,43	0,11	0,00	0,02	0,01	0,00	0,01	0,01	87,44
24	11,71	0,57	0,07	0,44	0,44	0,08	0,00	0,04	0,02	0,00	0,00	0,01	86,62
25	10,73	0,60	0,10	0,38	0,36	0,09	0,00	0,10	0,02	0,00	0,00	0,01	87,61

In order to obtain the Thermal Analysis (TA), samples with masses of approximately 300g ± 10g were poured into calibrates sand cups. Temperatures between 630 – 400 °C were recorded. The data for TA was collected using a high-speed National Instruments Data Acquisition System linked to a personal computer. Each TA trial was repeated three times. The cooling rate was 3°C/sec.

3. Results and discussion

From the solidification curves and derivatives were determined the Liquidus minimum and maximum (T_{Lmin}, T_{Lmax}), the Al-Si eutectic nucleation ($T_{AlSinuc}$), minimum ($T_{AlSimin}$) and maximum ($T_{AlSimax}$), the Mg_2Si precipitation (T_{Mg2Si}), and the Al-Si eutectic nucleation temperatures by the Si_{eq} method ($T_{AlSi\ Sieq}$), summarized on Table 2.

Table 2 Solidification Temperatures (ºC)

Ref.	TLmin	TLmax	TAlSi Si_{eq}	TAlSi nuc	TAlSi min	TAlSi max	TMg_2Si
1	590,38	591,67	593,75	586,56	563,59	565,31	552,73
2	604,43	605,94	601,65	596,05	564,79	569,15	550,26
3	600,01	600,06	596,09	590,50	564,14	565,36	549,92
4	586,55	588,09	584,00	582,76	561,70	563,53	549,47
5	584,11	584,11	584,71	571,11	555,74	560,10	546,68
6	594,82	595,07	590,42	587,13	562,57	567,17	539,26
7	597,61	598,45	593,74	591,10	563,40	564,72	541,13
8	593,64	593,72	584,97	585,73	558,64	560,10	543,58
9	592,16	592,33	589,08	588,51	561,60	562,14	548,25
10	577,18	577,18	575,15	570,89	561,55	564,19	549,62
11	584,00	584,00	580,84	575,49	561,08	563,66	549,04
12	576,80	576,80	570,85	571,55	564,11	565,33	543,36
13	572,79	573,39	568,88	570,47	562,51	563,50	549,63
14	575,46	575,46	575,34	569,22	560,73	563,92	548,99
15	570,82	572,16	571,97	567,85	562,16	564,63	547,33
16	576,60	576,60	569,43	571,96	563,00	564,12	540,32
17	568,71	570,62	567,02	568,18	564,27	565,81	541,33
18	583,20	583,97	585,05	578,81	563,71	566,42	552,37
19	584,53	584,99	582,61	576,63	563,35	564,61	548,88
20	579,20	580,55	582,55	575,31	561,34	563,48	553,05
21	586,03	587,68	583,48	582,89	564,72	567,86	553,59
22	578,67	579,62	581,03	574,73	563,59	565,50	553,42
23	580,97	581,91	578,91	576,42	563,11	565,95	552,75
24	575,04	575,55	573,47	572,00	564,78	567,34	551,21
25	580,92	581,74	580,15	576,34	563,19	565,71	552,09

By using linear regressions for the obtained values we obtain the equations from (1) to (6) in ºC. We can observe the linear regression coefficient (r^2) and the standard deviation (S_{ey}):

$$T_{Lmin}=665{,}72-7{,}65.Si-1{,}85.Mg-4{,}11.Fe+5{,}32.Cu -0{,}5.Ni+6{,}26.Cr-1{,}04.Mn+17{,}84.Ti-5{,}15.Zn +3{,}34.Pb-42{,}17Sn-32{,}79.Sr;\ r^2=0{,}978;\ S_{ey}=1{,}94 \quad (1)$$

$$T_{Lmax}=666{,}75-7{,}63.Si-1{,}29.Mg-3{,}4.Fe+2{,}66.Cu -2{,}58.Ni+5{,}54.Cr-1{,}95.Mn+11{,}99.Ti-10{,}34.Zn +5{,}95.Pb-30{,}75.Sn-16{,}05.Sr;\ r^2=0{,}978;\ S_{ey}=1{,}97 \quad (2)$$

$$T_{AlSiNuc}=649{,}16-6{,}54.Si-1{,}5.Mg+2{,}45.Fe+3{,}89.Cu -1{,}89Ni-5{,}79.Cr-4{,}1.Mn-1{,}15.Ti-11{,}68.Zn +7{,}08.Pb-59{,}0{,}Sn-5{,}7.Sr;\ r^2=0{,}947;\ S_{ey}=2{,}66 \quad (3)$$

$$T_{AlSimin}=561{,}16+0{,}73.Si-9{,}17.Mg-1{,}0.Fe+3{,}57.Cu -3{,}76.Ni-0{,}54.Cr-3{,}02.Mn-5{,}1.Ti-2{,}87.Zn +0{,}84.Pb-19{,}94.Sn+18{,}99.Sr;\ r^2=0{,}789;\ S_{ey}=1{,}33 \quad (4)$$

$$T_{AlSimax}=565{,}88+0{,}57.Si-7{,}56.Mg-3{,}19.Fe+4{,}91.Cu -0{,}53.Ni+4{,}06.Cr-4{,}0.Mn-6{,}69.Ti-5{,}14.Zn -0{,}32.Pb+7{,}5.Sn-8{,}47.Sr;\ r^2=0{,}794;\ S_{ey}=1{,}35 \quad (5)$$

$$T_{Mg2\ Si}=551{,}31+0{,}26.Si+5{,}22.Mg-5{,}47.Fe-8{,}84.Cu -15{,}94Ni+6{,}13.Cr-3{,}62.Mn+4{,}76.Ti-1{,}1.Zn +1{,}85.Pb-81{,}5Sn+8{,}05.Sr\ r^2=0{,}923;\ S_{ey}=1{,}72 \quad (6)$$

In order to determinate the accuracy of the Si_{eq} method, the obtained temperatures were plotted against calculated temperatures, as shown in figure 1:

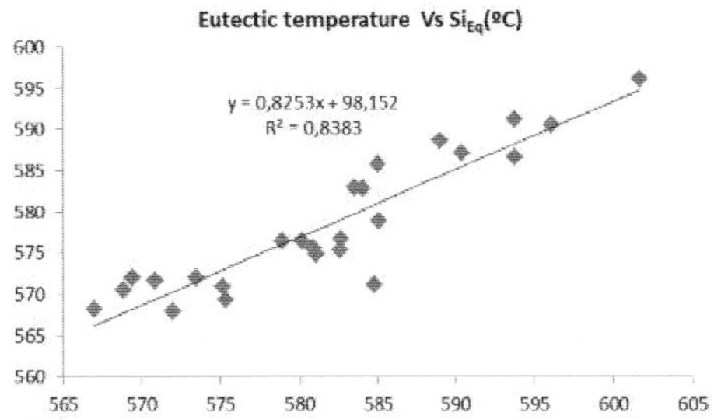

Fig. 1 Eutectic temperature vs. Si_{Eq} values

As it can be observed, the obtained values are less accurate than the obtained from developed equations.

4. Conclusions

A methodology based on Taguchi has been used to calculate the liquidus maximum and minimum, the eutectic and the Mg_2Si eutectic temperatures. The obtained results presented in this paper are more accurate than those found in the literature and the obtained by the Si_{Eq} method. The results show that the obtained equations allow to determinate with good accuracy the solidification parameters of any alloy.

Acknowledgements

This work has been partially funded by the Basque Government through the ETORGAI programme ER-2011/00002 and by the European Commission NIWE project (grant agreement 296024).

References

[1] M.B. Djurdjevic, I.Vicario. Rev. de metalurgia. 49 (2013) 161-171.
[2] G. Drossel. Giessereitechnik. 27 (1981) 7.
[3] R. Vijayaraghavan, N. Pelle, J. Boileau, J. Zindel, R. Beals: Scriptia Materalia. 35 (1996) 861.
[4] F.C. Robles, M.B. Djurdjevic, W.T. Kierkus, J.H. Sokolowski. Mat. Scien. and Eng. 396 (2005) 271-276.

Proceedings of the 72nd World Foundry Congress
21-25th May 2016, Nagoya, Japan

Effect of casting plan and alloying content on the hot tearing of Al-Si alloy

Hiroshi Tachibana [1], Shuuzi Koiwai [1], Mitsuru Adachi [1], Kazuma Murase[2] and Hiroshi Yamagata[3]
[1] KOIWAI Co., Ltd.
[2] Graduate School of Engineering, Gifu University
[3]Faculty Engineering, Gifu University

The effects of Si content and Sr addition on the hot tearing of Al-0%, 1%, 4%, 7%, 10% and 12.5% Si-0.3%Mg alloys were investigated using a T-shape product. Hot tearing was not observed in pure Al or in 12.5%Si-0.3%Mg alloy. The narrow solidification temperature range for the two alloys does not generate the high thermal stress that causes hot tearing. While hot tearing was observed in the 1%, 4%, 7% and 10% Si-0.3%Mg alloys, the crack length decreases with increasing Si content. The vertical bar portion of the T-shaped product is thick and undergoes delayed solidification. However, hot tearing takes place when solidification of the vertical bar portion is relatively fast. The addition of 0.1%Sr into the 7% and 10% Si-0.3%Mg alloys removes cracks with a filled eutectic phase through a healing mechanism, but the Sr addition does not heal the cracks in the 1% and 4% Si-0.3%Mg alloys.

Key words : *Hot tearing , Crack, Casting plan Sr addition, Thermal stress*

1. Introduction

JIS-AC4C alloy (Al-7%Si-0.3%Mg) is widely used for automotive components because of its superior castability. However, we found that hot tearing takes place when a particular casting plan is chosen. We designed four casting plans for a T-shape product and experimentally compared the influence of the casting plan on the hot tearing. It was found that the casting plan without a gate and runner, in particular, triggers hot tearing at the crossing corner portions between the horizontal bar and the vertical bar of the T-shape [1]. An elastic finite element method (FEM) stress analysis combined with a casting and solidification simulation revealed that high tensile stress causes hot tearing at the corners [2]. The high stress is generated by the temperature distribution associated with the T-shape. Hot tearing appears when the ratio of the solidification time for the vertical bar [ta] to that for the horizontal bar [tb] is less than 1.5. Also, it was found that a higher Sr concentration than 0.003% restricts the hot tearing by a healing process [3]. Based on previous research on the AC4C alloy, the purpose of this study was to investigate the effects of Si content and Sr addition on the hot tearing of Al-0% to 12.5% Si-0.3%Mg alloys.

2. Experimental procedures

A casting test was conducted using a top-pouring casting plan for the T-shaped product (**Fig. 1**). The casting plan without a gate and runner showed hot tearing in the AC4C alloy [1]

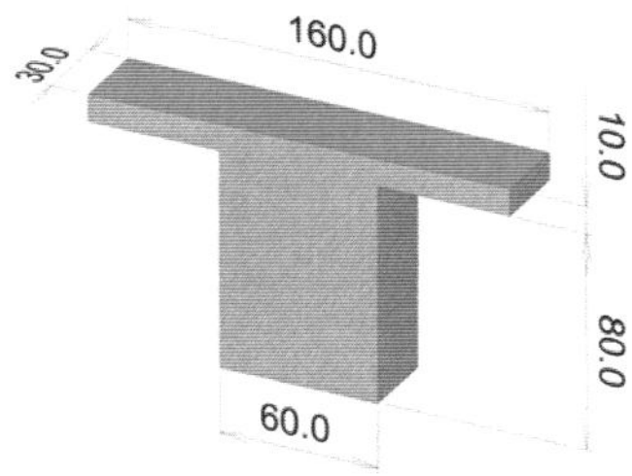

Fig.1 The casting plan.

A new silica sand mold was prepared with the addition of a 2% resin and a 0.4% curing agent. Al-0% to 12.5% Si-0.3%Mg alloys were used. Gas bubbling filtration (GBF) degassing reduced the hydrogen content to 0.2 cc/100 g Al. To investigate the effect of Sr, 0.01% Sr was added just before pouring. ADSTEFAN Ver. 2013 was used for the casting and solidification analysis. The parameters for the analysis were taken from the experimental conditions. The elastic FEM stress analysis using ANSYS Workbench 4.0 was used to calculate the stress distribution in the product.

3. Result and Discussion

Fig. 2 shows the cooling curves observed at three positions (horizontal bar, vertical bar and corner) in the 4%Si-0.3%Mg product. Solidification starts from the horizontal bar and proceeds through the corner to the vertical bar. The order was roughly the same regardless of the Si content of the alloys.**Fig. 3** shows the effects of Si content and Sr addition on the crack length. Pure Al and 12.5%Si-0.3%Mg alloy do not show any hot tearing. It was thought that the narrow solidification temperature range for the two alloys

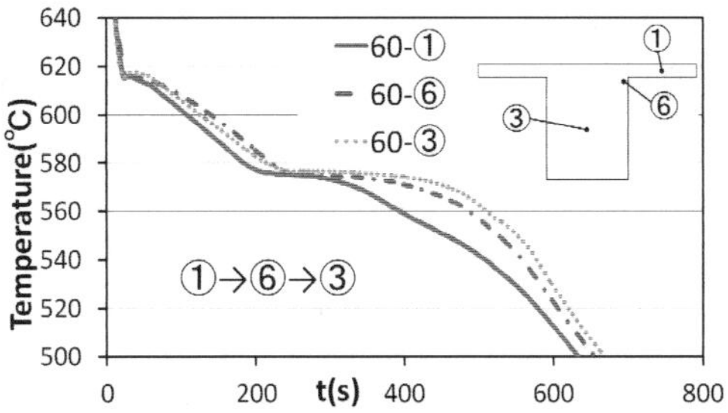

Fig.2 Cooling curves at the horizontal bar, vertical bar and corner.

does not generate the high stress that causes hot tearing. With increasing Si content, the crack length decreases. High stress is generated by the temperature distribution when the ratio (ta/tb) of the solidification time for the vertical bar (ta) to that for the horizontal bar (tb) is below 1.5. Fig. 3 also includes ta/tb lines. It is observed that the crack length decreases with increasing ta/tb. A ratio of unity indicates homogeneous solidification. If the thick vertical portion of the T-shape solidifies quickly when the corner portion is in the semi-solid state, the quick solidification generates high thermal stress on the crossing corner portions.

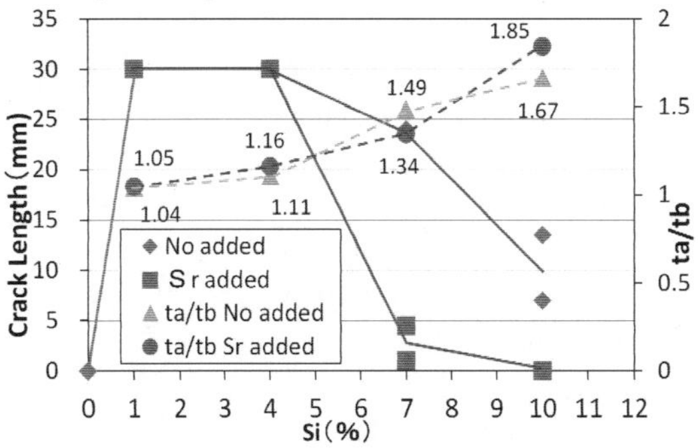

Fig.3 Effects of Si content and Sr addition on the crack length and ta/tb.

The addition of Sr into 7% and 10% Si-0.3%Mg alloys restricts cracking. However, even with the addition of Sr, hot tearing occurs in 1% and 4% Si-0.3%Mg alloys.

Fig. 4 shows the corner microstructures of 1%, 7% and 10% Si-0.3%Mg alloys with and without Sr. The Sr addition to the 7% and 10% Si-0.3%Mg alloys removed cracks through a healing process. However, the Sr addition to the 1%Si-0.3%Mg alloys did not remove cracks. It was thought that the Sr addition increases the permeability of the eutectic phase through a channel between the dendrites and enhances the healing of the alloys due to the higher amount of the eutectic phase.

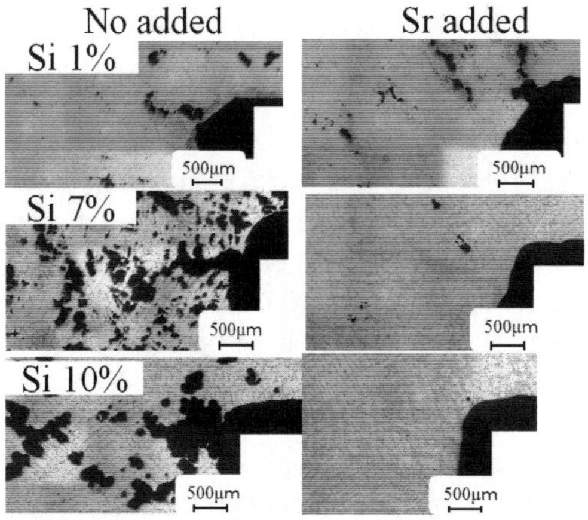

Fig. 4 Corner microstructure of Al-1%, 7% and 10%Si-Mg alloys with and without Sr.

4. Conclusion

The effects of Si content and Sr addition on the hot tearing of Al-0%, 1%, 4%, 7%, 10% and 12.5% Si-0.3%Mg alloys were investigated. The effects are summarized as follows.

(1) Hot tearing was not observed in pure Al and in Al-12.5%Si-0.3%Mg alloy. The narrow solidification temperature range does not generate a high thermal stress and restricts hot tearing.

(2) With decreasing Si content, the crack length increases. The solidification at the vertical bar portion of the T-shape will likely be delayed. Hot tearing occurs in the alloys when the solidification at the vertical bar portion is relatively fast and generates high tensile stress at the corner portion.

(3) The Sr addition into Al-7% and 10% Si-0.3%Mg alloys restricts cracking through a healing process, but it does not restrict hot tearing of Al-1% and 4% Si-0.3%Mg alloys. The healing is controlled by the amount of the eutectic phase and Sr itself.

References

[1] M. Adachi, S. Koiwai, H. Tachibana, K. Murase and H. Yamagata: J. JFS, 87(2015) pp. 531-537

[2] H. Yamagata, K. Murase, H. Tachibana, S. Koiwai and M. Adachi: J. JFS, (2015) in press

[3] M. Adachi, S. Koiwai, H. Tachibana, K. Murase and H. Yamagata: J. JFS, 87(2015) pp. 545-551

Study of low pressure die casting AlSi11 alloy solidification

Zdenka Zovko Brodarac[1], Josip Kalinić[2], Tamara Holjevac Grgurić[1]
[1] University of Zagreb Faculty of Metallurgy, Aleja narodnih heroja 3, Sisak, Croatia
[2] Jajce Alloy Wheels Ltd, Divicani bb, Jajce, Bosnia and Herzegovina

Aluminium alloys possess an excellent mechanical, physically-chemical and technological properties which have found an extensive application in the automotive industry. Alloy AlSi11 belongs to a group of eutectic alloys, which are characterized by high castability, low pouring temperature and narrow solidification interval which makes it suitable for low pressure casting technology (LPDC).

In addition to chemical composition, the structural and mechanical properties of alloys depend on many factors that act during solidification. Study of AlSi11 alloy solidification sequence has been accompanied by calculation of equilibrium phase diagram, and simultaneous thermal analysis, all in correlation to microstructure development. Thermodynamic calculation revealed solidification sequence with corresponded temperatures in equilibrium state. Differential scanning calorimetry enables determination of exact temperatures of phases' transformations. Metallographic investigation comprehends fractography, as well as optical and scanning electron microscopy. Microstructural examination performed on scanning electron microscope equipped by EDS revealed development of following constituents: primary aluminum (α_{Al}), high temperature intermetallic phases on the iron basis (Al_5FeSi and/or $Al_{15}(Fe,Mn)_3Si_2$), binary eutectic ($\alpha_{Al}+\beta_{Si}$) and secondary eutectic (Mg_2Si and $Al_5Mg_8FeSi_6$) as a last solidifying phases'.

Synergy of performed thermodynamic and microstructural investigation, as well as obtained mechanical properties enables an evaluation of material quality and management of solidification process.

Keywords: *AlSi11 alloy, solidification sequence, mechanical properties, microstructure.*

1. Introduction

The automotive industry is forced to apply advanced materials and technologies in order to overcome the mutual competition, but also for compliance with environmental regulations requiring reduction of emissions, and fuel consumption. One of main component for achieving this goal is weight saving by downsizing of components using advanced materials and production technologies. Average car containing 140 kg of aluminium exceeds a driving distance of 200.000 km (15.000 km / year). Aluminium component is on average 40% lighter than the replaced component (direct weight saving) and that 25% additional weight reduction is obtained by downsizing other components (weight savings) [1]. The widest production of aluminum component refers to wheels as the most safety-critical part, which has been produced by low pressure die casting technology. "Eutectic" AlSi11 alloy (EN AC 44000) is characterized with relatively low melting point and narrow solidification interval which both brought to the uniformly distributed eutectic microstructure indicating superior mechanical and technological properties [2]. Silicon is one of the most important alloying elements which comprehend to good castability of aluminium alloys. Addition of silicon improves resistance to hot cracks and feeding capability [2,3]. Synergy of influenced alloying and trace elements effect comprehend to different intermetallic phase evolution. The content of secondary alloying elements (Mg, Cu) and trace elements (Fe, Mn, Cr, Zn) significantly influence on solidification manner of an AlSi11 alloy [4-6].

2. Materials and methods

Investigation of an AlSi11 alloy comprehends modelling of equilibrium phase diagram (CALPHAD method-TCW 5.0), simultaneous phase analysis (Netzch STA-DSC) and microstructural investigations performed: optical (Olympus GX51) and scanning electron microscope (Tescan, JEOL).

3. Results and discussion

Chemical composition of produced sample, given in Table 1, corresponds to those prescribed by norm (EN 1706:2010, [7]).

Table 1. Chemical composition of AlSi11 alloy (wt.%).

Si	Fe	Cu	Mn	Mg	Al
10.6276	0.0905	0.001	0.0017	0.1808	Bal.

Modelling by CALPHAD method enables overview of solidification sequence AlSi11 alloy in polythermal section of phase, as shown in Figure 1.

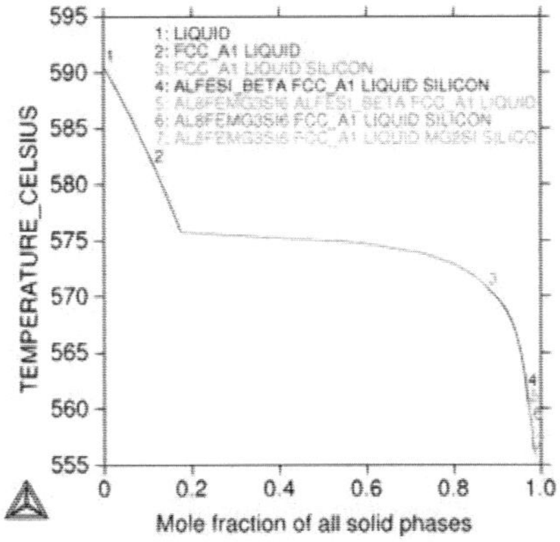

Fig. 1. Solidification sequence of AlSi11 alloy modelling by CALPHAD method.

Since liquid is stable till 566 °C, and secondary eutectic phases evaluate below this temperature which indicate limited sources and positions for its precipitation: evaluation of α_{Al}, precipitation of high temperature intermetallic phases on iron base, silicon (primary and eutectic) evaluation, and precipitation of complex secondary eutectic intermetallic phases $Al_8FeMg_3Si_6$ and Mg_2Si.

Simultaneous thermal analysis performed at heating / cooling rate of 10 (K/min) enabled exact identification of characteristic temperatures of phase transformations and precipitation. Comparison of CALPHAD (TCW) and STA temperatures is shown in Table 2.

Table 2. Comparison of TCW and DSC phase transformation temperatures

Reaction No.	Temperature, °C		Reaction
	TCW	DSC	
1	590	589,2	Dendrite network development
2	575	-	Primary silicon evolution
3	572	564,6	Precipitation of Al_5FeSi phase
4	502	563,2	Eutectic reaction
5	439	558,1	Precipitation of $Al_8FeMg_3Si_6$ phase
6	394	525,9	Precipitation of Mg_2Si phase

Metallographic analysis resulted in following microstructural constituents: primary aluminium α_{Al}, main eutectic $\alpha_{Al} + \beta_{Si}$, high temperature iron phase and intermetallic phases in form of secondary eutectic Mg_2Si and $Al_8FeMg_3Si_6$, as shown in Figure 2.

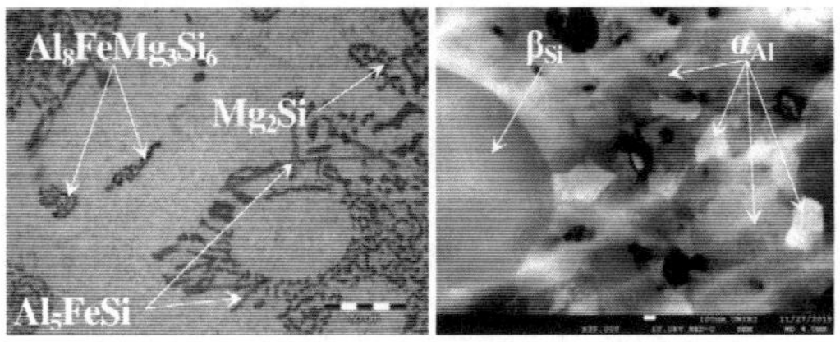

Fig. 2. Microstructure analysis of AlSi11 alloy.

Higher magnification enabled an overview of primary aluminium nucleation correlated with dislocation movement, as well as twinning mechanism of eutectic silicon.

4. Conclusions

An AlSi11 alloy, as a commonly used alloy in automotive applications reveals specific solidification sequence due to amount of secondary alloying and trace elements. Also thin-wall casting geometry and LPDC ensures rapid cooling and solidification. Modelling of equilibrium phase diagram, simultaneous phase analysis and microstructural investigations resulted in determination of manner and solidification sequence of AlSi11 alloy.

Acknowledgements

Company Jajce Alloy Wheels Ltd for samples and University of Rijeka, Department of physics for help are greatly acknowledged. Investigation was done in the frame of Financial support of investigation of University of Zagreb (5.10.1.1, TP1.79, TP167)

References

[1] EAA-Aluminium-in-Cars-Unlocking-the-light-weighting -potential, 2013, www.european-aluminium.eu

[2] J. R. Davis, Aluminim and Aluminum Alloys (ASM Specialty Handbook, Ohio, USA, 2002)

[3] J. Campbell, Metal quality studies in secondary remelting of aluminum, Technical Paper, 47(2004), 78-81.

[4] K. Krone, Aluminiumrecycling (Vereinigung Deutscher Schemlzhütten, Düsseldorf, 2000)

[5] H. Baker, Alloy Phase Diagrams, ASM Handbook – Vol. 3, (ASM International Ohio, USA, 1992.)

[6] L. Bäckerund, G. Chai, J. Tamminen, Solidification Characteristics of Aluminium Alloys, Foundry Alloys, Vol. 2, (AFS/Skanaluminium, Stockhlom, 1999).

[7] EN 1706:2010 Aluminium and aluminium alloys – Castings – Chemical composition and mechanical properties

O-35

HOW TO SOLVE SHRINKAGE POROSITY IN THE ALUMINUM GRAVITY DIE CASTINGS?

(A shop floor experience & case studies).

Author: Muthiah thirugnanam, Chennai, India. Email: callvmthiru@gmail.com,

Abstract:-Part 1:Many Aluminum castings are being produced all over the world, using gravity die casting process. Shrinkage porosity is one of the major problems faced. In part 1 theoretical aspects of shrinkage are illustrated. How to identify, analyze & find root causes for Macro & Micro shrinkages & develop solutions? are explained. In part 2:Case study 1: In a critical casting" Body", shrinkage occurred inside bore & solved by reducing wall thickness & re designing core. Case study 2: Micro shrinkage was occurred in a critical casting, "Valve body". By re designing the "casting shape a little" & providing copper chilling it was solved.

Key words: Micro & Macro shrinkage, feeding, chilling.

Introduction:- Shrinkage occurs in an Alum casting due to the "uneven shape of the casting. But it can not be changed by the casting maker. Only minor changes can be obtained from customer. If the casting has thick sections, shrinkage may occur. They are produced by *"Non homogeneous* variations of volume of metal ,occurring between the beginning & end of solidification. Shrinkages are two types: 1) Macro 2) Micro shrinkages, which are two different manifestations of same effect, with different sizes. To solve shrinkage, casting maker can *manipulate* only 1)the processing of metal ,but can *not* change the alloy.2) pouring conditions 3) design of the die.4) die coating & cooling methods.

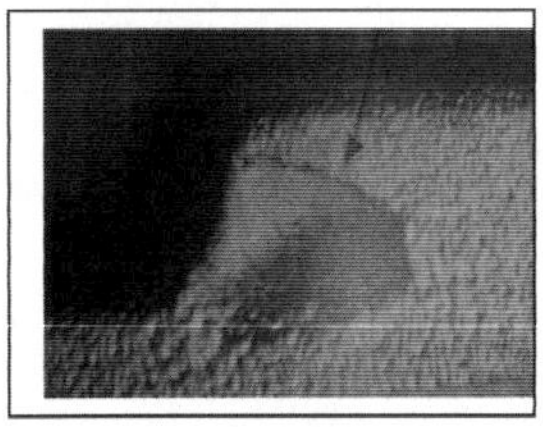

Fig 1. Macro shrinkage

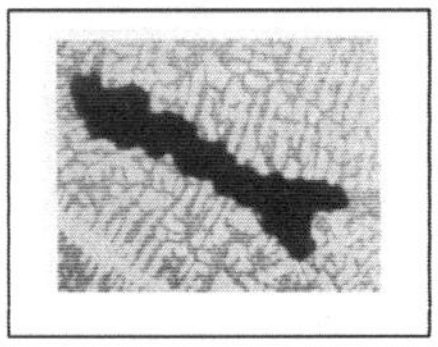

Fig 2. Micro shrinkage

Theoretical aspects of the causes:- (1)Macro shrinkage cavities are caused by massive volume reduction due to lack of feeding & when the alloy is of 'Short freezing range". It can be external or internal. The volume of the solid crystals is less than that of the liquid from which they grew. The solid mass occupy lesser volume than the liquid. Hence the difference in volume is made up of cavities of various forms.(2) Micro shrinkage occur between the dendrites, where the crystallization of the last liquid to solidify. They are noticed in the internal sections. The shape is always *'odd'*.

The factors that influence shrinkage cavities are :- (A)Factors depending on the casting alloy spec: (A.1)*Solidification range, (A.2) % Shrinkage during solidification,(A.3)Gravity (A.4)Evolution of gases* (B)Factors depending on the conditions of pouring:-(B.1)*Pouring temperature.(B.2)Rate of pouring (B.3)Feeding systems (Risers)*(C) Factors depending on the die (C.1)*Initial temperature of the die (C 2) Conductivity of the die* (By providing copper chills & cooling many surprising improvements can be achieved) (D) Factors depending on the casting *(D1)Heavy section thickness (D.2) Complexity of it's Shape* .(E) Factors depending on the shell sand cores *(E1)Type of metal entry in to the die(Eg: Over heating the ingate area) (E 2)* Resin content. (F)Factors depending on the molten Metal Quality:- *(F1)*Inclusion may encourages the nucleation of shrinkage .*(F2)* Calcium content (if calcium is > 0.0008%) micro shrinkage cavities may occur. (F3)Modification of molten alum:- If Sodium based flux are used , micro shrinkage may occur.

<u>Part 2:- Case Study.1</u> *(1)Housing casting:- Micro Shrinkages* occurred in the areas as shown in the Fig3&4.Air leaked in the pressure test

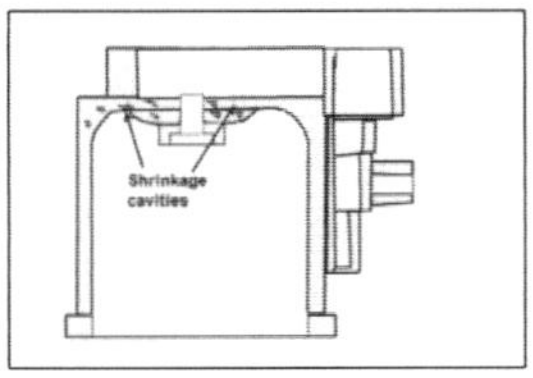

Fig.3 Shrinkage areas

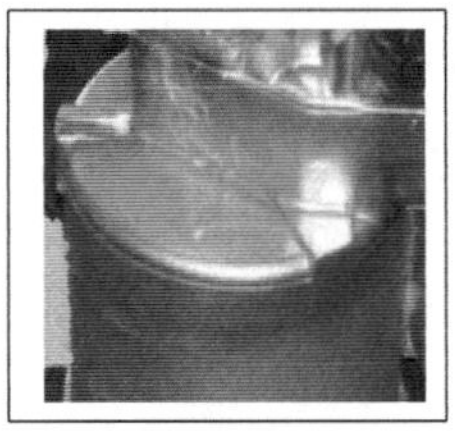

Fig.4.Body casting

<u>Causes</u>:-In the die, over heating of the big bottom core; 150 dia ;due to the direct hitting of molten metal(>400C) (Ref fig.5) & lack of feeding to this heavy Section thickness area. It became a hot spot. Unable to provide feeder due to die design problems.

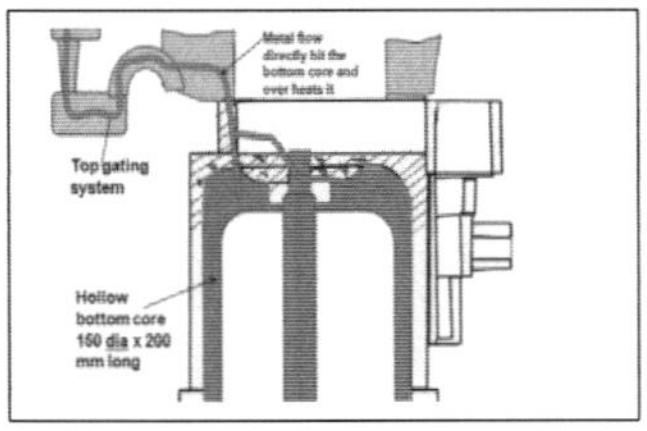

Fig.5Direct hitting of metal on the big bottom core

<u>Corrective actions taken</u>.(1) In the hot spot area ,wall thickness was reduced & made uniform ,7mm through out instead of 12 mm, with customer acceptance. (2) Cooling was provided below the over heating area.

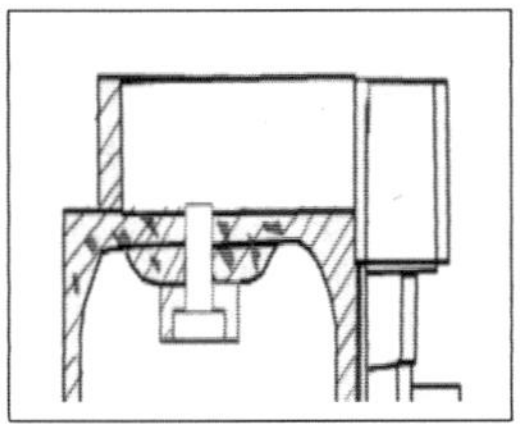

Fig.6.Wall thickness:12mm

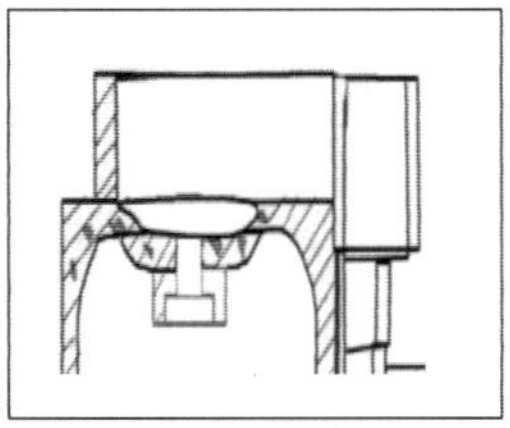

Fig.7 Wall thickness reduced to 7 mm after

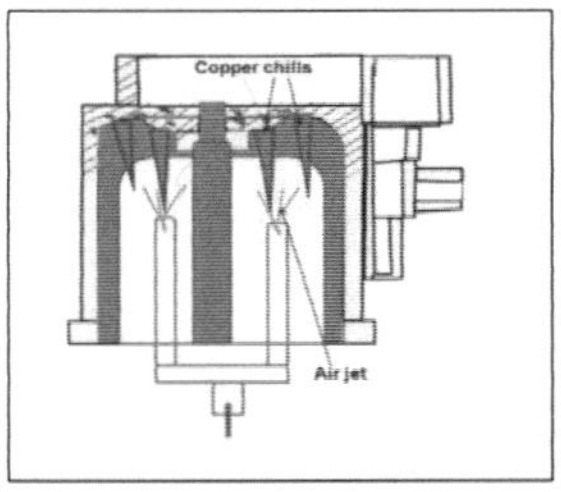

Fig 8. Copper chills with air jet cooling was provided, inside the 'Hollow core'.

<u>Result:</u>-The shrinkage was *eliminated.* But *air started leaking* through the *copper chills.* Also the shop became very ***<u>noisy</u>*** due to the discharge of compressed air &expensive. This was ***Not accept able to the management.*** Find another way**.** Core was studied & noticed that '*air* in 'hollow inner area of the core 'made it a poor conductor of heat. It was decided to try a solid core (ref Fig.9*)* expecting *sufficient conduc - tivity* for dissipating all the heat from the hot spots. <u>The result</u> was " <u>no shrinkage"</u> .

Fig.9. Solid metal bottom core,150 mm dia, conducts heat better & faster.

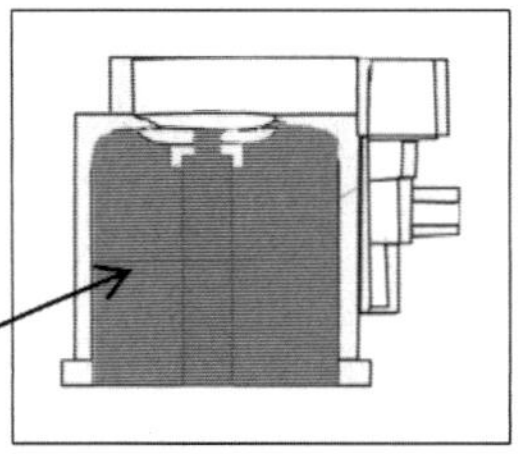

<u>Case study 2</u>: In the casting shown below micro shrinkage was occurring on valve seat.

Fig 10 Full shot of the casting, Valve Body

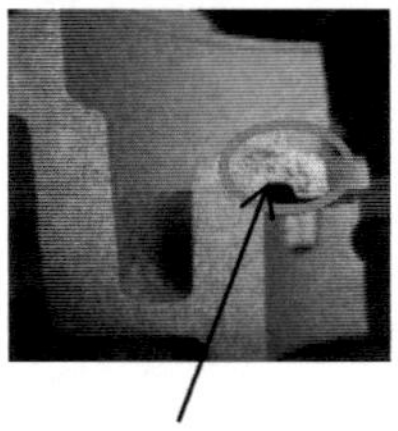

Fig.11Micro Shrinkage on Valve seat

<u>Corrective actions taken</u> (1) Machining allowance was reduced on the valve seat from 4.5 mm to 1.5 mm. (2) copper chills were provided & water cooling was done on the chills. Ref Fig 12. **<u>Result</u>**: *No shrinkages.*

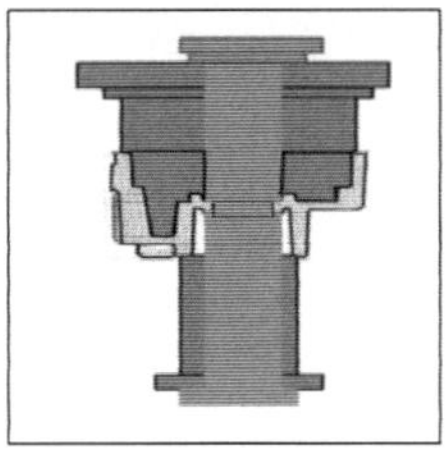

Fig12. Big copper chills were provided through the top & bottom cores

<u>Conclusions</u> *(*1) Try to provide risers (2) Explore for reducing wall thickness in the areas.(3) Provide copper chills with air or water cooling (4) Reduce machining allowance..

Product designers should consider foundry process capability& avoid frequent section changes

References:-

[1] T.P.Fisher Technology of Gravity Die casting- .

Proceedings of the 72nd World Foundry Congress
21-25th May 2016, Nagoya, Japan

Color Metallography of A356 Aluminum Alloy Castings using Weck's Reagent

Shinji Kumai, Yohei Harada and Li Gao
Department of Metallurgy and Ceramics Science, Tokyo Institute of Technology
2-12-1 O-okayama, Meguro-ku, Tokyo 152-8552, Japan

A color etching method using Weck's reagent (3.81mol%$KMnO_4$, 0.15mol%NaOH water solution) was developed for both the optical microscopy and micro-segregation characterization of A356 aluminum alloy (Al-7wt%Si-0.35wt%Mg) castings. The coloring mechanism of Weck's reagent was also elucidated. In this study, colored microstructures of A356 alloy (with 0.1wt% Ti added) were compared with the solute distributions measured by EPMA analyses to make clear the relationship between color and segregation. It was also evidenced that the tracking the distribution of Ti in Al phase during the spheroidization of Al phase at semi-solid state made possible. These examples prove that color etching with Weck's reagent is an efficient way of characterizing micro-segregations in Al alloy castings. The accurate evaluation of solid fraction of A356 alloy slurry using Weck's reagent was also achieved. With the help of Weck's reagent, the grain growth happened during rapid cooling of the alloy from semi-solid state to room temperature was clearly visualized. Thus the overestimation of solid fraction by conventional method was avoided by excluding the area of grain growth. Additionally, the spheroidization process of dendritic Al phase was observed in detail using Weck's reagent and EBSD analysis and the spheroidization mechanism was provided. Furthermore, the etched specimen surface was characterized by various observation and analysis methods to study the coloring mechanism of Weck's reagent.

Keywords: *Color etching, A356 alloys, semi-solid casting, micro-segregation, spheroidizing mechanism.*

1. Introduction

The present study is concerning the characterization of A356 alloys processed at semi-solid state through color metallography by using Weck's reagent. Applications of Weck's reagent were summarized, and then the coloring mechanism of the etching was studied by analyzing the etched surface using various kinds of advanced analyzing techniques.

2. Correlation between the micro-segregation in Al phase and color revealed by Weck's reagent

The detailed correlation between color and micro-segregation in A356 alloys was studied. EPMA mapping mode was applied to characterize the solute distribution and compared to optical micrograph after etching with Weck's reagent. The results show that micro-segregation of Ti, Si and Mg in Al phase can all lead to a color difference (Fig. 1(a-d)). Besides in dendritic Al structure, micro-segregation also exists in spherical Al grains obtained by semi-solid process (Fig. 1(e-f)). The spherical grain growth occurred when cooling it from semi-solid state to room temperature can also be visualized by Weck's reagent. [1]

3. Accurate solid fraction evaluation of A356 alloy at semi-solid state by Weck's reagent

As shown above, grain growth during quenching from semi-solid state could be visuallized by Weck's reagent. This finding can help to evaluate solid fraction at semi-solid state with a better accuracy by exclusing this part when calculating the area fraction of solid particles in the optical micrograph. The result for A356 alloy using this method were compared with both the traditional method without excluding the grain growth during queching and the results from lever rule referring to Al-Si binary phase diagram. It is shown that the results obtained with exclusion of the grain growth during water quenching agree well with the results calculated from Al-Si binary phase diagram. The results obtained without consideration of the grain growth showed an overestimation over the above two methods. [2, 3]

4. Spheroidization mechanism of Al phase at semi-solid state by compression and partial re-melting process

The microstructural evolution of Al grains from dendritic to spheroidal was focused. Since the materials were deformed and heated to semi-solid state, recrystallization occurred during heating. Al grains were refined compared to the specimen without deformation before heating. Weck's reagent finds its application again in this study. By etching the specimen which is transforming from dendritic to spheroidal, compressed dendritic Al grains are found separated and when the eutectic structure starts to melt, liquid phase penetrates into those separation sites and spheroidized Al grains are formed (Fig. 2). According to EBSD analysis, high angle grain boundaries are found inside Al phase when the compressed specimen is heated. This suggests that Weck's reagent is also able to reveal grain boundaries in Al phase clearly. [3]

5. Coloring mechanism of Weck's reagent studied by characterization of the etched surface

Coloring mechanism of Weck's reagent was studied by various characterizations and analyses of the etched surface. Surface observation by SEM and laser microscopy found that the locational change of color is related to the different surface roughness which is in the order of submicron. More detailed observation by TEM, STEM-EDS and SIMS showed that an amorphous film contains MnO_2 is formed on the Al surface via the conversion coating occurred during the etching with Weck's reagent. Depending on the different concentration of solute in Al (micro-segregation), the MnO_2 thin film with different thickness and roughness is formed on the specimen's surface. The higher concentration of Ti can promote the film growth, which is confirmed by the surface observation. It is concluded that the color of the film is basically brown. But the brightness of brown is strongly influenced by the film's thickness and the roughness of the interface between the film and substrate. As a result, reflection of light from both the film surface and interface between film and substrate (Al alloy) interfered, causing the different colors observed by optical microscopy (Fig. 3). [4]

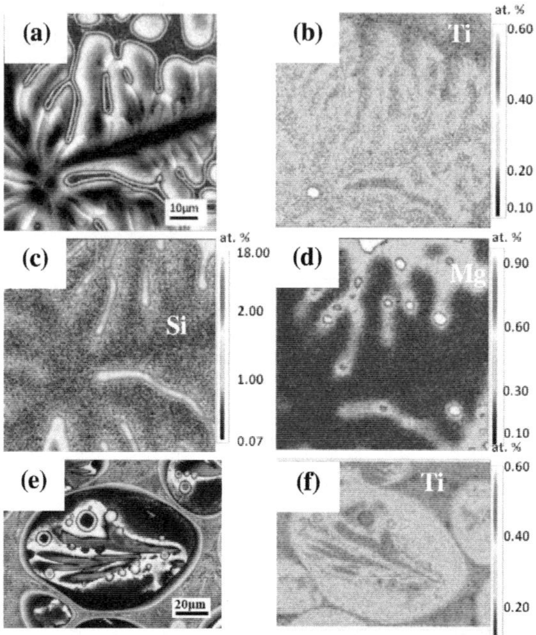

Fig. 1 Optical micrograph and EPMA mapping for solute distribution.

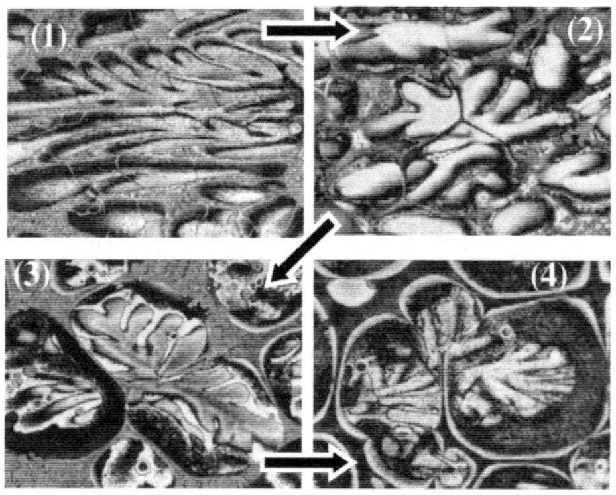

Fig. 2 Optical micrographs of Al grains showing spheroidizing process.

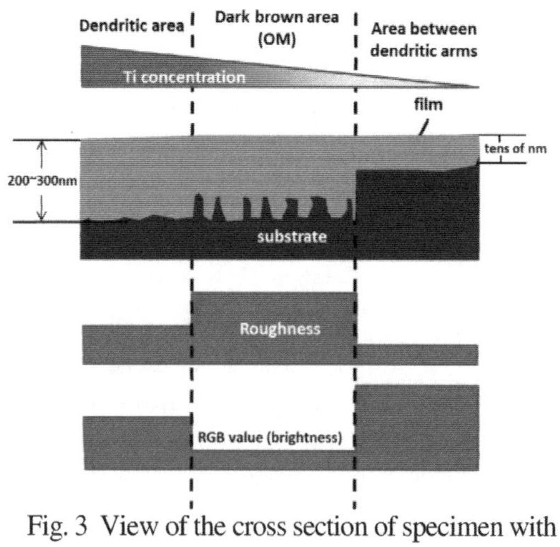

Fig. 3 View of the cross section of specimen with correlation to change of Ti concentration.

References

[1] Li. Gao, Y. Harada and S. Kumai: Materials Science Forum, 794-796, pp. 9-14, 2014.

[2] Li. Gao, Y. Harada and S. Kumai: Journal of Materials Science, 49(3), pp. 1286-1296, 2014.

[3] Li. Gao, Y. Harada and S. Kumai: Materials Characterization, 107, pp. 426-433, 2015.

[4] Li. Gao, Y. Harada and S. Kumai: Materials Characterization, 107, pp. 434-452, 2015.

Proceedings of the 72nd World Foundry Congress
21-25th May 2016, Nagoya, Japan

Microstructure Analysis of Al-5%Mg Alloy Semi-Solid Slurry by Weck's Reagent

Kamolwat Prapasajchavet, Yohei Harada and Shinji Kumai
Department of Metallurgy and Ceramics Science, Tokyo Institute of Technology, Tokyo 152-8552, Japan

Spheroidization of Al grains is required for the production of semi-solid slurry. In this research, Al-5%Mg aluminum alloy with dendrite structure was plastically deformed with 33% compression and partially melted in a semi-solid state in order to obtain globular solid grains. The microstructure evolution during partial melting was investigated by the quenched microstructure etched with Weck's reagent. EPMA analyses by X-ray Wavelength Dispersive Spectroscopy (WDS) were carried out to evaluate Mg content in globular solid grain. Imaging with secondary electrons by SEM was applied to characterize the etched surface. In the as-etched optical micrograph, the microstructure was revealed with different color and contrast. During microstructure evolution of compressed specimen, the deformed grain changed gradually to the spheroidal shape due to the partial melting at the semi-solid temperature range. The original shape of the globular Al grains at the semi-solid state could be visualized clearly in the quenched microstructure by eliminating the growth layer of Al grains formed during water quenching. The solid fractions were measured for the entire semi-solid temperature range. SEM images showed that the surface topology of the specimen is corresponding to the variation in color and contrast in the optical micrograph.

Keywords: *Aluminum alloys, Semi-solid process, Solid fraction, Color etching, Weck's reagent*

1. Introduction

Al-Mg alloy has the high strength and good corrosion resistance, however, this alloy shows a poor castability. By using conventional casting process, casting defects and solidification cracking often formed. The effective method to produce sound products is necessary.

Semi-solid metal process is the forming technique of the cast products from the semi-solid state at temperature between solidus and liquidus line. The semi-solid slurry consists of spheroidal solid grains suspended in the liquid phase at semi-solid temperature during forming [1]. The present process can decrease shrinkage porosity due to reduced processing temperature and improve the castability for this alloy system.

In the present research, Weck's reagent (Distilled water: 100mL; $KMnO_4$ 4g; NaOH 1g) is utilized to visualize the microstructure of Al-Mg alloy semi-solid slurry obtained by recrystallization and partial melting process for understanding the microstructure evolution during partial melting. The solid fractions in the semi-solid temperature range were also precisely evaluated.

2. Experiment

Al-5%Mg billet was plastically deformed at room temperature with 33% compression and partially melted to semi-solid temperature range: 580-630°C. The polished surface of as-cast, as-compressed and semi-solid specimen were etched with Weck's reagent for approximately 12s (±1s) at room temperature to examine the microstructure. Microstructure was observed by optical microscope. To investigate the microstructure evolution during partial melting, the compressed specimens were heated to various temperatures from the below solidus line to the temperature range of semi-solid state. The solid fractions were measured for the entire semi-solid temperature ranging from 585°C to 625°C. Imaging with secondary electrons by SEM was applied to characterize the etched surface. EPMA analyses by X-ray Wavelength Dispersive Spectroscopy (WDS) were carried out to evaluate Mg content in solid grain.

3. Results and Discussion

The optical micrograph observation showed the as cast specimen exhibited dendritic structure. For the as compressed specimen, dendritic branches were deformed and changed gradually to the spheroidal shape due to the partial remelting at the semi-solid temperature range [2] shown in Fig.1.

The dark brown area and the white area corresponded to the globular solid grain and liquid phase at the semi-solid state, respectively. While, the light brown area indicated the solid growth layer during water quenching shown in Fig.2.

By excluding the growth layer of Al grains formed during water quenching, the original shape of the globular Al grains at the semi-solid state could be visualized clearly [3]. The entrapped liquid inside the globular solid grains was considered as the solid phase at semi-solid state in order to evaluate the precise solid fraction which comparable to the actual situation during dies filling: red solid fraction curve shown in Fig.3.

As shown in Fig.4, SEM images showed that the surface topology of the specimen is corresponding to the variation in color and contrast in the optical micrograph. The dark brown area and the light brown area in the optical micrograph corresponded to the rough surface and smooth surface in SEM image, respectively.

Optical micrograph showed a variation in color and contrast, light brown and dark brown in globular solid grain. In contrast, the Mg content obtained by EPMA analysis showed the homogenous Mg distribution in the globular solid grain. This might suggest that Weck's reagent possibly has better sensitivity in detecting the local Mg content than EPMA analysis. In order to investigate this topic, coloring mechanism and precise investigation concerning the relationship between Mg composition and color (contrast) of the color-etched optical micrograph is necessary [4].

4. Conclusion

By etching Weck's Reagent, the original solid phase and liquid phase at semi-solid state were distinguished clearly with different color and contrast. The microstructure evolution during partial melting could be investigated. The present color metallography was effective for correct estimation of the solid-liquid ratio in Al-5%Mg alloy semi-solid slurry.

Acknowledgement

The authors would like to thank Nissan Motor Co., Ltd for supplying the starting material and technical supports in this research.

References

[1] Flemings MC (1991) Metall Trans A 22A:957

[2] L. Gao, Y. Harada, S. Kumai, J. Mater. Sci. 47 (2012) 6553-6564.

[3] L. Gao, Y. Harada, S. Kumai, J. Mater. Sci. 49 (2014) 1286-1296.

[4] L. Gao, Y. Harada, S. Kumai, Materials Characterization. 107(2015) 434-452.

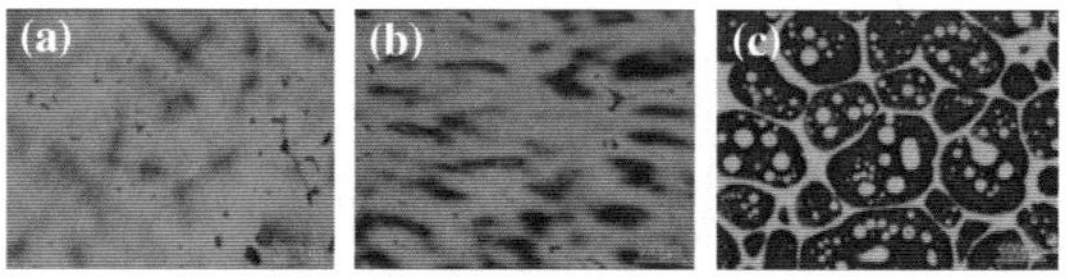

Fig.1 Microstructure of (a) as cast (b) as compressed and (c) semi–solid specimen

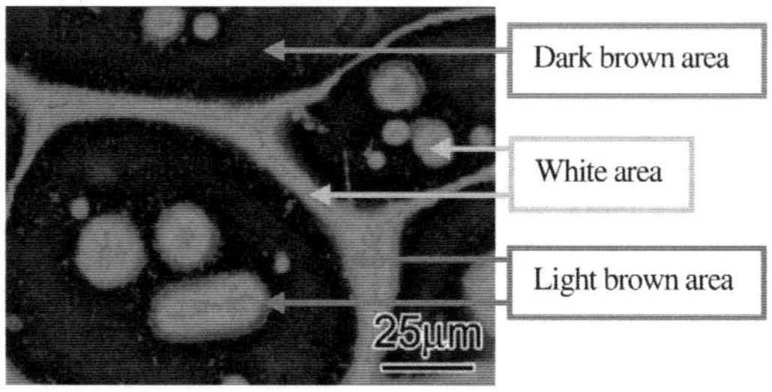

Fig.2 Detection of grain growth during quenching

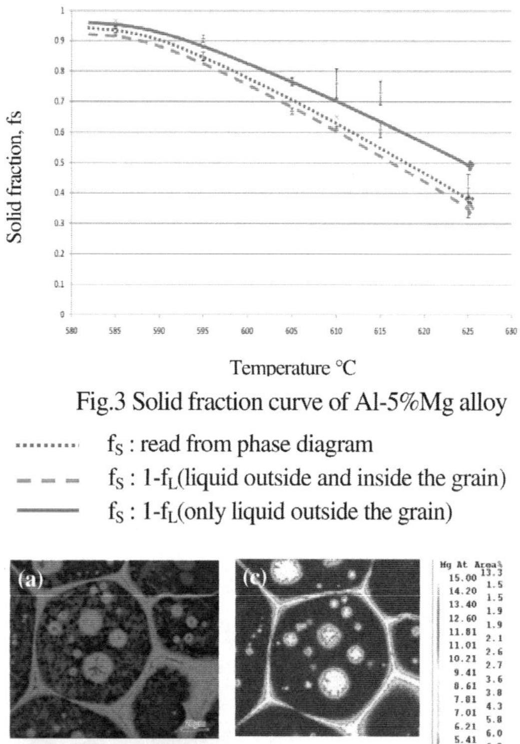

Fig.3 Solid fraction curve of Al-5%Mg alloy

········ f_S : read from phase diagram

– – – f_S : $1\text{-}f_L$(liquid outside and inside the grain)

——— f_S : $1\text{-}f_L$(only liquid outside the grain)

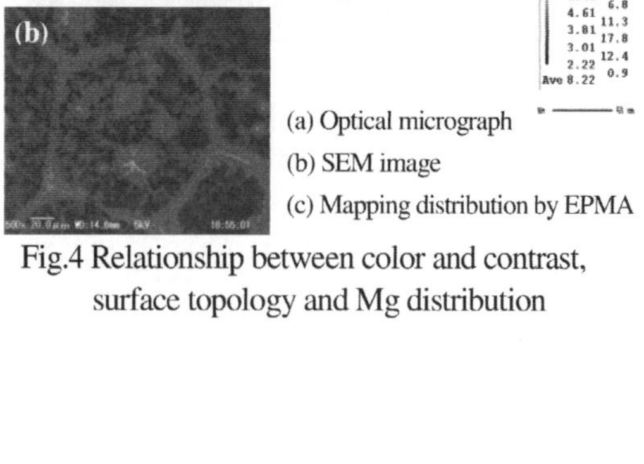

(a) Optical micrograph

(b) SEM image

(c) Mapping distribution by EPMA

Fig.4 Relationship between color and contrast, surface topology and Mg distribution

Proceedings of the 72nd World Foundry Congress
21-25th May 2016, Nagoya, Japan

Inorganic Technology – The End of Shell Sand?

Dr. Heinz Deters[1] and Kazuyuki Mase[2]

[1]ASK Chemicals GmbH, Reisholzstrasse 16-18, 40721 Hilden / Germany

[2]ASK Chemicals Japan Co., Ltd . 8th Floor, Bashamichi 450 Building, 50, Otamachi 4-Chome Naka-Ku, Yokohama 231-0011/ Japan

The Inorganic Technology (INOTEC) has proven to be especially suitable for aluminum casting applications, particularly for the manufacture of aluminum engine blocks and cylinder heads in permanent mold casting.
Odorless core production, no harmful emissions during casting, less cleaning of machines and tools, the resulting higher output and productivity, and the advantages in terms of casting, such as faster solidification due to reduced die mold temperatures – these benefits of INOTEC are already well-established when compared to classical organic binder technologies.[1-2] More and more foundries are converting from organic to inorganic binder systems – and there are several reasons for this trend.

Keywords: *INOTEC, Inorganic Binders, Shell sand, odorless, no emission, aluminum castings*

1. Introduction

Shell sand is a resin coated sand – typically silica sand but also more special ceramic sands.

INOTEC is an inorganic binder system which is based on two different components – a liquid INOTEC binder and a solid component – the so-called INOTEC Promotor. The latter component is not just an additive but rather a part 2 of the binder system as important properties are associated with it – such as initial strength, thermal stability, shake-out etc. Both of these components are mixed together with the refractory material in a mixing unit prior to their use.

2. Core manufacturing – similarities and main differences

Both binder systems are cured in a hot core box. Shell sand requires typically temperatures of higher than 250°C which are usually realized by electrical heating or gas burning heating.

INOTEC is typically cured in the temperature range between 150 and 210°C depending on the core geometry – significantly lower than that of shell sand, thereby offering the chance for energy and cost savings. Furthermore, the heating can be done electrically but, however, oil heating would be the method of choice due to the fact that the temperature only varies in a very narrow temperature range and therefore makes the process more stable. One crucial aspect for INOTEC is the hot air purging to accelerate the curing process. Such a purging is missing in the Shell-sand process.

There is a potential to decrease cycle times when using INOTEC compared to the traditional shell sand process. This may offer great opportunities to increase productivity in the core production.

Strengths values (both hot and cold) are high enough for the INOTEC cores to be handled. Care must be taken only in regard to the brittleness which is typically higher than that of shell sand cores. In this way INOTEC is rather comparable to glass, while shell sand is comparable to plastic. Once this is known in the core shop, core handling should not be a problem. Additionally, it must be noted that inorganic cores have – by nature – a higher affinity to water. Thus storage stability is more critical than in case of shell sand cores. Inorganic cores need to be protected from long exposure to high humidity. However, by recent product developments, ongoing efforts in research and development as well as the choice of proper storage conditions, significant improvements could be made to improve the humidity resistance of inorganic cores.

Shell sand has – by nature of this material – probably the best flowability properties of all sand/binder systems – there are almost no limitations in geometry. However, complex core geometries like water jackets of cylinder heads are also successfully produced in series production every day using the INOTEC binder system. This is thanks to both continuous product optimization and to substantial technical adjustments in terms of core box design but also in the core print design which is nowadays especially suited for the needs of inorganic binders systems.

One thing is without any doubt: INOTEC cores do not lead to any smoke or odor formation during core manufacturing which is in great contrast to shell sand cores.

3. Casting production

In case of shell sand there is a significant amount of smoke and gas that is generated during the casting process. Such a smoke formation is a result of the thermal decomposition of the organic resin leading to all different kinds of toxic organic substances which are emitted to the environment. Such a gas/smoke evolution could also generate some gas defects inside the casting if such gases are not properly driven out by vacuum systems. Furthermore the generated smoke leads to a tar build-up not only on the die molds but also in the air cleaning systems – thereby leading to additional cleaning efforts.

In case of the INOTEC binder system there would be no smoke formation as the nature of the binder can be entirely inorganic. As a consequence there is no tar formation on the die molds leading to less cleaning efforts and consequently less costs. However, it has to be noted that inorganic cores also generate gas to some extent during the casting process. This is related to the fact that inorganic binders still contain some residual water content to keep the silicate gel structure and therefore the binding properties "alive" and is further related to the core gas that will expand when exposed to heat. In particular for water jackets in cylinder heads such gas needs to be driven out by a proper vacuum system – similar to the shell sand but significantly less gas formation.

Due to the missing condensate build-up the die mold temperatures may be reduced leading to reduced solidification times of the liquid aluminum. All this can lead to an increase of productivity in the casting process – so there is a potential to produce more castings within the hour in case of INOTEC.

Furthermore, it has been shown that – as a result of faster solidification – secondary dentrite arm spacings are reduced which leads to better strength properties of the castings. Thus, inorganic binder systems offer great opportunities for future developments.

Just recently, a German automotive company revealed a new central feeding concept which is only possible when using inorganic binder technology because of the missing smoke and condensate formation. [2]

Shell sand is usually known for its high thermal stability leading to dimensional accuracy of the casting. INOTEC basically offers the chance to adjust the thermal stability according to the specific needs of the cores because the thermal stability is directly dependent on the chemical composition.

Last but not least, in case of inorganic binders no air cleaning systems and no after-air treatments are necessary due to missing smoke formation during the casting process. This is a great advantage to protect the environment and the employees working in the foundry.

4. Conclusion

"INOTEC – the end of shell sand?" is quite a provoking question. Certainly, INOTEC cannot outperform shell sand in every regard, but for sure in many ways. Given that the general trend towards green technologies holds on and governmental restrictions regarding emissions become stricter, more and more foundries will be converting from organic to inorganic binder systems.

But is not just the environmental aspect but also other potential benefits like increased productivity during core manufacturing and the casting process as well as new technological possibilities.

INOTEC has certainly proven to be more than just an alternative in aluminum die mold applications – such as the manufacture of cylinder heads.

References

[1] "Tomorrow's Cylinder Head Production - Ecology, Economy and Material Enhancement Brought in Line", Emmerich Weissenbek, Thomas Kautz, Jörg Brotzki, Jens Müller, MTZ06/2011, Volume 72, 484–489

[2] "Inorganic Innovation for the New Top-of-the-Range Diesel Engines in the BMW M550xd: Design and Casting technology of the Aluminum Crankcase", Emmerich Weissenbek, Bernhard Zabern, Andreas Fent, Johann Stastny, Christian Högl, Giesserei-Praxis 5/2013, 175–181

Proceedings of the 72nd World Foundry Congress
21-25th May 2016, Nagoya, Japan

Benchmarking the High Production Sand Core and Mold Binder Systems and Most Recent Advances

T. Homma ASK Chemicals Japan Co., Ltd. Yokohama, Japan
J. J. Archibald ASK Chemicals L.P., Dublin, Ohio, USA

Starting with the shell molding process introduced in the '50's, some of heat cured sand core and molding binder systems which can be applicable to high production foundries were developed. In late '60's, after the first cold box system, the amine cured phenolic urethane binder was developed, number of cold box systems were developed and introduced to the foundry industry. Each system was developed based on different chemistries. Motivations of the development were sand core and mold productivity, reduction of casting defects and environmental pressure. For their different chemistries and concepts of invention, it is important to benchmark the binder systems. In this paper, benchmarking the high production sand core and mold binder systems are conducted in terms of core and sand making property, productivity, casting quality, the cost-in-use and emissions.
Furthermore, future visions of high production sand binder systems including both organic and inorganic systems are discussed.

Keywords: *Cold box processes; sand binders; equipment; environmental compliance; productivity and cost*

1. Introduction

Today over 70% of high production sand casting operations use a CB process for producing cores and molds. The expansion of CB use has been driven by numerous factors including the need for more through-put and reduced work in process in the core room, improved dimensional accuracy, improved casting quality and reduction of emissions during core making and pouring, cooling and shakeout (PCS). The primary attribute of CB – productivity - is predicated on the fact that blowing sand into vented tooling and rapidly curing it enhances yield per tool, thus drives the return on investment and takes cost out of the process chain of producing complex castings like cylinder-heads. For the purpose of this paper we will discuss the four primary CB chemistries:

- Phenolic urethane CB (PUCB)
- Epoxy acrylic SO_2 cured ($EASO_2$)
- Ester- and CO_2-cured alkaline phenolic (ECAP)
- Acrylic urethane - Amine cured (AU-Amine)

2. Discussion

2.1 Core making properties of cold box binders

Among the cold box binders the epoxy acrylic system demonstrates by far the longest mixed sand bench life. The binder itself does not react under ambient conditions of temperature and humidity unless exposed to SO2 (fig.1)[1-3].

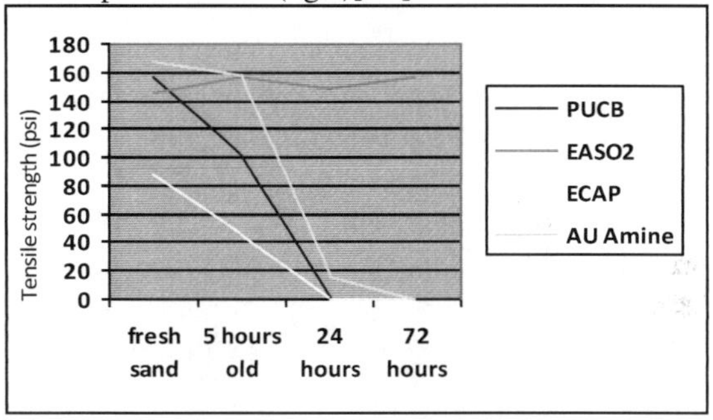

Figure 1. This chart compares mixed sand bench life vs. tensile strength for different CB binders.

It is important to note that over the years the other CB binder systems, especially the PUCB system, have also been significantly improved in mixed sand bench life performance, but more importantly the equipment including on-demand sand supply and easy cleanup has been engineered to reduce the impact of this property[4].

2.2 Advancement of PUCB in the environmental aspects

Primarily, the PUCB process historically used hydrocarbon solvents. New binders and base resins with improved solubility have resulted in a reduction up to total elimination of hydrocarbon solvents in some formulations. Ester-based and/or low evaporative solvent packages to reduce VOC's are now commonly employed (Tables 1)[5].

Table 1. PCS Emissions

CERP Data

PCS[1] emissions data generated at CERP / Technikon in 2000 with Phenolic-Urethane Cold Box Binders[2] relative to Baseline

	Baseline	PUCB Reduced HC	PUCB All Ester
HC:E Ratio[3]	63:37	52:48	0:100
VOC		-31%	-32%
HAP		-17%	-30%
Total HC[4]		-57%	-52%

1) PCS = Pouring, Cooling, Shake-out
2) CERP Reports RV100032CM; RV100061BP
3) Ratio of hydrocarbon-ester-solvent in the binder
4) Measured as undscans

2.3 Organic vs inorganic

Inorganic binder technology, which is synonymous with ultra-low emissions performance, is viable in high productivity semi-permanent mold applications for light metal automotive castings.

Compared to cores made with "traditional, organic binders" the sharply reduced VOC emissions from cores made with inorganic binder result in exceptionally clean-running SPM tooling with no tar buildup, which in turn drives productivity and casting quality[6].

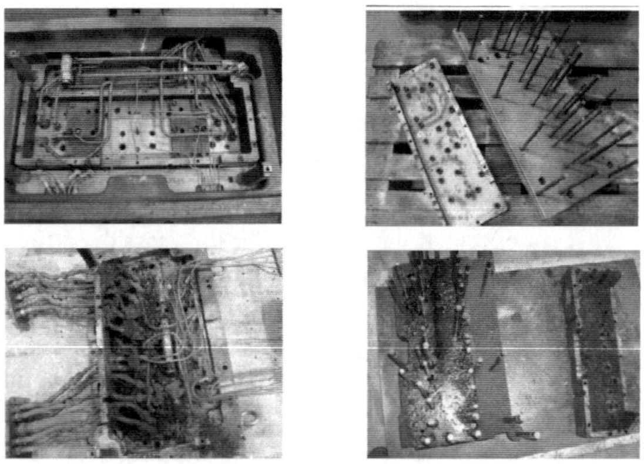

Figure 2. These pictures show SPM tooling used with inorganic core packages, prior to cleaning. There is no indication of tar formation on the tooling (Courtesy of BMW AG, Germany).

3. Conclusion

Technical, business and operational goals, as well as environmental and regulatory mandates will continue to guide and shape the development and improvements needed to advance metal casting manufacturing processes further. The information provided in this brief paper demonstrates that by combining experience, virtual simulation tools, the best available "consumables", processes and equipment, ingenuity and knowledge about the cost at each point in the metal casting manufacturing chain make up the recipe for success when designing and developing a casting. There are opportunities to remove cost from the process while improving the process itself, the return on investment, the work environment and last but not least, the performance of the final casting in service.

Acknowledgements

I would like to thank the many people and organizations who helped me write this technical publication. The Laempe Reich Company and Loramendi for updated automation and equipment illustrations. Numerous foundries and engineers we have worked with over the years including Amsted Industries, BMW Landshut, John Deere, Caterpillar, GM Powertrain, Ford Motor Company, Nemak, Hyundai Motor Company and numerous others. Friends, industry colleagues and co-workers at ASK Chemicals worldwide for their support and contributions.

References

[1] ASK Chemicals L.P.'s Sand School presentations (September 2011).
[2] ASK Chemicals L.P.'s Laboratory Reports.
[3] Traeger, W.R., Hysell, G.R., "The Technology of the Ester-Cured Phenolic Resin System and Its Comparison to Other Cold Box Systems," AFS Transactions (1985).
[4]Archibald, J., Warren, D.W., "Productivity and Environmental Aspects of Gas Cured Binder Systems," AFS Transactions (1989).
[5] CERP / TECHNICON Emission Reports (1999-2000).
[6] Weissenbek, E., Kautz, T., Brotzki, J. and Mueller, J. "Tomorrow's Cylinder Head Production Ecology, Economy and Material Enhancement Brought In Line" MTZ Vol. 72 48-52(June 2011).

O-40

Inorganic core production and casting application

Dipl-Ing. Klaus Löchte,
Hüttenes-Albertus GmbH, Wiesenstraße 23/64, 40549 Düsseldorf, Germany

1. Introduction

Inorganic binders for core production are known in the foundry industry for more than 50 years. Their heyday was in the 1950s and 1960s. With the upcoming of the Cold-Box-Process, the Waterglass CO_2-Process stepped back more and more. There are manifold reasons, just to mention a few: flowability of the moulding material as well as the bending strength of the cores were far behind the CB-Process.

Beginning of the year 2000 stricter environmental regulations came into force – particularly the protests by residents were decisive for the foundries to pose the question, if there is a less odour stressed core production process. The first inorganic systems were developed by binder supplier – these systems had taken a rapid development within the last 10 years.

Nowadays the leading german automotive manufacturers with own foundries – but also automotive suppliers – implement inorganic systems for core production.

Figure 1: Core Package taken to the die by manipulator

2. Basics of the IOB Systems

The Cordis Binder System is a complete inorganic two-component system.

Cordis:

- modificated silicate solution
- solvent water => frost product
- low viscosity (20 – 50 mPa·s)
- Density approx. 1,1 – 1,4 g/cm³

Anorgit:

- Systemic inorganic powder
- Bulk density 400 – 600 g/L
- Store under dry conditions

3. Chemistry of process

The above mentioned substances Cordis and Anorgit will be added to the silica sand und mixed with it. This moulding material will be blown into the core box, the hardening take place by the hot core box (150° - 200°C) while the solvent water is escaping/leaking. However, the more effective way to extract the water is purging with warm air (140°C). The curing takes place by a drying process. The Anorgit contains a reactive part, this cause a chemical hardening reaction. The sand grains will be linked three-dimensional - like the organic systems.

4. Equipment

4.1. Dosage and mixer

Each customary dosing component can be used for dosing the liquid component Cordis. Attention should be paid that Cordis is a highly alkaline liquid! Added amount is 2,0 – 2,6% referring to the silica sand.

For dosing the powder component Anorgit customary dosing screws with upstreaming weighing cell can be used. Please note that an amount of 0,6 – 1,6% must be added referring to the silica sand.

4.2. Core shooter

It is possible to modify already existing core shooter for the inorganic process. In most cases it is better to buy a new core shooter already constructed for the inorganic core production process.

If already existing core shooters are modified, it is necessary to pay attention to protect machine components against thermal radiation. It begins with movable machine components and ends with the storage container for the moulding material. The container should also be water-cooled as well as the shooting nozzles – this prevents the drying of the moulding material during storage. Several manufacturer of core shooter offer an additional moistening of the moulding material from above, others offer slide controller for the container. If it is possible regarding to the core box design, no shooting nozzles (dipping nozzles) should be used with a smaller diameter than 6 mm. Furthermore the weight of the core box used for the Inorganic is important – this is often higher than the one of a core box used for CB-Process.

4.3. Core boxes

The core box for the Inorganic should be manufactured with a heat resisting material that also withstands the mechanical stress during the core shooting, f.e.
Core box material
Tool Steel Types C45 W3
Material No 1.1730 or X38CrMoV51

The core box design should ensure running temperatures up to 200°C and more as well as a good and quick passage of warm and hot air to achieve short cycle times. This means an even temperature of approx. 140°C during the whole air purging cycle.
The following possibilities are available for heating the core boxes:

- Electric heating
- Steam heating
- Heat transfer liquid

5. Core production

During the core production is to be noted – as in the case of the other hot curing processes – that the thickness of the cured border area is sufficient for transport and moving of the cores. It depends on the used binder – but most of all of the interaction between core box temperature and length of purging time with warm air.
During the core production the thickness of the builded border area should be proved from time to time. If the area isn't thick enough, the core will break while removing of the core box. A sufficiently strong border area contains a part of not cured moulding material in the inner core. If the core is casted shortly after production, the not cured moulding material could lead to gas defects on the casting. If the core is stored (24h) the curing take place.
The curing will be accelerated by conduction of hot air. Thick-walled cores won't be cured during the machine cycle time, otherwise the productivity will suffer. The binder bridge of an inorganic core points a very brittle and glassy look. This brittleness also appears with the Warm Box Systems, which contains a high content of furfuryl alcohol. With these brittle systems a good maintenance of the core shooter is absolutely necessary.

6. Core storage

The inorganic core gets his bending strength by extraction of moisture. We are talking about a physical hardening. This factor has an influence on the core storage during the different seasons. During the winter the air can't store as much moisture as in the summer. So one could abandon a climatic core storage in the winter. During summer the inorganic cores should be stored under climatic conditions that means RH < 50%, better < 30% with a temperature of 20° - 25°C. Under these conditions and with the modern Cordis / Inorganic Systems the storage for five working days is guaranteed.
With the implementation of the Inorganic in the foundries the operating procedures also changed. For example the total linked inorganic core production for the cylinder head production at VW (Article Gießerei 07/2014).
During the past the work in this foundry was done like in any other foundry: the cores were produced and stored up to the time of need. The Inorganic changed this procedure. Today the core shooters are situated close to the casting table.
The cores will be removed by manipulators, another manipulator deburrs the cores. At the end of this process the core package will be manually assembled by employees. The tablet with the core package will be moved to the casting table where it is inserted into the die by manipulator and casted. There is no traditional core storage with this procedure. It is a just-in-time-production or a total linkage.
The above mentioned way of core production without core storage is not necessary to use the advantages of the Inorganic. Most of our customers are working with the climatic core storage.

7. Advantages of the Inorganic

- No emission => smaller extraction units
- No condensation in the die
- Less tool wear
- Extended die lifetime

O-41

Innovative Binder and Refractory Coating Solutions for Highly Complex Castings

Dipl-Ing. Peter Gröning[1]
Dr. Klaus Seeger[2]
[1]Hüttenes-Albertus GmbH, Wiesenstrasse 23/64, 40549 Düsseldorf, Germany
[2]Hüttenes-Albertus GmbH, Hansastrasse 1, 30419 Hannover, Germany

1. Introduction

A key element for the reduction of exhaust gas emissions is the so-called downsizing concept that enables to increase the specific performance of combustion engines along with reduced volumes. Light-weight, compact and small aggregates are growing ever more complex and pose new challenges to the foundry engineer. Today's engine components, such as cylinder crankcases and cylinder heads, enable several functions to be combined that previously needed additional aggregates. The new challenges in the manufacture of castings that have become very core intensive (Figure 1) have led to a specific increase in gasifiable substances and, hence, to the risk of gas-related casting defects.

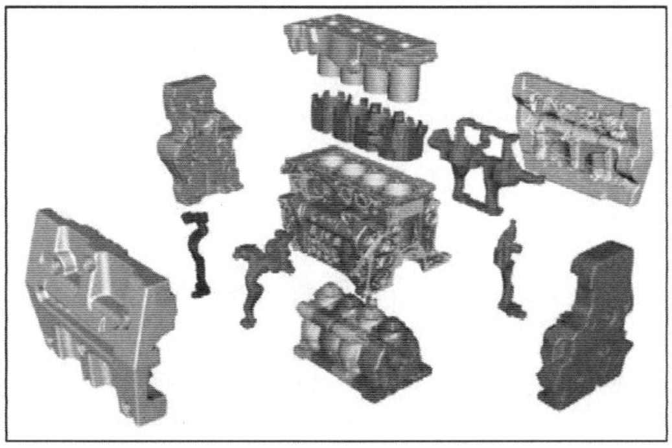

Figure 1: Core Package setting thin wall casting Eisenwerk Brühl

With the binding agents and coatings used, these new castings need new concepts. The conventional polyurethane process that is traditionally used in production lines due to its high productivity, however, fulfils the new requirements only very insufficiently and therefore requires a fundamentally new concept.

2. New Binder Concepts

The development of a new, innovative binder concept that incorporates inorganic components in its molecular structure and therefore contains less carbon provided an adequate solution to this problem. On the one hand, the result of this modification reduces the viscosity and, consequently, the quantity of solvents used. On the other hand, the new binder has a higher moleculare weight than the previous formulations, which results in a high thermal stability of the cores. This feature is especially important for the need to cast increasingly fragile cores without deformation. In addition, due to its structure, application of the new binder leads to a significant decrease in CO_2 emissions. In 1999, the first Cold Box system with silicium-based solvents (TEOS) has been used in foundries. Compared to the aliphatic and aromatic solvents that had been used in the Cold Box process until then, the use of tetraethyl orthosilicate (Figure 2) was the first step towards an partly inorganic Cold Box system.

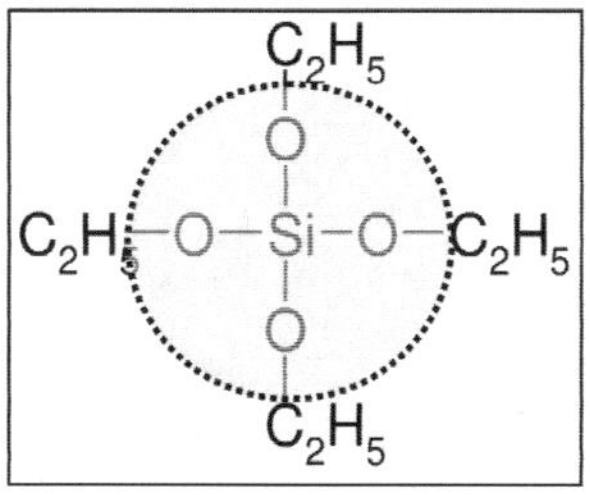

Figure 2: Formula TEOS (tetraethyl orthosilicate)

Now the objective of research was to gradually increase the fraction of these inorganic components in the Cold-Box system. To this day, a new generation of binder has been developed on the basis of tetraethyl silicate. In this new generation of binder the silicium units are not only contained in the solvent of the Cold-Box resin but are also integrated in the moleculare structure of the resin. Hüttenes-Albertus new, patented solution is a milestone in the development of novel Cold-Box systems. For the first time, it combines the advantages of both the Cold-Box process and certain properties of inorganic cores. The integration of inorganic silicon dioxide units is achieved through a substitution reaction during which the hydroxyl groups of the resin molecules are made to react with ethyl silicates. In the process, the size of the resin molecules increases, so that the amount of activator can be reduced in relation to the resin content. The viscosity of the base resin is reduced despite the higher molecular weight. As a result, significantly less solvent is required than with the Cold-Box binding agents used previously. With these new binder concepts it is possible to reduce the gas evolution (reducing gas related defects), the emissions (pollutants, smell and fume) and also the amount of condensates.

3. Coating Concepts

The manufacture of thin-walled, complex castings using the core package process requires the use of special coatings that besides good gas permeability, provide improved protection against sand expansion defects with reduced coating layer thickness. The novel coatings from Hüttenes-Albertus are designed with innovative refractory combinations that allow for controlled gas permeability to avoid gas defects. At the same time, their high ductility during pouring prevents the formation of expansion defects.

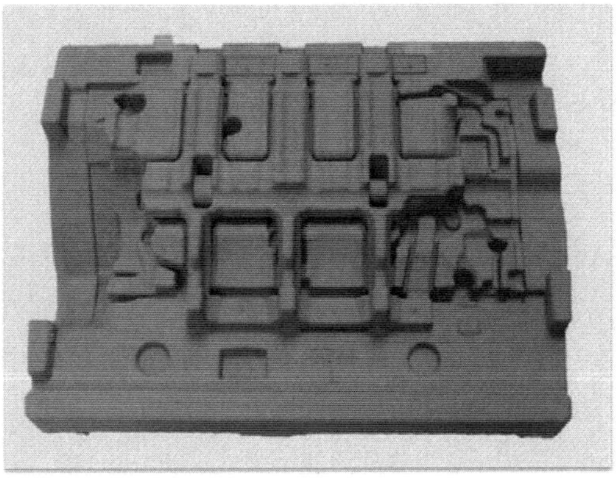

Figure 3: Coating Application core package process Eisenwerk Brühl Germany

The thin wall casting technology creates a lot of new challenges. The wall thickness of a conventional engine block in the past was 4 – 5 mm with a tolerance of ± 1 mm. The engine blocks of today have a wall thickness of 3 mm with a tolerance of ± 0,5 mm ! And the trend to reduced wall thicknesses is continuing. To manufacture thin walled engine blocks with a constant quality the system silica sand - sand additive – binder - coating has to fit the conditions. A big challenge is to avoid expansion and gas related defects (pinholes, scabbing etc.). Very important to avoid gas related defects is the gas pressure in the system sand – binder – coating. If a sand additive is used, this is also a very important substance to consider. If the gas pressure out of the core is higher than the metalostatic pressure the risk of gas defects is high. For the development of binder, additives and coatings or the examination of casting defects it is important to have a tool to examine the gas evolution and the gas pressure of the core sand system. In our technology center in Hannover we use a device to measure the gas pressure under conditions close to reality.

4. System - Sand-Additive-Binder-Coating

In order to reach the best results it is important to understand the crosslinking relations between the silica sand –sand additive-binder and the coating. The reason for using Sand additives is to minimize the effect of silica expansion and contraction behaviour. To choose the right sand additive it is important to know and understand the differences and the functionality of additives in the casting process in detail. There are different types of sand additives available, organic additives (burnout), inorganic additives and mixed forms. Organic additives are the traditional way of reducing the expansion rate of a core sand mix. These additives are relatively cost effective and easy to use. But organic additives create a lot of emissions like pollutants and smell. They increase rapidly the gas pressure in the core or core package and therefore can create gas related defects. That is the reason why the use of inorganic additives has become more in common, especially for thin wall applications. Inorganic additives create no additional gas evolution and no additional emissions. The impact on the core strength level is relatively small. Typical addition rates are 3 – 5 %. The inorganic additives reduce the sintering point of the silica core sand mix. It is basically a controlled manipulation of the sintering behaviour. The functionality of the sand additives also have a big impact on the performance of the coating. The unique properties especially of the inorganic additives can additionally open up new ways of designing a new generation of coatings.

Proceedings of the 72nd World Foundry Congress
21-25th May 2016, Nagoya, Japan

Evaluating a High Production Eco-friendly Core Binder System for Aluminum

S. Ramrattan[1], P. Patel[1], R. Shah[1], T. Aoki[2], Y. Kato[2] and H. Makino[2]
[1] Western Michigan University, Kalamazoo, MI, USA
[2] Sintokogio Ltd., Aichi, Japan

A sustainable polysaccharide sand binder system was developed for a modified injection molding process. The injection molded cores are designed for aluminum casting technology. The injection molding process affords complex core shapes and allows extremely thin sections at fast cycle times. Cores from conventional processes by means of cold-box and hot-box technologies were compared to those produced by the injection molding process. This paper relates the mechanical, physical, and thermo-mechanical properties of disc-shaped core specimens. Testing of disc-shaped core specimens included density, abrasion, impact strength, permeability index, thermal distortion. The test data is valuable to identify core quality and useful for process control in core production. Aluminum casting trials were conducted and the results show tremendous as-cast finish at the core/metal interface as-well-as shakeout benefits for the injection molded.

Keywords: *Chemically Bonded Sand, Cold-Box System, Core Binder System, Disc-shaped Specimen, Hot-Box Shell, Thermal Distortion*

1. Introduction

A popular medium for the production of cast powertrain metal parts is sand. Sand has many pluses, but it is far from perfect. Problems with chemically bonded sand systems arise from variation in materials and processes. This can come from many sources such as grain size, grain shape, chemical composition, binder level, additives, work-time, strip time, pouring temperature, metallostatic pressure, etc. [1]. Thus, chemically bonded sand has many potential sources of variation; but it is still subject to the pressures of delivering near-net shaped castings. Understanding those variations is a key issue for achieving good process control, and there have been several studies toward that end [1-3].

At the 71st World Foundry Congress a "New Core Binder System for Aluminum Casting Based on Polysaccharide" was offered [4]. This paper explains the core making process detailing the polysaccharide sand binder mixing process as well as the fast injection molding technology. The new core making process is commonly referred to as LYTE CORE (LC).

In order to produce a high quality casting, the purpose of this paper was to test the mechanical, thermal and physical properties of the new sand binder system. Additionally, aluminum casting trials were conducted for all sand binder systems tested.

2. Experimental procedure

LC specimens (two binder levels 0.35% and 0.70%) were prepared using an artificial sand made of alumina, ESPEARL #60. The LC disc-specimens (50 mm dia., 8 mm thick) were prepared by injection molding [4]. Curing temperature within tool was 220°C for a curing time of 180 sec. prior to ejection.

The 50mm diameter, 8 mm thick disc-shaped LC specimens were used in a variety of physical, mechanical, and thermal tests. The following five non-standard tests were conducted with a sample size of eight specimens for each test: 1) specimen weight, 2) impact, 3) modified permeability, 4) abrasion and 5) thermal distortion test (TDT) [5-6].

Specimens of LC, Hot-Box Shell and Cold-Box were compared using an experimental pattern to perform a casting trial where an aluminum alloy (356) was poured from 22.9 cm head height at 700 ± 3°C.

3. Results and discussion

3.1 Mechanical and physical properties of LC

The LC specimen weights are shown in Table 1 and it should be noted that the $LC_{0.70}$ specimens were more consistent. LC specimens were especially abrasion resistant. Moreover, the $LC_{0.70}$ specimens were tougher and show better venting.

Table 1 Summary of test results of LC samples

Test	$LC_{0.35}$		$LC_{0.70}$	
	Mean	σ	Mean	σ
Specimen Weight (g)	26.77	0.65	26.80	0.37
Impact Strength (J)	0.43	0.05	0.56	0.12
Permeability Index (#)	182.5	0.00	187	4.80
Abrasion (% loss)	0.00	0.00	0.00	0.00

3.2 Thermo-mechanical properties of LC

The thermal distortion curves (TDCs) for LC cores tested showed undulations that indicate thermo-mechanical and thermo-chemical changes in the binder system at elevated temperature. The longitudinal distortion (D_L) curves showed plastic deformation (downward movement of a TDC) (Fig. 1). Plastic deformation occurred for ~ 30 seconds and then leveled-off for the duration of the test. The radial distortion (D_R) indicated an expansion trend for 30 seconds followed by bending (Fig. 1).

All LC specimens had similar TDCs and there was no significant difference between $LC_{0.35}$ and $LC_{0.70}$ (Fig. 1). This finding is supported by with data form heat transfer data (Fig. 2).

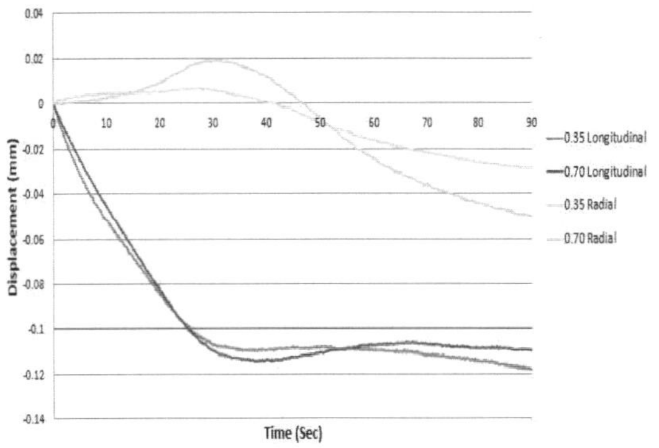

Fig. 1 Longitudinal and Radial TDC for LC 0.35% and 0.70 %

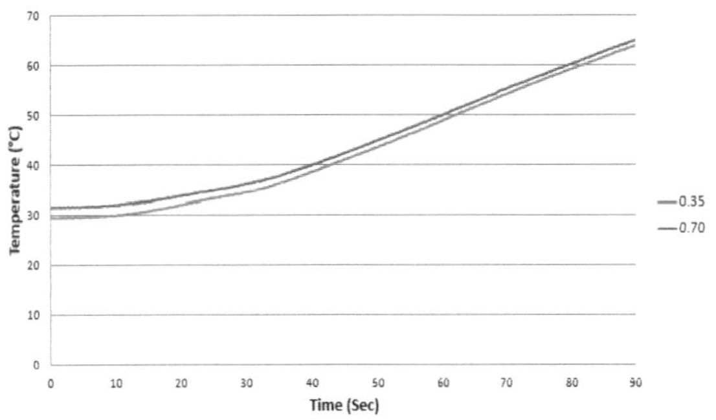

Fig. 2 Temperature versus time plots between the hot surface and the back of specimens

3.3 Observations from casting trial

All three sand binder systems were compared within the same molds using the same pouring temperature, alloy chemistry and metallostatic head pressure through solidification a shakeout comparison can be made. It is noteworthy to identify that the LC cores regardless of binder percentage collapsed from the core cavity without need for physical shakeout. Conversely, the removal of both Hot-Box Shell and Cold-Box systems required finishing equipment and aggressive shakeout.

4. Conclusions

This study has identified that thin section LC cores (8 mm thick) can handle at minimum 22.9 cm head height at 700 ± 3°C without incurring aluminum casting defects. The mechanical and physical properties of $LC_{0.70}$ are favorable to core handling in the foundry environment. Further, LC cores compares favorably with conventional Hot-Box Shell and Cold-Box systems at similar binder levels and sections thickness in aluminum casting trials. Apart from the established high production eco-friendly benefits of LC cores; this study has identified superior shakeout advantages.

Acknowledgements

The authors gratefully acknowledge Sintokogio, Ltd. for the research support. The authors thank Glenn Hall, Michael Konkel and Peter Thannhauser from Western Michigan University, for their technical support.

References

[1] American Society for Metals ASM, Metals Handbook, ISBN: 978-0-87170-711-6, vol. 15: Casting (2008).

[2] Iyer, R., Ramrattan, S., Lannutti, J., Li, W., "Thermo-Mechanical Properties of Chemically Bonded Sands," AFS Transactions, vol 109, pp. 1-9 (2001).

[3] Ramrattan, S. N., S. F. Cheah, and S. Zandarski, "Effects of Aluminium Fill Temperatures on the Distortion of Chemically Bonded Sand Systems," AFS Transactions, No. 03-154 (2003).

[4] Aoki, T., Y. Kato, T. Zenpo, N. Asano, "New Core Binder System for Aluminum Casting Based on Polysaccharide," 71st World Foundry Congress Proceedings, Bilbao, Spain (2014).

[5] Oman, A. J., S. N. Ramrattan, M. J. Keil, "Next Generation Thermal Distortion Tester", AFS Transactions, No. 13-1454 (2013).

[6] Ramrattan, S., A., M. Khoshgoftar, M. Konkel, J. Muniza and A. Pike, "Improvements to Disc-Shaped Specimens for Control of PUCB Sand Systems", AFS Transactions, No. 14-061 (2014).

Proceedings of the 72nd World Foundry Congress
21-25th May 2016, Nagoya, Japan

Improving the Moisture Resistance of Mold with Inorganic Binder

Hiroki Ameku[1], Kazuya Edane[1], Soichi nagai[1] and Yutaka kurokawa[1]
[1] Tsuchiyoshi Industry Co., Ltd, Technical Center, 1051-7 Wada Onan-Cho Ochi-Gun Shimane 696-0403, Japan

Abstract

The purpose of this study was to improve the moisture resistance of a mold with an inorganic binder. Some types of metallic salt were added to an inorganic binder to replace sodium ion in sodium silicate to metal ion, and produce an insoluble saline. In an experiment, a mold after shaping at 130 °C was stored at 30 °C and 90% relative humidity, and the mold strength was measured to compare the hygroscopic deterioration rate, which is defined as the strength ratio for fresh to regular hours of holding time, in order to find the deterioration by the absorption of moisture. Some types of alkali metal salt as an additive in the inorganic binder were found to reduce the hygroscopic deterioration rate to lower than 20% after 1 h of storing, where the storing time assumes a cycle time from molding to casting in the casting line. It is possible to use an inorganic binder in a mold in hot and humid conditions, such as those that prevail in Japan, by applying the additive of the alkali metal salt used in this study.

Keywords: *inorganic binder, moisture resistance. alkali metal salt, insoluble saline*

1. Introduction

An organic binder is used in the production of a mold, resulting in good productivity and mold quality. However, a mold with an organic binder might result in casting defects due to gases such as CO, CO_2, H_2, and N_2 owing to the heating during the casting process [1]. Inorganic binders have been actively investigated [2] for reducing casting defects and environmental concerns such as odor issues, and have become popular for use in the mold process in aluminum alloy casting. For the practical applications of inorganic binders with alkaline silicate in the molding process, the filling of sands with inorganic binder [3], moisture absorption of inorganic binders, and the collapse characteristics of the sand mold have been considered. An inorganic binder consists primarily of alkaline silicate and experiences a dehydration condensation reaction during the curing of the mold, which gives rise to the problem of hygroscopic deterioration of a mold prepared in hot and humid conditions. Hence, the effectiveness of spherical amorphous silica fillers and alkali metal salts, which are the agents added to inorganic binders to prevent moisture absorption for reducing the deterioration of the mold, was evaluated in this study.

2. Experimental method

2.1 Test pieces

Fig. 1 shows the optical and electron photomicrographs of natural silica sands and acid spherical amorphous silica fillers. The natural silica sands were coated by mixing them with 0.9mass% of spherical amorphous silica of acid (pH 3) fillers and 0.03 – 0.1mol of alkali metal salts for 60 s. Then, they were mixed for 60 s with 2% sodium silicate solution including surfactant agent. Next, test pieces were molded by curing with temperature of 130 °C and blow pressure of 0.1MPa.

Fig. 1 Optical and electron photomicrographs of natural silica sands and acid spherical amorphous silica fillers.

2.2 Measurement

The test molds were stored at 30 °C and 90% relative humidity, and the deflective strength was measured by supporting a 50-mm-long test piece at two points and applying a load to the test piece center. The hygroscopic deterioration rate, which is defined as the strength ratio for fresh to regular hours of holding time in order to determine the deterioration due to the absorption of moisture, was calculated using Eq. (1):

$$S = \{100-(A/B)\}\times 100 , \qquad (1)$$

Where,

S: Hygroscopic deterioration rate

A: Deflective strength after molding

B: Deflective strength after regular holding time at 30 °C and 90% relative humidity

3. Experimental Results

The moisture resistance was improved by adding acid spherical amorphous silica fillers in order to increase the gelation reaction and the filling rate during molding. Fig. 2 shows the hygroscopic deterioration rate for different cases, confirming the effectiveness of the use of spherical amorphous silica fillers. Lower hygroscopic deterioration rates indicate improved moisture resistance. The addition, the spherical amorphous silica fillers reduced the hygroscopic deterioration rate after 1 h of storage.

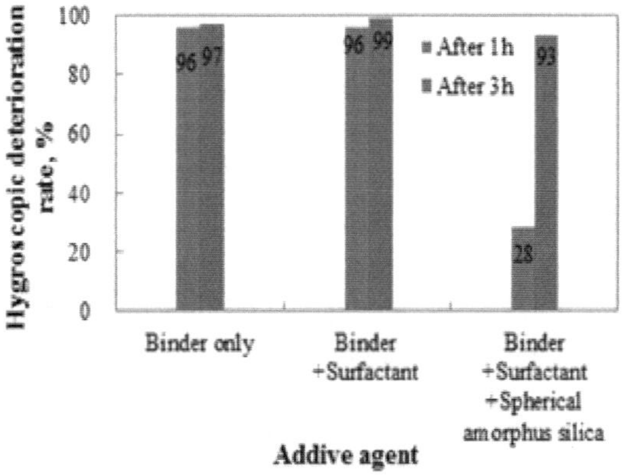

Fig. 2 Hygroscopic deterioration rate confirming the effectiveness of spherical amorphous silica fillers.

To further improve the moisture resistance, metallic salt was added because it is considered that the sodium ions in sodium silicate, which is an inorganic binder, are replaced by metal ions from the metallic salts, which results an insoluble saline compound , preventing hygroscopic deterioration. Fig. 4 shows a graph of the hygroscopic deterioration rate, which confirms the effectiveness of metal sulfate-s , obtained from metallic salts. A total of 0.003mol of each metal sulfates was added. Here, the blank consists of an inorganic binder, a surfactant agent, and spherical amorphous silica fillers, as shown in Fig. 3. The remaining additives were prepared by adding different metal sulfates to the blank. The alkali metal salt additive decreased the hygroscopic deterioration rate after 1 h of storage, which is sufficient for the mass production of sand molds with inorganic binders in hot and humid conditions. It is expected that this is a result of the insoluble saline compound partially generated by absorption or ion exchange between the inorganic binder and cations of the alkali metal salt.

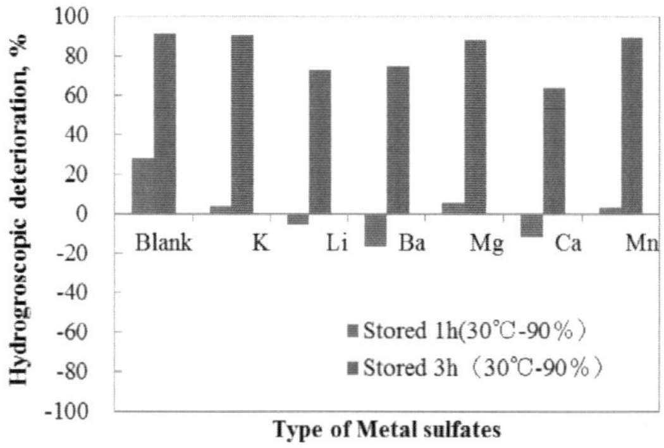

Fig. 3 Hygroscopic deterioration rate confirming the effectiveness of metal sulfates.

4. Conclusion

The moisture resistance of a mold with an inorganic binder was improved by using spherical amorphous silica fillers. The addition of alkali metal salts also proved to be effective for the same purpose. Hence, it is possible to use an inorganic binder in a mold in hot and humid conditions, such as those that prevail in Japan, by applying the alkali metal salts used in this study.

References

[1] Y. Shinada , O. Hideaki and Y. Ueda: Casting 57 (1985), 17-22

[2] A. Balinsky: Archives of Foundry Engineering. 13 (2013), 131-133.

[3] H. Ameku: Reports of the 164th JFS meeting (2015),76

Proceedings of the 72nd World Foundry Congress
21-25th May 2016, Nagoya, Japan

Post casting residual strength development of new generation inorganic binders using different South African silica sands.

Farai Chrispen Banganayi [1], Kasongo Nyembwe [2] and Hartmut Polzin [3]

[1] Metal Casting Technology Station, University of Johannesburg South Africa 37 Nind Street Doornfontein Johannesburg South Africa

[2] University of Johannesburg Faculty of Engineering and Built Environment Department of Metallurgy 37 Nind street Doornfontein Johannesburg South Africa

[3] Technische Universität Bergakademie Freiberg/Gießerei-Institut Bernhard –von-Cotta-Strasse Freiberg Germany

Abstract

The paper investigates the performance of two local silica sands using a new generation modified sodium silicate binder. The binder is a water glass based system. It is used as a one-component liquid binder in the range between 1.5 and 3.0 mass %. The hardening can be realized with CO_2 gassing, liquid ester hardeners or with heat and heated air. Three curing methods of the sodium silicate were considered which CO2, ester and heat are. The methodology followed consisted of sand characterization and residual strength measurement of strength specimen exposed to heat after curing. The test results revealed the profound influence of the sand properties on the residual strength of the sand. Because of the difference of properties of the two sands, they behave differently with the same sodium silicate binder.

Keywords: *Inorganic binder, residual strength, silica sands*

1. Introduction

The more understanding of the importance of environmental protection, the application of green products for foundry becomes one of the most urgent tasks [1]. Inorganic binders became unpopular in the foundry industry because of their poor shake out properties due to high residual strengths after casting. [2]. Despite sodium silicates' many favourable characteristics unmodified sodium silicate binders have one key problem which is the difficulty in shake out or inability to break down after metal solidification [3].Collapsibility is of particular importance in terms of recyclability of sodium silicate used sands. Poor collapsibility is due to high residual strength of used sand after the casting process. The Residual strength of sodium silicate used sand is a function of temperature reached by the sand during the casting process. The highest strength is developed in areas around the casting where the sand has been heated between 800^{0}C and 1000^{0}C. Lower strengths are developed in the regions where the sand reached temperature between 400^{0}C and 600^{0}C [4].

Despite the knowledge accumulated with regards to sodium silicate moulding technology, it is not always possible to predict how a new commercial binder will perform with untested refractory sands. This paper tackles this questions in the case of a single component new generation inorganic binder from Germany using South African local silica sands for foundry applications.

2. Experimental Procedure

The experiments related to this investigation were conducted according to figure 1 below:

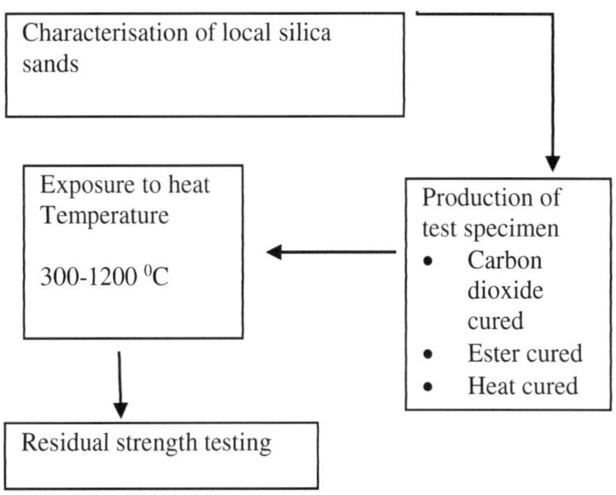

Figure 1 Experimental procedure

3. Residual strength results

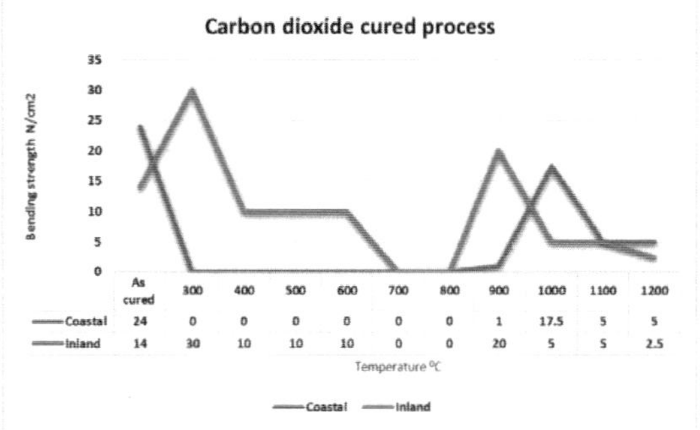

Figure 2: Residual strength of CO_2 cured sand at different temperatures of exposure to heat.

Figure 2 shows the residual strength results under the gas curing process. It can be seen that when exposed to heat, the inland and coastal sands differ in residual strength. For some of the heat exposure temperatures.

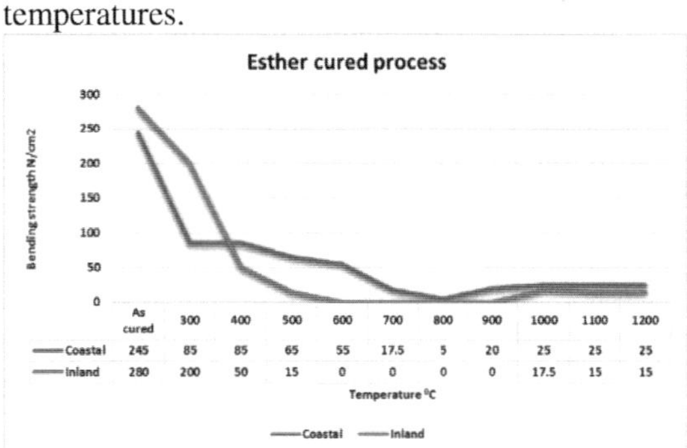

Figure 3: Residual strength of ester cured sand at different temperatures of exposure to heat.

Figure 3 shows the residual strength results under the ester curing process. It can be seen that the inland and coastal sands differ in residual strength for all of the heat exposure temperatures.

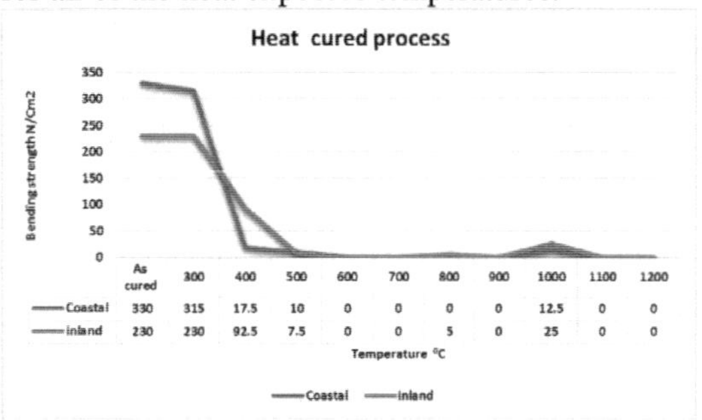

Figure 4: Residual strength of heat cured sand at different temperatures of exposure to heat.

Figure 4 shows the residual strength results under the heat curing process. It can be seen that when exposed to heat, the inland and coastal sands differ in residual strength. For some of the heat exposure temperatures.

4. Conclusion

The study investigated the effects of sand properties and the different curing methods on the residual strength of sodium silicate sands. It was found that;

- Each sand has its own characteristic behaviour with regards to the rate of drop in residual strength based on its properties.
  - The behaviour of the sands after the last peak or glassy phase was consistent with the type of curing process.
  - The residual strength after the glassy phase/last peak is ranked in the following order from highest to lowest
    - Ester cured
    - Carbon dioxide cured
    - Heat cured.

Acknowledgements

I would like to acknowledge the Metal Casting Technology Station staff for their assistance in the laboratory work

References

[1] J. Wang, Z. Fan, H. Wang, X. Dong and N. Huang, "An improved sodium silicate binder modified by ultra fine powder materials," *China Foundry Volume 4,* pp. 26-30, 2007.

[2] I. Ilzdebska-Szanda and A. Balisnki, "New generation ecological silicate binders," *Procedia Engineering,* pp. 887-893, 2011.

[3] J. T.Fox, FredS.Cannon, NicoleR.Brown, HeHuang and JamesC.Furness, "Comparison of a new,green foundry binder with conventional binders," *International journal of adhesion and adhesives Volume 34,* pp. 38-45, 2012.

[4] H. Polzin, Inorganic binders, Berlin: Schiele and Schon Gmbh, 2014.

Proceedings of the 72nd World Foundry Congress
21-25th May 2016, Nagoya, Japan

NEW COATINGS AND ADDITIVES CONCEPTS AS AN ENTIRE APPROACH FOR DEFECT AND RESIDUE FREE CASTINGS

Dr. Reinhard Stötzel[1], Ismail Yilmaz[2], Shengyin Liu[3], Fumihiko Kimura[4]

[1]ASKChemicals GmbH, Reisholzstrasse 16-18, 40721 Hilden/ Germany

[2] ASK Chemicals GmbH, Reisholzstrasse 16-18, 40721 Hilden/ Germany

*3 ASK-Chemicals (Zhenjiang) New Materials Technology Company Limited
Room 702 - 703; Block A, No 596 Middle Longhua Road, Xuhui District,
Shanghai 200032/China*

*4 ASK Chemicals Japan Co. Ltd. 8th Floor, Bashamichi 450 Building, 50, Otamachi 4-Chome
Naka-Ku, Yokohama 231-0011/ Japan*

There has been a lot of effort in the last years to eliminating coating process in the foundries. In some areas the target has been successfully implemented, e.g. for less demanded Ductile Iron castings. ASK Chemicals has been able to develop new type of additives which enable the foundries to reduce the casting defects significantly and thus pour more uncoated cores.

Engineered sand additives VEINO ULTRA and ISOSEAL are such developments, which have amazing anti-veining properties.

OEM's are forcing the foundries to deliver castings with a limited minimum amount of residue in their castings. A special coating based on good anti-veining and anti-penetration properties has been developed, which reduces the coating residue after pouring to a minimum.

***Keywords:** Mould and core coating, sand additives, foundry, performance, casting defect.*

1. Anti-Veining and Anti-Penetration Additive with the Scope to pour uncoated cores

There has been a lot of effort in the last years to eliminating coating process in the foundries. In some areas the target has been successfully implemented, e.g. for less demanded ductile iron castings [1].

ASK Chemicals has been able to develop new type of additives which enable the foundries to reduce the casting defects significantly and pour more cores uncoated.

VEINO ULTRA is such a development, which have amazing anti-veining properties.

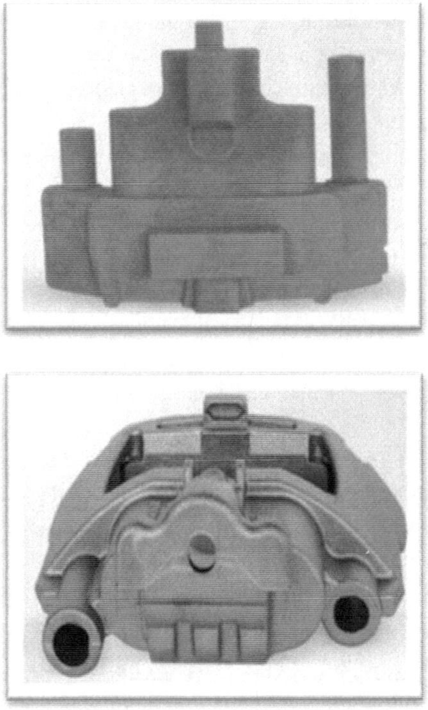

Fig. 1 Uncoated core and casting poured with uncoated core

2.1 Anti-Veining Coating

Special requirements concerning casting quality is to have no veining along the parting line of the core box.

For the production of the cores furthermore there is a strong demand to have no deformation of the cores during drying process .

Examples of the coating generation MIRATEC™ BD show that the coating eliminate the veining and by the improved application properties, the cores can be dipped in a reduced cycle which has been improved 100% without getting drops on the core.

2.2 High Gas Permeable Coatings against Scabbing

Scabs and gas defects belong to the sort of severe defects in serial production, because this leads to scrap in the foundry.

In a development project, ASK developed different coatings that are extremely gas-permeable and suppress these defects.

Castings, in which the cores are highly thermally loaded by a critical low mould filling on special areas, tend to suffer from scab formation. Based on the MIRATEC MB 501, a coating that shows good results (especially in the serial casting) to avoid veining and penetration, the feature "increased gas permeability" was implemented.

The new coating, which fulfils this profile, is the MIRATEC AH 501. By the implementation of this coating the scab defects could be removed both from a hydraulics casting part (core package Cold Box + shell sand) and from an Axle housing (Cold Box - core).

Furthermore the cycle time of the dipping process has being able to be reduced to the half with this coating in comparison to the initial coating.

2.3 Residue-free Castings with MIRATEC TS

Besides the primary function of the coating to prevent casting defects, the presence of residuals in cast components, especially automotive parts, has become increasingly significant and is determined by means of cleanliness analyses and limit controls carried out by OEMs in accordance with VDA 19 (ISO 16232).

The limiting of so-called residuals was originally necessary to prolong the maintenance intervals for motor vehicles. Residuals in the hollow spaces of the engine block caused the filters to clog early and lead to undesireable shorter maintenance intervals than required. On the other hand, it is quite difficult and expensive – in particular in a water jacket or oil channel – to shot-blast these areas in such a way that every spot is residue-free. Hence, special washing machines need to be installed in some cases, which increase costs and reduce productivity, however.

On the basis of the good protective properties of well known coatings against veining and penetration, a special coating MIRATEC™ TS has been developed which reduces the coating residue from casting to a minimum. The development was initially evaluated by means of a heating microscope (optical dilatometer) measurements and casting tests involving practical geometries, such as those of a water jacket. The casting tests sought to investigate whether or not the coating film can be removed from the cast part merely by blowing it with compressed air. Systematic series of developments have ultimately succeeded in producing such a "peel-off coating" that comes away from the metal on its own. SEM comparison (Scanning electron microscope) reveals that MIRATEC TS has these properties in comparison to a conventional coating.

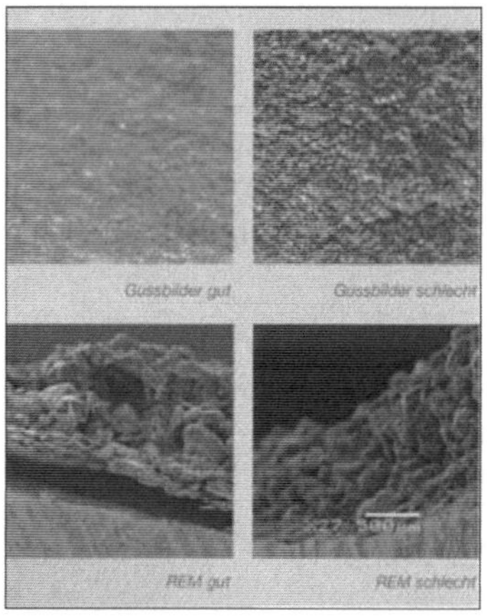

Fig. 2 SEM comparison of MIRATEC TS (left) and a conventional coating (right)

Summary

New coating and Additive developments enable foundries to produce flawless and residue free castings in a less complicated way with less side effects and even enable coating free casting production.

References

[1] Showman, R., Horvath, L., Clifford, S., Harmon, S., Lawson, E., "A Systematic Approach to Veining Control", FS Casting Congress Proceedings, 11-005.

O-46

Last developments on low-emission additives for molding sand

Thomas Engelhardt[1], Zefen He[2], Patrice Moreau[3]
[1]Clariant Produkte (Deutschland) GmbH, Ostenriederstr. 15, 85368 Moosburg, Germany
[2]Zefen He, Clariant Chemicals, Lucky Tower A 1306, No 3 North Road, Beijing, China
[3]Patrice Moreau, Clariant SE, Succursale France, 6 Rue Louise Michel, 94600 Choisy le Roi, France

In order to reduce emissions from molding sand systems, an innovative environmentally friendly foundry additive was introduced. Essential component of these additives are special natural graphites with which at least a part of the conventional lustrous carbon formers like sea coal and resins can be substituted. In the last years, the practicability of these system, named LE (low emission) technology, was proved in many foundries. In some foundries, the emission of volatile aromatics from the molding sand could be reduced by up to 80 %.

In addition to the reduction of the emission, graphite acts as a kind of lubricant within the molding sand. This leads to a more homogeneous compaction of the mold, related casting defects can be minimized.

Key words
Reduction emission, BTXE, graphite, molding sand, low emission technology

1. Introduction

Strict rules of the Goteborg protocol stipulate that, from the year 2020 onwards, a drastic country-specific emissions reduction of volatile organic compounds has to be followed. Therefore, the foundry industry has to intensify its efforts to lower its emissions, especially those of volatile aromatics. Thereby, the supplier industry is responsible for offering a technology to the foundry industry with which emissions can be reduced without increasing production costs. This concerns emissions from the lustrous carbon formers of the molding sand system as well as the today commonly used organic cold box binders for cores. Therefore it was the goal of a three year development project to develop an emission-free respectively low-emission alternatives for the lustrous carbon formers like sea coal and resins being in use today in the molding sand.

2. Sources of emission in a grey iron foundry

In a discussion about emissions from a grey iron foundry, it is important to focus on the lustrous carbon formers in the molding sand system as well as on the cold box binders for the cores. During pouring of liquid iron, the lustrous carbon formers in the molding sand are pyrolyzed and form low-boiling aromatics like benzene, toluene, xylene and ethyl benzene (BTXE). At the same time, polycyclic aromatic hydrocarbons (PAH) which have with a high boiling point are formed. Therefore, BTXE aromatics are a topic of gaseous emissions, while PAHs are adsorbed by the molding sand systems. These PAHs can create problems during waste disposal.

The cold box core binders also form low-boiling aromatics as well as PAHs during pouring. In addition, certain amounts of phenol are formed.

3. Technologies to reduce emissions from molding sand

Some different technologies are available to reduce the emissions from molding sand.

One technical solution is the adsorption of the aromatics in an external adsorption unit. Such external adsorption unit offers good efficiency, but additional investments are necessary. This leads to higher total process costs.

Other concepts follow the idea to adsorb the emissions in the molding sand system by adding special adsorber materials like zeolite to the molding sand. Such adsorber materials can offer also good efficiency, but transfer the emission problem from the gas phase to the waste disposal.

4. Low-emission (LE) technology

The most efficient way to reduce the level of emissions from a molding sand system is preventing the formation of toxic aromatics during pouring. The LE technology follows this concept. The challenge of such concept is to find a molding sand additive, which can overtake the function of classic lustrous carbon formers without formation of aromatics during pouring.

Graphite is a natural inorganic carbon-based material, which contains between 70 and 95 % pure carbon. This natural material is available with different morphologies and particle size distributions. All graphite grades are completely emission-free.

A special type of such natural graphite is the key component of the LE technology. This special type of graphite forms – like bentonite – very thin layers that can be separated from each other during the reprocessing of molding sand. By applying shear energy to the graphite agglomerates, the graphite can delaminate in very thin single sheets and can surround the silica grains like sea coal with a thin layer of carbon. So, the wetting characteristics of the silica grain by liquid iron is influenced in the same positive way like with sea coal.

While the morphology of graphite and bentonite is very similar, the surface characteristics are completely different. In contrast to bentonite, graphite has hydrophobic surface properties and therefore can be hardly be wetted by water. While bentonite is wetted very well, bentonite can swell in water very easily and forms well-dispersed bentonite layers which act as binders for the molding sand. Unmodified graphite is not wetted by water and swims on the surface in the form of coarse agglomerates. However, as the coverage of silica grains with graphite has to take place during molding sand processing in the mixer, the wetting of graphite by water has a special importance. When suitable dispersing agents are present, the dispersion will be successful like with bentonite.

The combination of special graphites with suitable dispersing agents is claimed in the patent application WO 2011/032668.

5. LE technology in foundry practice

By comparing the emission of benzene as most critical component from 60 different European foundries' molding sand systems, the huge differences in the benzene emission level are remarkable. These huge differences reflect the different levels of lustrous carbon formers in the molding sand systems as well as the different levels of core binders.

The comparison of the benzene emission of currently 15 European foundries processing LE technology shows very clearly that in most of the cases a reduction of the benzene emission by 50 % and even more can be realized (fig. 1). If a foundry is starting with a very high level of emission, a reduction of up to 80 % is possible.

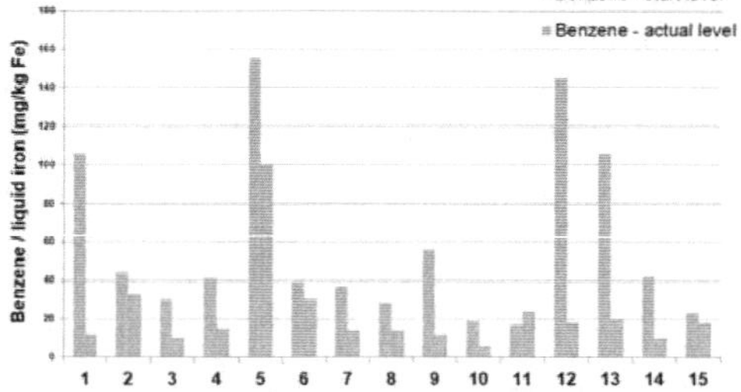

Fig. 1 Reduction of benzene emission by processing LE Technology in 15 European foundries

Proceedings of the 72nd World Foundry Congress
21-25th May 2016, Nagoya, Japan

Advancements in Materials for Three Dimensional Printing of Molds and Cores

J. Thiel, S. Ravi, N. Bryant
University of Northern Iowa, Metal Casting Center, USA

Abstract
Three dimensional printing of sand molds and cores is changing the way castings are produced. During the last five years major advances in equipment have allowed metal casters to realize castings designs at faster speeds than ever before. The elimination of tooling for mold and cores assemblies, has allowed the industry new flexibility in design optimization, reduced labor, increased dimensional accuracy and eliminated some defects associated with core assembly. While the equipment has been advancing, the materials used for printing have been very limited. The University of Northern Iowa has conducted new research into increasing the number of materials available for printing. These materials include regionally available resins, aggregates, and additives. These new materials have the ability to reduce the cost of printing and increase the number of applications while improving the casting quality. Veining defects in heavy iron or steel castings once prevalent on printed sand molds can be eliminated with engineered sand additives. Solidification rate and penetration defects can be addressed with specialty molding aggregates. This research has allowed the users of printed sand molds and cores to realize the potential of their sand printers and improve the quality of the castings they produced. The paper will detail the experiences with regionally available materials and compare their properties and performance with conventionally supplied materials.

Keywords: *Additive manufacturing, molding materials, 3D printing, bonded sands.*

1. Introduction

Three dimensional printing of cores and molds has come of age. Metal Casters are now able to drastically reduce lead times for pattern tooling while taking advantage of design freedom never before envisioned. Additive manufacturing for metal castings, AD4MC represents one of the most significant developments in metal casting technology in years.

The basic process introduced over twenty years ago has gained significant ground over the last two years. The technology once used exclusively for prototype low quantity is now being used for production cores and molds. The ability to produce very complex core assemblies has replaced a significant amount of manual labor not just in the core room but also in casting finishing operations. Single piece cores are replacing core assemblies that were manually glued in fixtures and mudded to seal joints.

90% of cores produced in North America utilize the phenolic urethane no-bake of cold box resin system. This robust process provides long working times and bench life along with good mechanical strength and compatibility with additives.

Almost all of the current three dimensional sand printers utilize the furan resin system catalyzed by sulphonic acid. Furan resins were very popular during the 70's and 80's in the foundry industry because of their high productivity and strength levels but price fluctuations and uncertain availability moved the foundry industry towards urethane and phenolic resins. The furan system is still used in foundry applications worldwide and has achieved a fair amount of market stability. It is available in several forms from several North America resin suppliers. It develops high tensile strengths while providing excellent shakeout and sand release from internal passageways.

While research continues on printing urethane or phenolic resins, furan systems remain the predominant systems in 3DP4MC applications. Recent testing has shown that although the furan sands are acid catalyzed, thermal reclamation effectively neutralizes the acidity of the sand making it compatible with phenolic urethane systems.

1.1 Equipment Technology

The equipment technology for AM4MC has been advancing at a tremendous rate. Developed by MIT researchers in the late 1980's it was licensed by three companies based in Germany. Zcorp had developed a resin system based on sodium silicate but ceased its sale and support of the printers in 2013. The remaining company split into what are the two largest manufacturers of sand mold printers. The first generation sand printers were introduced in the early 2000's are already reaching the end of their service life considered slow and inflexible. The second generation machines introduced in the late 2000's were two to three times as fast and twice as large. New equipment introduced in 2015 are touted to be twice as large again and four times as productive as the second generation equipment. North America suppliers have responded to the German made equipment with robotic arm printers promising to lower the cost of printed sand at a fraction of the price of larger units. The newest company to enter the market promises to combine very fast print times, low cost of equipment with freedom to use conventional sands and resins. This has the potential to lower the cost of printing sand to a point easily justifiable by even small foundries.

2.0 Materials Evaluated

Ten aggregates were evaluated for the binderjet printing process. A list of the materials evaluated is contained in Table 1.

Table 1 Aggregates tested

Sample	*Aggregate*
1	Silica M4000
2	Carbo Accucast ID50-K
3	Carbo Accucast LD70
4	ORC Chromite
5	Ceramsite
6	Cerabeads
7	Fused silica (crushed)
8	Incast 70
9	Incast 8-
10	Manley "Acti Sand"
11	Carbo 100/140

2.1 Mechanical Properties

Mechanical properties included Bulk density as defined by the AFS 1130-00-S [1], Tapped density per 1131-00-S, Surface area per AFS 1108-00-S, Grain shape or Coefficient of Angularity per AFS 1126-00-S, Base permeability per AFS 1119-00-S, and non-standard tests including angle of repose, material flow rate and penetration force. The penetration force was determined by measuring the resistance of a 1"(25mm) plate being pushed into a container of the material. The maximum force and penetration distance at the maximum force were recorded for all samples along with the sand displacement in pounds per inch.

Mechanical Properties Results

The untapped and tapped bulk density results are shown in table 2. With the exception of the specialty aggregates such as the ceramics and mullites, it can be observed that the untapped density of the silica sand samples are typically in the range of 1.4-1.6 g/cc. However, looking at the tapped density results, it can be observed that certain aggregates are arranged better upon tapping when compared to others. This largely depends on the grain shape factor of the aggregate. Silica M4000 has a tapped bulk density of 1.769 g/cc, which is the highest among the silica sand aggregates. Carbo 100/140 aggregate has a tapped density of 1.510 g/cc. The other silica sand aggregates are observed to range between 1.6-1.7 g/cc.

Table 2 Untapped and tapped bulk density results for all aggregates

Aggregate	*Bulk Density (g/cc)*	*Tapped Density (g/cc)*
Silica M4000	1.554	1.769
Carbo Accucast ID50-K	1.819	2.070
Carbo Accucast LD70	1.424	1.694
ORC Chromite	2.694	3.015
Ceramsite	1.839	2.007
Cerabeads	1.422	1.723
Fused silica (crushed)	0.989	1.256
Incast 70	1.516	1.673
Incast 80	1.445	1.620
Manley "Acti Sand"	1.438	1.623
Carbo 100/140	1.41	1.510

Acknowledgements

The authors wish to acknowledge the efforts of the student employees of the University of Northern Iowa Metal Casting Center for their assistance in the experiments required in the research.

References

[1] American Foundry Society, Mold And Core Test Handbook (2015)

Proceedings of the 72nd World Foundry Congress
21-25th May 2016, Nagoya, Japan

Evaluation of Ceramic sand produced by Ultrahigh temperature melting process

Soichi Nagai,[1], Hiroki Ameku[1], Kazuya Edane[1], Zizheng Huang[1] and Yutaka Kurokawa[1]
[1] Tsuchiyoshi Industry Co., Ltd, Technical Center, 1051-7 WADA ONAN-CHO OCHI-GUN SHIMANE 696-0403, Japan

Abstract

The recently ceramic sand is produced by our ultrahigh temperature melting process through advances in the melting technology for the ceramics and it is possible to produce the ceramic sand with the high refractoriness and dense particle. Against such a background, the characteristic of the ceramic sand produced by the ultrahigh temperature melting was evaluated by the particle compressive strength measurement and the burning resistant sand test in this study, which burning resistant sand test comes from that the mold wash is required for the molten metal in the cast steel due to the low refractoriness of the existing ceramic sand. As the result of this study, it was found that the ceramic sand produced by the ultrahigh temperature melting process had a few times of the particle strength as compared with the existing ceramic sand which showed the capable resistance for the stress in the casting line. For the sand burning, it was showed that the sand burning was encouraged by the low melting substance in the ceramic sand and the basic ceramic sand was effective to improve the burning sand.

Keywords: *ceramic sand, ultrahigh temperature melting, particle compressive strength, burning resistant sand*

1. Introduction

The usage of the ceramic sand as the environmentally-friendly foundry sand is increasing yearly because the factory waste and the casting defect rate can be reduced by it. For the action to prevent the casting defect by the ceramic sand, the reason to use it in the casting line comes from the low expansion for the deformation or the crack, the high refractoriness for the sand burning, the low usage of the binder with the good grain shape for the blow hole. In 2005, in order to estimate the amount of the factory waste, the authors investigated that the compressive strength of the particles in the casting sand was measured by the micro strength tester and the estimation was determined quantity in contrast with the stress of the particles in the casting sand damaged in the casting line. Against such a background, the characteristic of the ceramic sand produced by the ultrahigh temperature melting was evaluated by the particle compressive strength measurement and the burning resistant sand test in this study.

2. Experimental Procedure

2.1 Test Pieces

The ceramic sand as test pieces were prepared by the ultrahigh temperature melting process, in which ceramic materials are melted under plasma heat in atmosphere and the melted ceramic is gas atomized to form ceramic sands. The chemical composition of the ceramic sands was adjusted using four types of raw material including of SiO_2, Al_2O_3, MgO, and ZrO_2. As comparison samples, several types of marketed ceramic sand were provided with melting method using sintering heat or arc heat or LPG-O_2 flame heat.

2.2 Measurement

The obtained ceramic sands were evaluated by measuring the sand particle strength, sand burning resistant test for the melt cast steel and other characteristics as foundry sands.

3. Experimental Results

The sand particle strength and features of ceramic sands are presented in Table 1 with the purpose of comparing ceramic sand characteristics between ultrahigh temperature melting by plasma heat and existing melting methods by sintering heat or arc heat or LPG-O_2 flame heat. By evaluating the sand particle strength, it was found that the particle strength of the ceramic sand produced by the ultrahigh temperature melting process was a few multiples of that of the existing ceramic sand. As the chemical composition of the ceramic sand has been adjusted to the existing ones, the melting temperature of the plasma heat is considered to have caused an increase in strength.

Table 1. Sand particle strength and features of ceramic sands.

Sample#	Sand particle strength, MPa	Aspect ratio, -	Ceramic sand type
Test-1	2,627	0.944	Mullite
Exsiting-1	1,170	0.851	
Test-2	2,653	0.946	Corundum
Exsting-2	1,163	0.881	
Test-3	2,347	0.964	Corundum-Zirconia
Existing-3	1,466	0.845	
Test-4	3,193	0.957	Corundum
Exisitng-4	2,175	0.941	
Test-5	899	0.958	Periclase
Existing-5	222	0.940	

The cross sections of ceramic sands are shown in Fig. 1 to illustrate the reason behind the increase in strength; the images are of one of the comparison examples listed in Table 1. A comparison of the cross sections shows that the air bubble of melting by plasma heat (Test-1 to 5) is less than that from the existing melting ones (Existing-1 to 5). Additionally, there is an improvement in the aspect ratio defined as shortest diameter / longest diameter, as seen in Table 1; hence, it is expected that the higher temperature of plasma heat reduces the viscosity of the melting ceramic material and contributes to the reduction in the air bubble size and improves the aspect ratio during gas atomizing.

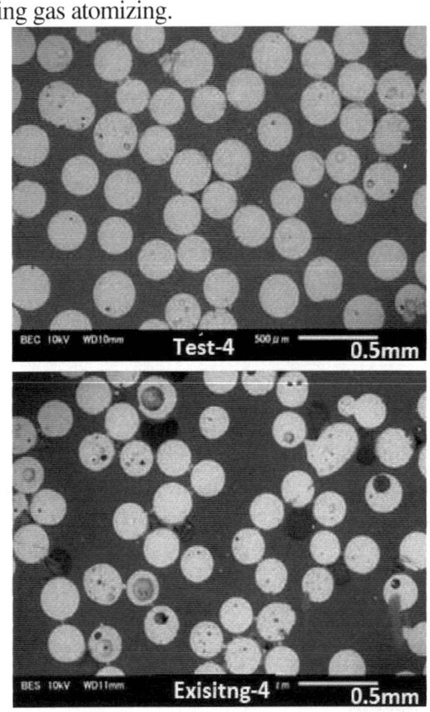

Fig. 1 Cross section of ceramic sands (Compositional image in BES mode)

The basic oxide of MgO in periclase is known to have burning resistance because the weakly-basic oxide of ferrous oxide or stainless steel is unreactive with MgO. On the other hand, it is difficult to melt the ceramic raw material containing a large amount of MgO because of its high refractoriness. However, as it is possible to melt raw material with more basic oxide using the plasma heat of the ultrahigh temperature melting process, the burning resistance was tested by reaction with melt stainless steel. The scanning electron micrograph of the burning resistance test with the stainless steel is shown in Fig.2; the images are of one of the comparison examples presented in Table 1. As seen in these images, there was little chemical burning in the ceramic sands with a ceramic sand type of periclase including MgO although there was chemical burning in other ceramic sand types such as mullite, corundum, and zirconia.

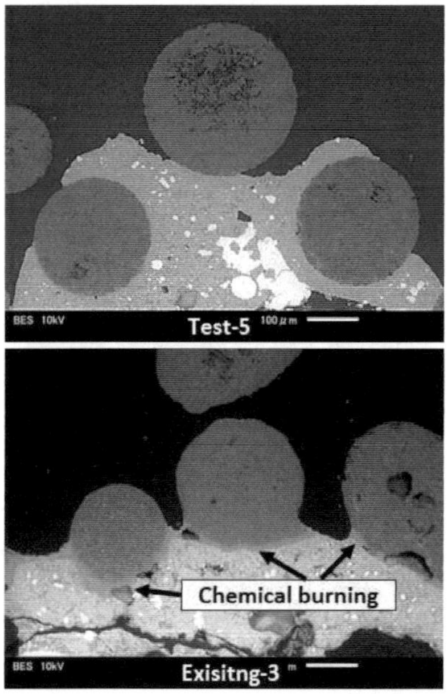

Fig. 2 Scanning electron micrograph of the burning resistance test with stainless steel (Compositional image in BES mode)

4. Conclusion

It was found that the particle strength of the ceramic sand produced by the ultrahigh temperature melting process was a few multiples of that of the existing ceramic sand, which implies capable resistance under stress in a casting line. It was also shown that sand burning was encouraged by the low melting substance in the ceramic sand and the basic ceramic sand was effective in improving the characteristics of burning sand.

O-49

Future Foundry Sand

Zhu Jianxun
Shengquan Group
Diaozhen Industrial Development Zone, Zhangqiu, Jinan, PRC

Sand has been widely used in big quantity not only for sand-mould casting, but also in lost foam , full mold, V-process, low pressure and metal mold gravity castings. Except the big consumption amount, it has significant influence on casting quality, working environment, environmental protection and production cost.

Deep analysis, trials and discussion on a variety of foundry sands, including quartz, chromites, zircon, olivine, bauxite, artificial ceramic sand have been done in this paper. The tendency of foundry sand has also been revealed: high quality and precision, functionalization, wide spread application of non-quartz and artificial spherical sand, finer granularity and non coating sand, new reclamation technology and 100% and non-hazardous recycling of used sand, 100% separation and reclamation technology of special sand, globalization of the foundry sand resource, variation of test methods and the new "3D" sand application.

Key Words: *Foundry sand, eight trends*

1. Introduction

The annual requirement of new sand will be 42 million tons, which also means 42 million tons of waste sand is disposed every year[1,2], which brings serious damage to the resource and environment.

Foundry sand possesses important influence on casting quality (e.g. dimension precision, surface finish) and some casting defects (metal penetration, burrs, shrinkage cavity). 50% of the casting defects are related to the sand [3, 4].

2. Future foundry sand will be high quality and more fine sand.

Future high quality sand will be higher SiO2 content, more rounded grain shape, lower clay content and the amount of fines and narrower AFS range. There will be new silica sand surface treatment technology. Development and application of finer and denser silica sand without coating and application of low expansion calcine.

Higher requirement and narrower control on sand specifications, which means sand strength should be improved by at least 20%, sand quality influence on casting rejection rate and quality problems should be reduced by at least 30%, and sand reclamation rate will be increased by at least 5%.

3. Artificial ceramic sand will be the next major casting sand in future.

Manufacture of artificial spherical sand.

Basically spherical artificial sand is resourced from rich Al-Si-based ore, and made by four processes.

They are mechanical granulation and sintering, spray granulation and sintering, fused spray granulation and Plasma arc granulation.

Advantages of artificial spherical sand.

Chemical composition is stable and adjustable, high refractoriness and wide applicability. Low thermal expansion coefficient. High bulk density and good thermal conductivity. High hardness, round grain shape and good heat resistance. Round shape and good surface finish. Improved sand reclamation rate. Improved casting yield and casting quality. Lower production cost.

Future development of artificial spherical ceramic sand.

It will gradually replace the silica sand and other sands. Its consumption may be even more than the amount of quartz sand in future. Artificial spherical ceramic sand will develop non-silicon system (such as MgO-SiO2), pure silica system, and pure Al2O3 system containing Fe3O4 system, lower Al2O3 system so as to meet different needs and reduce manufacturing costs. Spherical ceramic sand for inorganic binder system will also be developed.

4. The functionalization of foundry sand .

High permeability to solve porosity defects. Low-expansion sand brings higher casting dimensional accuracy, and solves problems of broken cores and burr defect. Insulation sand improves production yield. High-strength sand. High refractoriness, high fineness and high bulk density foundry sand which does not need coating.

5. "100%" Reclamation and Re-utilization of Foundry Sand.

This concept is consisted of three ideas. First, "100%" refers to rate of 99% or higher than 99.5%. Second, treatment will be pollution-free. The prevail thermal reclaiming method has disadvantage of low burning temperature (<700℃). Organic binder on the used sand is not able to be fully decomposed by combustion. Harmful gas such as dioxin may be released to the environment. Remained solid dust like fine powder and sand from mechanical or wet reclamation method contains organic and inorganic binding agent. Pollution-free treatment is brought to light to solve these problems. The third idea is re-utilization of used sand as resource. After pollution-free treatment, unrenewable used sand and solid powder will be reutilized as resource.

New wet reclaiming method for green sand. New hot reclamation method. New Mechanical Reclamation. New inorganic binder reclamation technology.

New development of special sand separation and recycle technology。

Gravity separation and reclaiming technology. Grain size screening separation and reclaiming technology.

6. New development of 3D casting sand.

The silica sand which is finer (AFS＞100), rounder (Circularity＞0.9),and with higher stacking density (wider grain size distribution) will be developed.Modification to 3D foundry sand surface, which helps improve sand wettability and adhesion , increase the strength of 3D cores.

7. New test method of the foundry sand.

Application of new test method for foundry sand particle shape. Rapid test on the foundry sand particle distribution. Test on the flow ability and bulk density. Test of high temperature performance.

8. Internationalization of foundry sand market.

In future, with development of casting technology and casting quality, also the development of high-quality foundry sand, and significantly increased sand reclamation rate (> 95% ---> 97%), foundry plants will use better quality sand, which will then reduce cost and bring the environmental benefits, the foundry sand will cross-border and sell to the international market, and promote the development of the global foundry industry.

References

[1]Foundry Sand Reclamation—Theory and Industrial Practice, R.Dańko~1 J.Dańko~2 M.Holtz

[2]Analysis of casting defects and identification of remedial measures - A diagnostic study, D.N.Shivappa~1,Rohit~2 and Abhijit Bhattacharya~3

[3]A.Z.Zook,"Low-Emission Core Binders Core Sand Dilution".71th World Foundry Congress(2014).

[4]A.J.Oman."Next Generation Thermal Distortion Tester".71th World Foundry Congress(2014)

O-50

Basic investigation into water cooling behavior of the high temperature green molding sand

Yasuhiro Maeda[1] and Hiroyasu Makino[2]

[1] Dept. of Mechanical Engineering, Daido University
Takiharu-cho 10-3, Minami-ku, Nagoya 457-8530, JAPAN

[2] R&D Center, Sintokogio LTD
Honohara 3-1, Toyokawa, 442-8505, JAPAN

The sand cooling process before the mulling of green sand is very important process in order to obtain the green molding sand with comfortable characteristics. In this study, the high temperature molding sand with more than 100 degree C (Celsius) is investigated, then the water cooling behavior of the molding sand is investigated by fundamental experiment. The components of the molding sand are the silica sand/the artificial mullite sand and Na-type bentonite. The compactability index which is one of the properties of molding sand is controlled by contents of water. Temperature and moisture content during sand cooling are studied to clarify the difference cooling method of by air and water. Further, theoretical analysis try to estimate the cooling the water cooling behavior by using a heat transfer and an evaporation equations. It is clear that the difference of cooling degree between more than 100 degree C of sand temperature and less than 100 degree C of ones.

Keywords: *green sand, moisture content, cooling behavior, heat transfer.*

1. Introduction

The temperature of return sand becomes higher because of the change of the component of green molding sand and the speedup of the molding line. The sand cooling process before the mulling of green sand is very important process in order to obtain the green molding sand with comfortable characteristics [1]. It is necessary to adjust the value of the sand temperature and moisture contents of the molding sand.

The high temperature molding sand with more than 100 degree C is investigated in the present study, then the water cooling behavior of the molding sand is investigated by fundamental experiment. Further, theoretical analysis try to estimate the cooling the water cooling behavior by using a heat transfer and an evaporation equations.

2. Experiments

2.1 Experimental setup

New silica sand and ceramic sand (synthetic mullite beads) with 10% sodium bentonite are used. The properties of molding sand are controlled by the compactability index. The experimental conditions are shown in Table 1.

Table 1 Experimental Conditions.

Molding sand	Sand 90% +Bentonite 10% +Water
Sand	Ceramic sand (Mullite beads), Silica sand
Bentonite	Na-Type
Compactability	30, 35, 40 %
Initial Temp.	120, 100, 80℃
Weight of sample	100g
Cooling type	Natural(air), Watering

2.2 Cooling Experiments

The molding sands are pre-heated to the initial temperature of 120, 100 and 80 degree C in the heater. In the first experiment, the molding sand is cooled naturally by being left in the atmosphere. Then, the temperature and moisture content of the molding sand are measured during cooling. The second experiment, the molding sand is receiving the proper amount of watering, then it has been left in the atmosphere.

2.3 Cooling behaviors

Figure 1 shows the cooling behavior of molding sand after watering in the case of ceramic sand and initial temp. of 120 degree C. The temperature of molding sand is rapidly decreased immediately after watering, then it is gradually cooled. Figure 2 shows the cooling behavior of molding sand in the case of natural (air) cooling and watering, initial temp. of 120 degree C and ceramic sand.

In the later stage of cooling, the temperature drop for the water-cooling and the air-cooling are almost the same. In the case of the high temperature molding sand with more than 100 degree C, the evaporation cooling was mainly occurred just sprinkling the water, then the temperature was decreased in accordance with heat transfer behavior. It is the same tendency in molding sand with silica sand.

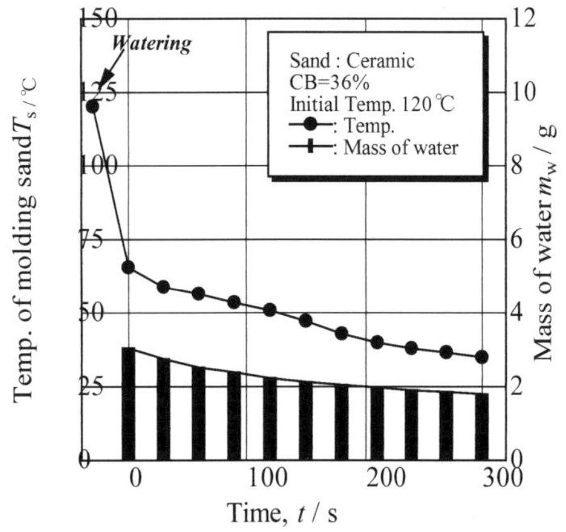

Fig.1 Cooling behavior of molding sand after watering in the case of ceramic sand and initial temp. of 120 degree C.

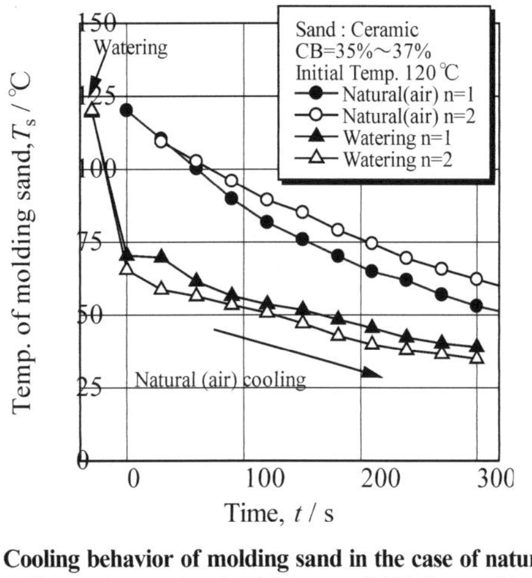

Fig.2 Cooling behavior of molding sand in the case of natural (air) cooling and watering, initial temp. of 120 degree C and ceramic sand.

3. Numerical Analysis

Theoretical analysis try to estimate the cooling the watering behavior by using a heat transfer and an evaporation equations. Referring the theoretical analysis of air-cooling for wet and dry sand by Ohnaka [2], the governing equations are following;

Heat transfer for wet sand:

$$\frac{dT_{sw}}{dt}=\frac{L_s}{C_s+w}\cdot\frac{dw}{dt}-\alpha\frac{\beta A_s}{\beta m_s(C_s+w)}\cdot(T_{sw}-T) \quad (1)$$

Heat transfer for dry sand:

$$\frac{dT_{sd}}{dt}=-\alpha\frac{\beta A_s}{\beta m_s(C_s+w)}\cdot(T_{sd}-T) \quad (2)$$

Where, T : Temperature [K], t : Time [s], L_s : Latent heat of vaporization [J/kg], C : Specific heat [J/kgK], w : moisture content [kg/dry.kg], α : heat transfer coefficient [W/m^{2}K], β : rate of wet sand, A : surface area [m2], m : Mass [kg], subscript: a : air, s : sand, sw : wet sand, sd : dray sand

Solving Eq.(1) and Eq.(2), it is possible to obtain the temperature of molding sand cooled by heat transfer. Figure 3 shows the comparison of cooling behavior between experiments and calculation in the case of natural (air) cooling, initial temp. of 120 degree Cand ceramic sand. The calculated values can be estimated well the experimental values.

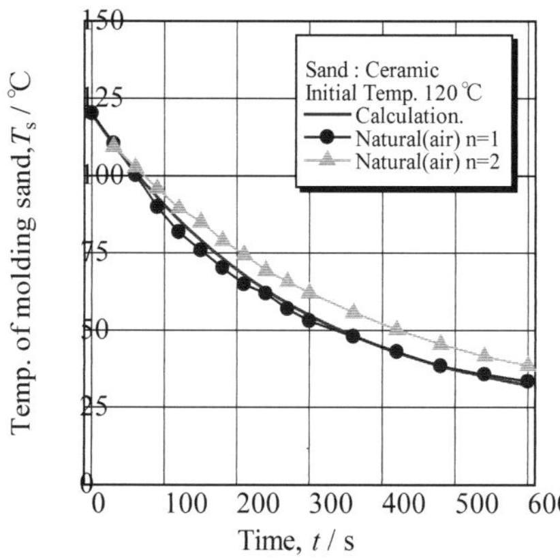

Fig.3 Comparison of cooling behavior between experiments and calculation in the case of natural (air) cooling, initial temp. of 120 degree C and ceramic sand.

4. Conclusions

It is clear that the difference of cooling phenomenon between more than 100 degree C of sand temperature and less than 100 degree C of ones. In the later stage of cooling, the temperature drop for the water-cooling and the air-cooling are almost the same. These results provide valuable information to produce casting stably.

References

[1] Japan Foundry Engineering Society : Report of Technical Session "Trend of Sand Mold Technics", No. 164 of JFS Meeting, Kyoto (2014) 5 pp.1-37.

[2] I. Ohnaka, T. Kamimura and K. Chijiiwa : J. of Japan Foundrymen's Society (IMONO), 43(1971) 10 pp.893-901.

Proceedings of the 72nd World Foundry Congress
21-25th May 2016, Nagoya, Japan

Dynamic Testing of Green Sand

S. Ramrattan[1], S. Butt[2] and H. Makino[3]
[1 & 2] Western Michigan University, Kalamazoo, MI USA
[3] Sintokogio Ltd, Aichi, Japan

Foundry engineers have long known that certain AFS standard green sand properties tests provide limited information for green sand control. This limitation is especially apparent when strict casting dimensional control and a superior casting surface is required. Dynamic green sand tests such as: thermal erosion test (TET), the wet tensile strength (WTS), the modified cone jolt toughness test (MCJT) and the friability test were able to differentiate among various compactability levels (moisture content) and clay levels. This paper focused on sodium Bentonite, calcium Bentonite and various blends. The dynamic tests provide more relevant data about these green sand systems. The results obtained in this study are very encouraging, and presents the opportunity to develop new green sand control tools using data from both room and elevated temperature tests in order to produce high quality castings.

Keywords: *thermal erosion test, wet tensile strength, modified cone jolt toughness, friability*

1. Introduction

The majority of casting defects in sand casting facilities are sand related, but many metal casters fail to recognize that molding sand quality is as important as metal quality [1]. Many foundries incorporate different bond formulations to produce the green properties they feel they need to produce acceptable castings. The green sand molding process brings with it some potential weaknesses for casting parts of various shapes, sizes, and alloys. Measurement and control of green sand properties is critical to the green sand foundry's success. The typical properties that are monitored are green compression strength, compactability, moisture, methylene blue clay percentage and wet tensile strength (WTS).

Casting quality is directly related to mold quality and mold quality is directly related to sand control. To produce world-class castings competitively and to meet the ever-increasing demands for higher quality and dimensional reproducibility, sand control is essential [2,3]. A good understanding of essential process requirements is needed to deal with day-to-day problems. Over the years, several noted papers on "green sand system control" have been published; however the common theme for success in green sand control is sticking to the basics [2-5].

Foundries carry out green sand testing as an essential part of process/quality control. Foundries depend on the AFS sand tests to help manage their new sands, green sand systems, chemically bonded cores and molds, sand additives and reclaimed sands. The testing procedures are defined by the AFS [6].

In order to produce high quality castings, the purpose of this paper is to determine whether a foundry can monitor changes in sand properties caused by small changes in clay and moisture levels when using non-standard green sand testing such as a thermal erosion test (TET) and a modified cone jolt toughness test (MCJT)

2. Dynamic Testing

Fig. 1 shows a picture of the thermal erosion test TET as one of dynamic test. This machine is developed at Western Michigan University (WMU) and allows dynamic testing of a modified two by two specimen against a rotating hot surface where eroded sands are measured in real time.

The modified cone jolt toughness test measures the sand's bulk brittleness, and is related to difficulty in pulling deep pockets in a pattern. The jolt/displacement curve affords ultimate strength, toughness, stiffness and plastic green sand property data. The modified cone jolt toughness test is directly related to compactability and measures the green sand system's ability to absorb energy [7].

When thermal erosion data is examined as compactability increases for the same level of clay in the green sand system there is generally a decrease in the percent green sand loss at room temperature (Fig. 2). However, there are more losses in Ca Bentonite green sand system for every compactability level tested but especially at the lower compactability levels.

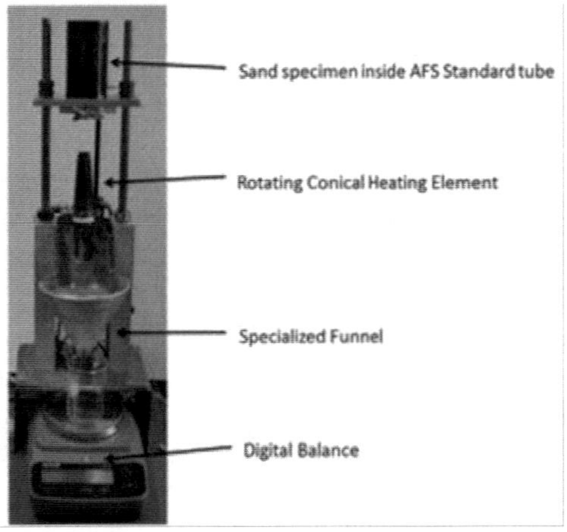

Fig. 1 TET and Data Acquisition System.

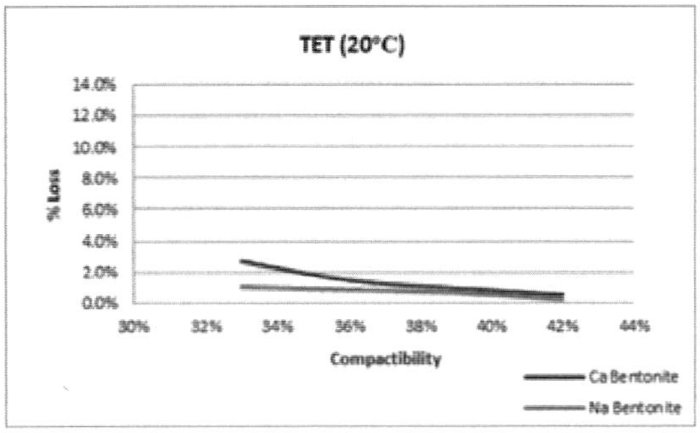

Fig. 2 Percent loss (abrasion) at room temperature as percent compactability is increased.

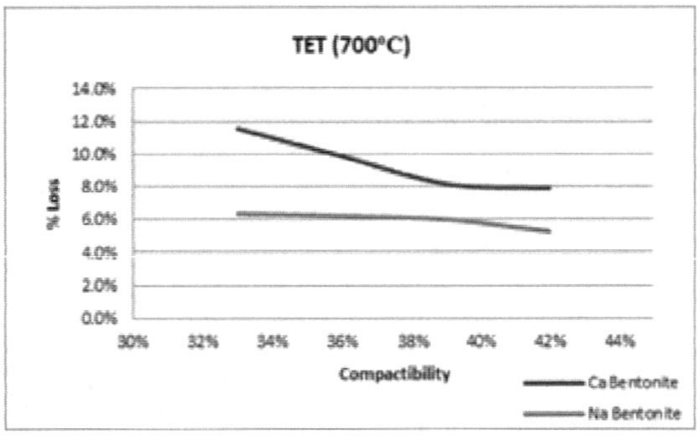

Fig. 3 Percent loss (abrasion) at elevated temperature of Na and Ca Bentonite Green Sand Systems as percent compactability is increased.

When Na and Ca Bentonite green sand systems were tested at an elevated temperature the same general trend was seen as tested at room temperature. As compactability was increased, there was less loss in both systems. However, the major difference is that a greater percent loss occurs at all the compactability levels but the losses are less drastic with Na Bentonite green sand system (Fig. 3).

3. Conclusions

The dynamic green sand tests such as: thermal erosion test (TET), the wet tensile strength (WTS), the modified cone jolt toughness test (MCJT) and the friability test were able to differentiate among various compactability levels (moisture content). Friability and TET (both ambient and elevated temperature) indicate that green sand system with 100% Na Bentonite had superior abrasion resistance. The results are very useful to define a good green sand to produce a high quality casting.

Acknowledgements

The authors gratefully acknowledge Sintokogio, Ltd. for the research support. The authors gratefully acknowledge Abraham Poot (posthumously), Glenn Hall, Peter Thannhauser, Michael Banion, Andrew Foster, and Ross Henry from Western Michigan University, for their technical support.

References

[1] Aycardi, M., M.B. Krysiak, and D. Martin, "Sand Lab to the Rescue," Modern Casting, (July 2009).

[2] Ramrattan, S., A. Khoshgoftar, H. Makino, M. Hirata, S. Takasu, and H. Miyazaki., 2013, "A Quick Technique for Comparing Aeration to Gravity Filled Green Sands" AFS Transactions, No. 13-1451, (2013).

[3] Krysiak. M.B., Ramrattan, S.N., Cheah S.F., "Thermal Distortion of Green Sand and Chemically Bonded Sand at Cast Iron Fill Temperatures," AFS Transactions, V 110 Paper No 02-037 P 651-664, (2002).

[4] Krysiak. M.B., "Reducing Casting Defects: A Basic Green Sand Control Program," 2 parts, Modern Casting, (Apr., May.1994).

[5] Krysiak, M.B., Pedicini, L.J., "Sand Testing Design", 3 parts, Modern Casting, (Feb., Mar., Apr. 1990).

[6] Mold and Core Test Handbook, 3rd. Edition, AFS (2001).

[7] Ramrattan, S. N., J. Rodriguez and A. Choudhury, 2010, "Evaluation of a Modified Cone Jolt Test on Green Sand Properties" No. 10-046, Transactions, American Foundry Society.

Proceedings of the 72nd World Foundry Congress
21-25th May 2016, Nagoya, Japan

Development of the New style Green Sand Molding Machine

Tokiya Terabe[1], Yutaka Hadano[1] and Koichi Sakaguchi [1]
[1] Sintokogio, LTD. Toyokawa, Aichi 442-8505, Japan

In the field of the green sand molding technology, various molding methods, such as jolt-squeeze, blow-squeeze and SEIATSU, etc. were developed in the 20th century. Thereafter, in the beginning of the 21st century, a new molding method with the new sand filling mechanism was developed[1,2]. That is the aeration sand filling mechanism with low pressure air. Through this technology, the sand filling to the mold was dramatically improved, the quality of mold has been stabilized and high performance molding has been carried out. At the same time, this molding method contributed to the environment issue by the large reduction of compressive air and decrease of the noise and vibration.

This innovative green sand molding technology is used as a core technology in our various molding machines including tight flask and flaskless ones. On the other hand, since foundrymen have a wide variety of requirements, the authors have developed two types of molding machines employing a different aeration sand filling technology to further satisfy their needs.

A sensing technology for visualization of operating conditions and preventive maintenance, etc. is applied in various forms to these molding machines from the view point of more stable quality cast and productivity improvement.

Keywords: *Aeration, Sensing technology, Visualization, Reduction of defects, Quality stabilization*

1. Introduction

The authors have a wide selection of molding machines with aeration sand filling[1,2] as a core technology which achieves uniform sand filling and saving energy. Futhermore, The authors have developed two types of molding machines employing a different aeration sand filling technology to further satisfy their needs.

2. Development of the New style Green Sand Molding Machine

2.1 New type tight flask molding machine

The molding method of tight flask molding machine employing aeration sand filling technology have two advantages: uniform mold strength by segment squeeze where each segment foot moves vertically as shown in Fig.1, a sand compaction ratio is preset as a/A=b/B at the time of completion of squeeze, mold sand is filled by aeration into the mold and then the mold sand is squeezed mechanically; and plane squeeze where the back face of the mold is flat when squeeze is completed.

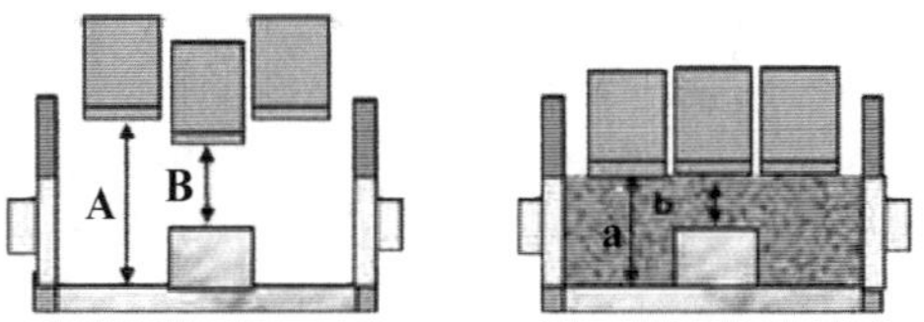

Fig. 1 Preset squeeze

Molding conditions such as aeration and squeeze, etc. can be registered for each pattern, which includes setting of pre-molding conditions. However, sometimes preset functions are not set properly and sometimes customers prefer simple segment squeeze due to troublesomeness of setting.

In this paper, the authors have developed a tight flask molding machine where segment cylinders are arranged on the whole surface of the squeeze board.

This made it possible to fill sand by aeration into every hole and corner of patterns without preset pre-molding and produce high quality molds by segment squeeze onto the overall sand. Furthermore, this made it possible to produce higher-quality molds in combination with preset pre-molding.

2.2 New type flaskless molding machine

The flaskless molding machine has various sand filling methods. The methods used for our flaskless molding machine are vertical aeration, side aeration and top & under blow sand filling.

In case that a molding machine is replaced, a failure of sand filling occurs sometimes due to a change in the sand filling method. Especially, the top & under blow type molding machine has a feature that sand filling is superior just under the blow nozzle. We applied this superiority to different sand filling types of molding machines. However, sand filling was sometimes unsatisfactory in specific points.

The new flaskless molding machine developed by us this time is the successor of top & under blow type of molding machine which uses a sand filling method of top & under aeration. As a result, mold quality and productivity are more improved.

Consumption of compressed air could be reduced by about 20% by filling sand at a lower pressure as compared with a conventional molding machine. We have also found that proportional valve can prevent sand from being compressed in the sand introduction pathway of aeration tank, etc.

Other features can be summarized as follows. Pattern drawability from the mold and use frequency of in-use match plates were improved by employing squeeze balance control and slightly adjusting squeeze force of the cope and drag.

2.3 Approach to visualization of operating conditions of green sand molding machine

To achieve visualization of the operating conditions of a molding machine, a system for monitoring them is installed as a standard in the newly developed molding machine. For that various sensors are set at the points to be monitored and the operating conditions can be understood through the data measured in real time and on line by the sensors.

Previously, aging deterioration due to operation was judged and cause investigation on the abnormal operation was performed based on monthly or annual inspection work at regular intervals. Accordingly, effective quantitative judgment based on numerical data provided big progress for preventive maintenance.

Moreover, there was great progress in the field of quantitative assessment of sand filling process which was indirectly assessed based on molds actually produced. Fig. 2 is a graph showing behavior of air pressure in the aeration sand filling process and pressure in the sand tank extracted from the measured data.

It is possible to estimate sand injection time and check if the function of sand filling is operating properly by understanding a fluctuation of injection time for the same pattern. Additionally, assessing differential pressure between the pressure in the chamber for supplying aeration air and the pressure in the sand tank leads to visualization of a status of clogging of the aeration filter, contributing to effective preventive maintenance against decreased function of sand filling.

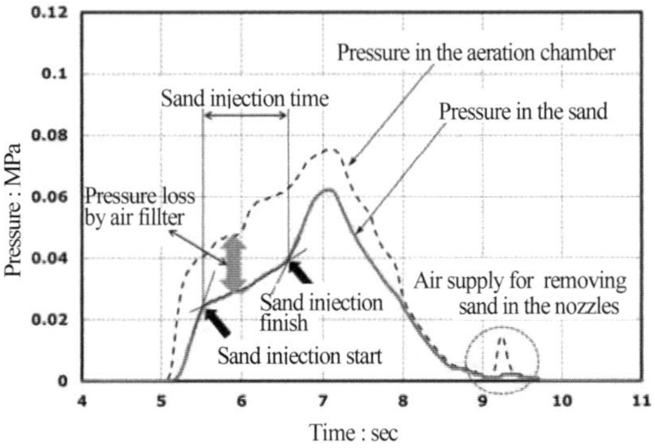

Fig. 2 Pressure behavior in sand filling process

These instances demonstrate that visualization of the operating conditions of the device using sensing technologies which have been used until now for the molding machine is very effective for stabilizing the operation of the molding machine and securing the quality of casts in combination with know-how about analysis of the measured data.

3. Conclusion

2 types of molding machines developed this time are added to a product group employing various types of aeration sand filling methods. It is expected that this will really contribute to response to a variety of foundrymen's needs, proposal about better production of casts and enhancement of the operating rate of equipment.

It seems that IoT (Internet of Things) technologies will increasingly advance in the manufacturing industry in the future and a mass of data will be accumulated and shared in connection with development of information and communications technologies. Bearing in mind enhancement of IoT technologies in the field of casting equipment, we will also utilize the data obtained through various sensing technologies and provide technologies and services based on the concept of ensuring the product quality and "preventing equipment from stopping unexpectedly".

Reference

[1] M. Hirata., "Eco-Conscious Molding System, Featuring Aeration Sand Filling, Preset Squeeze And Mold Height Feedback Control," AFS Transactions, Paper 02-081, Page 1 of 8, (2002).

[2] H. Makino, M. Hirata and Y. Hadano, "Computer Simulation and Optimization of Sand Filling Using the Distinct Element Method", WFO Technical Forum on GIFA2003, Düsseldolf, session 5.1

Proceedings of the 72nd World Foundry Congress
21-25th May 2016, Nagoya, Japan

Results of new developments for efficient and intelligent equipment in vertical molding machines

Borja García[1], Ander Eguidazu[1], Raul López[1],Patxi Rodriguez[2],Oier Gereñu[2]
[1]LORAMENDI, S. Coop., Alibarra Kalea, 26, 01010 Vitoria-Gasteiz (Araba), Spain
[2] FAGOR EDERLAN S. Coop. Torrebaso Pasealekua, 7, 20540 Eskoriatza (Gipuzkoa) Spain

ABSTRACT

The competitiveness in the current foundry market pushes the companies to deploy cost saving programs over all activities. Among the points to be technically reinforced in the molding area, we have: foundry process optimization, energy consumption, resources consumption and environmental protection. According to current data in terms of energy usage share in an average foundry, 12% of energy consumption relates to molding equipment, and 60% of the cost relates with materials.

Increasing prices of electric power and raw materials together with demand for new technical casting solutions force the foundry to find out new ways of production where high efficient process, high accuracy and highly enviromental friendly are challences to face.

The present paper presents results and production data collected during two-year production of the new concept of electrical Vertical Molding Machine (e-VMM) proving its high energetic efficiency, high reliability, high accuracy on its performance and environmental friendliness.

Keywords: vertical moulding, Electrical, Efficiency, Accuracy Smart Casting.

INTRODUCTION

When we started with the project of installing an electric moulding machine at one of the Fagor Ederlan foundry plant we set up some technical targets, which gather the new technology advantages.

We initially defined those targets between both teams, from Loramendi and Fagor Ederlan, based on trials carry out in advance during the prototype phase.

From the foundry point of view, it was a challenge to produce complex steering knuckles in a vertical line and, on the other hand, from the machine builder point of view it was a challenge to reach expectations for energy efficiency and high accuracy of the new technology.

Additionally, those common targets were defined in a simple way, which will later on allow us to measure, record them, and therefore recognise success of the project. That finally means proving the advantages of the new technology as per expectation.

EQUIPMENT AND TARGETS DEFINITION

A hydraulic moulding machine has been replaced by a new LORAMENDI electric vertical moulding machine, namely, e-VMM 5070B© together with an e-AMC© 18 m in length.

Speed of such a model was 400 molds per hour, although today a 500 molds per hour are already in production at other foundries.

Fig. 1. Electric moulding machine at Foundry

We have needed two years before we can present data and say that targets of the project have been reached by far in a stable way and therefore that this new technology offers to foundries new advantages the hydraulic one can not.

Targest for the new equipment were as follows:

- Energy saving of 60%
- Cooling water saving. No water required

- Hydraulic oil saving. Not oil required. No oil waste management costs
- Less maintenance time. Machine Uptime > 95%
- Higher accuracy & casting quality. Mismatch < 0.2mm
- Lower noise level
- Cleaner work enviroment

RESULTS AFTER A TWO-YEAR OPERATION

After a two-year operation and 2,000,000 moulds produced, we have recorded parameters and average values shown as follows:

Energy saving of 60%
Energy consumption per mould has been recorded and actual values are 0.182 KWh/mould for the hydraulic version against 0.064 KWh/mould for the electrical version, resulting a 65% energy saving

Cooling water saving
Due to "servo-engine-configuration" machine instead of "hydraulic-unit" one, there is no cooling water required and therefore a 4,800 litre/hour saving has happened. This point is of great importance, as cooling water consumption in an industrial environment needs special treatment before sending back to the network or, alternatively, be used within a water closed circuit

Hydraulic oil saving
Here again, due to "servo-engine-configuration" machine instead of "hydraulic-unit" one, there is no hydraulic oil involved. That means a 7,000 litre/year (e.g. Renolin type) and 880 litre/year (e.g. HLP46 type) savings have happened. Once again, waste oil management becomes of great importance within an industrial environment following European rules for the environment after oil usage.

Less maintenance time. Machine uptime > 95%
After recording machine stops during these two years, the average of uptime for the electrical moulding machine has been of 96.12%. Here we consider the percentage of the machine available time when it was operational, excluding from this value stops that may have happened due to peripheral equipment such as pouring unit, sand plant unit, cores supply, finishing line or any other.

Higher accuracy & casting quality. Mismatch < 0.2mm
The already reported mould hardness homogeneity in this LORAMENDI equipment (see article form 71st WFC Bilbao 2014, "A new concept in efficient vertical moulding lines") due to unique system for chamber filling and the squeeze of the mould, now it is incorporated the mismatch values of < 0.145mm.

Unlike hydraulic technology, electric one allows us to perform "à la carte" programs on speed & force. Higher precision on position, speed and accelerations when dealing with moulds squeezing, sorting out and conveying as well as performance concerned to the core setting are a reality today, resulting on a more accurate castings production and new capabilities to reduce flash and finishing operations on castings.

Lower noise level
Today, noise level of the machine is lower than 80 dB and further reduction are on going today.

Cleaner work enviroment
The fact of avoiding hydraulic oil and its inherent leakage control and repair is giving place to a cleaner work environment.

Spare parts consumption
Although there is no enough working hours to assess precisely consumption of spare parts in such an electrical machine, preliminary data show huge potential reduction on them. Less number of parts and robust mechanism design (rack and pinion assembly and so on) has wear out extremely little. Therefore, they keep performance without replacement up today.

Future developments
Today, Industry 4.0 concept consists of big data collection, fast processing and feeding back to equipment to reduce scrap and anticipate maintenance jobs. The electrical vertical moulding technology is in the best condition to perform diagnosis and welcome several sensors, which later on will supply required data for the mentioned concepts.

AKNOWLEDGEMENTS
The authors would like to thank to the foundry maintenance people from FAGOR EDERLAN ESKORIATZA for their commitment and strong support.

Proceedings of the 72nd World Foundry Congress
21-25th May 2016, Nagoya, Japan

A successful machine and tooling concept to produce inorganic sand cores

Borja García, Gorka Paramo
LORAMENDI, S. Coop., Alibarra Kalea, 26, 01010 Vitoria-Gasteiz (Araba), Spain
AURRENAK, S. Coop., Bitorialanda Kalea, 15, 01010 Vitoria-Gasteiz (Araba), Spain

Keywords: inorganic, sand cores, binders,tooling, equipment

ABSTRACT

A decade ago both the society and the government's regulations were demanding to the foundry industry solutions to respect our environment. Concerning the aluminium casting production, a new technology appeared to produce sand cores in such a way that respected the environment inorganic cores.

That basic concept behind was that new inorganic binders were used to shape the sand grains into a specific form given by a tooling. The process was so sensitive to so many factors that finding an industrial solution, cost efficiently for a mass production of those cores becomes a real challenge where core machine was main concern, together with sand preparation and core handling.

Ten years ago, Germany was leading this technology and the rest of the world was looking forward to knowing the real potential results of the new process. Nowadays, the inorganic process has been successfully exported to other countries as Spain, Italy, Russia, USA, Mexico, China and Korea, where it has become a reality.

INTRODUCTION

The journey was not easy. There have been many factors which affect in a process where day to day there are new steps and improvements thanks to all involved parties: foundrymen, machinery manufacturers, tooling makers and raw material suppliers.

These new discoveries determined the equipment design and manufacturing and today we already have the required knowledge to carry out also this process in new destinations as Brazil and Japan.

EQUIPMENT AND TARGETS DEFINITION

The start point was an anthropological work to detail all influential factors, which affect directly the inorganic process. From the raw materials to their processing machines, following with the cores delivery into the fusion lines per gravity or low pressure.

After listing all those factors, we need to know how they influence directly or indirectly in the whole process together or separately. Finally, it should be defined the maximal and minimum and maximum limits of those factors in order to redesign the installation equipment.

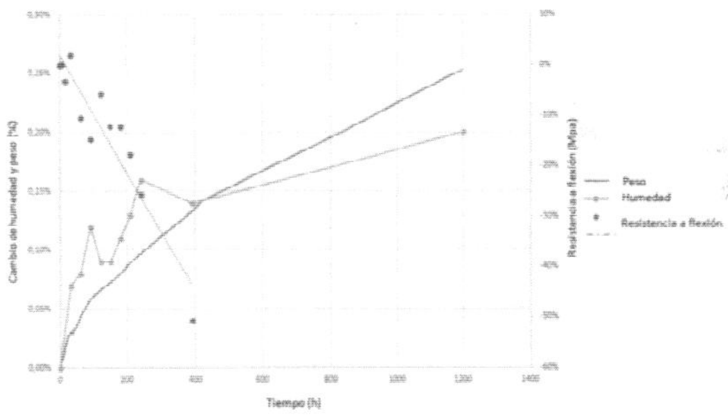

Fig. 1. Properties of test pieces subjected to increased humidity

Knowledge of all these data and their influence are the way to implement this new manufacturing process in new countries.
Moreover, the sensing equipments to directly know the current state of parameters is necessary to act upon the request, which is why this process industry 4.0 has a prominent role in the new developments of Loramendi and Aurrenak.

RESULTS AFTER A DECADE OF OPERATION

After working closely with prescribers of inorganic process as BMW and VW have been, Loramendi and Aurrenak have defined the most significant parameters of the inorganic process being the following:

 # Sand, solid Additive, Binder:
Particle size, shape, moisture and acidity of sand as well as sand additive and resin temperature,

 # Mixing of components:
Percentage of each components, sequence, dose rate of components and mixer speed at different steps of mixing.

 # Maintenance of sand mix before blown in core box:
Control of temperature and humidity of sand mixture as well as contact with the outside environment.

 # Cores Production:
Time and blow pressure, temperature of corebox, pressure, drying temperature, humidity and air flow, drying time, temperature and flow of tooling and machine cooling system.

 # Core Box:
Shooting nozzle diameter and length, types of air vents, tooling heatingsystem, air inlets types for cores drying.

Cores Storage:
Temperature, humidity and storage time.

Loramendi and Aurrenak have been working in those fields concerning the industrial solution and came out to a machine concept with the required features that are able to handle the process parameters within the narrow range they must be to produce high quality cores in the industrial scale and costly efficient in any developed industrial place of the world.

CONCLUSIONS AND NEXT STEPS

Many parameters forming the inorganic process and its individual control and adjustment function of the rest parameters, so the Industry 4.0 concept consists of big data collection, fast processing and feeding back to equipment to readjust and reduce scrap levels, is a key issue in the design of new equipments. That's why Loramendi and Aurrenak is in the best condition to perform diagnosis and welcome several sensors, which later on will supply required data for the mentioned concepts and reajust of them to get high production and quality levels as it expected by the inorganic process.
Now the new challenge presented is to successfully implement the inorganic process in countries like Japan where conditions are unique and specific to the knower who have given us the work done during the last decade in the inorganic process.

Proceedings of the 72nd World Foundry Congress
21-25th May 2016, Nagoya, Japan

A Contribution to the Understanding of the Combined Effect of Nitrogen and Boron in Grey Cast Iron.

Knud Strande[1], Niels Tiedje[2], Ming Chen[3]
1 Dansk Udviklings Formidling ApS, Bygaden 60, DK 2630 Taastrup, Denmark.
2 Technical University of Denmark, DTU, Mechanical Engineering, DK 2800 Lyngby, Denmark.
3 Technical University of Denmark, DTU, Energy Conversion and Storage, DK 4000 Roskilde, Denmark.

Abstract

This investigation deals with the combined effect of nitrogen and boron and how it in a simple way is possible to utilize this effect to enhance material properties in heavy grey iron castings.
It is shown that controlled addition of nitrogen and boron in practice can be used to control microstructure of heavy grey iron castings.
A plausible theory for formation of boron nitride nuclei effective for graphite growth is presented.

Keywords: *Nitrogen, Boron, Control, Properties, Grey Iron, Heavy Castings.*

1. Introduction

Inoculation is an essential part of controlling material properties in grey cast iron. Inoculation practice has for decades been based on the addition to the melt of small amounts of elements with strong affinity to O (and S) just before casting takes place. This method is proven - both in theory and in practice - to be effective in most cases. But it has the disadvantage that the nucleation effect fades away over time. Especially in heavy castings (slow cooling) this effect may cause non uniform and unacceptable material properties in some parts of the casting. Nitrogen is also known to influence grey iron microstructure. Both graphite flake formation and matrix formation are influenced. However, the obtained effects differ considerably and unpredictable in practice.

2 Experimental

In an attempt to quantify the effect of N and B on grey cast iron properties the following experiment was conducted.

2.1 Test casting

The test casting is a vertically cast solid cylinder Ø180/200 mm x 460 mm poured in a sodium silicate bonded sand mould with an Ø200x200 mm insulating feeder on top. Total weight ~130 kg.
Slices approximately 35 mm thick were cut and used for different heat treatments and determination of material properties and microstructure. From each slice it is possible to machine 3 tensile test pieces - ISO 185 type B with test diameter Ø16 mm. Those test pieces are also used for determination of material structure and hardness.
The casting was cooled to temperature (T) < 300^{0}C before shake out took place.
Besides the test casting itself the following test items were made from the melt for control purposes:

- Two white solidified coupons for chemical analysis.
- 2 pcs. separately cast Ø30 mm tensile test pieces (ISO 185).
- One Quick Cup for thermal analysis.

2.2 Melt treatment

The charge was melted in an acid lined 150 kg medium frequency induction furnace. The charge material was made up of (in the following order): pig iron, foundry return, steel plate, and carburizer. When all melted, T was raised to ~ 1500^{0}C and kept constant for about a 20 min. carbon boil. Thus the melt could be considered as clean and relatively poor of nuclei. The chemical composition (except N) was adjusted by adding the necessary amount of different ferroalloys. The target N content was achieved by adding calcium cyan amide ($CaCN_2$) in bottom of the pouring ladle during metal transfer.
No conventional inoculation was performed.
The test casting was poured at 1300^{0}C.

2.3 Material investigations

Determination of material properties and microstructural analysis were performed on one as cast slice (designated A2) and on one stress relieved slice (designated A3).

For reference the two separately cast Ø30 mm test pieces were treated likewise.

Stress relieving cycle: Heating to 600°C, 6 h; holding at 600°C for 10 h; cooling in furnace until T < 300°C.

3 Results and discussions

3.1 Chemical composition

Table 1. Achieved metal composition (w %).

C	Si	Mn	P	S	Cu
3.25	1.0	0.44	0.25	0.08	0.44
B	N	Cr	V	Ti	Al
0.026	0.013	0.06	0.017	0.007	0.005

3.2 Mechanical properties

Table 2. Mechanical properties of test casting.

Speci men	Rm [MPa]	Atot [%]	E-modul [GPa]	Hardness HBW 10/3000	Hardness, HV25 g. Ave. of 5 impressions in pearlite structures of A2-2.
A2-1	203	0.73	122	170	
A2-2	200	0.62	146	165	
A2-3	200	0.71	161	NA	
A2	*201*	*0.69*	*143*	*168*	*272*
A3-1	214	0.65	161	164	A3-2
A3-2	204	0.61	153	165	
A3-3	215	0.74	140	NA	
A3	*211*	*0.67*	*151*	*165*	*303*
Ø30	*230*	-	-	*230*	-
Ø30H	*230*	-	-	*180*	-

A2 represents values obtained on an as cast test slice.

A3 are values obtained on a stress relieved test slice.

Ø30 represents values obtained on an as cast Ø30 mm tensile test piece.

Ø30H represents values obtained on a stress relieved Ø30 mm tensile test piece.

3.3 Microstructure

Table 3. Characterization of microstructure - Ø30 mm test pieces.

Specimen	Microstructure
Ø30 as cast	Fine lamellar graphite in a matrix of pearlite surrounded by hard phase steadite . Hard phase share ~ 15 %.
Ø 30 stress relieved	Fine lamellar graphite in a matrix of pearlite. Hard phase share is now reduced to 3- 4 %. The cementite part of the steadite has been broken down to graphite and perlite during the low temperature heat treatment.

Microstructure of the test casting was evaluated using a MAN B&W standardized IAS procedure.

Table 4. Microstructure of test casting. Average ferrite content in both items is measured < 1 %.

Item	Graphite type, distribution, size		Gra-phite	Gra-phite	hard phase [%]	
	Type	Size	no/mm^2	area %	Av.	Max.
A2-2	I A 3/4/5	3.7	91	8.3	3.4	5.5
A3-2	I A 3/4/5	3.7	85	8.8	3.3	5.3

The high hardness seen in the as cast Ø30 mm test piece indicates that the state of inoculation has been inadequate for the actual combination of metal composition and cooling rate.

In the slow cooling casting the state of inoculation seems to be perfect and the stress relieving heat treatment causes an increase in matrix hardness/ material strength.

4 Conclusions

1. Provided controlled content of vital trace elements is enforced small amounts of boron and nitrogen causes uniform microstructure as well as good material properties all over the cross section of heavy castings.
2. Graphite growth conditions are running in "nitrogen" mode creating short and compact flakes.
3. Dissolved nitrogen - not tied up by hexagonal boron nitride (and TiN, AlN) formation - gives rise to age hardening of the pearlite when heating to > 560°C.
4. Some sort of "over aging" may be a consequence when time at elevated temperature is extended.
5. In phosphorous alloyed wear resisting materials (cylinder liners) it is possible to achieve a specific, constant amount of hard phase evenly distributed in all part of the casting.

It is possible to implement those findings in practical foundry operation without big fuss. The first step will be a routine control of trace elements in the melt. Today's analysing equipment can do the job together with motivated melting shop operators.

Acknowledgement

This work has been made possible by the support from:

1. MAN Diesel & Turbo, Copenhagen, Denmark.
2. Vald. Birn A/S, Holstebro, Denmark.
3. Brdr. Christensen ApS, Skuderløse, Denmark.

Proceedings of the 72nd World Foundry Congress
21-25th May 2016, Nagoya, Japan

Mechanism of the molten metal surface pattern generating in gray cast iron

Toshitake Kanno[1], Yuki Iwami[1], Taiji Fukuo[2], Natsuki Miyai[2], Yuichi Hiramoto[1], Hideo Nakae[1]
[1] Kimura Chuzosho Co., Ltd., 1157 Shimizu-cho, Shizuoka-ken, Japan
[2] Nagaoka unibersity of Technology, 1603-1 Kamitomioka Nagaoka, Niigata-ken, Japan

The surface patterns called Bamboo-leaf type, Pine needle type or Hexagonal type appear at around 1623 K on the molten iron. The origin of the surface patterns is a surface SiO_2 film which is formed by the oxidation of Si. The difference in radiation rate between SiO_2 and molten iron makes the patterns visible. The addition of 0.02% S reduces the surface tension remarkably. As a result, a complex Marangoni convection occurs beneath the surface and it generates the complex surface patterns. It has been said that the surface pattern can be shown only in a cupola. However, the reason is that Marangoni convection occurs due to sulfur coming from cokes in a cupola. Therefore, even in an electric induction furnace, if above conditions are satisfied, the surface pattern can be shown.

Keywords: *molten metal, surface pattern, gray cast iron, sulfur, silicon, surface tension.* .

1. Introduction

In a cupola molten metal, a surface pattern of Bamboo leaf type, Pine needle type, or Hexagonal type appears at around 1623 K, as shown in Fig.1. Sometimes the quality of molten gray iron metal was judged by those patterns.

As for the surface patterns, reported that it is an oxidation film of FeO-SiO_2 reported that it is not shown in an electric induction furnace [1-5]. However, the mechanism of the surface pattern is still unknown definitely. This research aims to reveal the formation mechanism of the patterns.

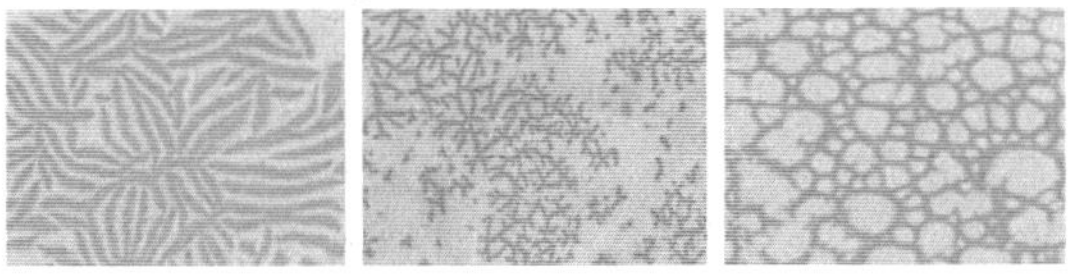

Bamboo-leaf Pine needle Hexagon

Fig. 1 Molten cast iron surface pattern in gray iron.

2. Experimental method

We observed the surface of molten metal during increasing temperature with heating(power on) and during decreasing temperature with cooling(power off) in a high-frequency induction furnace(output 50kW and 3000Hz, internal diameter of the furnace 200mm, melting capacity 50kg, refractory lining with 23mass%Al_2O_3 and 72mass%MgO (hereinafter referred to as %)). In this experiment, S and Si content were varied, based on the composition of FC300 (Fe-3.2%C-0.75%Mn-0.07%P). To investigate the effect of Si, Si content was varied form 0.05% to 3.5% while S content was set at 0.16%.

To identify the substance of the patterns, air was blown on the melt's surface at 1773 K where the surface pattern was not shown (we call it patternless surface) and Ar gas was blown on the melt's surface at 1623 K where the surface pattern was shown (We call it patterned surface).

3. Results of the experiments and discussion

3.1 Effect of S and Si

Fig.2 shows the relationship between S content and molten cast iron surface pattern. Molten iron with S content of over 0.02% starts to form a pattern, which develops in the order of bamboo-leaf, pine-needle, and hexagonal type as the S content increases.

Fig.3 shows the relationship between Si content and molten cast iron surface pattern. Molten iron with over 0.5%Si forms the pattern. As Si content increases, pattern develops in the order of bamboo leaves, large hexagons, and small hexagons. Also as Si content increases, the temperature rage at which the patterns appear widens.

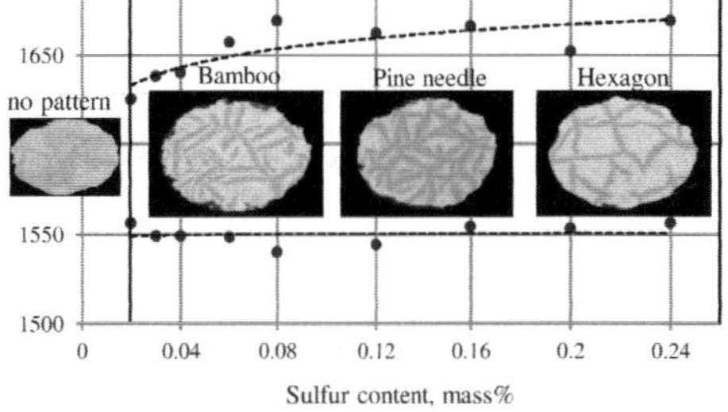

Fig. 2 Relationship between S content and surface pattern (Decreasing temperature).

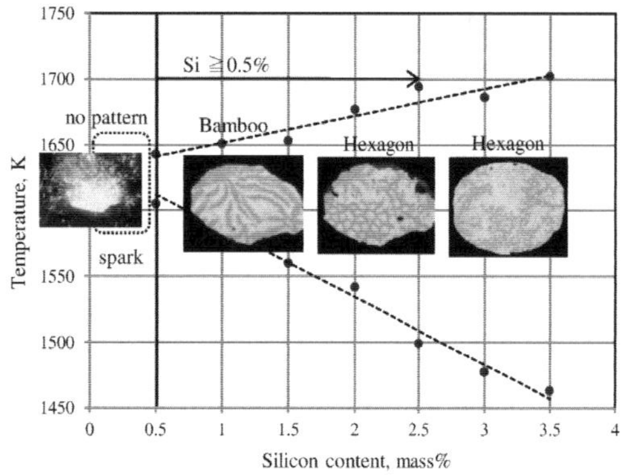

Fig. 3 Relationship between Si content and surface pattern (Decreasing temperature).

3.2 Effect of air or Ar on SiO_2 film

As shown in Fig.4, a pattern suddenly appears when air is blown on the patternless melt surface at 1773 K. On the other hand, the pattern disappears at the part where Ar gas is blown at 1623 K. This result shows that a surface pattern can be visualized, because SiO_2 film forms based on an reaction equation of $SiO_2+2C=Si+2CO$, happening in cast iron molten metal with much quanntity of C and Si. It is thought that a surface pattern appears in more than 0.5%Si, because Si is necessary for this reaction.

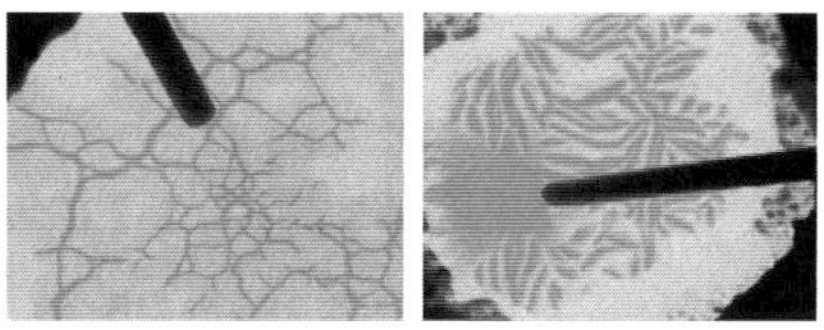

Fig. 4 a) Air flow. b) Argon flow.

3.3 Visualization due to difference in a radiation rate

As shown in Fig.5, because a radiation rate of the molten metal is different from SiO_2 film, the surface pattern is visualized. The molten metal's temperature becomes lower directly under the SiO_2 film. It is thought that the temperature difference causes the surface tension difference.

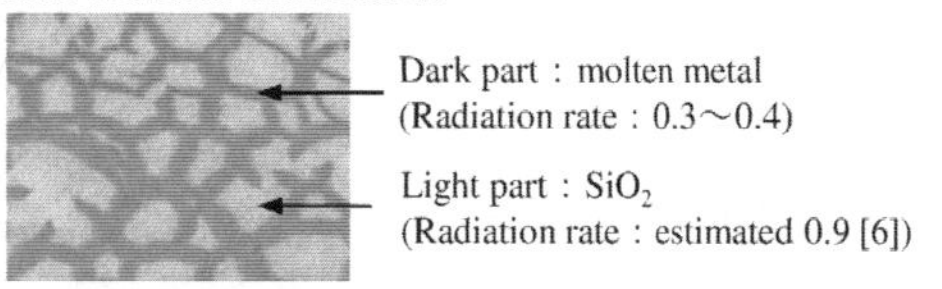

Fig. 5 Visualization due to radiation rate difference.

3.4 Effect of S on Marangoni convection

Surface tension is known to decrease when S is added in cast iron molten metal. Thus, it is thought that a surface pattern appears when S content becomes more than 0.02%. However, the complicated phenomenon such as the surface pattern does not occur only by the fallen surface tension.

Fig.6 shows the effect of surface tension of 18-8 stainless steel (Calculated results) [7]. There is a flexion point temperature in the relation between temperature and surface tension. The flexion point temperature and above mentioned radiation rate difference causes the surface tension difference It is thought that the complicated Marangoni convection appears due to the surface tension difference.

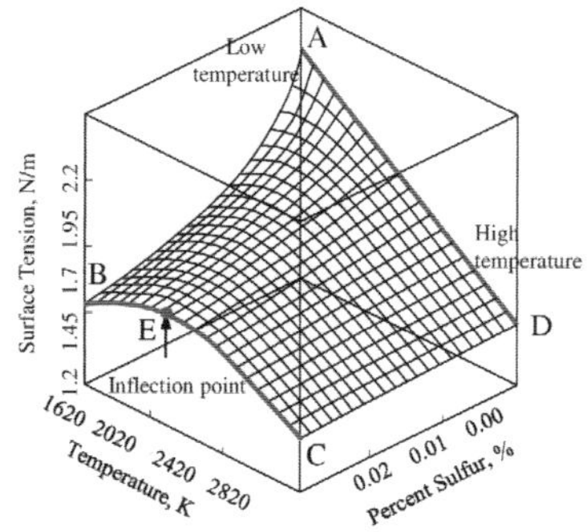

Fig. 6 Effect of concentration on the surface tension of molten 8Ni-18Cr-Fe Stainless steel (Calculation results) [7].

4. Summary

When SiO_2 film occurs based on the reaction of $SiO_2+2C=Si+2CO$, the surface pattern is visualized. S content of over 0.02% reduce surface tension . In addition to, there is a inflection point temperature in the relation between temperature and surface tension. As the result, the complicated Marangoni convection occurs and surface patterns appear. Therefore, the required conditions for pattern formation are as follows; (1) over 0.02%S, (2) over 0.5%Si, (3) appropriate amount of C, and (4) forming SiO_2 film.

References

[1] T. Kusakawa: *the Cupola Handbook* (Nihon Imono Kyokai) (Maruzen) (1957) 211

[2] T. Ishino: *the Handbook on Melting Cast Iron* (Nihon Imono Kyokai) (Maruzen) (1983) 185, 222

[3] Y. Saito: *Foundry Engineering* (Marauzen) (1965) 174

[4] O. Madono: Giessenei, 21 (1957) Nov. 5. 714

[5] E. Fuji: *Surface Patterns and Qualities of Molten Cast Iron* (Japan Foundry Forging Society) (2001) 104

[6] The Iron and Steel Institure of Japan: *the Handbook on Melting Welding Properties* (1972) 310

[7] W. Pitscheneder, T. DebRoy, K. Mundra and R. Ebner: 75Welding Journal March (1996) 71

Proceedings of the 72nd World Foundry Congress
21-25th May 2016, Nagoya, Japan

Dynamic measurements of the load on GCI castings and the contraction of castings during cooling in green sand molds

Kazuki YOSHIDA[1], Haruki KASHIMURA[1], Seigo UENO[1] , Hiroyasu MAKINO[2] and Makoto YOSHIDA[3]
[1] Department of Modern Mechanical Engineering, Graduate School of Waseda University, 3-4-1 Shinjyuku-ku Okubo, Tokyo 169-8555, Japan
[2] Research And Develop Center, SINTOKOGIO LTD., 3-1, Honohara, Toyokawa Aichi 442-8505, Japan
[3] Kagami Memorial Laboratory for Materials Science and Technology, Waseda University, 2-8-26, Nishi-Waseda, Shinjyuku-ku, Tokyo, 169-0051, Japan

For cylinder head of diesel engine and bed of machine tools which are made of sand casting, residual stress and deformation are serious problems. These defects are known to be due to the restraint by sand mold of heat shrinkage during casting. In previous studies, Motoyama et al. [1-4] conducted dynamic measurement of the aluminum alloy casting the resulting sand mold reaction force and the casting shrinkage, but studies in iron castings has not been performed so far. A continuous measuring device of sand mold reaction force and casting shrinkage of the green sand was newly developed. In this study, using the newly developed device, the authors conduct continuous measurement of the sand mold reaction force and casting shrinkage that occur in iron casting in green sand mold. The measured results provide valuable information to produce high quality casting.
Keywords: *sand mold, sand casting, gray cast iron, dynamic measurements*

1. Introduction

Residual stress and warpage have been severe defects of sand castings, such as cylinder heads or machine tool beds. The authors developed a dynamically measuring device of restraint force from green sand mold and contraction of a GCI. This device can apply squeezing pressure to green sand mold.

2. Experimental procedure

The table 1 shows the properties of the green sand. Fig.1 shows the squeezed mold. The green sand mold was controlled to the compactability ranging from 40 % - 45 % and squeezed in metal molding box at 0.8 MPa. Fig.2 shows schematic illustration of the device. An upper mold and a lower mold were formed through the same operation. When the molds were assembled, a quartz rod and the cast part were set. Then, tie bar, load cell (Maximum load 100 kN) and rigid plate were assembled.

The dimensions of the flange were 90 mm square. The casting was made from gray cast iron JIS FC 300(almost same as ASTM 45). The end of the casting opposite side of the flange was fixed by a cast part after pouring. Longitudinal contraction of the casting was obtained dynamically by measuring displacement of the quartz rod by using laser displacement gauge. Simultaneously, restraint force reacted on the flange of the casting was measured dynamically by using a load cell.

Table 1. Properties of green sand

Moisture (mass %)	Bentonite (mass %)	Active Bentonite (mass %)	Compactability (%)
3.2	10.8	7.6	38

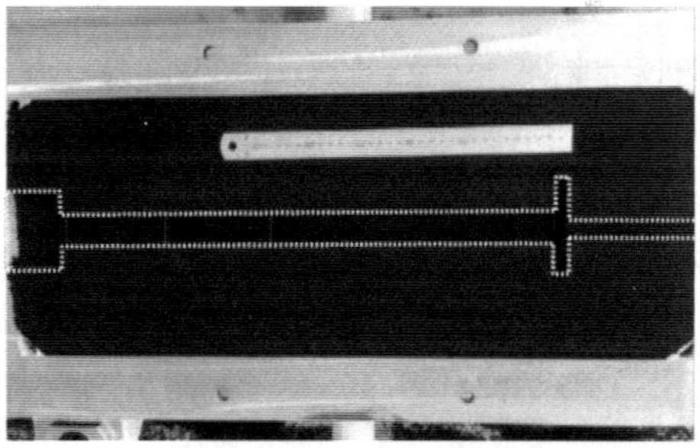

Fig. 1 The squeezed mold
(The dashed lines show the edge of the mold)

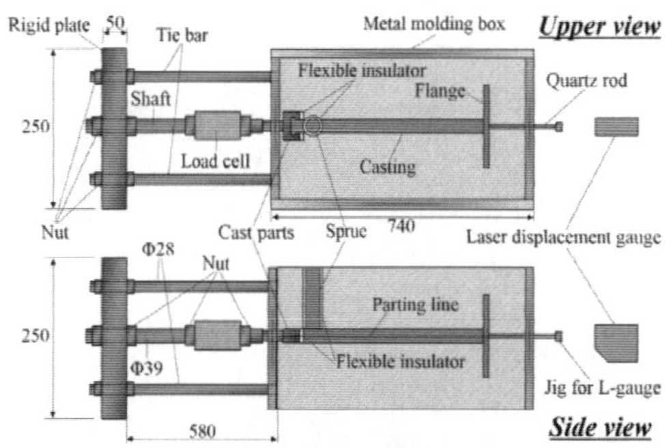

Fig. 2 Schematic illustration of the device

3. Result

Fig.3 shows the dynamic measurements of the load on the casting and the contraction of the casting. The horizontal axis shows the temperature of the center of the casting.

The load on the casting reached -0.3 kN after beginning of solidification (1210 °C). This means that the casting expands. Then, the load on the casting increased with cooling. After reached 2.5 kN at 730 °C, it decreased to 0.7 kN at 700 °C. After that, the load on the casting increased linearly with cooling again. It stayed constant under 650 °C.

The contraction of the casting reached -0.4 mm after beginning of solidification. This result indicates that casting expands in this temperature. Then, the contraction increased with cooling, but it stayed constant between 730 °C and 700 °C. After that, the contraction increased linearly with cooling again. The contraction increased while the load of casting stayed constant.

Summarizing the above, the contraction of the casting or the load on the casting, dropped or stayed constant in the following temperature range.

Range 1: 1210 °C - 1150 °C

Range 2: 730 °C - 700 °C

Range 3: 650 °C - 500 °C

Range 4: 500 °C - 200 °C

The phenomena in the four temperature ranges will be discussed in next section.

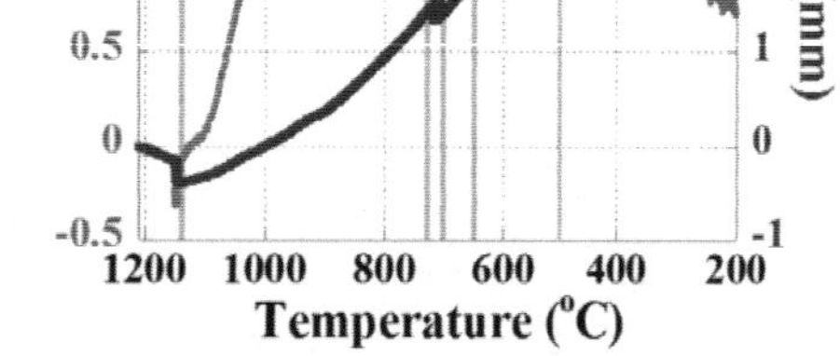

Fig. 3 The dynamic measurements of the load on the casting and the contraction of the casting

1. Discussion

Range 1;

The casting expands between 1160 °C - 1150 °C by crystallization of graphite during the solidification. Thus, these experimental results indicate that the cast iron casting expanded while it receiving the reaction force from the sand mold during the solidification.

Range 2;

This is caused by the expansion of the cast iron casting during A1 transformation. The experimental results indicate that the reaction force on the cast iron casting from the sand mold is relieved by the expansion during the A1 transformation.

Range 3;

In this range, the load on the casting stayed constant. This should be caused by the fracture of the sand mold around the flange. The load when the sand mold stayed constant was approximately 2.5 kN. Since the dimensions of the flange were 90 mm square, the sand mold would fracture in approximately 0.3 MPa compressive stress. The mold started to fracture in this stage.

Range 4;

The fracture of mold progressed in this stage. Therefore the load on casting was decreasing and the contraction of the casting was still increasing.

2. Conclusion

The authors developed the device which can measure dynamically both the restraint force from the green sand mold and the contraction of the GCI casting. By using the device, changing in the restraint force and contraction of the casting, which provide valuable information to produce high quality casting, were successfully obtained.

References

[1] Y. Motoyama, H. Takahashi, Y. Inoue, K. Shinji, M. Yoshida, 2012: Journal of Materials Processing Technology 212(6), pp.1399-1405.

[2] Y. Motoyama, H. Takahashi, Y. Inoue, K. Shinji, M. Yoshida, 2012: Journal of Materials Processing Technology 213(2), pp.238-244.

[3] Y. Inoue, Y. Motoyama, H. Takahashi, K. Shinji, M. Yoshida, 2013: Journal of Materials Processing Technology 213(7), pp.1157-1165.

[4] Y. Motoyama, Y. Inoue, G. Saito, M. Yoshida, 2013: Journal of Materials Processing Technology 213(12), pp.2270-2277.

[5] H. Polzin, W. Tilch, M. Strehle, 2013: International Foundry Research 65(2013) No.4, pp.18-24

[6] A. Baghani, P. Davami, 2014.: Metallurgical and Materials Transactions B 45, pp.1157-1169.

O-58

Numerical Value Evaluation of Graphite Morphology and Mechanical Property of Flake Graphite Cast Iron

Sadato Hiratsuka[1], Hajime Ichimura[2] and Makoto Yoshida[3]

[1] Iwate University, Japan

[2] Nissan Motor Co.,Ltd.,Japan

[3] Waseda University, Japan

The variations in graphite morphology have been classified, together with the length of the flakes, by standards that have been utilized for many years. K-FGI(KIRIU Flake Graphite Structure Index)of graphitic structure using an image analyses was proposed to estimate the graphite shape and the size of flake graphite cast iron newly here. The obtained supposition graphite area is divided by 100μm of supposition graphite length. The calculated value was defined as TTR(Thick Thin Ratio). Using K- FGI, the influence of the graphite shape on the thermal conductivity, the tensile strength and the hardness of flake graphite cast iron was investigated. The mechanical property and the thermal property were influenced by the graphite morphology and distribution of flake graphite cast iron.

Keywords: *Cast Iron, Graphite morphology,*

1. Introduction

The mechanical property, the physical properties and the electric properties of the cast iron are governed in part of by the graphite shape, size amount and distribution. Up to now, the graphite shape and the distribution are only applied to the numerical value to the graphitic structure of cast iron by graphite spheroidization ratio about simple spheroidal graphite cast iron and compacted vermicular graphite cast iron relatively. A method for evaluating graphite flake distribution and size in given in ASTM A 247. There are five graphite classifications from type A to type E graphite about the size of the graphite and the distribution by flake graphite cast iron. The large flakes are desirable in applications requiring high thermal conductivity and damping capacity.

Small flakes, because they disrupt the matrix to a lesser extent, are desired when maximum tensile properties. But graphite shape is not estimated numerically.

In this paper, estimation of the graphite shape (the length of the graphite and the thickness) by the numerical value was studied, and the mechanical property and thermal properties of the cast iron based on the flake graphite structure index were suggested.

2. Experimental procedure

2.1 Melting of cast iron

All of the iron samples were produced in a high frequency induction furnace (10kHz, 20kW). The charging materials consisted of high purity pig iron, Fe-75%Si, Fe-72%Mn and Fe-50%S. The target chemical composition is 3.1-3.6%C, 1.8-2.3%Si, 0.5%Mn, 0.02%P and 0.02%S. Each molten iron temperature was raised to 1753K and inoculated at 1723K by Fe-75.5%Si inoculant. After this, the molten iron was poured into CO_2 mold at 1673K.

2.2 Measurement of the number of graphite

A picture of graphitic microstructure of 100 times of magnification is indicated on the monitor with the resolution with 1920 X 1080 dots using CCD camera more than 1,500,000 numbers of pixels. 1000 X 1000 dots of measurement area are set on the indication monitor screen. The number of graphite of more than 5μm of average diameter in the area is measured. The number of measured graphite was defined as KFGI (Kiriu Flake Graphite structure Index).

2.3 Definition of Thick Thin Ratio

A picture of graphitic microstructure of 100 times of magnification is indicated on the monitor with the resolution with 1920 X 1080 dots using CCD camera more than 1,500,000 numbers of pixels. 1000 X 1000 dots of measurement area are set on the indication monitor screen. The length of the graphite and the area are measured about all graphite from 50 to 150μm in the area. The area of the graphite is checked about supposition graphite of 100μm length near the median length of the graphite in the area. The obtained supposition graphite area is divided by 100μm of supposition graphite length. The calculated value was defined as TTR (Thick Thin Ratio).

3. Result

Fig.1 shows a relation between graphitic micro-structure with the various shapes and the number of graphite. It was possible to indicate A type graphite in KFGI of 40-250. The E type graphite is 251-350 KFGI. D type graphite is more than 351KFGI.

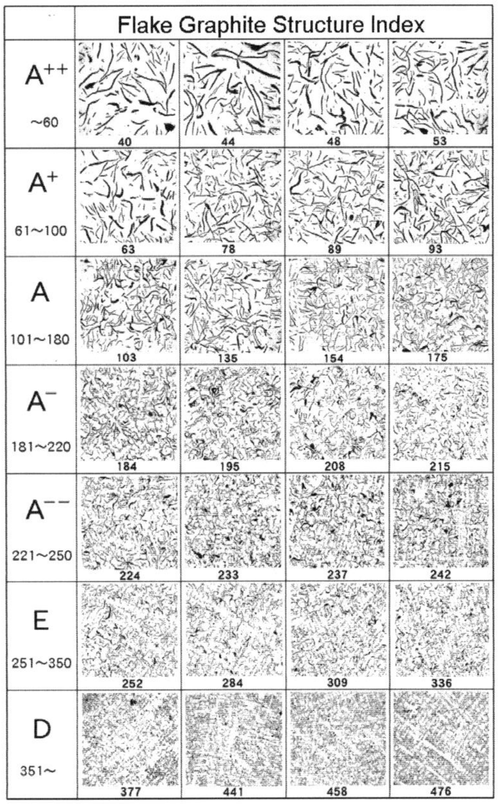

Fig.1 Relation between graphitic microstructure and KFGI.

Fig.2 shows Thick Thin Ratio(TTR) of samples. The number of graphite by each graphitic microstructure was 90KFGI, but respective Thick Thin Ratio (TTR) was different. When the thickness and thinness of the graphite are converted into rectangular latitude of 100μm of length, the size of the flake graphite is the index to show the tendency of the thickness of the graphite in spite of the size of the graphite and the relative length of the graphite. It is possible to make it the means to compare graphitic microstructure.

Fig.3 shows a relation between thermal conductivity and KFGI.

Fig.4 shows a relation between thermal conductivity and Thick Thin Ratio. As Thick thin ratio is increased, thermal conductivity is increased.

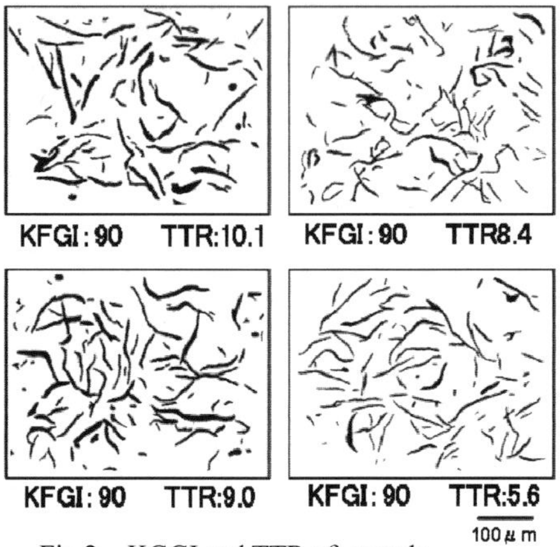

Fig.2 KGGI and TTR of sample

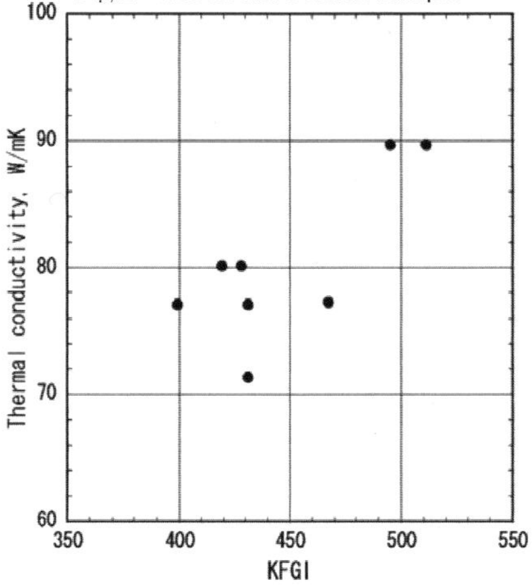

Fig.3 Relation between thermal conductivity and KFGI

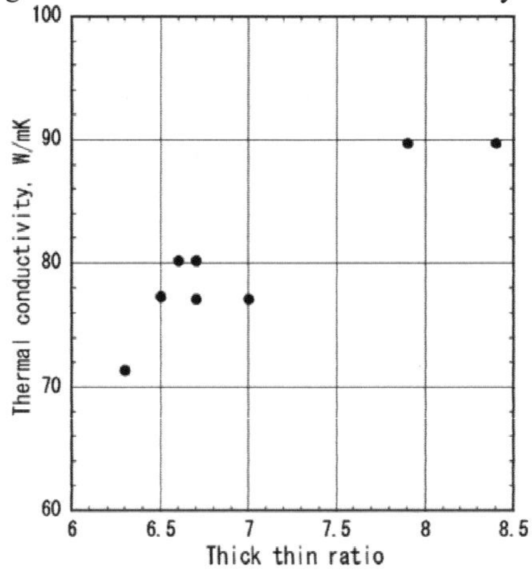

Fig.4 Relation between thermal conductivity and TTR

4. Conclusion

1) It was possible to show graphite shape at KFGI and Thick Thin Ratio.
2) A type graphite was indicated by 40-250KFGI.
3) As thick thin ratio is increased, thermal conductivity is increased.

Proceedings of the 72nd World Foundry Congress
21-25th May 2016, Nagoya, Japan

A consistent model and a new continuous reference row for the graphite structure in grey cast iron alloys

Prof. Dr.-Ing. Gotthard Wolf[1], Dr.-Ing. Claudia Dommaschk[1] Dipl.-Ing. Benjamin Aurich[1]
[1] TU Bergakademie Freiberg, Foundry Institute, Germany

Currently, we merely dispose of a visual, idealized rating standard to distinguish the single graphite structures of GJL. In a research project, we undertook to elaborate a model for a quantitative and continuous description of the graphite structure of EN-GJL. It was intended to vary the main influential factors and quantify them by measurements. The variable graphite formation was to be described by applying the definition of a structural parameter. Graphite description proceeded by determining the specific line length (LA in mm-1) between the graphite particles and their surrounding matrix. An objective determination of graphite precipitations was achieved by applying this parameter. We were also able to quantify relations with the chemical composition and the chill rate and elaborate a classification chart proposal for a continuous differentiation of graphite structures.

Keywords: *Grey Iron, Graphite structure, Cooling rate,*

1. Introduction

The development of the graphite structure in cast-iron alloys is decisive for the mechanical and physical properties. In case of GJL, the shape, dimension, quantity and arrangement of the graphite lamellae play a decisive role. The solidification mechanisms of cast iron alloys are very complex and are influenced by the chemical composition, the nucleation conditions and the cooling rate. The growth of the austenite and graphite phase of the eutectic system, in turn, depends on the undercooling, the local temperature gradient and cooling rate dT/dt [1]. The classification chart to determine the structure of graphite (DIN EN ISO 945 [2]) is currently a merely phenomenological assessment standard relating to the formation of the respective lamellar structure. It is known that a uniform formation mechanism of the graphite phase exists, however, the respective forms of graphite are not subject to a continuous transition in dependence of specific process parameters. The current classification chart according to DIN EN ISO 945 [2], used to for structural analysis, is merely a visual, idealized rating standard.

2. Graphite description

In a research project, we undertook to elaborate a model for a quantitative and continuous description of the graphite structure of EN-GJL. It was intended to vary the main influential factors in an adequate fashion and quantify them by measurements. The variable graphite formation was to be described by applying the definition of an appropriate structural parameter.

The formation of graphite structure was controlled by the chemical composition and variation of solidification conditions.

The development of various graphite morphologies was achieved by using a temperature-controlled chill. By being cast against a chill, the sample specimen was subject to a linear-directed dissipation of heat. The local solidification conditions were measured with thermocouple elements.

Graphite description proceeded by determining the specific line length (LA in mm-1) between graphite particles and their surrounding matrix. An objective determination of graphite precipitations was achieved by applying this parameter.

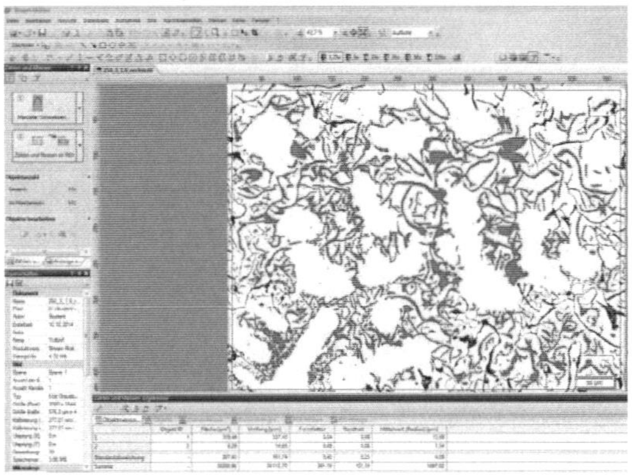

Fig.1 Example of structure analysis of grey iron

$$L_A = \frac{\sum U_G - U_R}{S_{effective}} \quad (1)$$

3. Results

Graphite fining and an interdendritic arrangement resulted in an increase of L_A, whereas a decrease of L_A signalized graphite coarsening. A regular graphite precipitation (in accordance with DIN EN ISO 945, Graphite Structure A) results in a specific line length of $L_A < 80mm^{-1}$.

We determined that a critical cooling rate of $dT/dt_{crit} = 6$ K/min. modified the graphite structure. Depending on the chemical composition, this was found to be equivalent to a solidification time ranging between 17.5 and 23 minutes. Below dT/dt_{crit}, no substantial variation of graphite formation could be observed, a coarse A graphite developed. It followed from the experiments that the formation of microstructures strongly depended on the chemical composition once dT/dt_{crit}, had been exceeded. There was a strong dependence on the cooling rate when the carbon equivalent CE increased. Strongly under-eutectic compositions displayed a distinctly lower cooling-rate dependence. In addition, it could be demonstrated that the given C/Si ratio also exerts a strong influence on the microstructure when the CE is the same. A high C/Si ratio enhances the dependence on the cooling rate.

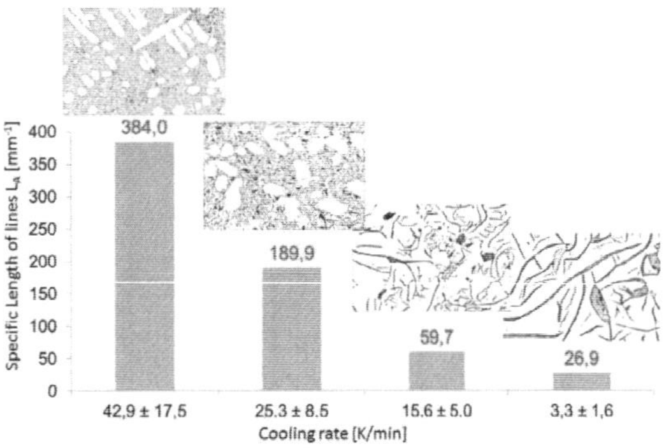

Fig.2 L_A depending on the cooling rate (nearly eutectic composition)

4. Conclusion

The results we obtained allowed for a revision of the previous, purely phenomenological classification charts that applied to GJL. Based on the data now available, we elaborated a continuous model to differentiate graphite structures. The figure 3 below shows a classification chart proposal [3].

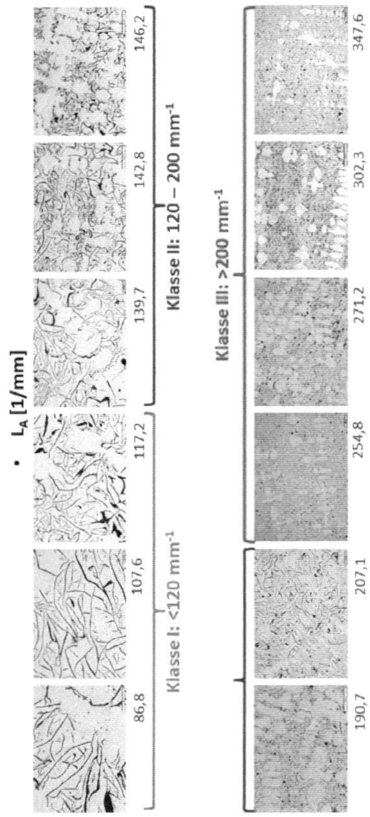

Fig.3 Classification chart proposal

References

[1] W. Kurz und D. J. Fisher: *Fundamentals of Solidification, Bd. 3*, Trans Tech Publications, 1989

[2] DIN Deutsches Institut für Normung e. V., Beuth Verlag GmbH, 2010.

[3] G. Wolf; C. Dommaschk; B. Aurich.: GIESSEREI 102, 06/2015, 102-108

Streamlining the design, development and manufacture of grey iron brake discs through Computer-Aided Design, Manufacturing and Engineering

Richard Sims BEng (Hons) CEng Prof MICME [1] and Dr. David Eggleston[2]
[1] Senior Production Engineer, Technical Dept., EURAC Poole Ltd, Mannings Heath Rd, Poole, Dorset, UK
[2] Head of Technical Quality, EURAC Poole Ltd, Mannings Heath Rd, Poole, Dorset, UK

It is now commonplace for foundries to be supplying more than just a raw casting. Machining, assembly, painting, etc. can all add profitable value when the core competence lies in the finished product. Casting buyers will often require the foundry to take responsibility for the design, development and prototyping of components, driving the process from initial concept to series production. Customers' SQA departments generally prefer pared-down supply chains where a single supplier can demonstrate competence at all stages leading to the delivery of finished parts.

In the procurement of OE brake disc rotors, design and development is often delegated to a nominated foundry as the first tier supplier.
The current method of producing a finished brake rotor from simple calculations to a fully tested and approved brake rotor takes the best part of a year and beyond. The sometimes laborious interactions between numerous independent companies or departments, demarcated by specialisation; design, engineering, foundry prototyping, testing, series foundry production, machine shop, paint shop, etc. increases the time-to-market. The "silo mentality" is a well-documented phenomenon. However, all of these stages can be completed effectively through a single, lean department when using well-trained, multi-skilled and motivated personnel. This presentation sets out to describe how the application of state-of-the-art software and technology (CAD, CAM, CAE and Rapid Prototyping) to the design, development and manufacture of brake rotors can be accomplished in a fraction of the time needed via more traditional routes.

Keywords: *Brake disc, simulation, agile, rapid prototyping, DISAmatic.*

1. Introduction

This paper summarises and highlights some of the steps involved in designing, developing and producing disc brake rotors for the OE marketplace, contrasting, where appropriate, those methods exercised at EURAC to reduce time-to-market based on a standardised *New Product Introduction* (NPI) process focusing on simulation, rapid prototyping and the skills profiles of the engineers involved.

2. Functional Test Correlation *(Benchmarking)*

Considerable effort has been expended at EURAC to enable the accurate prediction of braking performance using fully-coupled, thermo-mechanical and elastoplastic mathematical models. Such simulations have proven capable of predicting maximum bulk rotor temperatures to within 3 % when compared with actual results from several Grossglockner alpine descents. Given the repeatability of the thermal model, there is no need for complex, frictional models incorporating the contact regime between rotor and stator.

The brake rotor is a safety-critical automotive braking component which cannot be permitted to fail on the vehicle such that computer simulation does not negate the need for legislated, industry standard vehicle testing, but it does predict the test's successful outcome. This process reduces the time-to-market and prototyping costs, as only one design iteration is needed to produce the prototype component.

Figure 1. VW Touran wheels instrumented for road testing.

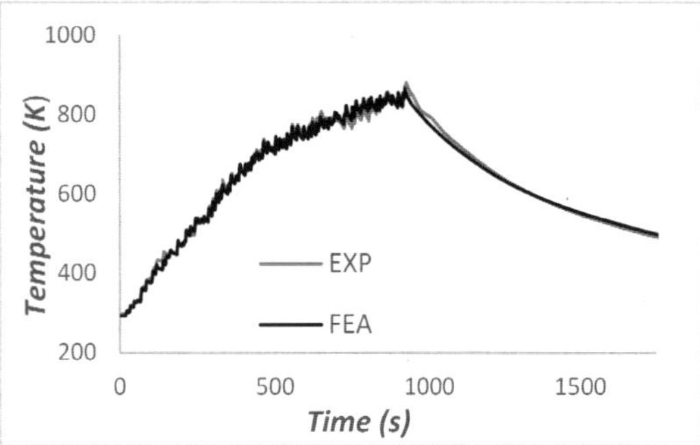

Figure 2. Correlation between experimental vehicle testing and FEA data VW Touran [1]

3. Rapid Prototyping of Castings

The *Rapid Prototyping* process can be used to directly print moulds and cores. These processes build the cores and moulds layer-by-layer by laser-fusing either polymer-bonded sand or using a wide-area inkjet to bond the sand [2].

Pre-production samples will be cast from these printed cores and moulds and will be designed around the intended patterns and runner system. These rapid-prototyped castings are produced on the same *DISAmatic* production line such that the resulting castings follow the existing foundry process and the desired in-specification material grade for production parts.

Once pre-production parts have proven successful, the foundry tooling is ordered and manufactured and will rarely require subsequent modification. The rapid prototyping of castings is the main factor in reducing lead times where there is no waiting for foundry tooling to be manufactured before samples can be cast, which would typically take in the order of 12 weeks.

Figure 3. Photograph of 3D printed core with as-cast holes in brake faces.

4. *Agile* multi-skilled Engineering Department

Many different engineering functions are involved in the various stages of the design and manufacture of an OE brake rotor. Mechanical Design Engineers, Foundry Method Engineers, Machine Shop Production Engineers and other skill bases are called upon to add value along the NPI process. This is in contrast to the engineering department at EURAC, where, although specialisms still exist, each individual is provided with the expertise and competence to lead projects effectively through each of the required engineering functions, across multiple disciplines.

This cross-disciplinary competence has been achieved by simplifying processes, designing standard work procedures, training people in these and ensuring these approved methods are being adhered to by robust audit procedures. Today, one highly-trained engineer can design a brake rotor to produce machining CNC code for the finished part and complete all the engineering tasks for each step in-between.

The benefits to the employee are vast such that they can take ownership of the whole NPI process with less fatigue and a more varied work content [3]. However, this style of working is also beneficial to the business because projects do not have to be planned to specific individuals and the customer benefits from much shorter lead times. In many ways, this agile, holistic approach brings lean manufacturing full circle.

5. Results, Conclusions and Recommendations

As a result of time saving initiatives, a 42 % improvement has been realised. A further direct benefit is project cost savings of up to 23 %. Future considerations should be given to product lifecycle management software, where all development work could be administered through a single software application.

Acknowledgements

Stephen Borthwick, www.eurac-group.com & www.matholdingsinc.com

References

[1] S Wren, R Sims & F Suarez, EURAC Poole Ltd, UK: *FEA Simulations for Comparison to Thermal Data Obtained from Vehicle Testing* (Unpublished Company Report 2015)

[2] K Woods & S Ravi, Univ. of Northern Iowa, Cedar Falls, Iowa: *Design Considerations for 3-D-Printed Cores and Molds* (Modern Casting, American Foundry Society May 2015) pp. 24-29.

[3] K Zheltoukhova: *HR: Getting smart about agile working* (CIPD 2014) pp.25-28

Proceedings of the 72nd World Foundry Congress
21-25th May 2016, Nagoya, Japan

Is it all Chunky Graphite that you see or could it be something else?

Cathrine Hartung[1], Lena Magnusson Åberg[1], Doug White[2], and Jacques Lacaze[3]
[1] Elkem AS, Kristiansand, Norway
[2] Elkem Materials Inc. Pittsburgh, USA
[3] CIRIMAT, Toulouse, France

Chunky graphite has been a phenomena encountered and studied since the invention of ductile iron in 1948. Still chunky is confused or mixed with other graphite forms such as compacted graphite and degenerated graphite. The paper will show why chunky can be confused with compacted graphite and why degenerated graphite can easily be called chunky graphite and why it can be difficult to find a good cure.

Keywords: *Chunky graphite, Compacted Graphite, Degenerated graphite, Cooling rate, Silicon level.*

1. Introduction

Graphite is one of the essential components in cast iron. Controlling the formation and growth of graphite is important to acquire the desired properties in the final casting. What makes it complicated is that the graphite can come in many different shapes, and trying to fix one can create unwanted graphite forms. Some of the shapes are very difficult to keep apart and they can appear in the same sample and even side by side.

2. Compacted to Chunky

2.1 Compacted Graphite

Compacted graphite has similar appearance to flake graphite only with thicker and more curved randomly oriented flakes. Closer examination reveals the branch ending in a half sphere. Compacted graphite can be a wanted graphite shape, as it gives strength and thermal properties intermediate to flake and spheroidal graphite [1]. Most common causes for unwanted compacted graphite are insufficient residual Mg level in relation to base iron S-level; too high S level in relation to residual Mg or contamination with Ti, Zr or Al. [2, 3]

2.2 Chunky Graphite

Chunky graphite is often found in the areas last to freeze and looks like small chips or fragmented compacted graphite finely and randomly distributed, either over large areas or confined to cells. [4, 5]

Reviewing literature, the most common causes listed for chunky are too high level of Cerium, high Si-level or slow cooling. [4, 6]

2.3 Is it possible to mix them up?

Yes, as compacted and chunky share similarities in growth [4, 6], the type you get depends on cooling rate, composition and difference in nucleation potential.

In figure 1, transition from spheroidal to compacted to chunky graphite is observed in samples from varying section size. This illustrate the effect cooling rate has on graphite formation and growth. With decreasing cooling rate, the risk for chunky graphite increases [5, 6]. The example shows how the structure can transform from spheroidal to compacted to chunky in the same casting with decreasing cooling rate. As a result, it can make it difficult both to classify the graphite correctly and to select best strategy to overcome the problem since the problem is different in different casting sections.

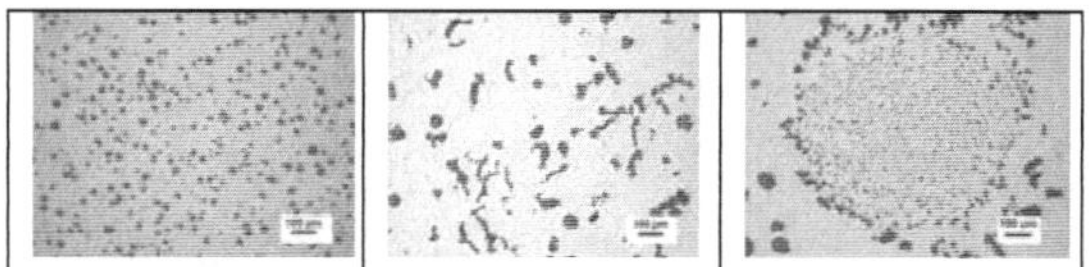

Spheroidal	Compacted	Chunky
Decreasing cooling rate →		

Fig. 1 Transition in structure caused by decreasing cooling rate.

By changing the composition and especially the Si-level it is possible to transform graphite in the same iron from compacted to chunky. In figure 2 it can be seen how a structure with compacted graphite can be improved to more spheroidal by increasing the Si-level.

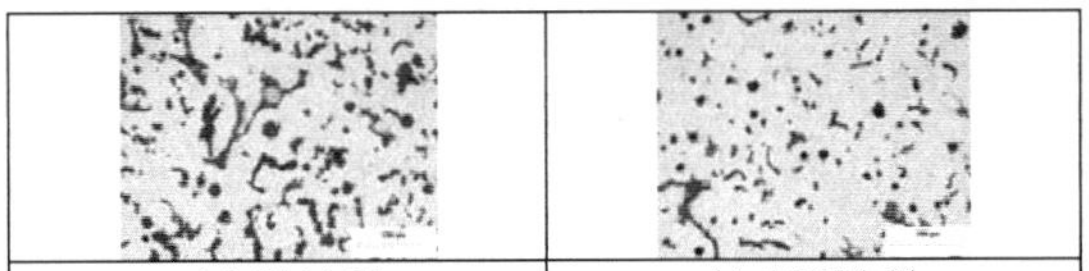

a) 3.81% Si	b) 4.28% Si

Fig. 2 Transition in structure caused by increase in Si-level.

Inoculation is listed as something that can both help prevent chunky [4] and can cause chunky [7].

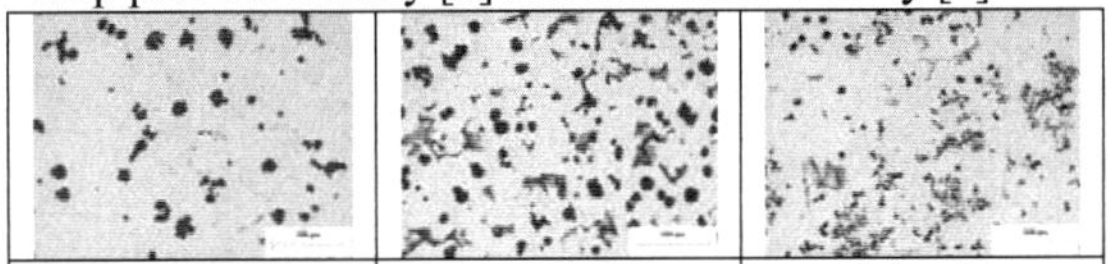

No inoculation	1.44 wt% inoculation	2.0 wt% inoculation

Fig. 3 Transition in structure caused inoculation and increase in Si-level.

Figure 3 illustrate how inoculation can help improve a structure by increasing the nodule count, but also how inoculation can cause chunky graphite if not controlled. Inoculation introduces Si, and increasing the Si-level increases the risk for chunky graphite, but at the same time it is recommended to increase nodule count to avoid chunky graphite.

3. Degenerated Graphite

Unwanted graphite forms are often referred to as degenerated graphite forms [8], but behind this term, there can be many graphite forms with very different causes and cures. Hoover [9] actually use the term chunky or coupled vermicular, while Murthy et al. [10] use compacted graphite in connection with both chunky and exploded graphite. The imprecise term degenerated graphite along with mixing of chunky, compacted and exploded graphite can generate confusion.

With the recent boom in wind energy it has become more common to make large ductile iron castings. In these large and thick castings, chunky or degenerated graphite is more encountered and unwanted.

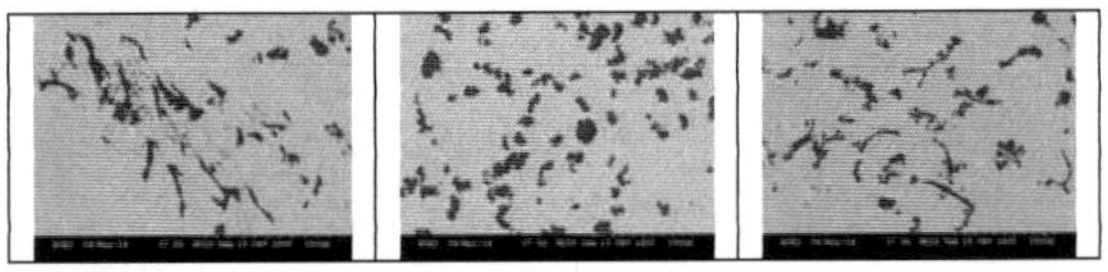

Fig. 4 Variation in graphite shape observed within a small area in one sample.

Figure 4 shows how the graphite morphology varies within the same sample, from a heavy casting, over an area of 4*4 mm. This is an example of the difficulty in classifying the different graphite morphologies that can arise and how conditions for graphite growth and nucleation can vary within a very small area.

4. Conclusion

Is it all chunky graphite that you see or could it be something else was the question.

The answer is yes and no.

It is difficult to correctly identify the various graphite shapes as they in many cases share both structural similarities, growth similarities and co-exists within the same casting. This problem is more pronounced in larger castings with variation in section size as effect of cooling rate, segregation and fading of Mg and inoculation overlaps.

Combined with the imprecise term degenerated graphite and the mixing of graphite morphologies chunky can easily be confused with compacted graphite and graphite nucleated on small inclusions can be classified as chunky graphite.

It is therefore important to analyze the conditions to take the correct actions and avoid introducing another defect.

References

[1] S. Dawson, T. Schroeder "Compacted Graphite Iron: A Viable Alternative" SinterCast Publication (2000).

[2] S. Hasse: *Guss und Gefügefehler* (Schiel & Schön, Berlin, 1999) pp. 93-108

[3] J. Zhou: *Colour Metallography of Cast Iron* (*China* Foundry, 2011), Vol. 8, issue 1, pp 154-155

[4] K. Hamberg, et. Al: *Chunkygrafit i segjern – en litteratur- genomgång* (Gjuteriforeningen 2003) pp. 36-37

[5] M. Gagne, 101 Casting Congress AFS, paper 97-034

[6] J. Lacaze, L. Magnusson Åberg, J. Sertucha, Keith Millis Symposium 2013, 23 pp. 1-11

[7] I. Asenjo, et.al , International Journal of Cast Metals Research 2007, Vol. 20, No 6, pp. 319-324

[8] A. Javaid, C. R. Loper Jr , AFS Transaction 95-49, pp. 135-150

[9] H. W Hoover Jr., AFS Transactions 86-102, pp. 601-608

[10] V. S. R Murty et.al, Journal of Metals, 1986, Vol.38, No 12 pp. 24-28

Proceedings of the 72nd World Foundry Congress
21-25th May 2016, Nagoya, Japan

On the structure of ausferrite in ductile iron

Dawid Myszka
Warsaw University of Technology, Faculty of Production Engineering,
ul.Narbutta 85, 02-524 Warszawa, Poland
mail: myszkadawid@wp.pl

Matrix of the heat treated ductile iron can be distinguished into several different types of ausferrite - mixture of austenite and bainitic ferrite. It may contain fine or feathery ferrite plates, mechanically unstable austenite grains, martensite, nanoausferrite, etc. Each of these structures will affect the properties of ductile iron. The article described these structures and their impact on selected properties.

Keywords: *Ductile Iron, Ausferrite, Microstructure, Nanostructure, Austempering.*

1. Ausferrite matrix of ductile iron

Detailed examinations of the microstructure morphology of an ausferritic matrix in ductile iron show several characteristic features, which can be discussed as dependent on the time of isothermal transformation. With short time of this transformation (A point on Fig. 1), the ductile iron tends to form an almost fully martensitic matrix, containing only a very small amount of the precipitates of lamellar ferrite. This is due to complete transformation at ambient temperature of the thermally unstable austenite.

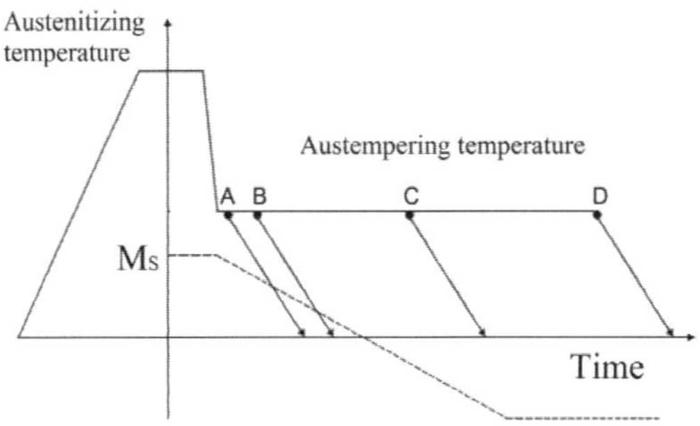

Fig. 1 Schematic representation of time-related changes in temperature during heat treatment shaping the ductile iron matrix (changes in the M_S temperature during isothermal transformation).

Slightly longer time of the transformation increases ferrite content in the form of packets of plates separated by carbon-rich austenite. The continuation of transformation leads to austenite stabilization. This makes the matrix of ductile iron fully ausferritic with austenite present in two forms, i.e. in the form of layers separating the plates of ferrite and in the form of blocks located between the ferritic-austenitic packets. Very long time of the isothermal transformation usually leads to the formation of typically bainitic microstructure or, in term of very low temperature of austempering, could leads to the nanoausferritic structure (D point on Fig. 2) [1,2].

Fig. 2 Nanoausferritic matrix of ductile iron.
phot. Emilia Skołek, PhD, DSc.

Studies show that with the progressing time of isothermal transformation, austenite acquires its thermal stability and the M_S temperature is decreasing [3]. The schema neglects the time and temperature of the isothermal transformation, but the results of experiments clearly indicate that changes in the M_S temperature will follow a general course similar for all types of the ausferritic ductile iron.

2. Ausferrite transformation during subzero treatment

In classic austempered ductile iron [4], the time of isothermal transformation usually ends in the region between points B-C (Fig. 1). From Figure 1 it can be concluded that in this range the M_S temperature will drop below room temperature. This indicates the condition in which austenite can be stable at ambient temperature, because its M_S temperature will be slightly lower than room temperature. However, even a small change in the temperature, e.g. during subzero treatment, can trigger the martensitic transformation. This theory has been verified experimentally [5] by treatment of ausferritic ductile iron samples at the temperature of liquid nitrogen (-196°C). Microscopic examinations of areas containing austenite showed the presence of a phase with quasi-martensitic morphology appearing in those places after the process of freezing. In the same microstructure, areas of stable austenite located between the ferrite plates were traced, which means that these are the areas heavily oversaturated with carbon migrating from the regions of growing ferrite plates.

3. Mechanical instability of austenite

Analysis of the freezing process indicates that austenite transformation into martensite at temperatures below M_S occurs only in some specific areas of the microstructure, i.e. in the blocky austenite. These places are also privileged for the deformation-induced transformation [6-8]. As a result of such transformation, hard, carbon-rich, martensite is formed. This effect can be easily observed during microscopic examination of samples after hardness measurements, when the indenter permanently deforms the tested material, leaving a ring of plastic deformation around the indentation. Martensite formed under a pressure of the penetrating indenter of the hardness tester also creates problems in finding a clear relationship between hardness and austenite content in ausferrite [9]. The same problem appears in numerous other tests, e.g. in the static tensile test, when martensite is traced on fractures of the specimens after failure [7]. Martensite of this type is also present in samples subjected to machining or abrasion wear. Therefore, using TEM images of the microstructure and diffraction patterns, it is difficult to tell if the examined martensite is a result of the transformation caused by technological process or transformation due to wrong sample preparation. The phenomenon of martensitic transformation is related with proper level of stress or strain which, arising in the examined material, can trigger TRIP effect [6,8].

4. Effect of high temperature on ausferrite transformation

Ausferritic ductile iron is known to suffer a drastic drop of mechanical properties when subjected to the impact of high temperature. This is extremely important for the technological process and possible high temperature applications. The cause of such changes are transformations occurring in the ausferrite. Deeper analysis of this problem indicates that high temperature provides energy necessary to activate and complete the bainitic transformation. At a temperature above 400°C, different for the different types of ausferrite, intense exothermic reaction begins. It is related with the formation of a mixture of ferrite and carbides originating from the austenite.

The decomposition of ausferrite to form a bainitic structure under the effect of high temperature carries a specific activation energy associated with this transformation. From calculations made in [10] it follows that to activate the transformation in ductile iron treated in a high range of the isothermal transformation temperatures (370°C and 400°C) it is necessary to supply the energy of 200kJ/mole. For a lower range of the transformation temperatures, i.e. 300°C and 320°C, this energy increases to about 280kJ/mole. From the studies it also follows that with the higher heating rate more of the activation energy is needed to trigger the bainitic transformation and the temperature at which this transformation starts must be higher, too.

Acknowledgements

This work was partially supported by the National Centre for Research and Development NCBiR project no PBS3/B5/45/2015.

References

[1] D.Myszka, K.Wasiluk, E.Skołek, W.Świątnicki, Materials Science and Technology 7(2015) 31 829-834.
[2] D.Myszka, E.Skołek, A.Wieczorek, Archives of Metallurgy and Materials 59 (2014) 1227-1231.
[3] K.B. Rundman, J.R. Parolini, D.J. Moore, AFS Transactions 05-145 (2005) 1-15.
[4] www.ductile.org
[5] D.Myszka, T.Borowski, T.Babul, 5th WSEAS Int. Conf. on Heat Transfer, Greece 2007.
[6] Srinivasmurthy Daber, P.Prasad Rao, Journal of Materials Science 43 (2008) 357-367.
[7] J.L.Garin, R.L.Mannheim, Journal of Materials Processing Technology 143–144 (2003) 347–351.
[8] M.Naili Ahmadabadi, Metallugical and Materials Transactions A 28A (1997) 2159-2162.
[9] D.Myszka, Archives of Metallurgy and Materials 52 (2007) 475-480.
[10] R.E.Campos-Cambranis, at all., Scripta Materialia 38/8 (1998) 1281-1287.

Proceedings of the 72nd World Foundry Congress
21-25th May 2016, Nagoya, Japan

Austempered Ductile Iron (ADI)
Alternative material for high performance applications

G. Artola[1,4], I. Gallastegi[1], J. Izaga[1], M. Barreña[2] and A. Rimmer[3]
[1] IK4-Azterlan, Aliendalde Auzunea 6, 48200 Durango (Bizkaia), Spain
[2] Adilan Group, Pol. Ind. Santa Apolonia 2-2, 48215 Iurreta (Bizkaia), Spain
[3] ADI Treatments Ltd., Doranda Way, West Bromwich B71, UK
[4] Tecnun (University of Navarra), Paseo Manuel Lardizabal 13, 20018 Donostia (Gipuzkoa), Spain

ADI grades are standardized and the requirements of current international standards (EN 1564-12 / ASTM A897-15) are given in terms of conventional mechanical properties.
Nevertheless, these properties don't show the real potential of the ADI grades. In order to promote the use of ADI parts in place of other materials, this work proposes a comparison between ADI 900 and ADI 1200 grades, both in terms of conventional and advanced mechanical properties, employing stress intensity factors and critical CTODs (Crack Tip Opening Displacement).
This study is completed with mechanical fatigue testing, so that it can be shown that the service life of ADI parts is comparable to that given by other heavier and more expensive options.

Keywords: *Austempered Ductile Iron, ADI, fracture mechanics, fatigue, alternative materials.*

1. Introduction

Conventional mechanical testing is adequate for structural integrity calculations with ductile materials under plane stress conditions. When designs involve high strength materials, plane strains are more likely to occur and both fracture mechanics and fatigue become more relevant. This is the case for ADI and in this paper the characterization of two representative grades has been investigated.

2. Experimental method

All the materials in the tests planned have been processed in industrial facilities and under production conditions. It has been checked if the "as cast" materials are capable of reaching GJS-900-8 and GJS-1200-3 grades and, after austempering, a broad mechanical characterization has been performed.

All tests have followed European EN standards, except for fracture testing (BS) and uniaxial fatigue (ASTM).

3. Results and discussion

3.1 "As cast" condition

The study materials have been obtained from 22 one-inch thickness type Y2 keel-blocks from the same heat. The "as cast" condition fulfils the requirements for a proper austempering treatment.

Table 1 Chemical composition (%)

C	Si	Mn	Mg	Ni	Mo	Cu
3.69	2.28	0.19	0.037	2.45	0.20	0.79

Table 2. Mechanical properties "as cast".

Hardness (HB10/3000W)	273					
Tensile testing	$R_{p0.2}$ (MPa)		R_m (MPa)		A (%)	
	625		852		4.7	
Impact testing	KV (J)			Unnotched (J)		
	5	5	5	39	34	37

Table 3 Microstructure.

Nodularity (%)	Graphite density (nod./mm^2)	Ferrite/pearlite ratio
>90	200	20/80

3.2 Austempering

The samples are machined to obtain two sets of specimen pre-shapes, with a minimum machining clearance of 2 mm. Austempering is performed under controlled atmosphere with a carbon potential of 0.8 during austenitization and inert gas protection during the transfer to the isothermal bath. After heat treatment, pre-shapes are machined to standard testing specimens and it is verified that one set of specimens has been obtained for each grade GJS-900-8 and GJS-1200-3.

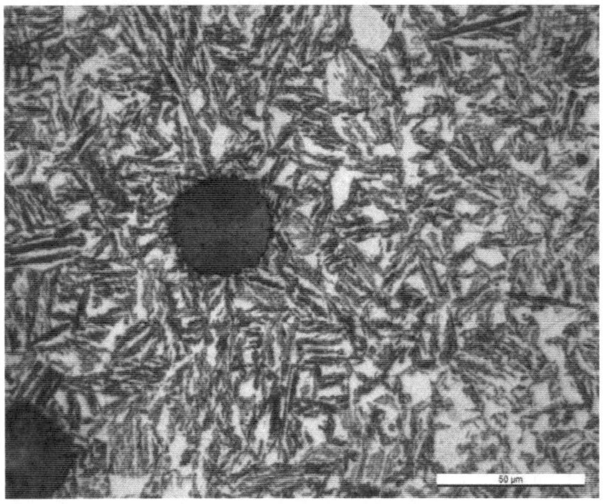

Fig. 1 GJS-900-8 microstructure.

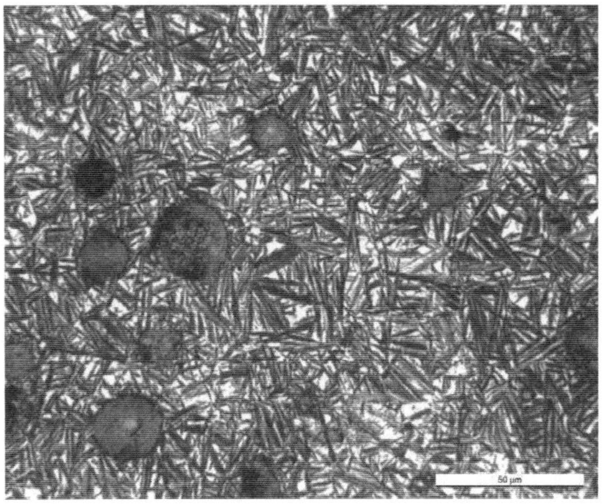

Fig. 2 GJS-1200-3 microstructure.

Table 5 GJS-900-8 mechanical properties.

Hardness (HB10/3000W)	314					
Tensile testing	$R_{p0.2}$ (MPa)		R_m (MPa)		A (%)	
	622		963		10.4	
Impact testing	KV (J)			Unnotched (J)		
	8	9	9	105	99	106

Table 6 GJS-1200-3 mechanical properties.

Hardness (HB10/3000W)	397					
Tensile testing	$R_{p0.2}$ (MPa)		R_m (MPa)		A (%)	
	1035		1260		9.8	
Impact testing	KV (J)			Unnotched (J)		
	7	7	8	104	102	94

BS7448-1 B(E) specimens with a 12.5x25 mm section have been employed for toughness testing. All fractures have shown an elastic-plastic behavior, despite the plasticity being small.

Table 7 Toughness testing.

Grade	K_Q (MPa·m$^{1/2}$)	δ_c (mm)
GJS-900-8	50.3	0.05
GJS-1200-3	63.7	0.05

Fatigue curves have been built for an average stress σ_m equal to half $R_{p0.2}$ of the grade tested. Loading amplitude $\Delta\sigma$ ranged from 50% to 100%·$R_{p0.2}$.

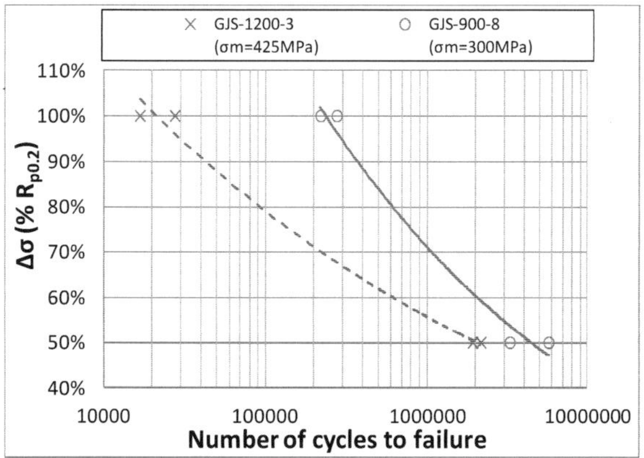

Fig. 3 S-N fatigue curves.

4. Conclusions

GJS-900-8 and GJS-1200-3 grades develop plastic crack tip blunting for the studied thickness.

The obtained fatigue curves point to service lives that are higher than the reference values given in the European Standard.

The range of applications where ADI grades become an alternative material to current solutions is extended when design calculations take into account fracture and fatigue criteria.

Acknowledgements

The authors most sincerely thank to Furesa S.Coop., member of ADILAN Group, and to ADI Treatments Ltd. for their invaluable support in this work.

References

[1] A.Iglesias, I.Gallastegi, G.Artola, M.Muro et al.: *71st World Foundry Congress* (2014).

[2] S.K. Putatunda: *Materials Science and Engineering A315* (2001) 70-80.

[3] M.G. Hafiz: *AFS Transactions* (2009).

[4] M.Kuna, M.Springman, M.Mädler, P.Hübner and G.Pusch: *Engineering Fracture Mechanics 72* (2005) 241-253.

Proceedings of the 72nd World Foundry Congress
21-25th May 2016, Nagoya, Japan

Scanning electron microscope analysis data for casting defects in iron foundry

Kazuya Edane[1],Aki Kanno[1],Soichi Nagai[1],Hiroki Ameku1,Yutaka Kurokawa [1]
[1] Tsuchiyoshi Industry Co., Ltd, Technical Center, 1051-7 WADA ONAN-CHO OCHI-GUN SHIMANE 696-0403, Japan

Abstract
Visual and microscopic observations of the defect sites are conducted after analyzing the macroscopic information to identify the cause of cast iron defects. In addition, analysis by scanning electron microscopy (SEM) with energy dispersive X-ray analysis (EDS) has been extensively used, because SEM-EDS identification of the cause of casting iron defects. However, SEM-EDS only provides two-dimensional data in a micro area and the analysis data is affected by diffusion of the electron beam along depth of the defect sample, which may lead to a difficulty in the identification of the cause of defect formation. In this study, causes of defect formation in iron casting were investigated from several perspectives, and the samples were analyzed using metallographic observation, X-ray computed tomography (X-ray CT) observation, wavelength dispersive X-ray spectrometer (WDS), and other types of analytical equipment. Metallographic observation was used to determine the causes of the defect formation -for the graphite precipitation film in a pinhole defect, the thickness of an oxidized membrane defect, and the metallographic structure around a leak defect. Additionally, three-dimensional analysis using X-ray CT and WDS made it feasible to determine the cause of defect formation by X-ray CT and WDS by raising the assay precision of EDS.

Keywords: *Casting Iron defect, scanning electron microscopy, energy dispersive X-ray analysis*

1. Introduction

A cast iron product is required to pass checks such as visual examination, non-destructive test, microstructure observation, and leakage test. Because there are several kinds of defects and their corresponding causes, we require severak checks, wichi complicates the identification of the cause the defects, because the causes of cast defects casting plan, molding, and melting or those reciprocal actions.The effect of cast defects on factory management is substantial, therefore it is necessary to identify the cause that resulted in a cast defect. There are physical factors and chemical factors responsible for cast defect. Cast defect caused by physical factors are often revealed by the visual inspection of the outward appearance, but cast defects caused by chemical factors are minute, and therefore are difficult to identify based on the outward appearance. It is therefore effective to analyze cast defects caused by chemical factors by using SEM/EDS and taking measures. This study considered the possibility of investigating cast defects by adding microstructure observation, metallic gas analysis, and WDS to SEM-EDS.

2. Experimental Procedure

The cast defect that occurred at a cast iron manufacturing factory was used as the experimental sample. The sample was cut such that no heat was produced by friction, and the sample was analyzed using SEM-EDS. Additionally as the need arises, Microstructure observation, metallic gas analysis and WDS were conducted.

3. Results and discussion

The cast defect that occurs is different in the the material such as flake graphite cast iron and spheroidal graphite cast iron. Restrictions on the number of pages, only the results that considered dross inclusion and porosity are indicated.

Classification of dross inclusion and the results of the analysis are indicated in Fig.1. Dross is classified into oxide dross, graphite dross, and one in which the oxide and graphite dross are interminingled. Line dross and Massive dross are indicated as classification types of dross in Fig.1(A).

The shape of Line dross resembles a thin crack, as observed using SEM and metallography, and Mg, Al, and Si detect clacks on the basis of EDS, Line Dross destroyed eutectie call, as observed using metallography. On the basis of this data, it was judged that Line dross had flowed. Next, we analyzed cast defects using X-ray CT and obtained a three-dimensional image. Subsequently, the image was processed in to an image of only defect.This obtained image illustratede that Line dross resemble sheet. Line dross occurred by the oxidization of dross on the melt surface.When sheet dross is cut, it is seen as line. Line dross is flowed sheet dross made on the melt surface, therefore measures to control surplus dross ingredient and prevent melt from oxidation are imperative.

Fig1(A-2) indicates Massive dross inclusion. The shape of the dross is massive and Ba detect dross on the basis of

EDS.The dross was influenced inoculamt. Graphite was observed in flakes around dross, and around the graphite flakes were surrounded by eutectic cells of Spheroidal graphite. Consider a case where dross and iron oxide flowed, and subsequently solidify. An effective measure would be to prevent oxide formation than controlling the surplus dross ingredient.

Fig1(B) indicates graphite dross. In metallography, flowed graphite exists and poor nodularity dose no't occur under castin surface.Consider that graphite equivalent was high or temperture was low when casting.

Fig1(C) indicates the intermingled type of oxide dross and graphite dross.Dross was detected on casting sarface by SEM-EDS.Flake graphite and flowed graphite exists under the casting surface. Because, this defect is intermingled with

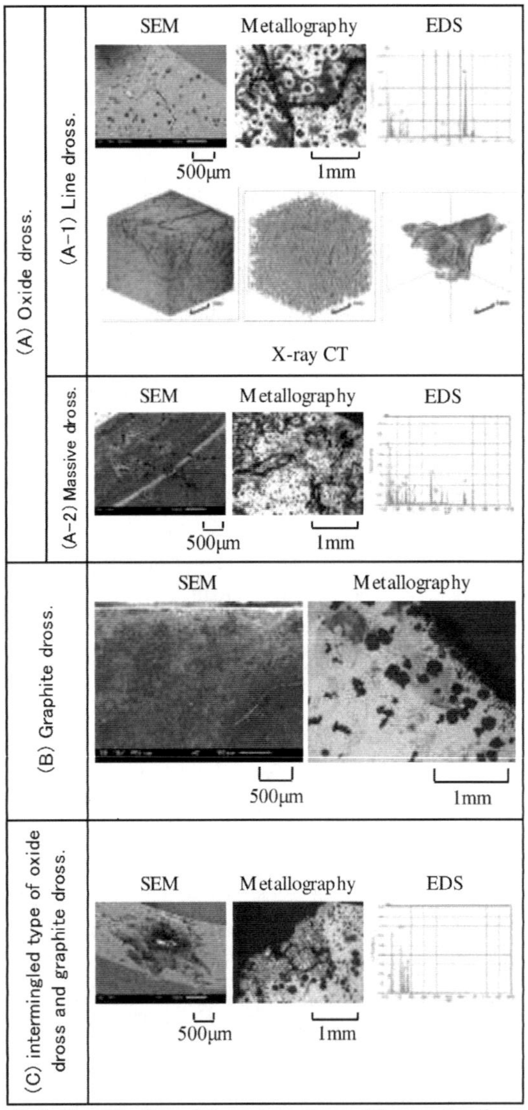

Fig1. Classification of dross inclusion and the results of the analysis

oxide dross and graphite dross, we need to take suitable measures to prevent oxide dross and graphite dross.

Fig2 indicates porosity defects.Fig2(A) is illustrates micro shrinkage, while Fig2(B) indivate micro porosity.

Fig2(A), the defect occurred inside the sample, around core and thickly. Dendrite and graphite were present and Mg, Ca, and O were detected to have penetrated into the defect. This porosity was surrounded by eutectic cell, wichi are a feature of porocity.Three-dimensional image indicated that porocity occurs between eutecite cells.

The defect shown in Fig2(B) occurred under the riser. The sample was analyzed to distinguish defects and non defects using metallic gas test. Large amounts of oxide were detected, and Nitrogen and Hydrogen were also present in incremented quantities in the part of defect that was analyzed.

4. Conclusion

The study showed that causes of cast defects can be identified effectively by adding data obtained from microstructure obsernation, metallic gas analysis and WDS to SEM-EDS.

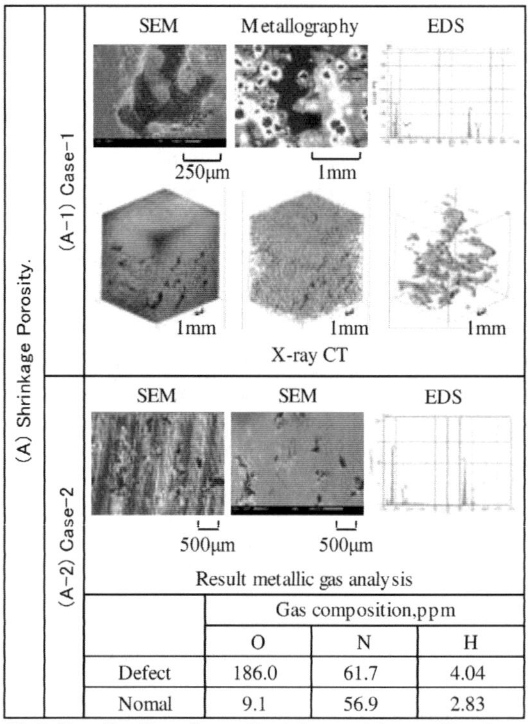

Result metallic gas analysis

	Gas composition,ppm		
	O	N	H
Defect	186.0	61.7	4.04
Nomal	9.1	56.9	2.83

Fig2. Classification of porosity defect and the results of the analysis

Proceedings of the 72nd World Foundry Congress
21-25th May 2016, Nagoya, Japan

Improved Numerical Methods for Accuracy of Filling and Shrinkage Prediction in Cast Iron and Steel Castings

Sung-Bin Kim, Young-Hoon Yim, Joong-Mook Yoon, Se-Ho Ahn and Jung-Soo Baek
AnyCasting Co., Ltd.
583 Yangchun-ro, Gangeso-gu, Seoul, South Korea

A hybrid method incorporating cut-cell and porous media approaches is proposed to improve the filling accuracy of castings made using sand casting process. The retained melt modulus and surface methods were coupled with the probability distribution model for the prediction of shrinkage positions and area sizes, and the evaluation of the accuracy of shrinkage prediction. They highly depend upon the shrinkage prediction model and the critical solidification fraction.

Keywords: *hybrid method, filling accuracy, shrinkage model, critical solidification fraction, accuracy of shrinkage prediction*

1. Introduction

Effective numerical algorithms are very crucial both for the mold filling at between cavity and mold interface and for the calculation of shrinkage defects in solidification stage. A hybrid method incorporating cut-cell[1-2] and porous media approaches[3-4] is proposed to improve the filling accuracy of castings made using sand casting process. The method was verified using a Plexiglas-water model and compared with a simulation using the general finite difference method(FDM).

Lots of efforts have been made for the prediction of solidification shrinkage defects by numerical approaches[5-7], and they can be divided into two groups, the tracing of retained melts during solidification for gravity and macro porosity prediction, and appropriate criterion by the balance of temperature gradient and solidification speed in mushy zone for micro porosity. The retained melt modulus(RMM) and surface(RMS) methods in the first group were coupled with the probability distribution model for the prediction of shrinkage positions and area sizes with the variation of critical solidification fraction(CSF).

2. Filling Accuracy

As shown in Fig. 1, an appropriate number of sub-mesh was used to trace the curved boundary of the mold and cavity using the cut-cell method. With this sub-mesh structure, the information for the interface cell, including the volume, advection area, and direction of fluid, was calculated. The fluid volume and advection area of the interface cell were integrated into the momentum, pressure, and free surface transport equations (equations (1), (2) and (3), respectively).

$$V_f \frac{\partial \phi}{\partial t} + u_j \frac{\partial (A_j \phi)}{\partial x_j} = \frac{\mu}{\rho} \frac{\partial}{\partial x_j}\left(A_j \frac{\partial \phi}{\partial x_j} \right) - \frac{V_f}{\rho} \frac{\partial p}{\partial x_j} + V_f g_j \quad (1)$$

$$\frac{\partial}{\partial x}\left(A_j \frac{1}{\rho} \frac{\partial p^{n+1}}{\partial x_j} \right) = \frac{D^n}{\Delta t} = \frac{1}{\Delta t} \cdot \frac{\partial}{\partial x_j}(A_j u_j) \quad (2)$$

$$\frac{\partial}{\partial t}(V_f F) + \frac{\partial}{\partial x_j}(A_j u_j F) = 0 \quad (3)$$

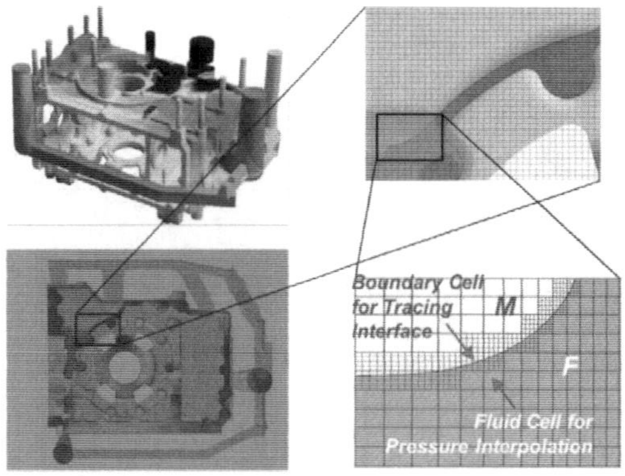

Fig. 1 Trace of the curved boundary of mold and cavity interface using sub-mesh.

As shown in Fig. 2, the Plexiglas-water experiment and simulation results using the FDM and hybrid methods were compared for a sand casting process. The simulation using the hybrid method resulted in a filling pattern similar to that shown in Fig. 2(c).

(a) (b) (c)

Fig. 2 Filling accuracy in comparison with the Plexiglas-water experiment.

3. Shrinkage Prediction Accuracy

A probabilistic model has been developed for the prediction of shrinkage area size based upon total shrinkage contraction during solidification, and considering both micro and macro feeding by riser and cast as shown in Fig. 3. Two shrinkage models, retained melt surface and modulus have been compared for the accurate prediction of shrinkages with real castings of ductile cast iron diffuser case housing(33 kg) and cast steel barrel(4.6 ton). The accuracy of shrinkage defects prediction was highly depend upon the selection of these shrinkage models and the critical solidifiaciton fraction, as show in Fig .4 and 5.

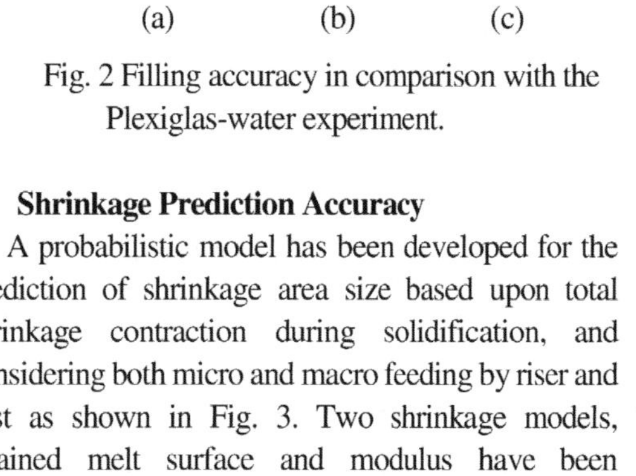

Fig. 3 Prediction of size for shrinkage area using probabilistic model.

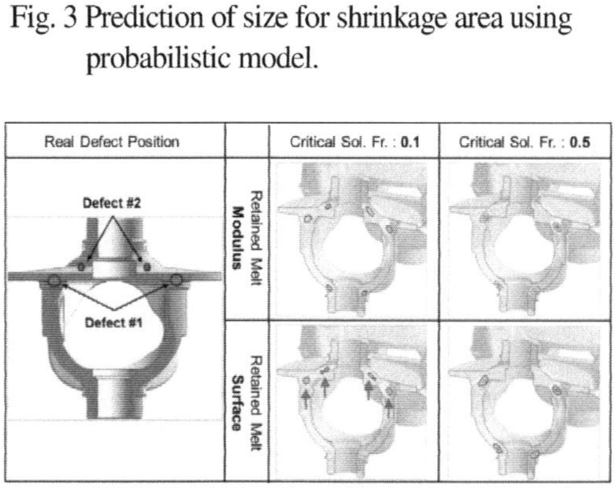

Fig. 4 Shrinkage prediction accuracy according to model and CSF in ductile cast iron casting.

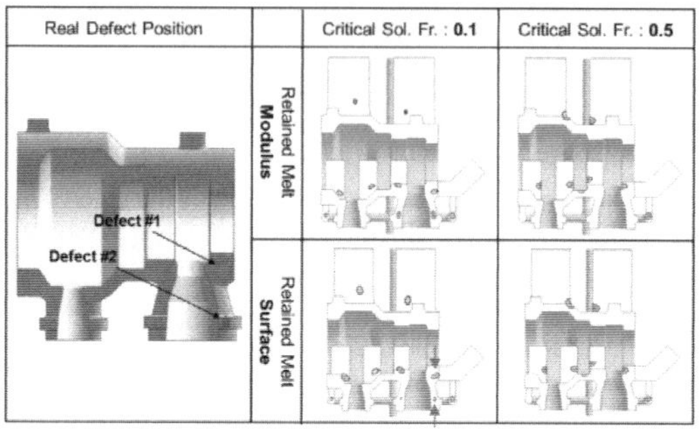

Fig. 5 Shrinkage prediction accuracy according to model and CSF in cast steel casting.

4. Concluding Remarks

For the improvement of shrinkage prediction in the castings of cast iron and steel, three kinds of researches were performed as follows.

a. improve the filling accuracy by the hybrid method

b. propose a probabilistic model for the prediction of shrinkage area size.

c. effect of the selection of shrinkgae prediction model and criticial solidification fraction upon the accuracy for the analysis of shrinkage positions and sizes

References

[1] T. Ye, R. Mittal, H. S. Udaykumar and W. Shyy, Journal of Computational Physics, 156 (1999) 209.

[2] P. G. Tucker and Z. Pan, Applied Mathematical Modelling, 24 (2000) 591.

[3] P. Lin, K.A. Chang and T. Sakakiyama, J. of Waterway, Port, Coastal and Ocean Engineering, 11 (1999) 322.

[4] P. D. Hieu and K. Tanimoto, Ocean Engineering, 33 (2006) 1565.

[5] S. Viswanathan and V.S. Sikka, MCWASP VI, 1998, p285.

[6] E. Niyama, T. Uchida, M. Morikawa and S. Saito, AFS Int. Cast Metals Journal, 7 (1982) 52.

[7] Y.W. Lee, E. Chang and C.F. Chieu, Metall. Trans. B, 21B (1990) 715.

Proceedings of the 72nd World Foundry Congress
21-25th May 2016, Nagoya, Japan

Prediction of Graphite Nodule Count and Shrinkage Tendency in Ductile Cast Iron, with 1 Cup Thermal Analysis

Toshitake Kanno[1], Taiji Fukuo[1], Ilgoo Kang[1], Yuki Iwami[1], Naoaki Kikuchi[1], Yuichi Hiramoto[1]
[1] Kimura Chuzosho Co., Ltd.,1157 Shimizu-cho,Shizuoka-ken,Japan

We have succeeded in investigating the graphite eutectic temperature (TEG) and cementite eutectic temperature (TEC) in ductile cast iron, from the chemical composition of the molten metal, as follows;
TEG=1149.1(℃)+4.7Si%-4.0Mn%-44P% etc.
TEC=1142.6(℃)-11.6Si%-0.75Mn%-46.2P% etc.
We added various kinds of inoculants into the molten metals, and then cast them at various fading time into 1 cup. Using the cooling curve of 1 cup and above equations, we calculated eutectic graphitization ability (EGA).The graphite nodule count (N) of CE cup T.P. is determined by (DT1/DTE), as follow; N = 6.13 x (100 x DT1/DTE) - 126 In addition to, when EGA becomes over 70%, shrinkage tendency (θ) becomes low as near 60°. Consequently, 1 cup thermal analysis is a good system to predict graphite nodule count and shrinkage tendency (θ) correctly in ductile cast iron, before pouring it.

Keywords: *1 cup thermal analysis, ductile cast iron, eutectic graphitization ability, molten metal's quality*

1. Introduction

T. Kanno developed the 3 cups thermal analysis system[1)]. It consists of 3 cups (First: Inoculated, Second: Base melt, Third: Tellurium added). Using this system, the eutectic graphitization ability (EGA =DT1/DTE) can be investigated and the graphite ability, chilling tendency and tensile strength can be predicted, in gray cast iron. Here, DTE means the difference between the graphite eutectic temperature (TEG) and cementite eutectic temperature (TEC). DT1 means the difference between the supercooling turning point temperature (TSC) and cementite eutectic temperature (TEC).

However, this 3 cups thermal analysis system cannot be used in ductile cast iron. In the ductile iron, as austenite solidifies ahead of graphite, TEG cannot be shown even with inoculated cup (a). In addition, as Te bonds to make MgTe, so it cannot work for making chill and TEC cannot be shown, even with Te added cup (c).

In this study, using 1 cup thermal analysis system, we calculated EGA and investigated the relationship between EGA and the molten metal's quality in ductile cast iron.

2. Experimental procedure

2.1 Principle of 1 cup thermal analysis system

Fig.1 shows the 1 Cup Thermal Analysis System. With the cooling curve of 1 cup (no addition), the supercooling turning point temperature (TSC) can be searched.

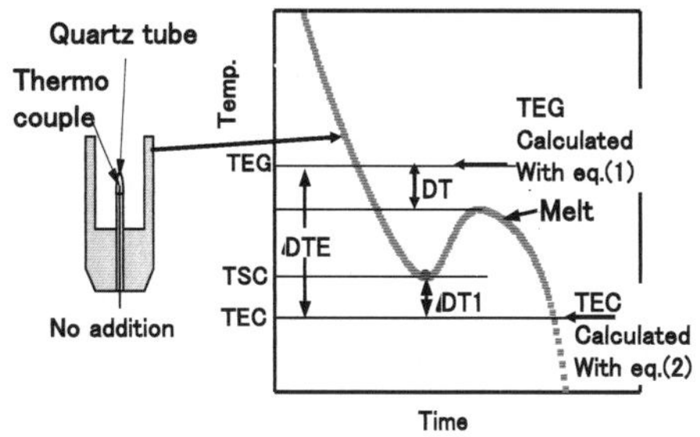

Fig.1. 1 Cup Thermal Analysis System

TEG and TEC can be calculated from the following principle. There is a report about the effect of element on eutectic temperature TEG and TEC[2)]. From this data, we can calculate TEG and TEC in ductile cast iron, from the chemical composition of the molten metal, as follows;

TEG=1149.1(℃)+4.7Si%-4.0Mn%-44P%+2.7Cu%+1.0Ni%+1.8Co%+13.9Al%-17.7Mo%-10.5Cr%-9.3Sn%-5.2Sb%-6.1W%-3.7Nb-14.8V%-80.3B% (1)

TEC=1142.6(℃)-11.6Si%-0.75Mn%-46.2P%-1.4Cu%-1.1Ni%-0.7Co%-1.8Al%-14.5Mo%+5.9Cr%-6.0Sn%-5.1Sb%-2.8W%+0Nb%+3.3V%-26.0B% (2)

In Ductile cast iron, S bonds Mg to makes MgS, so we do not consider its effect. Mg amount is small enough to ignore. As these results, EGA can be calculated. (EGA = 100 x (TSC - TEC) / (TEG - TEC)

2.2 Pouring test

Melting was done in a silica lined high frequency furnace of 60 kg, 3000 Hz. The basic composition was as follows: 3.7mass% C (hereafter simply shown as %), 2.15% Si, 0.3% Mn, 0.02% P, 0.01% S and 0.05% Mg. After Mg spheroidizing treatment as sandwich method, the molten metal was returned to furnace and was held at 1350 ℃. We added various kinds of inoculants (Graphite type, Ba type, Ba-Ce type, Ca-Si type and Bi type) into the molten metals, and then cast them at various fading time (0, 1, 2.5, 5.0, 7.5, 10, 15 and 20 min) into 1 cup (D30xH50mm, 250g). Using the cooling curve of 1 cup and above equations, we calculated EGA and investigated the relationship between EGA and the molten metal's quality of ductile cast iron.

2.3 Principle of investigating shrinkage tendency

Fig.2 shows the principle of investigating shrinkage tendency (θ), with 1 cup thermal analysis system. When shrinkage cavity forms, thermal conductivity becomes smaller, owing to the cavity in the casting, and cooling speed becomes slower and the tendency (θ) becomes larger. That is, the larger the θ is, the larger the shrinkage is.

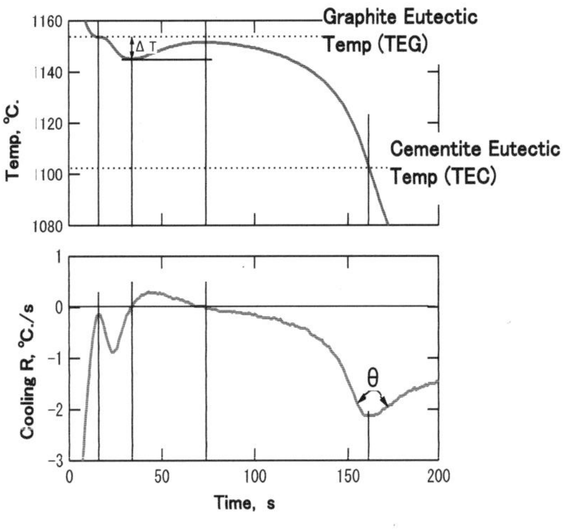

Fig.2 Principle of investigating shrinkage tendency (θ), with 1 cup thermal analysis system.

3.1 Prediction of graphite nodule count

Fig.3 shows the relationship between (DT1/DTE) and the graphite nodule count (N) of CE cup. T.P. The graphite nodule count (N) is determined by (DT1/DTE), as follow.

$$N = 6.13 \times (100 \times DT1/DTE) - 126 \qquad (3)$$

1 Cup thermal analysis is a good system to predict graphite nodule count correctly.

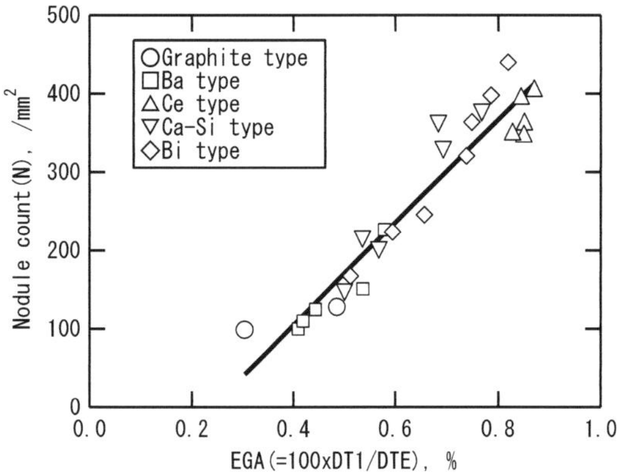

Fig.3 Relationship between EGA and the graphite nodule count (N) (T.P.: CE cup, D30xH30mm, 250g)

3.2 Prediction of shrinkage tendency

Fig.4 shows the relationship between EGA and shrinkage tendency (θ). When EGA becomes over 70%, shrinkage tendency (θ) becomes low as near 60°. 1 Cup thermal analysis is a good system to predict shrinkage tendency (θ) correctly.

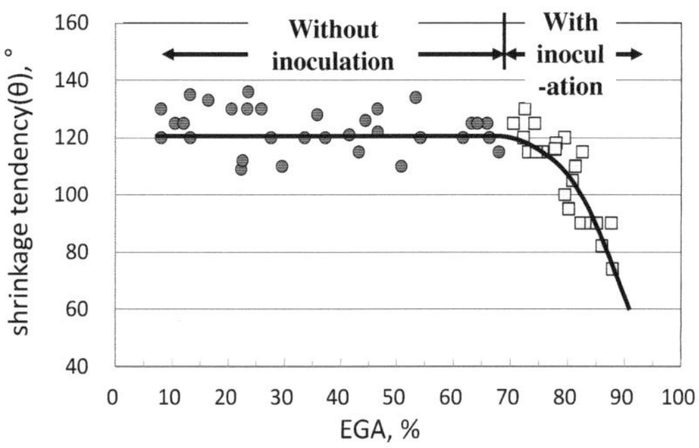

Fig.4 Relationship between (EGA=100 x DT1/DTE) and shrinkage tendency (θ).

Conclusion

1 cup thermal analysis is a good system to predict graphite nodule count and shrinkage tendency (θ) correctly in ductile cast iron, before pouring it.

Reference

[1] T. Kanno, M. Morinaka and H. Nakae: J. Japan Foundry Society, 70 (1998) 773-778.

[2] T. Kanno, M. Morinaka and H. Nakae: J. Japan Foundry Society, 70 (1998) 465-470.

Proceedings of the 72nd World Foundry Congress
21-25th May 2016, Nagoya, Japan

Application of the forced cooling in the mold for prevention of the occurrences of metal penetration

Jun Kannari[1], Makoto Masuyama[2], Misao Okino[3], and Yoshinao Fukuma[4]
Daisuke Tomigashi [5], Hiroaki_Kudo [6]
[1, 2, 3, 4] The Japan Steel Works, LTD, 4 CHATSU – MACHI MURORAN, HOKKAIDO, JAPAN
[5, 6]NIKKO MEC Co. , LTD, 2-1 CHATSU – MACHI MURORAN, HOKKAIDO, JAPAN

Among the steel castings by using the sand mold, particularly of heavy thickness and large mass, *metal penetration* often occurs on the product surface toward the core mold. Generally, in order to prevent the occurrences of *metal penetration*, the thick coating of mold wash (known as *Surfacer*) and/or the high refractory sand has been applied in our foundry shop. But, it is not so effective as an Extra-high temperature portion, so called as *Hot Spot* where a core is surrounded by melted steel, and the occurrence has not been prevented. It takes a large amount of worktime to remove the zone of *metal penetration*, and it makes the productivity stay low.

We'd like to report herein on the prevention of *metal penetration* by using the *forced cooling* at *Hot Spot*.

Keywords: *metal penetration, Hot Spot , Forced Cooling*

1. The occurrences of *metal penetration*

We are manufacturing heavy thickness and Large mass steel castings by using the sand molds. The *metal penetration* often occurs on these large items due to the high pressure head of melted steel. Appearance of *Metal Penetration* occurred in our foundry shop is as shown in Fig.1. This Figure shows the typical *metal penetration*, melted steel Penetrating into the void between sand particles.

It takes a large amount of worktime to remove the zone of *metal penetration*, and it makes the productivity stay low. Therefore, we need a new method to prevent the occurrences of *metal penetration*.

Generally, in order to prevent the occurrence of *metal penetration*, the thick coating of mold wash (known as *Surfacer*) and/or the high refractory sand has been applied in our foundry shop. But, it is not so effective as an Extra-high temperature portion, so called as *Hot Spot* where a core is surrounded by melted steel, and the occurrence has not been prevented.

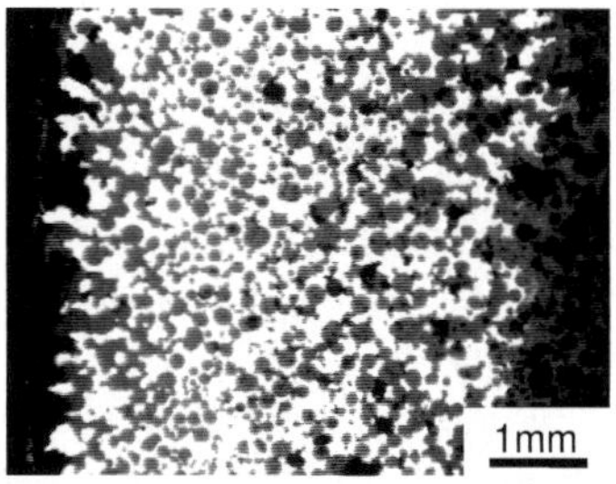

Fig.1 Appearance of *Metal Penetration* (Enlarged)
(The left side is product, The opposite side is Mold)

2. The estimated cause of the *metal penetration*

The *metal penetration* seems to occur on the *Hot spot*, therefore we focused on the strength of the casting mold. The higher the temperature of the mold arises, the extremely lower mold strength becomes, due that the resin (as binding material) will be decomposed or burned. After the pouring into the mold, mold strength becomes lower and cracks will happen on the contact surface. Then, unsolidified melted steel penetrates into the cracks.

We estimated these phenomenon causes the *metal penetration*.

3. Method of prevention of the *metal penetration*

It is supposed that the appropriate cooling will keep the molding temperature lower and it helps to maintain the mold strength. We planned the *forced cooling* method by aeration pipes set in the core mold at expected *Hot Spot*.

4. Test plan

The planned method was applied to the actual production. Aeration pipe was set in the expected *Hot Spot*. The applicable part is as shown in Fig.2. Concept of molding around the applicable part is as shown in Fig.3. As the aeration pipe, stainless steel pipe with 20(mm) of internal diameter was used. Also, 2 Thermo-couples were set to measure the mold temperature. Thermo-couple 1 is to measure the mold temperature distant from the aeration pipe, while thermos-couple 2 is closed to aeration pipe. Compressed air was used for the *forced cooling*. It is supposed to take 6(Hr.) from the end pouring to complete solidification, the planned aeration duration is 6(Hr.).

5. Result

The result of the mold temperature measurement is as shown in Fig. 5. As comparison of Curve 1 and 2, the rising rate closed to aeration pipe is slower than those distant from aeration pipe. The ultimate temperature of Curve 2 is lower than Curve 1. And, the Appearance of the applicable part after Sand Cleaning is as shown in Fig. 4. There was not any *metal penetration*. The result indicates the *forced cooling* of the mold is effective to prevent the occurrences of crack on the mold surface until the complete solidification even where direct contact with melted steel.

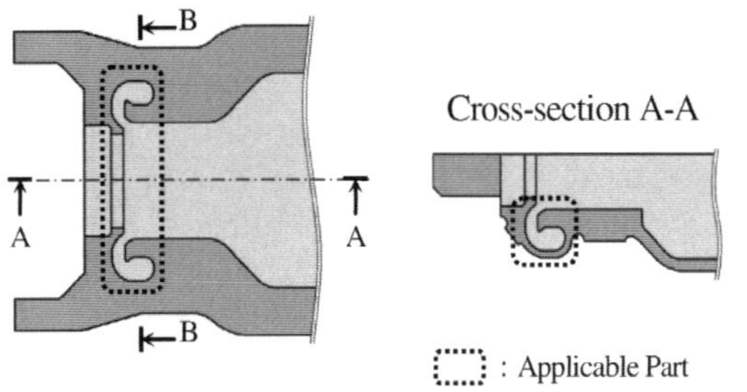

Fig.2 Applicable part of *forced cooling*

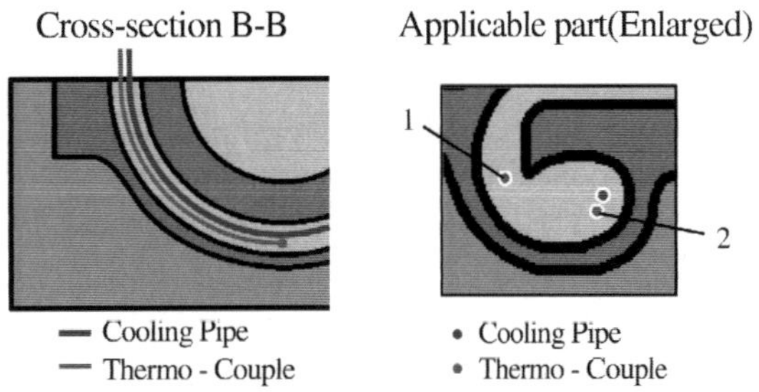

Fig.3 Concept of molding around the applicable part

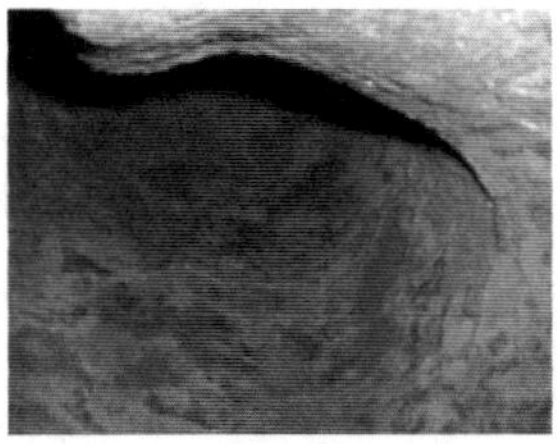

Fig.4 Appearance of the applicable part after Sand Cleaning (After the shot-blast)

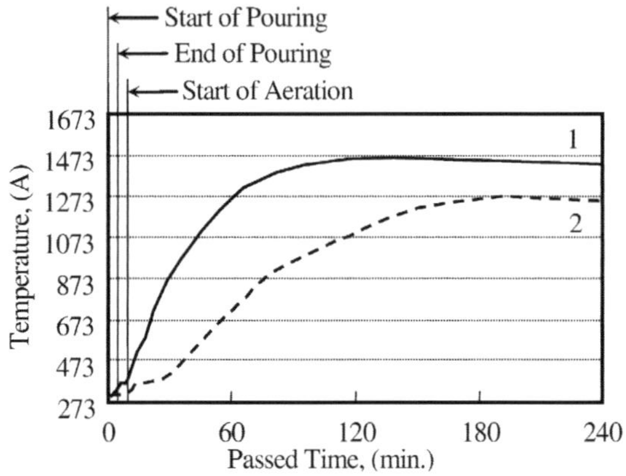

Fig.5 The result of the mold temperature measurement

6. Effect

The effect of prevention of *metal penetration* on *Hot spot* by *forced cooling* is as shown in Table1. Although the setting of aeration pipe is needed in the molding procedure, surface shaping worktime is reduced. We succeeded to improve the total productivity by prevention of the occurrences of *metal penetration* on the *Hot Spot*.

Table1 Effect of *forced cooling* application

process	Effect
Molding	Relaxation of sand grade (High refractory Sand ⇒ Normal Sand)
	Increase of core molding time (+1.8(%))
Shaping	Reduction of the time to gouging (−95(%))
	Reduction of the time to grinding (−58(%))

7. Summary

(1) By application of *the forced cooling* in the mold the rising rate and the maximum temperature of the mold temperature could be lowered.
The *metal penetration* on the *Hot Spot* which has often occurred this discontinuity Could be prevented.

(2) As a result of the prevention of the occurrences of *metal penetration* on the *Hot Spot*, the total productivity was improved.

Proceedings of the 72nd World Foundry Congress
21-25th May 2016, Nagoya, Japan

Online Tracking for Single Unit in Investment Casting of Jet Engine Critical Components

Michal Kwiatkowski[1], Adam Sury[2], Daniel Pociecha[1], Rafal Cygan[3], Marek Wojtas[2] and Krzysztof Zaba[1]
[1] AGH University of Science and Technology, Faculty of Non-Ferrous Metals, A. Mickiewicza 30 Av., 30-059 Cracow, Poland
[2] Infoster sp. z o.o., Gorna 5A Str., 30-094 Cracow, Poland
[3] Rzeszow University of Technology, Faculty of Mechanical Engineering and Aeronautics, Powstancow Warszawy 8 Av., Rzeszow 35-959, Poland

The paper shows the concept and implementation of a dedicated, computer module of production unit tracking – wax patterns and assemblies, moulds and single casts in production line, which data are collected automatically supplemented by the information entered manually by the operator. Currently there is no production unit's position tracking in the individual processes. The essence of this module is full control of the production unit location in each stages. It is assumed that by monitoring the location of the unit in each device can be identified by extraordinary variability arising from the state of individual devices in all stages. The module allows specifying the location of individual parts in casting system. Innovative dedicated software was developed, that allows to determinate the defects and generates detailed statistical analysis based on multiple filters.

Keywords: *superalloys, investment casting, expert system, tracking*

1. Introduction

Manufacturing processes of the jet engine critical components using lost-wax investment casting are classic examples of the so-called special processes. Due to the complexity of each single processes and their collection, strong emphasis is put on their computer-aided realization.

Diagram of the manufacturing process, under the tracking system shown in Fig. 1.

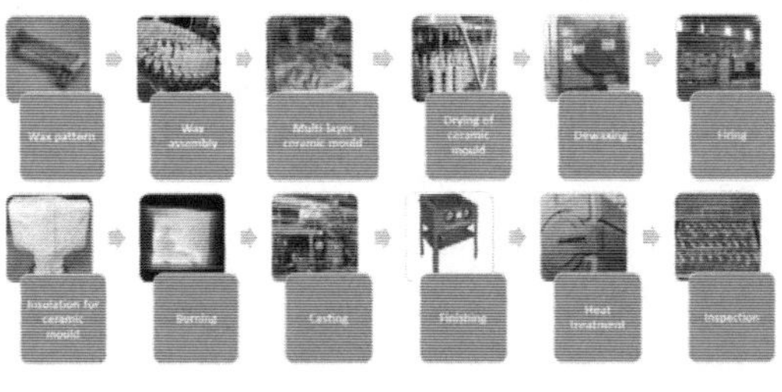

Fig. 1 Investment casting process scheme

2. Concept of expert system

2.1 Production unit "Online tracking" module

Presented functionalities are part of the computer expert system to ensure comprehensive supervision for all technological and organizational processes.

The essence of this module is full control of the production unit location in injection molding machines, producer, autoclaves for dewaxing, furnace for moulds firing, burning, pouring and cores leaching autoclaves. It is assumed that by monitoring the location of the unit in each device can be identified by extraordinary variability arising from the state of individual devices in the stages: wax patterns and wax assemblies making, moulds making and drying, dewaxing, moulds firing and burning, casting, ceramic cores leaching and inspection.

The figure below shows the main window tracking system.

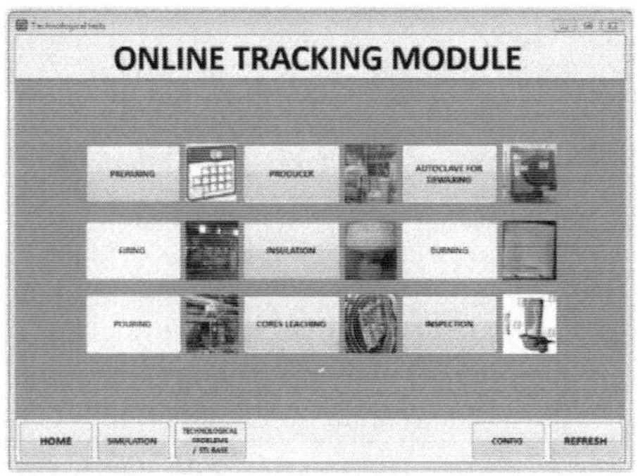

Fig. 2 "Online tracking module" main window

The screen gives the possibility move to the specific stage of the production process and viewing information that is collected automatically and entered by operators. For each of the individual steps may be accompanied by documentation (audio, video, *.txt files). The location of each patterns in a single mould and device, is tracked by sample.

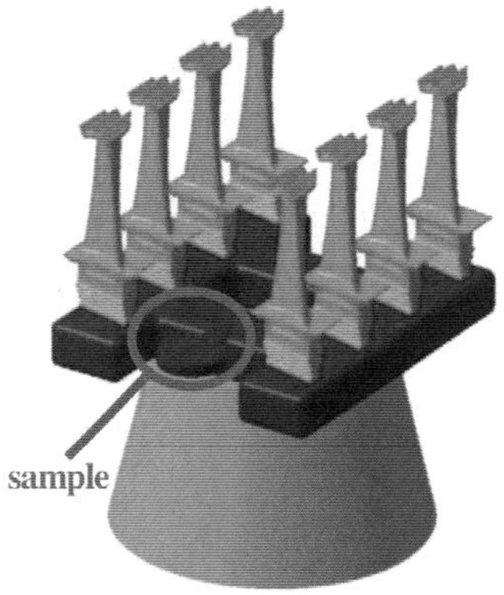

Fig. 3 Wax assembly with sample

Selected and presented examples of screens for the unit operations – coating and drying moulds in producer and ceramic cores leaching in an autoclave.

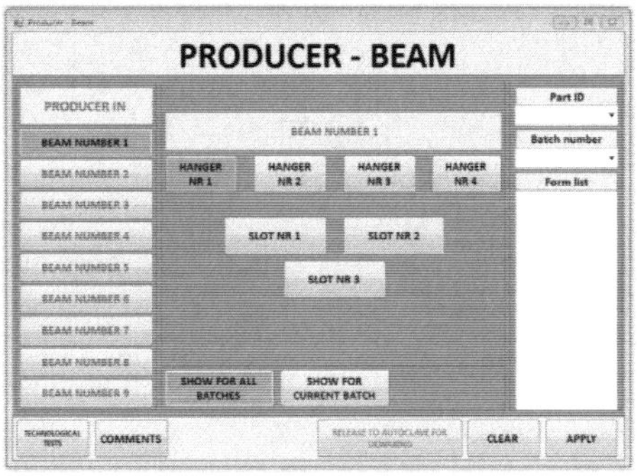

Fig. 4 “Producer” window

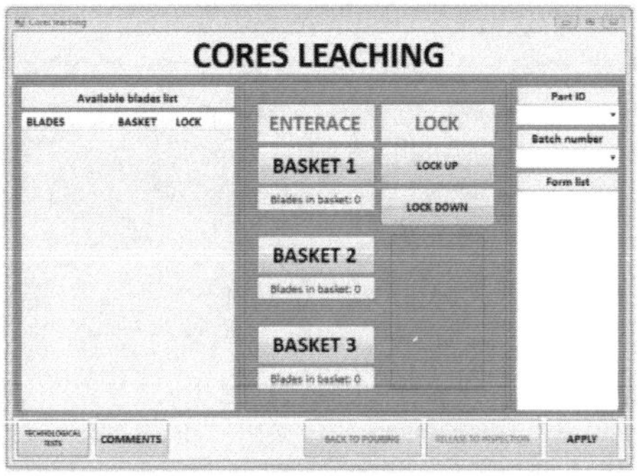

Fig. 5 “Cores leaching” window

Screen presented on Fig. 4, it is possible to mark the information on the location of the mould on the beam, hanger and place. The module collects data about the coating process parameters, humidity, temperature and drying time.
Another exemplary screen (Fig. 5) shows the tracking of a single blade in an autoclave for leaching the cores. The screen allows the operator to identify the location of individual compontents in one of three baskets along with the record information on the direction of their arrangement. The module collects information about the leaching, including the composition and temperature of the electrolyte, pressure in the autoclave and leaching time.

2.2 "Interactive map of defects" module

The module allows specifying the location of individual parts in casting system, and then mark certain defects on models parts (divided into stages of quality control). Innovative dedicated software was developed, that allows to determinate the defects and generates detailed reports and statistical analysis based on multiple filters, areas of casts, type of alloy, type of casting furnace and other.
The areas are defined in the *.stl file using a recording color information for each the singe triangles. On loaded model into module applied can be selected identified defects in inspection stages. The database and thereby to the module analysis are recorded location coordinates and information about the defect. An example screen presented in Fig. 6.

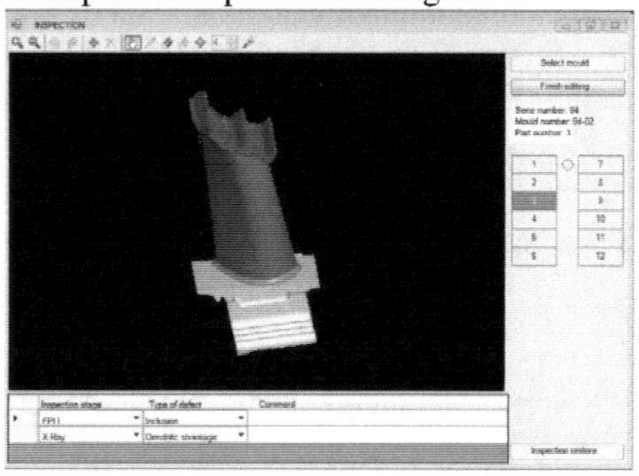

Fig. 6 “Interactive maps of defects ”

3. Conclusions

Presented concept has unique character, it is possible to implement in different casting processes.

Acknowledgements

Financial support of Structural Funds in the Operational Program – Intelligent Development (ID OP) financed from the "INNOLOT – Innovative aviation", Project "Innovative investment casting technologies", No. "INNOLOT/I/8/NCBR/2013 – is gratefully acknowledged.

References

[1] A. Sury, M. Wojtas, M. Kwiatkowski, D. Pociecha, K. Zaba, R. Cygan: *Expert systems in foundry industries*, 13th Asian Foundry Congress, Hanoi, Vietnam (2015).

Proceedings of the 72nd World Foundry Congress
21-25th May 2016, Nagoya, Japan

Enhancement of porosity and mechanical properties of ceramic shell in investment casting process by the addition of camphor and needle coke/wax powder- A comparison

Khyati Tamta[1], D. Benny Karunakar[2] and Wankhede Umesh Pralhad[3]
[1] Mechanical and Industrial Engineering Department, Indian Institute of Technology Roorkee, India
[2] Mechanical and Industrial Engineering Department, Indian Institute of Technology Roorkee, India
[3] Mechanical and Industrial Engineering Department, Indian Institute of Technology Roorkee, India

In the investment casting process, molten metal is poured into a ceramic shell, which is created around wax pattern by the repeated application of slurry/stucco. Because of the fine ingredients used, the ceramic shell often exhibits poor permeability and leads to porosity defects. In the present study, an attempt was made to enhance the permeability of ceramic shell by mixing certain fillers like camphor, needle coke and wax powder to the ceramic slurry. In the first investigation, camphor was added to the inner coat slurry and needle coke was added to the outer coat slurry in small proportions. In the second investigation, camphor was again added to the inner coat slurry and wax powder was added to the outer coat slurry in small proportions. During the firing of the shell, these fillers (camphor, needle coke / wax) were burnt and pores were created inside the shell, which enabled escape of undesirable gases. A comparison was made between the performances and properties of the shells obtained in both the cases.

Keywords: *Investment casting, Ceramic shell, Permeability, Camphor, Needle coke and Wax powder.*

1. Introduction

Investment casting process is known for its good surface finish and excellent dimensional accuracy. Excellent surface finish of the cast component is achieved by the excellent surface finish of the inner coat of the shell, due to the very fine grain size of the refractory powder used. This very fine grain size of the ceramic powder leads to poor permeability of the shell and consequently gas porosity defects. Poor selection of ingredients could lead to poor permeability of the shell and also poor mechanical properties of the shell. To overcome these problems, organic fibres were added to the slurry ingredients to improve the mould thickness at the round edges and sharp corners. These fibres required very less resistance and forces while removal from the matrix. Further addition to this, Wang et al. have improved the thickness of the shell by adding needle coke to the outer layer of the shell. The shell, thus produced, had improved mechanical properties than the conventional shell [1].

In the present work, camphor, needle coke / wax powder were added to the inner and outer coat slurries respectively. At room temperature and pressure, camphor sublimates from solid to gas without turning to liquid [2]. Needle coke is a special type of petroleum product prepared with coal and crude oil. It is known for high strength, good resistance to thermal shock [3]. Montan wax is hard and is most resistant to oxidation which will improve the mechanical properties of the shell [4].

2. Experimental work

Ceramic shells were prepared using the slurries made up of aluminum silicate refractory powder and colloidal silica binder. In the first investigation, camphor was added to the inner coat slurry and needle coke was added to the outer coat slurry in small proportions (around 2%). In the second investigation, camphor was again added to the inner coat slurry and montan wax powder was added to the outer coat slurry in small proportions. Both the modified slurries were stirred for 24-48 hours. Six layers of coatings were given around the pattern using these modified slurries. During firing, camphor sublimated leaving micropores in the inner coating while needle coke / montan wax powder got burnt leaving macropores in the outer coatings.

2.1 Permeability measurement test

Permeability is the ability of ceramic shell to allow hot gases to escape through it. For the measurement of permeability, a sample has been prepared using silica tube of 300 mm long and 8 mm diameter, as shown in Fig.1.

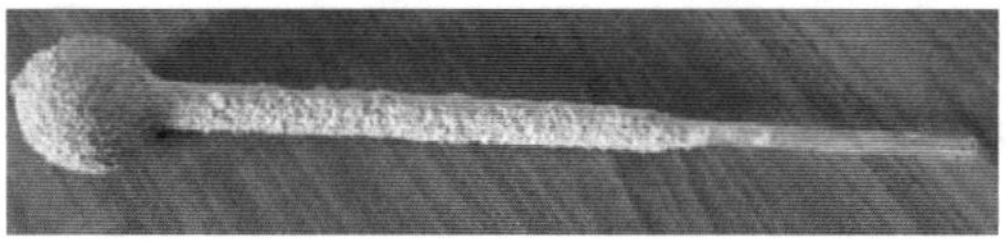

Fig. 1 Sample prepared for permeability measurement.

2.2 Shell strength test

For shell strength measurement, flexural strength tests was conducted on three- point bending over a span of 90 mm using test pieces of 70 mm long 10 mm width and 10 mm depth. Instron 5800R, UTM was used for the testing machine at a constant load rate of 1 mm/min until failure. The MOR, σ_{Max}, was calculated using the following equation.

$$\sigma_{max} = \frac{3P_{max}L}{2WH^2}$$

Where P_{max} is the fracture load, L is the test span, W is the width of test piece and H is the height of test piece.

3. Results

Permeability measurement test was carried out on both the modified shells and the relation between pressure and flow of air has been measured at the constant temperature. From Fig. 2, it can be seen that modified shell 2 (with needle coke) is more permeable than the modified shell 1 (with montan wax). As the pressure was increasing, flow of air also increased.

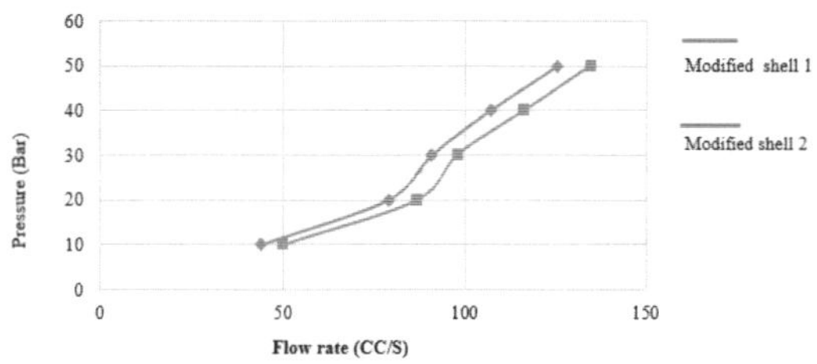

Fig. 2 Sample prepared for permeability measurement.

For the mcechanical testing, five samples were tested and MOR was calculated, as shwon in Table 1. It can be seen that modified shell 2 (with needle coke) has higher MOR (σ_{max}) for all the samples than that of the modified shell 1 (with montan wax powder). From Fig. 3, it can be seen that, after firing camphor got sublimated leaving micropores behind. Image was captured using a LIO, 435VP scanning electron microscope.

Table 1 MOR calculated for both modified shells.

Sn.	Modified shell 1	Modified shell 2
	σ_{max} (MPa)	σ_{max} (MPa)
1	8.834	9.713
2	8.528	9.640
3	8.319	9.935
4	8.480	9.404
5	8.720	9.773

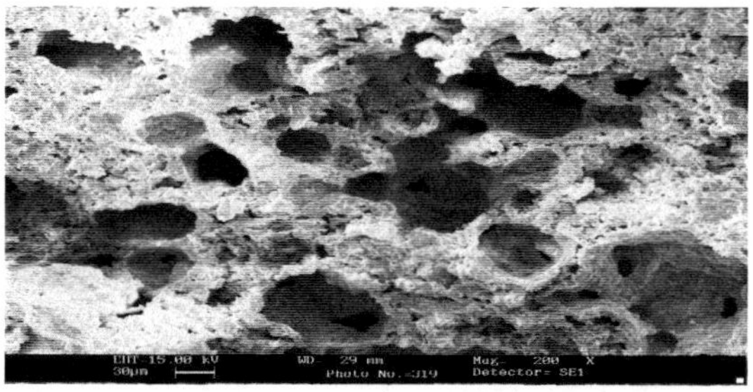

Fig. 3 Microstructure of inner coat of the shell after adding camphor.

4. Conclusions

- Needle coke modified shell has shown more permeability than montan wax powder modified shell.
- Wax modified shell has a maximum MOR of 8.834 MPa while the needle coke modified shell has a maximum MOR of 9.713 MPa.
- Modification of the ceramic slurry by the addition of camphor has improved the permeability of the shell by creating micro-pores inside the inner part of the shell.

References

[1] F. Wang, F. Li, B. He, and B. Sun.: Ceramics International. 40 (2014) 479-486.

[2] U. Bayram, S. Aksoz, and N. Marasli: Thermochimica Acta. 531 (2012) 12-20.

[3] S. Birghila, I. C. Carazeanu, and A. Dumitru: *Study on Physical-Chemical Properties of Petroleum Cokes,* (Romania Journal of Physics, 2011) pp. 976-982.

[4] O. Bemblage and D. B. Karunakr: *A Study on the Blended Wax Patterns in Investment Casting Process,* (Proceedings of the World Congress on Engineering 2011) pp. 721-727.

Proceedings of the 72nd World Foundry Congress
21-25th May 2016, Nagoya, Japan

Novel Bayesian Inference Based Approach to Identify Critical Parameters Affecting Mechanical Properties of Investment Castings

Amit Sata[1], B. Ravi[2]
[1]Associate Professor, B H Gardi College of Engineering and Technology, Rajkot - 361162 (India)
[2]Institute Chair Professor, Mechanical Engineering Department, IIT Bombay, Mumbai - 400076 (India)

In addition to shape fidelity and internal soundness, mechanical properties have also become critical acceptance criteria for investment cast parts. These properties, including ultimate tensile strength, yield strength and percentage elongation, are driven by the chemical composition of cast alloy as well as process parameters related to wax making (time taken for injection, press room temperature and humidity) and other steps in investment casting: shell making, dewaxing, and melting. It is however, difficult to identify the most critical parameters and their specific values resulting in the mechanical properties. This is achieved in the present work by employing Bayesian inference to compute posterior probability values for each input parameter. This is demonstrated on real-life data collected from an industrial foundry. Controlling the identified parameters within the specific range of values resulted in improved mechanical properties. Unlike computer simulation, artificial neural networks and statistical methods explored by earlier researchers, the proposed approach is easy to implement in industry for controlling and optimizing the parameters to achieve the desired range of mechanical properties.

Keywords: *Bayesian Inference, Investment Casting, Mechanical Properties, Process Parameters.*

1. Introduction

Investment castings used in critical applications like aerospace and biomedical need to be free of internal defects as well as possess the desired mechanical properties like ultimate tensile strength, yield strength, and percentage elongation to provide the required functionality and service life. The casting quality and mechanical properties are greatly affected by the flow, solidification and microstructure of cast metal in the ceramic shell. These phenomena in turn depend on the chemical composition of the alloy and various process parameters. There is a need to identify the most critical parameters and their range of values to enable controlling and optimizing these parameters, thereby minimizing quality problems and achieving the desired mechanical properties. This is however, a challenging task since there are a large number of parameters involved in investment casting process. Moreover, it should be possible to implement the proposed approach in industrial foundries.

In the past, several researchers have explored different techniques for predicting the effect of alloy composition and selected process parameters on mechanical properties, though mainly for sand and die casting processes. These include computer modeling and simulation of metal flow and solidification [1-3]; statistical methods (mainly multiple regression analysis) [4-5]; and artificial neural networks [6-8]. There is very little work to identify critical parameters and their specific range of values to achieve the desired range of mechanical properties.

A relatively new approach called Bayesian inference can be explored for the above purpose. It relies on the Bayes' rule that estimates the probability of occurrence (prediction) based on the evidences (observations). The basic methodology is presented next, followed by testing in an industrial foundry.

2. Methodology

The basic methodology is shown in figure 1. Input data related to process parameters (wax making, shell making, melting and pouring) and alloy composition is collected, and stored in Microsoft Excel spreadsheet. This data is processed using Bayesian inference to identify critical parameters and their specific range of values affecting mechanical properties. This involves computation of posterior probabilities (programmed in Excel), briefly explained next.

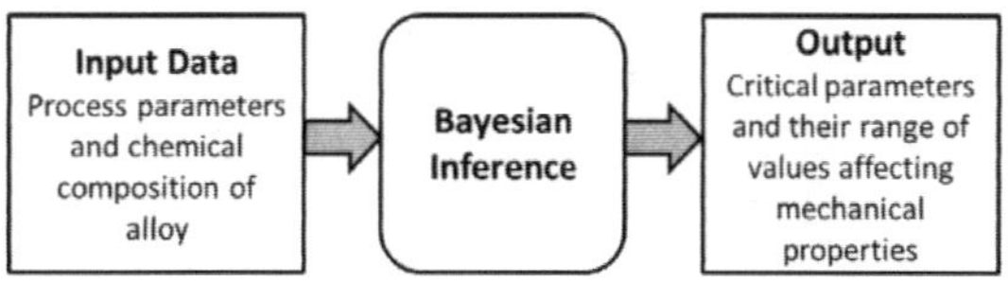

Fig. 1 Basic methodology

The total range of each input parameter as well as mechanical property is divided into four ranges (i = 1 to 4), and the observed value of each input is placed in one of the four ranges. The input parameter addend (P_{IP}) and mechanical property addend (P_{MP}) are calculated. The values of posterior probability for the upper range of mechanical properties (R_{MP4}) are computed for all input parameters from an input data. This in turn requires computing the values of *local probability* (*LP*), *prior odd* (O_{prior}), *joint probability* (*JP*), *conditional probability* (*CP*), *likelihood ratio* (*LR*), and *posterior odd* (O_{post}) based on input data. The relevant equations are given in table 1.

Table 1 Computation of posterior probabilities [9]

Parameters	Equation
LP_{IPRi}	N_{IPRi}/N_{IP}
$O_{prior(IPRi)}$	$LP_{IPRi}/(1- LP_{IPRi})$
$JP_{MPRi/\ IPRi}$	$N_{IPRi/MPRi}/N_{MPRi}$
$CP_{IPRi/MPRi}$	$P_{MPRi}* JP_{MPRi/\ IPRi}/ LP_{IPRi}$
$LR_{MPRi/\ IPRi}$	$CP_{IPRi/MPRi}/(1- CP_{IPRi/MPRi})$
$O_{post(MPRi/(IPRi)}$	$O_{prior(IPRi)} * LR_{MPRi/\ RIPi}$
PR_{IPRi}	$O_{post(MPRi/(IPRi)}/(1+ O_{post(MPRi/(IPRi)})$

The values of posterior probabilities are used to identify the critical parameters affecting mechanical properties. An input parameters and its specific range of values are considered to be critical if its value of posterior probabilities is high.

3. Testing

The input data of more than 450 heats related to process parameters and chemical composition of alloy was collected from an industrial foundry. The part is a steel valve body used in automobiles (figure 2). The foundry measured the mechanical properties (ultimate tensile strength, yield strength and percentage elongation) for each batch, by casting sample bars along with the castings in each batch, and testing each sample bar on an universal testing machine as per ASTM A370 (figure 2). The total data set comprised 30 input parameters and three output parameters.

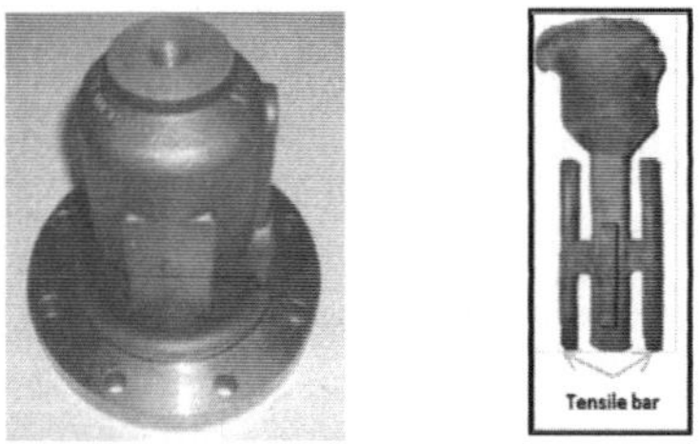

Fig. 2 Industrial casting & tensile bar

The input data was entered into the system in the form of Microsoft Excel sheet. This data was analyzed using Bayesian inference as described earlier, to compute posterior probabilities. Posterior probability of metal preparation time (one of the input parameters) for the desired ultimate tensile strength (for 4th range) is shown in table 2.

Table 2 Posterior probability of metal preparation time

Parameters	Values
LP_{IPR1}	0.96
$O_{prior(IPR1)}$	29.80
$JP_{MPR4/\ IPR1}$	1
$CP_{IPR1/MPR4}$	0.11
$LR_{MPR4/\ IPR1}$	0.12
$O_{post(MPR4/(IPR1)}$	3.58
PR_{IPR1}	0.78

The optimal range of other critical parameters (posterior probability more than 0.5) were similarly identified: viscosity of primary slurry (19 to 24 sec); pH of primary slurry (9.0 to 9.1); metal preparation time (100 to 139 minutes); Molybdenum (0.002 to 0.010 %). Using these values, 10 heats were melted, from which 22 castings were poured and tested. The mechanical properties were found to be in the desired range for 15 castings.

To summarize, usefulness of Bayesian inference methodology has been successfully demonstrated for determining the optimal range of process parameters and alloy composition, to improve quality assurance of industrial investment castings. The relative ease of implementing and using the proposed methodology make it valuable for practical application.

References

[1] S. Seifeddine, M. Wessen and I. Svensson: Metall Sci. Tech. 24 (2006) 26-32.

[2] M. Schneider et al.: Mat Sci and Engg. 33(2012) 1-8.

[3] Guo et al.: Mfg. Engg. and Perfor. 16(2007)680-684.

[4] N. Costa, N. Machado and F. S. Silva: Journal of Fatigue. 32(2010)988-995.

[5] Pucher et al.: Proceeding of MFPCM by Metal and Material Society(2011).

[6] M. Perzyk and A. Kochanski: Journal of Proc Tech. 109(2001)305-307.

[7] L A Dobrazanski and M Krol: Arch of Comp. Mat. Sci. and Surface Engg. 2(2010)181-188.

[8] Amit Sata and B. Ravi: Journal of Mat. Engg. and Perf. 23(2014)2953-2964.

[9] Amit Sata: PhD Thesis: IIT Bombay (2015).

Proceedings of the 72nd World Foundry Congress
21-25th May 2016, Nagoya, Japan

Critical Analysis of Rapid Prototyping assisted Investment casting for Medical Implants

Dheeraj S. Bhiogade[1], Sanjay Randiwe[1], Dr. A. M. Kuthe[2]
[1] Ph.D scholars, Department of Mechanical Engineering (CAD-CAM), India
[2] Professor, Department of Mechanical Engineering, India
Visvesvaraya National Institute of Technology, Nagpur, Maharashtra, India,

Medical applications are some of the most interesting applications of the rapid manufacturing using rapid prototyping (RP) technologies. Produce RP model of the organ structure, which is very useful for diagnosis, surgery planning, training, and development of custom implants. The paper presents the rapid manufacturing process i.e. Casting using ABS polymer pattern & investment casting process for SS316L alloy femur implant. Simulation programs (AutoCAST-X) are used to achieve sound, high quality castings in first attempt. Also the procedure for design of casting system for any intricate shapes has been elaborated, thus foundry engineer facing difficulty. RP pattern based investment casting overcomes the conventional wax pattern process along with validated & optimized process.
With the use of computer software the mold filling and solidification has assured the quality of castings with a higher degree of confidence, and reduces the cost of rejects & major improvements being realized in the areas of controlling shrinkage porosity defects.

Keywords: *Rapid prototyping, Investment casting, Rapid manufacturing, Custom implants, Simulation.*

1. Introduction

The product development from the conceptualization stage to the finish product, numbers of stages involved such as CAD, analysis, casting simulation, optimization, pattern/mold making, melting & pouring of metal, inspection & testing. A dominant technology for producing tailor made product is rapid prototyping (RP). Fused deposition modeling (FDM) is an additive technology suite for producing prototypes and final parts with complex geometries using ABS (Acrylonitrile Butadiene Styerne) [1, 2]. RP has proved be a cost-effective and time-efficient approach for development of products [4, 5]. Worldwide, metal called stainless steel is extensively used, now it also used in human body as prosthesis (implant) [3].

In the one of the study indicated regarding the shell cracking during the burnout process of investment casting, when the invested shell is not strong enough to resist the exerted forces due to thermal expansion. From FEA, the predicted rupture temperature is 350° C, at which the induced stress exceeds the Modulus of rupture (MOR) of the ceramic and thus fractural cracking occurs [6]. In this study procedure for avoiding the mold cracking during burn out process was demonstrated.

2. Methodology of rapid-prototyping

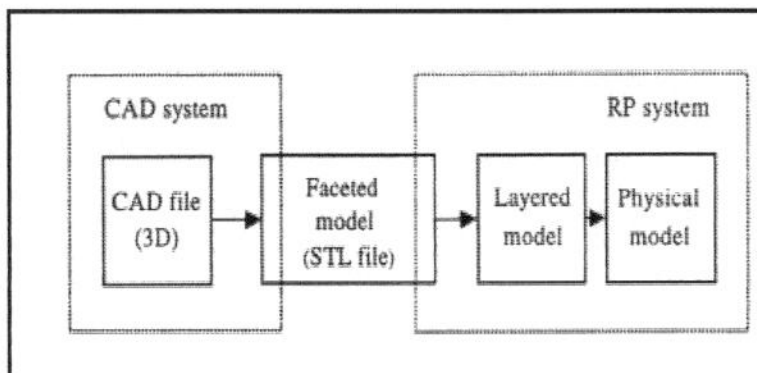

Fig. 1 Rapid Prototyping approach

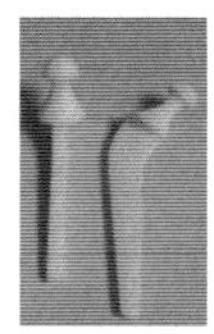

Fig. 2 pattern

Fig. 1 shows the block diagram of pattern development approach, & finished and cleaned RP pattern shown in fig.2.

3. Methodology of rapid manufacturing/Rapid casting

3.1 Steps involved in rapid casting process are

a. Pattern was developed using RP technology & used for investment casting.
b. Baking of mould & complete evaporation of RP pattern was inside of muffle furnace.
c. Metal melting & pouring.
d. Finished casting after solidification.

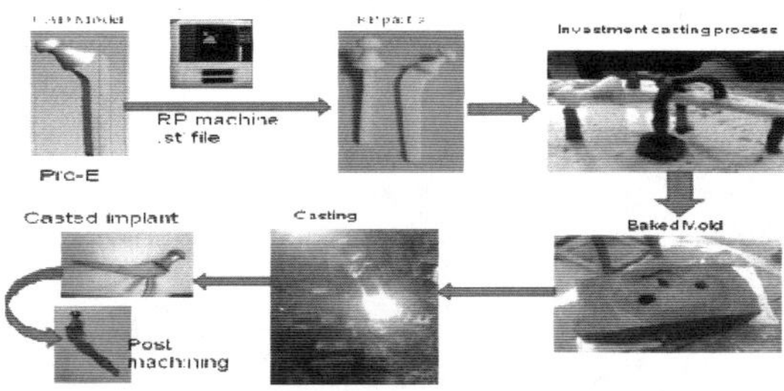

Fig. 5 Rapid manufacturing process for Implants

4. Experimentation

The investigated temperature increasing pattern inside muffle furnace for complete evaporatiton of RP pattern as shown in Table 1:

Table 1: The investigated temp. increasing pattern

Sr. No.	Temperature in (^{0}c)	Status of ABS-Plastic Material
1	At 100 ^{0}Cat 15 min	Solid form
2	At 127 ^{0}C at 15 min	Start to melt
3	At 200 ^{0}C at 15 min	Melting start
4	At 300 ^{0}C at 15 min	Melting continue
5	At 350 ^{0}C at 15 min	Melting complete
6	At 400 ^{0}C at 15 min	Smoke out start
7	At 500 ^{0}C at 15 min	Vaporization start
8	At 500 ^{0}C at 30 min	Vaporization continues at 30 min. Smoke continue
9	At 500 ^{0}C at 40 min	Evaporates completely

5. Mould making

Materials and accessories required for Molding are; vibrator, zirconium sand (biosint extra powder), biosole liquid, muffle furnace, RP-ABS polymer part.

Table 2: Technical data for zirconium sand

Mixing ratio	100 g powder :15 ml liquid
Total expansion	0.95 % - 1.65 %
Mixing time under vacuum	60 second
Processing time span	3 – 5 min.
Compression strength	According to concentration of the expansion liquid between 15 and 20 MPa

Formula for number of packets required for making mould (found from small experiment)

From 1 packets made =287cc volume of net block

Then for 1 cc volume of block = (1/287) packets

Then we can say for any volume of $(\pi/4*d2*h)$ block

Net Volume of sand material= Volume of block-Volume of pattern (sample)

$= \pi/4*d2*h- \pi/4*D2*H$ ... if sample were cylinder.

No. of packets required to be is

$$(N) = \frac{\text{Net Volume of sand material}}{287}$$

A ceramic block, which shape is depends upon the RP part, makes a block with minimum sand material and optimum shell thickness, made by poured coat of agitated ceramic slurry (zirconium sand slurry) & allowed to dried under controlled conditions within 30 min. After the blocking process completed baked in muffle furnace for complete burnout of pattern by ramping the temp. up to 500°C according to investigated temp. increasing pattern as per table 1. Backing process sinters the ceramics block and cause invested RP model to be burned out.

5.1 Block Washing

To remove ash and ceramic dust, the block was washed with a forceful air by blower & by water. The water stream is allowed to enter one gate and exit through another gate or a vent or riser. During washing, the water is agitated by shaking the block vigorously.

5.2 Casting

The completed block is now ready to receive the molten alloy. The block is preheated, and the alloy is cast into it according to the foundry's operational procedures.

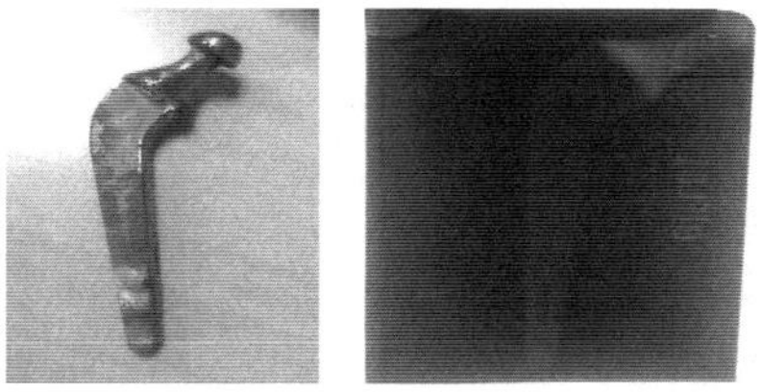

Fig.6 Actual femur implant casting along with radiograph

Conclusion

- Complete vaporizion of ABS pattern obtained by varying the temp. up to 500 ^{0}C in stages.
- Changes in time & temperature profile of baking was susceptible to crack, residue or half burn.
- If shell thickness reduces, mould may damage due to the forcees of molten metal.
- With the help of simulation data the casting process was improved & ease in manufacturing of complex shapes.
- The study is much helpful for tailor made implants casting in order to overcome the defects.

References

[1]R. Patil, S. M. Kumar, E. Abhilash :IJEIT, Volume 1, Issue 4, April 2012

[2] A. Pranjal, B. Prof. A. M. Kuthe :Proceedings of the 5thMESI Conference – Zaragoza – June 2013.

[3] Roger A :Dairy food & environmental sanitation ,vol.20,No.7,Pg. no. 506-517.

[4] S. Jauhar, K. M. Asthankar & A.M. Kuthe :(AMEA) 181 Vol. 2, No. 3, 2012, ISSN 2167-6380 Copyright © World Science Publisher, United States

[5] D.S. Ingole , A.M. Kuthe: Emerlad Group Pub. Ltd', 2009.

[6] W.L.Yao, Ming C. Leu: MCB University Press',1999.

O-73

Deflection simulation of a die installed in an HPDC machine to predict the die life

Hiroshi Yamagata
Faculty of Engineering, Gifu University, 1-1 Yanagido, Gifu City, Japan

This research proposed a method to predict heat-checking. The analysis of the die surface for a flat part revealed that heat-checking occurs at the portion causing plastic deformation. High stress concentration during curing causes the Mises stress to be above the yield stress. The diagram illustrating the maximum value of the stress difference (= Mises stress- yield stress) well predicted the crack positions observed in the die after 120,000 shots. The surface strain analysis revealed that the initiated crack widens with repeated expansion and contraction. The mode II mechanism of fracture mechanics well explains the crack initiation and propagation.

Keywords: *High-pressure die-casting, Heat-checking, Thermal distortion, Stress simulation.*

1. Introduction

Heat-checking (cracking) and gross-cracking typically shorten the durability of an HPDC die. In the previous study [1-3], we found that heat-checking occurs where the Mises stress exceeds the yield stress during curing. The surface strain analysis revealed that the initiated crack widens with repeated expansion and contraction. The coincidence of the stress analysis with the actual observation demonstrated that the mode II mechanism of fracture mechanics well explains the crack initiation and propagation.

In the present study, the author uses a similar approach to reproduce the three elements: the die-locking force, the casting pressure, and the thermal stress. The casting and solidification simulation revealed the temperature distribution of the die. The FEM stress analysis evaluated the stress causing heat-checking. The detailed mechanism is demonstrated and the method to enable precise prediction of the heat-checking is proposed.

2. Simulation

First, an FDM flow and solidification simulation calculated the temperature distribution of the die insert. Second, the thermal deflection of the die insert in the die holder was calculated from the temperature distribution. Finally, the deflection of the die insert caused by the additional die-locking force as well as casting pressures was simulated. The die and machine were modeled in accordance with the actual installation. The HPDC operation used an oil-based die-release agent. The injection conditions and additional parameters for the simulation were selected so as to follow the actual HPDC operation. Typically, the initial temperature of the molten metal JIS-ADC12 was 580°C. The slow injection speed was 0.3 m/s, and the high injection speed was 1.7 m/s. The curing time was 5.5 s. One shot of the actual casting process comprises five stages. A total of 20 shots reached the steady state temperature distribution of the die. The die-locking force was applied from the toggle of the movable platen with a load of 2,450 kN. A casting pressure of 60 MPa was vertically and homogeneously applied on the cavity surface in order to simulate the squeezing pressure. The detail of these installations was illustrated in the previous paper [1-3].

3. Results and discussion

Fig. 1 shows a detailed variation of cavity surface temperature at points 1 through 12 during the 20th shot. The point number corresponds to the numbers in Fig. 2. The horizontal axis shows the time from the first shot as well as the time from the onset of this shot. The curing after the injection starts from 560.5 s. The timings discussed are indicated by the vertical lines in the diagram.

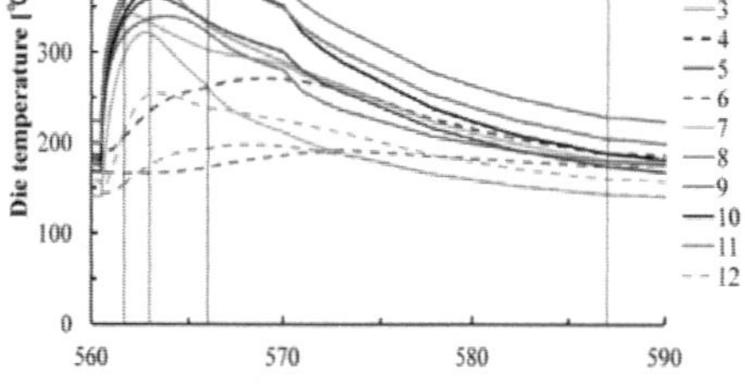

Fig. 1 Variation of temperature at points 1 through 12 during the 20th shot. The injection ends at 560.5 s (0.5 s).

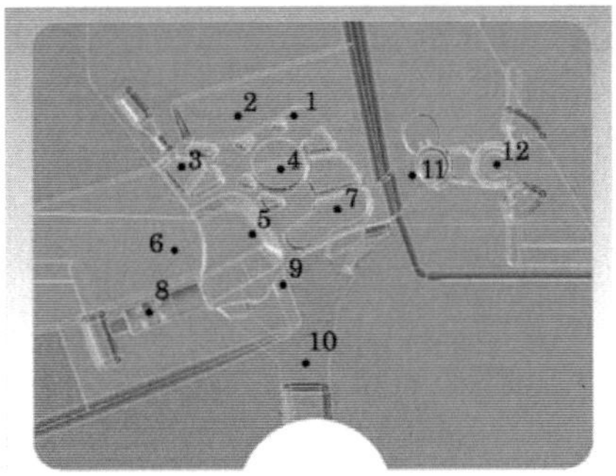

Fig. 2 Points 1 through 12 used in stress calculation for the movable-die surface.

The surface temperature increases rapidly after injection and shows a peak, but the timing showing the peak is slightly different from peak to peak. In the diagram, the curing finally ends at t3. The timings t1 = 561.7 s and t2 = 563.0 exist during curing, followed by the timings at the end of die opening and at the end of part release from the die. The end of spray-air blow after parting the die is also indicated at 587.0 s. The highest thermal stress at the area of focus on the die surface takes place during curing. In the previous study [1, 2], we used timing, typically t2, for the judgment (Mises stress> yield stress) at the analyzed position. However, the temperature-peak time differs from point to point, as shown in Fig. 1. The difference (= Mises stress- yield stress) will be proportional to the degree of plastic strain. In order to avoid this issue and to best evaluate the place of heat-checking, the stress difference at t1, t2 or t3 was separately calculated at each point and the maximum value of the stress difference was plotted as shown in Fig. 3.

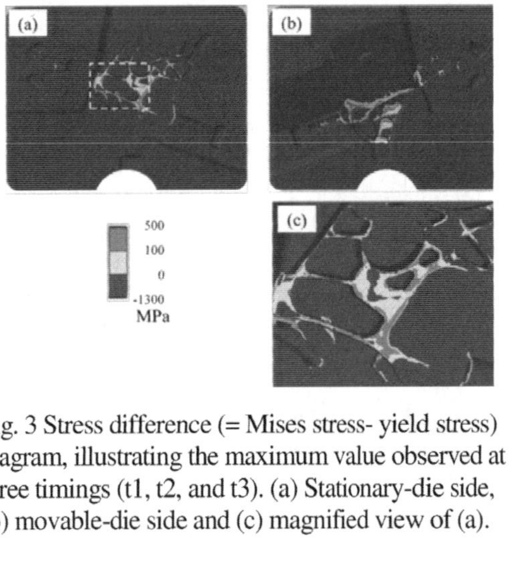

Fig. 3 Stress difference (= Mises stress- yield stress) diagram, illustrating the maximum value observed at three timings (t1, t2, and t3). (a) Stationary-die side, (b) movable-die side and (c) magnified view of (a).

This diagram was found to be valuable in determining whether heat-checking has occurred. Fig. 4 shows an actual heat-checking which corresponds to the enlarged view of the square area enclosed by the dotted line in Fig.3 (c). The coincidence confirmed the validity of the mode II mechanism. Fig. 3(b) shows the similar stresses on the stationary-die side. The high stress difference appears just beyond the gate. Thus, typical heat-checking just beyond the gate is predictable.

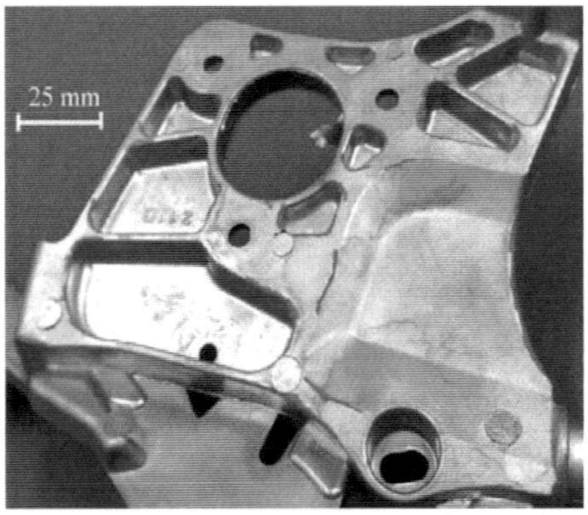

Fig. 4 Transcribed heat-checking observed in the part on the movable-die side after 120,000 shots. Enlarged view of the square area enclosed by the dotted line in Fig.3 (a).

4. Conclusion

FEM stress analysis of the die revealed that heat-checking occurs at the localized portion where plastic deformation takes place during curing. The diagram showing the maximum value of the stress difference during curing well predicted the crack positions observed on a flat part of the die after 120,000 shots. The coincidence confirmed the validity of the mode II mechanism.

References

[1] H. Yamagata, T. Kimura and S. Tanikawa: Transaction of JDCATEC (2014) JD14-15.

[2] H. Yamagata and T. Kimura: Proceedings of the Advances in Materials and Processing Technologies (2015) Madrid, 294-295.

[3] H. Yamagata: Visualization of die-technology of high-pressure die-casting using computer simulation, (Kallos Publishing, Tokyo, 2015) 41-52.

Proceedings of the 72nd World Foundry Congress
21-25th May 2016, Nagoya, Japan

Confirmation of the heat check deterrent effect by the shot peening

Yuji Kobayashi1, Akinori Matsui1
1 SINTOKOGIO, Ltd.,Toyokawa, Aichi Pref.441-1205, Japan

Die cast method is often used for automotive parts that can make high strength and high precision parts. Parts are molded in high temperature and high pressure. Then, high strength is required on die casting die. Most common destruction forms of die cast die are called heat checking which generate casting defects.
In this study was used specimens of the same material, but submitted to different treatments. The cyclic heat stress was simulated on the specimens and were compared the results among the different treatments.

Keywords: *shot peening, die cast, heat checking.*

1. Introduction

Die cast method is often used for automotive parts that can make high strength and high precision parts.

Parts are molded in high temperature and high pressure. Then, high strength is required on die casting die.

Most common destruction forms of die cast die are called heat checking which generate casting defects. Heat checking is occurred on the grain boundary by cyclic heat stress due to molten and release agent. Therefore, it will occur flat surface easily.

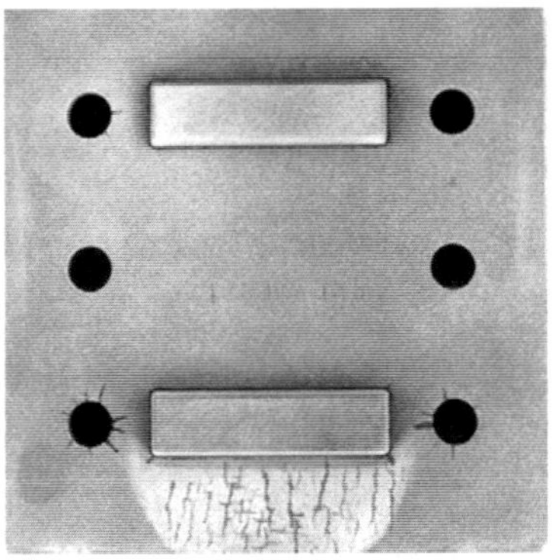

Fig.1 Heat checking

Fig.1 shows the typical heat cheking that occurred at the die surface [1] [2] . Shot peening is a method of cold working. Residual stress is generated by shot peening near the metal surface by hitting small spherical metal on the surface in order to improve fatigue strength. Mainly, it is used for transmission gear or valve spring.

In this study, authors will report the shot peening deterrent effect to heat checking on die casting die.

2. Experiment method

Hot tool steel that is called SKD61 is prepared for specimens. Table1 shows chemical composition of SKD61. After machining, we carried out various kind of surface treatment. Table 2 shows condition of heat treatment.

Table1 Chemical composition of SKD61 (%)

C	Si	Mn	Cr	Mo	V
0.4	1.0	0.4	5.2	1.2	0.8

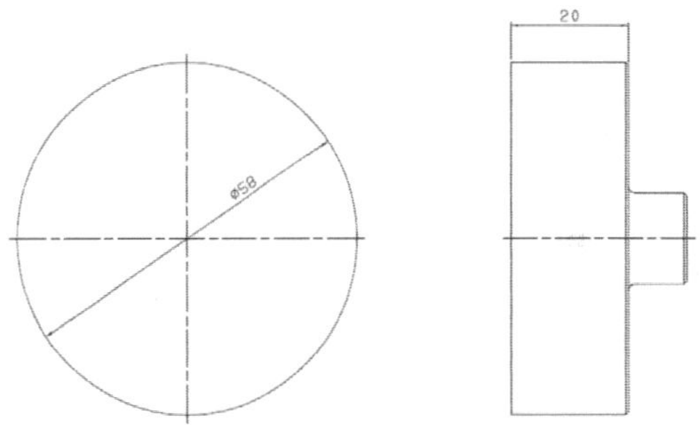

Fig.2 Shape of specimen

Table2 Condition of heat treatment

Heat treatment	Number of cycle
Quenching +Tempering	15000, 100
Salt bath nitriding	10000
Repeening	25000 (5000/Repeening)
Laser peening	100

Fig. 2 shows the shape of specimens. Fig.3 shows the method of experiment. Surface of specimen was put on the heater that was heated in approximately 853K. After this, the specimen was dropped into cold water. Then, the specimens were submitted to cyclic heat stress. In this cycle was simulated the cyclic heat stress that occurs on die casting die.

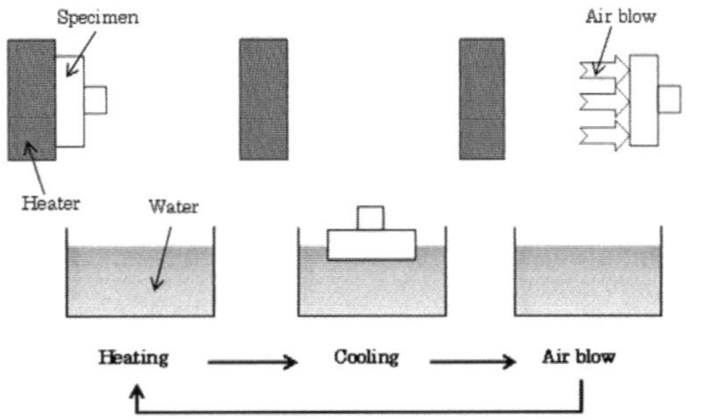

Fig.3 Method of experiment

Basically, heat checking was beginning to occur after 100cycle. But, authors observed heat checking after 1000 ~ 25000 shots. In one specimen, the shot peening was applied again every 5000cycles in order to expand the life of die. Residual stress, FWHM and 3D shape were investigated.

3. Result

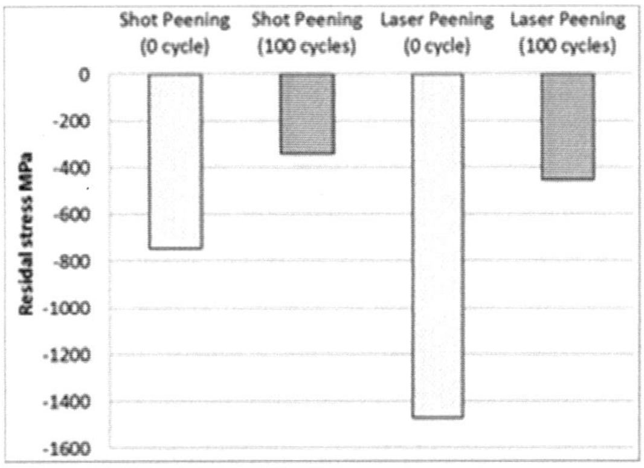

Fig.4 Surface residual stress

The Fig.4 shows the residual stress in the shot peened and laser peened specimens before the cyclic heat stress (0 cycle) and after the cyclic heat stress (100 cycles). The residual stress in the laser peened specimen is 97% higher than the residual stress in the shot peened specimen, but after 100 cycles the residual stress decreases considerably in both cases and now the residual stress in the laser peened specimen is only 31% higher than the residual stress in the shot peened specimen.

In the Fig.5 is shown the photograph of heat checking in a shot peened specimen before the cyclic heat stress, after 100 cycles and a laser peened specimen after 100 cycles. In the first picture, regarding to shot peened specimen before the cyclic heat stress, was not found any heat checking. In the second picture, regarding to shot peened specimen after 100 cycles, almost there are no cracks due the

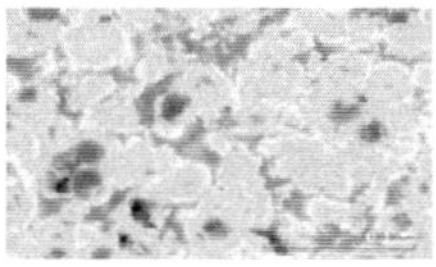

Shot Peening - Before cyclic heat stress

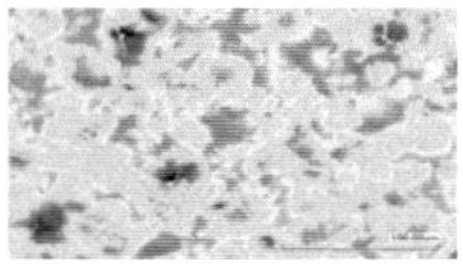

Shot Peening – After 100 cycles

Laser Peening – After 100 cycles

Fig.5 Comparison of heat checking between Shot Penning and Laser Penning

heat checking effect. And, in the third picture, regarding to laser peened specimen after 100 cycles, there are a representative quantity of cracks due the heat checking effect. Although the laser peening process have better remaining residual stress, the shot peening process presents better results for heat checking deterrent effect, because the plastic deformation is much higher in the shot peening process.

Shot peening process is more effective than the laser peening process, beyond being cheaper.

4. Summary

Heat checking deterrent effect by shot peening is confirmed. The laser peening process have better remaining residual stress, but the shot peening process is more effective for heat checking deterrent effect than the laser peening process, beyond being cheaper.

The results provide valuable information to reduce casting defects.

References

[1] T. Kitagawa and Y. Hiraoka and Y. Kobayashi, Complex surface treatment, improving heat-checking – resistance for die casting mold " AMINIT DS", Die and Mould Technology , Vol. 25 No.27(2010), pp 118-129.

[2] S. Nakahama, Technologies to improve the casting die life , Die and Mould Technology , Vol. 25 No.4(2010), pp 48-51.

Increasing the Lifetime of Die Casting Molds by Additive Remanufacturing Applying Electron-Beam Deposit Welding with a Local Process Integrated Heat Treatment

T. Schuchardt, K. Dilger and H. Pries
Institute of Joining and Welding, Braunschweig University of Technology, Braunschweig, Germany

High pressure die casting molds are constantly exposed to high thermal, mechanical, chemical and tribological cyclic loads during the operation. These loads might cause different defects on or in the mold and thereby reduce the die lifetime significantly. In the case of an unexpected production stop caused by critical defects in the mold, repair welding by means of deposit welding is often the only way to reinstate the casting tool contemporary to the production.

Hence, the primary objective is the development of the electron-beam welding technology for an economic regeneration of locally damaged die casting tools.

Keywords: *HPDC, additive remanufacturing, electron-beam welding, permanent mold*

1. Introduction

The profitability of each die casting process is significantly influenced by the achievable lifetime of the die casting die.

Moreover, the surface quality of the cast component is directly correlated with the surface quality of the permanent mold. The process-related damages, like stress cracks, heat checkings and erosion increase constantly with every casting cycle and map onto to the so-called functional surfaces of the cast component. Damages have to be reworked manually, which is time- and costintensive. Due to the high costs for the production of a new mold insert, which can amount to at least 50 % of the initial cost of a new mold insert, repair welding instead of new manufacturing is a preferred approach. [1]

2. Repair welding of casting molds

Currently, the TIG or plasma welding processes are mainly used for such repair welding routines, because of their relatively low-cost system technology. Repair welding is also increasingly performed by using the semi-automated laser surface cladding. [2] Nevertheless, these conventional welding techniques manifest an insufficient process-reliability and therefore achieve an unsatisfactory increase in the lifetime.

The mold material mostly used in aluminum die casting is the hot-work steel H13. Due to its relatively high carbon, molybdenum, vanadium and chromium contents, this steel is not very well weldable. To avoid failures like hardening and cracking it is essential to heat the mold insert before welding to above the martensite start temperature. However, preheating temperatures should be avoided above 450 °C, as this may lead to embrittlement of the base material. [3]

As a consequence, insufficient preheating of the die or a too long cooling prior to performing the repair welding cause a very high hardness in the weld microstructure. This may also lead to macro cracks in the weld deposit. Tempering reduces the hardness in the weld structure within certain limits, by welding martensitic materials. Apart from the outstanding hardness peaks, also areas with hardness values in the heat-affected zone below the base material are noted. These are caused by annealing effects due to the heat input during the welding process. Because of the high hardness there is still a metallurgical notch in the heat-affected zone. Despite the final machining of the surface, this notch might quickly lead to a renewed crack and a failure of the permanent mold in the environment of the repair weld during the casting process.

3. Electron-beam welding

3.1 EB-technique

In addition to the laser beam method, electron-beam technology represents the second beam technology for the processing of thermal materials. Thereby, almost massless and thus inertia-less electrons for the thermal material processing are accelerated up to 2/3 of the speed of light by means of an acceleration voltage of 120 kV from an electron cloud in front of the cathode. The focusing and highly dynamic deflection of the electron beam is effected by a number of electromagnetic lenses. If the electrons impact on the workpiece surface, they will be braked. At this deceleration process, the kinetic energy of the electrons is converted into thermal energy and thus creates the possibility of melting the metal base or filler material. [4]

3.2 Researches on deposit welding by electron-beam welding

In the first part of the researches suitable filler materials were identified and tested under vacuum conditions. An under-alloyed, a similar and an over-alloyed material were tested. For possible buffer layers an additional filler material should be examined for Cobalt basis and on the basis of maraging steels with the standard designation 1.2709. Each filler material is fed as a filler wire with a diameter of 1.6 mm into the molten bath with an in the electron-beam system built-in wire conveyor. It has been shown that filler wires are not weldable in vacuum without a massive formation of weld spatters. That's why only massive wires were used for the further investigations.

In the main part of the investigations build-up tests were performed in extra pre-milled swales. The milled swale should simulate the real situation in the repair welding process in this connection, since any possible cracks in the die are milled and then back-filled with the filler material. The basic material is the hot work tool steel H13 (ESR-quality, annealed to 45 HRC), using SzFe3 as massive filler material.

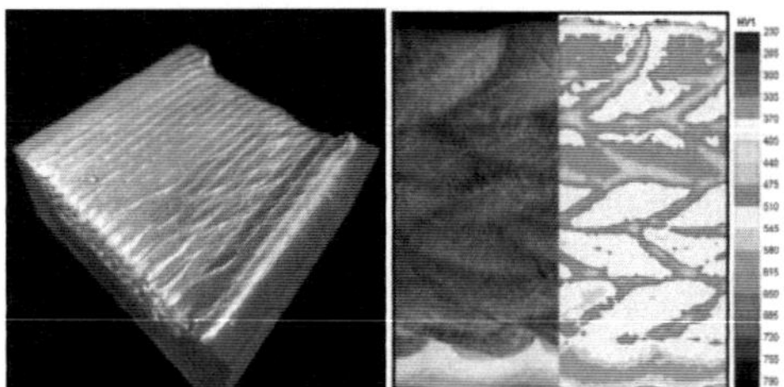

Fig. 1 Deposit welded filled test piece (swale)/ Cross-section polish and a hardness allocation of an EB-welded swale (without tempering).

It was welded with a torch directed toward the finished part of the weld, at which the sample was rotated to 180° by means of clamping. This was the reason that at the end of each track, a drop height of the weld took shape. A rotation after each layer compensated this situation. A total of five layers are required to fill the 10 mm depth of recess. Different heat treatment strategies were applied by adaption of the beam, simultaneously. Based on these experiments, it was found that a generative structure using the electron-beam system and an additional material is generally possible, as the test piece in figure 2 shows.

Furthermore, it could be shown that the connection between filler material and base material is well constituted with a relatively small heat-affected zone. Welding defects such as hot or cold cracking did not occur at all surfacings. Besides, figure 2 left shows clearly that the typical metallurgical notch in the heat-affected zone is less distinct than in conventional welded examples.

4. Conclusion and prospect

The conducted experiments show that the electron beam welding method is suitable for a generative development of metallic materials. For this purpose, multi-layer deposit weldings were realized with a typical hot work steel and suitable filler materials. Possible build-up strategies for three-dimensional bodies were demonstrated.

These results form the basis for an additive remanufacturing of permanent casting molds. A pre-and post-heating of the weld metal is hence essential in this or any case to avoid heat cracks.

Due to its multi-beam technology, the electron-beam welding technique allows in contrast to conventional welding methods a process-integrated pre-and post-heating. The next step of will be the application of the achieved results to real damaged die casting dies.

Acknowledgements

The authors would like to thank the Federal Ministry for Economics and Technology (project no. AiF 17.843 N) for the funding of this research project.

References

[1] Wanke, P.: Leichtbaupotentiale im Automobilbau durch Gießtechnik, Landshut, 27.10.2010.

[2] Müller, S.; Pries, H.; Dilger, K.; Ocylok, S. et al. (2011): Applying Functionally Graded Materials by Laser Cladding, Proceedings of the 18th CIRP International

[3] Garthoff, Chr.; Müller, S.; Dilger, K. (2011): Reparaturschweißen von Aluminiumdruckgießformen

[4] Schultz, H. (2000): Elektronenstrahlschweißen, Fachbuchreihe Schweißtechnik, Band 93

Proceedings of the 72nd World Foundry Congress
21-25th May 2016, Nagoya, Japan

Development of Cooling System for Small Core Pin in Aluminum HPDC and Its Applications

Suguru Takeda[1], Shin Orii[1], Nobuyuki Sakai[2],Shinji Sannakanishi[3]
[1] Ahresty Corp. Production Eng. Div.; 1-2 Nakahara Mitsuya-cho Toyohashi, Aichi, Japan
[2] Ahresty Mexicana S.A. de C.V.; Calle Industria Automotriz #20 Complejo de Naves Industriales la Zacatecana Guadalupe, Zacatecas C.P.98600, Mexico
[3]Ahresty Corp. Advanced Production Eng. Dev.; 1-2 Nakahara Mitsuya-cho Toyohashi, Aichi, Japan

This paper reports the technical aspects of the cooling system in tackling soldering as well as its application in metal saver. As a result of the experiments, no soldering is observed on the small core pin in the Aluminum alloy die casting with using thin walled core pin and greater quantity of cooling water without special die surface treatment.

Keywords: *Aluminum Alloy Die casting, Thin Wall Core Pin, Heat Flow, Surface Temperature, Soldering,*

1. Introduction

To cope with the ever increasing demand for compact engine lay-out, energy efficiency, cost savings, etc., automobile parts and hence die casting die shapes are getting more complex than ever before. Inconsistent part thickness, slender cored holes, deep part shapes have become common features, which in turn needs the use of long and thin core pins and intricately shaped deep insert blocks. Core pin in thick part section often suffers from soldering due to insufficient cooling. Porosity and lower productivity due to longer machine down time caused by core and die polishing and changing broken core pins have become two of the most severe quality and cost concerns, respectively.

In order to address these problems, authors think that thermal approach would provide a better solution to these problems. Thus they have studied [1],[2],[3],[4] the thermal behavior of die castings, heat flow ratio and die surface temperature pattern. And then they have developed the concept of cooling core pin with high pressure water jet with subsequent purging of remaining water enabling real time control of heat flow, and then successfully used the new system in mass production.

2. Thermal behavior of core pin in various thermal conditions

Thermal behaviors of core pin in various conditions were measured. As shown in Fig.1, there are four patterns depending on the casting volume and cooling conditions. In the beginning, just after the metal fills in the cavity, the temperature and heat flux patterns are not so different. However, after reaching the peak value, the thermal behaviors are quite different in each case. In case of low cooling water pressure and large casting volume, counter heat flow (heat flow from core pin to casting) is observed[4)]. There's very little heat removal form the cavity to core pin which causes defects such as soldering, shrinkages, and porosities.

The high efficient heat removal form the casting to core pin is achieved with a combination of high flow rate and thin wall of the core pin. That reduces the peak temperature and increases heat flux of the core pin which prevents soldering of core pins. The wall thickness should be no more than 15% of the outer diameter of the pin.

As for example, a core pin with outer diameter of 7mm, the wall thickness will be only 1 mm, which may apparently be thought to cause core pin failure. However, it has been established by computer simulations as well as experiment that the durability of the thin walled core pin against bending force is almost same as that of solid core pin[55]].

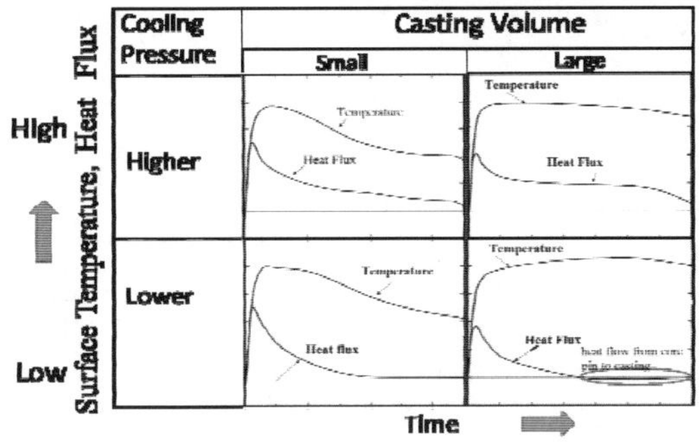

Fig.1 Thermal Behavior of Core pin in various casting conditions

3. Controlling method and its apparatus for preventing defects caused by thermal effects

From the experimental result of thermal behavior of core pin, a new cooling system has been developed as shown in Fig.2. There are four main sections such as "High Pressure Water Supply Unit" capable of supplying large amount of cooling water into small channel of core pin; "Air Purging System" that removes remaining water to control the core pin temperature; "Water purifier" that purifies the cooling water preventing from scale and; "Water Cooler" to chill the returning water from core pin and dies. In addition to the cooling system, core pin and cooling pipe have also been developed.

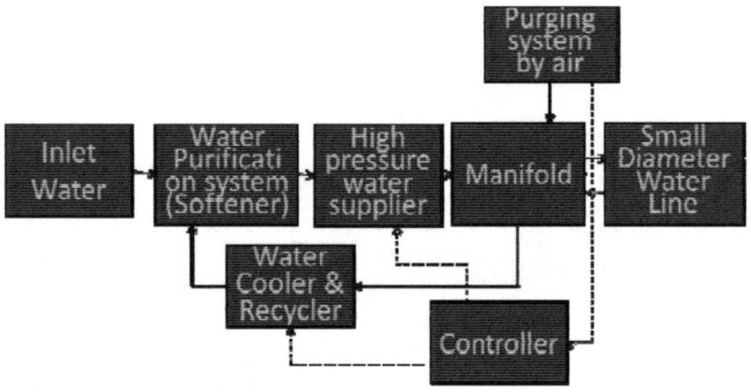

Fig.2 Block diagram of developed cooling system

4. Applications using the Cooling System

The developed apparatus works not only improving soldering but also reducing another defect such as porosity. The typical examples are shown in Fig.3 and Fig .4. As shown in Fig.3, there is no soldering of the core pin after 5000shots by applying this cooling system, and core pin polishing time was extremely reduced to zero which contribute productivity and safety.

Fig.3 Prevention of soldering using this system

Moreover, with conventional cooling system, shrinkage porosity appeared around core pin due to overheat as shown in Fig.4(a), but when the newly developed cooling system was applied, the shrinkage porosity disappeared as shown in Fig.4(b). Hence, they improves the leakage problem in the machined surface.

Fig.5 shows the result of applying the cooling system to draft angle zero core pin. After 100shots, slight sticking is observed in case of rougher surface (Rmax=3.9micro-meter), however no sticking and no soldering are observed in case of polished pin (Rmax=1.1micro-meter). This shows the possibility of eliminating machining after casting.

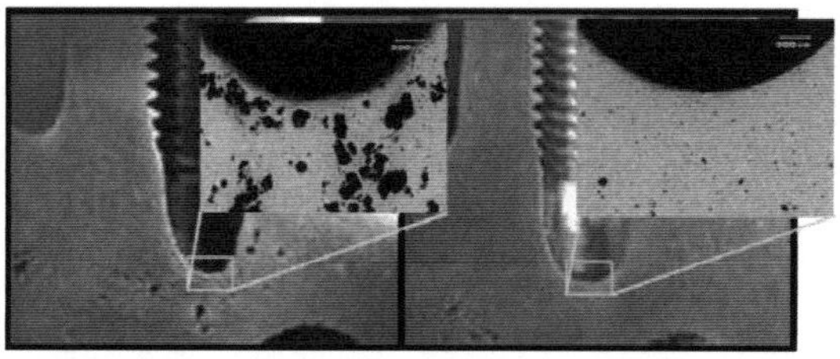

a) Conventional Cooling b) Developed Cooling

Fig.4 Reducing porosity near the core pin surface with this system

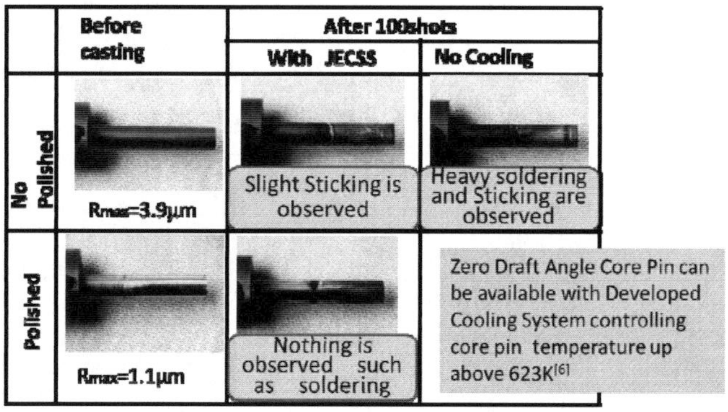

Fig.5 Trial approach of draft angle zero core pin using this system

5. Conclusions

By determining the thermal behavior of core pins, new cooling system has been developed. This system is proved to be effective for small core pin preventing defects caused by thermal problem such as soldering, shrinkage and leakage. Moreover this system contributes to the net-shape manufacturing by HPDC, which eliminate machining to a large extent.

References

[1] K. Shirahige, Y. Hatamura, S. Takeda, et al. ;Trans. Japan Die Ccasting Congress, (1986), 123-130

[2] Y. Hatamura, K. Shirahige, S. Takeda, Y. Mochiku; Trans. 15th International Die Casting Congress, (1989), paper No. G-T89-023, 1-7

[3] S.Takeda, S.Orii; J. Japan Inst. Light Metals, 65(6), (2015), 1-5

[4] S. Takeda, S. Orii; 128th of JILM (May 2015, Sendai), 11-12

[5] S. Takeda, S. Sannakanishi, et al.; 166th of JFS (May 2015, Tokyo), 84

[6] K. Tashimo, K. Sakamoto; JFS 60 (1988)747

Proceedings of the 72nd World Foundry Congress
21-25th May 2016, Nagoya, Japan

Effect of Manganese on Intermetallic layer Morphology between Dies and Al-Si alloy

D. Ferdian[1], Vika Rizkia[2], Bambang Suharno[1], Sri Harjanto[1].

1 Dept. of Metallurgy and Materials Engineering, Universitas Indonesia, Kampus Baru UI Depok 16424, Indonesia

2 Dept. of Mechanical, State Polytechnic Jakarta, Kampus Baru UI Depok 16424, Indonesia

Die soldering phenomena has been a problem encounter in die casting industry. Several methods already developed to mitigate and reduce their detrimental effect. Manganese was known for their capability to suppress the harmful effect of iron rich intermetallic phase. In this work, the effect of Manganese to the intermetallic layer morphology was investigated. The experiment was carried out by immersion of H13 into molten Al-Si alloys. The alloy was added 0.1-0.7% Mn with immersion time carried out for 10 to 60 minutes. The immersion held at various holding temperature of 680-740°C. The intermetallic layer was then characterized using micro hardness, scanning electron microscopy (SEM) and energy dispersive spectroscopy (EDS). The result showed that compact and broken intermetallic layer was formed during immersion process. Observation also showed that Manganese reduced intermetallic layer thickness without reacted with the layer. Furthermore, mechanical properties by hardness method showed no significant changes of intermetallic layer hardness value in relation to the manganese addition.

Keywords: *Die soldering, intermetallic layer, manganese, Aluminium-silicon alloy.*

Die soldering is one of the major causes of die failure and occurs directly as a result of interactions between the die steel and the injected molten aluminium alloy [1]. Numerous researches already conducted to study die soldering phenomena. Sticking of the casting alloy to the die, as a direct result of die soldering, can produce defective castings, hinder ejection of the casting from the die and shorten the useful die life [2]. Several researchers studied the effect of element on capability to reduce the detrimental effect of soldering phenomena. Xiaoxia et al. [3] studied that the existence of Si could reduce that interaction between die and molten aluminium. Nevertheless, another work by Springer et al. [4] showed different result where the addition of Si accelerate reaction layer in solid-semisolid interdiffusion.

Manganese were known as effective element for neutralize the detrimental effect of iron rich intermetallic phase, especially β-$Al_9Fe_2Si_2$ into less harmful of α-Al(MnFe)Si. Studied by Chu et al. [5] showed that the ability of molten metal to be ejected from the die will increase by at least 0.4% manganese addition. Their studied showed the Mn also increase the intermetallic layer thickness. However, different result was obtained by Shankar et al. [6,7] which observed that the addition of manganese into molten aluminium slows the growth of the intermetallic layer. Preliminary studies on Al 12%Si by Harjanto et al. [8] showed that manganese addition could reduce the formation of compact layer at about 39%.

In general, the intermetallic layer consists of compact intermetallic and broken intermetallic layer, which will be refers as compact and broken layer in this paper. The presence work aims to investigate further effect of manganese addition into Al-7%Si and Al-12%Si alloy on the intermetallic layer morphology of H13 surface with different holding time and temperature.

The materials studied in this work were master alloy of Al-7%Si and Al-12%Si which added 0.1, 0.3, 0.5 and 0.7% Mn in the melting furnace. No modifier or grain refiner was added to the alloys. Half cylinder sample of annealed H13 tool steel with 30 mm in diameter and 4 mm in height was prepared. H13 steel samples then polished to 1000 grit prior the immersion into the molten alloys. The H13 sample then dipped into molten Al-7% and Al-12%Si alloy for 10-60 minutes at holding temperature of 680-740°C. The temperature of the melt was measured using a k-type thermocouple. After the dipping process, the samples were air cooled to room temperature. The cross sections of the dipped samples were prepared for metallographic observation using Scanning electron microscopy and Energy Dispersive X-Ray Spectrometer (SEM/EDX) were used to measure and characterize the intermetallic layer.

The immersion test was conducted to investigate the relation between Manganese additions to the intermetallic layer growth at H13 contact surface. For Al-7Si, the morphology of compact layer seems influenced by the Mn content and the holding temperature. In Fig 1, the value appear shifted to higher Mn content in positive parabolic curve as the holding time increases for holding temperature of 700°C. The difference between holding time is not clearly seen at low Mn content.

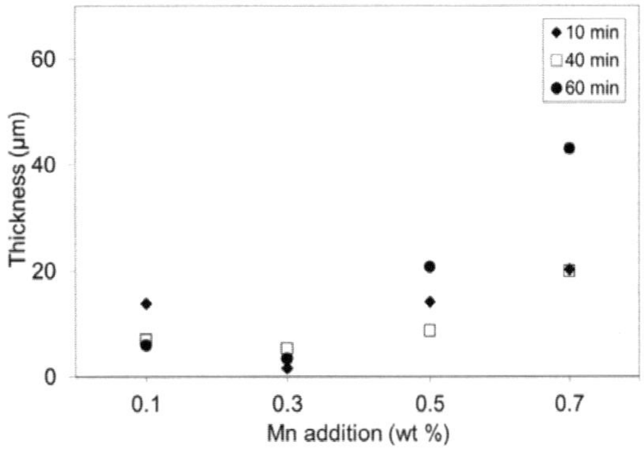

Fig. 1 Effect of Mn addition into Al-7%Si alloy at temperature 700°C on the thickness of the compact layer

For Al-12Si, as seen in Fig. 2, it shown that the Mn content and holding temperature also affected the compact layer morphology, where the compact layer morphology appeared reduced as the Mn and holding temperature increases. Generally, the compact layer has higher iron content than broken layer. EDS analysis of the layer as seen in showed the chemical composition range for binary phase and ternary phase such as $FeAl_2$, η-Fe_2Al_5 and α-$Al8Fe_2Si_2$. Analysis shows no significant effect of Mn content level to the intermetallic layer chemical composition. Furthermore, XRD analysis did not show any appearance of quarternary intermetallic phase.

Addition Mn to Al-Si alloy could reduce the thickness of the compact layer. The layer thickness also influenced by the holding temperature. The Mn seems effectively reduced the formation of compact layer without reacting with the intermetallic layer phases. Although, the phenomena in lower holding temperature for Al-7Si requires further investigation.

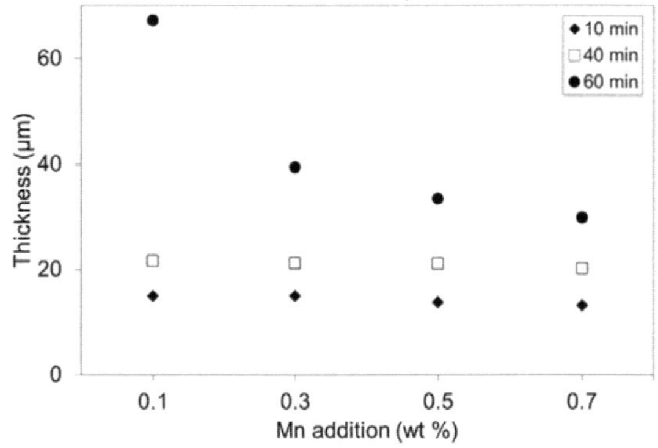

Fig. 2 Effect of Mn addition into Al-12%Si alloy at temperature 720°C on the thickness of the compact layer.

Acknowledgements

Partial financial support to this research from Faculty of Engineering and Directorate for Research and Public Services-Universitas Indonesia is highly acknowledged.

References

1. B. Zabala, A. Igartua, F. Pagano, M. Conte, H. Eibisch, R. Heid, On the nature of high presure die casting failure, 71st World Foundry Congress, Bilbao, Spain, 2014
2. S. Shankar and D. Apelian, The Role of Aluminum Alloy Chemistry and Die Material in Die Soldering, NADCA Transactions (1999) 281–290.
3. H. Xiaoxia, Y. Hua, Z. Yan, P. Fuzhen: Materials Letter. 58 (2004) 3424-3427
4. H. Springer, A. Kostka, E.J Payton, D. Raabe, A. Kaysser-Pyzalla, G. Eggeler: Acta Materialia. 59 (2011) 1586-1900
5. Y.L Chu, P.S. Cheng, R. Shivpuri: Transactions of North American Die Casting Association (1993) 360-371
6. S. Shankar and D. Apelian: Metall Mat Trans B, 33(1993), p. 465-476.
7. S. Shankar, and D. Apelian.: Intl.Journal of Cast Metal Res. 15 (2002), 103-116
8. S. Harjanto, V. Rizkia, B. Suharno, Effect of manganese addition into Al-12%Si alloy on die soldering phenomenon, International Conference on Material Processing Technology, Bangkok, 2010.

Proceedings of the 72nd World Foundry Congress
21-25th May 2016, Nagoya, Japan

Development of Crack Prediction Simulation Technology

Naoki Nishimura[1], Kazuteru Ishihara[1], Tatsuyuki Kobayashi[1] and Tsutomu Ishikawa[1]

80 Kowari, Minaminakane-cho, Nishio, Aichi 445-0810, Japan

In aluminum die-casting, we predict by simulation (hereinafter referred to as "CAE") risks to productivity including cavities, leaks, burrs, and soldering marks, and incorporate the results in die design and casting condition settings to make quality improvements. However, prediction technology by CAE has yet to be established for shrinkage cracks, which that is a known defect. Under these circumstances, we focused on shrinkage cracks that would have a significant effect on quality and successfully established related prediction technology. This paper reports on that technology.

Keywords: *die-casting, CAE*

1. Introduction

Of the cracks seen in aluminum die-casting, there are cracks caused by mechanical force such as die opening force and those caused by casting stress and inhibited thermal contraction. The latter includes hot cracks and shrinkage cracks. Shrinkage cracks occur in the final section to solidify in the solidification process, such as a thick-wall, due to inadequate supply of molten metal in aluminum die-casting [1]. So, we focused on the solidification process to work on the development of prediction technology.

2. Shrinkage cracking tests

2.1 Test method and conditions

We conducted casting tests using the I-beam test mold shown in Fig. 1, in order to assess the shrinkage cracking phenomenon with a simple model. We composed a mold of four separate sections using SKD61 hot work tool steel as a mold material, and poured molten JIS ADC12 aluminum alloy at 923K into the mold. Table 1 shows the test conditions. We conducted the test with and without heat insulator to change the solidification rate, and assessed the temperature of casting up on the occurrence of cracks.

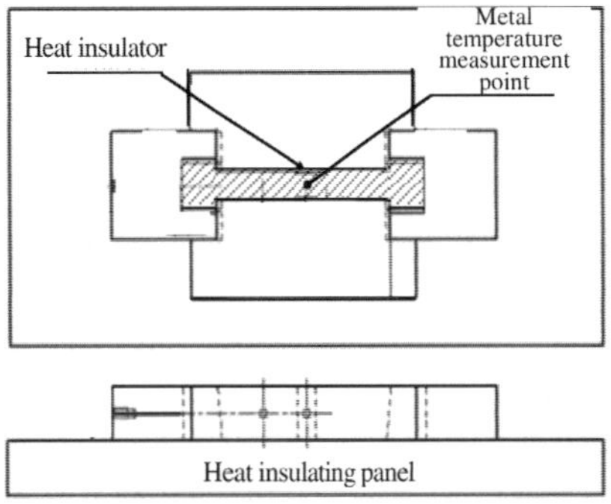

Fig. 1 I-beam test mold

Table 1 Test conditions

	Condition 1	Condition 2
Heat insulator	Provided	Not provided
Pouring temperature	923K	
Initial mold temperature	473K	

2.2 Test results

In tests, a crack occurred under Condition 1 that delayed solidification (Fig. 2). The temperature of the casting in the center of the test piece where the crack occurred was about 773K. This proved that the crack occurred in the solidification process (Fig. 3).

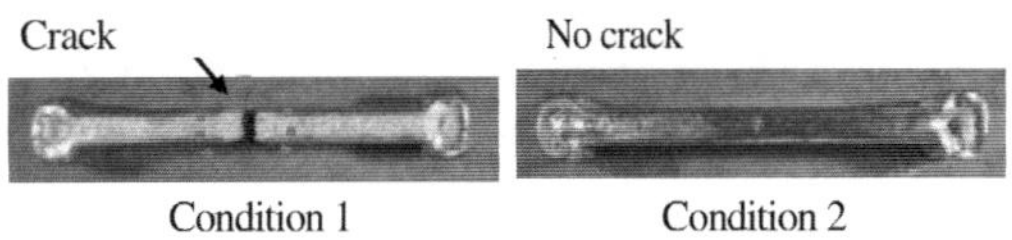

Fig. 2 Test results

Molten metal temperature	893K	843 to 773K	773 to 673K
Appearance	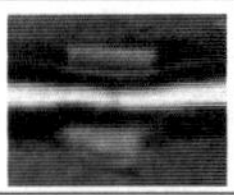	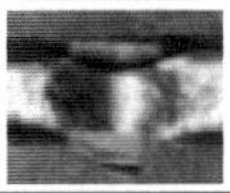	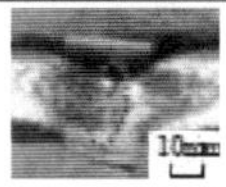
Remarks	Right after molten metal was poured	External shrinkage occurred	Crack occurred

Fig. 3 Cracking process (Condition 1)

3. Prediction of shrinkage cracking

3.1 CAE analysis model and results

We prepared a CAE analysis model like the model used for shrinkage crack assessment tests and analyzed the effects of changes made to the solidification rate. We measured the maximum principal strain in the center of the test piece in which the shrinkage crack occurred during tests. As a result, the maximum principal strain in the solid-liquid zone under Condition 1 was larger than that under Condition 2, as shown in Fig. 4. Consequently, we prepared an index to assess the risks of shrinkage cracks based on the maximum principal strain, and assessed them.

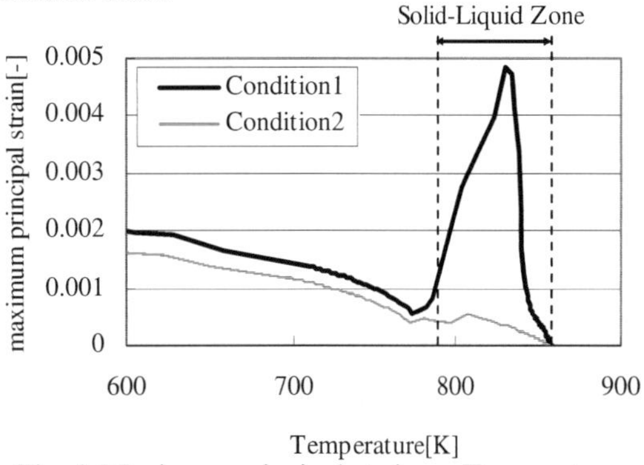

Fig. 4 Maximum principal strain vs. Temperature

As a result, Condition 1 showed a higher risk of shrinkage cracks, on the other hand Condition 2 showed a lower risk, as shown in Fig. 5. Consequently, we considered it possible to predict shrinkage cracks using CAE analysis and established the shrinkage crack prediction technology.

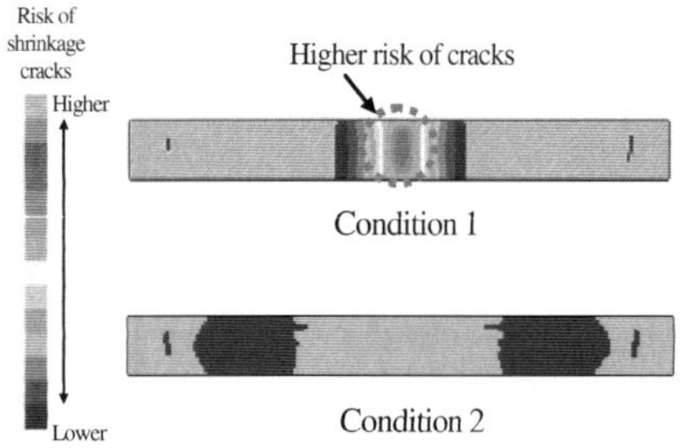

Fig. 5 Results of prediction of shrinkage cracks

3.2 Application to aluminum die-cast components

We verified the shrinkage crack prediction technology using an actual aluminum diecast product based on the results of CAE analysis conducted with the use of the I-beam test mold.

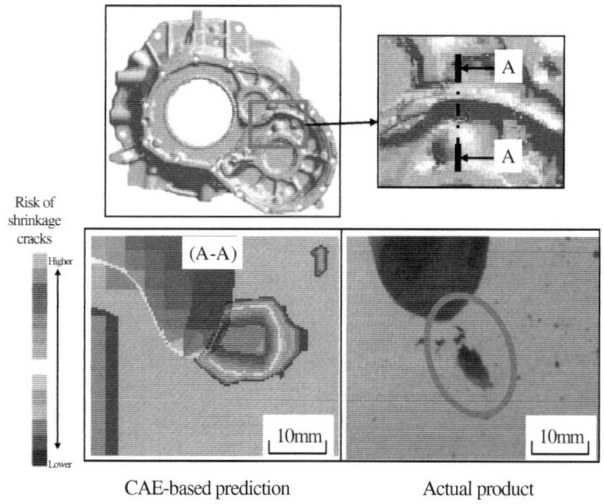

Fig. 6 Comparison between actual product and CAE-based prediction

As shown in Fig. 6, the location of the crack in the actual product was consistent with that predicted by CAE analysis. This proved that this shrinkage crack prediction technology, which takes the maximum principal strain assessed using the I-beam test mold as a threshold, was applicable to aluminum die-cast components.

4. Conclusion

We successfully established shrinkage crack prediction technology applicable to aluminum die-cast components by examining the mechanism that causes shrinkage cracks and extracting the maximum principal strain in the solidification process of casting by CAE.

[1] The Japan Foundry Engineering Society: *Case Examples of Casting Defects/Faults in Die Casting and Countermeasures* (2000) pp.43-44

Proceedings of the 72nd World Foundry Congress
21-25th May 2016, Nagoya, Japan

Direct measurements and process monitoring for High Pressure Die Casting (HPDC) process

I.Vicario[1], I. Crespo[1], J.C. Garcia[1], E. Arratibel[2], I. Aguillo[2], D. Val[3], D. Cao[3]
[1] Tecnalia Research & Innovation, Derio, Spain
[2] Fundiciones inyectadas Alavesas, S.A., Nanclares de la Oca, Spain
[3]Grupo Antolín S.A., Burgos, Spain

Process variables have a direct impact over High Pressure Die Casting (HPDC) quality parts. The measurement of metal temperature and the intensification pressure directly from the injected part, the vacuum pressure and flow and other parameters permits to determinate if the injected part is under the desired range on the controlled parameters.
This work demonstrates how it's possible to have a better control of the HPDC process increasing final part quality by combining different sensors and actuators.

Keywords: *High pressure die casting, squeeze pin, sensor, vacuum casting, process parameters.*

1. Introduction

High Pressure Die Casting (HPDC) is the most usual method to produce aluminum and magnesium castings, due to the excellent balance of cost and properties for high production volumes. Nevertheless, conventional HPDC process presents certain limitations in terms of final mechanical properties, due to the intrinsic porosity of HPDC castings, the alloy composition and all the parameters than can interfere in the quality of an injected part. More than 400 parameters can have influence over the final part quality. The real values of process parameters in the cavity are an important criterion for the consistent quality of these parts [1-4]. In the case were there is a tendency to promote shrinkage porosity, squeeze pins are employed to reduce the porosity by a local pressure multiplication pin [5]. The results from the temperature control and monitoring of the die, machine, vacuum and sensors in direct contact with the injected metal can achieve minimum scrap rates and ensure constant quality.

2. Experimental conditions

In order to obtain the results, a base AlSi9Cu3Fe1 alloy has been employed. The base alloy composition is in table 1.

Table 1 Composition of base alloy (mass %).

Si	Fe	Cu	Mn	Mg	Cr	Ni	Zn	Ti	Al
8.4	0.7	1.56	0.13	0.03	0.03	0,04	0.94	0,04	Bal.

In order to determinate the temperature directly from the injected part, a replaceable insert was designed to place the thermocouples (T1, T2 and T3). To measure also pressure (P1, P2) and metal contact (C1, C2), the same external shape was employed for all the different sensors. Also two bayonet RS 621-2221 K type thermocouple (T4, T5) were placed in the die. Sensor placements on the die are shown on figure 1.

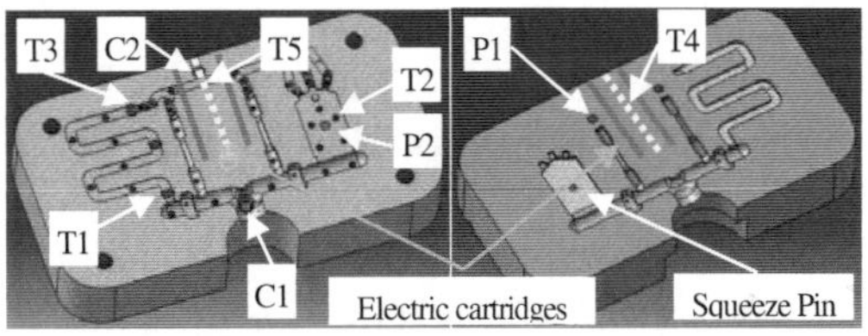

Fig.1 Sensors placements on the die.

To determinate directly the specific pressure of the injected metal, a Kistler cavity pressure sensors Type 6175A2. To determinate the contact of metal with the cavity, a FKS-03 sensors and for vacuum, humidity, gas and air quantity a MASS multi-sensor from Electronics. All the data were registered with a FieldLogger 512K data logger, a Siemens Simatic S7-300 PLC and an Advantech PCI-1711 PCI card.

For achieving an accurate temperature measurement over the injected part, a new sensor design was developed. Due to the HPDC short cycles times related with a rapid cooling, the thermocouple has been designed with a wire diameter of only 0.15 mm embedded in a H13 tool steel insert with 42 HRC hardness and with the two thermocouple wires separated 2 mm one from the center of the other welded by laser to the contact surface. The two thermocouple wires are isolated with a no conductive ceramic material along the insert.

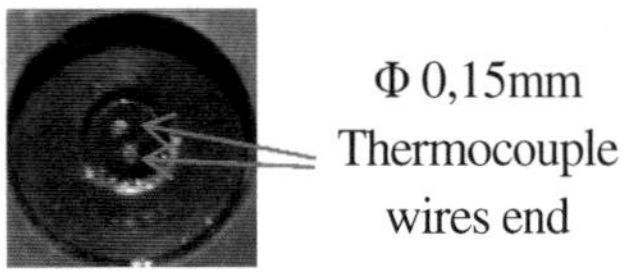

Fig. 2 Detail of direct contact thermocouples.

The thermal control of the die was obtained by a combination of water and oil (2x16kW Tool Temp - TT-388/2) tempering circuits and four Aloña 800W electrical heating cartridges. The 10 water and 2 oil circuits were equipped with individual flowmeters, water exit temperature thermocouples and actuator. The flow of lubing agent, air pressure and lubricating time was registered.

Trials have been made at Tecnalia's foundry pilot plant, with a 950 Tn HPDC machine equipped with a VDS Controller ProVac PLC-250 and vacuum valve ProVac Standard 1000 and injection parameters diagram (Distance, speed and pressure). 400 kg of aluminum alloy were melted and casted, with a continuous registration of metal temperature.

3. Results and discussion

The system is capable of determining the real temperature of the injected part in different points, as shown in figure 3. The quick response of temperature thermocouples permits to obtain the complete solidification curve, what it gives the solidification curve that can be employed to determinate the real solidification parameters by thermal analysis. We can observe how the maximum temperature variates from T1 = 620ºC, T2= 605ºC and T3=602 ºC, well above the nucleation temperature. Monitoring these temperatures can help us in the determination of part quality and to adjust thermal treatment temperatures with the obtained solidification temperature for the analyzed injected part. In the case of P2 installed opposite to the squeeze pin, a 200 Mpa value is obtained. The squeeze pin is operated by controlling the injected part temperature by T2 or by time. Depending on the solidification curve of an alloy obtained from T2, the temperature set point to apply the local intensification pressure in the semi-solid state is adjusted. In this way, a correct squeeze pin use is guaranteed. In fig. 3 we can also observe the main the vacuum parameters values with a minimum vacuum of 86 mbar and flow of 6.7 l/s. The temperature in a point of the die is smart controlled by regulating the water or oil flow or by the actuation time of the electrical cartridge, with a variation of only ±2ºC.

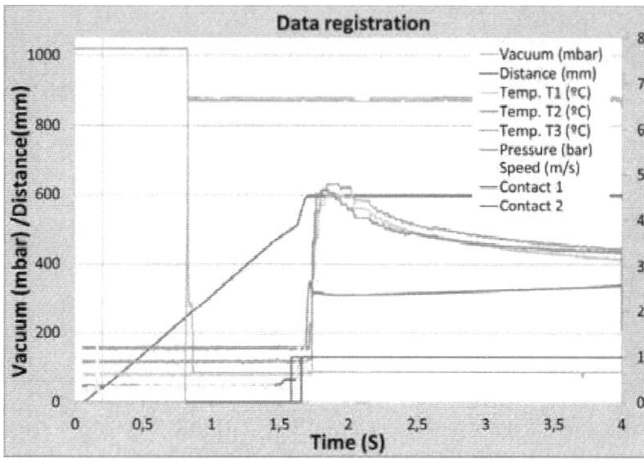

Fig. 3 Parameters control diagram.

4. Conclusions

The determination of HPDC parameters can lead in a better control of part quality. The alloy composition that promotes differences in the fluidity and solidification of alloys can be controlled by adjusting process parameters using temperature, contact and pressure sensors in direct contact with injected metal. The quality control system can be optimized for a determined alloy composition range, obtaining an optimal temperature and pressure value in a determinate point, a maximum filling time and use them also to regulate the use of squeeze pins.

Monitoring the vacuum values can lead to an automatic control of injected parts quality, especially by controlling the vacuum and flow values.

Quality defective rates have been reduced up to a 60% by acting over the process parameters with the obtained data.

Acknowledgements

This work has been partially funded by the Basque Government through the research projects ETORGAI ER-2011/00002 and by the European Commission to EFEVE's project with grant agreement Nº: 314582.

References

[1] A. Long, D. Thornhill, C. Armstrong and D. Watson: Applied Thermal Engineering. 44 (2012) 100-107.

[2] K.K.S. Tong, B.H. Hu, X.P. Niu and I. Pinwill: Journal of Materials Processing Technology. 127 (2002) 238–241.

[3] G. Dour, M. Dargusch, C. Davidson: International Journal of Heat and Mass Transfer. 49 (2006) 1773–1789.

[4] Matthew S. Dargusch, G. Dour, N. Schauer, C.M. Dinnis, G. Savage: Journal of Materials Processing Technology. 180 (2006) 37–43.

[5] E.S. Kim, K.H. Lee, Y.H. Moon: Journal of Materials Processing Technology. 105 (2000) 42-48.

Proceedings of the 72nd World Foundry Congress
21-25th May 2016, Nagoya, Japan

Reduction in Die-casting Burrs by Shot Profile Control

Kazuya Inoue [1], Akihiro Teshima [2], Hideo Tana [3], Tatsuyuki Kobayashi[4] and Tsutomu Ishikawa[5]

80 Kowari, Minaminakane-cho, Nishio, Aichi 445-0810, Japan

Burrs formed on aluminum die-castings had resulted in increased man-hours for deburring and the suspension of production due to the flashes of molten metal, and had eventually become major factors affecting productivity. To eliminate these factors, we had taken conventional countermeasures, such as reducing casting pressure and reducing shot speed immediately before completely filling the die cavities with molten metal. However, these countermeasures did not stabilize the external and internal quality of products. Under these circumstances, we successfully developed technology that reduces burrs without having any adverse influence on quality, by controlling shot pressure before and after completely filling the die cavities with molten metal. This paper reports on that technology.

Keywords: *burrs, surge pressure, intensification profile*

1. Introduction

We mounted a displacement sensor on the parting line of the dies to check the clearance between the die halves during casting, and conducted practical casting operations. This caused the gap of die to be affected by surge pressure upon completion of filling of molten metal into die cavities and shot pressure after the completion of that, as shown in Fig. 1. Furthermore, as shown in Fig. 2, the gap of die and burr thickness were found to have a strong correlation. Consequently, we worked on improving these two items concerning the gap of die.

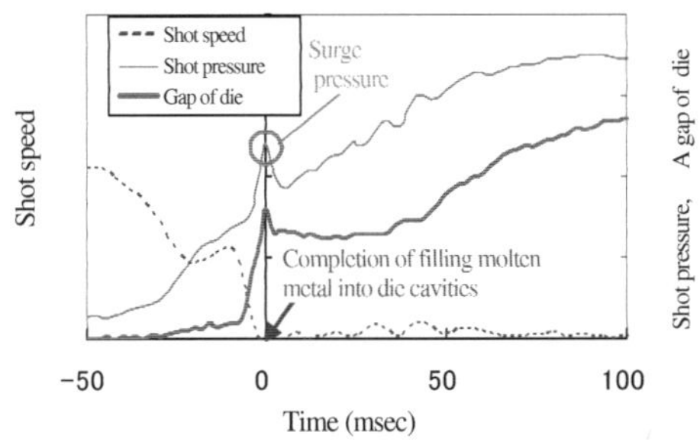

Fig. 1 Typical shot pressure and a gap of die during casting

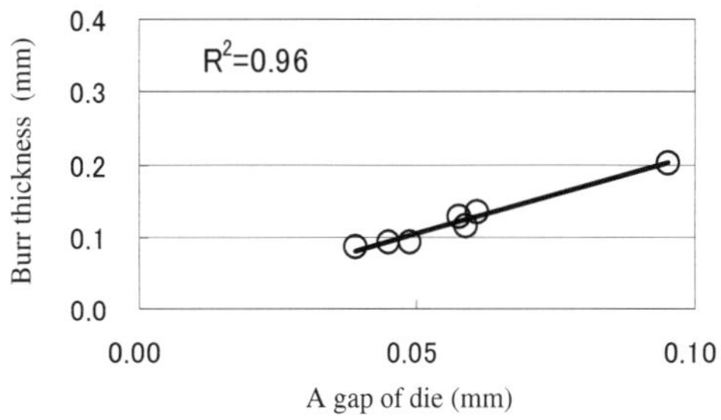

Fig. 2 Correlation between a gap of die and burr thickness

2. Reduction in surge pressure upon completion of filling of molten metal into die cavities

Figure 3 shows a diagram of the start of intensification under shot position control. In die-casting, starting intensification after completely filling the die cavities with molten metal results in quality degradation. Therefore, intensification is set to start at a point where it can begin without delay even if the pouring amount increases.

In the meantime, we found that a time lag from the start of intensification to the completion of filling (i.e., intensification time lag) varied according to changes in pouring amount and this affected surge pressure upon the completion of filling (Fig. 4).

So, we examined ways to start intensification after judging the filled amount of molten metal into die cavities. As a result, we found that making this judgment according to shot speed provided higher reproducibility.

Figure 5 shows the intensification time lag when the intensification start control factor was changed from the shot position to the shot speed. This change reduced the variation in intensification time lag to 1/9, thus making it possible to suppress variations in intensification start timing and control surge pressure upon completion of molten metal filling into the die cavities.

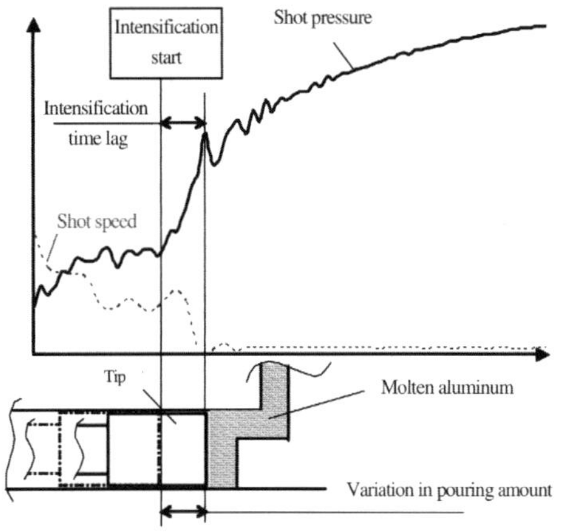

Fig. 3 Diagram for intensification start under shot position control

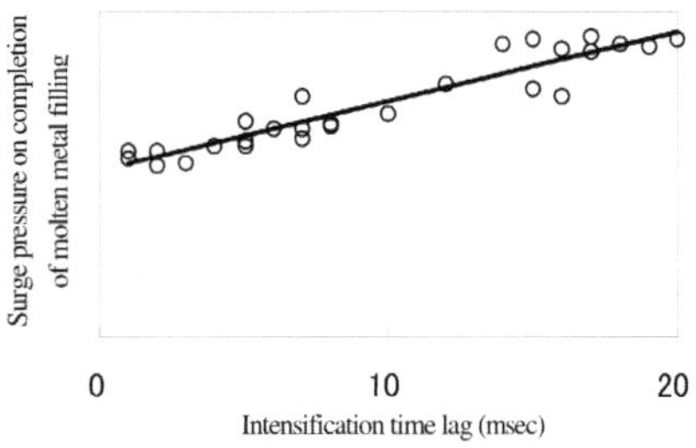

Fig. 4 Correlation between intensification time lag and surge pressure upon completion of molten metal filling into die cavities

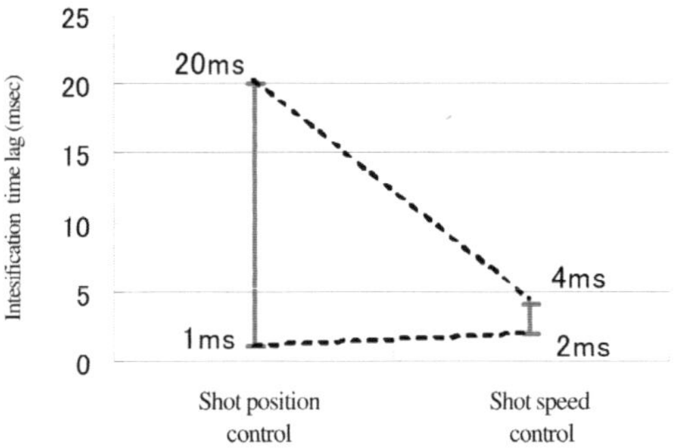

Fig. 5 Intensification time lag values

3. Pressure build up profile change

As shown in Fig. 1, the gap of die for intensification increases with the rise in pressure. However, it is important to "add the necessary pressure when needed" in order to reduce burrs and ensure quality.

Therefore, we conducted tests to verify the correlation with quality by replacing the intensification control valve with one of excellent responsiveness and changing the pressure build up pattern with this valve.

As a result, we found that shot pressure served as follows in each period.

1. Initial intensification: Completely fills molten metal up to the ends of die cavities and builds in appearance quality.
   (Solidified ratio: Not more than 35%)
2. Final intensification: Builds in internal quality after the outmost surface of the product solidifies.
   (Solidified ratio: Not less than 45%)

Intensification in the solidified ratio range of 35 to 45% has no effect on quality. However, continuing to increase pressure in this range will result in increased a gap of die as shown in Fig. 1. Therefore, we controlled the pressure build up profile so as to suppress intermediate intensification without changing the initial intensification or final shot pressure.

4. Verification of effect

As a result of using surge pressure control and intensification profile control, we successfully reduced a gap of die during casting by 36% and a significant reduction in burrs while maintaining the quality of castings at the conventional level (Figs. 6 and 7).

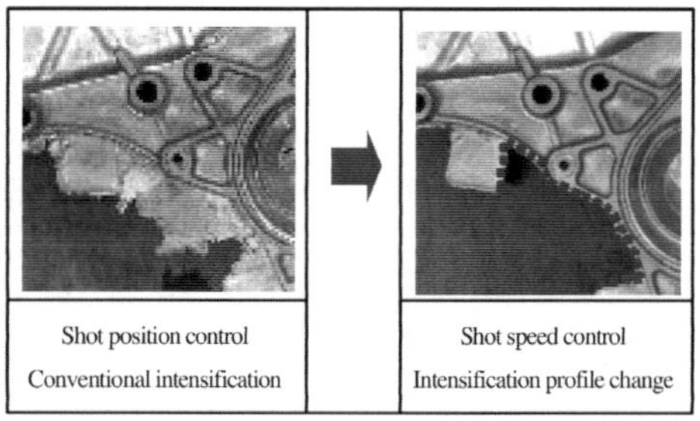

Fig. 6 Comparison of development of burrs

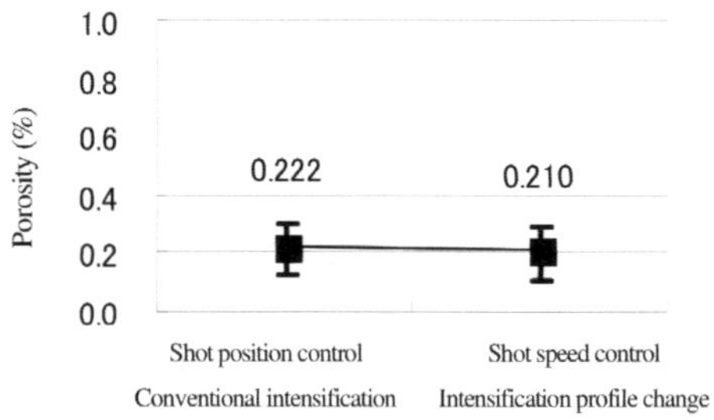

Fig. 7 Comparison of porosity of castings

Effect of Vibration Conditions and Shear Rate on the Shape of Solid Particles in JIS AC4CH Aluminum Alloy Slurry Made by Applying Mechanical Vibration

Yuichiro Murakami[1], Kenji Miwa[2], Masayuki Kito[3], Takashi Honda[3] and Shuji Tada[1]

[1] National Institute of Advanced Industrial Science and Technology, 2266-98, Anagahora, Shimo-Shidami, Moriyama-ku Nagoya, Japan

[2] Aichi Science and Technology Foundation, 1-157-1, Onda-cho, Kariya, Aichi, Japan

[3] Aisan Industry Co., Ltd. 1-1-1, Kyowa-cho, Obu, Aichi, Japan

The shape of solid particles affects on fluidity in the semi-solid die-casting process, so to clarify the effects of shear stress and initial shape of solid particles on deformation are an important issue. The slurry made by applying mechanical vibration under different acceleration amplitude was injected into the spiral shape cavity with different shear rate at the gate. The solid particles became fine and globular in the case of the high acceleration amplitude and high shear rate. Moreover, at the high acceleration amplitude slurry, fraction solid after the injection process became higher than low acceleration amplitude. This result suggested that the slurry made with high acceleration amplitude can be formed stably and high fluidity.

Keywords: *Semi-solid slurry, high pressure die-casting, mechanical vibration, shear stress, solid particle*

1. Introduction

Semi-solid process is a method of forming at the solid-liquid coexistence phase. The semi-solid metal has a higher viscosity and lower solidification shrinkage ratio than the liquid metal, so the casting defects caused by air entrainment or shrinkage porosity are expected to be decreased. However, the semi-solid process has the disadvantages of formability in low fluidity, and a difficulty of temperature control to make a semi-solid slurry.

We have developed the method to prepare the slurry by applying mechanical vibration during cooling from liquid to semi-solid temperature [1]. The fluidity of the slurry could be improved by applying shear stress to the slurry at the gate of mold. In this method, the solid particles in the slurry were deformed to fine and globular by applying shear stress, and fluidity of slurry was increased with fine and globular solid particles [2-4]. Therefore, deformation of solid particles is thought to have effects on the fluidity.

In this study, to clarify the effect of the shear stress and the initial shape of solid particles on deformation, the fluidity tests for the semi-solid slurry made by applying various mechanical vibration conditions were carried out.

2. Experimental procedure

The slurry was prepared as follows [1]. In this study, an Al–7wt%Si–0.3wt%Mg alloy (AC4CH in the JIS, equivalent to A356 in the ASTM standard) was used. The molten AC4CH alloy at 650 °C was poured into the horizontal vibrating a stainless steel container at the room temperature. The vibration was applied for 20 s. The frequency of vibration was constant at 50 Hz, and the acceleration amplitude was varied as 117.6 m/s^2 (12 G), 147.0 m/s^2 (15 G) and 166.7 m/s^2 (17 G).

The slurry was injected into the spiral shape cavity (Fig. 1) by high pressure die-casting machine. Moreover, to vary the shear rate on the slurry, the thickness of the gate at the entrance of the cavity was set at 1.0 mm or 4.0 mm. The shear rate γ is 7.5×10^5 /s or 4.7×10^4 /s, respectively. The specimens were cut, and polished, then were subsequently etched with 0.5% HF solution. Then the microstructure at vertical section to flow direction was observed.

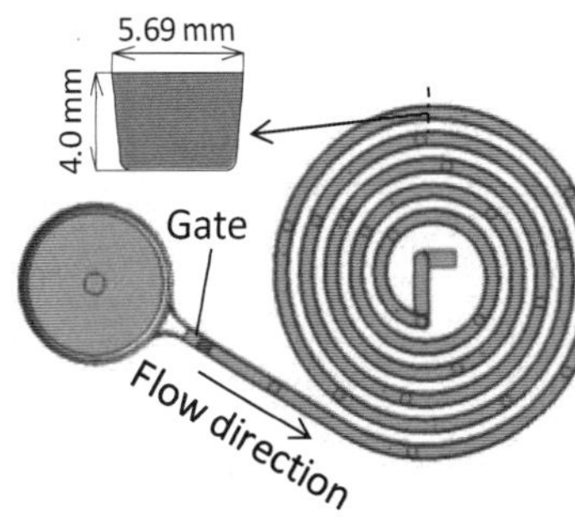

Fig. 1 Schematic illustration of fluidity test sample

3. Results and discussion

Fig. 2 shows the microstructure of the specimens. The mean particle diameter and roundness were calculated by these microphotographs, and then the effect of the acceleration amplitude on both these values was shown in Fig. 3. The roundness was defined as the ratio of measured circumference and equivalent circle circumference [3]. Therefore, if roundness is 1, the particle has the true circle shape, and this value increases if the particle deviates from the circle. The solid particles became fine and globular by increasing the acceleration amplitude of the mechanical vibration before injection. Moreover, the solid particles became finer and more globular after injection. Meanwhile, at low acceleration amplitude, the coarse solid particles could be seen. Fig. 4 shows the relation between acceleration amplitude and fraction solid. The fraction solid was decreased after injection in low acceleration amplitude. By the

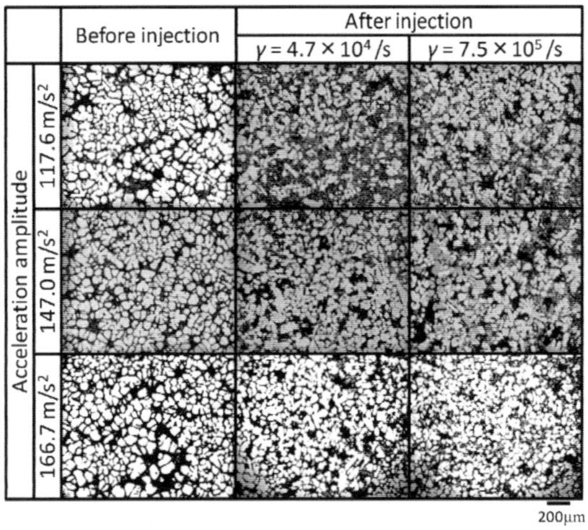

Fig. 2 Microstructure of specimens

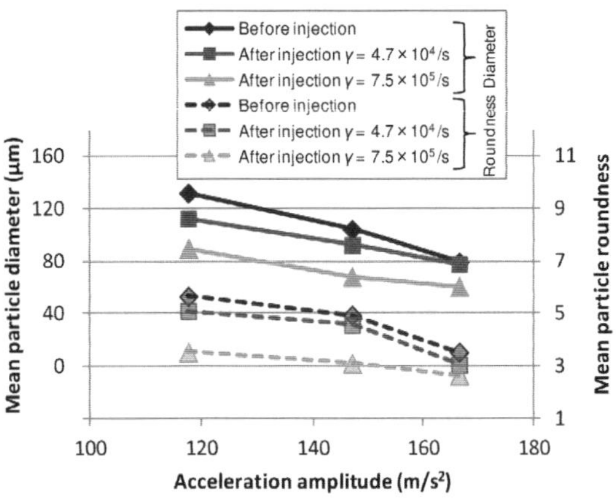

Fig. 3 Effect of acceleration amplitude on the shape on solid particles.

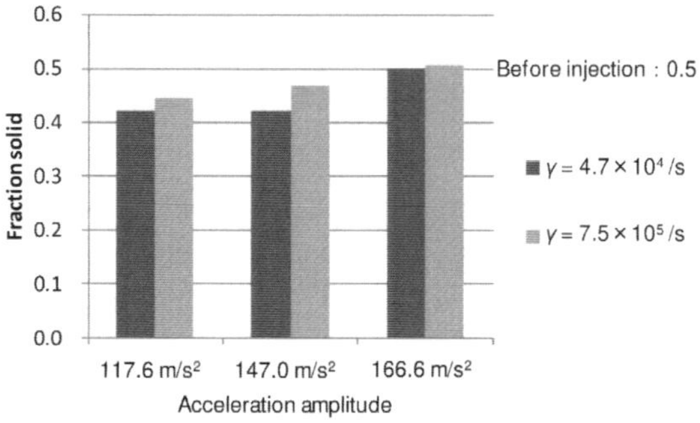

Fig. 4 Effect of acceleration amplitude on the fraction solid after injection process

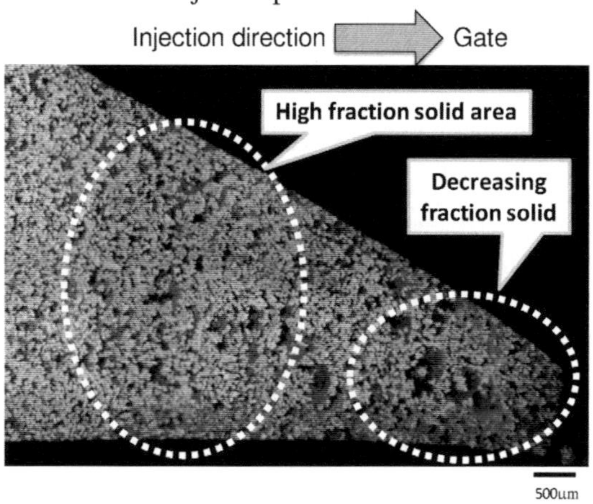

Fig. 5 Microstructure at the gate (acceleration amplitude 117.6 m/s^2, $\gamma = 7.5\times10^5$ /s)

observation of the microstructure at the gate (Fig. 5), the high fraction solid area was seen. Then the gate was considered to be clogged in low acceleration amplitude. This is considered as the cause of the decreasing of fraction solid. This phenomenon is also thought to be the reason for the high ratio of refinement, because coarse solid particles could not flow into the cavity by clogging. Meanwhile, the fraction solid after injection was same as before the injection at high acceleration amplitude. Therefore, the slurry made with high acceleration amplitude is expected to be formed stably.

References

[1] Y. Murakami, K. Miwa, M. Kito, T. Honda, N. Kanetake, S. Tada : J. JFS 86 (2014), 728-733 (in Japanese).

[2] Y. Murakami, K. Miwa, M. Kito, T. Honda, K. Yorioka, N. Kanetake, S. Tada : *Light Metals 2012*, (2012), 297-301.

[3] Y. Murakami, K. Miwa, M. Kito, T. Honda, N. Kanetake, S. Tada : Mater. Trans. 54 (2013), 1788-1794.

[4] Y. Murakami, K. Miwa, M. Kito, T. Honda, N. Kanetake, S. Tada, J. JFS 86 (2014), 773-780 (in Japanese).

Proceedings of the 72nd World Foundry Congress
21-25th May 2016, Nagoya, Japan

Squeeze Casting Technology in Production of High-Strength Monolithic Aluminum Alloys and Aluminum Matrix Composites

Piotr Dudek, Jerzy J. Sobczak, Paweł Darłak, Piotr Długosz
Foundry Research Institute, 73 Zakopianska St., Krakow, Poland
piotr.dudek@iod.krakow.pl, jerzy.sobczak@iod.krakow.pl, pawel.darlak@iod.krakow.pl, piotr.dlugosz@iod.krakow.pl

The paper presents the results of investigation of monolithic and composite aluminum-based cast and materials with wrought alloys: AW 2024, AW 6061, AW 7075 in accordance with EN 573-3 and cast alloys AC 42100 according to EN 1706. Castings were produced using VSC 500 stand at a high squeeze pressure of 150 MPa and, for comparison, by gravity casting into permanent mold. Composite castings were performed using both commercially available ceramic preform (based on Al_2O_3 – Saffil® fibers) and preforms produced by Institute of Foundry using high energy ball milling and mechanical alloying of ceramic substrates. The preforms are made from shot Saffil® fibers and colloidal silica (LUDOX®) with fillers increasing porosity. Castings made of aluminum wrought alloy after heat treatment have shown almost double increased mechanical properties. In research characterized the material properties, including the influence on pressure for microstructure, defined physical properties, namely density, electrical conductivity and thermal conductivity, coefficient of thermal expansion, as well as basic mechanical characteristics.

Keywords: *squeeze casting, aluminum alloys, metal matrix composites.*

1. Introduction

Squeeze casting technology allows producing castings with a very accurate representation of their shape and surface. Components made by this method are characterized by a fine microstructure with reduced porosity, which allows applying the heat treatment, welding and resistance welding [1]. It can be thus produce a variety of composites reinforced by ceramic phase, e.g. dispersoids (particules) and/or short fibers or whiskers, long fibers and use porous preforms [2,3]. Low speed of shot plunger and high hydrostatic pressure provide excellent combination of ceramic and metal materials in the uniform homogeneous structure.

Research of squeeze casting and infiltration process has been performed on hydraulic complex VSC 500.

Metal was melted in the resistance furnace with a SiC crucible. The temperature of the die was kept stable by means of two heating/cooling devices and maintained at 200°C. All squeeze castings were solidified under the external pressure of 150 MPa. Moreover, a series of gravity reference castings (solidifying under the atmospheric pressure) were created.

2. Squeeze infiltration of ceramic preforms

In order to fabricate preforms with high open porosity the following methods were applied:
1) introducing different sizes of particles and
2) addition of organic substances.

The highest quality of preforms has been reach using of sawdust as the addition. The result is surprising because during burning sawdust, besides gases, releases pollution which may cause the degradation of the structure of preforms. Ceramic preforms were characteristic of open porosity contribution within measured range from 24 to 75%.

For infiltration preforms were selected with open porosity above 50%. The preheated preforms were infiltrated with 7075 liquid aluminium alloy with the use of the so-called "bottom" infiltration (Fig.1). The external pressure was also posed directly on the liquid metal surface (direct squeeze casting) with the use of a piston. The pressure equaled 150 MPa, and time was 60 seconds.

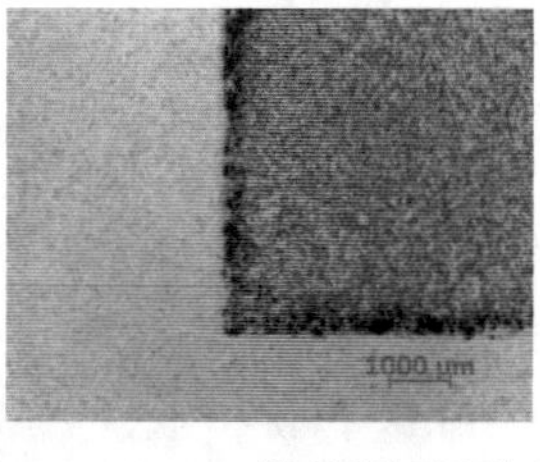

Fig.1 Microstructure of MMC 7075 alloy with Saffil® fibers

3. Squeeze casting

For tested alloys the shot speed was respectively 60, 50, and 40 mm/s, melt temperature: 750, 730, 720 and 690°C, mold temperature: 240, 220, 200°C and a squeeze time 25 sec. The differences of applied parameters are caused by the different alloys technological properties used in attempts.

It was found out those positive changes of microstructure castings solidifying under external pressure caused by the operation of mechanisms that can be divided into mechanical and thermodynamic ones. The first of these cause a significant increase in the rate of cooling of the solidifying metal, by reduction of the gap between the casting and the wall of the mold, thereby increasing the cooling rate of castings. Second increase the rate of nucleation of the solidifying metal by the increase of supercooling (Fig. 2).

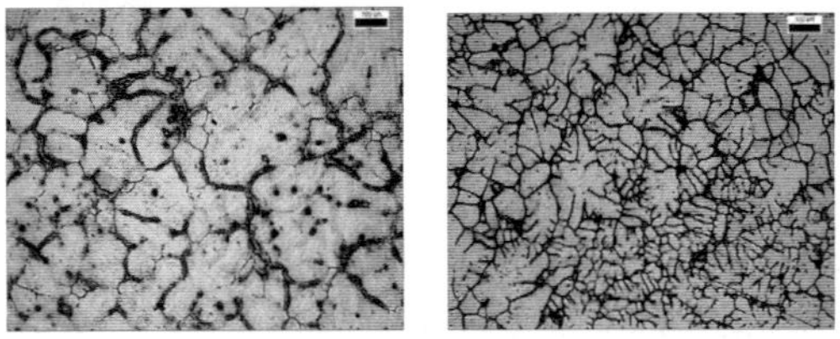

a) b)

Fig. 2 Microstructures of the 7075 alloy gravity (a) and squeezed cast (b)

The mechanical properties of as-cast samples F, solution T4, the aged artificially T5 and precipitation treatment T6 were performed.

An advantage complex of mechanical properties was obtained regarding to castings made by squeeze casting method compared to standard castings solidified under atmospheric pressure (Fig. 3).

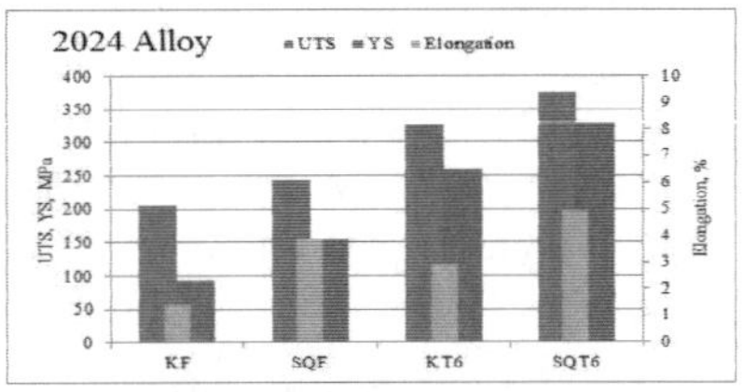

a)

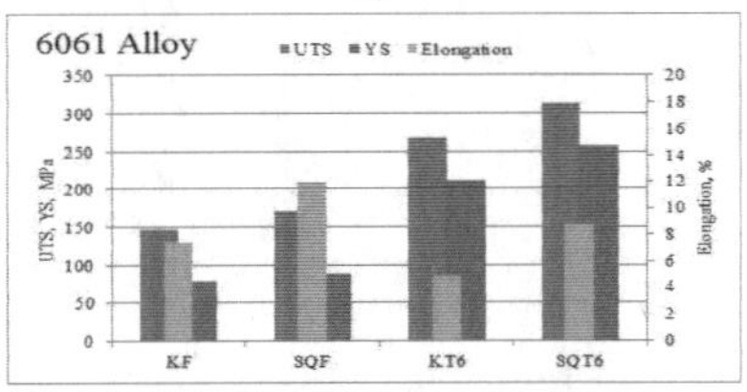

b)

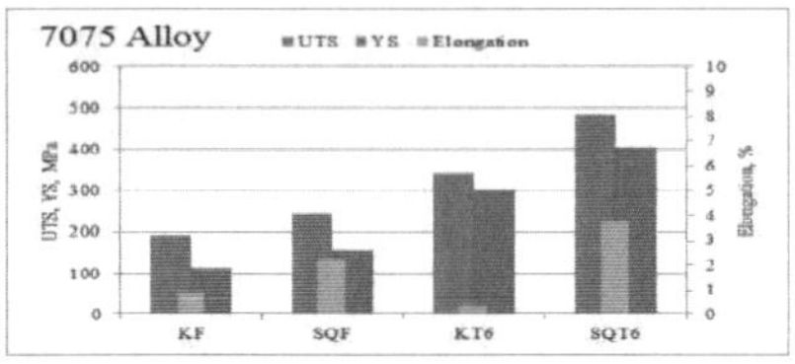

c)

Fig. 3 Mechanical properties of gravity (K) and squeezed cast 150 MPa (SQ) in the as-cast state (F) and after heat treatment to T6 condition: a) 2024, b) 6061, c) 7075 alloys

4. Conclusions

1. The highest porosity and compressive strength were achieved with the application of sawdust as the pore-creating substance, for producing preforms.
2. The wrought alloys can be successfully applied for squeeze casting technology.
3. The increase of exploitation characteristics for investigation materials can be improved by optimizations of squeeze casting parameters and heat treatment conditions.
4. Maximum level of mechanical properties were achieved for squeeze casted 7075 aluminium alloy (UTS = 483 MPa, YS = 405 MPa with elongation of almost 4%).
5. Metal matrix composites obtained by squeeze casting characterized by higher mechanical properties than monolithic alloys. The highest values of mechanical properties obtained for composites made from 7075 alloy fibers Saffil® after heat treatment T6 (UTS = 487 MPa, YS = 385 MPa, elongation – 7%).

Acknowledgements

Studies were performed as part of strategic project No. POIG.01.03.01-00-015/09 entitled "Advanced Materials and Technologies for Their Manufacture".

References

[1] J. J. Sobczak: Teoretical and practical backgrounds of squeeze casting of non-ferrous metals. Transaction of the Foundry Research Institute, Special Issue 41, Krakow, 1993

[2] R. Ashiri, F. Karimzadeh, B. Niroumand: Materials Science and Technology, 2014, vol. 30, no 10.

[3] T. Reguła, A. Fajkiel, P. Dudek, K. Saja: Transactions of Foundry Research Institute, Volume LIV, Year 2014, Number 4.

Proceedings of the 72nd World Foundry Congress
21-25th May 2016, Nagoya, Japan

Multiple Spot Electron Beam Welding of Aluminium Die Castings

Fabian Teichmann[1], Helge Pries[1] and Klaus Dilger[1]
[1] Braunschweig University of Technology, Institute of Joining and Welding, Germany

When welding aluminium high pressure die casting (HPDC) parts metallurgic pore formation is a main challenge. In particular welding of thin-walled, pressure-proof housing parts, such as heat exchangers for vehicle application, is mainly performed using laser or electron beam welding (EBW). A novel approach is the application of multiple spot EBW to achieve good degassing conditions through remelting in a low pressure environment. Moreover high speed beam deflection enables the generation of multiple melt pools at spatially divided locations. The present investigation focuses on the influence of the number of melt pools, the distance between them and power distribution (PD) on porosity. In addition the results are related to the HPDC process and its influence on EBW.

Keywords: *Electron beam welding, multiple spot, aluminium casting, aluminium high pressure die casting*

1. Introduction

The demand for thin-walled aluminium die castings for lightweight and medical applications is steadily increasing. In many applications these parts need to be welded to fulfil their purpose. The main challenge in welding aluminium HPDC parts is porosity. The applied welding technique, the surface condition, the total hydrogen content and the casting porosity are key factors regarding the formation of weld porosity. According to literature hydrogen bonded in hydrides and hydrogen dissolved within the casting material is carried into the melt pool by diffusion during the welding process. After solidification the hydrogen is trapped inside the weld metal and forms porosity. An opportunity to avoid metallurgical pore formation is multiple melting with different penetration depths. At each run the weld metal melts again and trapped porosity, which was formed during a former run, can rise to the surface of the melt pool where is leaves the thaw. In state of the art welding this is achieved by repeating the welding process several times. In order to increase process efficiency the long term goal is to use the capability of EBW for multi spot welding to achieve remelting during one run. Subsequently the current investigation aims to find basic relations between weld bead quality, power distribution and the distance between the melt pools. [1,2]

2. Materials and Methods

2.1 Materials

All trials performed were carried out on material casted at the institute for joining and welding (ifs) on a Bühler B53 Evolution HPDC machine. Alloy AlSi10MnMg was casted as sheets with dimensions of 260 x 110 x 4 mm. The HPDC process was performed using the wax free release agent HERA EU DC-120918A, plunger lubricant ChemTrend Power-Lube 766. All plates were casted with vacuum assistance (150 mbar, evacuation after shot sleeve filling 1.5 s)

2.2 Methods

The welding process was performed on a pro-beam K-26 electron beam welding machine. Welding speed and acceleration voltage were kept constant at 60 mm/s and 60 kV in order to ensure production related welding speeds and avoid high voltage discharge. The energy input per unit length was set to 162/234/288 J/m for 2/3/4 spot welding. The focal position, deflection pattern, number of melt pools and the power distribution between the melt pools were varied (see Table 1).

Table 1 Welding parameters alternated

Parameter	Range
Focal position	+20/+10/0/-10/-20
Pattern [-]	Spot/Circle
Frequency [Hz]	1667/1111/1000
Number of melt pools [-]	2/3/4
Melt pool distance [mm]	10/20/30/40

All weld paths were created as 3 mm bead on plate welds of 210 mm length, containing a 10 mm slope in/out region. Moreover stop action welds were produced for each set of parameters, in order to distinguish the influence of each weld pool. Photographs were taken from each weld and longitudinal as well as cross sectional metallurgic sections were taken from chosen weld beads. The stop action welds were additionally CT-scanned and analysed for porosity.

3. Results and discussion

3.1 Melt pool distance

The weld paths performed with small gaps between the melt pools showed a strong tendency to surface defects. These defects were identified as overheating driven surface porosity. The occurrence of these defects was further observed as dependent on the number of melt pools. Thus a lower bound for the melt pool distance was defined for each number of melt pools. For the current conditions a lower bound of 5 mm melt pool distance was found. It was further observed that the penetration depth of the following melt pools drops with increasing gaps between the keyholes, as expected. Fig 1 gives an example for this observation. The metallurgic cross-section each show two nail head formed welds, where the upper one is caused by the second keyhole and the lower one by the first melt pool. As a result of the increasing gap about 20 mm the penetration depth drops as shown.

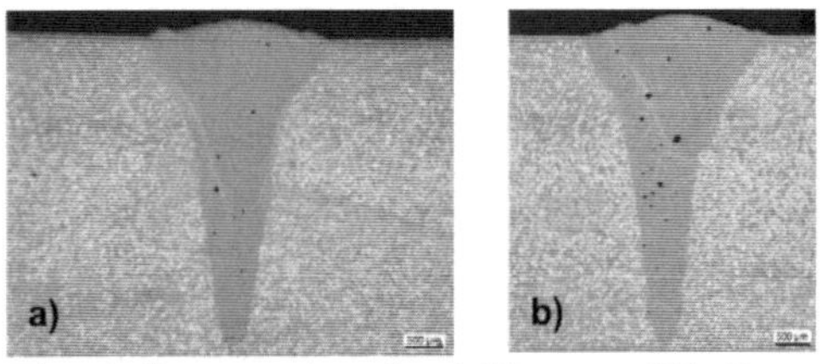

Fig. 1 Cross-sections for a) 20 mm and b) 40 mm melt pool distance for a two spot weld path

As an additional outcome the penetration depth must be carefully set by beam power, distance between the melt pools and power distribution to avoid melting base material at the repeating runs.

3.2 Power distribution

The influence of power distribution was observed by analyzing the stop welds performed. Fig 2 shows a longitudinal section taken from the stop action welds performed. The areas highlighted areas show the target location of the different melt pools. Area (A) for example shows the area where only one melt pool passed the joint. As the comparison of area A, B, and C shows, the amount of porosity decreases after each melt pool passed a specific place within the weld path. From the areas highlighted in Fig 2, it is apparent that the power distribution should be decreased slightly from the first to the last keyhole. The weld bead shown in Fig 2 was welded with three spots and a 78/17/6 % power distribution, a frequency within the deflection pattern of 1111 Hz and 3900 W beam power. As Fig 2 shows, the decrease of power used for this specific example was too sharp. Area A is much bigger than area C but C is the area with the lowest porosity content.

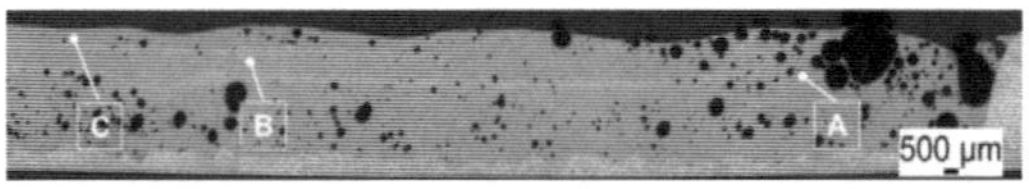

Fig. 2 Longitudinal section, three spots, PD 78/17/6 %

3.3 HPDC process influence on welding

The trials although confirmed that the casting quality is a key factor for the quality of the resulting weld bead. Vacuum assisted casting, the use of wax free release agent as well as low density indices were confirmed to be a good casting strategy for weldable HPDC Parts.

4. Summary and conclusion

In this investigation, the aim was to assess the influence of power distribution and melt pool distance for multiple spot EBW of aluminium HPDC parts. It was shown that the melt pool effects penetration depth and porosity content and is effected by the power distribution used. It was confirmed that the power distribution has to be decreased counter to the direction of welding slightly.

Acknowledgements

The current research has received funding by the Dobeneck-Technologie-Stiftung. The authors gratefully acknowledge the received funding.

References

[1] S. Wiesner: *Wirtschaftliche Herstellung von gasarmem, schweißbarem Aluminium-Druckguss.* (Shaker-Verlag, Aachen, 2003)

[2] R. Winkler: *Porenbildung beim Laserstrahl-schweissen von Aluminium Druckguss.* (Herbert Utz Verlag Wissenschaft, München, 2003)

Proceedings of the 72nd World Foundry Congress
21-25th May 2016, Nagoya, Japan

A Chemical Pre-treatment Process for Adhesive Bonding of High-Pressure Die Cast Aluminum Alloys

Jianmin Shi, Torsten Schuchardt, Helge Pries, Elisabeth Stammen and Klaus Dilger
Institute of Joining and Welding, Technische Universität Braunschweig,
Langer Kamp 8, 38106 Braunschweig, Germany

High-pressure die cast (HPDC) aluminum components are widely used in the assembly of vehicle structures for weight-reduction. Joining of HPDC aluminum castings with other structural parts using adhesives is still challenging due to the inherent contamination of casting surfaces by die lubricants. In this paper we report on the development of a universal chemical pre-treatment process for the reliable fabrication of adhesively bonded HPDC aluminum joints. The effectiveness of the three-step chemical pre-treatment process, consisting in degreasing in alkaline solutions, deoxidation in fluoride-containing sulfuric acid solutions and dipping in hexaflurotitanic acid solutions, was verified by the evaluation of the durability of epoxy-bonded HPDC plates after aging in 5 wt % NaCl solution at 70 °C.

Keywords: *High-pressure die casting, aluminum alloys, surface pretreatment, adhesive bonding, durability.*

1. Introduction

High-pressure die casting (HPDC) of Al alloys is an attractive manufacturing process for light-weight structural parts in the automobile industry. Integration of Al die castings into hybrid vehicle-structures using adhesives is still challenging due to the inherent contamination of the casting surfaces by residual lubricants and the inhomogeneous distributions of alloying elements. Our previous work on the adhesion properties of HPDC Al alloys showed that the adhesive bonding property of an HPDC Al casting varies significantly at different locations on the casting surface [1,2]. Based on the surface chemical analysis and the evaluation of bonding strength, we have developed a chemical process for the pre-treatment of HPDC Al alloys to fabricate adhesively bonded structures using HPDC Al components. This 3-step process was optimized with respect to chemicals used in the second and third steps [1,2]. In this paper we report on the applicability of the optimized pre-treatment process to other HPDC alloys cast with different die lubricants.

2. Materials and experimental methods

Two Al alloys, EN AC-AlSi9Cu3(Fe) and EN AC AlSi10MnMg were fabricated into plates (260×150×4 mm^3) using HPDC machine with polysiloxane-based emulsions as die lubricants.

The chemicals used for the pre-treatment of HPDC Al alloys were selected from commercially available industrial products based on their main ingredients. Surface pretreatment was carried by degreasing in NaOH solutions followed by deoxidation in F^--containing H_2SO_4 solutions and deposition of Ti-based conversion coatings. The surface morphology and chemical compositions before and after chemical treatment were examined by using scanning electron microcopy (SEM) and X-ray photoelectron spectroscopy (XPS), respectively.

A structural adhesive based on epoxy was used to fabricate wedge specimens according to the German standard, DIN 65448 and cured at 180°C for 30 min. The initial crack length and crack extension were measured before and after aging at 70°C in 5wt% NaCl solutions. Finally, the wedge specimens were opened for examination of fracture surfaces.

3. Results and discussion

3.1 Changes in surface morphology and chemical compositions after chemical treatment

The change in surface morphology after the chemical treatment of HPDC AlSi10MnMg is shown in Fig. 1. The relatively featureless surface morphology of the as-cast surface changed into a rough pattern with exposure of intermetallic phases and Si particles and formation of pitting holes at the grain boundaries, due to etching effects of chemicals used. The surface chemical composition was characterized using XPS survey scans and high-resolution spectra of Si2p, respectively, before and after chemical treatment.

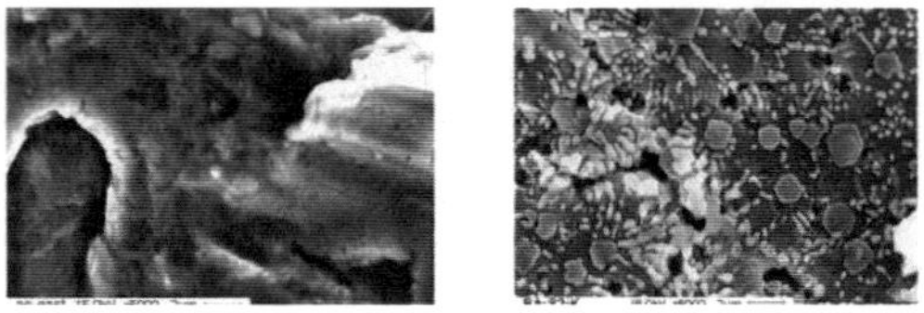

(a) (b)

Fig. 1 SEM micrographs of the surface of HPDC AlSi10MnMg plates: (a) as-cast and (b) after chemical treatment.

After chemical treatment, the concentration of C-containing species decreased from about 65 at% on the as-cast surface to about 40 at% on chemically treated ones and is accompanied by an increase in atomic concentration of oxygen and aluminum. The deposition of Ti-oxide conversion coating was confirmed by the appearance of Ti signals (about 0.5 at%) in the survey scan of the chemically treated surface. On one hand, the decrease in Si concentration after chemical treatment is another indicator for removing residual lubricants. On the other hand, the high-resolution Si2p spectrum of the as-cast surface is dominated by organic Si species resulting from the residual lubricants while that of chemically treated surfaces is dominated by inorganic Si, i.e., Si and SiO_2. The results of surface characterization confirmed the effectiveness of chemical treatment for the removal of residual lubricants and the deposition of corrosion-resistant Ti-oxide coatings.

3.2 Initial bonding properties and the durability of adhesive Joints

The evaluation of adhesion properties using wedge specimens show that longer initial crack length and crack extension were observed for the as-cast state than for the chemically treated specimen. It is also noticed that the alloying elements in HPDC alloys influence only the durability of adhesive joints fabricated in the as-cast state. This result suggested that the chemical pretreatment process is suitable for the pre-treatment of different HPDC alloys, thus being a universal process for adhesive bonding of HPDC Al alloys.

Representative fracture surfaces after opening the wedge specimen fabricated with and without chemical treatment are shown in Fig. 2 for the two HPDC alloys. The fracture surfaces of adhesive joints in the as-cast state exhibit a failure mode dominated by adhesive failure while those of chemically treated specimens show a cohesive failure mode for both HPDC alloys. From the surface characterization and the comparison of bonding properties of the as-cast and the chemically pretreated specimens, one can conclude that the developed chemical pre-treatment process can substantially improve the durability of adhesive joints involving HPDC Al alloys and is applicable to various HPDC Al alloys for adhesive bonding.

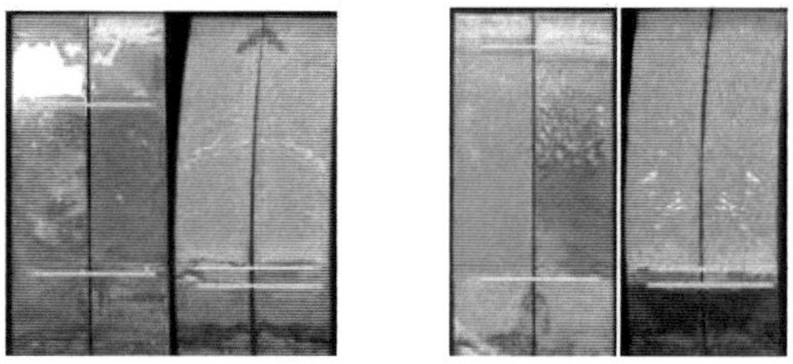

as-cast chem. as-cast chem.

(a) (b)

Fig. 2 Fracture surfaces of wedge specimens fabricated in the as-cast and chemically treated (chem.) states of (a) HPDC AlSi10MnMg and (b) HPDC AlSi9Cu3(Fe) alloys. The crack extension is marked by the green lines in each photograph.

4. Conclusions

A chemical pretreatment process for the reliable fabrication of adhesive joints involving HPDC Al alloys is demonstrated in this paper. The chemical pre-treatment process consists in degreasing in NaOH solutions, deoxidation in fluoride-containing sulfuric acid solutions and deposition of Ti-oxide coatings. The effectiveness of the chemical treatment process was verified by surface characterization and durability tests of adhesive joints. The optimized chemical process is found to be a universal pre-treatment process for HPDC Al alloys of various chemical compositions and cast with different die lubricants.

Acknowledgements

Financial support from the German Ministry of Economics and Energy (BMWi) through Industrial Collective Research (IGF) is greatly acknowledged.

References

[1] J. Shi, H. Pries, K. Dilger, M. Martin and F. Walther: Surface chemical inhomogeneity of high-pressure die cast Al-Si alloys and its effects on corrosion behaviors and adhesive bonding capabilities, in Proc. of the 71st World Foundry Congress. 2014, Bilbao, Spain.

[2] J. Shi, H. Pries, E. Stammen and K. Dilger: Int. J. Adhes. Adhes. 61 (2015) 112-121.

Proceedings of the 72nd World Foundry Congress
21-25th May 2016, Nagoya, Japan

Solidification Manner of Overlay Alloy in Al-Mn/Al-Si Alloy Clad Strips Produced by Vertical-Type Tandem Twin-Roll Casting

Yusuke Takayama, Yohei Harada and Shinji Kumai
Department of Metallurgy and Ceramics Science, Tokyo Institute of Technology, 2-12-1, O-okayama, Meguro-ku, Tokyo, 152-8552, Japan

Three Al-Si alloys with different Si compositions, 10 %Si, 12.6 %Si, 14 %Si were provided as the overlay alloy to clad the A3003 base strip by vertical-type tandem twin-roll casting. The influence of the Si composition on the overlay solidification manner and the bonding condition of interface were investigated. Regardless Si composition, the microstructure of overlay surface region exhibited fine eutectic structure, and the mid-thickness region of overlay consisted of α-Al dendrites and eutectic structure. Globular α-Al grain structure was also observed near the base/overlay interface region for 10 %Si. No un-bonded area was observed for 10 %Si. In contrast, a limited number of local un-bonded areas were observed at the interface for 12.6 %Si and 14 %Si. Bonding condition at the base/overlay interface depended on the growth manner of solidifying shell and the location where solidification finished. For 10%Si, the final solidification area was located at the boundary of growth front of solidifying shell and the globular grain band.

Keywords: *twin roll casting, clad strip, interface*

1. Introduction

Clad plates are widely used for both structural and functional material applications. However, the fabrication process of the clad sheets, such as hot-roll bonding, is complicated and involves the expenditure of a large amount of time and energy. In order to solve this problem, the present authors have developed a vertical-type tandem twin-roll casting method[1], which can produce the clad strips from molten metal directly by single step. This provides great time- and energy saving effects. The caster consists of two or three twin-roll casters arranged vertically. The base strip is produced by the upper caster and introduced to the roll gap of lower caster. When the base strip passes through the roll gap of lower caster, the overlay alloy melt is poured to the roll gap. The solidification of overlay strip alloy takes place and the clad strip is produced. In the previous studies, A4045/A3003 /A4045 clad strips, for automobile heat exchanger were produced by the present method and processed to the clad sheets[2]. The microstructure and mechanical properties were compared to those of the clad sheets fabricated by the conventional hot-rolling process. Twin-roll cast products exhibited refined microstructure and the improved mechanical properties than the hot-roll bonded products. Hypo-eutectic Al-Si alloys A4045 (Al-10 %Si) are generally used for the overlay strip of brazing sheets. This is partly because the hot-roll bonding is not suitable for the eutectic and hyper-eutectic Al-Si alloys due to the occurrence of fracture and de-bonding of Si particles during the repeated rolling process. In contrast, the present method is free from such a restriction. Therefore, in the present study, three Al-Si alloys with different Si compositions were provided as the overlay alloy to clad the A3003 base strip. The influence of the Si composition on the overlay solidification manner and the bonding condition of base/overlay interface was investigated.

2. Experimental procedure

2.1 Sample materials

Three layer clad strip was fabricated by vertical-type tandem twin-roll caster. Sample materials are the 3003 aluminium alloy (Al-1 %Mn) for the base strip and Al-10 %Si (corresponding to A4045), Al-12.6 %Si (Al-Si eutectic composition), and Al-14 %Si alloy for the overlay strip.

2.2 Vertical-type tandem twin-roll casting

Casting condition on the vertical-type tandem twin-roll casting is as follow. Solidification length of the upper and lower caster are 60 and 80 mm, respectively. Roll separating force are 2.2 and 4.4 kN. Roll rotating speed is 40 m/min. Casting temperature of molten metal is 15 ℃ higher than each liquidus temperature. In addtion, 50 ℃ higher pouring temperature experiment was also carried out to consider the effect of pouring temperature.

2.3 Microstructure observation

Optical microscopic observation was carried out. The polished cross-section was etched by Weck's reagent (4 g $KMnO_4$, 1 g NaOH, and 100 mL distilled water). SEM-SEI observation was also conducted.

3. Result and discussion

3.1 Microstructure of overlay strip

Regardless the Si composition, the microstructure of the overlay surface region exhibited fine eutectic structure, and the mid-thickness region of overlay consisted of α-Al dendrites and eutectic structure. Especially for 14 %Si, the microstructure was the mixture of dendrites and eutectic structure through the thickness. In contrast, in the case of 10 %Si, there were a large amount of globular grains near the base/overlay interface. The reason of this segregated globular grain is considered as follows. The growth front of overlay solidifying shell on the roll surface is a mushy layer consisting of dendrite branches and liquid phase. Then, the fragmentation of dendrite tip in the mushy layer occurs by melt convection. These fragments grow into the globular grains[3]. Eventually, these grains gather in the interface and forms a globular grain band.

3.2 Base/overlay strip interface

No un-bonded area was observed at base/overlay interface for 10 %Si. In contrast, a limited number of local un-bonded areas observed at the interface for 12.6 and 14 %Si (Fig. 1). In the case of both 12.6 and 14 %Si overlay strips, the defect such as solidification shrinkage and gas porosity was formed at the base/overlay interface, it is considered that overlay solidification took place from the roll surface and the growth front developed toward the base strip, finally overlay solidification finished at the interface.

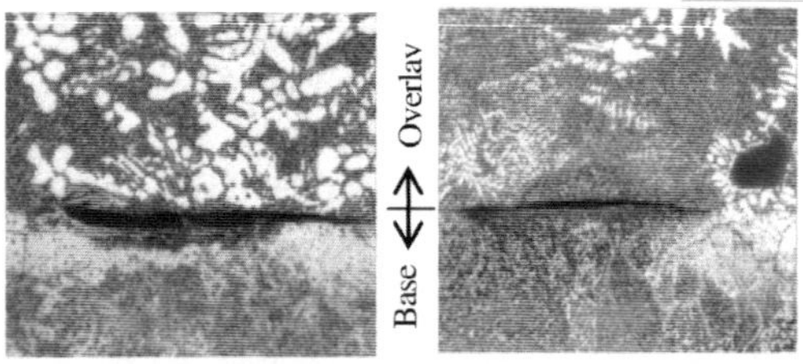

Fig. 1 Un-bonded zone of interface (12.6 %Si, 14 %Si)

3.3 Final solidification area for 10 %Si overlay

To examine the final solidification area for 10 %Si overlay strip, the experiment was conducted of 50 ℃ higher pouring temperature. Relatively higher pouring temperature can lead increasing of solidification defects. Left side of Fig. 2 shows overlay microstructure observed by OM. Continuous internal crack (black line) was observed at boundary area of overlay solidifying shell and globular grain band. Right side of Fig. 2 is SEM-SEI image of internal crack surface. Dendrite tips were observed and it indicates this internal crack was induced by high pouring temperature. Thus, it is investigated that the solidification finished at around this area, boundary of solidifying shell and globular grain band for 10 %Si overlay. Therefore, the defects were not observed at the base/overlay interface in the case of 10 %Si.

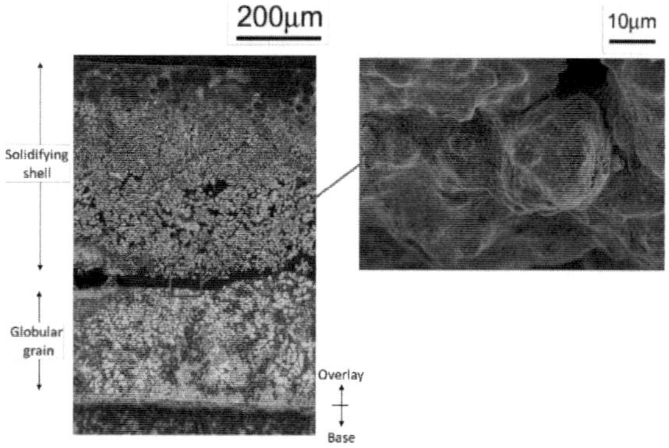

Fig. 2 OM image of 50 ℃ higher pouring temperature overlay microsturcure and SEM image of defect

4. Conclusion

Solidification manner was investigated for several Si composition overlay strip for 3 layer clad strips produced by vertical-type tandem twin-roll casting. For 10 %Si overlay strip, the overlay solidification finished at the boundary area of solidifying shell growth from roll surface and globular grain band. Local un-bonded areas were observed at the base/overlay interface for 12.6 and 14 %Si overlay strips. This indicates the overlay solidification finished at the base/overlay interface for 12.6 and 14 %Si. These microstructure difference in the final solidification area is considered to be due to the difference of solidification manner depending on the Si composition.

References

[1] T. Haga, R. Nakamura, S. Kumai, and H. Watari: Proc. 33rd Int. MATADOR Conf. 37 (2009) 117–124.

[2] R. Nakamura, et.al.: J. Japan Inst. Light Met 64 (2014) 399–406.

[3] M.-S. Kim, Y. Arai, Y. Hori, and S. Kumai: Mater. Trans. 51 (2010) 1854–1860.

Proceedings of the 72nd World Foundry Congress
21-25th May 2016, Nagoya, Japan

Grain Refinement of Cast Magnesium Alloy Containing Aluminum

Jun Ho Bae[1], Young Min Kim[1], Byoung Gi Moon[1], Nguyen Ngoc Ha[2] and Bong Sun You[1]
[1] Korea Institute of Materials and Science (KIMS), Changwon 642-831, Korea
[2] Korea University of Science Technology (UST), Changwon 642-831, Korea

Experiments were conducted to examine the grain refinement ability of AMC master alloy for Al containing Mg alloys. The addition of AMC master alloy shows excellent grain refinement ability in the AZ91 alloy and the Al-Mn morphology change to polygonal type cause by duplex nucleation hypothesis. Refining efficiency was not disappeared after re-melting of refined AZ91 alloy and its refining ability was maintained for 4 hours holding time in melt condition. Therefore, the AMC master alloy can be confidently contributed to make grain refined Mg-Al base alloys with commercialized casting processes.

Keywords: *Magnesium alloy, grain refinement, AZ91 alloy.*

1. Introduction

Grain refinement is an important technique to improve the mechanical properties of magnesium alloys. In general, refined microstructure in as-cast components leads to more uniform microstructure, secondary phases and microporosity on a fine scale. And also it can provide superior extrudability, rollability and forgeability, excellent resistance to hot tearing, good surface finish and so on, leading to considerable cost reduction for production of wrought magnesium alloys [1-3].

Currently, Al-free magnesium alloys can be grain refined simply by zirconium addition, but for Mg-Al based alloys which comprise the main of the magnesium industry, there are no commercialized grain refiners and refining process despite numerical researches of grain refinement such as carbon inoculation, Elfinal process, superheating, and so on [4, 5]. Although, carbon inoculation using C_2Cl_6 is a very effective way for grain refinement of Mg-Al based alloys, the emission of toxic gases causes environment problems and workplace, resulting in the restriction of its use [6-9]. Therefore, alternative grain refiners and refining technology, i.e. effective, reliable and easy to apply for industry, should be developed. Recently, KIMS obtained excellent refining efficiency by developing a new refiner, AMC master alloy composed of Al and Mn as main components and it improved refining ability by inoculation carbon through various ways. The purpose of this study was to evaluate microstructure change after re-melting and fading time in order to apply AMC master alloy to mass production process.

2. Experiments

AMC master alloy was added so that the final composition is the same as commercial AZ91D alloy. In order to measure fading time of AMC master alloy, the alloy melt including Mg, Al, Zn and AMC hold at 710°C for a period of time from master alloy addition using steel crucible and was poured in cylindrical mold that was 80mm in diameter preheated at 200°C. Cast billets were cut 20mm high from the bottom and mechanically polished and etched for microstructure analysis. To confirm the duration of refining efficiency of AMC master alloy after re-melting, grain refined AZ91 alloy by AMC master alloy was melted again at 710°C, and hold for 30 minutes, and then manually stirred for five minutes. It was then casted in the same methods before and microstructure was observed. The grain size was measured by linear intercept method.

3. Results and Discussion

Figure 1 shows the microstructures of commercial AZ91 alloy and grain refined AZ91 alloy by using AMC master alloy. The average grain size of not refined and refined alloy was measured at 400 ± 50 μm and 80 ± 5 μm, respectively.

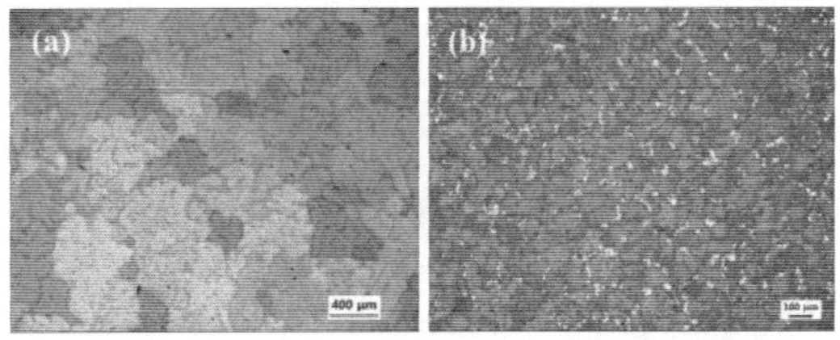

Fig. 1 Microstructures of the cast AZ91 alloy; (a) commercial and (b) grain refined by AMC master alloy.

Figure 2 show the SEM micrographs of commercial and grain refined AZ91 alloy which conducted heat treatment at 400 °C for 10h. Remined particles showing bright after heat treatment is confirmed Al-Mn phase. Two samples show a distinct difference in morphology of Al-Mn phase. The irregular needle shape existed in boundary is shown in case of commercial AZ91 alloy, while the refined AZ91 alloy is showing polygonal shape with the size of 5μm and exist inside the grains . This is because grain refinement using AMC master alloy accompanied with morphology change of Al-Mn phase due to the duplex nucleation mechanism reported by Kim and Wang [10-12].

In the case of AMC master alloy, it cannot be used in commercial AZ91 alloy directly due to the additional Al content in master alloy. So AMC master alloy should be added when AZ91 alloy making. In this case, grain refined AZ91 alloy should re-melt to make products in final manufacturing process. After re-melting, grain refinement efficiency can reduce or disappear during final manufacturing process due to fading phenomenon. Figure 3 shows change of refining efficiency of refined AZ91 alloy after re-melting and during holding in melt state. As shown in figure, the average grain size is 80 μm even after re-melting, and then grain size increased after about 4 hours. Although casting time in industrial fields may differ greatly based on the processes, 4 hours fading time is thought tobe sufficient time to produce grain refined product for use in low pressure die casting, sand casting, or direct chill casting process.

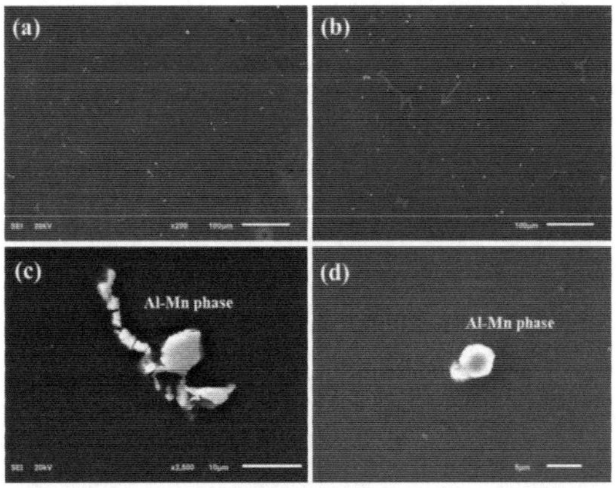

Fig. 2 Scanning electron micrographs of AZ91 alloy heat treated at 400 °C for 10h; (a), (c) commercial AZ91 alloy and (b), (d) grain refined AZ91 alloy by addition of AMC master alloy.

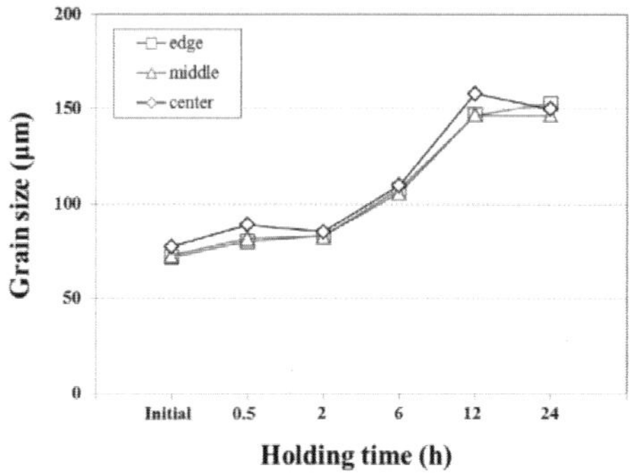

Fig 3. Grain size changes after re-melting and the holding time of refined AZ91 alloy.

4. Conclusion

AMC master alloy that was developed for grain refinement of Al containing Mg alloys is added during AZ91 alloy making process. It was proved that the AMC master alloy addition leads to considerable grain refinement and refining efficiency maintains even after re-melting and 4 hours holding time.

Acknowledgements

This research was supported by the Fundamental Research Program of the Korea Institute of Materials Science (KIMS)

References

[1] C.E. Nelson : Trans American Foundrymen's Soc. 56 (1948) 1-23.

[2] Y.C. Lee, A.K. Dahle and D.H. StJohn : Metal. Mater. Trans. 31A (2000) 2895-2906.

[3] E.F. Emley : *Principles of Magnesium Technology* (Oxford: Pergamon Press, 1966) pp. 200-211.

[4] A.K. Dahle, et al. : J. Light Metal.1 (2001) 61-72.

[5] M. Bamberger : Mater. Sci. Tech. 17 (2001) 15-24.

[6] Q. Jin, et al. : Scripta Mater. 49 (2003) 1129-1132.

[7] Q. Jin, et al. : Scripta Mater.52 (2005) 421-423.

[8] L. Lu, A.K. Dahle and D.H. StJohn : Scripta Mater. 54 (2006) 2197–2201.

[9] L. Lu, A.K. Dahle and D.H. StJohn : Scripta Mater. 53 (2005) 517-522.

[10] Y.M. Kim, C. D. Yim and B.S. You : Scripta Mater. 57 (2007) 691-694.

[11] Y. M. Kim, et al. : TMS Magnesium Tech. (2007) 121-126.

[12] Y. Wang, et al.: Intermetallics 18 (2010) 1683-1689.

Proceedings of the 72nd World Foundry Congress
21-25th May 2016, Nagoya, Japan

Thermal Analysis as a Microstructure Prediction Tool in Aluminum Foundry: A Literature Review

Waleed Khalifa
Faculty of Engineering, Cairo University, 12613, Giza, Egypt

Thermal analysis is one of the promising microstructure prediction techniques, with a high potential of improvement and accuracy. This paper aims at reviewing the available literature on using the thermal analysis and cooling curves as prediction tools in aluminum foundry. This review covers the relationships between parameters of thermal analysis and the quality of melt preparation, and hence the use of these relationships for the prediction of the major microstructural characteristics, such as the grain-refining potential, the eutectic modification, the formation of porosity in the cast products.

Keywords: *Thermal analysis, cooling curves, grain refining, Si modification, porosity.*

1. Introduction

The quality and properties of Al-Si cast products; such as the mechanical properties, the hot tearing susceptibility, the surface finish, the machinability, and the pressure tightness of products; are based mainly on certain microstructural characteristics. Therefore, the assessment and prediction of such microstructural characteristics before pouring the melt is of prime interest. Of these characteristics is the level of melt quality including the extent of grain refining; and the level of modification [1]; and the level of dissolved hydrogen. In this brief review, thermal analysis parameters related to these microstructural characteristics are discussed.

2. Cooling curves

Thermal analysis using the cooling curves is called conventional thermal analysis. In this technique, the molten alloy is allowed to cool down slowly (0.1 -0.8^{o}Cs^{-1} [2]), and the data are collected form the thermal analysis curve, see Fig. 1. Three regions are of significance in this curve: region A indicating the extent of grain refining, region B indicating the level of modification attained, and region C which determines the end of solidification. The general shape of the cooling curve gives an indication on level of grain refinement and if the grain morphology is columnar or equiaxed [3]. The thermal analysis can be done using either one or two thermocouples. The first thermocouple is placed at the center of the cub mold and the second one is placed near the mold wall.

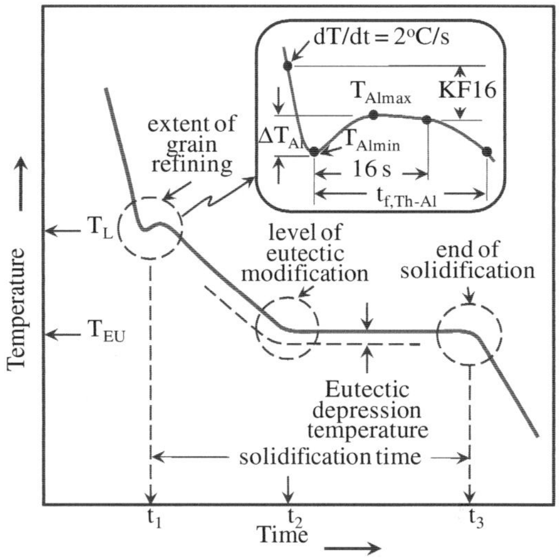

Fig. 1 Cooling curve obtained during thermal analysis and the main thermal analysis parameters.

3. Grain refinement

The grain refining process provides several advantages to alloy properties and processing [4], including better distribution of porosity and second phase particles, improved surface finish and machinability, greater fatigue strength, better pressure tightness, and reducing tendency of ingot cracking, and provides improved mechanical properties and several other beneficial effects. This is why monitoring of the extent of grain refining is necessary for quality assurance before pouring the melt.

Thermal analysis can be used as a technique to control the grain size in castings, and to determine the dendrite coherency point [5]. The relevant thermal analysis parameters used in the grain-size prediction (Fig. 1) of the hypo-eutectic Al-Si alloys [6], are (i) ΔT_{Al} is the recalescence, i.e. the difference between maximum T_{Almax} and minimum T_{Almin} temperatures,

(ii) KF16 is the difference between the temperature when the cooling rate is 2°C/s and the temperature recorded 16s later, and (iii) $t_{f,Th\text{-}Al}$ is the time difference between the moment corresponding to the minimum in temperature and the time at which that temperature is again reached after recalescence.

Several research works were carried to study grain refining potential using this technique [7]. The breaking points of cooling curves were calibrated against the growth temperature T_{Almax}. One the complexities that face this approach is that the growth temperature is a function of the cooling rate. Therefore, constant cooling rate should be experienced to obtain consistent data [5, 8].

When the nucleation undercooling, ΔT_{Al} is large, a relatively large grain size in obtained, and when this value is less than 0.3°C this indicates that there is no barrier for nucleation and the grain size is relatively fine [1]. Hence assessment and comparisons of the grain refining potential of the melts may be done. Grain refinement affects the other thermal analysis parameters such as nucleation and growth temperatures of various phases, *e.g.*, eutectic Si, the solidification range and total solidification time [9].

4. Si modification

During solidification of Al-Si alloys, the liquid phase ends with the eutectic reaction resulting in the formation of the eutectic phases. The unmodified alloy exhibits poor ductility leading to brittle fracture of castings, while the properly modified alloy by strontium or sodium shows ductile mode of fracture, with increased elongation and tensile strength [1].

The thermal analysis parameters related to the level of modification are the eutectic temperature depression and the time difference between eutectic nucleation and eutectic finish points (i.e., t_3-t_2 of Fig. 1). The amount of depression ΔT measured in degrees Celsius is a common indicative parameter for slow cooling rates. The larger the depression the finer and more fibrous is the eutectic structure. For example, well-modified structures occur at depression temperatures of more than 5.5°C in alloys A356.2 [1]. The eutectic temperature is a function of modifier level, and it is lowest at the optimum level of modifier, with over modified alloy showing smaller depression temperatures [10, 11]. The time difference parameter is useful at higher cooling rates [2].

5. Hydrogen level and porosity

The amount of dissolved hydrogen influences the cooling curve characteristics, so that certain parameters from the cooling curve can be used to assess the amount of porosity in as-cast aluminum products. These parameters are the characteristic nucleation temperatures: α-Al nucleation, the Al–Si eutectic and the Al–Si–Cu eutectic temperatures. The effect of different amounts of hydrogen on the Al–Si–Cu eutectic temperature is much more significant than the other two parameters, i.e., the dissolved hydrogen depresses the nucleation temperature of copper rich eutectic phases [12]. The depression reached 12.3°C for a melt containing 0.18 mL H_2/100 g Al [5].

Conclusions

Thermal analysis is a potential technique for the prediction of microstructure characteristics in Al-Si foundry alloys.

Acknowledgements

The financial support of the Egyptian Science and Technology Development Fund, grant No. 14902 is highly appreciated.

References

[1] D. Apelian, G. Sigworth, K. Whaler: *AFS Trans.* (1984) 297-307.

[2] N. Tenekedjiev, J. Gruzleski: *AFS Trans.* (1991) 1-6.

[3] L. Arnberg, L. Backerud, H. Klang: *Grain refinment in Castings and Welds,* Ed. By G. Abbaschian, S. David, 1983.

[4] W. Khalifa, Y. Tsunekawa: *71st World Foundry Congress* 2014.

[5] M. Johnsson: Thermochimica Acta 256 (1995) 107-121.

[6] J. Lacase, D. Ferdian, I. Lizarralde, A. Niklas, S. Eguskiza, and A.I. Fernadez-Calvo: *71st World Foundry Congress* 2014.

[7] M. Johnsson, L. Backerud, G. Sigworth: Metallurgical Transactions 24A (1993) 481-491.

[8] D. Gloria, J. E. Gruzleski,: *Light Metals* (1999), Met Soc., Ed. by M. Bouchard, A. Faucher.

[9] S. G. Shabestari, M. Malekan: Journal of Alloys and Compounds 492 (2010) 134-142.

[10] R. DasGupta, C. Brown, S. Marek: *AFS Trans.* (1988) 297-310.

[11] H. Beumler, A. Hammerstad, B. Wieting, R. DasGupta: *AFS Trans.* (1988) 1-12.

[12] A. Mitrasinovic, F.C. Robles Hern´andez, M. Djurdjevic, J.H. Sokolowski: Materials Science and Engineering A 428 (2006) 41–46.

Proceedings of the 72nd World Foundry Congress
21-25th May 2016, Nagoya, Japan

Influence of Mo content on Tungsten Heavy Alloys for Die Casting Applications

Rafael Cury[1], Laurent Dartus[1] and Nicolas Nuns[2]
[1] Plansee Tungsten Alloys, 466, avenue de Digues, Saint Pierre en Faucigny, France
[2] Université de Lille 1, Villeneuve D'Ascq, France

Tungsten heavy alloys containing Mo are widely used on the Al die casting industry for several decades. Obtained through solid-liquid sintering from a compacted mixture of prime materials at the powder state, some aspects of these alloys, notably with respect to the influence of a small variation on the Mo content are unknown. Present investigation evaluates the influence of Mo not only at the bulk properties of tungsten heavy alloys but most importantly at the surface level, where alloy will interact directly with the solidifying liquid cast. Usual investigations were performed and a complementary ToF-SIMS was performed, characterizing the very first few layers of the material.

Keywords: *tungsten heavy alloys, Mo content, sintering, mechanical resistance, Aluminiun die casting*

1. Introduction

Recent previous works [1-5] described correspondingly mechanical and thermal properties of two available trademarks from Plansee tungsten heavy alloys: Densimet D185 © and Densimet D2M © (the latter, alloy containing Mo as an alloying element). In the present study, efforts were concentrated on the Mo role of at Densimet D2M © material, notably at the assessed microstructure for variations on Mo concentration with some perspectives for an improved Densimet D2M ©. According to the literature [6], the presence of Mo will play a role on tungsten solubility at the binder. Therefore, the increasing concentration of Mo shall reduce the total solubility of W during solid liquid sintering [5]. Authors supposed that at higher concentrations (more than 10%), the kinetic growth of W grain will be slower and better mechanical properties can be then obtained by grain refinement. Such high Mo concentration can be though industrially critical because of the possible presence of brittle intermetallics, particularly the ones described on the Ni-Mo system.

The Mo concentration effect was then evaluated by a very specific surface analysis technique, the ToF-SIMS. Time-of-Flight Secondary Ion Mass Spectroscopy (ToF SIMS)[7] is a very sensitive analytical technique which uses a pulsed ion beam to remove both atoms and molecules from the sample surface (< 4 atomic layers). ToF SIMS machine offers the opportunity to perform chemical mapping of element or molecules with a high lateral resolution (< 100 nm). By combining depth profile with 2D imaging, it is possible to reconstruct 3D images.

2. Experimental

2.1 Sample preparation

Standard Densimet D2M© and 4 prototypes were elaborated according to the alloying composition listed on Table 1. (cf. Table 2).

Table 2- Standard Densimet D2M and prototype composition

Alloy	Ni/Fe	Mo	W
Densimet D2M©	2	4	balance
Prototype A	0.5	4	
Prototype B	2	2	
Prototype C	3	2	
Prototype D	2	6	

They were then isostatic pressed in cylinders in a cylindrical shape with approximately 12 mm of diameter. Sintering step took place under pure H_2 atmosphere, aiming the chemical reduction of oxide layers. Samples were then machined, cut and polished using SiC paper of different mesh and then finished using ordinary alumina polishing. They were cleaned on US bath and dried on desiccator before usual and TOF-SIMS analysis

2.2 Results

All density values were superior to 17,0 g/cm^3. In terms of microstructure analysis, the first striking point was the perceptible presence of intermetallics on Prototypes A, B and D. Indeed, the presence of various intermetallic phases associated with the intense variation on the Mo content and the different Ni/Fe ratio might generate brittleness for those compositions.

Rather interesting results were obtained on prototype C. This relatively low density alloy (around 17.3 g/cm^3) showed an overall consolidated structure, similar to the classical Densimet D2M © alloy. Samples were then deeply investigated on ToF-SIMS analysis, notably prototype C, as shown at the images below (Figure 4 and Figure 5). A representative surface of $100\mu m^2$ from the sample was scrutinized by the device, with respect to each one of the constitutional elements. This chemical mapping enables the observation of rich or depleted zones on the material according to the elements, including W and Mo distributions.

As previously observed [2], Fe and Ni will be mainly restricted to the binder (cf. Figure 4). Some past results [6,8] pointed out the possibility that the levels of Ni evaluated inside the W phase are higher than the ones predicted by the Ni-W phase diagram. It is though not observed any trace of either Fe or Ni on the nodular domains. The nodular domains are mainly evaluated with respect to Mo and W. Figure 5 shows some surprising results with respect to the nodules: some of the: are showing depleted Mo zones. Those results are in contrast with previously obtained microstructure in which Mo core surrounded by W were observed by some authors. Also, in small scale on the binder, some zones are also Mo deprived.

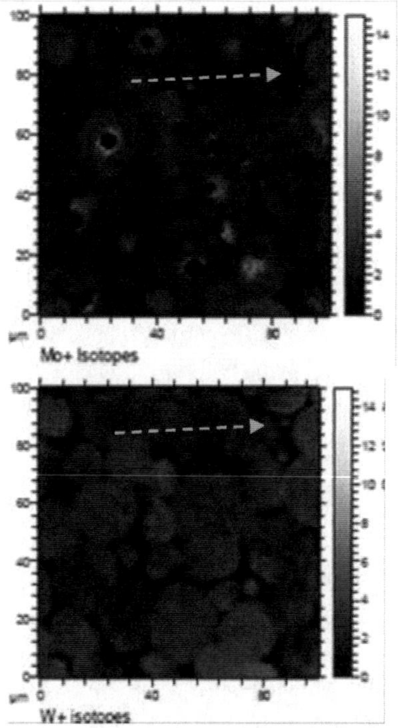

Figure 5- Images obtained with ToF-SIMS analysis (prototype C) with W and Mo mapping.

The chemical mapping is related with W and Mo occurrence. The arrows on Figure 5 indicate the existence of a depleted Mo zone. Also, several nodular phases are showing Mo depleted areas in their core.

3. Conclusions and perspectives

Prototype C has the highest Ni/Fe ratio and less Mo than other compositions. In a first observation, it showed similar microstructure with respect to the standard industrial alloy. ToF-SIMS analysis though put in evidence some depleted Mo zones in that alloy. Observed microstructure is similar to the existent industrial alloy.

Even though some results showed drawbacks in t terms of homogeneity, tungsten heavy alloys containing Mo remain as the ideal material for die casting application. A better control on Mo distribution will undoubtedly lead to an optimized alloy which must be verified in application.

References

[1].German, R.M., *Sintering: From Empirical Observations to Scientific Principles, Elsevier, First Edition,* 2014.

[2].Cury, R., Dartus, L. Schroeder, J. *Properties and possible improvements of Tungsten Heavy Alloys for Die Casting Application*, NADCA 2014 Die Casting Conference and Tabletop.

[3].Kemp, P.B., German, R.M., *Grain growth in liquid-phase-sintered W-Mo-Ni-Fe alloys, J. of the Less-Commom Metals*, 175, p.353-368, (1991)

[4].Lin; K.H, Hsu, C.S., Lin, S.T., *Precipitation of an Intermetallic Phase in Mo-alloyed Tungsten Heavy Alloys, Met.Trans.* vol. 44, No. 3, p. 358 (2003).

[5].Guillermet, A.F., Ostlund, L., Met. Trans. A, 17 (1986)

[6]. Cury, R. "Etude métallurgique des alliages Ni-W et Ni-W-Cr ", PhD thesis, Université Paris, (2007).

[7].http://uccs.univ-lille1.fr/index.php/ressources/2-non-categorise/55-regional-platform-of-surface-analysis, Université de Lille, published on website on May 2015.

[8]..Edmonds, D.V., Jones, P., *Interfacial embrittlement in liquid-phase sintered tungsten heavy alloys, Metall. Trans. A*, vol. 10A, p.289, (1985)

Proceedings of the 72nd World Foundry Congress
21-25th May 2016, Nagoya, Japan

High temperature interaction between liquid Gd-containing alloys and selected oxides

N. Sobczak[1,2], P. Turalska[1], M. Homa[1], A. Siewiorek[1], G. Bruzda[1], R. Nowak[1], A. Kudyba[1], I. Kaban[3], N. Mattern[3], J. Eckert[4,5]

[1]Foundry Research Institute, 73 Zakopianska Str., 30-418 Krakow, Poland
[2]Motor Transport Institute, 80 Jagiellonska Str., Warsaw, Poland
[3]IFW Dresden, Institute for Complex Materials, P.O. Box 270116, 01171 Dresden, Germany
[4]Erich Schmid Institute of Materials Science, Austrian Academy of Sciences, 8700 Leoben, Austria
[5]Department Materials Physics, Montanuniversitat Leoben, Jahnstrase 12, 8700 Leoben, Austria

The sessile drop method was applied for investigation of high temperature interaction of liquid Gd, Ti, binary Gd-Ti and ternary Gd-Ti-Zr alloys in contact with dense polycrystalline Y_2O_3 and ZrO_2+3%Y_2O_3 substrates. During high temperature tests performed under flowing Ar, the contact angles were determined from the images of drop/substrate couples recorded with CCD camera. The effect of metal chemistry on its melting, wetting behaviour and reactivity in contact with oxide substrates are analysed.

Keywords: *wettability, reactivity, Gd-alloys, oxides*

1. Introduction

Gadolinium (Gd) is widely used in advanced alloys to improve their functional and utility properties. However, high melting temperature, high oxygen affinity and chemical aggressiveness against most available refractory materials makes difficult melting, casting as well as high-temperature testing of Gd-containing alloys using container-assisted methods (e.g. calorimetric studies, measurements of surface tension of liquid alloys) [1,2].

The paper is focused on investigation of high temperature interaction between liquid Gd-containing alloys and selected oxides.

2. Experimental

The materials were pure Gd (99.95%), Ti (99.98%) and their binary and ternary alloys produced by arc melting (Table 1). Dense polycrystalline oxide substrates were made from Y_2O_3 and ZrO_2 powders by HT-HP synthesis and polished up to a roughness of about 120 nm.

Real-time melting, wetting, spreading and solidification behaviour of selected Gd/oxide systems were examined by the sessile drop method [3,4] using experimental facility described in [4,5].

During high temperature tests, the selected metal/substrate couples were contact-heated to the test temperature under protective atmosphere (flowing Ar) and their images were recorded with a high-speed high-resolution CCD camera. The test temperatures were about 50 to 100°C above the melting points of the metal samples.

The structure and chemistry of solidified couples were examined both on the top surfaces and on cross-sections using optical microscopy (OM) and scanning electron microscopy (SEM) coupled with energy dispersive X-ray analysis.

3. Results

Wetting phenomenon was observed between Y_2O_3 substrate and pure Gd immediately after its melting (1312°C) showing θ~65°. Upon 2 min heating to 1360°C, a contact angle degreased to a final value of 33°. Similar wetting behavior was observed for $Gd_{60}Ti_{40}$ alloy at 1473°C.

Table 1. Contact angles of selected Gd/oxide couples

Metal	Substrate	Temp. °C	θ,°
Gd	Y_2O_3	1360	33
$Gd_{60}Ti_{40}$	Y_2O_3	1473	33
$Gd_{30}Ti_{70}$	Y_2O_3	1500	157
Ti	Y_2O_3	1760	77
Gd	ZrO_2+3%Y_2O_3	1362/1412	68
Ti	ZrO_2+3%Y_2O_3	1720/1770	122
$Gd_{60}Ti_{40}$	ZrO_2+3%Y_2O_3	1697	92
$Gd_{40}Ti_{60}$	ZrO_2+3%Y_2O_3	1760	80
$Gd_{40}Ti_{30}Zr_{30}$	ZrO_2+3%Y_2O_3	1650	104
$Gd_{40}Ti_{30}Zr_{30}$	ZrO_2+3%Y_2O_3	1660	101
$Gd_{40}Ti_{40}Zr_{20}$	ZrO_2+3%Y_2O_3	1690	95
$Gd_{40}Ti_{20}Zr_{40}$	ZrO_2+3%Y_2O_3	1700	97
$Gd_{40}Ti_{20}Zr_{40}$	ZrO_2+3%Y_2O_3	1717	94

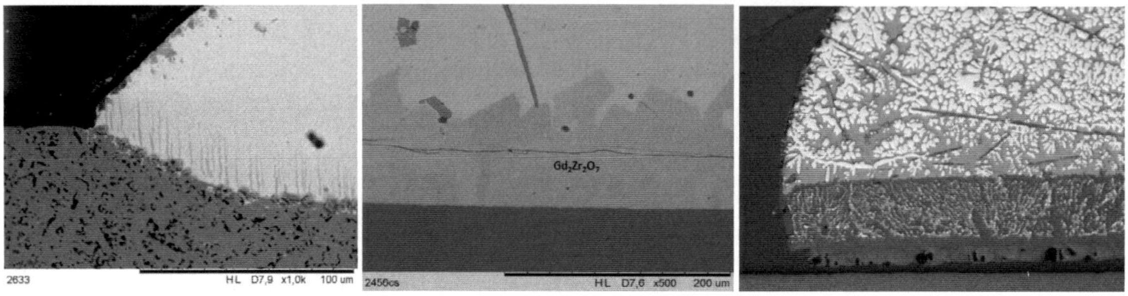

Fig. 1 Microstructures of cross-sectioned sessile drop couples:
a) Gd/Y_2O_3; b) Gd/ZrO_2+3% Y_2O_3; c) $Gd_{40}Ti_{40}Zr_{20}$/ZrO_2+3% Y_2O_3; a,b - SEM; c - OM

Y_2O_3 was also wetted by pure Ti, contrary to the high Ti-containing alloy $Gd_{30}Ti_{70}$ showing non-wetting behavior and a very high contact angle.

The ZrO_2 substrates containing 3%Y_2O_3 exhibited much worse wettability with both pure metals and their alloys. These substrates were not wettable by pure Ti at the experimental temperatures of 1720°C and 1770°C, contrary to pure Gd showing a good wetting at 1362°C and 1412°C. Alloying Gd with 40% Ti caused significant increase in the contact angle and non-wetting behavior at 1697°C. Further increase of the Ti content to 60% made ZrO_2+3%Y_2O_3 substrate wettable at 1760°C. For Gd-Ti-Zr alloys with 40% Gd, when Ti was partially replaced by Zr, no wetting was observed at selected temperatures.

Structural characterization of solidified couples revealed significant difference in the mechanisms of high temperature interaction in the studied systems. In the Gd/Y_2O_3 couples, the interaction is dominated by a strong dissolution of the substrate in the liquid metal resulting in the formation of a deep crater in the substrate, alloying Gd drop with Y and the growth of directionally solidified interfacial region during cooling (Fig. 1a). On the contrary, the interaction of both pure Gd and selected Gd-containing alloys with ZrO_2-3%Y_2O_3 substrates is controlled by the formation of continuous interfacial reaction product $Gd_2Zr_2O_7$ (Fig. 1b).

Additionally in the Gd-Ti-Zr/ZrO_2-3%Y_2O_3 couples, a Gd-rich layer is evidenced in the Gd-Ti-Zr drop near the reactively formed interfacial layer while the other part of the drop is almost free of Gd (Fig. 1c). This effect is related to the liquid metal de-mixing since the Gd-Ti-Zr alloys exhibit miscibility gap in the temperature rage studied in this work.

4. Summary

Dissolutive wetting mechanism is responsible for a good wetting of liquid Gd and Gd-rich alloys on Y_2O_3 substrates.

Wetting phenomenon in the Gd/ZrO_2+3%Y_2O_3 couple is dominated by the formation of continuous and wettable layer of a new phase $Gd_2Zr_2O_7$ at the interface.

Alloying Gd with Ti or Ti+Zr reduces reactivity of liquid metal with both Y_2O_3 and ZrO_2+3%Y_2O_3 substrates thus resulting in higher contact angles.

Acknowledgements

This study was done in the frame of collaboration program between the German Academic Exchange Service DAAD and the Ministry of Science and Higher Education of Poland. Financial support from the National Science Centre of Poland (program HARMONIA, project No. 193624) and the German Academic Exchange Service DAAD (Project No. DAAD-56269397) is acknowledged.

.

References

[1] I. Kaban, R. Nowak, O. Shuleshova, B. Korpala, G. Bruzda, A. Siewiorek, J.H. Han, N. Sobczak, N. Mattern, J. Eckert: J. Mater. Sci. 47/24 (2012) 8381-8386

[2] N. Sobczak, R. Nowak, A. Siewiorek, B. Korpala, G. Bruzda, I. Kaban, O. Shuleshova, J.H. Han, N. Mattern, J. Eckert: Proc. 71st World Foundry Congress, Bilbao, Spain (2014).

[3] N. Sobczak, M. Singh, R. Asthana: Curr. Opin. Solid State Mater. Sci., 9/4 (2005) 241-253

[4] N. Sobczak, J. Sobczak, R. Asthana, R. Purgert: China Foundry 7/4 (2010) 425-437

[5] N. Sobczak, R. Nowak, W. Radziwill, J. Budzioch, A. Glenz: Mater. Sci. Eng. A495 (2008) 43-49

High temperature behaviour of liquid Ni-based alloys with oxide ceramics

J.J. Sobczak[1], N. Sobczak[1,2], R.M. Purgert[3], R. Asthana[4], R. Nowak[1], G. Bruzda[1], P. Turalska[1], M. Homa[1], A. Kudyba[1], A. Siewiorek[1]

[1]Foundry Research Institute, 73 Zakopianska Str., 30-418 Cracow, Poland
[2]Motor Transport Institute, 80 Jagiellonska Str., Warsaw, Poland
[3]Energy Industries of Ohio, Cleveland, OH 44131, USA
[4]University of Wisconsin–Stout, Menomonie, WI 54751, USA

High temperature interaction between liquid Ni and its alloys (NiCo10, NiCr10, NiW10, Inconel 740, Haynes 282) and refractory oxide-based ceramics containing Al_2O_3, MgO, $MgAl_2O_4$, Mg_2SiO_4, ZrO_2 was examined by a sessile drop method at 1500°C in argon. A specially developed procedure for *in situ* opening of the metal/ceramic interfaces directly at the testing temperature was used. Structure and chemistry of interfaces were examined using SEM+EDS combined with scanning probe microscopy analysis. The results are discussed in terms of chemical interactions in relation to the role of alloying elements and substrate chemistry on reactivity and interface formation in selected metal/oxide systems.

Keywords: *wetting, reactivity, Ni alloys, oxides*

1. Introduction

Information on high temperature behaviour of refractory materials in contact with Ni-based alloys, particularly the effect of alloying additions on wettability and reactivity between an alloy and a refractory, is of great practical importance for metal casting applications [1-6].

The paper is focused on applicability of oxide-based ceramics for making foundry appliances needed for melting and casting of the new generation of Ni-based superalloys in order to manufacture cast boiler and turbine components for Advanced UltraSuperCritical Power Plants (A-USC).

2. Experimental procedure

Wettability and reactivity in selected metal/oxide couples were examined by a sessile drop method allowing the direct measurements of contact angle (θ) [7]. The materials used are shown in Table 1. Oxides were in the form of single crystals (sc) of a roughness R_a<1 nm or dense polycrystalline (pc) substrates produced by HT-HP sintering from high-purity powder and polished up to R_a<150 nm.

The tests were performed in an experimental system specially designed for high temperature studies [8]. Its details together with testing procedures are described in [3-8]. The following testing conditions were used: high-purity flowing Ar (5N), 1500°C, contact heating with a rate of 10 K min^{-1}. Real time behavior of an alloy on a substrate was recorded by a high-speed high-resolution CCD camera. In order to get the drop/substrate interfaces unaffected by cooling history, the interfaces were exposed at 1500°C directly in the UHV chamber by drop pushing from the 1st to 2nd position on a substrate [4-6]. Structural analysis of solidified sessile drop samples was performed by scanning electron microscopy (SEM), energy-dispersive X-ray spectroscopy (EDS) and scanning probe microscopy (SPM).

3. Results

The results from a comparative investigation of the effects of alloying additions to Ni and oxide type on wettability in selected metal/oxide couples are summarized in Table 1. Liquid Ni and its binary alloys do not wet (θ>90°) any of the selected oxide substrates. An exception is NiCr10/MgO couple since, after melting, liquid NiCr10 wets and spreads over the substrate but after a few seconds it again de-wets it. The wetting-dewetting phenomenon takes place several times during heating up to 1500°C and it is accompanied with drop movement. Most probably, this phenomenon is related with periodic formation and disappearance of the wettable interfacial phase ($MgCr_2O_4$ [6]) due to its dissolution in the alloy and further transfer of Cr to flowing gas. Real time observations of melting, wetting, spreading and solidification of Cr-containing alloys (NiCr10, IN740, H282) on MgO-containing substrates (MgO, $MgAl_2O_4$, Mg_2SiO_4) evidenced drop movement, a larger scatter in contact angle measurements and a greenish color of the substrate surface that had been in contact with or in the vicinity of the drop.

Table 1 Contact angles of alloy/oxide couples at 1500°C
("→" corresponds to the tests using drop pushing from the 1st to 2nd position on the substrate at 1500°C)

Substrate	Contact angle (degree) of selected alloys (chemical composition in wt%)					
	Ni (99.99%)	NiCo10	NiCr10	NiW10	IN740[a]	H282[b]
Al_2O_3 (sc)	107	118	114	112	107	84
MgO (sc)	111→111	124	62-118*	128	91±5	80-95*→85*
$MgAl_2O_4$ (sc)	93	110→110	148→110±10	135±5→130±5	95±5	85*→87*
Mg_2SiO_4 (pc)					154→154	91*
ZrO_2+MgO (pc)						81→86
ZrO_2+3%Y_2O_3 (pc)	100±10*	120→110	100→100		94±5	82→81
$ZrO_2+Y_2O_3{}^{np}$ (pc)	110				147	134→109
$ZrO_2+ZrO_2{}^{np}$ (pc)	120				127	131→106

[a] INCONEL®740: 20Co, 25Cr, 0.5Mo, 0.9Al, 1.8Ti, 2Nb, 0.7Fe, 0.3Mn, 0.5Si, 0.03C (Special Metals Corp., USA)
[b] HAYNES®282: 10Co, 20Cr, 8.5Mo, 1.5Al, 2.1Ti, 1.5Fe, 0.3Mn, 0.15Si, 0.06C, 0.005B (Haynes International Inc., USA)
* apparent values due to drop movement or strong secondary substrate roughening

IN740 and H282 alloys do not wet polycrystalline Mg_2SiO_4, however, the formation of an Al_2O_3-rich phase was noted to occur. Especially with H282 alloy, Al_2O_3 was evidenced both at the interface and on the substrate surface near the drop, forming a continuous network along substrate grain boundaries. The worst wetting and the smallest reactivity were observed for IN740 and H282 on ZrO_2 substrates containing additions of Y_2O_3 or ZrO_2 nanoparticles (np). SEM, EDS and SPM characterization of interfaces showed that initially smooth substrate surfaces became rough due to a high temperature interaction with all alloys. Secondary roughening is particularly pronounced for MgO^{sc} after wettability tests with NiCr10, IN740 and H282 alloys showing two orders of magnitude increase in drop-side substrate surface roughness.

4. Summary

Due to high melting temperature and the presence of reactive alloying elements, the interaction between A-USC candidate alloys (IN740, H282) and oxide-based ceramics is accompanied with substrate dissolution in the alloy and the formation of interfacial reaction products. It results in secondary substrate roughening, a decrease in the contact angle values and an increase in adhesion of the liquid alloy to the oxide substrate. Unusual wetting behavior of liquid Cr-rich alloys on MgO-rich substrates, accompanied with the drop movement, is caused from continuous change in the substrate surface chemistry. The above phenomena may contribute to the formation of casting defects responsible for a significant reduction in mechanical properties of products.

Acknowledgements

The research is a part of a supplier development effort initiated by the State of Ohio in support of the U.S. DOE/Ohio Coal Development Office Advanced UltraSuperCritical Materials Program and performed by the Energy Industries of Ohio in cooperation with the Foundry Research Institute in Poland. Financial support from the Ministry of Science and Higher Education of Poland in frame of projects NN507450634 and 721/N-NICKEL/2010 is acknowledged.

References

[1] R. Asthana, S.T Mileiko, N. Sobczak: Bull. Polish Acad. Sci., Technical Sciences 54/2 (2006). 147-166
[2] F. Valenza, R. Nowak, N. Sobczak, A. Passerone, M. Di Foggia, M.L. Muolo: Adv. Sci. Techn. 70 (2010) 130-135
[3] N. Sobczak, J. Sobczak, R. Asthana, R. Purgert: China Foundry 7/4 (2010) 425-437
[4] N. Sobczak, R. Nowak, R. Asthana, R. Purgert: Scripta Mater. 62 (2010) 949-954
[5] N. Sobczak, R.M. Purgert, R. Asthana, J.J. Sobczak, M. Homa, R. Nowak, G. Bruzda, A. Siewiorek, Z. Pirowski: Ceram. Eng. Sci. Proc., 36, 8 (2016) 309-321
[6] N. Sobczak, R.M. Purgert, J.J. Sobczak, R. Asthana, R. Nowak, M. Homa, J. Morgiel: J. Mater. Eng. Perform. (2016)
[7] N. Sobczak, M. Singh, R. Asthana: Curr. Opin. Solid State Mater. Sci., 9/4 (2005) 241-253
[8] N. Sobczak, R. Nowak, W. Radziwill, J. Budzioch, A. Glenz: Mater. Sci. Eng A495 (2008) 43-49

Proceedings of the 72nd World Foundry Congress
21-25th May 2016, Nagoya, Japan

Evaluation of the α-case with Titania Mold for Titanium Investment Casting

Seul LEE[1] and Young-Jig KIM[1]

School of advanced materials science and engineering, Sungkyunkwan University, 2066 Seobu-ro, Jangan-gu, Suwon, Gyeonggi-do 16419, Korea.

The brittle α-case must be removed by chemical milling because it reduce the mechanical properties. The aim of this study is to evaluate and prevent α-case formation as a mold materials. Titania which was most of the α-case phase was generated by addition of titanium powder or simply added on the Al_2O_3 mold for preventing of α-case. The titanium casting surface was investigated by using optical microscopy (OM). Composition and morphology of mold was confirmed by using X-ray diffraction (XRD) and field emission scanning electron microscope (FE-SEM). The castings using titanium added mold (Al_2O_3+xTi) was effectively reduced the formation of α-case. However, the chemical stability of titania add mold (Al_2O_3+x TiO_2) that has almost identical composition to titanium added mold had different effect of α-case formation.

Keywords: Titanium, Investment casting, Titiania added mold, titanium added mold, α-case

1. Introduction

Because of the poor machinability and workability, investment casting was widely used for producing complex and near net shape titanium product. Investment casting processes has excellent reproducibility of castings within close dimensional limits [1]. However, the titanium was extremely reactive with mold material, resulting in α-case. α -case was formed by the diffusion of oxygen and metallic element from the mold to titanium. And the α-case were composed of titanium oxide and intermetallic compound [2]. The formation of the α-case on casting degraded the mechanical properties such as ductility by acting as cracks initiate site [3]. Therefore the α-case usually removed by chemical milling before use. But the milling process was expensive and limits the dimensional tolerance. In this study was to suppress the α-case by generating or adding titania on the mold. The formation of titania in the mold side was induced during the mold firing process or simply added to slurry.

2. Experimental procedures

Alumina, titanium and titania were used for primary coating. In order to produce TiO_2 in the mold, the 50 % of titanium or titania were added to the alumina plus colloidal SiO_2 slurry respectively. Primary coating process was repeated twice. Backup coating underwent three times with chamotte slurry over at intervals of 4hour. The mold was dewaxed in an autoclave at 150 MPa and fired in furnace at 950°C. Titanium (ASTM B-348, grade 2) was melted in the water-cooled copper crucible. The titanium, weighing 120g, was placed on top of the copper crucible. And the mold was located inside the copper crucible for the drop of titanium melts. The phase identification of molds was analyzed by x-ray diffractometer (M18XHF-SRA, Mac Science). The thickness of the α-case was determined using a micro Vickers hardness (MVK-H2, Mitutoyo) and OM (PME 3, Olympus). In our work, the micro Vickers hardness test was performed with 50 μm from the casting surface to 1000 μm, applying a 100 g load. Macrostructure and composition of before and after mold surface were examined by FE-SEM (JSM-7600F, JEOL).

3. Result and discussion

3.1 Analysis of mold composition

Figure 1 shows the XRD spectra of various mold materials. The composition of Al_2O_3 and Al_2O_3+ 50TiO_2 mold was not changed, indicating that the Al_2O_3 and TiO_2 were stable phase after mold firing condition. While, Al_2O_3, TiO_2 and Ti_5Si_3 were found in Al_2O_3+50Ti mold. TiO_2 was caused by reaction between the atmospheric oxygen and titanium during mold firing. Also, the reaction between the titanium and colloidal silica resulted in formation of Ti_5Si_3.

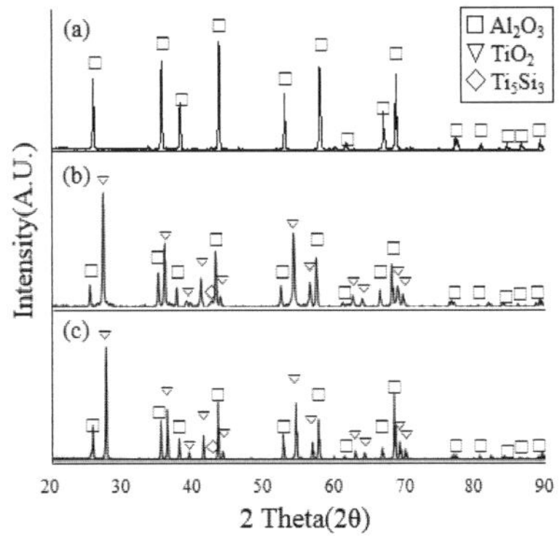

Fig. 1 XRD results of refractoty mold (a) Al_2O_3, (b) Al_2O_3+50Ti and (c) Al_2O_3+50TiO_2

3.2 Thickness of α-case

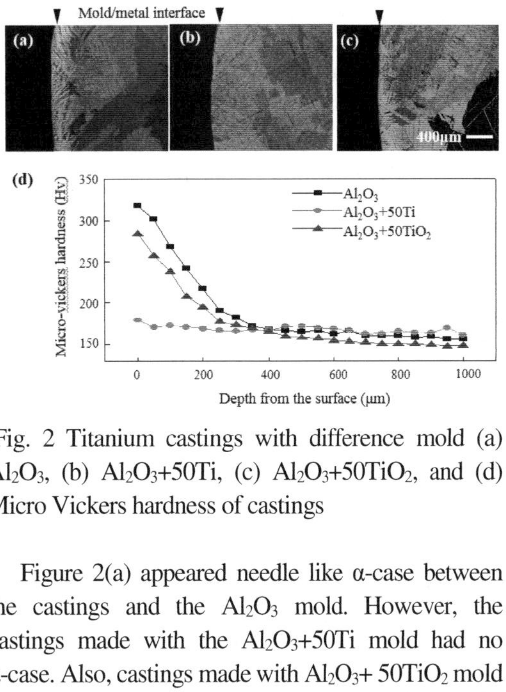

Fig. 2 Titanium castings with difference mold (a) Al_2O_3, (b) Al_2O_3+50Ti, (c) Al_2O_3+50TiO_2, and (d) Micro Vickers hardness of castings

Figure 2(a) appeared needle like α-case between the castings and the Al_2O_3 mold. However, the castings made with the Al_2O_3+50Ti mold had no α-case. Also, castings made with Al_2O_3+ 50TiO_2 mold has small amount of α-case, compared to Al_2O_3 mold. Figure 2(d) exhibit a micro Vickers hardness profile in titanium castings depending on the depth from the surface. Castings made with Al_2O_3 mold had 350μm α-case depth. However, the α-case was only 50μm with Al_2O_3+50Ti mold. Also, α-case was 300μm when Al_2O_3+50TiO_2 mold was used.

3.3 Effect of TiO_2 on α-case formation

Both Al_2O_3+50Ti and Al_2O_3+50TiO_2 mold of composition was nearly identical. However, effect of α-case was different. Mold surface morphology was analyzed by FE-SEM. Figure 3 shown the FE-SEM image of mold surface before and after casting. Before casting, Al_2O_3 and Al_2O_3+50TiO_2 mold surface contained homogeneously distributed Al_2O_3+SiO_2 and TiO_2+SiO_2. In Al_2O_3 +50Ti mold, TiO_2 phase shaped all pointed, in conjunction with Al_2O_3+SiO_2 phase, were observed. After casting, Al_2O_3+SiO_2 was not observed and residues in mold was confirmed AlO. TiO_2 phase was also changed into the TiO because oxygen was diffused into melts. Oxygen solubility of the titanium contact with TiO_2 phase was 1.6% at 1770℃. However oxygen solubility of the titanium was changed into 0.14% when titanium contacted with TiO phase at 1750℃ [4]. As a result, TiO_2 phase on the mold reduced α-case formation. However, the effect of TiO_2 phase in the Al_2O_3+50Ti and Al_2O_3+ 50TiO_2 mold was apparently contradictory result. That was supposed to be distinguished morphology between Al_2O_3+50Ti and Al_2O_3+50TiO_2 mold.

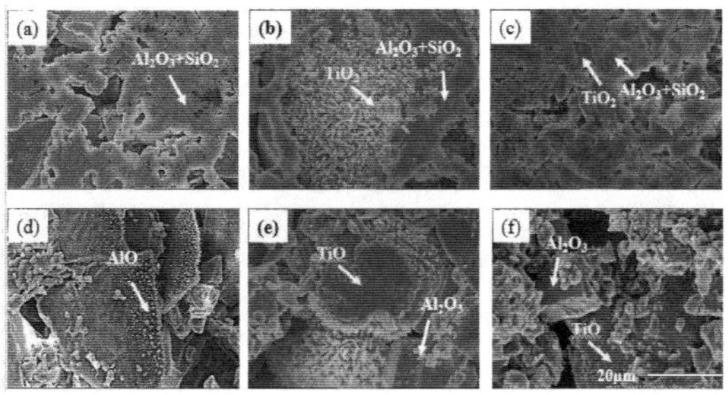

Fig. 3 FE-SEM image of mold surface (a) Al_2O_3, (b) Al_2O_3+50Ti, (c) Al_2O_3+ 50TiO_2 before casting, (d) Al_2O_3, (e) Al_2O_3+50Ti, (f) Al_2O_3+ 50TiO_2 after casting,

4. Conclusions

Composition of Al_2O_3+50Ti and Al_2O_3+50TiO_2 mold was nearly identical because of the titanium oxidation. While, effect of α-case formation was different. That was supposed that the TiO_2 had different shape on the mold.

References

[1] Peter R. Beeley, Robert F. Smart: *Investment Casting,* Ed. By Peter R. Beeley (The institute of materials, London, 1995) pp. 23-24

[2] Si-Young Sung, Bong-Jae Choi, Beom-Suck Han, Han-Jun Oh and Young-Jig Kim, Journal of Materials Science & Technology 24 (2008) 70-74

[3] Kun-Fung Lin, Chien-Cheng Lin, Journal of material of materials Science 34 (1999) 5899 – 5906

4] L. N. Belyanchikov, Russian Metallurgy, 2010 (2010) 1156-1163

Proceedings of the 72nd World Foundry Congress
21-25th May 2016, Nagoya, Japan

Improvement of Creep Resistance in Wrought Nickel Based Superalloys by Addition of Trace Elements

Byung-IL Kang, Jong-Cheon Yoon, Jeong-IL Yoon, Young-Jig Kim*
School of Advanced Materials Science and Engineering, Sungkyunkwan University, Korea
2066, Seobu-ro, Jangan-gu, Gyeong gi-do, Korea

The alloys used for fossil fuel power plants are transitioning from stainless steels that operate below 600 ℃ to nickel based superalloys that can operate up to 700 ℃. The typical requirement is a creep rupture life of 100,000 hr at 700 ℃ under a stress of 100 MPa.

In nickel based superalloys, the approach has been to strengthen the matrix and to increase the volume fraction of γ'. In addition, the grain boundaries are reinforced by carbide precipitation and by the use of minor addition of boron and zirconium. Conversely, the mechanism of the trace elements on grain boundary reinforcement is not fully studied.

In this study it was observed that the addition of trace elements influenced not only the morphology of grain boundary but also the creep resistance. As a result of the evaluation on microstructure and the creep test we have to clarify the effect of the trace elements.

Keywords: *Nickel based superalloy, Boron, Zirconium, Creep resistance, Grain boundary*

1. Introduction

It is well known that the grain boundaries (GBs) structure is mainly influenced mechanical properties at elevated temperature such as creep resistance in nickel base superalloys. There are commonly investigated several phase and structure in GB including carbides, gamma prime and so on. The carbides are classified primary MC type carbide and precipitation carbides such as M_6C, $M_{23}C_6$ type [1].

The effect of trace elements is discovered improving the creep rupture time in nickel base superalloy by interaction of melts and crucible and this phenomena is caused that the boron and zirconium are segregated in GBs, and disturbed formation of micro crack and coarsening of carbides in GBs [2]. Many researchers studied the mechanism of trace elements, but it is not clarify about below mentioned.

1) the changes of morphology of GBs by addition of trace elements,
2) the effects of forming carbides by addition of trace elements
3) the effects of creep resistance of GBs morphology

In this study, we want to study the influence of addition of trace element between the creep resistance and microstructure in GBs. So we added trace elements in our designed alloy LESS 1 (Ni - 22Cr - 20Co - 2W - 2Nb - 1.5Ti - 1.5Al - 0.03C), evaluated microstructure in GBs and creep resistance.

2. Experimental procedures

2.1 Materials and processes

The experiments were performed on vacuum induction melting (VIM) LESS 1 alloy having the nominal composition (LESS 1) and addition of trace elements such as boron and zirconium. After the casting, the specimens are conducted homogeneous treatment (1200 ℃, 16 hr), forging at 1100 ℃, solution treatment (1200 ℃, 30 min) and aging treatment (800 ℃, 16 hr). After the heat treatment and forging, the specimens are prepared for metallography and small punch creep test.

2.2 Metallography

The metallographic specimens were ground using 200 – 2000 grit abrasive SiC papers. Immediately after the last grinding step, the specimens were washed with water, cleaned ultrasonic in cleaner with ethanol and dried. The fine polishing was conducted using 6 and 1 ㎛ water based diamond compounds. After the polishing, the specimens were washed, cleaned and dried in the same method as the above mentioned.

The polished specimens were performed on electrical etching by mixture of perchloric acid, propionic acid and ethanol. The etched specimens were evaluated the microstructure on grain boundary using SEM.

2.3 Small punch creep test

The small punch creep test specimen were cut 10 x 10 x 0.6 ㎜ by wire cutting machine and ground 200 – 2000 grit abrasive SiC papers both sides of the specimen to avoid the influence of heat affected surface during cutting stage. The last grinding step, the size of specimens were 10 x 10 x 0.5 ㎜.

The prepared specimens were elevated up to 700 ℃ in inert atmosphere using Ar gas, and kept at 700 ℃ for 1 hr because of thermal maintenance. After then it was loaded 50kgf and evaluated the rupture time of specimens by small punch creep test. After creep test we observed the rupture surface by SEM.

3. Results and discussion

The composition of trace element in fabricated specimens are listed in table 1.

In this study, the significant changes by addition of trace elements are morphology of carbides in GBs.

It is observed the similar results that MC and $M_{23}C_6$ carbidse are distributed on GBs in LESS 1 and LESS 1 (Zr) alloy. However in the case of LESS 1 (B), the morphology of GBs is changed that the MC carbides could not be obseved, only γ' film with carbides were formed in GBs. This resluts was represnted at Fig. 1.

It is clarify that the effect of trace elements is affected to form the carbide, zirconium is stabilized MC carbide and boron is formed γ' film with carbides during aging treatment.

Fig. 2 shows the results of creep test, the creep rupture time is similar between LESS 1 and LESS 1 (Zr), the rupture time of LESS 1 (B) alloy is significantly improved.

After the test we observed the rupture surface by FE-SEM and it is shown at Fig. 3. It is very interested that the surface of LESS 1 and LESS 1 (Zr) alloy is shown traditional fracture of intergranular, the case of LESS 1 (B) is similar observed intergranular fracture but remained precipitation on the surface.

This results show that formation of γ' film with carbides is more effective about creep resistance than MC and $M_{23}C_6$ carbide in GBs

Table 1 The composition of trace elements (ppm).

Alloy	B	Zr
LESS 1	10	30
LESS 1 (Zr)	10	110
LESS 1 (B)	90	30

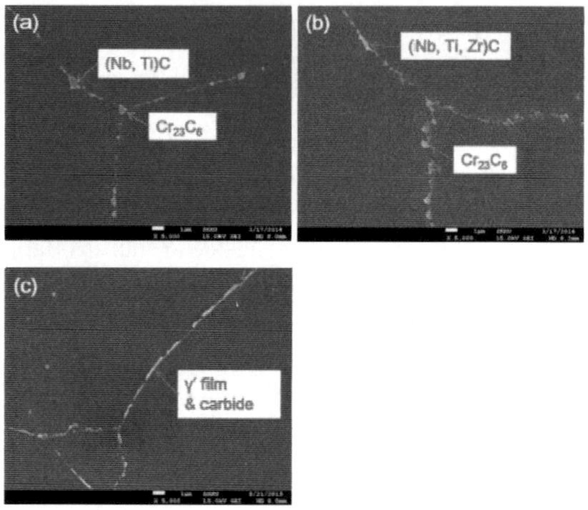

Fig. 1 FE-SEM image of the morphology on GBs (a) LESS 1, (b) LESS 1 (Zr) (c) LESS 1 (B)

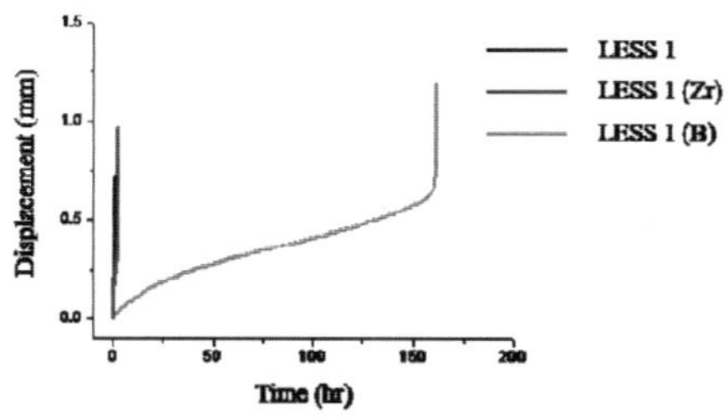

Fig. 2 The results of small punch creep test

Fig. 3 FE-SEM image of the rupture surface (a) LESS 1 (Zr) (b) LESS 1 (B)

4. Conclusions

The effect of trace elements is summarized below

1) the effect of trace elements on GBs is formation of carbide,
2) a point of view about creep resistance, it is more effective boron than zirconium because of formation of γ' film with carbides.

References

[1] C. T. Sims and W. C. Hagel, *The Superalloy*, (John Wiley & Sons, Inc, London, 1972) pp.52 – 58.

[2] S. Floreen and J. M. Davidson; Metall. Trans. A, 14 (1983) 895-901.

Proceedings of the 72nd World Foundry Congress
21-25th May 2016, Nagoya, Japan

Defect susceptibility of tensile properties to microporosity variation in as-cast Al-xSi alloys

Young-Ki Lee[1], Jeong-il Youn[1], Choong-Do Lee[2*], Young-Jig Kim[1]
[1]School of Advanced Materials Science and Engineering, Sungkyunkwan University, Suwon, Korea
[2]Department of Metallurgical and Materials Engineering, Inha Technical College, Incheon, Korea
[*]Corresponding author: cdlee@inhatc.ac.kr

The variability in the defect susceptibility of tensile properties to microporosity variation in as-cast Al-xSi (x=2,5,8,11) alloys was investigated in terms of the effective void area fraction which includes the fraction of micro-voids and the damage evolution of eutectic Si particles. The tensile properties of Al-xSi alloys depend fundamentally upon the microstructural change, with a remarkable deviation by microporosity variation. The defect susceptibility of UTS and elongation is practically varied with the area fraction of eutectic Si particle. Also, the contribution of effective void area fraction to the defect susceptibility was estimated by the constitutive model for tensile instability.

Keywords: Defect susceptibility, Microporosity, Al-Si alloy, Tensile property

1. Introduction

Tensile property of Al-Si castings depends practically upon the porosity variation [1-8], and the UTS and elongation are empirically described in terms of power law relation between the defect susceptibility to microporosity variation and maximum value of a defect-free condition as following equations [3-6]:

$$e = e_o [1 - f]^a \quad (1)$$

$$s = s_o [1 - f]^b \quad (2)$$

where e and s are the elongation and UTS of a material with a microporosity f, e_o and s_o are the elongation and UTS of a defect-free material, and a and b can be defined as the defect susceptibility of the elongation and UTS to microporosity variation, respectively.

2. Experiments

The Al-xSi binary alloys (x=2,5,8,11) used in the present study were fabricated with addition of 0.02%Sr, via a gravity-casting process. A melt controlled at 710 °C was poured into 5-step metallic mold (width = 120 mm, thickness = 5 ~ 25 mm, length = 50 mm per thickness) heated to approximately 200 °C.

The tension test was carried out at room temperature under strain rate of 2.78×10^{-4} s^{-1} using an extensometer. The microporosity was measured by SEM fractography analyses on the fractured surface of the entire specimen. Also, the individual size of the micro-voids and the spacing between micro-voids on the fractured surface were measured via image analysis of SEM observations.

3. Result

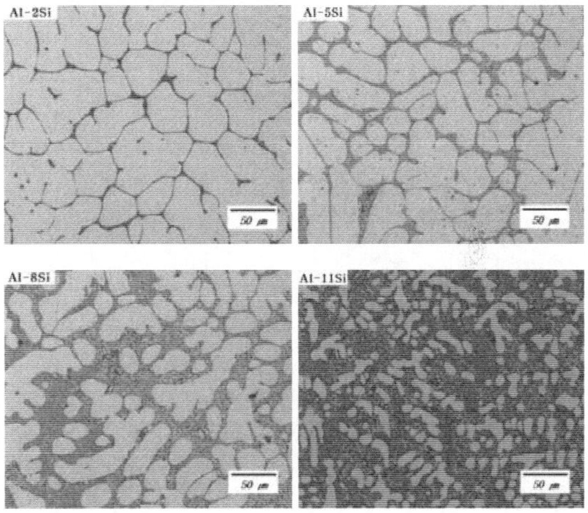

Fig. 1 Optical images of as-cast Al-xSi alloys.

Table 1 Microstructural features of Al-xSi alloys.

Alloys / Microstructure	2Si	5Si	8Si	11Si
Grain size (μm)	675	524	427	393
Nominal fraction of Si-colony (%)	6.2	18.3	34.0	50.8
Fraction of Si-particle (%)	2.7	6.4	11.9	17.8
Width of pro-eutectic α-Al(μm)	76.9	55.9	33.4	20.7
Thick. of Si-colony (μm)	5.9	13.7	34.6	42.8

Figure 1 shows typical microstructural images of Al-xSi alloys. The area fraction of Si-particle and SDAS are remarkably varied with nominal Si-content.

Typical values of various microstructural features are listed in Table 1.

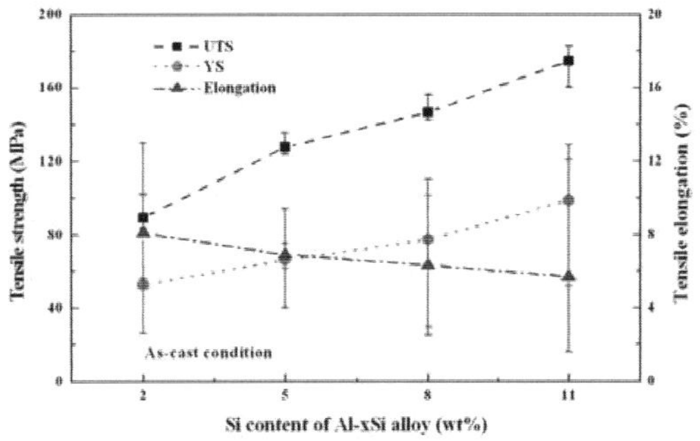

Fig. 2 Variation of tensile properties on Si-content.

Figure 2 shows the variability in the tensile properties on Si-content. The UTS and YS are remarkably increased with Si-addition, whereas the tensile elongation decreased with a considerable deviation for each alloy condition.

The overall dependence of elongation (a) and UTS (b) on microporosity variation was described via the defect susceptibility are shown in Figure 3. The nominal value of defect susceptibility increases with Si-content. And, the difference at a given microporosity is mainly arisen from the microstructural changes, i.e. the fraction of Si-particle and SDAS.

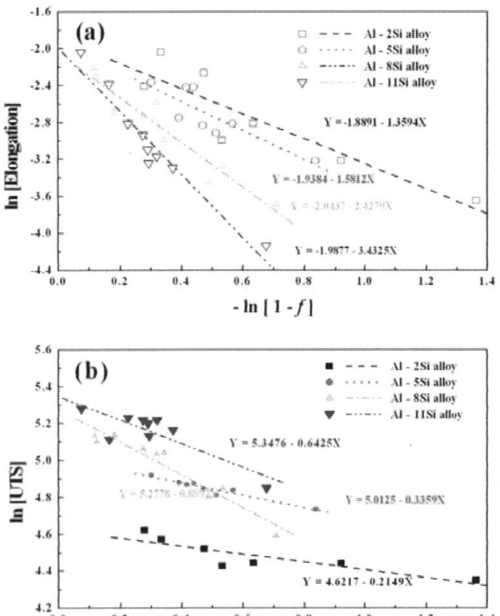

Fig. 3 Tensile properties via defect susceptibility.

Also, the maximum value of UTS achievable in a defect-free condition increases with Si-content, while the elongation is slightly decreased even rarely at similar level.

Figure 4 shows the comparison of theoretical prediction and experimental result. The prediction by modified model [9] show good agreement with experiments and the difference between both approaches means the maximum contribution by micro-failure (cracking mode) of eutectic Si-particles.

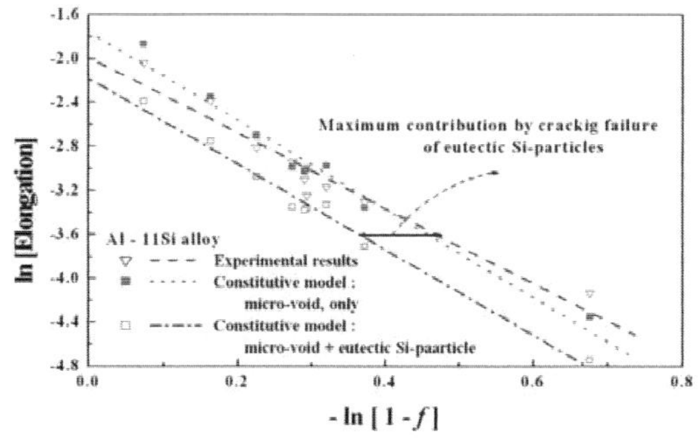

Fig. 4 Typical example of constitutive prediction for contribution of cracking failure of Si-particle.

Acknowledgements

This research was supported by the Development Program for Industrial Core-Technology through the Korea Evaluation Institute of Industrial Technology (KEIT) funded by the Ministry of Trade, Industry and Energy (10048817).

References

[1] C.H.Cáceres: Scripta Metall 32(1995) 1851-1855.

[2] C.H.Cáceres, B.I.Selling: Mat Sci Eng A 220 (1996) 109-116.

[3] A.M.Gokhale G.R.Patel: Mater Char. 54 (2005) 13-20.

[4] A.M.Gokhale, G.R.Patel. Mat Sci Eng A 392 (2005) 184-190.

[5] A.M.Gokhale, G.R.Patel: Scripta Mater 52 (2005) 237-241.

[6] S.G.Lee, G.R.Patel, A.M.Gokhale, A.Sreeranganathan, M.F.Horstemeyer: Scripta Mater 53 (2005) 851-856.

[7] C.D.Lee: Mat Sci Eng A 527 (2010) 3144-3150.

[8] C.D.Lee: Mat Sci Eng A 488 (2008) 296-302.

[9] C.D.Lee, K.S.Shin, Y.G.Kim: J Mat Sci (submitted)

Proceedings of the 72nd World Foundry Congress
21-25th May 2016, Nagoya, Japan

Experimental Investigation of Microstructural Characteristics and Tribological Properties of Al-Si Alloys Modified under the Influence of Mechanical Mold Vibration.

Vardhaman Mudakappanavar[1] and H M Nanjundaswamy [2]

[1]Department of Mechanical Engineering, B M S College of Engineering, Bangalore 560019, India
mvardhaman@gmail.com
[2] Department of Industrial and Production Engineering, P E S College of Engineering, Mandya, India
hmnanjunda@gmail.com,

Dry sliding wear behavior of Al-Si alloys depends on the size, shape of the silicon particles and size and distribution of the α-Al grains in the interdendritic region. This paper highlights the effect of mechanical mold vibration during solidification on the microstructure and tribological properties of Al-12% Si alloy. The size of eutectic silicon, grain size and wear resistance were investigated.

Keywords- Mechanical mold vibration, dry sliding wear test.

1. Introduction

Aluminum-Silicon alloys are the most commonly used foundry alloys because of their good thermal conductivity, excellent cast ability, high strength to weight ratio, wear and corrosion resistance etc. Therefore they are well suited for automotive and air craft components. The mechanical properties of Aluminum-Silicon alloys are related to the size and shape of eutectic silicon which in turns depends on the solidification process.

In the present work, mold material (A), mechanical mold vibration (B) and pouring temperature (C) were considered as variables during the solidification process.

Imposition of mechanical vibration on liquid Al-Si alloy during solidification has shown improvements like grain refinement [1], reduction in shrinkage pipe [2], fragmentation of dendrites and transition of eutectic structures from flakes to fibrous [3], reduction in average size of silicon needle [4] resulting in improved properties.

2. Experimental

A simple mechanical vibrator set up was used to subject the mold to vibration. Molten commercial grade LM6 alloy was solidified under the influence of vibration with frequency and amplitude maintained at 25 Hz and 0.05 mm respectively. The vibration was maintained until the melt was completely solidified. After solidification the cast specimen were prepared for testing.

Experiments were conducted as per the design matrix [5] of full factorial design as shown in Table 1. The notations are as follows: For mold material, cast iron mold is designated as -1 and graphite mold as 1, absence of vibration as -1 and presence of vibration as 1 and pouring temperature of 700°C as -1 and 800°C as 1.

3. Results and Discussion

The results of microstructure examination as per ASTM-E407 standard of samples cast as per Table 1 are shown in Fig 1. The microstructure for cast specimen in static mold shows presence of dendritic growth but for the cast specimen in vibrated mold the aluminum dendrites have broken. Inducing vibration fragments the dendrites thereby resulting in less dendritic growth.

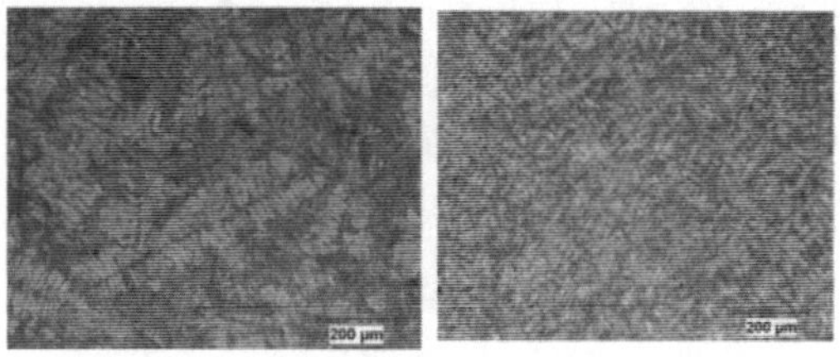

(a) (b)

Fig.1 Microstructure of cast specimen poured in graphite mold at 800°C under (a) static condition (b) vibrated condition at 100x magnification.

Fig. 2 and Fig. 3 shows the variation of average silicon size and grain size and it is observed that the average size of silicon needles and average grain size for cast specimen designated as *abc* i.e vibrated graphite mold with alloy poured at 800°C is least.

Table 1 Design Matrix

		Factors		
Experiment Number	Label	A	B	C
1	(1)	-1	-1	-1
2	a	1	-1	-1
3	b	-1	1	-1
4	c	-1	-1	1
5	ab	1	1	-1
6	bc	-1	1	1
7	ac	1	-1	1
8	abc	1	1	1

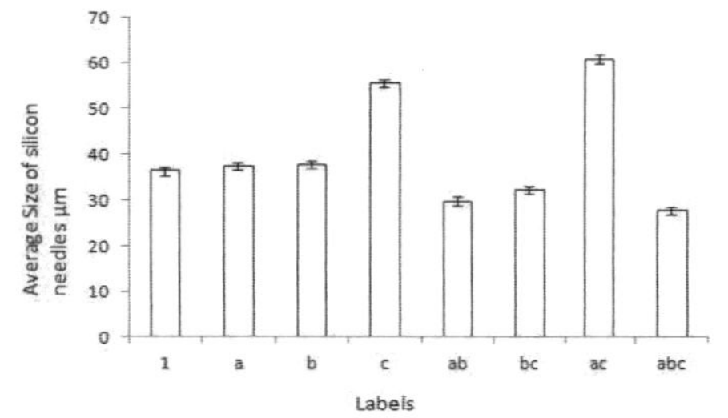

Fig. 2 Average size of silicon needles of the cast specimen designated with labels as per the design matrix table 1.

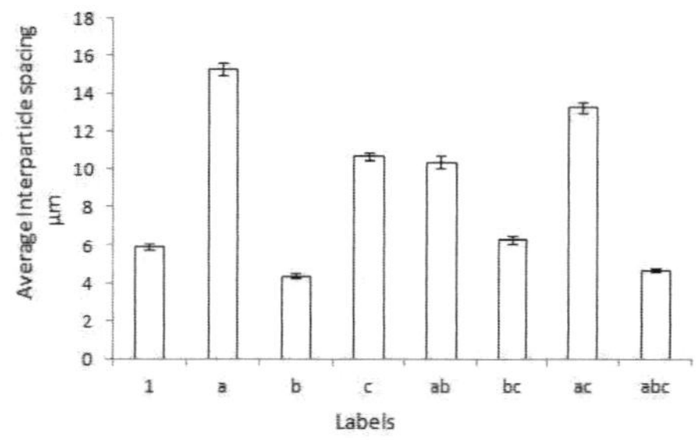

Fig. 3 Average grain size number v/s Label as designated in design matrix table 1.

The energy of mechanical vibrations promotes microstructure refinement during solidification. Also vibration induces a higher heat transfer along the mold walls due to the alternated movement of liquid molten metal. This could be the reason for change in the silicon particle size and grain size.

Fig. 4 shows the variation of volume loss with normal load at a constant sliding speed of 5.495 m/s and a sliding distance of 9891 m. It is observed that there is an improvement in the wear behavior of the cast specimen in vibrated mold as compared to the wear behavior of cast specimen in static mold.

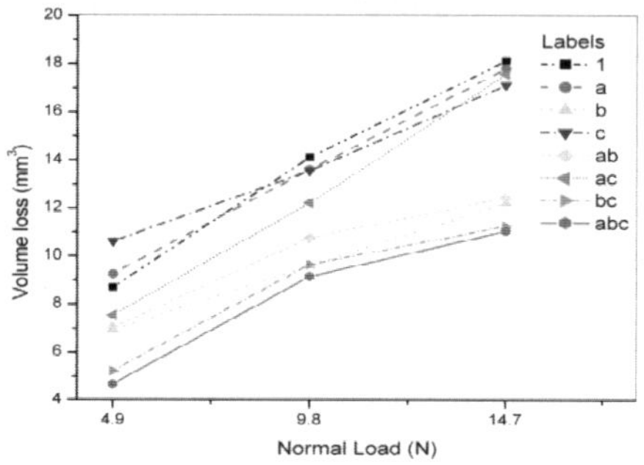

Fig. 4 Volume loss vs. normal load

4. Conclusion

Applying vibration to the solidifying LM6 alloy leads to microstructural changes to both the dendritic structure and eutectic silicon which results in improved properties. The technique currently in use is addition of grain refiners like master alloys of Al-Ti or Al-Ti-B and modifiers like Sodium to the melt, but efficiency is limited due to fading effect and is not environment friendly. Hence the technique studied in this paper is hoped to be beneficial to the industries producing Al-Si alloy casting.

Acknowledgements

The work reported in this paper is supported by the B.M.S College of Engineering, Bangalore, India.

References

[1] N. R. Pillai, "Effect of low frequency Mechanical vibration on structure of Modified Al- Si eutectic", Metallurgical Transaction, vol 3, 1972, pp. 1313-1316.

[2] K. Kocatepe, and C. F. Burdett, " Effect of low frequency vibration on macro and micro structures of LM6 alloys", Journal of Material Science, vol 35, 2000, pp 3327-3335.

[3] N. Abu-Dheir, M. Khraishseh, K. Saito, and A. Male, "Silicon morphology modification in the eutectic Al – Si alloy using mechanical mold vibration", Material Science and Engineering A, vol 393A, 2005, pp 109-117.

[4] U. Pandel, A. Sharma , and D. B. Goel, "Study on the effect of vibrations during solidification on Cast Al-Si alloy", Indian Foundry Journal,vol 51, No. 2, 2005, pp 42-45.

[5] D.C. Montgomery, Design and Analysis of Experiments V[th] edn., John Wiley & Sons (Asia) Pte Ltd., Singapore, 2007.

O-95

Microstructure Refinement in Al-7%Si Alloys Solidified by an Electromagnetic Stirring Technique

Mingjun Li, Naoki Omura, Yuichiro Murakami, Isao Matsui and Shuji Tada

National Institute of Advanced Industrial Science and Technology (AIST), Structural Materials Research Institute, Nagoya, Aichi 463-8560 Japan

In the present study, we poured the Al-7%Si aluminum melt into a ceramic crucible that was then set in a rotating electromagnetic field. Solidification of the melt was accomplished while the rotating magnetic field was imposed and thus electromagnetic stirring (EMS) could be produced. The structure development versus reverse EMS imposition time, *t*, has been observed, revealing that the longer imposition time will result in the finer primary solid solution phase. The refinement mechanism is discussed when considering fluid flow and uncoupled movement behavior between the primary solid and the remaining liquid.

Keywords: *Aluminum alloy, solidification, structure, electromagnetic stirring, grain refinement.*

1. Introduction

A grain-refined structure is always desired during casting as the processing cost of an aluminum alloy billet with a refined structure can be substantially reduced in the subsequent treatment and moreover, it can improve the performance of the alloy. Although TiB_2 compound has been proved to be quite effective in refining Al alloys [1], these two rare-metal elements increase the production cost. Moreover, the unstable supplying market may make the Al industry subject to a high risk. Therefore, it is critical to develop new techniques, by which Al billets with grain-refined structures can be produced while no refining additives are required. In the present study, we employed the EMS technique to yield grain refined structures for a bulk aluminum alloy ingot, by which no rare metals are externally added. The microstructure development as a function of processing parameters has been observed and the refinement mechanism is discussed when considering the EMS-induced melt flow and the displacement behavior between the primary solid and remaining liquid.

2. Experimental procedure

To generate a rotating electromagnetic field, a three-phase induction motor was used and the rotor was removed to produce a hollow space. The molten Al-7%Si aluminum alloy was placed into the space to solidify while EMS was imposed. Here it should be noted that by switching over the sequence of the alternating current, one can readily alter the rotating direction of the magnetic field and thus change that of Lorentz force accordingly. The movement direction of the crystallizing melt in the crucible can also be changed by switching over the phase sequence of the three-phase alternating current, i.e., from clockwise (CW) to counterclockwise (CCW) and vice versa. For the present study, we employed an inverter to change the phase sequence of the three-phase alternating current. The EMS imposition time could be well controlled with a timer. Fig. 1 indicates the EMS imposition pattern in the present study.

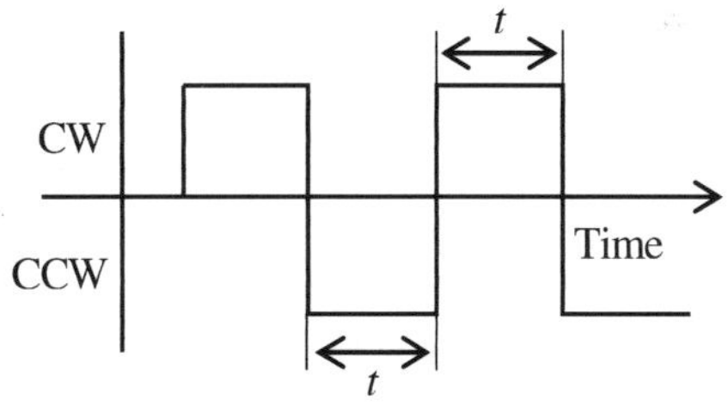

Fig. 1 The EMS imposition pattern versus time

The polarized optical microstructures were observed along the traditional metallographic method after the electrolytic corrosion was carried out.

3. Results and discussions

In order to reveal the influence of EMS on structure formation, we first show the microstructure solidified under the usual casting state without EMS, as depicted in Fig. 2. One can see that it consists of developed coarse dendrites. Fig. 3 shows two microstructures of the alloy solidified at (a) t = 0.2 seconds and (b) t = 0.5 seconds, respectively. It can be discerned that in comparison with that solidified at the casting state, the structure solidified at t = 0.2 seconds

Fig. 2 Microstructure solidified without EMS

consists of some fine particles. When the reverse imposition time is further extended to $t = 0.5$ seconds, as revealed in Fig. 3(b), the microstructure almost consists of fine equiaxed grains and no traces of complete dendrites can be identified. One can readily become aware that using the EMS processing can refine the structure and the longer imposition of the reverse time can result in equiaxed fine grains.

As far as the refinement mechanism is concerned, two principal reasons are considered. The first is that EMS can induce fluid flow that has been well documented [2]. The primary mechanism concerning refinement is that fluid flow can lead to thermosolutal convection that originates fragmentation by local remelting of higher order of dendritic arms and thus resulting in grain-refined structures.

The second reason concerns the movement behavior of the solid and liquid during EMS processing. The electromagnetic torque of a conductor in a magnetic field is reversely proportional to its electrical resistivity. For simplicity, the electrical resistivities of the Al solid solution and the remaining liquid can be approximated to that of pure solid and liquid aluminum, respectively, which are 110n Ω m for solid and 242nΩ m for liquid [3]. Hence, one can deduce that the electromagnetic torque produced upon the solid is about twice of that upon liquid. Considering that there is little difference in density between the solid and liquid, it is clear that the body force upon the solid is about twice of that the liquid. Therefore, the solid can be driven to move with a larger acceleration and thus covers a longer distance in comparison with the liquid. So, uncoupled movement occurs between the mobile leading solid and the sluggish remaining liquid, making solute distribution boundary layer unstable and thus difficult to produce constitutional undercooling zone to yield dendrites.

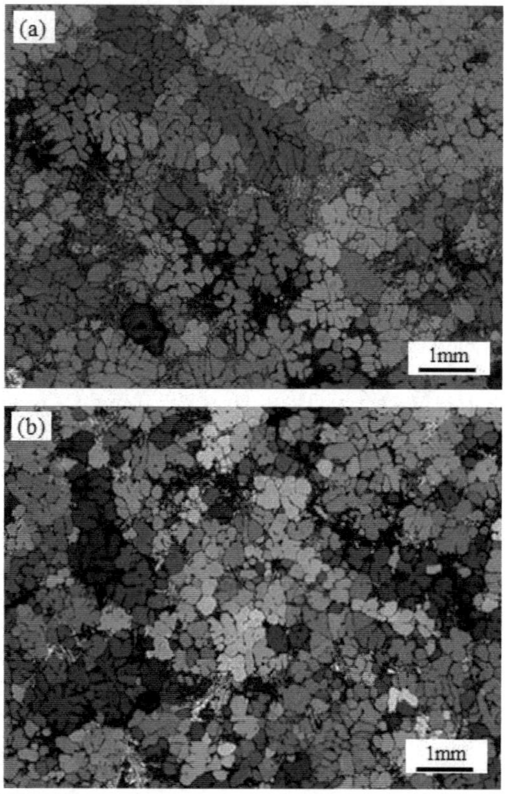

Fig. 3 Two microstructures solidified with EMS at the imposition time of (a) $t = 0.2$ and (b) $t = 0.5$ seconds, respectively.

Meanwhile, the longer imposition time may lead to more severe turbulent fluid flow and furthermore, yield considerable uncoupled displacement behavior, which may result in finer equiaxed grains.

4. Conclusions

The EMS technique has been proved to be effective in refining the microstructure of the Al-7%Si alloy. The refinement can be attributed to fluid flow and uncoupled movement between the solid and liquid, and thus promoting the fragmentation of dendrites to yield refined structures during EMS processing.

References

[1] T. Quested and A.L. Greer: Acta mater. 52 (2004) 3859-3868.

[2] A. Hellawell, S. Liu and S.Z. Lu: JOM 49(3) (1997) 18-20.

[3] J.R. Davis: *ASM Specialty Handbook Aluminum and Aluminum Alloys*. The materials international society (The materials park OH, 1993) p.641.

Proceedings of the 72nd World Foundry Congress
21-25th May 2016, Nagoya, Japan

Novel Grain Refining Process using Ti Clusters generated by Ultrasonic Cavitation

Jeong IL YOUN[1], Young Ki LEE[1], Young Jig KIM[1]*,
Jeong Wook PARK[2] and Woo Chun KIM[2]
[1] School of advanced materials science and engineering, Sungkyunkwan University,
2066 Seobu-ro, Jangan-gu, Suwon, Gyeonggi-do 16419, Korea.
[2] R&D Center, DR AXION Co., Ltd., 55 Sannam-gil, Ulju-gun, Ulsan 45010, Korea.

Grain refinement is the essential process to improve mechanical properties of castings. Nucleation enhanced ultrasonic melt treatment (NEUMT) proposed in this study increases refining efficiency through heterogeneous nucleation by the formation of proper intermetallic compounds in the complete melt. Ti clusters were supplied by the cavitation erosion on the surfaces of the sonotrode, and were simultaneously mixed uniformly into the melt. These particles act as potential nuclei and refine the structure.

Keywords: *Grain refinement, Cavitation, Ultrasound, A356 Al alloy, AZ31 Mg alloy.*

1. Introduction

Research to improve mechanical properties of Al alloys has been performed, and it is well accepted that grain refinement is a very effective method for improving structural uniformity, strength and ductility [1]. Especially, the removal of dissolved hydrogen from the Al alloys melt is also critical to the production of high quality casting [2]. The study to apply the ultrasound in casting process has been studied for the last a decade [3,4]. The application of ultrasound in casting process has been focused on not only the degassing but also the grain refinement, and especially the grain refinement has been mainly studied on the point of view the mechanism that the dispersion of the broken particles of dendrite resulting from the ultrasound injection during solidification or the state which is the coexistence of solid and liquid phase of metal [5]. However, this makes effectiveness low to adapt in industry, and in addition the degassing by ultrasound never performs well because the fluidity of melt extremely dropped. Based on this research back ground, we applied a new processing concept, *Nucleation Enhanced Ultrasonic Melt Treatment* (NEUMT), which is based on the novel paradigm of the sonotrode, to use ultrasound during normal Al casting processes. This paper discusses the effect of NEUMT on grain refinement of Al alloys, especially A356 and A390, and degassing of the alloys. The mechanism of the NEUMT based on the analysis of the chemical composition and the microstructure of the alloy were involved in this paper.

Element	A356	A390
Si	6.9	16.9
Cu	0.2	4.3
Mg	0.4	0.5
Mn	0.1	0.1
Fe	0.1	0.1
Al	Bal.	Bal.

Table 1 Chemical composition of A356 and A390 aluminum alloy. (mass%)

2. Experimental procedure

The experimental system for NEUMT which can generate and inject the ultrasound in metal melts. The frequency of the system is 20 kHz and the maximum power output is about 1900 W/cm^2. A356 and A390 Al alloys were selected as the material to be investigated in this study, and the chemical compositions are shown in Table 1. The about 500 g alloy ingot was melted at 700 °C for A356 and 750 °C for A390 alloy. After the ingot melting, Ultrasound was injected for 0~600 seconds, and the melt is poured into the steel mold. The sonotrode was rod shaped with a length of 360 mm and a diameter of 40 mm, and was made of commercial purity Ti.

3. Results and Discussion

Fig. 1 shows the the grain refinement mechanism of NEUMT. The process makes full use of cavitation effect to supply the nuclei in the Al alloys melt without a separate process for refiner feeding. The atoms and

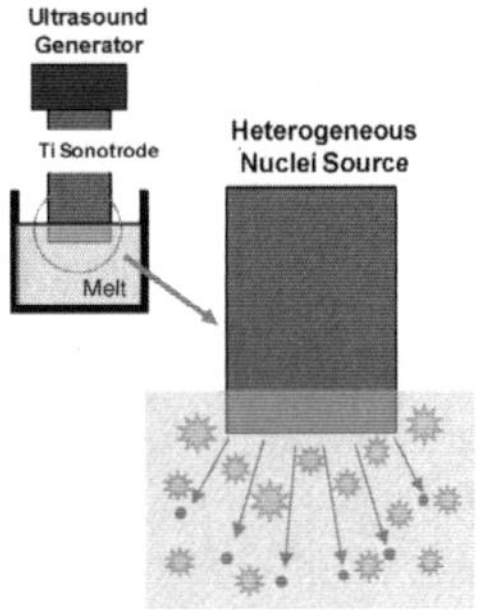

Fig. 1 Grain refinement mechanism of NEUMT process by Ti sonotrode.

clusters parted from the sonotrode by the cavitation erosion are uniformly distributed in the melt by acoustic streaming, and react with the melt and form proper intermetallic compounds that can act as heterogeneous nucleation sites. Fig. 2 shows the macrostructure of A356 alloy after NEUMT for 300 seconds and as-cast without NEUMT. The result shown in figure suggests that NUMT is very effective to obtain very fine grains of A356 alloy in a short time. The grain size of A356 and primary Si size was measured by the image analysis to characterize NEUMT effect as the size with time, and the results was represented at Fig. 3. The grain size of A356 without NEUMT was about 760 μm, however, the size decreased with increasing the process time and reached to about 100 μm. The size was maintained after the time in spite of increasing of NEUMT time. The effect of NEUMT on the refining was clear to the

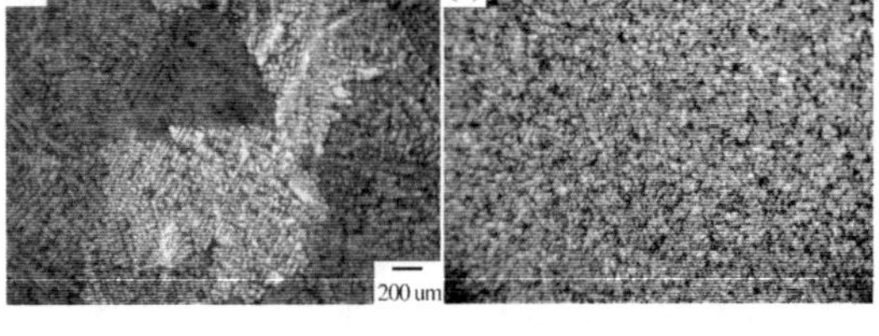

Fig. 2 Macrostructure of A356 alloy; (a) without NEUMT and (b) NEUMT for 300 seconds.

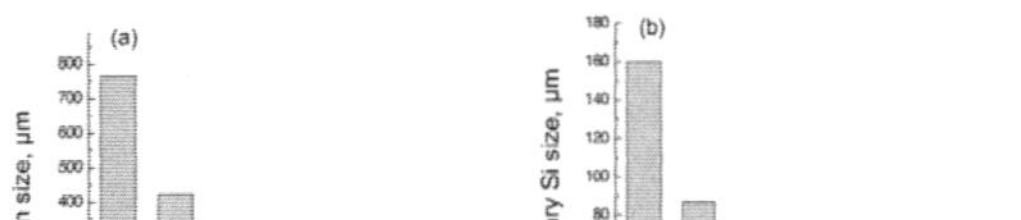

Fig. 3 (a) Grain size of A356 and (b) primary Si size of A390 alloy as NEUMT time.

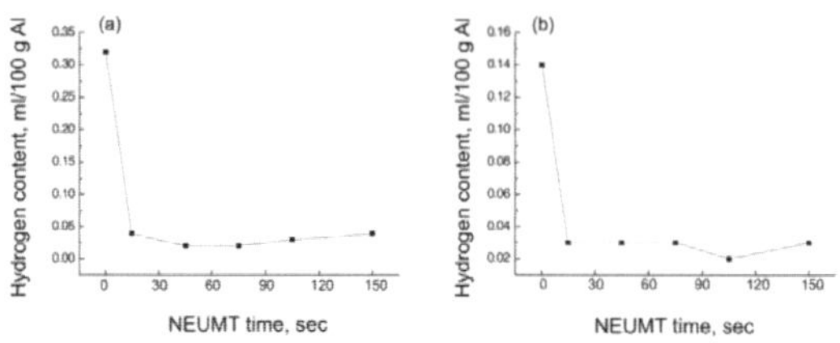

Fig. 4 Hydrogen content in melt with NUMT time: (a) A356 and (b) A390.

hypereutectic alloy, A390 as well. With NEUMT, the hydrogen contents decreased sharply with increasing process time in just few seconds and then reached a plateau hydrogen content, which corresponds to the steady-state hydrogen concentration. The results shown in Fig. 5 suggest that NUMT time for degassing of aluminum melt was extremely short. Degassing in the aluminum melt by NEUMT can be explained by the cavitation effect. Cavitations are made by the formation of tiny discontinuities or cavities in the melt, known as cavitation bubbles. In a melt containing dissolved gases, hydrogen atoms diffuse into the cavitation bubbles at the bubble/melt interface during the nucleation and growth stage of the cavitation bubbles where they form molecular hydrogen. The hydrogen molecules and/or bubbles float to the melt surface due to their buoyancy and are then removed.

4. Conclusion

The effect of NEUMT was normalized by the grain size and hydrogen content at various experimental conditions, and these data could be the guideline of the process. The grain size was decreased to a half by NEUMT for just dozens of seconds, and the more fine structure has shown with the more process time. When the initial hydrogen concentration was 0.32 ml and 0.15 ml at A356 and A390 without degassing, respectively, the hydrogen contents decreased dramatically to 0.02 ml/100 g Al with increasing NEUMT time in a few seconds.

Reference

[1] N. Hansen: Scr. Mater. 51 (2004) 801-806.

[2] D. Talbot: Int. Metall. Rev. 20 (1975) 166-184.

[3] G. I. Eskin: Ultrasonic Treatment of Light Alloy Melts, (Gordon and Breach, Amsterdam, 1998) pp. 245-280.

[4] M. K. Aghayani, and B. Niroumand: J. Alloy. Compd. 509 (2011) 114-122.

[5] J. Campbell: Int. Met. Rev. 26 (1981) 71-108.

Proceedings of the 72nd World Foundry Congress
21-25th May 2016, Nagoya, Japan

The microstructure and refinement performance of Al-Ti-C master alloy smelted via improved SHS-approach

Kateryna Svynarenko[1], Tingju Li[1]

[1] Laboratory of Special Processing of Raw Materials, School of Material Science and Engineering, Dalian University of Technology, Dalian, 116024, China

The reduction of TiC-containing refiners smelting temperature below 1250 °C inevitably leads to the deterioration of their performance. In this work, an improved combustion route is used to prepare Al-5Ti-0.2C master alloy. The separate smelting of Al-10TiC and Al-Ti followed by the intermixing enabled to better kinetic conditions and accelerate the formation of TiC phase. As a result, the smelting temperature was decreased down to 1050 °C while the refinement efficacy of the alloy was improved. Comparative analysis of the master alloy microstructure is given with relation to the results of refinement tests.

Keywords: *Al-Ti-C master alloy: SHS-approach; Grain refinement; Smelting Temperature.*

1. Introduction

Chemical inoculation of aluminium-based alloys has been widely used industrial practice for decades. The master alloys of Al-Ti-C system possess several advantages compare to refiners of other systems. However, difficulties related to the high production temperatures and consequent high energy and cost consumption have always led to the restriction in extending their usage scope [1, 2].

The Self-propagating High-temperature Synthesis (SHS) is considered the most promising in terms of energy consumption and cost reduction [3]. With great potency of the SHS-approach to facilitate the TiC formation at lower temperature, it is of special importance to adapt the smelting process and utilize maximum benefits from the combustion reaction. Thus, the production route for Al-Ti-C refiners may be improved and production expenses may be reduced.

2. Experimental procedures

Starting materials for master alloy preparation: commercial purity Al (99.8%, hereafter – CPAl) and elemental powders of Al (99 % purity), Ti (99 % purity) and graphite (99% purity). All compositions quoted in this work are in wt.% unless otherwise stated. The Al-5Ti-0.2C master alloy was smelted in a resistance furnace at 1050 °C and 1250 °C.

The main feature of the SHS approach presented in this work resides in the separation of raw materials before the master alloy smelting. In this case, the Al-10TiC was smelted first, and then mixed with Al-Ti to obtain Al-Ti-C master alloy of desired composition. The holding conditions for Al-10TiC smelting and following Al-10TiC and Al-Ti mixing are 10 min and 20 min, respectively.

The Al-5Ti-0.2C master alloy was also smelted via conventional combustion approach, i.e. a green preform of Al, Ti and C powders was inserted into molten Al and then it was held for 30 min at the experimental temperature.

A Zeiss Supra55 FE-SEM was used for the master alloy microstructure observation and analysis. Refinement tests were performed by addition of 0.25% Al-5Ti-0.2C master alloy into CPAl melt at 740 °C and subsequent holding for 5 min. The mean linear intercept method was applied to determine the average grain size of CPAl.

3 Results and Discussion

The macrostructures of CPAl in initial and refined with 0.25% Al-5Ti-0.2C master alloy conditions are presented in Fig. 1. Although the master alloy smelted via conventional SHS method at 1250 °C refines CPAl effiently and the grain size decreases from 2380 μm (Fig. 1a) to 370 μm (Fig. 1b), its refinement efficiency deteriorates significantly when the master alloy smelting temperature is 1050 °C (Fig. 1c). In contrast, the performance of refiner smelted via improved method is superior even at 1050 °C smelting temperature (Fig. 1d), the CPAl grain size is 330 μm.

In Al-Ti-C master alloys designed to refine aluminium, the main phases are TiC and Al_3Ti. The TiC particles are considered to provide nucleation sites for Al grains, while the Al_3Ti phase dissolves fast into the Al melt and supplies it with solute Ti.

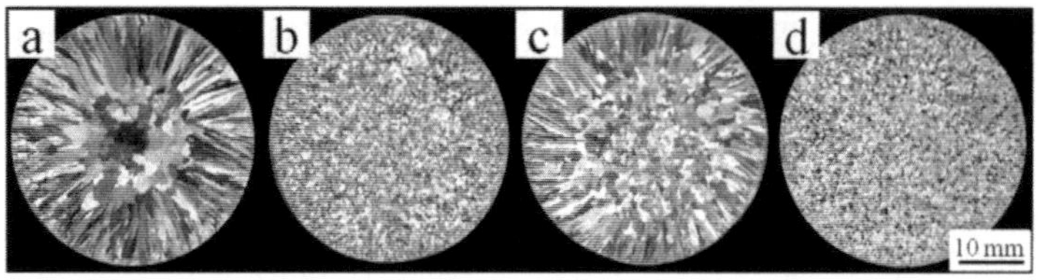

Fig. 1 Macrostructure of CPAl without (a) and with (b-d) addition 0.25% Al-5Ti-0.2C master alloy; b) $SHS_{conv.}$, 1250 °C; c) $SHS_{conv.}$, 1050 °C; d) $SHS_{impr.}$, 1050 °C

Fig. 2 SEM micrographs of Al-5Ti-0.2C master alloy smelted via conventional (a, b) and improved (c) SHS approach; a) 1250 °C; b), c) 1050 °C; × 30000

One of the factors affecting the performance of refiner is the size of particles which are claimed to act as heterogeneous nuclei [4]. Fig. 2 represents the TiC-rich region in the microstructure of Al-5Ti-0.2C master alloy smelted via conventional (Fig. 2a, b) and improved (Fig. 2c) SHS approach. In the case of conventional smelting ($SHS_{conv.}$), the average TiC particle size decreases from 0.35 μm to 0.27 μm at 1250 °C and 1050 °C, respectively. Separation of raw materials appears to accelerate the growth of carbide particles even at lower temperature (0.45 μm). Moreover, the master alloy smelted by $SHS_{conv.}$ method possess TiC particles of inconsistent size (Fig. 2a, b) compare to the one smelted by $SHS_{impr.}$ method (Fig. 2c).

The application of SHS-approach to the preparation of different materials is caused by the potential of combustion phenomena to accelerate considerably the interaction between system constituents. However, low content of TiC in Al-5Ti-0.2C master alloy is suggested to impede utilization of benefits provided by combustion wave during the SHS-reaction. Diffusion processes during the holding rather than combustion itself, appear to dominate the formation of TiC [5]. Therefore, the smelting temperature above 1200 °C is required to ensure the satisfactory performance of Al-5Ti-0.2C master alloy. The separation of raw materials during the smelting leads to an improvement in kinetic conditions, thus accelerating the formation of desired TiC phase even at lower temperature.

Acknowledgements

The authors gratefully acknowledge the support of the Natural Science Foundation of China (No. 51271042, 51471042), and the Fundamental Research Funds for the Central Universities of China.

References

[1] Landry K, Kalogeropoulou S, Eustathopoulos N: J. Mater. Sci. Eng. A 254 (1998) 99-111.

[2] Pastukhov E.A., Chentsov V.P., Kiselev A.V., Bodrova L.E., Dolmatov A.V., Popova E.A., Petrova S.A., Zakharov R.G. (2006): in Proc. 4 th Int. Conf."Mathematical modeling and computer simulation in materials technologies".

[3] Vallauri D, Atías Adrián IC, Chrysanthou A: J. Eur. Ceram. Soc. 28 (2008) 1697-1713.

[4] Quested, T.E., Greer A.L.: Acta Mater. 52 (2004) 3859-3868.

[5] Svynarenko K., Zhang Y., Jie J., Kutsova V., Li T: Int. J. Mater. Res. 106 (2015) E 1–9. DOI 10.3139/146.111328.

Proceedings of the 72nd World Foundry Congress
21-25th May 2016, Nagoya, Japan

Structure, Mechanical Properties and Erosion Resistance of HVOF Sprayed Cr_3C_2+NiCr Coating on the Surface of Castings Made from Al-Si alloy

M.Ksiazek[1], , L. Boron[1], A. Tchorz[1] , R. Grzelka[2]
[1] Foundry Research Institute, 30-418 Kraków, ul. Zakopiańska 73
[2] Plasma System S.A., 41-103 Siemianowice Slaskie, ul. Towarowa 14

The paper presents research of the microstructure and mechanical properties of HVOF (high velocity oxy-fuel process) sprayed Cr_3C_2–NiCr coating on the surface of castings made from Al-Si alloy in the presence of NiAl interlayer. The investigations were aimed at the adherence of the interlayer and the outer coating deposited by HVOF technique. The structure of *ceramic coating/interlayer/metallic substrate* system was examined using the methods of light (LM) and scanning (SEM) microscopies as well as the analysis of chemical composition in microareas (EDS).
The results were discussed in reference to examination of cyclic fatigue test and an assessment of wear resistance of the coating *(Cr_3C_2-NiCr)/NiAl/*substrate *(Al-Si alloy)* system in combination with the cracking and delamination analysis of the coating in the interface area. Good adhesion to the metal base with NiAl interlayer and durability in cyclic loads make carbide coatings more effective as protective coatings for castings made from Al-Si alloys operating under the conditions of heavy dust and variable loads.

Keywords: *high velocity oxy-fuel, Al-Si alloy, chromium carbide coatings, microstructure, erosive wear*

1. Introduction

One of the most recent methods allowing local change of surface layer properties is the hypersonic powder spraying at high speeds (HVOF- High Velocity Oxy Fuel). The process of hypersonic spraying with a relatively low spray stream temperature (about 2600°C) and short residence time of powder particles in the spray stream has significantly reduced adverse effects of variations in phase composition of coatings present in the conventional plasma spraying, including: carbide decay, drastic reduction of the degree of metallic and carbide material oxidation, which in turn has improved the performance characteristics of coatings [1-2]. The coatings applied by means of high-velocity technology HVOF have excellent wear resistance, reduced porosity and increased adhesion compared to conventional plasma spraying. The quality of coatings, their strength, hardness, density, adhesion, and related properties depend on the velocity of powder particles, which in this method exceeds that of sound. High collision speeds may cause very high pressure and severe plastic deformations, including occurrence of the rotational mechanism. Therefore, a unique feature of this technology is that, unlike all other thermal spraying methods, it ensures spraying of the coating with pressure stress. The pressure stress in a coating significantly increases sprayed material adhesion to the substrate and is also beneficial in terms of the fatigue properties of components being coated [3-4].

2. Materials and methods

The composite powder Cr_3C_2-20wt.% NiCr of a particle size range of 45-60μm used as feedstock. A commercial HVOF spraying equipment (HV-50 HVOF System in Plasma System S.A,), which uses kerosene and oxygen to generate a hypersonic combustion flame, was used to spray these powders onto aluminium alloy substrates of following chemical composition: AK9 (AlSi9Mg) 10.5% Si, 0.3% Mg, 0.35% Mn, 0.8% Fe). Before application of coating, the substrate was precoated with an interlayer of NiAl (80% Ni and 20% Al), applied also by plasma technique. Prior to spraying, the surface of the aluminium alloy was sand–blasted using 20–mesh alumina grits to a surface roughness of about Ra=30 μm. The thickness of coatings was approximately 300 μm.
The microstructure of the system type *coating/substrate* was examined by light microscopy (LM) and scanning electron microscopy (SEM) with the measurements of the chemical composition in microareas (EDS). Micro-hardness measurements of the coatings were performed on the

cross sections of normal samples relative to their surfaces with Vickers method with the load 1N. The fatigue resistance was tested in 3-point bending test on INSTRON 8800M machine with hydropulse system. In a single test three specimens (of dimensions of 100x15x5 mm) were used and the fatigue resistance data were reported as the average values. The dynamic tests were carried out under variable bending loads of 1 Hz amplitude. Total number of the loading cycles in each tested specimen was 5000. The fractography of the coating surface after bending test was done under a scanning microscopy. Erosion test was performed on a special device for testing erosive wear resistance.

3. The results

The typical microstructure of the HVOF-sprayed chromium carbide coating on Al-Si alloy is shown in Fig. 1. The coatings consisted of lamellae elongated in the direction parallel to coating surface. The cross section and top surface of the both coatings exhibited dense microstructure with high cohesion without cracks, and very fine Cr_3C_2 particles embedded in NiCr alloy matrix, coming to the size of nanocrystalline due to the high impact velocity of the coating particles. Darker regions are identifiable, consisting of micrometric Cr_3C_2 grains, on the other hand brighter Ni-Cr containing regions exist. HVOF sprayed Cr_3C_2-NiCr coatings, exhibit low porosity (approx. 4.2±0,2% measured on microscopic images) and high hardness. An average microhardness of the Cr_3C_2-NiCr coating is 1300HV0.1, for NiAl interlayer is 1000 HV0.1, and then decreases to the value of 310HV0.1 for the substrate.

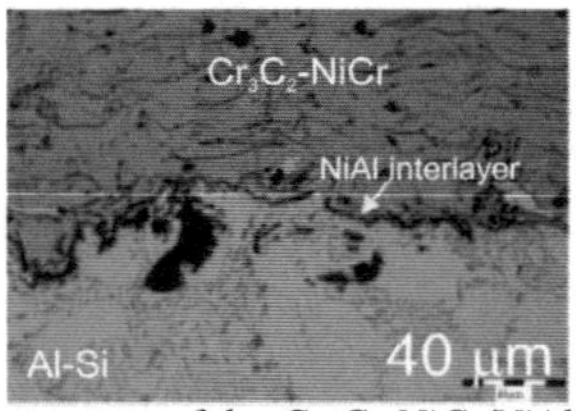

Fig. 1. Microstructure of the *Cr_3C_2-NiCr/NiAl/Al-Si* interface

Table 1. Test results of the cyclic bending

Sample	Bend stress [MPa]	Deflection [mm]
Al-Si	230	8
Cr_3C_2-NiCr/NiAl/Al-Si	290	29

For *Cr_3C_2-NiCr/NiCr/Al-Si alloy* samples observed increase in the maximum bending stress after cycles (Table 1). This can be explained by mechanism increase adhesion of the ceramic coating samples under the load (effect of strengthening during reciprocating motion). For this system, the damage occurs along the NiAl interlayer-substrate interface, with affecting it (Fig.2).

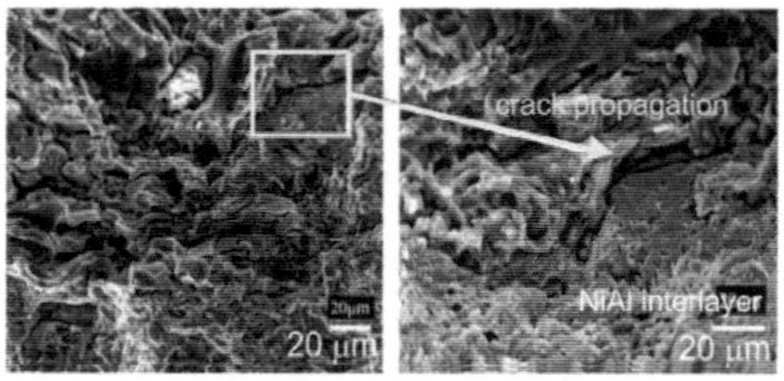

Fig. 2. Scanning micrographs of the fracture surface of the *Cr_3C_2-NiCr/NiAl/Al-Si* system after bending test.

4. Summary

The carbide coating Cr_3C_2-NiCr deposited on Al-Si alloy, using NiAl interlayer by HVOF technique is characterised by low porosity, dense structure, good adhesion and high hardness. In the structure of the coating there are very fine Cr_3C_2 particles embedded in NiCr matrix, coming to the size of nanocrystalline. The structure of this coating provides good resistance to cracking. The failure occurs along the *interlayer–substrate* interface and do not propagate to the substrate material. The carbide coating Cr_3C_2-NiCr on aluminium alloy has good erosive wear resistance associated with the effect of low porosity and high hardness as well as the phase composition of coating.

References

[1] A. Kulkarni, J. Gutleber, S. Sampath, A. Goland, W.B. Lindquist et al, Studies of the microstructure and properties of dense ceramic coatings produced by high velocity oxygen fuel combustion spraying, *Mater. Sci. Eng.,* 2007, **A369**, p 124-127

[2] Y.-Y. Wang, C.-J. Li, A. Ohmori, Examination of factors influencing the bond strength of high velocity oxy-fuel sprayed coatings, *Surf. Coat. Technol.*, 2006, **200**, p 2923-2928.

[3] W. Tillmann , E. Vogli , I. Baumann , G. Kopp, C. Weihs, Desirability-Based Multi-Criteria Optimization of HVOF Spray Experiments to Manufacture Fine Structured Wear- Resistant Cr3C2-25(NiCr20) Coatings, *J. Therm. Spray Technol.*, 2010, **19**, p 392– 408.

[4] J. Mateos, J.M. Cuetos , R. Vijande, E. Ferna´ndez, Tribological properties of plasma sprayed and laser remelted 75/25 Cr3C2/NiCr coatings, *Trib. Int.* 2001, **34**, p 345–351

Proceedings of the 72nd World Foundry Congress
21-25th May 2016, Nagoya, Japan

Development of Complete Gravity Die Casting Production System for Aluminum Castings

Yukiyoshi Funakoshi[1], Keishiro Kaneda[1]
[1] SINTOKOGIO,LTD. Toyokawa, Aichi, 442-8505, Japan

According to the world population explosion, it is expected the demand for automobile will still go up at least until 2050. It is required for manufacturing equipment to improve the productivity and quality under above-mentioned situation. Especially, the improvement of the aluminum casting equipment is required greater since aluminum castings are important components in order to reduce weight of automotive parts. To meet this demand, it is very important to optimize into complete system as well as improving each device in each process.
Gravity die casting method is popularly utilized for light metal castings especially to get high strength and pressure tightness. Authors have developed a completely new gravity die casting machine, which achieves energy saving and downsizing and also allows producing quality castings with above mentioned advantages. The authors also offer the complete system from melting to finishing in order to deliver the following benefits for foundries with this completely new gravity die casting machine as core device.

Keywords: *gravity die casting, high strength, pressure tightness, energy saving, downsizing ,*

1. Introduction

The developed gravity die casting machine has construction and feature which could produce high quarity castings besides low energy, downsized and also improves maintenance. Figure 1 shows the outside of equipment.

2. Feature

2.1 Producing high quality castings

The developed machine enables to pour molten metal into cavities without turbulence since the center of matching-die rotating axis is about the same, it reduces ladle tilting distance.

Application of the servo motor for die tilting mechanism allows molten metal flow without turbulence since shock and oscillation are reduced.

Additonaly,pouring movements is up to each type of dimensional shape cavity and pouring condition is managing by patterned. When flame exchanging actualized pouring high reproducibility and quality stabilization by pattern chooseing. Figure 2 shows structural drawing.

Fig.1 Development machine

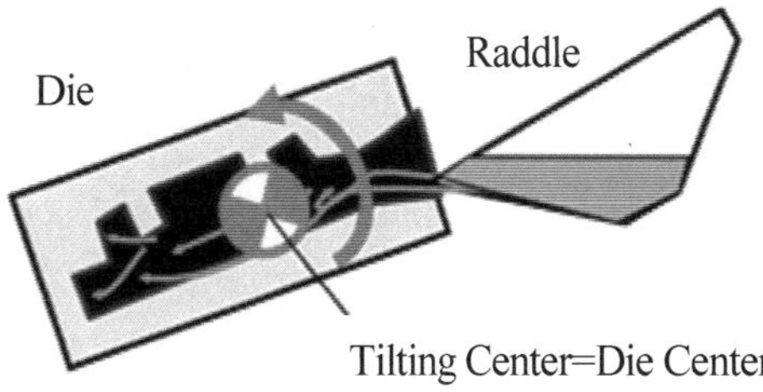

Fig.2 Tilting method

2.2 Energy saving

Figure 3 shows the comparison of the electric power consumption taking conventional machine as a base of 100.

As shown in this figure, the developed machine saved energy 80% compared to the conventional machine. The main reasons are following: (1) the number of actuators was reduced because one actuator operates multiple actions (2) Actuators power was reduced because the rotation center is set near the die center.(3) The machine weight is reduced. (4)

Optimizing the hydraulic unit control. These features reduced the machine running cost.

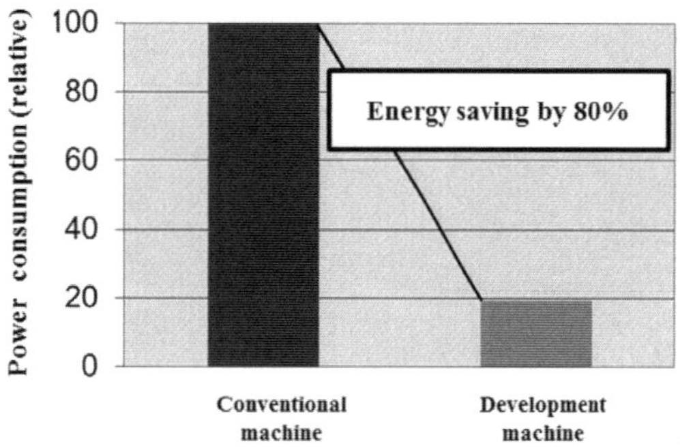

Fig.3 Effective save energy

2.3 Downsizing

Figure 4 shows the machine size comparison with (a) cope up-and-down method, (b) cope flip-up method and (c) developed machine taking (a) cope up-and-down method as a base of 100.

This comparison shows that new die casitng machine realized the significant downsizing with new die opening system.The developed machine reduced the size 40% compared to up-and-down method and 50% to core flip-up method.

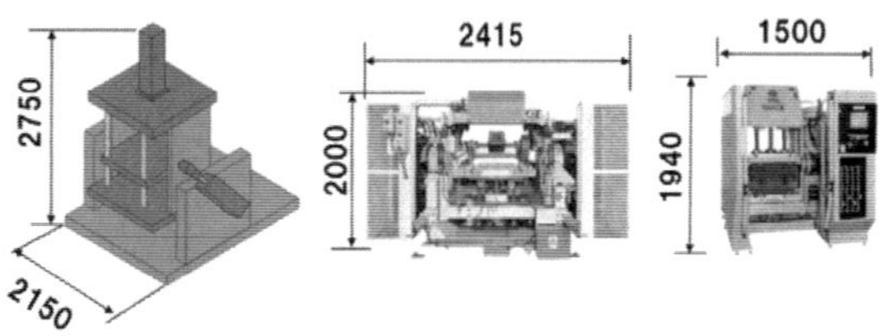

(a) Cope up-and-down method (b) Cope flip-up met (c) Development machine

(a) 100 : (b)85 : (c) 52

Fig.4 Comparison of machine size

Therefore the developed machine is easy to upgrade from the existing machine and install more machines thanks to the flexibility of the layout.

2.4 Total traceability

This developed machine monitors various data such as die temperature, cooling flow, tilting pattern, cycle time,etc to utilize them for total traceability. In addition, production data in entire die casintg system are monitored and integrated as a database with combing with core making machine or pouring machine.

Figure 5. shows the example of total traceability in entrire system. This system is able to clarify the points to be improved for quality aluminium casting and it leads to reduce the casting defects.

3. Conclusion

Authors have developed a completely new gravity die casting machine, which achieves energy saving and downsizing and also allows producing quality castings with reconsidering the conventional die casting machine. Moreover, application of the servo motor for die tilting mechanism improves the casting quality.

Authors believe that this developed die casting machine contributes to produce higher quality die casting.

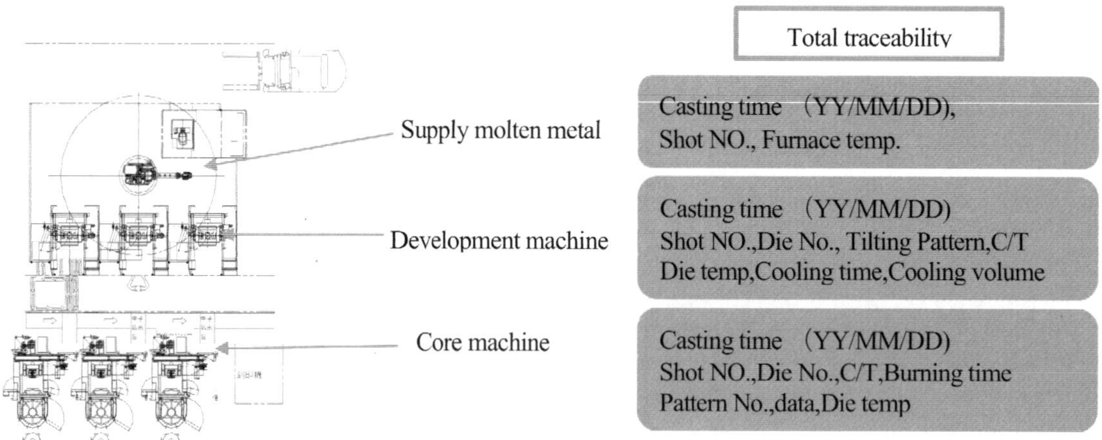

Fig.5 Total traceability(ex.)

Proceedings of the 72nd World Foundry Congress
21-25th May 2016, Nagoya, Japan

Semi-Solid Processing of Aluminium A201 – expectations and fulfilment

Platon Kapranos[1]

[1] Department of Materials Science & Engineering, The University of Sheffield, Sir Robert Hadfield Building, Mappin Street, Sheffield, S1 3JD, U.K. E-Mail: p.kapranos@sheffield.ac.uk;
Tel.: +44-114-222-5509; Fax: +44-114-222-5946.

Semi-Solid Metal Processing (SSM) or Thixoforming is the near net shaping of metals in the semi-solid state, i.e. within the freezing or melting range between the fully solid and fully liquid states. A201 is a copper containing casting alloy with additional small quantities of magnesium, silicon and silver. Although this alloy is difficult to cast, it has a particularly high response to age-hardening and therefore offers mechanical properties close to the wrought 2014 alloy. Alloy development experiments carried out at the University of Sheffield on A201 alloy have shown that feedstock of this alloy can be produced having uniform non-dendritic microstructures amenable to thixoforming, exhibits thixotropic properties under these conditions, can be shaped into complex near net-shape parts and develops impressive mechanical properties after appropriate heat treatment. A201 alloy billets thixoformed in the semi-solid state, heat treated with a T7 followed by T6 heat treatment before being mechanically tested with properties approaching these of wrought alloys. Here follows a review of work done to date on shaping the A201 aluminium alloy in the semi-solid state describing the hopes, expectations and fulfilment of researchers in the field.

Keywords: *A201 aluminum alloy; Non-dendritic feedstock; Thixoforming; Semi-Solid Metal Processing; Mechanical Properties*

1. Introduction

Thixoforming, the near net shaping of metals in the semi-solid state, relies on material that has a non-dendritic microstructure that exhibits thixotropic properties, i.e. when undisturbed the material remains stiff and holds its shape so it can be readily handled, but rapidly thinning and behaving like a liquid when put under shear. This behavior is the key to the thixoforming process where material flows as a semi-solid slurry into a die, as in conventional die-casting. This behavior is exhibited in all metal alloys that possess a microstructure that consists of metal spheroids e.g. α-Al in aluminum alloys, surrounded by a contiguous layer of eutectic liquid when heated to the semi-solid state.

The thixoforming process offers a number of advantages over both the casting and forging processes, such as fine, uniform and virtually free of porosity microstructures, products that may be heat-treated to give mechanical properties superior to those of casting, reduced energy consumption and reduced die thermal shock and therefore longer die-lives, due to the lower heat content of the semi-solid material. In addition, higher melting point alloys, such as hypereutectic aluminum-silicon alloys with very high silicon contents (~25-40%), superalloys or tool steels, which cannot be easily be die cast, may nevertheless, be thixoformed.

However, even though the potential of the process is clear to see and through its long history had numerous attempts to full commercialization, it has only been successfully exploited in a limited number of cases [1], using mainly conventional aluminum casting alloys of the AlSi7Mg type (A356 and 357).

At Sheffield we looked at the A201 copper containing casting alloy with additional small quantities of magnesium, silicon and silver and although this alloy is difficult to cast, it has a particularly high response to age-hardening and therefore offers mechanical properties close to the wrought 2014 alloy, therefore making it an idea candidate for thixoforming applications with potential to be used in aerospace.

1.1 Work on Aluminum A201

Although A201 is castable, it suffers from hot-shortness. However, better casting procedures and process control approved by the Casting Technology International has improved the situation. A201 is sand cast, permanent mold and investment cast into parts such as structural castings members, aerospace housings, electrical transmission line fittings, insulator caps, truck and trailer castings, and applications requiring highest tensile and yield strengths with moderate elongation. A201 castings also find applications in gasoline engine cylinders heads and pistons, turbine and supercharger impellers, rocker arms, connecting rods, missile fins, and other applications where elevated temperatures are important, as well as structural gear housings, aircraft landing gear castings, ordnance castings, pump housings and other applications where high strength and high energy- absorption capacity are required.

A factor that has kept the A201 cast alloy products from being widely used in aerospace applications has been its low elongation and the lack of fatigue property data. Thixoforming A201 has provided some evidence of good mechanical properties, almost doubled the elongation over conventional castings and encouraging data on fatigue behavior [2,3] showing that the alloy holds considerable promise if the data can be reproduced and the manufacturing route optimized both for its technical as well as its commercial aspects. The fact that the fatigue data were impressive, triggered thixoforming trials on the A201, conducted at the University of Sheffield's Advanced Manufacturing Research Centre producing a quite favorable comparison between the mechanical test results for the thixoformed and heat-treated A201 and a similar wrought alloy, 2014. These results were encouraging and showed that the target values for ultimate tensile strength and elongation of the standard as cast A201 (480 MPa UTS and 7% E), have been matched and bettered by the thixoformed alternative (480 MPa UTS and 13% E) and that these values are quite well compared to those of the wrought 2014 at 480 MPa UTS and 13%E [3].

2.1 Thixoformed products

Successful thixoforming of aluminum alloys has been carried out over the years in a number of applications, especially of automotive parts and these have been publicized in the wider literature [1, 4].

A commercial application of thixoforming the high strength casting alloy 201 into impeller shapes has been recently published [5]. The same authors have shown that heat treatment trials on material cut from the impeller were carried out to compare T7, T71 and T71* conditions and their hardness results suggest that T71 with an ageing time of just under 3 hours give the optimum behavior and their results also show an enhancement relative to the ASM Handbook recommended value for T7 treatment.

3. Discussion & Conclusions

Work carried out on the high strength A201 casting alloy has provided us with a lot of useful information as well as data of mechanical properties that are very encouraging.

- ✓ A201 aluminum alloy can be shaped into near net-shapes by Semi-Solid Metal Processing.
- ✓ The mechanical properties of thixoformed A201 alloy parts are better than conventional castings and can approach those of wrought alloys if the process and heat treatments are optimized and appropriately controlled.
- ✓ Consistency of mechanical properties is dictated by large extent by the feedstock material used.
- ✓ The suitability of feedstock material and in effect the commercial success of this process will be cost and market driven.

Acknowledgments

The author would like to thank all his colleagues and students over the years that contributed in filling the various gaps in our knowledge on shaping this alloy in the semi-solid state.

References

[1] P.Kapranos, *Semi-Solid Metal Processing – A Process Looking for a Market*, Solid State Phenomena, Trans Tech Publications, Switzerland, Vols. 141-143 (2008), pp 1-8.

[2] P. Kapranos, *Thixoforming: from automotive to aerospace*, APT Aluminium, Process & Product Technology, Vol. 5, Issue 1, February 2008, pp 39-44.

[3] P. Kapranos and M. Farnsworth, *'Semi-Solid Processing of A201 Alloy*, 2nd 2nd JSMEA/ASME, Proc. Int. Conf. on Materials Processing, Seattle, USA, 19th-22nd June 2005

[4] M.N. Mohammed, M.Z. Omar, M.S. Salleh, K.S. Alhawari, and P. Kapranos, *Semisolid Metal Processing Techniques for Nondendritic Feedstock Production*, The Scientific World Journal, Hindawi Publishing Corporation, Volume 2013, Article ID 752175, pp 1-16. http://dx.doi.org/10.1155/2013/752175

[5] Z. Qiang, S.P. Midson, W.M. Chang, and H.V. Atkinson, *'Casting and Heat Treatment of Turbocharger Impellers Thixocast from Alloy 201'*, Solid State Phenomena, (Volumes 192 - 193), Semi-Solid Processing of Alloys and Composites XII, Chapter 5: Industrial Applications, Editors Möller, H. and Govender, G., pp 556-561, October, 2012

Proceedings of the 72nd World Foundry Congress
21-25th May 2016, Nagoya, Japan

Accurate Controlling and Monitoring Sand Properties by sensing in Green Sand Treatment System

Yuichi Ogura[1], Katsuhito Sato[1], Tsuyoshi Sakai[1]
[1] Sintokogio, LTD. Toyokawa, Aichi 442-8505, Japan

In a casting process, green sand cannot always keep its stable properties due to loss of additives, water evaporation and change in the Oolistics layer, etc. caused by pouring heat.

On the other hand, to achieve "production of superior castings", it is desired that sand of high quality and stable properties is fed to a molding machine, which leads to prevention of defects. The sand after pouring and cooling is adjusted as appropriate in sand treatment equipment[1]. Since sand properties in a line depend on cooling time, number of cores and sand to metal ratio, etc. sand conditions are not always the same. Sand is required to be adjusted according to such conditions. This is not easy. Under these circumstances, the authors have developed a new feedback system using sensing technologies for sand treatment equipment in order to stabilize mold sand properties. The feedback system enables molding sand to be supplied in a state of stable properties.

This system can contribute to increased improvement of the quality assurance of cast products from the perspective of sand and equipment. Furtheremore, a foundry can become competitive by constantly managing the data collected by a sand control system.

Keywords: Sensing, Feedback, Quality assurance

1. Introduction

This paper explains accurate controlling and monitoring sand properties by sensing in green sand treatment system. This system can supply the sand of high quality and stable properties.

2. Sand treatment control system

2.1 Overall structure of control system

A sensor required of each process of sand treatment equipment[1] consisting of shake-out equipment, cooling device, sand muller and sand feeder to the molding machine are arranged in this system. The measured data by each sensor manage if the equipment is in proper operation and if the sand is adjusted and adapted to its appropriate properties in each relevant stage. Fig. 1 shows an outline of sand treatment control system. To measure sand properties, a sensor is fitted before and after each device and the sand situation in each process can be confirmed with the measurement results.

The development of this control system focuses on sand temperature and moisture. The first consideration is the sand temperature. In the shake-out equipment, heat-affected sand is mixed with sand not affected by heat. And the whole temperature of sand is lowered by spraying water on the recovered sand. In the cooling equipment the sand temperature is more lowered by latent heat of vaporization of water sprayed on the sand. Proper sand treatment following these basic principles ensures that molding sand is supplied at an appropriate temperature.

The next consideration is the sand moisture. Sand water affected by pouring heat is vaporized and discharged outward and the moisture level decreases in the whole mold. After that, the moisture level increases by spraying water on the sand in the shake-out and cooling equipment. In the sand muller, the moisture level is adjusted to reach its target value. After that, the moisture is slightly vaporized in the conveyance process between the sand muller and the molding machine.

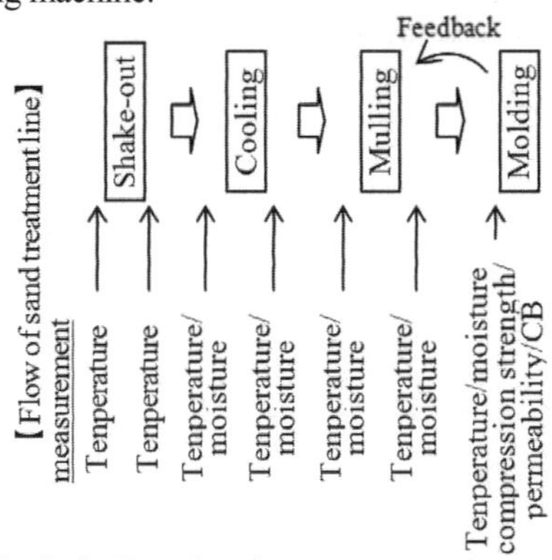

Fig. 1: Outline of sand treatment control system

Most foundries control sand temperature and moisture by using a factor of compactability (CB) closely correlating with the moisture. However, this system stabilizes the properties of molding sand by measuring the CB just before sand feeding to the molding machine and feeding it back to the controller of the sand muller.

2.2 Control results in each equipment

(1) Shake-out equipment

The shake-out equipment is equiped with a rotary-type mold cooler. The mold cooler measures the temperature of sand and products at its inlet and sprays water on them according to their temperature. Measuring the temperature of products at the inlet of the cooling equipment of the downstream process makes it possible to confirm the cooling effect in the shake-out equipment.

(2) Hydrous cooling equipment

The sensors arranged on the belt conveyor measure the sand temperature and moisture and the cooling equipment sprays water on the sand accordingly. After water spraying, the sand is fed to the sand cooler and cooled by laten heat of vaporization in it. The sand temperature and moisture is measured at the outlet of the sand cooler, which makes it possible to confirm the cooling effect in the cooling equipment. Thus, measuring and controling the sand temperature and moisture at the inlet and outlet of the cooling equipment lead to preventing sand inclusion due to insufficient cooling equipment.

(3) Sand muller, molding machine

In the sand mixing process the temperature and moisture of sand fed into the sand muller is measured and the sand is moistened accordingly and mixed. Eventually, the CB controller fitted on the sand muller measures CB. The CB controller feeds the sand to the molding machine after checking that the CB is within its target. The properties of sand to be molded are not yet automatically confirmed although the sand properties change due to external factors until the sand is supplied to the molding machine. Some foundries take by hand and test sample sand at regular intervals. In this case, the test is carried out at most several times per day. Therefore, it is difficult to understand the then sand properties and take appropriate measures against defective molding even if defective molds are produced due to the sand. In this context a device measuring the sand properties automatically is fitted just prior to the molding machine. The device provides data about the properties of sand molded actually and the operating conditions of the molding machine and such data can be used for traceability and measures to be taken against defective molding.

Moreover, comparison was made between the measured data by the sand property measuring device just prior to the molding machine and the measured data by the CB controller fitted on the sand muller. As a result, it was found that there was a close correlation between the CB just prior to the molding machine and the temperature of mixed sand. The measured reults are shown in Fig. 2. Correlating the data of sand properties just prior to the molding machine with various factors made it possible to feed back the data to the targeted CB of the sand muller. The authors are challenging to construct a system introducing such sensing technologies, control the upstream process with the measured results just prior the molding machine and accordingly stabilize the sand introduced into the molding machine.

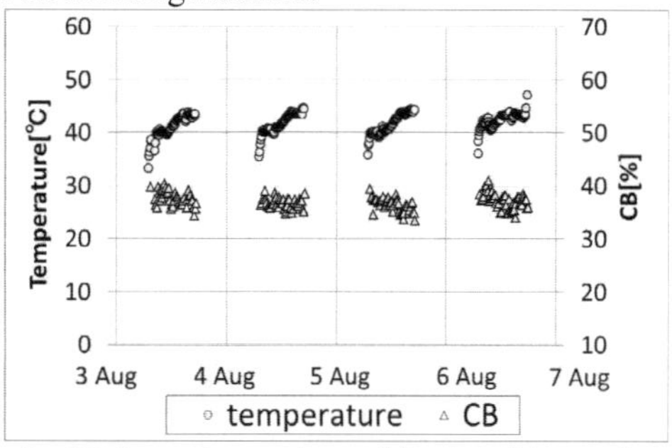

Fig. 2: CB just prior to the molding machine and sand temperature in the sand muller

3. Conclusion

Sand control in the past depended on the experiences and/or intuition of a skilled worker. However, this development ensures that the sand control is “visualized“ with sensing technologies and that the sand properties can be understood in real time. In addition, the feedback function between equipment with sensing technologies ensures that sand of more stable properties can be produced automatically, not by a skilled worker.

Furthermore, the sand treatment control system can be used as a tool for measures against defective molding by analyzing the sand properties in each process, the operating conditions of equipment and a trend of defect occurrence due to the sand. The authors are expecting that use of the data for “visualizing“ the entire foundry with sensing technologies ensures increased improvement of the product quality important for the operation of a foundry.

References

[1] K. Hashimoto, M. Fukihara and H. Makino, “Automation Technology in Green Sand Processing Equipment” International Journal of Automation Technology 2 (2008) No.4

Proceedings of the 72nd World Foundry Congress
21-25th May 2016, Nagoya, Japan

In-situ Measurement of Green Sand during Molding Process

Dr.-Ing. habil. Hartmut Polzin[1], Dr.-Ing. Matthias Strehle[1], Prof. Dr.-Ing. Gotthard Wolf[1]
Dr. Hiroyasu Makino[2], Shuji Takasu[2] and Hiroyuki Miyazaki[2]
[1] TU Bergakademie Freiberg Germany, Foundry Institute
[2] Sintokogio Ltd, Aichi, Japan

A green sand molding process depends on a high number of variable influence factors. For stable mold strength at pattern plate, a useful sensor and control system is needed. Following, different sensors will be shown, which allow to measure and control sand filling and squeeze process in-situ. This technical paper shows valuable knowledge to understand green sand molding process and to achieve high quality mold with low energy.

Keywords: *sensor, green sand molding, aeration sand filling, squeeze*

1. Introduction

A high quality of casting product is depending on quality of sand mold. Generally, the green sand molding process consists of sand filling process and squeeze process. In order to produce and control high quality mold stably, it is necessary to know this processes in-situ.

Parameters of sand preparation, sand filling and squeeze are possible to measure and control. Industrial production always needs to produce with stable conditions by the same machine parameters. Measurements in foundry show that all parameters fluctuate in a range. The influence of each parameter is known by statistic investigation. With this measured parameter, the knowledge of influence and also a control system, it is possible to get high mold quality. This is the main reason for in-situ control of mold quality at pattern plate with sensors. As last step in mold production, squeeze is used. Here, the mold quality is influenced by the filling time, air pressure and also squeeze pressure. To have the right setting for this parameter, the knowledge about mold strength at pattern plate is needed. Until now, there was no measurement device that can be fixed on pattern plate, for the registration of sand filling and squeeze.

2. Sensor devices

2.1 First sensor

The first sensor, which was developed by Prof. Bast for air-flow (Seiatsu, Impact) and squeeze process, shows mold strength or mold hardness in-situ.

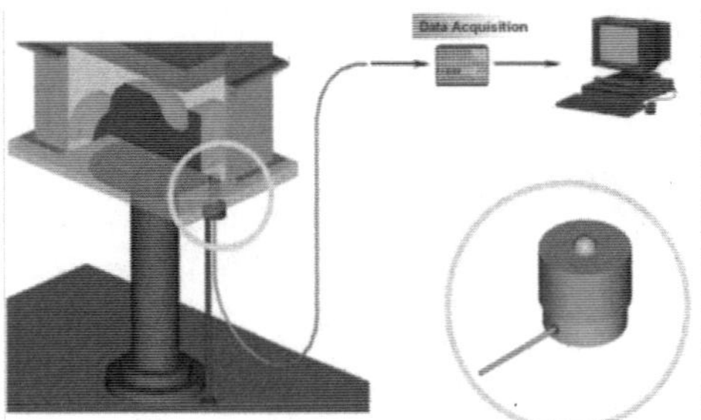

Fig. 1: First Sensor device

This sensor is useful to save energy and shorten cycle time for high quality casting. The construction of this sensor allows stable measurements for air-flow of Seiatsu/ Impact/, gravity sand filling and squeeze process. But for molding processes which use a very low air-pressure for sand filling, such as aeration sand filling, it is not possible to use the first sensor because of sensitivity [1].

2.2 Sensor for observation of sand filling

The second sensor, shown in Fig. 2, was developed to detect sand filling, aeration sand filling, high pressure blow sand filling and gravity sand filling. Actually, it was impossible to observe in-situ sand filling behavior inside tight flask at actual foundry shop. It is necessary to understand how sand flows inside flask and to use this sensor signal for squeeze control.

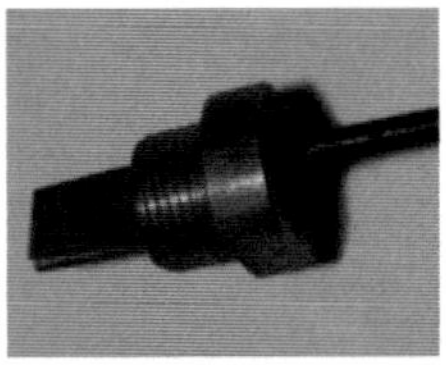

Fig 2: sensor for sand detection

Experiments were conducted at university and in foundry for practical trials. This experiments show, how sand flow inside the flask and fill the flask [2]. For this experiment, sensors were fixed at different position on the pattern plate and in sleeve with different high (0, 70, 90, 110mm). Now, it was possible to understand how sand flow inside the flask, sand filling time and also the influence of pattern shape (represented by different sleeve high).

2.3 Final sensor for all sand molding processes

The third sensor is shown in the Figure 3 and developed for all sand molding processes in-situ.

Fig 3: sensor for in-situ measurement of mold

The third new sensor combines all advantages, such as using for molding processes and a short reaction time for precision mold machine control. The new Sensor shows a high accuracy in experiments. It was possible to see the low air pressure, such as from aeration sand filling, in detail. Different settings from air pressure at molding machine will be measured by the sensor reliable. Also different squeeze pressure will be detected easily. Figure 4 shows an example of measured sensor voltage curve during aeration sand filling and squeeze. Three sensors were fixed at the bottom of sleeves on the pattern plate. When green sand contact the sensors during aeration sand filling, sensor voltages increased a little. Then, sensor voltages went up rapidly during squeeze.

With this new sensor, a valuable knowledge of mold quality at the area of pattern plate is available. Furthermore, this value of strength can use on one side for documentation the machine settings and mold quality for customer. But on the other side, this measured value is perfect for mold machine control. This possibility is given, because of the knowledge, when the maximum or needed strength is achieved at pattern. Without this sensor, it was not possible to know the right squeeze pressure for switch of hydraulic, by fluctuating properties of mixed sand and weight. In addition to the high and stable quality, it is also a tool for saving energy and protect molding machine, because of early stop of aeration and squeeze.

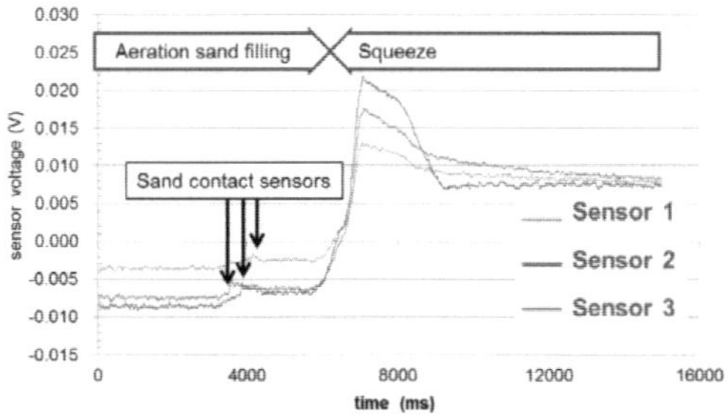

Fig 4: Measured sensor voltage curve during aeration sand filling and squeeze

3. Conclusion

Different sensors are developed for sand filling and squeeze process. The final sensors are tested in foundry and can be used such as follows:

- In-situ measurement of sand filling and squeeze for all technologies
- Using for machine control in a control cycle.
- Saving energy and time by knowledge of mold properties at pattern plate.
- Increase quality of mold, by compensation of fluctuating sand properties and sand weight.
- Using sensor as a tool for online documentation of molding properties for customer.

Acknowledgements

The authors gratefully acknowledge Sintokogio Ltd. for the research support.

References

[1] J. Bast, T. Malaschkin and T. Hahn: *Optimizing Molding Parameters for Green Sand Compaction*, AFS Transactions, Paper 05-008 (2005) pp. 309-318

[2] M. Strehle, H. Polzin, H. Makino and H. Miyazaki: *Sensor for the Detection of Sand-Filling-Process in Aeration Molding Machines*, Presentation at 71st World Foundry Congress in Bilbao 2014

Proceedings of the 72nd World Foundry Congress
21-25th May 2016, Nagoya, Japan

Porosity Discrimination System for Aluminum Alloy Die Castings Using X-ray CT Imaging

Eito Tanaka[1], Ken'ichi Kanazawa[1], Ken'ichi Yano[1], Ryusei Kawatani[2] and Jun'ichi Ogura[2]
[1] Department of Mechanical Engineering, Mie University, Tsu, Japan
[2] Yamaha Motor Co., Ltd., Iwata, Japan

In die casting, it is necessary to decrease porosity defects by taking some of counter measures. Such defects can be observed by using an X-ray CT scanner, which is a non-destructive measurement device. In this study, we propose a discrimination system for porosity defects between gas and shrinkage porosity using blister tests and X-ray CT images. We demonstrate the effectiveness of the system in discriminating between gas and shrinkage porosity defects by applying it to actual die casting.

Keywords: *X-ray CT, blister test, gas porosity, shrinkage porosity, discrimination.*

1. Introduction

Porosity defects in a casting are roughly divided into two types: shrinkage porosity and gas porosity. There are different types of counter measure used for the different types of porosity defects. Porosity can be visualized using an X-ray CT scanner, but the different types cannot be quantitatively discriminated. The blister tests make gas porosity, but not shrinkage porosity, expand remarkably. However, it is difficult to compare the volumes of the porosity. The castings positions and postures in X-ray photography cannot be adjusted exactly before and after blister tests. In addition, the blister tests cause casting products to deform due to thermal stress. Therefore, we propose a method of discriminating porosity defects by comparing volumes before and after blister tests.

2. Discrimination system

Our proposed system uses image registration of X-ray CT images before and after blister tests. Furthermore, the system estimates deviations of a pair of porosities, which cannot be adjusted only the registration. The positions and volumes of the porosity are calculated by image processing. The appropriate discrimination is achieved by calculating the volume ratio between identified porosity before and after blister tests.

An image registration uses homography. The matrix to adjust the castings before and after the blister test is calculated according to feature points of the castings, which are selected on 2-D projection images by the user. The registration is then repeated with each of the tomographic images, operating step by step on the x-y, y-z, and x-z planes.

CT values are given in grayscale intensity, and are normalized from 0 to 1 in this system. Porosity is then extracted from a binary image based on Morphology Reconstruction [1]. However, the precision of extraction has been improved in the proposed system to reflect dynamic changes in thresholds. The volume of the porosity can be calculated based on the method described by Hishida [2]. Each of the points of the porosity is calculated using a specific moment of CT value as the gravity point.

The extracted porosity cannot be exactly identified before and after blister tests. Deviation is estimated by calculating deviations in the cross-section images of the casting. Cross-section images before the blister test are made of the x-y, y-z, and z-x planes, which cross the gravity point of the porosity before the blister test. The size of images is resized as an $l \times l$ square, whose center is the gravity point and l is calculated by the following formula:

$$l = 2\sqrt{\frac{S_1+S_2+S_3}{3}} \tag{1}$$

The cross-section areas of the x-y, y-z, and z-x planes are S_1, S_2, and S_3, respectively. The cross-section areas are calculated using binary images. Cross-section images after the blister test are made of the x-y, y-z, and z-x planes, whose positions are same as before the blister test. Position vectors between the two gravity points before and after blister tests are calculated from the binary images of the cross-sections. Let each of the vectors on the x-y, y-z, and z-x planes be $\boldsymbol{k}_1 = (k_{1x}, k_{1y})$, $\boldsymbol{k}_2 = (k_{2y}, k_{2z})$, and $\boldsymbol{k}_3 = (k_{3z}, k_{3x})$, respectively, with $\boldsymbol{k}$ calculated by the following formula:

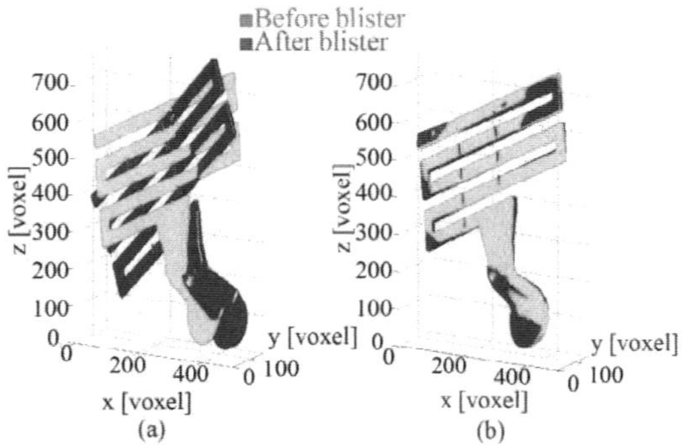

Fig. 1 Comparsion of the forms before and after registration. (a) Before. (b) After.

$$\boldsymbol{k} = \frac{1}{2}\begin{pmatrix} k_{1x} + k_{3x} \\ k_{1y} + k_{2y} \\ k_{2z} + k_{3z} \end{pmatrix}. \quad (2)$$

A sphere of radius is calculated by $k = \beta \, \| \boldsymbol{k} \|$. The center is the gravity point of the porosity before the blister test. The corrective coefficient for allowance is β. Porosity after the blister test existed in the sphere of radius k and the porosity before the blister test regarded as a pair of porosity. If i porosities after the blister test exist in the sphere, let the vectors between the porosities before and after the blister test be $\boldsymbol{r}_i$, with identical porosities then calculated using the following formula:

$$\arg\min_i \| \boldsymbol{k} - \boldsymbol{r}_i \|. \quad (3)$$

The expansion coefficient of the porosity between before and after the blister test is defined as V_{after}/V_{before}. If the coefficient is less threshold α or porosity before the blister test have non-target to identical that after the blister test, the porosity is determined to be shrinkage porosity. If the coefficient is equal to or greater than α or porosity exists only after the blister test, the porosity is determined to be gas porosity. Threshold α is set as a coefficient of the total volume of porosity before and after blister tests between $1 < \alpha < \alpha_{total}$.

3. Demonstration of the system

The test piece to which this system was applied was made from aluminum (ADC12) and produced by a die casting machine (DC350I-MS). For imaging, we used a SOMATOM Emotion 6 X-ray CT scanner

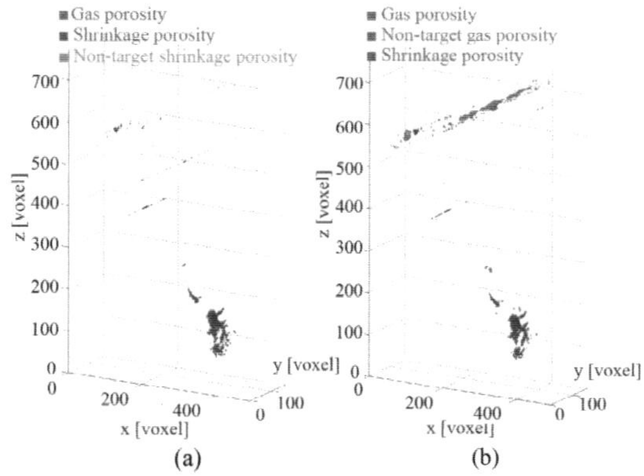

Fig. 2 Results of the porosity discrimination.
(a) Before the blister test. (b) After the bliste test.

manufactured by Siemens Japan K.K. The space resolution was 0.7871^3 mm^3 per voxel.

The thresholds and coefficients in this system were set to obtain stable results by multiple trials: $\beta = 1.5$, $\alpha = 1.3$ (setting in $1 < \alpha < \alpha_{total} = 2.032$).

Figure 1 shows 3-D images of the test piece before and after the blister test. The results of discrimination are shown in Fig. 2. The volumes of the porosities are shown in Table 1. This system was verified to be able to discriminate between two types of porosity even if mixed and unclear.

Table 1 Volume of porosity before and after the blister test

Porosity	Volume (mm^3)	
	Before	After
Shrinkage	3.484×10^3	3.207×10^3
Non-target shrinkage	52.27	–
Gas	5.238	2.985×10^3
Non-target gas	–	1.088×10^3
Total	3.541×10^3	7.199×10^3

4. Conclusion

In the conventional evaluation of porosity defects, it is possible to calculate volumes, but not possible to discriminate between the two types of porosity. Our proposed system, however, can not only calculate volumes but also identify the type of porosity in various positions.

References

[1] P. Soille: *Morphological Image Analysis*, Springer-Verlag, Italy (1999) pp. 208-210.

[2] H. Hishida, T. Michikawa, H. Suzuki and Y. Ohtake: Precision Engineering Jounal, 76 (2010) pp. 960-965.

Proceedings of the 72nd World Foundry Congress
21-25th May 2016, Nagoya, Japan

Technological conversion applicable for manufacturing elements from Nickel superalloy H282

Z. Pirowski [1], K. Jaśkowiec [1], A. Tchórz [1], I. Krzak [1], J. J. Sobczak [1], R. M. Purgert [2]
[1] Instytut Odlewnictwa, 30-418 Kraków, ul. Zakopiańska 73,Polska
[2] Energy Industries of Ohio, Cleveland, OH, USA

Nickel alloys are ones of the most resistant materials designed to work in extremely harsh conditions at high temperature. The majority of working elements, made from those alloys is being manufactured using very expensive erosive treatment, based on material that has been pre-plastically reworked. The application of metal casting processes for manufacturing Ni alloys poses a number of difficulties, the scale of which is rising equally to expected increase of performance of those alloys.

The aim of the proposed study was to optimize the casting process of manufacturing of Ni alloy with properties comparable to alloys manufactured by squeezing method with associated plastic forming. The study also involved the execution of experimental utility castings based on the example of Haynes superalloy 282 (H282).

Keywords: *nickel alloys, computer simulation, computed tomography, technological conversion*

1. Introduction

Nickel alloys plays the key role, where materials are designed for elements with the largest thermal load, including also stress and corrosion conditions. Such alloys, including those crawling resistant, are used among others in energy sector facility, chemical and petrochemical industry, aviation and space. Many of those alloys are used as materials plastically fabricated and elements made from such alloys are manufactured by costly erosive treatment. In the following study the trials of melting and casting of H282 has been presented. Actually the Haynes superalloy H282 is being manufactured only using the plastic fabrication method. It was assumed that the result of the study will be manufactured experimental utility castings.

2. Material investigation

For the purpose of experimental work the H282 superalloy has been chosen, additionally reinforced precipitation of phase γ'. This superalloy is an outstanding material designed to work at high temperature, in very aggressive environment, including large loads [1-5]. Up to now it is only used and manufactured by plastic fabrication method. The attempt to use such material as casting encountered the number of problems. Those problem has been identified in the area of both – melting processes (high reactivity of the melt with ceramic shells and the atmosphere appeared above metal bath), and the casting process itself (low fluidity, large casting shrinkage).

The trials of melting and casting of H282 superalloy in laboratory conditions were focused on optimization of melting technology in the open induction furnace, as well as the evaluation of acquired properties using the method of gravity casting in sand moulds.

Casting alloys are characterized by decreased mechanical properties at ambient temperature, than their plastically fabricated equivalents. This is due to the lack of reinforcement of the structure by squeeze of the casting alloys. Those differences are being decreased while the temperature of investigation is going higher. And in sufficiently high temperature, the level of strength properties of alloys often exceeds the appropriate level for materials manufactured forging, rolling, and drawing techniques. Appointed in the following study temperature for conversion of H282 is equal to about 920°C (Fig. 1).

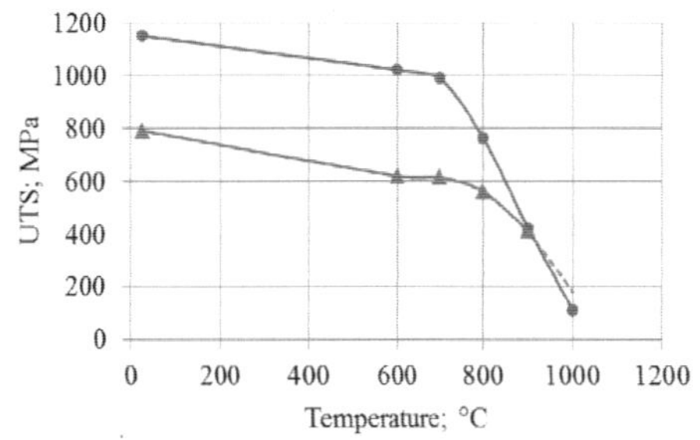

Fig. 1. Dependence UTS=f(T) for alloy H282:
▲ - casting alloy (own invesigations),
● - plastically fabricated (literature data) [6, 7]

3. Prototype castings

After carrying out the computer simulation of the technological process, there has been made utility castings of burner nozzle for ladle preheating. The quality of manufactured castings was compared to the results of numerical analysis. Estimated potential distribution of porosity in the casting nozzle flooded by the riser was confirmed in the experimentally-made casting, and further proved by computed tomography (CT). The prevalence of similiar porosity was not found in the case where metal was introduced into the mould through the bottom of the gating system (Fig. 2). The results has confirmed that numerical analysis was stated correctly.

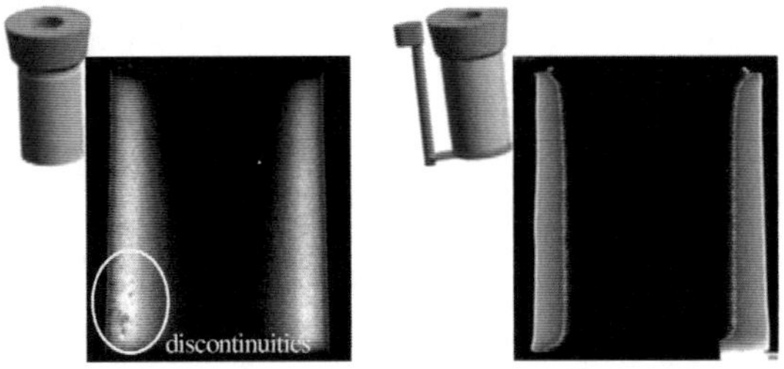

Fig. 2. Computed tomography of burner nozzle casting made from H282 in different melting technologies

Additional experiments attempting to perform experimental casting showed that from H282 superalloy shaped casting can be manufactured with satisfactory quality of the surface, flared shape, and a low degree of occurrence of internal discontinuities. The working nozzle, mounted in the burner is shown on Fig.3. Measured temperature of its working conditions locally exceeds 800°C.

Fig. 3. Working burner nozzle during the heating ladle: left – overall view, right – thermal picture

4. Summary

Knowledge related to level of properties: mechanical, thermo-physical, technological, functional, as well as process development of macro- and microstructures of H282 allows to propose the melting technology of H282 in an open crucible induction furnace. The technology of gravity casting has been performed, allowing to manufacture of good quality castings to operate in extreme conditions, i.e. in the temperature that significantly exceeds 700°C. The basic condition necessary to obtain the output of casting alloys, without defects is the strict adherence to the developed manufacturing technology.

Acknowledgements

The following study is the effect of international collaboration in the framework of international project called Advanced Ultra-Supercritical NICKEL (A-USC NICKEL)

References

[1] Z. Pirowski: Nickel alloys as a modern casting material designed to work in extreme conditions; Professor' dissertation; Instytut Odlewnictwa; Kraków 2013.

[2] R. Schwant, C. Shen, M. Soare: New Materials Enable Unprecedented Improvement in Turbine Performance; Advanced Materials & Processes; ASM International 2013.

[3] M. Santella, J. Shingledecker, B. Swindeman: Materials for Advanced Ultra-Supercritical Steam Boilers; 24th Annual Conference on Fossil Energy Materials; Pittsburg; RPA; 2010.

[4] H. White, M. Santella, E. D. Specht: Weldability of HAYNES 282 alloy for new fabrications and after service exposure; Energy Materials Materials Science and Engineering for Energy Systems; 2009; 4 (2: Welding and microstructure evolution in power generation plant); pp.84-91.

[5] N. Sobczak, Z. Pirowski, R. M. Purgert, W. Uhl, K. Jaśkowiec, Ł. Boroń, S. Pysz, J. J. Sobczak: Castability of HAYNES 282 alloy; Workshop "Advanced Ultrasupercritical Coal-fired Power Plants"; EVN Forum Maria Enzerdorf Vienna; Austria; 19-20 September 2012.

[6] HAYNES®282® alloy; materiały reklamowe firmy Haynes International, Inc.; Kokomo Indiana USA; 2008; www.haynesintl.com.

[7] L. M. Pike: Development of a fabricable gamma-prime (γ') strengthened superalloy; Superalloys 2008 (Eleventh International Symposium); TSM 2008; pp. 191-200.

Revert Alloy Impact on Mechanical Properties of Nickel Based Superalloys Used in Production of Aircraft Engine Critical Components

Rafal Cygan[1], Pawel Zywicki[2], Tomasz Szczech[3] and Michal Kwiatkowski[4]
[1] Rzeszow University of Technology, Faculty of Mechanical Engineering and Aeronautics, Powstancow Warszawy 8 Av., Rzeszow 35-959, Poland
[2] Warsaw University of Technology, Faculty of Materials Science and Engineering, Woloska 141 Str., Warsaw 02-507, Poland
[3] Pratt & Whitney Rzeszow S.A., Hetmanska 120 Str., 35-078 Rzeszow, Poland
[4] AGH University of Science and Technology, Faculty of Non-Ferrous Metals, A. Mickiewicza 30 Av., 30-059 Cracow, Poland

The requirements of jet engine critical components investment casting process determine the operational strength of cast components and are strictly enforced by the engine final manufacturers. One of the main factors affecting technological process of casting is the charge alloy, provided by so-called approved source of supply. The quality of supplied charge material is of fundamental importance for quality of castings as all bad characteristics of the alloy transfer naturally into castings. Charge melts are ordered in form of rolls having a diameter of 60 to 90 mm in batches having a mass of 3 to 5 tons. In further steps they are cut into batch portions of 5 to 15 kg weights suitable for casting furnaces capacity. Supplier produces alloys in so-called "Virgin" state - without revert alloy, or with revert alloy of the same type, which is recovered in a form of gating systems and defective components that have been rejected in the quality control process. The weight percentage of revert alloy that is added to the production process can affect the quality and properties of resulting material. In the article compares the effect of the amount of revert material on the properties casting parts.

Keywords: *investment casting, virgin and revert superalloys, tensile and creep strength*

1. Introduction

Elements of aircraft engines, including high and low pressure turbine blades and vane clusters, made of nickel and cobalt based superalloys, during long-term operation at elevated temperatures, are exposed to mass forces as well as corrosive agents (Ref 1). These facts force very high quality requirements for castings, both in terms of dimensional requirements, surface condition, as well as requirements for micro- and macrostructure. Castings quality is dependent on a complex, multi-step production process (Ref 2). One of the main factors affecting technological process of casting is the charge alloy, provided by so-called approved source of supply. The quality of supplied charge material is of fundamental importance for quality of castings as all bad characteristics of the alloy transfer naturally into castings. Charge melts are ordered in form of rolls having a diameter of 60 to 90 mm in batches having a mass of 3 to 5 tons. In further steps they are cut into batch portions of 5 to 15 kg weights suitable for casting furnaces capacity. Along with charge material obtained from the manufacturer, a certificate of chemical composition analysis is provided, which corresponds to the material standard for a given alloy type and, as was proven by previous experience, certificate coincides with analysis results carried out in the quality control laboratory.

Supplier produces alloys in so-called "Virgin" state - without revert alloy, or with revert alloy of the same type, which is recovered in a form of gating systems and defective components that have been rejected in the quality control process. The weight percentage of revert alloy that is added to the production process can affect the quality and properties of resulting material. Therefore, the maximum number of elements in the revert alloy has to be approved by preparing and examining test series of the castings.

2. Material and experimental procedure

2.1 Scope of the research

Since it was found that content of revert alloy affects the quality of resulting material, in the present study an analysis of the test alloys with various contents of the revert alloy is presented. The analysis was performed for IN713C alloy, four different masterheats were analysed, that were delivered at different times and manufactured in different melting processes. Among tested alloys there were ones in "Virgin" state and ones produced using revert alloy.

For the purposes of this study, the following designations of the amount of revert alloy was adopted. INC713C was subjected to strength tests according to the materials standards.

Table 1 List of investigated alloys.

Alloy	Virgin	Revert	Designation
IN713C	100	0	A
IN713C	60	40	B
IN713C	50	50	C

2.2 Preparation of specimens for strength tests

Before any taken action on the alloy, supplier sends a portion of the material in order to conduct study and to qualify the material for production. For each batch of a material received from the supplier, a separate sample is prepared. Test samples are casted to a ceramic mould according to standardized, repeatable procedure.

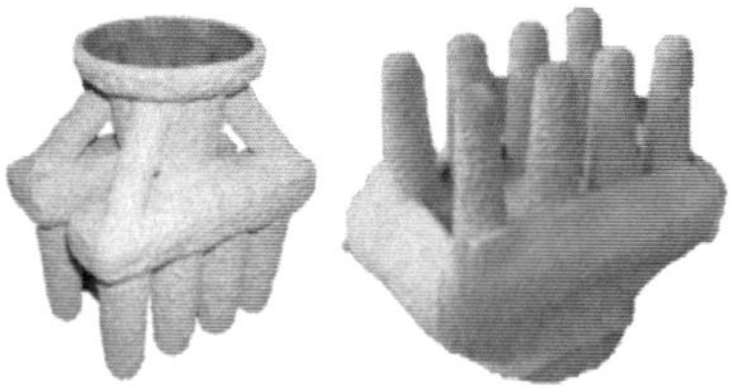

Fig. 1 Ceramic mould

The casting of the sample - the so-called. "Carrot" - has shape of a truncated cone, which is machined to obtain a rod of a diameter required by standards. The ends of the sample are machined to appropriate diameter and threaded to allow assembly in the testing machine.

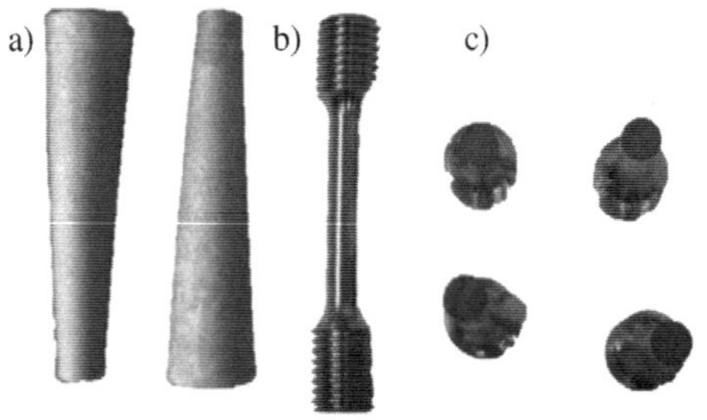

Fig. 2 (a) Casting samples prior to machining. (b)) sample after machining (c) sample after the strength test

Materials standard specify parameters: creep strength, time to sample rupture subjected to 152 MPa (22 kpsi) load at a temperature of 982°C (1800 F) and hardness.

3. Results and discussion

The following fig. presents the example results for INC713.

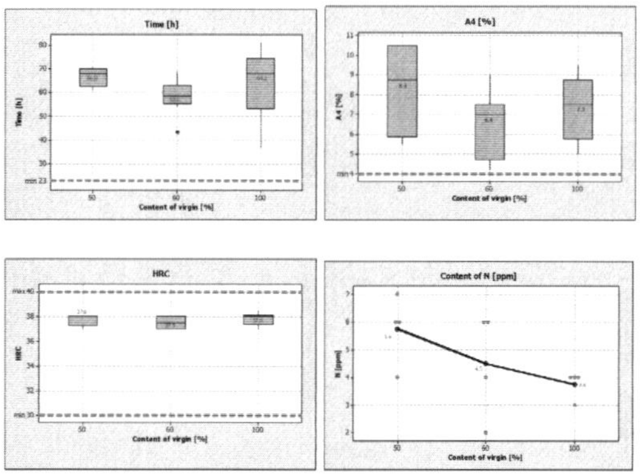

Fig. 3 Impact of pure elements in the IN 713C alloy on: (a) creep strength (b) elongation (c) hardness (d) nitrogen content

4. Conclusions

IN713C - tested materials show no dependence of revert alloy content on mechanical properties at elevated temperature. 100V alloy is characterized by low nitrogen content. Conducted study confirms that the procedure applied by the manufacturer for cleaning, preparation of revert alloy for remelting and casting allow the use of remelted material in the investment casting process. All parameters of tested INC713 superalloy are within the limits of the standards. In most cases, it is difficult to determine the dependence of the contents of the revert alloy on parameters of final material. Revert alloy does not affect the properties of the resulting alloy in studied range.

Acknowledgements

Financial support of Structural Funds in the Operational Programme - Innovative Economy (IE OP) financed from the European Regional Development Fund - Project "Modern material technologies in aerospace industry", Nr POIG. 01.01.02-00-015/08-01 is gratefully acknowledged.

References

[1] H. Matysiak, M. Zagorska, A. Balkowiec, B. Adamczyk-Cieslak, R. Cygan, J. Cwajna, J. Nawrocki, K. J. Kurzydłowski: Journal of Materials Engineering and Performance. 23(9). (2014) 3305-3313.

[2] A. Ges, H. Palacio, R. Versaci: J. Mater. Sci. 29 (1994) 3572-3576.

Proceedings of the 72nd World Foundry Congress
21-25th May 2016, Nagoya, Japan

Semi-automatic data linkage and acquisition process for the preparation of Austempered Ductile Iron

D. Wilk-Kołodziejczyk[1,2], I. Olejarczyk-Wożeńska[1], W. Adrian[1], K. Regulski[1], G. Rojek[1], S. Kluska-Nawarecka[3]
1AGH University of Science and Technology,
st. A. Mickiewicza 30, 30-059 Krakow, Poland, (e-mail: dwilk@agh.edu.pl)
[2]Foundry Research Institute, Zakopianska 73, 30-418 Krakow, Poland

Austempered Ductile Iron (ADI) is the result of heat treatment carried out on the cast iron with nodular graphite. Various parameters of the manufacturing process such as temperature and time in different stages of the process, and also the chemical composition are responsible for the mechanical properties of final material. Data concerning processes of ADI manufacturing derived from experiments could be find in the Web. The paper describes methodology of collecting experimental data concerning the process of ADI manufacturing as well as Web exploration with use of semantic techniques and the framework developed in the School of Information Technology and Electrical Engineering at The University of Queensland, named OGDL: Ontology Guided Data Linkage.

Keywords: *Austempered Ductile Iron, thermal treatment, properties, data integration, ontology.*

1. Introduction

Austempered Ductile Iron (ADI) is the result of heat treatment carried out on the cast iron with nodular graphite. Excellent combination of properties obtained in ADI, including the strength, toughness and fatigue behavior, makes this material a successful substitute for steel or aluminum alloys. ADI has a high fatigue strength, higher than aluminum, is resistant to abrasive wear like steel, but most of all – its use can significantly reduce the cost of production. An important advantage for the automotive industry is also high damping capacity, as a matter of fact, by 40% better than that of steel [1-2].

The technologically demanding process of the ADI is a good example of how important the accurate information about various process parameters really is. As described above, various parameters such as temperature and time in different stages of the process, and also the chemical composition exert a strong influence on the final properties of this material.

The aim of the project was to propose a methodology supporting the task of collecting the comparative data on studies of the mechanical properties of ADI. Collecting of research data is an important step in the process of finding the optimum design solutions for newly made products - experimental data allow us properly calibrate the manufacturing process of ADI to let the final product achieve the required properties. Parameters of the ADI production process, i.e. the time and temperature of austenitizing and austempering, as well as the alloying elements added to ductile iron affect the ADI properties. The design process can use research data collected, among others, from the Web [3-4].

The process of data acquisition can be supported by semantic technologies, including ontologies which are descriptive logic formalism. The authors propose the use of a genuine proprietary framework developed in the School of Information Technology and Electrical Engineering at The University of Queensland, named OGDL: Ontology Guided Data Linkage which is also presented in the paper. The framework reveals hidden relationships between data sources towards patterns discovery at different levels of data abstraction.

2. Ontology-Driven Data Acquisition

The aim of the first stage of the work was to collect comparative data on the manufacturing process of ADI. Foundry Research Institute in Cracow has an experienced team of engineers conducting research on this material. Studies of this material are also conducted in nearly all parts of the world. The diversity of the experiments results in the development of new variants of the heat treatment, and thus in new ways to obtain the required properties of this material.

Ontology in computer science is a formal record of the definition of a certain part of the knowledge expressed in the form of a taxonomy of related classes. In other words, ontology is a semantic model of defined field of knowledge. Ontologies are computerized implementation of description logic [5-7].

Based on the literature, one can create a very accurate and reliable knowledge base. Semantic model of relationships existing between the data could reduce required experience of the technologist to evaluate the data found, and most of all – time needed. The authors propose the use of a genuine proprietary framework developed in the School of Information Technology and Electrical Engineering at The University of Queensland, named OGDL: Ontology Guided Data Linkage [8].

The framework produces an ontology based on the data model of the heterogeneous data sources. An ontology is capable of describing the properties of data objects from different application domains. Clusters of object are connected with the domain ontology of ADI and the semantic transformation of the relations among the sources is proceeded.

Having the data classified with the use of the domain ontology, knowledge based data search among the data sources is possible.

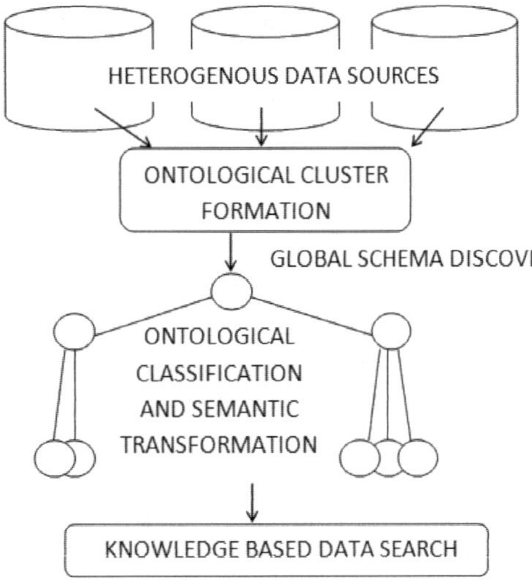

Fig. 1 Simplified schema of process flow with Ontology Guided Data Linkage framework.

3. Conclusions

Presented methodology enables the use of a modern approach to data integration, which is the Agile Data Integration system. Collecting of research data and searching with the use of knowledge based systems is an important step in the process of finding the optimum design solutions for newly made products - experimental data allow us properly calibrate the manufacturing process of ADI to let the final product achieve the required properties. The design process can use the research data collected, among others, from the Web. As indicated in the article, the process of data acquisition can be supported by the semantic technologies.

Acknowledgements

Financial support of The National Centre for Research and Development LIDER/028/593/L-4/12/NCBR/2013 is gratefully acknowledged.

References

[1] Skoczylas, P., Krzyńska, A., Kaczorowski, M.:Archives of Foundry Engineering 11 (2011) 123–126.

[2] Olejarczyk-Wożeńska, I., Adrian, H., Mrzygłód, B.: Archives of Metallurgy and Materials 59 (2014) 613-617.

[3] Glowacz, A.: Measurement Science Review 14 (2014) 257-262.

[4] Opalinski, A., Turek W., Cetnarowicz K..: Proc.Conf. FedCSIS (2013) 1261–1267.

[5] Wilk-Kołodziejczyk, D., Dajda, J., Macura, M., Regulski, K., Kluska-Nawarecka, S.: Archives of Metallurgy and Materials 59 (2014) 744–746.

[6] Kluska-Nawarecka, S., Regulski, K., Krzyżak, M., Leśniak, G., Gurda, M.: Archives of Metallurgy and Materials 58 (2013) 927–930.

[7] Kluska-Nawarecka, S., Śnieżyński, B., Parada, W., Lustofin, M., Wilk-Kołodziejczyk, D.: Archives of Civil and Mechanical Engineering 14 (2014) 25-31.

[8] Gollapalli, M., Li, X., Wood, I., Governatori, G.: Lecture Notes in Computer Science 7121 (2011) 252-265.

Proceedings of the 72nd World Foundry Congress
21-25th May 2016, Nagoya, Japan

Technology transfer through industry-University-government collaboration

Meera K. Joseph[1] and Farouk Varachia[2]
[1, 2] University of Johannesburg, South Africa

This study examines technology transfer through collaboration between Government, industry and Universities. The Metal Casting Technology Station at the University of Johannesburg (MCTS-UJ) has been a reputed Technology Station and center of excellence for trouble shooting foundry problems and conducting research in South African foundries. The Department of Science and Technology in South Africa funds the Technology Stations Programme (TSP) through the Technology Innovation Agency (TIA). There are eighteen technology stations across the country at various universities and three of these focus in supporting the foundry industry. The mandate of the Technology Station is to develop human capital and support localization through technology and skills transfer. In this paper we discuss two case studies one on human capital development and the other on localization of technology with a partner university in Germany and a local manufacturer in South Africa. Both projects, Research and Innovation in Foundry Technology (RIFT) and inorganic binders for environmental friendly moulding material has benefited graduate students in acquiring the necessary skills and industry with new technology through the use of inorganic binders for environmental friendly moulding material which helped to reduce toxic fumes while mould making. This is an innovative idea used while mould making through MCTS-University of Johannesburg and Technische Universität Bergakademie Freiberg, Germany collaboration.

Keywords: *Technology transfer, RIFT programme, use of inorganic binders, Government-Industry-University collaboration.*

1. Introduction

"Technology transfer is the putting of information (technological innovation) to use, and the technology transfer process usually involves moving a technological innovation from an R&D organization to a receptor organization" [1]. Techakanont and Terdudomtham [2] opines technology transfer is of two types namely product and process transfer. In this paper we discuss how technology transfer happened in the context of RIFT programme and how the collaboration with industry led to the use of inorganic binders in one of the projects. Sodium silicate binders are used as inorganic binders in foundries as the curing time is low and it is environmentally beneficial [3]. Silicate binders are used to produce moulds and cores especially for castings [4].

2. Research methodology

2.1 Research strategy

We use case study research method as part of the methodology where it covered multiple cases and drew a single set of cross-case conclusion [5]. Convenience sampling [6] was used as it was easy to access and inexpensive to study. Open coding [7] was used to open up case study data related to technology transfer to uncover meanings.

2.2 Case studies

We discuss case studies that covered technology transfer through industry-University-government collaboration namely RIFT programme for localization and the Inorganic binder project. The two cases are discussed in the next section.

2.3 Data Analysis and Results

Case study 1: Technische Universität Bergakademie Freiberg (Germany)-MCTS-UJ (SouthAfrica) –Industry partner : **Inorganic binder project** . Industry partner (IP) initially used the hot box process and they were having problems with binder suppliers and environmental issues. MCTS-UJ collaborated with consultants from Technische Universität Bergakademie Freiberg, Germany to replace the hot box binder. The binder being used after consultation for the project trials is a sodium silicate based binder with slow and fast curing time hardeners. The binder worked well with a slightly increased curing time from the normal time of 10s up to a set up temperature of 160 degree Celcius. There is growing environmental regulations in South Africa. The binder helped to reduce toxic fumes and assisted in producing environmentally friendly moulding material. Examples of a product using Inorganic binders is shown in figure 1.

Fig. 1 Product using Inorganic binder project

Case study 2: RIFT programme – The Department of Science and Technology (DST), South Africa identified technology gaps and provided assistance to MCTS-UJ to provide an opportunity to four female students to study at the Technische Universität Bergakademie Freiberg, Germany. The students successfully completed their Dipl.-Ing in Foundry Technolgy in Germany at the end of 2014. In 2015 the graduate engineers were integrated into MCTS-UJ reporting structures that enabled MCTS to enhance its offerings in assisting foundries with skills and new technologies. One of the students on completion of the degree / RIFT programme has been assigned to the Sand Focus Area within the MCTS-UJ, where she contributes to routine checks and tests. She also assisted some local SA foundries with green sand analysis and helped to compare the quality of moulding sand before and after installation of their new sand equipment. One of the other graduates assisted the foundry in the localization of the striker bar casting, simulation and problem solving. The graduates have been involved in developing course material and training activities at the foundry industry.

3. Findings and discussions

The industry-University collaboration led to the use of inorganic binders in the project and helped the industry partner to reduce toxic fume. It also assisted in producing environmental friendly moulding material. The inorganic binder also brought some cost savings to the project. The Government-Industry-University collaboration based on RIFT programme in turn helped in the localization. After completing the degree as part of the RIFT programme the graduates were incorporated into the MCTS structures to strengthen the technical knowledge of local foundries and to provide additional capacity. The graduates used their knowledge and competence they gained in Germany to assist larger SME foundries with troubleshooting and technology transfer.

4. Conclusion and future recommendations

In this paper we discussed technology transfer through industry-University-government collaboration. RIFT for localization and the use of inorganic binders for environmental friendly moulding material. MCTS-UJ and Technische Universität Bergakademie Freiberg, Germany collaboration led to the use of inorganic binders to reduce toxic fumes and assisted in producing environmentally friendly moulding material. Through the RIFT programme the MCTS staff were able to set up a sand control system to improve the environment. In both programs the technology transfer through industry-University collaboration benefited the staff. The graduates used the knowledge they got in Germany to assist the local South African SME foundries with troubleshooting and technology transfer.

Acknowledgements

We are thankful to Mr. Farrai Banganayi-Station Engineer, Metal Casting Technology Station for the valuable information on inorganic binders.

References

[1] Rogers, E.M., Takegami, S., Yin, J. Lessons learned about technology transfer, (Technovation, 2001, 21) pp.253-261.

[2] Techakanont, K. and Terdudomtham, T. Evolution of inter-firm technology transfer and technological capability formation of local parts firms in the Thai automotive industry, (Journal of Technology Innovation, 2004, 12(2)) pp 151-183.

[3] Rabbii A. Sodium Silicate Glass as an Inorganic Binder in Foundry Industry, (Iranian Polymer Journal 10(4), 2001) pp. 229-235.

[4] Izdebska-Szanda, I. And Balinski A. New Generation of ecological silicate binders (Procedia Engineering 10, 2011) pp. 887-893

[5] Yin, R. K. (Case Study research design and method", 2009, Vol 5, Sage publication, Fourth edition) pp. 20.

[6] Patton, M., Designing qualitative studies (Qualitative Evaluation and Research Methods, 2002, Third Edition, Sage publication), pp. 241-242.

[7] Given L.M Open coding (The sage encyclopedia of qualitative research methods, 2008, Sage publications, Vol 2)

Proceedings of the 72nd World Foundry Congress
21-25th May 2016, Nagoya, Japan

Industry 4.0 and what it means to the foundry industry

Mr Mark Lewis
Omega Foundry Machinery Ltd, Morley Way, Peterborough, UK

The first thing to understand about Industry 4.0 is its not one technology but a combination of modern technologies combined to create a 'Smart Factory'. The 4.0 stands for the fourth industrial revolution which at first sounds extreme but when you start to look at the possibilities it's easy to see how these technologies can become real game changers.
Industry 4.0 is the brain child of the German government, and the train of thought is to create smarter more efficient manufacturing through the use of Smart Factories in the not too distant future. This will be achieved by various technologies communicating in a way that allow autonomous running of the factory and processes.
The big question is how can we utilise these new technologies within the foundry industry and what are the benefits?

Keywords: *Industry 4.0, Technology, Internet of Things, Automation, Cloud*

1. Introduction

In our everyday lives we are becoming more and more reliant on technology, with smarter cars keeping us safe through to smart phones keeping us connected. If you consider the things we take for granted in our everyday lives like streaming music or films, saving documents to the cloud, or remotely connecting to the office these are all using state of the art technology with one important link, the internet. The high speed internet of today is allowing a lot more data to be transferred remotely and giving us a lot more control over various aspects of our lives. And this is where industry will start to see massive leaps forward within the workplace.
Businesses are starting to utilise this connectivity in many ways, from automatic material ordering through to cloud based software control. The premise behind Industry 4.0 is to take this one step further by connecting not just one machine but all the factory so it communicates as one entity.
To achieve this there's one more key element needed and that is the Industrial Internet of Things and this boils down to creating smart devices/machines that communicate with each other and the outside world.

2. Foundry Applications

If we were to take all these technologies and look at how they can be utilised in a foundry for example one using silica sand monitored by a smart system. When the sand drops below the reorder level the Smart Factory automatically places an order on the sand supplier for the required quantity of sand. So far this is simple but its reactive not proactive so to take it to the next level if that same system was tied into the production control system within the foundry and using data from material consumptions could predict the sand, chemical, and consumable requirements for the coming week or month it could then have orders ready with suppliers for when its needed. Of course whilst all this is happening the relevant person within the organisation is kept informed via notifications and can easily see what is happening via any device with a web browser and internet connection from anywhere in the world.
This is a very simple example of what could easily be achieved and if the rest of the foundry was automated and connected you start to get an understanding of how far reaching Industry 4.0 can truly be.

3. Today's Foundry

We may be some years away from a truly automated foundry but the technology is already available to achieve a lot of the benefits we will see in the future.
As an example you can already monitor your machinery in your foundry remotely via cloud based control systems giving complete access to the data on the machine and if needed remote control of certain elements. Also using technologies like RIFD we are able to automate control of various machines for example on sand mixers its possible to deliver the exact sand recipe and quantity along with fully automatic filling sequence, this level of control can reduces wastage and improves overall casting quality. Because this process is automated it becomes easier to

record production information and material usage as it is automatically collated and stored.
Add the ability to then access this data remotely on PC, table or phone from anywhere in the world and we can see the future foundry is not so far away.

4. Benefits and Future advantages

With less time spent doing the mundane work and by removing the guess work from the equation it's easy to see the efficiency gains possible. In Germany they are talking about average productivity gains of 5-8% with some sectors seeing up to 20% and potential of Industry 4.0 adding over $14 trillion to the global economy in the next 15 years.
Foundries of the future will need to be reactive to the changing market place and by investing in Industry 4.0 will give them competitive edge. Being more efficient and improving productivity but at the same time being able to be more reactive to their customer needs as these systems will give huge flexibility allowing more affordable short production runs.

5. Pitfalls and Cybersecurity

Obviously there are disadvantages to any system and Industry 4.0 doesn't come without its issues. Firstly the systems are very dependent on connectivity and the internet, if the factory were to lose its internet connection it would have no means of communicating with the outside world. Secondly the risk of cybercrime and hacking become even more apparent when the whole plant is connected to the internet.
These issues are easily overcome with clear planning and preparation, the plant must be able to continue operating if connectively is lost and the systems also need to have robust security and protection. When undertaking the task of installing a smart foundry it's important to understand all the limitations and minimize their impact.
Another point worth considering is the supply chain around the foundry, there is no point creating an automated process if the current supply chain are not on board or capable of working with Industry 4.0. There is nothing stopping businesses implementing Industry 4.0 in small sections of the foundry as this gives a clear and steady path to implementation, but again planning is the key element and choosing the correct partners to work with will be paramount.

6. What's next

It will be many years before smart foundries become common place but that does not mean it's not important now to understand what the benefits are and what can be done to prepare for the future. It is possible to retro fit smart technology to old plant so we don't have to wait for new factories and equipment to take advantage of the Industrial Internet of Things. As devices and equipment in our factories get smarter so must we on how we use the connectivity made available to us.
The possibilities are endless and by simply integrating smarter open technologies now it will make the future of these foundries easier to upgrade to the Industry 4.0 ethos.

7. Final Goal

The final goal is a Foundry where the customer orders are placed via a centralised control system and using integrated MRP/ERP systems the foundry manages its supply chain and production needs automatically. Machines communicate to each other and the supply chain placing orders for raw materials and planning production needs to meet lead times. The equipment then works together in the most efficient manor to achieve the customers requirements.
This doesn't mean the end of human involvement but does necessitate a different skill sets, so it's important to have a work force able to understand and deal with this advance in technology.
As technology has changed our everyday lives away from work it is now time to see how it can improve our working environments too. We all need to get a better understanding of what can and cant be done with Industry 4.0 so we can make the transition as smooth as possible.

Proceedings of the 72nd World Foundry Congress
21-25th May 2016, Nagoya, Japan

Development of Al-Li Alloy Castings Excellent in Formability.

Seiji Saikawa[1], Susumu Ikeno[2] and Koichi Komai[3]

[1] Graduate School of Science and Engineering for Research, University of Toyama, Toyama-shi, Toyama 930-8555, Japan

[2] Hokuriku Polytechnic College, Uozu-shi, Toyama 937-0856, Japan

[3] Tanida Ltd., 28-2, Higashikagatsume-machi, Kanazawa-shi, Ishikawa 920-0209, Japan

The suitability of coated material for a casting mold used for producing Al-Li alloy excellent in formability was investigated. Visual inspections of the casting surfaces of Al-Li alloys indicated that there were large gas bubble defects due to melt-mold interaction when those were cast into conventional molds consist of SiO_2 sand and phenol resin binder. XRD analysis showed that the main product of the interaction was Li_2O and Li_2CO_3. Except metal mold castings, the quality of Al-Li castings was order of CaO, Al_2O_3, TiO_2, BN, MgO and no-coated sand mold. This order mostly agreed with the standard generation free energy of oxide, except in MgO coated mold.

Keywords: Al-Li alloy, casting, mold, reaction, formability

1. Introduction

In the aerospace , automobile and motorcycle fields, high-strength aluminum alloys are widely used for body, structural, and engine components, where weight reduction is needed to improve movement performance and fuel efficiency[1]. Al-Li alloys have excellent characteristic with high specific strength, high modulus and lightweight property, it is used for aerospace-related fields in particular[2].

For casting of high quality Al-Li alloy excellent in formability, it is important to control the process of melting , refining, molding and casting . However, since the Al-Li alloy are highly active and hard to cast, there has been limited research on those process of Al-Li alloy for casting[3, 4].

In this study, the suitability of molds used for producing Al-Li alloy castings excellent in formability was investigated.

2. Experimental procedure

Pure Al and ingots of other elements were melted at 973K in an electrical furnace. When the temperature of molten metal was up to 1003K, pure Li added into molten metal under innert atmosphere. After all elements were melted, molten metal was degassed by argon gas bubbling. Fig.1 shows schematic drawing of the mold for casting. Temperature of molten metal was contorolled to 973K, it was cast into various molds with a Y-shaped cavity.

Table 1 shows detail of molds using for Al-2.5%Li casting. Five kinds of oxide material, such as BN, TiO_2, Al_2O_3, MgO and CaO coated to inner surface of sand molds, and then those molds were heated at 473K for 4.8ks before cast .

The soundness and quality of castings were evaluated by visual inspection of cross section of castings, and amount and kind of reaction products occurred at the surface of the castings using by X-ray Diffraction method(Cu-K,40kV,40mA).

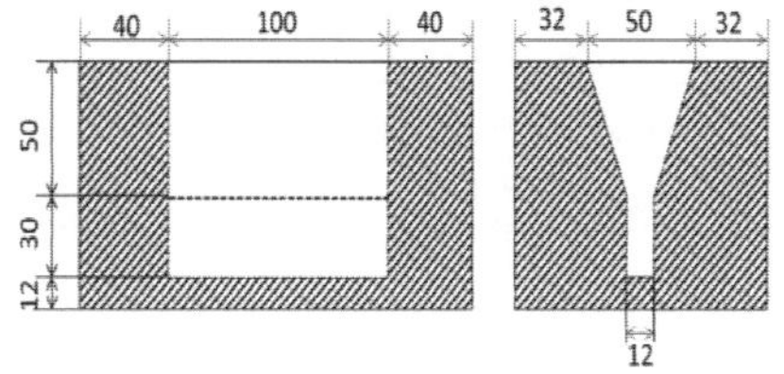

Fig.1 Schematic drawing of the mold for casting.

Table 1 Detail of molds using for Al-Li casting.

Code	Mold Material	Binder	Coated Material	Heating of Mold before casting
Sand-mold	sand	Phenol Resin	---	473K, 4.8ks
BN-Coated			BN	473K, 4.8ks
TiO_2-Coated			TiO_2	473K, 4.8ks
Al_2O_3-Coated			Al_2O_3	473K, 4.8ks
MgO-Coated			MgO	473K, 4.8ks
CaO-Coated			CaO	473K, 4.8ks
Metal-mold	metal	---	BN	453-473K

3. Results and discussion

The sectional appearance of the central part of Al-Li castings are shown in Fig. 2. Many porosity defects occured in the inside of all castings except metal mold one's. MgO coated, sand mold and BN coated mold have lowest casting quality, because those had many porosity defect. CaO, TiO_2 and Al_2O_3 coated molds had relatively slightly good quality, because those defects were rather less.

Fig.3 shows X-ray diffraction result obtained from the castings surface. Peaks from aluminum matrix was detected by all castings. In addition to these, diffraction peaks of reaction products, such as Li_2O, Li_2CO_3, $LiOH\text{-}H_2O$ and γ-$LiAlO_2$ peaks were detected in all castings. In reference to peak strength and the kind of the product, we evaluate the soundness of each castings. As above results, the quality of the castings was order of CaO, Al_2O_3, TiO_2, BN, MgO and no-coated sand mold. This order mostly agreed with the standard generation free energy of oxide, except in MgO coated mold.

Fig.4 shows appearance of Al-2%Li-2%Cu -1.5%Mg-0.10%Zr alloy plate hot rolling in 90%. Before of the rolling, slab was cast by ceramic coated mold. The rolling temperature was 473-573K, and rolling passes was 20 times. Because the crack was observed in neither part, it was proved that Al-Li alloy cast by before rolling was superior in a formability.

4. Conclusions

Good quality Al-Li alloy castings are obtained when molten metal are refined by Ar degassing, and then pouring (cast) into molds which surfaces are coated with cermics materials, such as CaO, Al_2O_3 and TiO_2.

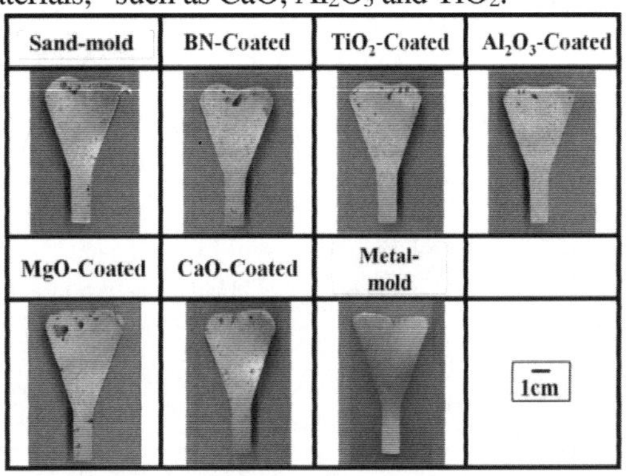

Fig.2 X-ray diffraction result obtained from the casting surface of Al-2.5%Li alloys.

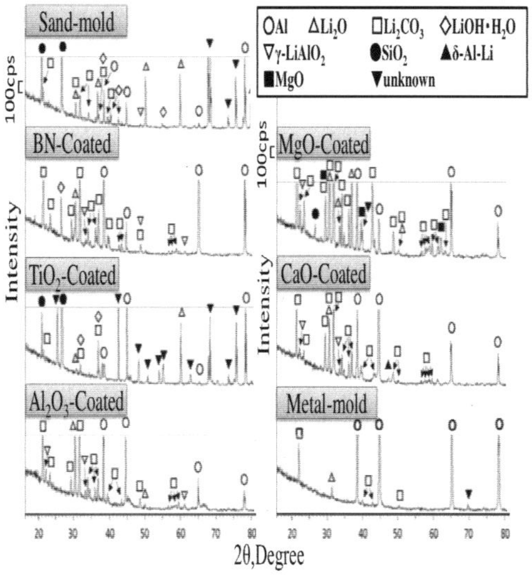

Fig.3 X-ray diffraction result obtained from the castings surface of Al-2.5%Li alloys.

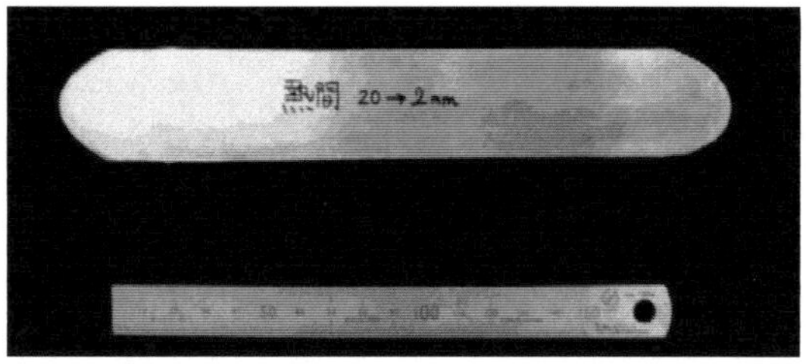

Fig.4 Appearance of Al-2%Li-2%Cu-1.5%Mg -0.10%Zr alloy plate hot rolled in 90%.

Acknowledgements

This work was supported by KAKENHI (15K06523).

References

[1]T.Ishikawa: Japan Institute of Light Metals. 62(2011), 469–474.

[2]T. Nakamura: Japan Institute of Light Metals. 62 (2012), 249–256.

[3]S. Saikawa, S. Sugioka, K. Nakai, Y. Sugiura and A. Kamio: Japan Institute of Light Metals. 41 (1991),596-601.

[4]]S. Saikawa, S. Sugioka, K. Nakai, Y. Sugiura and A. Kamio: Japan Institute of Light Metals. 43 (1993), 630-634.

O-111

New concept cracking tubes for ethylene plant in a petrochemical industry

Makoto Hineno[1], Kunihide Hashimoto[2], Shigemoto Matsueda[3], Yohei Enjo[3]
1 Materials Development Group, Materials Center, KUBOTA Corporation, Japan
2 Engineering & QA, Kubota Saudi Arabia Company, LLC, Kingdom of Saudi Arabia
3 Materials Technology Dept., Materials Division, KUBOTA Corporation, Japan

Alumina-forming materials are next generation materials used in petrochemical industry for ethylene cracking tubes. Carbon deposited on internal surface of the cracking tube in the process of thermal cracking of hydrocarbon during operation causes material degradation problem due to significant carburization and blockage caused by decrease of the tube inner diameter. Therefore, authors developed the technology to form dense alumina film with high purity on the internal surface of cracking tubes by alloy composition and optimizing of the generation of oxide film for material protection. This material has the ability to generate a stable aluminum film at high temperatures, it also possesses far greater resistance to oxidation and carburization. Tests performed in the recent years by domestic and international petrochemical manufacturers with actual furnaces confirmed excellent performance of the material and it is focus of attention from the industry.

***Keywords**: Ethylene, Cracking tube, Carbon, Alumina, Thermal decomposition, Oxide film*

1. Introduction

The cracking tubes for the ethylene plant in a petrochemical industry are used in high temperature condition of more than 1000°C. During the operation, coke formation occurs at tube inside by thermal cracking of hydrocarbon. This coke formation at high temperature causes carburization at tube material. Therefore, the heat-resistant alloys that have good oxidation resistance and carburization resistance are applied as cracking tube material. Generally, a chromium oxide film to be formed on the surface during operation prevents carburization of base metal, but a chromium oxide film isn't stable in the severe high temperature. Therefore, we focused on an aluminum oxide film which is more stable at the severe condition. Based on the results of the experiment, we have developed the new concept cracking tube that generate a high purity alumina oxide film on the inside surface.

In addition to the investigation of the oxide film produced on the test material described above, its oxidation resistance and carburization resistance were evaluated and good results were observed.

2. Experimental method

2.1 Material under test

The composition of conventional cracking tube material is 35Cr-45Ni. On the other hand, development material is 25Cr-35Ni-Al. Authors tested materials 35AF and 45A shown in the table 1. Samples were manufactured by using a centrifugal casting process, and test sample was taken for analysis from the thick central part of the casted body. Chemical composition analysis was performed by using a spark optical emission spectrometer.

Table 1 Component of the material(mass%)

	C	Cr	Ni	Al	Fe	other
Development material (35AF)	0.4/0.6	22/28	29/37	2.0/4.0	Bal.	Add
Conventional material (35Cr-45Ni:45A)	0.4/0.6	30/35	40/46	<0.5	Bal.	−

2.2 Oxide film analysis

Analysis of the components and composition of the oxide film produced on test material was performed by observation of its cross-section with using EPMA. To identify internal structure of the sample it was analyzed from surface to the center using a XDR method.

2.3 Oxidation resistance test

To evaluate the oxidation characteristics sample was subjected to 10 hours cyclic oxidation tests. During the cyclic oxidation test sample was heated from room temperature to 1050°C in air atmosphere, subjected to 10 hours hold and cooled to the room temperature using air cooling. This process was repeated 50 times. Evalution based on weight change measurement and surface observation.

2.4 Carburization resistance test

Carburizing properties were evaluated using gas carburizing tests.During the gas carburization test sample was held in a mixed CH_4 and H_2 atmosphere in a gas ring furnace at 980°C for 200 hours, and carbon weight increase on outer surface of the sample was measured in arbitrary points.

3. Test results and considerations

3.1 Oxide film analysis

Film observation results are shown on the Figs 1, 2. The result of mapping analysis by EPMA conducted to determine the composition of the oxide film showed that oxide film on the surface of the conventional material based from Cr oxide, while the oxide film on the developed material contained Al oxide (Fig. 1). Further results of XRD measurements showed that the oxide film of the new material composed from α-alumina (Fig. 2).

It can be surmised that for each of the oxide films the generated oxide film is grown by internal diffusion of oxygen. Thus, we can suppose that because α-alumina oxide is denser than Cr oxide, it suppresses the internal diffusion of oxygen, and alumina oxide film became thinner.

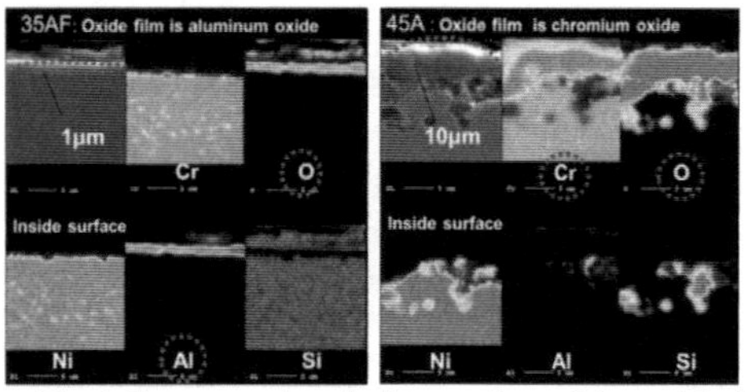

Fig.1 Cross-sectional observation of the oxide film

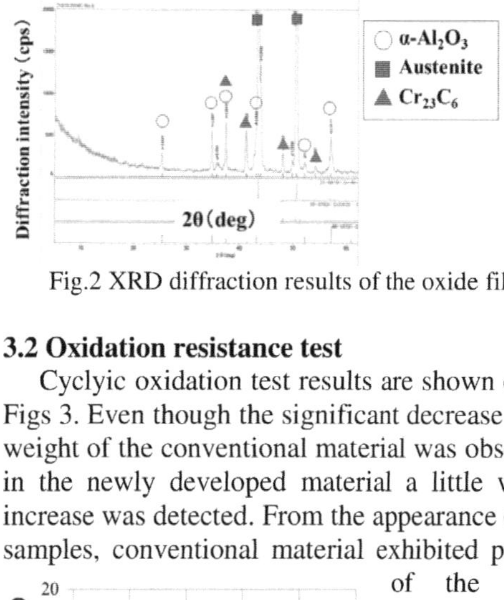

Fig.2 XRD diffraction results of the oxide film

3.2 Oxidation resistance test

Cyclyic oxidation test results are shown on the Figs 3. Even though the significant decrease in the weight of the conventional material was observed, in the newly developed material a little weight increase was detected. From the appearance of test samples, conventional material exhibited peeling of the oxide film, but new material showed almost no peeling. We considered the peeling mechanism as follows. First, the oxide film is swelled by intermittent oxidation. Next, on cooling, a shock on the oxide film is given due to a difference in thermal stress of the base material. These occurs repeatedly, and the oxide film is peeling.

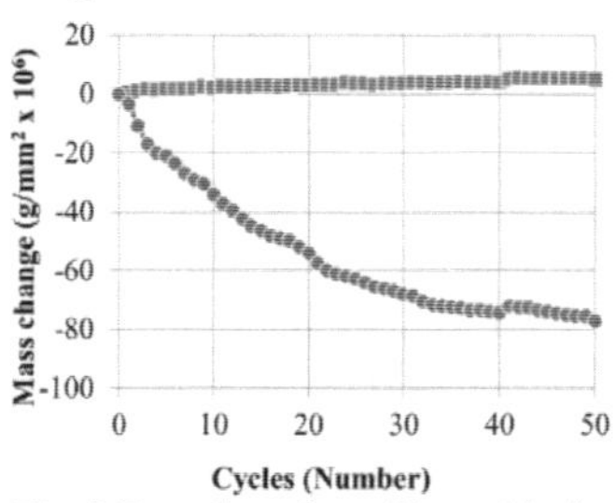

Fig.3 Results of cyclic oxidation test

3.3 Carburization resistance test

Carburization resistance test results are shown on the Figs 4. From the results of gas carburizing test, while the conventional material showed carburization from the surface, in the developed material almost no carburization was found (Fig. 4). It can be explained that because Cr oxide film generated on surface of the conventional material is less dense than the Al oxide film produced on surface of the developed material, carbon penetrates the Cr oxide which causes carburization. Because Al oxide prevents carbon penetration, carburization does not occur. Additionally, whereas in move severe conditions carburization in the conventional material accelerated, in new material carburization speed did not change.

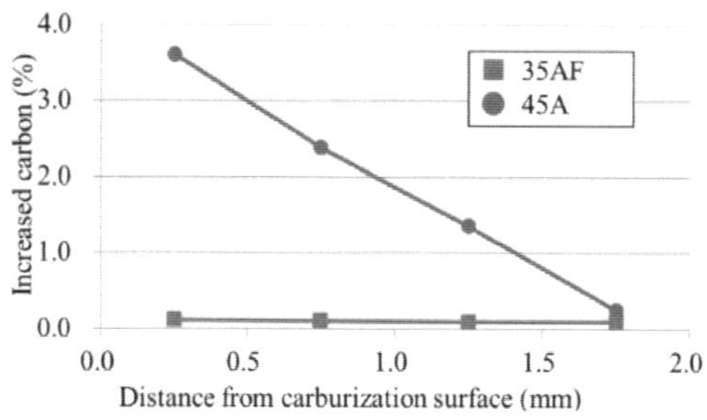

Fig.4 Results of gas carburization test

4. Conclusion

By analyzing the oxide film generated on the internal surface of cracking tubes and performing oxidation resistance and carburizing resistance tests, the following results were obtained.

(1) From the observation of the oxide film generated on the developed material, the film was identified as α-alumina by XRD measurements.

(2) Oxidation test results showed that Cr oxide film on conventional material becomes unstable at the high temperature exhibiting swelling and peeling. On the other hand, Al oxide film on the developed material showed almost no peeling and kept stability at the high temperature.

(3) Carburization test results showed that Al oxide film produced on the developed material performs well at the test temperature and indicates much less carburization comparing to Cr oxide film generated on the conventional material.

Proceedings of the 72nd World Foundry Congress
21-25th May 2016, Nagoya, Japan

A vertical full-continuous casting method for aluminum alloy billets

Kenji Niwa[1], Takeshi Fujita[1], Yuichi Hasegawa[1], Kiyoshi Katayama[1],
Shinya Mizuno[2], Mikiya Nozaki[2], Noriyuki Ueno[2]
[1]Nippon Light Metal Co., Ltd., Tokyo, Japan
[2]Toyota Motor Corporation, Aichi, Japan

The conventional methods used for casting aluminum alloy billets have been the vertical semi-continuous casting method and the horizontal continuous casting method. We have developed a vertical full-continuous casting method with the combined benefits of the aforementioned methods. Our method improves the production yield rate by approximately 15% compared to the vertical semi-continuous casting method and by approximately 10% compared to the horizontal continuous casting method.

Keywords: *Aluminum alloy, Continuous casting, High-speed casting, Billet for forging*

1. Introduction

The vertical semi-continuous casting method and the horizontal continuous casting method have been used to cast aluminum alloy billets. The vertical semi-continuous casting method suffers from poor productivity. Because the length of the cast billet is limited to the depth of the pit, the casting must be interrupted for each predetermined length. In addition, because the ratio of end materials in non-stationary regions during the initial and final stages of casting is large, the cutting yield of this method is poor. However, the peeling yield of this method is good because of the uniform casting structure of the billet surface, which is a consequence of the lack of any influence of gravity on the distribution of the cooling water. The horizontal continuous casting method features good productivity; the casting is not stopped whenever a billet having a predetermined length is cast because the cutting and casting of the billet are performed simultaneously. The peeling yield of this method is poor because the surface structure of the billet is likely to become heterogeneous; this heterogeneity is a consequence of the difference in the amount of cooling water at the upper and lower surfaces of the billet, which, in turn, is caused by the effect of gravity. Therefore, incorporating the advantages of both technologies into a vertical full-continuous casting method would result in improvements in the yield, quality, and productivity compared to the other two methods. However, realization of such a casting method has been thought to be impossible because the cooling-water facilities are complex and the counterplan of vibrations at the time of cutting is inherently difficult. In this work, we aimed to substantially improve the production of aluminum alloy billets by implementing the vertical full-continuous casting method by developing specialized equipment.

2. Development of the vertical full-continuous casting machine

A schematic of the vertical full-continuous casting machine is shown in Fig. 2. The machine comprises five primary components:

(1) High-speed casting mold: the casting speed when using this mold is 2–3 times more than that when using a general mold. Thus, this mold provides versatility in production and improves productivity.

(2) Guide for suppressing the bending: vibrations generated during cutting and pulling of the billet can be suppressed by using this guide, thereby suppressing the bending of the billet.

(3) Seal: this is placed at the bottom of the water tank in order to separate other equipment from the cooling water of the casting sections.

(4) Pulling machine: this device controls the casting speed.

(5) Synchronized cutting machine: this machine cuts the billet while moving synchronously with the casting speed to cut the billet exactly to the standard product length.

3. Production results

Since the beginning of production, our casting machine has produced ≥10,000 tons of billet in total. This technique can be used to cast billets stably and continuously for >100 h, which is a more than 50-fold increase compared to the continuous casting time achieved with the conventional vertical semi-continuous casting method. The production yield of this method represents a 15% improvement over that obtained using the vertical semi-continuous casting method and a 10% improvement over that obtained using the horizontal continuous casting method. Furthermore, in our developed casting method, variation of the mechanical properties of the billet is reduced. Presumably, this reduction is because of the microstructure of the billet becoming fine and homogeneous as a result of the stable high-speed casting. As an example, the Charpy impact values and variations for 6061 alloy billets cast using our method and each of the conventional methods are shown in Fig. 4.

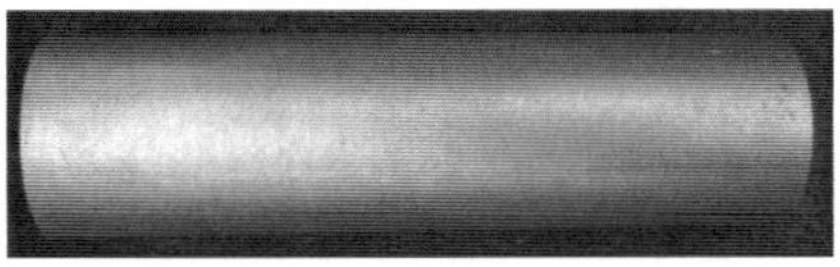

Fig. 1 An aluminum alloy billet.

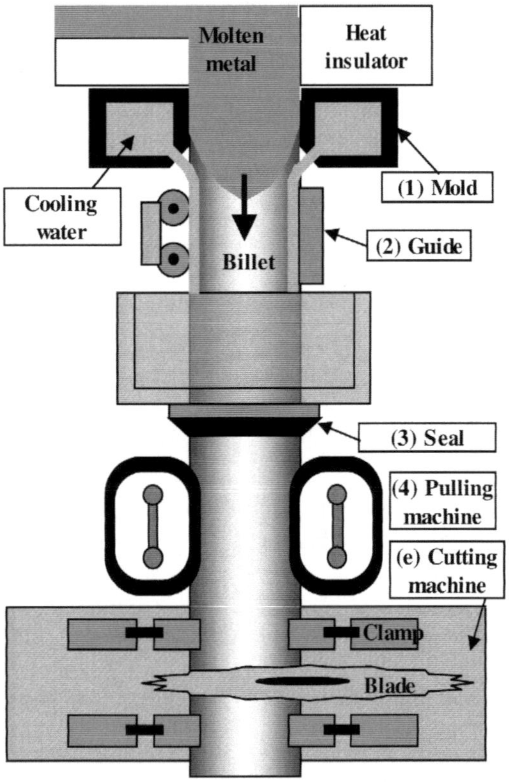

Fig. 2 Schematic of the vertical full-continuous casting machine.

Table 1 Comparison of the casting method developed in this work and the conventional casting methods.

Method	Cutting yield	Peeling yield	Quality	Productivity
Vertical semi-continuous	bad	good	good	bad
Horizontal continuous	good	bad	intermediate	good
Developed	good	good	good	good

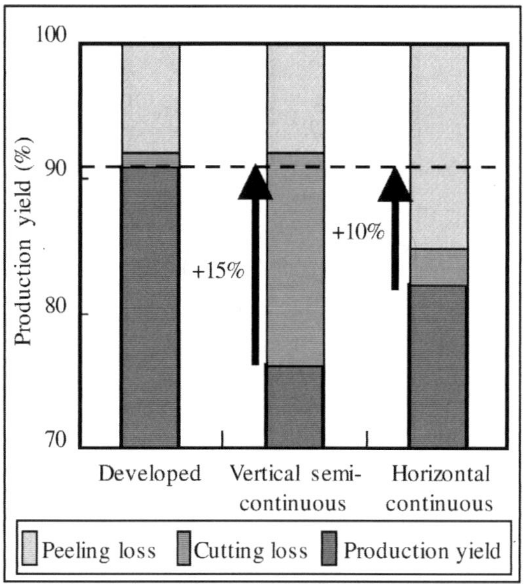

Fig. 3 Comparison of production yields of the casting method developed in this work and the conventional casting methods.

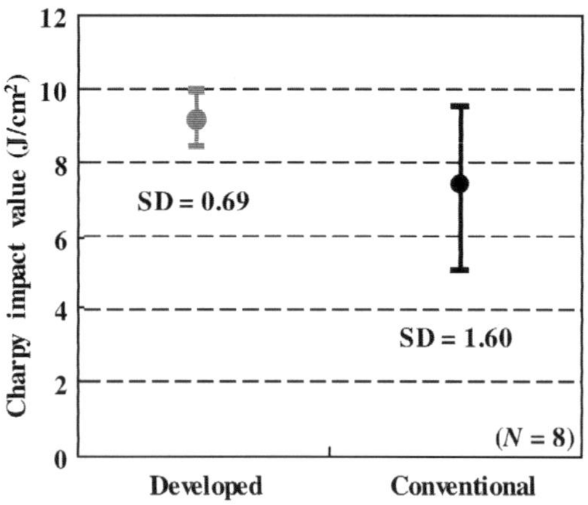

Fig. 4 Comparison of Charpy impact values for 6061 alloy.

Innovation of a roll caster for aluminum alloy

Toshio Haga[1]
[1] Osaka Institute of Technology, 5-16-1 Omiya Asahiku Osaka city 535-8585, Japan

In this study, we propose three prototypes of roll casters, each with innovative features: a single roll caster equipped with a scraper (SRCS), a vertical type high-speed twin roll caster for aluminum alloy (VHSTRC), and a vertical type tandem twin roll caster (VTTRC). The scraper of the SRCS improves the flatness of the free solidified surface of the strip, and the cast strip exhibits no center line segregation. The VHSTRC can cast Al-11%Si strip at a speed of 60 m/min. Furthermore, a cup test was conducted for this strip, which demonstrated that a limiting drawing ratio of up to 1.9 can be obtained. Finally, the VTTRC can cast three-layer clad strips. In this system, one VHSTRC is mounted on the other VHSTRC to cast the clad strip. The clad strips can be cast without mixing of the two kinds of aluminum alloys.

Keywords: *single roll caster, twin roll caster, clad strip, scraper*

1. Introduction

Although conventional twin roll casters for aluminum alloy (CTRCA) provide advantages such as rapid solidification and a straightforward casting process, their disadvantages are center line segregation and a low casting speed. A single roll caster equipped with a scraper (SRCS) and a vertical type high-speed twin roll caster for aluminum alloy (VHSTRC) were proposed to overcome these disadvantages [1,2]. A straightforward process is critical in the fabrication of clad strips. Therefore, a vertical type tandem twin roll caster (VTTRC) was proposed [2]. In this paper, the characteristics of the three innovative twin roll casters are shown.

2. Single roll caster equipped with a scraper

A schematic illustration of the SRCS is shown in Fig. 1 [1]. Semisolid metal on the free solidified surface is scribed by the scraper, and the scribed surface is shown in Fig. 2. As shown in the figure, the scraper produced a flat free solidified surface. The thickness distribution in the lateral direction of an as-cast 5182 strip is shown in Fig. 3, and was found to be neither convex nor concave. Cross-sections of the as-cast strip and the strip after cold rolling and annealing are shown in Fig. 4. No significant difference was seen in the microstructure between the roll contact surface and the scribed surface. In addition, no center line segregation was observed.

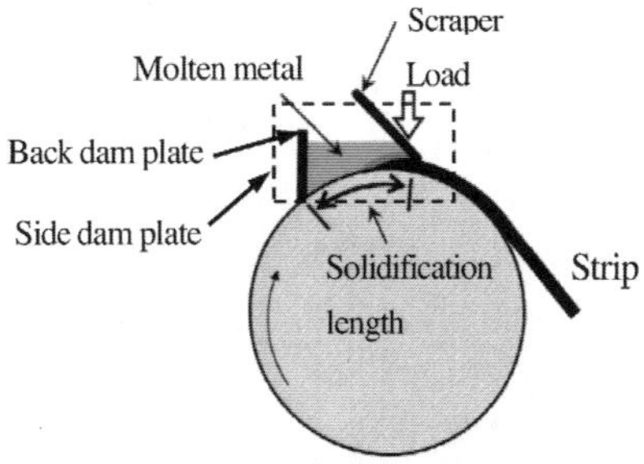

Fig. 1 Single roll caster equipped with a scraper

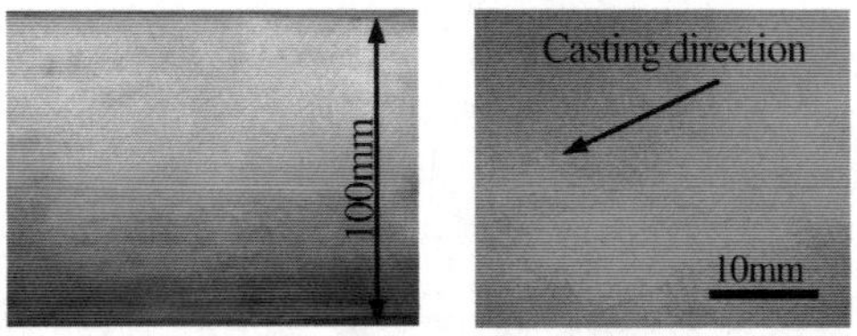

Fig. 2 Scribed surface of as-cast 5182 strip

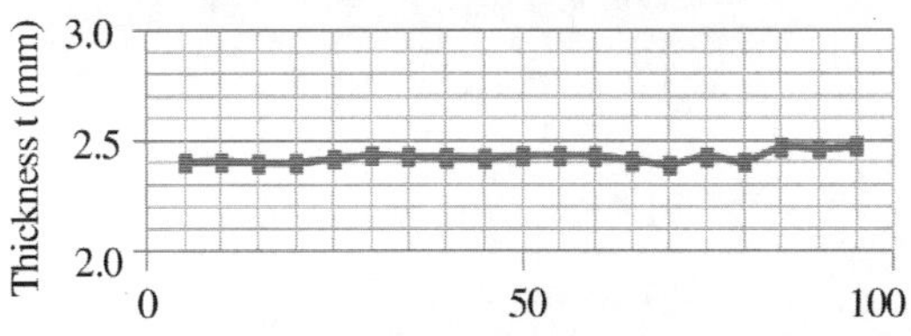

Fig. 3 Thickness distribution of as-cast 5182 strip

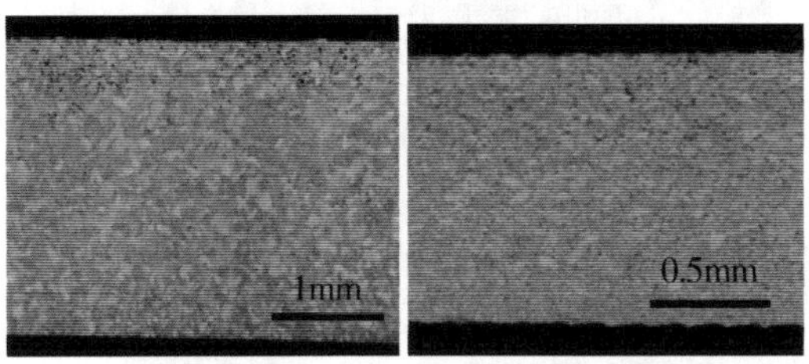

(a) as-cast strip (b) cold rolled and annealed

Fig. 4 Cross section of 5182 strip

3. Vertical type high speed twin roll caster

A schematic illustration of the VHSTRC is shown in Fig. 5 [1]. The VHSTRC uses hydrostatic pressure provided by a metal head to improve the contact between the molten metal and the roll. Additionally, copper rolls are used in this system. A very fine microstructure was observed for an as-cast Al-11%Si strip cast at 60 m/min, as shown in Fig. 6. The eutectic Si was globular and smaller than 2 μm, and rapid solidification was attained. A bending test and a deep drawing test were conducted on the cold rolled and annealed 1-mm-thick plate, and the result is shown in Fig. 7. Loop bending at a 0.5 mm radius of curvature was achieved. Additionally, we performed a deep drawing test, which demonstrated that a limiting drawing ratio (LDR) of 1.9 could be achieved.

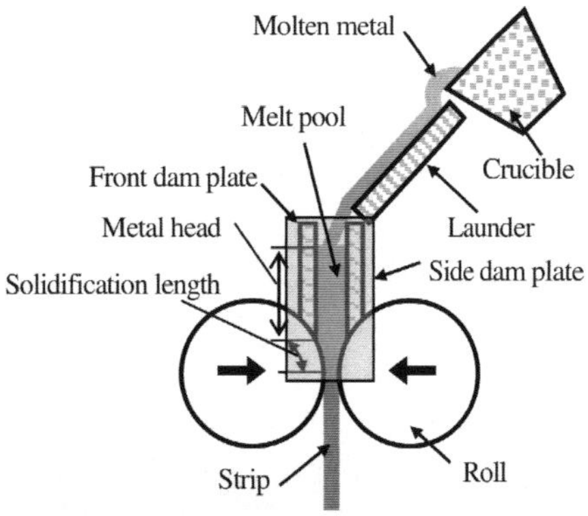

Fig. 5 Vertical type high speed twin roll caster

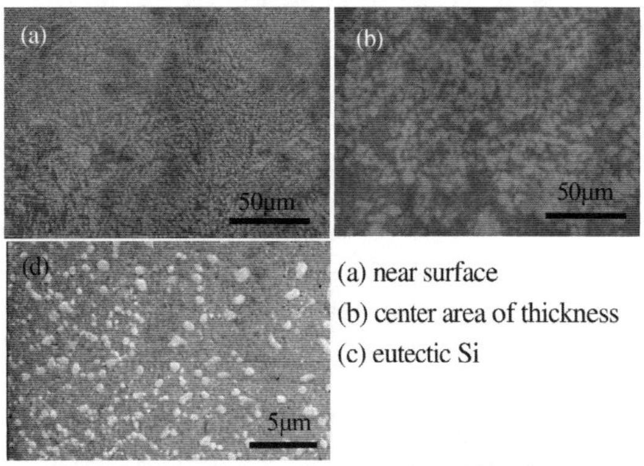

Fig. 6 Microstructure of as-cast Al-11%Si strip

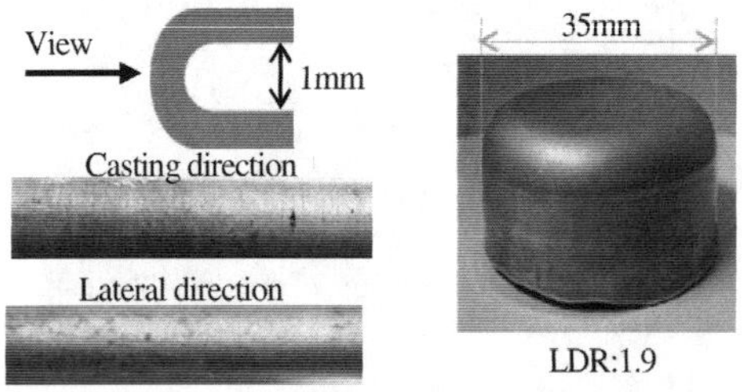

Fig. 7 Result of bending test and deep drawing test.

4. Vertical type tandem twin roll caster

A schematic illustration of the VTTRC for producing clad strips is shown in Fig. 8 [1,2]. A base strip and overlap strips are cast by the upper and lower twin roll casters, respectively. The cross-section of the as-cast three-layer clad strip cast at 30 m/min and the line analysis of Si at the interface are shown in Fig. 9. The interface between the strips was clear, and the Si in 4045 did not diffuse into 3003. A continuous bending test was conducted, as shown in Fig. 10, as an easy investigation method of the bonding conditions. It is clear from Fig. 10 that the overlay strip was strongly bonded to the base strip, because cracks did not occur at the interface between the 4045 and the 3003 alloys.

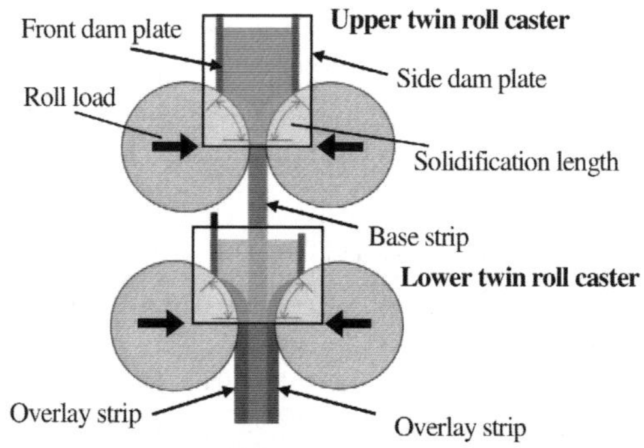

Fig. 8 Vertical type tandem twin roll caster for clad strip

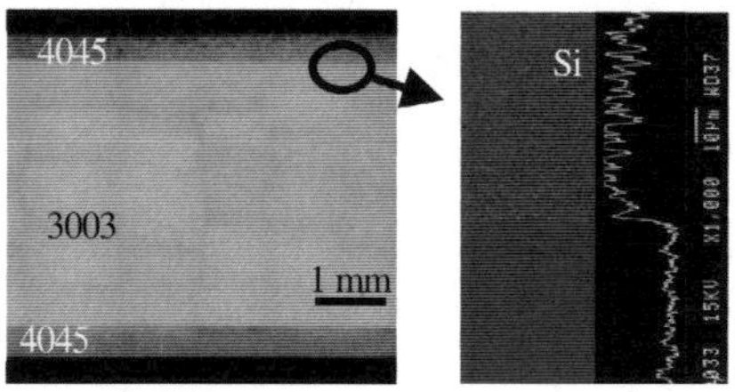

Fig. 9 Cross section of as-cast clad strip and result of line analysis of Si at the interface

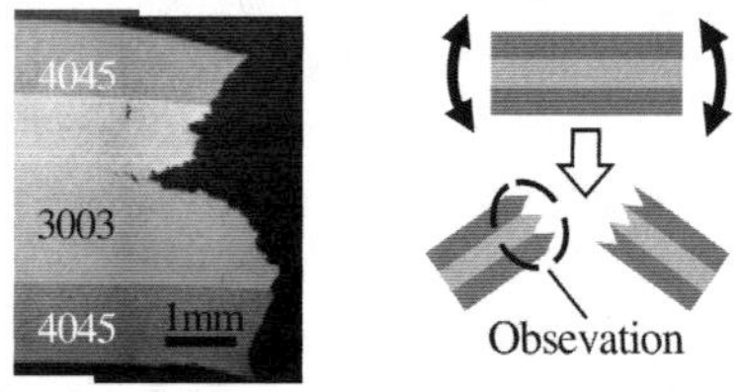

Fig. 10 Broken area after continuous bending test

5. Conclusion

Selected characteristics of innovative roll casters are discussed in this paper. These new characteristics are not features of the conventional twin roll casters typically used with aluminum alloys.

References

[1] T.Haga: Journal of Achievements in Materials and Manufacturing Engineering 43(2010) 393-402.
[2] T.Haga, R.Nakamura, S.Kumai and H.Watari: ibid. 62(2013) 36-44.

Proceedings of the 72nd World Foundry Congress
21-25th May 2016, Nagoya, Japan

Development of Lead Free Bronze with Sulfide Dispersion for Sliding Members

Tomohiro Sato[1], Yoshimasa Hirai[2] and Takeshi Kobayashi[3]
[1] Kansai University, Department of Mechanical Engineering,
3-3-35 Yamate-cho, Suita-shi, Osaka, Japan
[2] Kurimoto,LTD. Technology Development Division,
2-8-45 Shibatani, Suminoe-ku, Osaka, Japan
[3]Kansai University, Professor emeritus

Cu-alloy has been used for a bearing material for a long time because of the good running in properties for steels as the counter materials. These Cu-alloy were selectable for the applications. For example, high strength brass castings and phosphorus bronze castings were used for high strength domain, lead bronze castings was used for anti-seizure domain. For preventing scoring and seizure, it was often combine with a solid lubricant. In this study, tribological properties of the sulfide dispersed bronze were investigated. As a result, seizure was prevented only sulfide bronze under some test conditions. It was estimated that the dispersed sulfide restricted the transfer of the Cu-alloy elements to the mating steel surface.

Keywords: *Cu-alloy, bronze, Sulfide, Friction, Seizure*

1. Introduction

Lead was well-known solid lubricant and lead bronze was useful material for sliding bearings. However, to protect the environment and human health, various regulations have been enacted. These regulations or trends have an impact on industrial manufacturers. Many industries have developed new materials as substitutes for lead in solid lubricants. For example, lead bronze that has been used for slide bearings, has been replaced by a sulfide dispersed bronze (sulfide bronze). Tribological properties such as friction and wear, were investigated on casting [1] and sintering [2] materials comparing with those properties on lead bronze. As a results, sulfide bronze showed the superior friction properties than lead bronze. However, effects of sulfides on anti-seizure is not clarified.

In this study, tribological properties of the sulfide dispersed bronze were investigated. Especially, effects of sulfides on anti-seizure is focused. For the observation of effects of sulfide, sulfide bronze and phosphorous bronze were prepared.

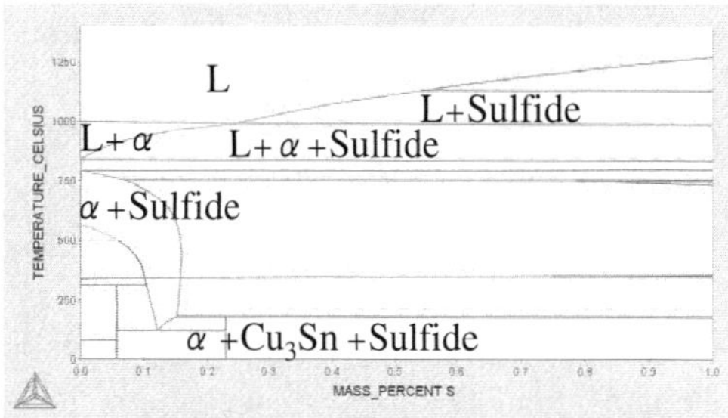

Fig. 1 Calculated phase diagram of sulfide bronze

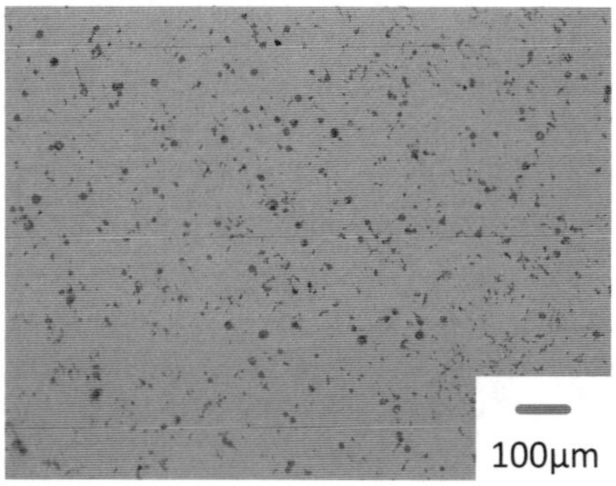

Fig. 2 Microstructure of sulfide dispersed bronze

2. Experiments

2.1 Material

At first, to determine the suitable sulfide contents, phase diagram ware calculated based on CALPHAD (Calculation of Phase Diagram) method on Cu-Sn-Fe-S system as shown in Fig.1. The content of Fe was determined under 1.3 mass% contents because of preventing crystallization of α'-phase of Fe in the Cu-Sn alloy. As a result, it was clarified that sulfide was dispersed in the Cu-alloy because sulfide under 0.25 mass% contents in the alloy was crystalized after crystallization of α-phase Cu. On the other hand, the sulfide remained in the matrix under 0.60 mass % which clarified experimentally. As shown in Fig.2, sulfide was well dispersed in the bronze matrix. Moreover, the sulfides were trapped among the

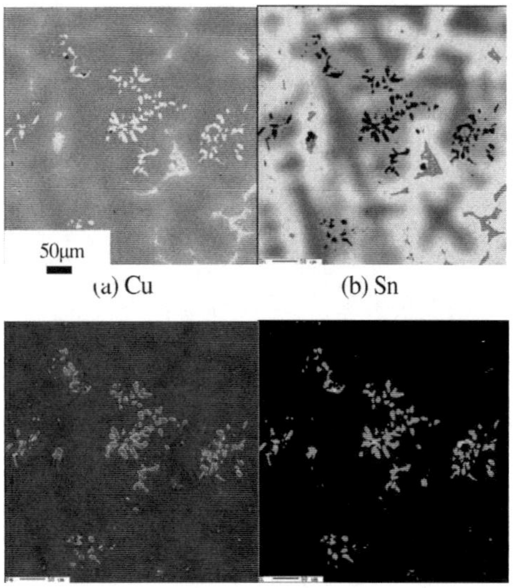

(a) Cu (b) Sn

(c) Fe (d) S

Fig. 3 Element mapping by EPMA

dendrites of Cu-Sn alloy. These results are observed by optical microscope and EPMA (Electron Probe Micro Analyzer) shown in Fig.3.

2.2 Experimental procedure

Effects of sulfide dispersion in the Cu-Sn alloy were experimentally evaluated using a 3 balls type ring on disc testing apparatus in dry condition. Only the effect of the sulfides was observed, phosphorus bronze so called CAC502C restricted in JIS (Japan Industrial Standard) was prepared for comparison. The matrix of phosphorus bronze was almost as same as the sulfide bronze. Details of compositions are shown in Table 1.

Table 1 Chemical compositions (mass %)

	Cu	Sn	S	Fe	P
Sulfide bronze	Bal.	10.00	0.55	0.15	0.02
Phosphorus bronze	Bal.	10.00	-	-	0.12

Mating materials were carbon steels so called S45C restricted in JIS. In the test, constant force 10 N was loaded and constant speed 1.0 m/s or 2.0 m/s were selected as the conditions.

3. Results and discussion

Fig. 4 shows results of the friction test. During the test, sulfide bronze did not occur scoring and seizure. On the other hand, Phosphorus bronze had seizure beginning on the test. As a result, seizure was prevented only sulfide bronze. As shown in Fig.5, Sulphur was detected on the friction tracks. It was estimated that the dispersed sulfide restricted the transfer of the Cu-alloy elements to the mating steel surface.

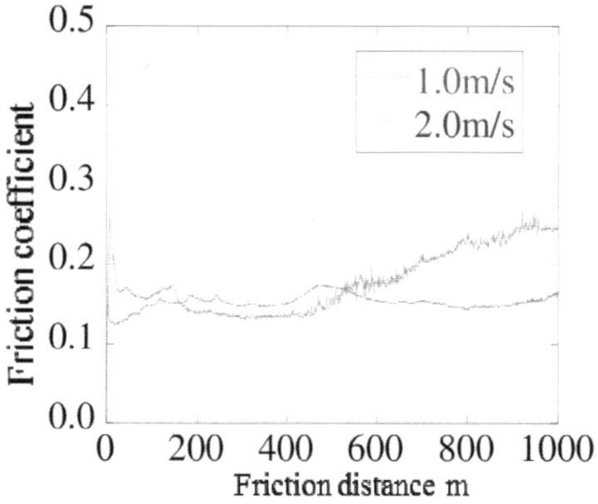

(a) Sulfide bronze

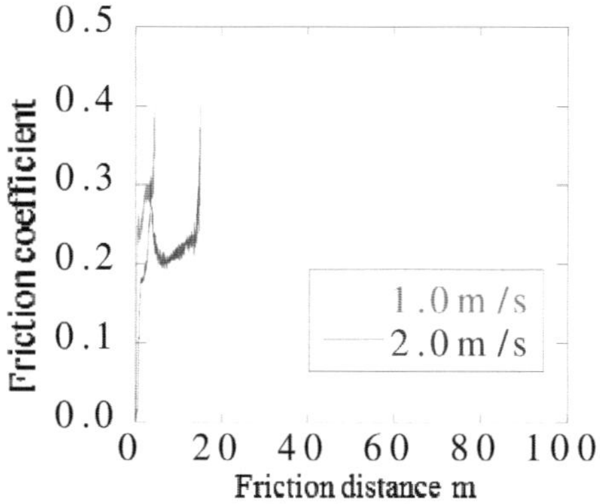

(b) Phosphorus bronze

Fig. 4 Results of friction test

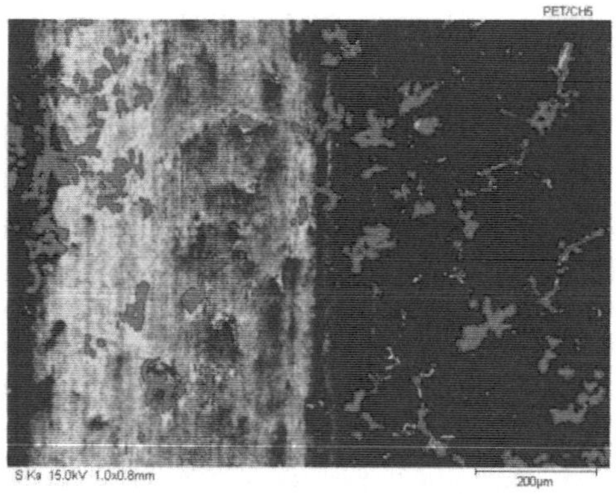

Fig. 5 Mapping of S by EPMA after the test

References

[1] Y.Hirai,T.Sato, T.Maruyama and T.Kobayashi: Proc. of fall meetings of JFS Kansai (2009)22-24. (in Japanese).

[2] T. Sato, Y. Hirai, T. Fukui, K. Tanizawa and H. Usami: Special Issue on ICMDT 2011- Advances in Manufacturing, Machine Design and Tribology 6 (2012) 158-164.

Proceedings of the 72nd World Foundry Congress
21-25th May 2016, Nagoya, Japan

Development of Energy Saving Type Vertically Grooved Crucible

Tamio Okada[1],Ryuun boku[2] ,Oka Nobuyuki[3], Shinnosuke Takeuchi [4]
[1] Nippon Crucible Co.Ltd Ebisu NR Bldg 21-3 Ebisu 1-Chome, Shibuya-Ku, Tokyo, Japan
[2] Nippon Crucible Co.Ltd Ebisu NR Bldg 21-3 Ebisu 1-Chome, Shibuya-Ku, Tokyo, Japan
[3] Nippon Crucible Co.Ltd Ebisu NR Bldg 21-3 Ebisu 1-Chome, Shibuya-Ku, Tokyo, Japan
[4]Nippon Crucible Co.Ltd Ebisu NR Bldg 21-3 Ebisu 1-Chome, Shibuya-Ku, Tokyo, Japan

1. Although the crucible furnace has various advantages, it has also disadvantage of lower thermal efficiency. However, higher thermal efficiency of the melting crucible itself has yet to be achieved. In this study on the development of a new crucible, the aim was to increase the thermal efficiency of the crucible itself by 5 to 10% compared which conventional melting crucibles by applying vertical grooves around the outer crucible surface which increases the thermal exchange surface area by 30%. We also performed investigations on the physical properties of the developed crucible and microscopic analysis of the melting crucible material in order to clarify the destruction mechanism of the crucible.

Key words: Vertical grooved crucible, Energy saving, Thermal exchange area, Thermal efficiency

2. Introduction

Crucible furnaces are widely used in non-ferrous foundry industry to melt and hold the molten metal. Although crucible furnace has various advantages, it has also disadvantage of lower thermal efficiency. There are various ways to improve its efficiency such as the reuse of combustion exhaust gas heat. However, the thermal efficiency of the melting crucible itself could not be improved so far.

This study describes the melting crucible thermal efficiency improvement of 5 to 10% compared with that of the conventional crucible. This improvement has been made by incresing the thermal exchange surface area of the crucible by 30%.

We also investigated the physical properties of the developed crucible materials and made the microscopic analysis of the melting crucible in order to clarify the destruction mechanism of the melting crucible.

2. Experiment test

2.1 Designing of the melting crucible materials

We have designed well balanced formulation of the crucible compositions with the cosideration of its strength, thermal conductivity, spalling resistance and oxidation resistance. Table 1 shows the crucible composition material formulation and grain distributions.

Table 1. Crucible composition formulation

	SiC	C	Carbon bond	Si	Low melting piont glass	Borides
Material formulation (mass%)	35~50	30~45	10~20	3~8	0~3	0~2
Particle size (μm)	0~250	150~840	liquid	0~325	0~325	0~325

2.2 Crucible manufacturing process

Graphite crucible is manufactured with following production process.

Raw material formulation – Mixing – Kneading – Mixed raw material adjustment – Forming by press – firing – Nondestructive inspection – Impregnation – Baking – Glazing – Coating – Baking – Packing & Shipping.

Raw material formulation is shown in the Table 1.

2.3 Vertically grooved crucible test at the customer

For obtaining the fuel consumption data of the conventional and developed crucibles, those crucibles are actually tested at the customer for 12 months.

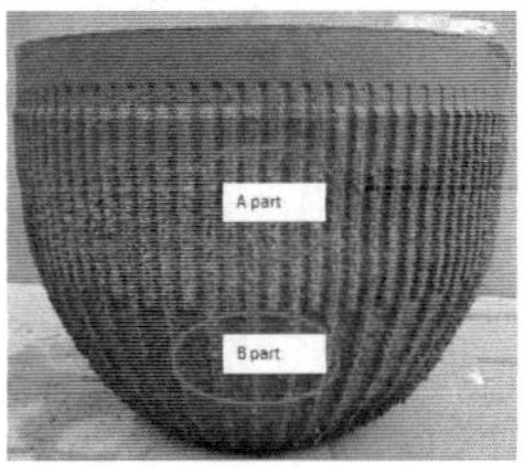

Fig1. Appearance of the crucible after 12 months use

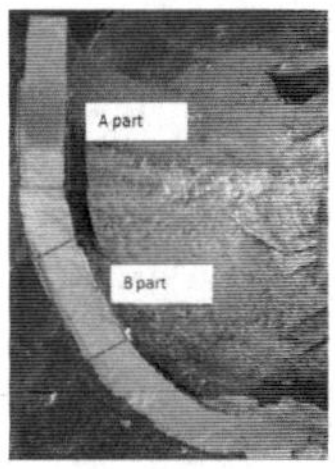

Fig2.Cut crucible photo

After 12 months operation, those crucibles were removed from the furnaces. Then samples were cut out from A part (Oxidized area) and B part (Less oxidized area) as shown in the below Fig. 1 and 2. Then we inspected the samples with optical microscope, SEM and X-ray diffraction. We also analyzed the contents of Si, SiO_2, SiC, B_2O_3 according to JIS-R2015 and JIS-R2011.

3. Result and consideration

3-1 Crucible fuel consumption and life

Thermal exchange surface area of the developed crucible has 30% more than that of the conventional one. As a result of melting test, it has been proved that developed crucible has 5 to 10% less fuel consumption than the conventional crucible. We also made follow-up survey of the life of 30 pcs. each of the developed and conventional crucibles. The result was almost the same as the above result.

3-2 Crucible destruction by composition change

Fig. 3. shows the photos of optical microscope composition of unused crucible and the sample pieces of "A" and "B" parts of the 12 months used crucibles.

Comparing the textures of "A" part (Photo b), "B"part (Photo c) and unused crucible with photo (a), it is distinctive that the amount of flake graphite is significantly reduced.

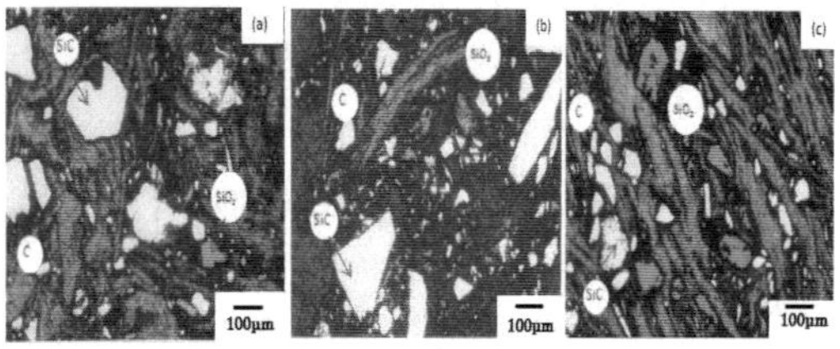

Fig.3.Optical microscope observation of crucible material and the result of phase analysis
(a)Unused (b) "A" part after 12 months usage (c) "B" part after 12 months usage.

Fig. 4 shows the photos of the result of X-ray analysis of the unused crucible, "A" and "B" positions of the sample pieces of 12 months used crucible. Unused crucible (a) and "B" position (c) are lower oxygen content and SiC and little amount of SiO_2 were detected.

On the contrary, higher oxygen content was detected in the severely oxidized "A" position (b), and SiC and the large amount of SiO_2 were also observed.

The large amount of SiO_2 is generated by SiC which starts to be oxidized from 800 degree C, and the chemical reaction of $SiC+2O_2$ to SiO_2+CO_2 forms SiO_2.

In addition to this, metalic silicon in the crucible composition will form beta type ultra fine SiC during firing process which will also form SiO_2 as shown in the above reaction formula. And during firing process, non-reacting metalic silicon will contact with oxigen and directly form SiO_2. So, there are three types of oxidizing reactions.

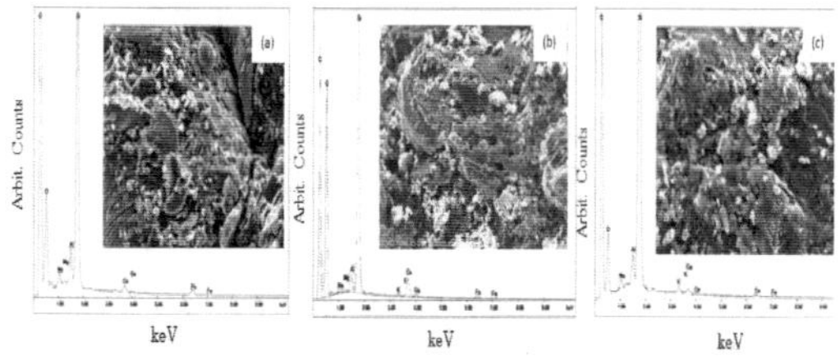

Fig.4. X-ray analysis result of crucible
(a) Before usage (b) "A" position after 12 months usage (c) "B" position after 12 months usage

Those results can also be justified by the result of the components analysis of Si, SiO_2, SiC and B_2O_3. When we compare Si content of un-used crucible component and part A and part B of 12 months used crucible, Si in the part A drastically reduced but SiO_2 content increased in stead.

Those SiO_2 generated as above will create glass film under working temperature which has effect to prevent oxigen to penetrate in the crucible component. This means that main component of flake graphite or SiC can be protected from oxigen. Actual field test result of longer crucible life was the mainly the result of the above effect. At the same time, vertically grooved crucible had the same effect and the same life as the normal crucible. This point was also the benefit of this study.

4. Conclusion

(1) Vertically grooved crucible can achieve thermal efficiency improvement of 5 to 10% by increasing 30% of the thermal exchange surface area.

(2) SiO_2 film generated by metalic silicon is highly effective for preventing oxidation of flake graphite which is the one of the crucible component. However, the film has lack of compactness.

(3) Life of the vertically grooved crucible has almost the same life as the normal crucible.

Proceedings of the 72nd World Foundry Congress
21-25th May 2016, Nagoya, Japan

Development of Aluminum Cylinder Head Castings for Diesel Engines

Ken Moizumi[1], Yoshinori Hase[1] and Kazumi Koyama[1]
[1] ISUZU Motors Limited, Tsuchi-dana 8, Fujisawa, Kanagawa Japan

An Al-Si-Mg-Cu alloy and its heat treatment process for the cylinder head castings of diesel engines have been developed. Developed alloy is not only superior on the mechanical strength, but also on the thermal fatigue durability comparing with the conventional aluminum casting alloys. This alloy allows increasing the engine output under the condition of higher combustion pressure. In order to improve durability of cylinder head castings, it has been studied that influences of precipitation behavior in various contents of alloy elements and heat treatment conditions. And it was founded out that precipitated phase and its microstructures influence to thermal fatigue durability. This study results in improving the both mechanical and thermal fatigue properties. In addition to optimizations of aluminum casting alloy, casting processes have been improved. Developed tilting casting process prevents generation of oxides. New cooling system of the mold contributes to increase solidification speed, and to decrease shrinkage porosities with lower riser weight.

Keywords: *Diesel engine cylinder head, Aluminum casting, fatigue, thermal fatigue.*

1. Introduction

In our company, it has produced a variety of commercial vehicles and diesel engines for light to heavy duty use. Diesel engines are required higher reliability because it is an important power source to support the operation of the logistics and also manufacturing in industry. As main components of diesel engines, cylinder head, cylinder block, connecting rod, crank shaft and cam shaft, those are called "5C" components, are also required higher reliability. In particular the cylinder head is a key component for the weight reduction and improving durability of the engine. In order to meet these demands, it is increased of which not only adoption of aluminum castings but also development of higher strength aluminum alloys.

In order to improve mechanical strength and thermal fatigue durability in simultaneously, we have studied chemical components of alloys, heat treatment procedures and other manufacturing processes of aluminum castings. And also it was studied that the mas-production process which does not generate inclusions and shrinkage porosities as a detrimental defect as high quality aluminum casting.

2. Development of aluminum alloy

In conventionally, Al-Si-Cu-based JIS AC2A and AC2B or the Al-Si-Mg-based JIS AC4C alloys are widely used in the cylinder head castings. These aluminum alloys are suitable for cylinder head because conventional alloys have a good cast-ability in which good fluidity and less shrinkage cavities in complex shape castings. In addition to above, mechanical strength is able to be increased by heat treatment. However, it is reported that thermal fatigue durability of alloys including Cu is not higher than the alloys not including Cu [1]. This is due to differences in precipitation microstructures. We have investigated changes of precipitation types and morphologies by chemical content and heat treatment conditions. And influence of precipitation microstructures on thermal fatigue durability was studied. In this study, it is found out that precipitation microstructures do not influence only in mechanical strength but also thermal-mechanical fatigue durability. In AC4C alloy including 0.3mass% of Mg and 7mass% of Si, it is known that Mg-Si binary β'' or β' phases precipitate as a reinforcement phase. These precipitates improve ductility of Al matrix during thermal fatigue deformation because it is easy to growth in high temperature environment. On the other hand, in alloys including Cu, Q', θ' and/or relating phases are precipitate. These precipitates increased mechanical strength but prevent ductility of Al matrix during thermal fatigue deformation. Less ductility of Al matrix reduces thermal fatigue leaves. In our series of studies, it was founded out that 1mass% of Cu and long term of over aging treatment moderately disperses Q' phase. Optimized precipitate microstructure has been improved both of mechanical fatigue strength and thermal fatigue durability. Fig.1 shows results of high cycle fatigue test of conventional and developed alloys. Fatigue strength of developed alloy increases +20MPa comparing with the conventional alloy. And also thermal fatigue leaves of developed alloy improved to almost 2 times of the conventional alloy as shown in Fig.2.

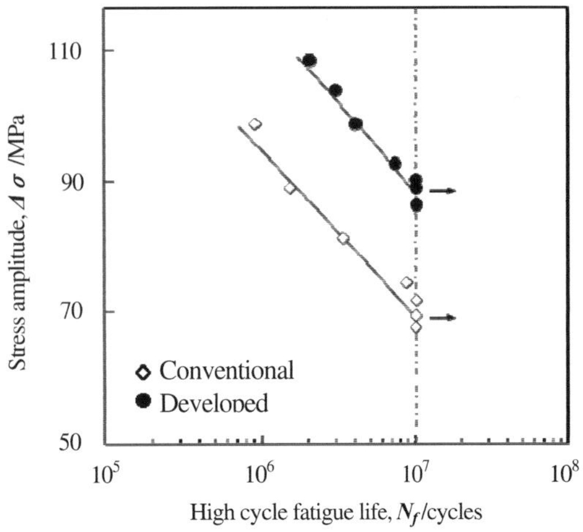

Fig.1 Result of Mechanical fatigue test in conventional and developed alloys.

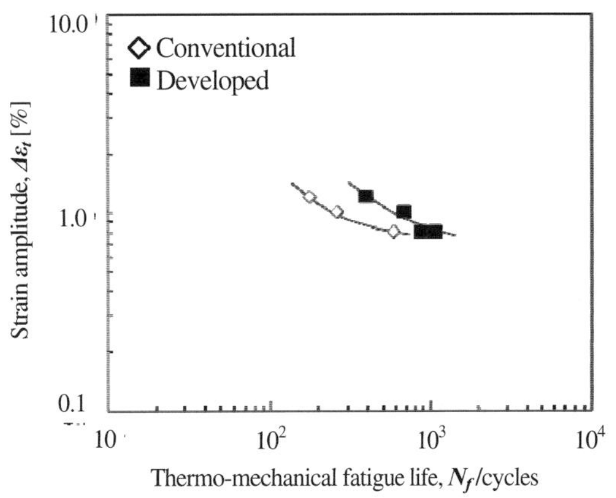

Fig.2 Result of thermal fatigue test in conventional and developed alloys.

3. Development of casting processes

3.1 Elimination of micro porosities

Microporosities locating casting surface are able to affects as initiation sites of fatigue cracks. In order to reduce of this affection, it was studied to improve the cooling method of the parmanent mold during solidification.

Fig.3 shows the structure of inserted chiller whitch made of Cu alloy on the lower face of parmanent mold. Inside of chillers have been cooled by water flow from the bottom side. Fig.4 shows the microstructures of the cylinder head castings. In the improved cooling method, higher solidification speed decreases microporosities of the casting surface.

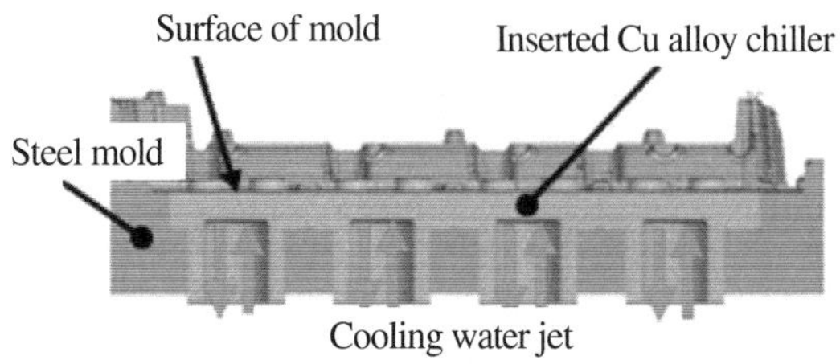

Fig.3 Sturucture of parmanend mold with Cu alloy chiller

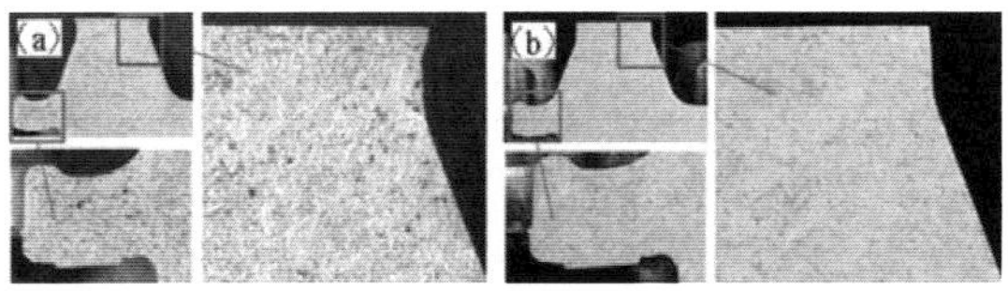

Fig.4 Microporosities in sections of (a) conventional and (b) improved castings.

3.2 Elimination of incrusions

It is also important to eliminate incursions such an aluminum oxide in order to prevent the affection for the fatigue crack generation. The pouring process has been also improved showing in Fig.5. A raddle handled with a robot tilts in synchronization with the mold. In this process, scooped molten metal from the crucible is poured directory into the mold. In this method, less to produce an oxide due to the absence for transfers of the molten metal.

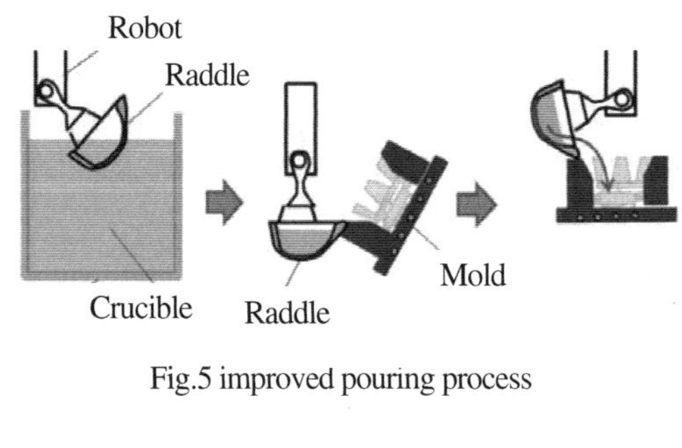

Fig.5 improved pouring process

4. Summary

Developed aluminum casting alloy, heat treatment procedure and casting processes totally improve the durability of cylinder head castings. It allows that the aluminum cast cylinder heads are able to be used for higher output and less fuel consumption of diesel engines.

[1] K. Moizumi, K. Mine, H. Tezuka and T. Sato: Material Science Forum 01/2002, 396-402 (2002), 1371-1376.

Proceedings of the 72nd World Foundry Congress
21-25th May 2016, Nagoya, Japan

Compacting of Greensand Control Equipment

Hiroaki Amano, Shinichi Chomatsuken, Yoshimasa Sawada
Casting Production Engineering Dept., Aisin Takaoka Co., Ltd.
1 Tennoh, Takaokashin-machi, Toyota, Aichi, 473-8501 Japan

Sand control equipment in conventional greensand molding lines requires much space and investment. With the aim of putting into practical use compact sand control equipment that ensures stable sand properties, a method for stabilizing sand properties while improving casting quality was devised. In the method, collected sand was processed after separating into thermally deteriorated and undeteriorated sand. Since undeteriorated sand could be processed more simply, the amount of sand to be processed was reduced, and thus the newly developed sand control equipment achieved 20% space reduction for sand processing as compared with conventional equipment.

Keywords: *greensand control equipment, compacting , separation and collection process.*

1. Introduction

In conventional greensand molding lines, sand control equipment requires much space and investment, making it difficult to develop compact greensand molding lines. Therefore, we have developed a sand control method that ensures stable sand properties even in compact equipment.

2. Concept of compacting greensand control equipment

Conventional sand control equipment necessitates much space for recycling all collected greensand and processing a larger amount of greensand than that required for molding. To achieve equipment compacting, we considered reducing the processing amount, focusing on the state of greensand mold after pouring molten metal, as shown in Fig. 1. We anticipated that it is possible to reduce processing amount by normally processing only sand that contains a reduced amount of activated clay as a result of bentonite degradation due to heat of molten metal around the casting (hereinafter referred to as "deteriorated sand"), and by performing simple agitation for moisture adjustment of recycling sand that contains no bentonite degraded by heat in the outer layer (hereinafter referred to as "undeteriorated sand"). To this end, we developed a method for separating a mold after pouring molten metal into deteriorated and undeteriorated sand at the water condensed layer where mold strength is lowered.

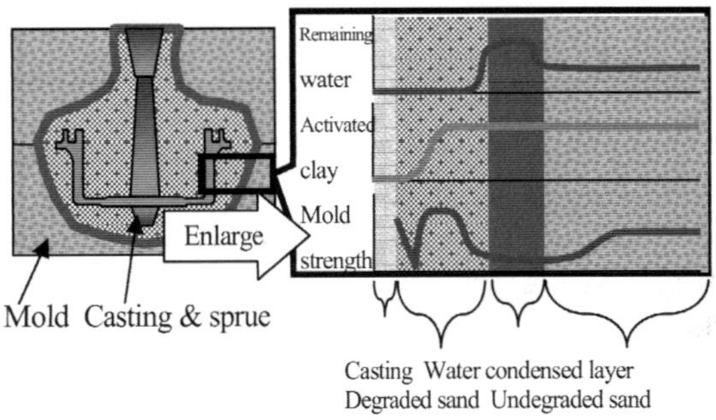

Fig. 1 Mold cross-section after pouring molten metal and properties of sand

3. Basic research

We investigated the possibility of separating sand by means of the water condensed layer, by examining changes in sand temperature and the degree of bentonite degradation due to heat of molten metal from the casting to the mold surface. Almost no changes in the activated clay content were noted in undeteriorated sand at 100 °C or higher temperatures where the condensed water layer was formed, whereas marked depletion of activated clay content was observed in deteriorated sand, as shown in Fig. 1. The properties of deteriorated and undeteriorated sand separated by the water condensed layer were hardly affected by differences in the type of casting, the sand-metal ratio, or the ratio of deteriorated to undeteriorated sand. These findings demonstrated that the separation into deteriorated and undeteriorated sand might have suppressed changes in sand properties even if they were separately processed in constant amounts, in contrast to conventional sand processing methods, where sand is processed in the same manner despite differences in the degree of thermal deterioration of sand found in different parts of the mold due to changes in the sand-metal ratio, causing variation in the properties of the sand.

Then, we developed a method for separating a mold after pouring molten metal into deteriorated and undeteriorated sand by means of the water condensed layer. The method facilitated the separation of a horizontal splitting mold by making vertical cuts at the center of its four horizontal sides to split it. We applied this method to various actual products to separate deteriorated and undeteriorated sand and examined the proportion of deteriorated and undeteriorated sand to estimate the capability of greensand processing based on proportions where the amount of deteriorated sand is larger.

Finally, we fabricated compact greensand control equipment based on the above results, and evaluated it in a mass-production environment.

4. Results of using the greensand control equipment

The newly developed compact greensand control equipment was used in a molding line for mass production. The process of separating and collecting deteriorated and undeteriorated greensand was introduced in the shake-out process. As a result, a 20% reduction in space and a 10% reduction in investment cost were achieved, as shown in Fig. 2 and Fig. 3, respectively.

In addition, as shown in Fig. 4, compared with the conventional method, sand properties were greatly stabilized by processing stable deteriorated and undeteriorated sand separately through the separation and collection process.

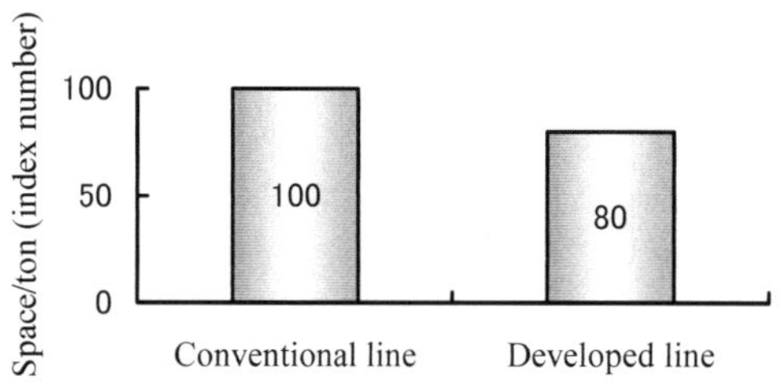

Fig. 2 Comparison of space per ton of sand processed

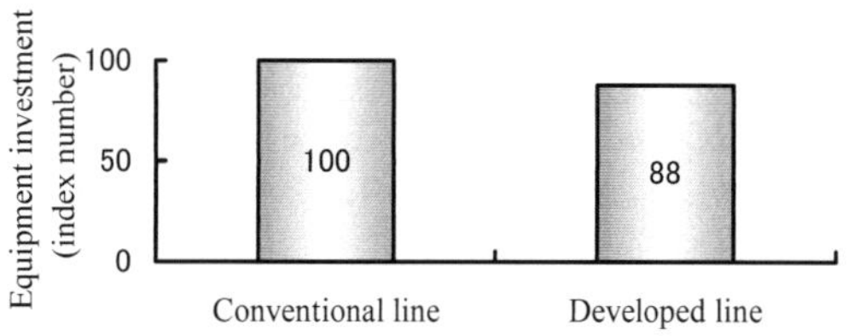

Fig. 3 Comparison of greensand control equipment costs

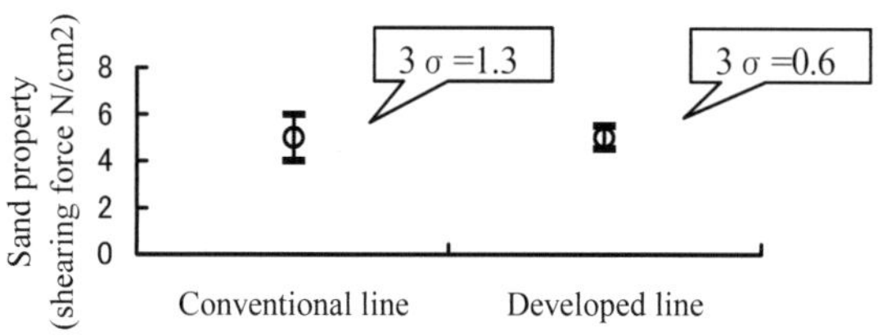

Fig. 4 Comparison of variation in sand properties

Proceedings of the 72nd World Foundry Congress
21-25th May 2016, Nagoya, Japan

Development of the latest small-size flaskless horizontal molding machine

Shuji Takasu [1], Toshikazu Morishita[1], Yutaka Hadano[1] Takayuki Komiyama[1] and Shuichi Tuzuki[1]
[1] Sintokogio, Ltd, Toyokawa, Aichi, Japan

In recent years, environmental protection against to global warming and deterioration of environment has been paid attention to. Authors have been developing molding machine not only increasing productivity but also achieving environmentally-friendly production, safety and improving working environment.
In that development, the key technology is "Aeration sand filling". In the conventional sand filling, the falling sand particles have directional property. It always caused problems such as uneven sand density even on the same pattern and insufficient sand filling into small bore pockets. These problems are solved by aeration sand filling technology. The compressed air with low pressure is introduced into the sand tank to fluidize the molding sand, and guide the sand into the molding flask. As the compressed air with low pressure is used, the air consumption is reduced, and the noise caused by compressed air is decreased. Aeration sand filling carries many other advantages. One of them is low noise and low vibration. Dimensional deviation of casting has been improved, thus finishing cost could have been reduced. Sand filling performance in the complex shape has been improved by the aeration sand filling. Therefore, defects of mold has been greatly reduced.
This technology was applied to tight-flask molding machine at first. Next, authors developed the small-size flaskiless horizontal molding machine with latest technology for the advanced of the small jobbing foundry.

Keywords: *Green sand mold, Aeration sand filling, Flaskless molding, Energy saving*

1. Introduction

In the recent domestic foundry industry, the obsolescence of technologies and the aging of the facilities have progressed gradually. Therefore, the expectations for the molding machine with the latest technology have increased especially from the small-scale foundry plants. On the other hand, in overseas plants mainly in China and Southeast Asia, has been changing from the semi-automatic molding machine with hand work to full automatic.

To answer such demand, authors developed small-size molding machine (type FDNX). This user-friendly molding machine achieved the stable operation for quality casting with the aeration sand filling technology [1-3] as the core technology and the latest control system.

2. The principle of aeration sand filling technology

Figure.1 shows a concept of aeration sand filling mechanism.

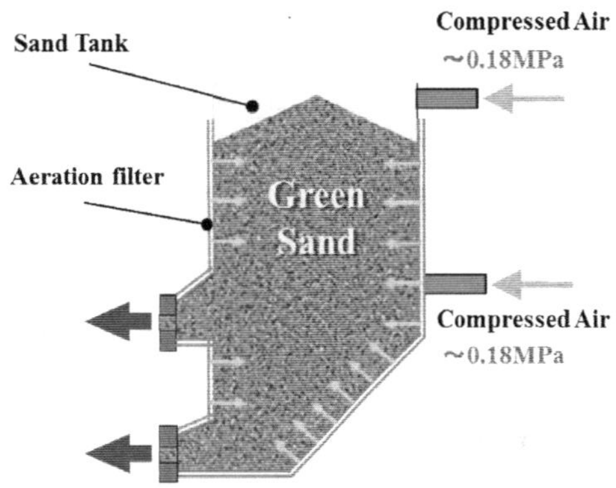

Fig.1 Aeration sand filling mechanism

The conventional blow type molding machine had required high pressure compressed air of more than 0.3MPa for the sand filling process.

In contrast, the newly developed molding machine adopted the aeration sand filling technology as the core technology. Uniform sand filling is realized with saving energy because aeration air in low pressure under 0.18MPa is led in through a special filter in the sand tank.

3. Test result

3.1 Mold strength

The mold strength distribution of the mold were compared by the conventional blow type molding machine and upgraded molding machine in a Japanese foundry plant. The results are shown in Figure 2. Compared to the conventional blow type molding machine, new aeration type molding machine has improved the mold strength, furthermore, achieved uniform sand filling with the aeration filling.

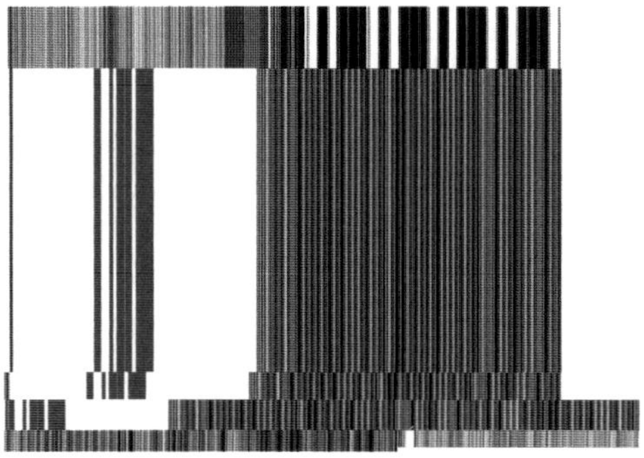

Fig.2 Mold strength distribution comparison

3.2 Casting dimensional accuracy comparison test results

Dimensional accuracy was also compared by the molding machine before and after updating as with the intensity distribution. The result is shown in Figure 3.

The developed molding machine achieved the improvement of the casting quality in terms of dimensional accuracy even in small plant.

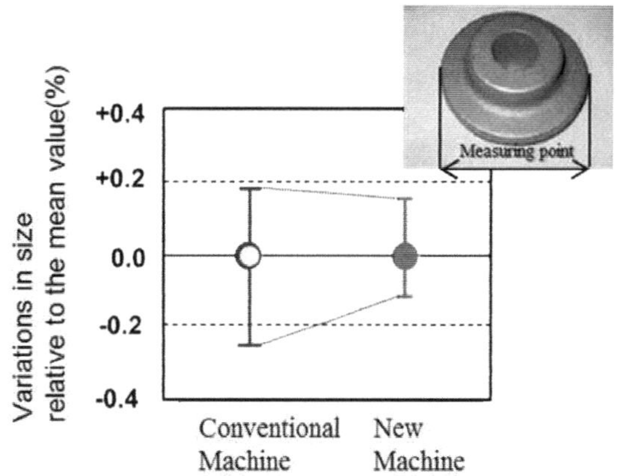

Fig.3 Casting accuracy comparison result

4. Conclusion

The improvement of casting quality is important task for the domestic small-scale foundry in the global competition. In addition, it also contributes to improve Japanese foundry technology.

Authors deveroped small-size flaskless molding machine.It has achieved uniform sand filling and the improvement of the casting quality in terms of dimensional accuracy with the aeration filling.

Semi-automatic molding machine with hand work is still widely used in emerging countries. On the other hand, the countries developing such as an automobile industry need to stably mass-produce high precision casting. New developed small-size flaskiless molding machine meets such demand.

References

[1] Hirata, M., "Eco-Conscious Molding System, Featuring Aeration Sand Filling, Preset Squeeze And Mold Height Feedback Control," AFS Transactions, Paper 02-081, Page 1 of 8, (2002).

[2] Hirata, M., Sugita, K., "New Sand Filling Method in Flaskless Molding and Its Controls," AFS Transactions, Paper 05-037, Page 1 of 8, (2005).

[3] Makino, H., Hirata, M. and Hadano, Y. "Computer Simulation and Optimization of Sand Filling Using the Distinct Element Method", WFO Technical Forum on GIFA2003, Düsseldolf, session 5.1

Proceedings of the 72nd World Foundry Congress
21-25th May 2016, Nagoya, Japan

Development of Mechanical Sand Reclamation System Using Roller Squeezing Mechanism

Takafumi Oba[1], Shigeaki Nakayama[2] and Yutaka Hagata[1]
[1] Sintokogio, LTD. Toyokawa Works, 3-1 Honohara, Toyokawa, Aichi 442-8505, Japan
[2] Sintokogio, LTD. Kyushu Works, 590-6 Furumon Kurate, Kurate, Fukuoka 807-1306, Japan

Sand reclamation is currently one of the most important technologies for nobake sand plants from the perspective of waste minimization and reuse of mold sand. Conventional sand reclamation technologies are becoming increasingly obsolete due to the increasing popularity of high adhesion nobake binders such as Phenol Nobake (PNB) process and the growing need for reusing waste mold sand due to tighter environmental standards. Consequently, advanced sand reclamation technology suitable for PNB was desired.
To meet this foundrymen's desire, authors have developed advanced sand reclamation technology named *Roller Squeezing Type Reclaimer*. This system enables to reclaim used sand completely with lower LOI and higher mold strength. These results mean this technology contributes casting quality improvement since lower LOI eliminates gas defects caused by binder and higher mold strength eliminates sand inclusion and achieves superior dimensional accuracy.

Keywords: *Sand Reclamation, Phenol Nobake, Squeeze, Ceramic Roller, Reuse*

1. Introduction

Recently invented binders, such as PNB, adheres tightly onto sand grain surfaces. The adhered binder requires hard scrubbing to be removed. Conventional mechanical reclaimers, such as pneumatic type, attrition type or grinding type are insufficient to be applied for PNB. Consequently, kinds of mechanical reclamation mechanisms had been tried [1].

Finally, the advanced mechanical reclamation mechanism, named *Roller Squeezing Type* was developed [1-3].

2. Principle and Mechanism

Roller Squeezing Type Reclaimer consists of a pair of ceramic rollers and a rotating drum. Sand fed into the reclaimer makes a layer on the inner wall of the drum and scrubbed by the attrition among sand grains.

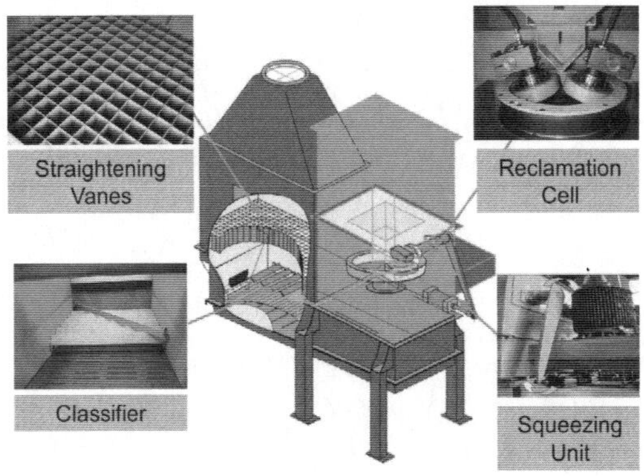

Fig. 1 Reclaimer Unit

Simultaneously, squeezing rollers press the sand layer. Combination of scrubbing and squeezing removes residual binder stronger than conventional methods.

Fig. 1 shows the whole unit of the reclaimer. The reclaimer consists of a reclamation cell, a fluidized bed type classifier, and a settling chamber. The settling chamber equips straightening vanes to separate sand grains and fines accurately.

3. Comparison Tests Against Conventional Reclaimer

A series of comparison test was conducted against rotary scrubbing type reclaimer, the most popular conventional reclaimer in Japan. Test conditions are given in Table 1.

Table 1 LOI Comparion Test Conditions

Binder	Ester-Cured Phenol Nobake (PNB)
Sand	Silica Sand (AFS GFN 39)
Motor Power	30kW/Pass
Throughput	5 MTPH

3.1 LOI Comparison

Fig. 2 gives the LOI comparison test result [2, 3]. As shown in Fig. 2, *Roller Squeezing Type* reaches the upper limit at second pass. However, conventional rotary scrubbing type requires four passes to reach the upper limit. This result means *Roller Squeezing Type* requires only half energy compared to conventional method for achieving same LOI level.

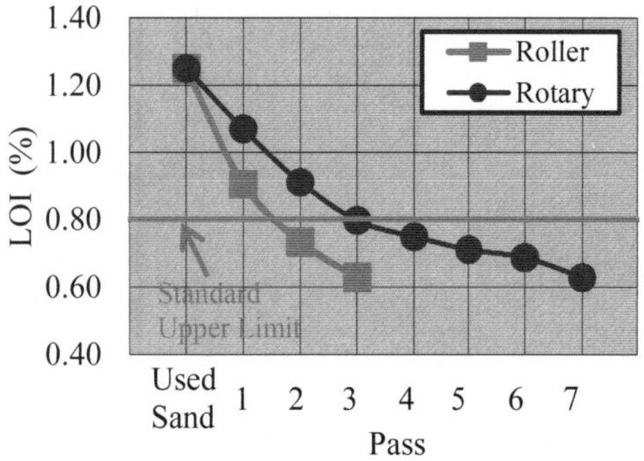

Fig. 2 LOI Comparison Result

3.2 Compression Strength Comparison

Using the same property sand, second pass for *Roller Squeezing Type* and fourth pass for rotary scrubbing type, compression strength of 50mm Dia. x 50mm Height test pieces was also compared. Table 2 gives the compression strength test conditions and Fig. 3 gives the test results [3].

Table 2 Compression Strength Comparison Test Conditions

Binder	Ester-Cured Phenol Nobake (PNB)
Sand	Silica Sand (AFS GFN 39)
Resin Content	1.4% Based On Sand (BOS)
Hardener Content	20% Based On Resin (BOR)

Fig. 3 shows that compression strength for reclaimed sand of *Roller Squeezing Type* is superior to that of rotary scrubbing type both after three hours and twenty four hours after molded. These results indicate that *Roller Squeezing Type* enables to remove firmly adhered binder without crushing sand grains. Besides, *Roller Squeezing Type* improves sand grain shape more rounded and compression strength is also increased by rounded sand grain shape.

4. Conclusions

Roller Squeezing Type Reclaimer achieves high efficiency and quality sand reclamation and reclaimed sand with this technology shows superior sand properties and mold strength against conventional one. These results mean this technology enables to improve reclaimed sand quality and then improved reclaimed

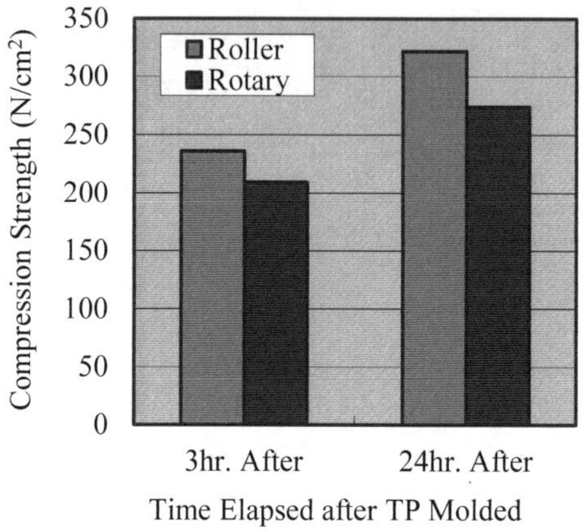

Fig. 3 Compression Strength Comparison Results

sand also improves casting quality. This technology was originally developed in Japan, however now spreads all over the world and contributes to sand quality improvement waste reduction. For example, it is reported that one Mexican steel foundry *Roller Squeezing Type Reclaimer* contributes to sand quality improvement [4]. In addition, *Roller Squeezing Type Reclaimer* and thermal reclamation system was compared in Europe and *Roller Squeezing Type Reclaimer* enables to replace thermal reclamation system [5].

Authors believe *Roller Squeezing Type Reclaimer* will change mind of foundrymen that mechanical reclamation is less powerful and reclaimed sand cannot be alternative of unused sand due to its low quality. And also, we expect this technology improves casting quality through high quality reclaimed sand.

References

[1] Oba, Iyoda, Iwasaki and Aoki: AFS Transactions 107 (1999) 89-93.

[2] Oba, Hagata and Nakayama: *Development of Mechanical Sand Reclamation Technology Using Roller Squeezing Method*, (159th Japan Foundry Society Meeting, Matsue Japan, 2011) pp12.

[3] Oba, Hagata and Nakayama: Journal of Japan Foundry Society 84 (2012) 533-538.

[4] Fett and Oyarzabal: AFS Transactions 113 (2005) 51-58

[5] Fenyes: *High Level Sand Reclamation by means of Secondary Attrition - A Viable Alternative to Thermal Reclamation*, (55th Indian Foundry Congress, Agra India, 2007) pp217 - 230

Proceedings of the 72nd World Foundry Congress
21-25th May 2016, Nagoya, Japan

Achieving a clean foundry plant with adopting artificial sand in an old plant

Tohru Mizuki[1], Taiji Fukuo[1], Seitaro Tsuchiya[2], Takanori Aizawa[2],
Toshinari Sudo[2], Yusuke Katayama[2] and Toshitake Kanno[1]
[1] R&D Dept., Kimura Chuzosho Co., Ltd. [2] Head quarter plant, Kimura Chuzosho Co., Ltd.

In this report, we report about improving our plant's environment with adopting artificial sand.
By changing all the silica sand to alumina artificial sand, free silica in the powder dust dropped from 9.7% to 3.0%. As the result, the lung disease risk caused by dust (pneumonia) decreased and the working environment of level 3 was improved to level 2.
But, there were several problems to use the artificial sand. We have solved the artificial sand's problems and brought out the artificial sand's merits.

Keywords: *artificial sand, working environment, thermal expansion, lung disease caused by dust (pneumonia), external shrinkage, foundry plant*

1. Introduction

As our head-quarter foundry plant was built about 60 years ago, many equipments were old and its environment was very bad. In addition to, the government revised strictly the environmental law, so we must improve our plants environment. Therefore, in order to achieve a clean foundry plant, we considered adopting artificial sand.

2. Technical problems to be solved.

In order to check the problems caused by changing the silica sand to the artificial sand, we cast test pieces, 300*200*H300mm. We varied the artificial ratio to silica sand and molded all the 8 test pieces with the ratio varied sand around 100mm from test pieces in the same flask. We cast the test pieces with the composition of ISO 185/JL/300 gray iron.

After shake-out, we investigated all the 8 test pieces. External shrinkages were formed on the cope surface. Fig.1 shows the effect of the artificial sand ratio on external shrinkage volume in gray iron. The external shrinkage volume increases with increasing the artificial sand ratio. As the countermeasure against the external shrinkage, we determine to use risers.

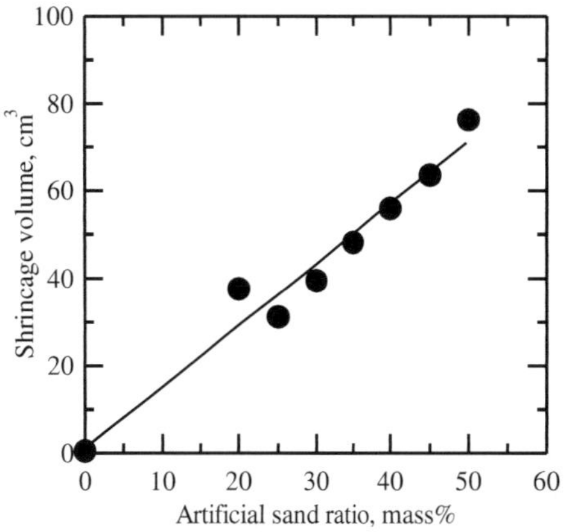

Fig.1 Effect of the artificial sand ratio on external shrinkage volume in gray cast iron.

3. Effect of changing sand

Table 1 shows the environmental level improvement with introducing artificial sand. Level 1 means good and level 3 means bad. In the case of level 3, the environmental law orders us to improve the place. The environmental level of shake-out had been 3. Following the law, we improved the shake-out with introducing artificial sand. As the result, the working environmental level was improved from 3 to 2. Table 2 shows the detail of the environment measurement data at shake-out room. Free silica in the powder dust dropped from 9.7% to 3.0%. The lung disease risk caused by dust (pneumonia) decrease. We could improve the working environment with introducing the artificial sand.

Table 3 shows the other effects of changing sand. Compared to the silica sand, the artificial sand is said not to be broken and can be used semi-permanently. To confirm it, we did friability test. As shown in table 3, the friability of artificial sand is 1.1, lower than that

Table 1 Environmental level with introducing artificial sand

	Before introducing	After introducing				
	2008 Mar	2008 Aug	2009 Sep	2010 Sep	2011 Sep	2012 Sep
Shake-out room	Level3	Level2	Level2	Level2	Level2	Level1
Moldig area1	Level2	Level1	Level2	Level2	Level2	Level2
Moldig area2	Level2	Level1	Level1	Level	Level1	Level1

Table 2 Detail of the environment measurement data at shake-out room

	Before introducing artificial sand	After introducing artificial sand		
	2008 Mar	2008 Aug	2011 Sep	2012 Sep
Free silica ratio (%)	9.7	3.6	3.0	3.0
Requirement for level 1	E_{A1}<0.45 and C_B<0.45	E_{A1}<0.96 and C_B<0.96	E_{A1}<0.66 and C_B<0.66	E_{A1}<0.66 and C_B<0.66
Requirement for level 2	E_{A2}<0.45 and C_B<0.675	E_{A2}<0.96 and C_B<1.44	E_{A2}<0.66 and C_B<0.99	E_{A2}<0.66 and C_B<0.99
Requirement for level 3	0.45<E_{A2} or 0.675<C_B	0.96<EA2 or 1.44<CB	0.66<EA2 or 0.99<CB	0.66<EA2 or 0.99<CB
1st Enviromental Assessment E_{A1}(mg/m^3)	1.95	1.78	1.50	0.62
2nd Enviromental Assessment E_{A2}(mg/m^3)	0.81	0.63	0.61	0.23
Measurement data at B point C_B(mg/m^3)	0.77	0.47	0.56	0.25
Enviromental Level	3	2	2	1

of silica sand (1.25). As the artificial sand does not break easily, the additive amount reduced from 1.2% to 0.25% and the waste amount reduced from 1.72% to 1.13%. However, in fact, the artificial sand breaks, to a certain extent.

Table 3 Other effects of changing sand

	Before introducing artificial sand	After introducing artificial sand
Friability	1.25	1.1
Additive new sand(%)	1.27	0.25
Waste rate (%)	1.72	1.13
Penetration	Existed	Nothing
Riser in gray iron	Nothing	Necessary
Bulk specific gravity(g/cm^3)	1.3-1.6	1.9-2.1

We checked where the artificial sand breaks and found it breaks during sand's reconditioned process. At that time, we used the sand reconditioning equipment, where the sand was collided to the wings with high-speed rotation. During this process, the sand broke. As the countermeasure against it, we introduced new method that the sand is reconditioned with mutual friction. With this method we could prevent the sand-broke.

Secondly, the artificial sand is said to have high penetration resistibility because of its high refractoriness. Fig.2 shows the penetration test result. We varied the artificial sand ratio to the silica sand and cast test pieces. When the artificial sand is less than 50%, the penetration is shown. When its ratio is more than 65%, the penetration reduced greatly. When its ratio is more than 80%, the penetration is not shown.

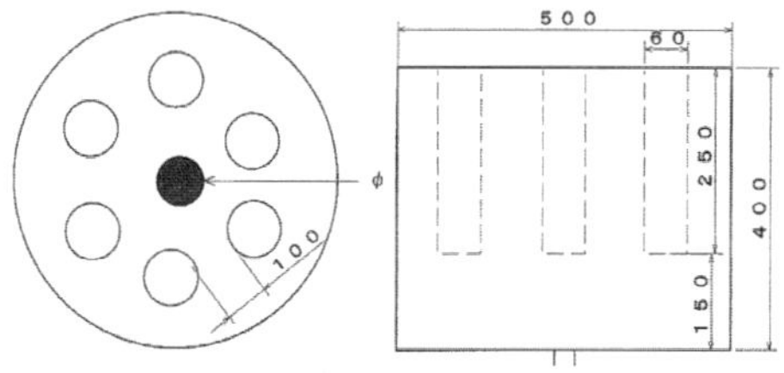

(a)Shape of Test piece, mm

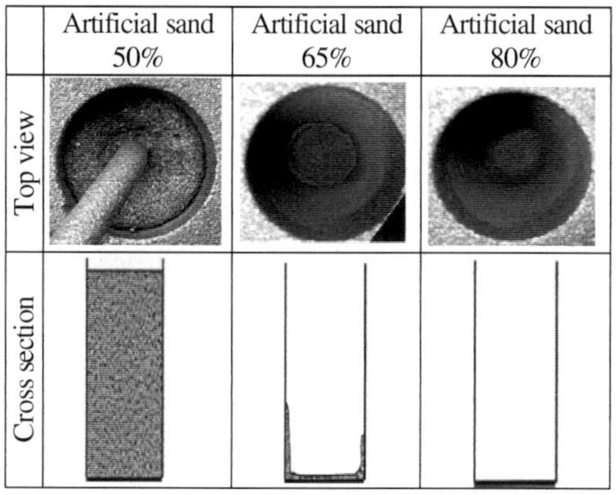

(b)Test resuluts

Fig.2 Penetration test result

By the way, there were several problems to use the artificial sand. Firstly, there was a problem of external shrinkage. In the past, we did not use risers in gray iron. But in the case of using artificial sand, we must use risers even in gray iron, in order to prevent the external shrinkage.

Secondly, there was a problem of reducing maximum product weight. Compared to the silica sand, the artificial sand shows higher apparent density. Because of it, the maximum casting weight which we could produce became smaller.

4. Conclusion

As described above, we have solved the artificial sand's problems and brought out the artificial sand's merits. Therefore, even though the plant is still old, we have achieved a clean foundry plant friendly to human.

O-121

Effect of Surface Condition on Eddy Current Evaluation of Ductile Cast Iron Matrix

Noritaka HORIKAWA[1], Takato HONMA[2]
*1 National Institute of Technology, Asahikawa College, 1-6,2-2 Shunkoudai, Asahikawa, 072-8142, Japan
*2 Advanced Course of Production System Engineering, National institute of technology, Asahikawa college

This study examined the effect of the surface conditions on evaluating ductile cast iron matrix with the eddy current method. Specimens having matrix with various pearlite ratio were prepared. The surfaces of the specimens are machined by face mill and shaping machine with various cutting conditions. Surface roughness of the test piece was measured. Eddy current testing was carried out with an eddy current flaw tester. Cutting condition affects the eddy current signals especially when a worn cutting chip was used. Difference of matrix can be evaluated by using eddy current signals, but the surface machining has unignorable effect.

Keywords: *Ductile cast iron, Eddy current, Nondestructive testing, Cutting condition*

1. Introduction

Thin wall ductile cast iron has been developed to reduce weight of the casting parts and it is proposed to apply to automobile parts. Since microstructure and mechanical properties are varied by thickness of the casting, nondestructive evaluation method on the microstructure and mechanical properties of cast iron is required to apply it to structural parts. An eddy current method is one of the techniques to detect material properties, and there are several attempts to adopt the method for evaluating cast iron matrix. [1] In that case, effect of surface condition, such as casting surface, machined surface, ground surface, should be concerned. The effect of the machining condition of the ductile cast iron on eddy current evaluation was examined.

2. Experiment

2.1 Specimens

Ductile cast iron specimens with various matrixes were prepared and the microstructures are shown in Fig.1. All test pieces were cut from Y-block. Test pieces of CBR and FCD were machined by a face mill. Cutting speed and feeding rate were changed and they are listed in Table 1. The surface of CA and CB were cut with a shaping machine.[2] Surface roughness was measured and the results were also shown in Table 1.

2.2 Experimental procedure

Eddy current testing was performed with an eddy current flaw detector and it is shown in Fig.2. The testing frequency was 10, 25, 50kHz. A polyethylene sheet was put on between the testing probe and a test piece. Test piece having the cut surface by an abrasive wheel was used as the standard test piece. Eddy current signals Vx, Vy are output of the eddy current flaw detector. It compares the test piece to be evaluated and the standard test piece. Differences in Vx and Vy have correlation with the impedance of the coil in the testing probe. They will change depending on the electromagnetic properties of the test piece, or contact state between the coil and test piece.

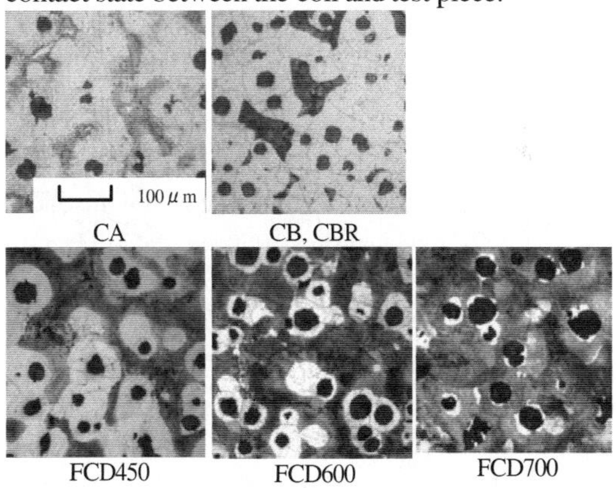

Fig.1 Microstructures of the test pieces

Table1 Cutting conditions and surface roughness of the CBR series test pieces

Test piece	Cutting speed (m/min)	Feeding rate (mm/min)	Ra(μm)
CBR1	100	50	2.76
CBR2	192		3.13
CBR3	345		3.55
CBR5			7.14
CBR4		150	2.30
CBR6			1.50
CBR7	100		6.72
CBR8, FCD	100	100	6.16
CBR9,FCD	192		2.42
CBR10, FCD	345		0.89

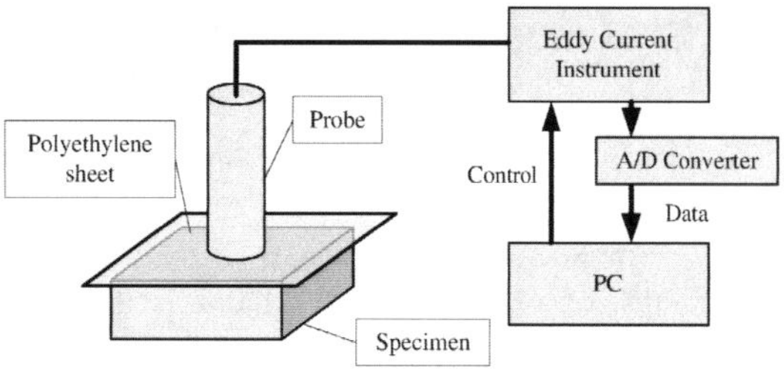

Fig.2 Schematic illustration of eddy current testing

3. Results and discussion

3.1 Effect of cutting condition

Eddy current signals of CBR series test pieces with the testing frequency of 25kHz are shown in Fig.3. For CBR1 ~ 7, both Vx and Vy increase with larger cutting speed and they are in line. Author reported that the difference in microstructure and residual stress appeared in Vy and surface roughness was strongly affected to Vx with same testing equipments and testing condition.[2] Since relation between the signals and the surface roughness is not clear, the change of the signal was mainly caused by the residual stress induced on the surface of the test pieces. On the other hand, the signals of CBR 8~10 showed opposite tendency and smaller change compared with CBR1~7. When cutting the test pieces of CBR8~10, new cutting chips were used. New cutting chip may induce smaller residual stress on the surface of them, and lead smaller change on the eddy current signals. Residual stress of these test pieces are to be measured.

3.2 Signal changes with the microstructure and effect of the surface

Fig.4 shows the eddy current signals of the test pieces with various microstructures and surface condition. Signals of the test pieces of FCD450 indicate similar value with CBR8~10. Both the Vx and Vy of FCD600 and FCD700 are larger compared with other test pieces. Difference of matrix can be distinguished by using eddy current signals, but the effect of the surface machining cannot be ignored. Some of the plot of CA and CB, including the annealed test pieces, distribute right area of the line which the signals of CBR consist. Those test pieces of CA and CB were machined on its surface by a shaping machine, and they have rough surface (Ra > 10μm). It suggests that the signals of Vx is large when the roughness of the test piece is large even with same microstructures.

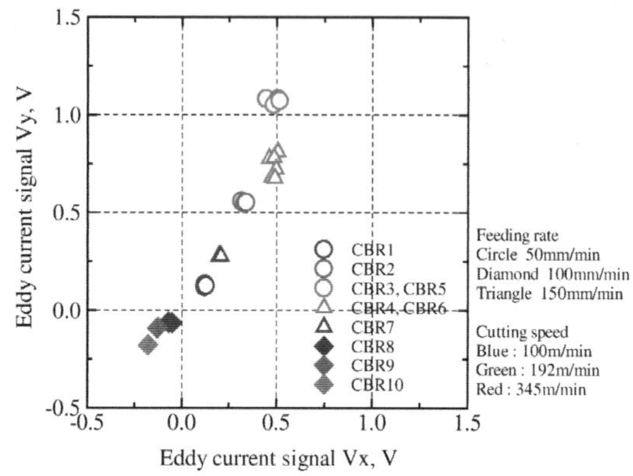

Fig.3 Distributions of eddy current signals of the test pieces with various surface cutting condition

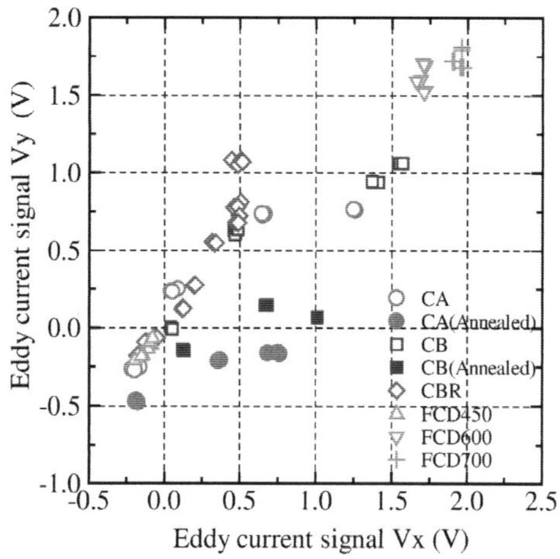

Fig.4 Distributions of eddy current signals of the test pieces with various microstructure and cutting condition (25kHz)

4. Conclusion

1) Effect of the residual stress and the surface roughness on the eddy current signal can be distinguished.
2) Difference of matrix can be evaluated by using eddy current signals, but the effect of the surface machining should be considered.

Acknowledgements

This research was supported by Japan Society for the Promotion of Science Grant-in-Aid for Young Scientists (B)

References

[1] S. Konoplyuk, T. Abe, T. Uchimoto, T. Takagi, M. Kurosawa, NDT&E Int. , vol.38, 623–626(2005)

[2] N. Horikawa, Proc. of the 71st world foundry congress, (2015)

Proceedings of the 72nd World Foundry Congress
21-25th May 2016, Nagoya, Japan

Solidification sequence of ductile cast iron - In-situ observation and modeling -

Hideyuki Yasuda[1], Akira Sugiyama[2], Tomoya Nagira[3], Kohei Morishita[1], Masato Yoshiya[3], Tomohiro Nishimura[1] and Koshiro Yamane[4]
[1] Department of Materials Science and Engineering, Kyoto University, Japan
[2] Osaka Sangyo University, Japan
[3] Department of Adaptive Machine Systems, Osaka University, Japan
[4] IHI Master Metals, Japan

Radiography was performed to examine solidification and melting behaviors in hypereutectic cast iron specimens with / without Mg addition. The specimens without Mg addition showed that graphite particles exhibited spherical shape at the beginning and then transformed to the flake-like shape. Growth of γ-Fe dendrites was followed by nucleation of graphite particles ahead of the eutectic front. In contrast, growth of graphite particles and γ-Fe dendrites occurred nearly at the same time in the specimens with Mg addition and the spherical graphite particles were engulfed by γ-Fe or eutectic. The addition of 0.05mass%Mg significantly reduced the temperature range in which the graphite particles grew as the primary phase. Image-based analysis of melting behavior clearly proved that even 0.05mass% addition modified the phase equilibrium of the liquid, γ-Fe and graphite phases.

Keywords: *Radiography, X-ray imaging, spheroidal graphite, cast iron*

1. Introduction

Small addition of Ce or Mg to hypereutectic cast iron changes morphology of graphite particles from a flake shape to a more compacted / spheroidal shape [1-2]. The solidification sequence in the cast iron is rather complicated [3]. The shape of graphite particles transforms from lamellar graphite in a coupled eutectic structure to compacted graphite, and from compacted graphite to spheroidal graphite in the divorced eutectic structure with increasing Mg or Ce content in the melt [3-5]. The results show that the divorced eutectic growth is normally selected in the melt containing Mg or Ce. However, it is still unclear how the solidification proceeds in the melt with / without Mg and the spheroidal graphite particles form.

In recent years, radiography using synchrotron radiation X-rays has enabled to the observation of solidification in metallic alloys. The technique has been developed for observing solidification of Fe-based alloys [6]. The solidification of the cast iron with hypoeutectic composition [7] and the hypereutectic compositions were observed [8-10].

This paper demonstrates the solidification sequence in the hypereutectic cast iron with and without Mg [8-9]. Second, the influence of Mg on the phase equilibrium (liquid phase, γ-Fe and graphite) is presented [10].

Table 1 Compositions of specimens (mass%).

Sample	C	Si	Mn	Mg	S	CE
0.05Mg	3.73	2.57	0.45	0.05	0.004	4.59
0.002Mg	3.69	2.71	0.45	0.002	0.004	4.59

2. Nucleation and growth of graphite

Table 1 lists initial compositions of the specimens used for the radiography. The carbon equivalent values, which is calculated by [mass%C] + [mass%Si] / 3, are 4.59 for both specimens. Figures 4 (a) and (b) show the solidification structure in 0.002Mg and 0.05Mg at 0.17K/s [9], respectively.

In 0.002Mg (Fig. 1(a)), graphite particles were spheroidal just after the nucleation events. The transition to the flake-like shape was observed at t>-99s. γ-Fe dendrites were immediately followed by the eutectic solidification (γ-Fe and graphite. referred to as the eutectic). In addition, a number of graphite particles nucleated ahead of the eutectic front even though the primary graphite particles were pre-existing.

Different solidification structures were observed in 0.05Mg (Fig.1(b)). Nucleation of graphite particles and dendrites of γ-Fe occurred nearly at the same time and they grew independently. Nucleation events of the primary graphite particles continuously occurred until the eutectic solidification commenced. The graphite particles with spherical shape grew for 20s and then were engulfed by the eutectic front. In 0.05Mg, the temperature range in which the graphite particles grew

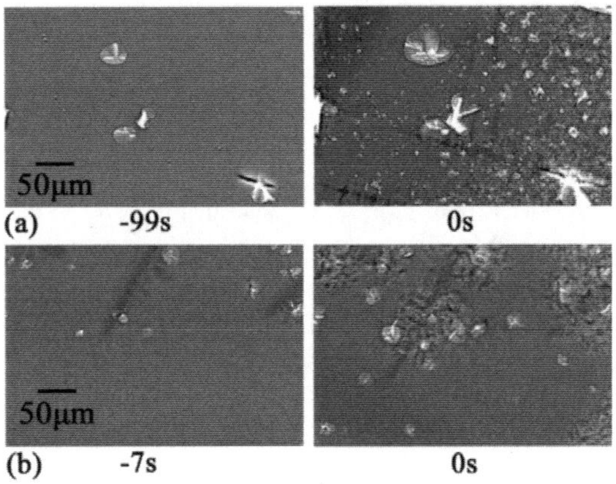

Fig. 1 Solidification (a) in 0.002Mg at 10K/s and (b) in 0.05Mg at 10K/s [9]. Time is designated 'zero' when the eutectic solidification front reached the center of observation area.

independently was only 3-4K. In addition, the nucleation frequency of graphite particles in 0.05Mg was approximately three times larger than that in 0.002Mg.

3. Influence of Mg on phase equilibrium

There are two possible explanations for the decrease in the temperature range of the primary graphite particles; one is modification of phase equilibrium (thermodynamic influence) and the other inhabitation of graphite nucleation (kinetic influence). However, the inhibiting conflicts with the promotion of graphite nucleation (Fig.1(b)).

A thermal image analysis [10] was performed to examine the Mg influence. During the heating procedure (0.194K/s), a part of the specimen was locally melted but the melting region was not extended only in the 0.05Mg specimen. Melting of matrix was immediately followed by melting of graphite particles in 0.05Mg. On the other hand, melting of the matrix started and melting of graphite was hardly observed in 0.002Mg. The addition of 0.05mass%Mg slightly decreased the eutectic temperature (6K) and largely decreased the liquidus temperature (45K). The difference between the liquidus and the eutectic temperatures was estimated to be 56 K for 0.002Mg and 17 K for 0.05Mg, respectively. The differential image analysis proved the thermodynamic influence of Mg largely reduced the difference between the liquidus and the eutectic temperatures.

4. Summary

The addition of 0.05mass%Mg significantly caused a significant increase of the graphite nucleation events and decreased the temperature range in which the primary graphite particles grew independently. The solidification sequence in 0.002Mg is 1) "L » Gr", 2) "L » Gr " + "L » γ" (divorced eutectic), 3) "L » Gr" + " L » γ" + "L » Gr" (divorced eutectic) and 4) "L » γ + Gr " (coupled eutectic). For 0.05Mg, 1) "L » Gr" or "L » γ", 2) "L » Gr" + "L » γ" (divorced eutectic), 3) "L » Gr" + "L » γ" + "L » Gr" (divorced eutectic) and 4) "L » γ + Gr" (coupled eutectic). The differential image analysis demonstrated the influence of Mg on the phase equilibrium. The addition of 0.05mass% Mg decreased the eutectic temperature slightly and the liquidus temperature remarkably.

Acknowledgements

This study was supported by a Grant-in-Aid for Scientific Research (S) from MEXT (No.24226018). The synchrotron radiation experiments were performed as general projects at beamlines of BL20XU and BL20B2 at SPring-8 (JASRI).

References

[1] H. Morrogh, J. W. Grant, J. Iron and Steel Inst., 155 (1947) 321.
[2] A. P. Gagnebin, K. D. Millis, N. B. Pilling, The Iron Age, Feb. (1949) 77.
[3] D. M. Stefanescu, Metall. Mater. Trans. A, 38A (2007) 1433.
[4] Y.X. Li, B.C. Liu, C.R. Loper, Jr., AFS Trans., 98 (1990) 483.
[5] G. Rivera, R. Boeri, J. Sikora, J. Adv. Mater. Res., 4-5 (1997) 169.
[6] H.Yasuda, T. Nagira, M. Yoshiya, N. Nakatsuka, A. Sugiyama, K. Uesugi K. Umetani, ISIJ Int., 51 (2011) 402.
[7] A, Sugiyama, H. Yasuda, T. Nagira, M. Yoshiya, K. Uesugi, K. Umetani, I. Ohnaka, J. Japan Foundry Eng Soc, 82 (20111) 131.
[8] K. Yamane, H. Yasuda, A. Sugiyama, T. Nagira, M. Yoshiya, K. Uesugi, K. Umetani, C. Ushigome, A. Sato, J. Japan Foundry Eng Soc, 85 (2013) 760.
[9] K. Yamane, A. Sugiyama, T. Nagira, M. Yoshiya, H. Yasuda, Y. Tanaka, A, Sato, K. Uesugi, A. Takeuchi, Y. Suzuki, H. Honda, K. Sato, J. Japan Foundry Eng Soc, 86 (2014) 461.
[10] K. Yamane, H. Yasuda, A. Sugiyama, T. Nagira, M. Yoshiya, K. Morishita, K. Uesugi, A. Takeuchi, Y. Suzuki, Metall Mater Trans A, 46A (2015) 4937.

Proceedings of the 72nd World Foundry Congress
21-25th May 2016, Nagoya, Japan

Effect of Copper and Nickel on the Crystallization, Microstructure and Properties of Vermicular Cast Iron

Grzegorz Gumienny[1], Tomasz Giętka[2], Barbara Kurowska[1], Tadeusz Pacyniak[1]
[1] Department of Materials Engineering and Production Systems, Lodz University of Technology, Stefanowskiego 1/15 Street, 90-924 Łódź, Poland
[2] Material Engineering Unit, University of Technology and Life Science in Bydgoszcz
7 prof. S. Kaliskiego Str., 85-796 Bydgoszcz, Poland

Abstract

The paper presents the results of studies of the effect of copper and nickel concentration on the solidification process, microstructure and selected properties of cast iron with vermicular graphite. The vermicularizing treatment of cast iron was done by an Inmold process. The effect of alloying elements on the temperature of eutectic crystallization and on the temperature of the austenite transformation was discussed. The relationship between the concentration of Cu and Ni, and the temperature of phase transformation were given. The effect of aforementioned elements on graphite shape and the cast iron matrix microstructure was tested. It was demonstrated that Cu and Ni are perlite-forming elements. A comparison of the intensity of their impact was showed. The conditions were given for which in vermicular cast iron with nickel or/and copper an pearlitic matrix is achieved. The effect of the above mentioned elements on the hardness and microhardness of the matrix components was given.

Keywords: *vermicular cast iron, Inmold, crystallization, copper, nickel*

1. Introduction

Vermicular cast iron is one of the modern and the developmental materials with high potential for utilitarian.

Compared to nodular cast iron, cast iron with vermicular graphite offers the following advantages: lower coefficient of thermal expansion, higher heat conductivity, lower modulus of elasticity, lower thermally induced internal stress level, better thermoshock resistance and lower tendency to distortion, better damping capacity, better pouring properties (shrinkage, mold-filling, etc.).

Very interesting complex of properties offered by this cast iron is the object of intensive research and numerous publications [1-6]. The standard specifies five grades of this cast iron with a tensile strength of 300 to 500 MPa at an elongation from 2.0 to 0.5%, respectively.

The increase in the mechanical properties of vermicular cast iron is associated with heat treatment of castings or the change of chemical composition.

There are very few publications, however, which would discuss the effect of alloying elements on the microstructure and properties of cast iron with vermicular graphite [7, 8]. So, the aim of this work was to study the effect of copper and nickel on the solidification behaviour, microstructure and hardness as well as microhardness of vermicular graphite cast iron.

2. Results

The chemical composition of tested cast iron is presented in Table 1.

Table 1
The chemical composition of tested cast iron

Chemical composition, mass %					
C	Si	Mn	Mg	Cu	Ni
3.28-3.83	2.23-2.48	0.27-0.34	0.014-0.025	0-2.92	0-5.21

The study of the crystallization process carried out by derivative thermal analysis method (DTA) indicated that copper (as a graphitizing agent) increased the eutectic transformation temperature of about 6°C per 1% Cu. The addition of Cu decreases the temperature of austenite transformation start of about 7°C per 1% concentration, and the end of about 13°C.

Nickel increases the eutectic transformation temperature of about 4.5°C per 1% of Ni. This element causes decrease in the temperature of austenite transformation start of about 23°C and finish 26°C per 1% Ni.

Figure 2 shows the microstructure of vermicular cast iron containing about 3% Cu.

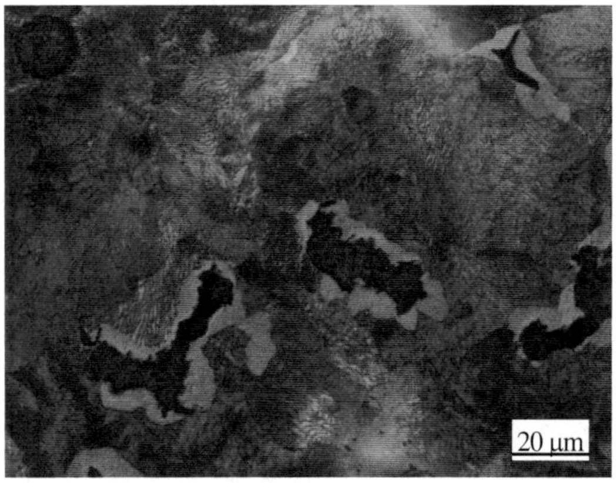

Fig. 2 Microstructure of vermicular cast iron containing about 3% Cu: vermicular graphite, pearlite, ferrite

Fig. 2 indicates that copper in concentration of about 3% caused an almost complete pearlitic matrix of tested cast iron. The small separation of ferrite remained around of graphite due to the formation of the barrier by Cu atoms impeding diffusion of carbon atoms.

In Figure 3 the microstructure of vermicular cast iron containing about 3% Ni is presented.

Fig. 3 Microstructure of vermicular cast iron containing about 3% Ni: vermicular graphite, pearlite, ferrite

The study showed that the surface fraction of ferrite in the cast iron containing about 3% Ni amounts to 35%. It is smaller than in unalloyed cast iron wherein the surface fraction of ferrite was 70%. The increasing of the nickel concentration to about 5% resulted in a further reduction of ferrite and the occurrence of martensite.

Figure 4 presents hardness of vermicular cast iron vs Cu and Ni concentration.

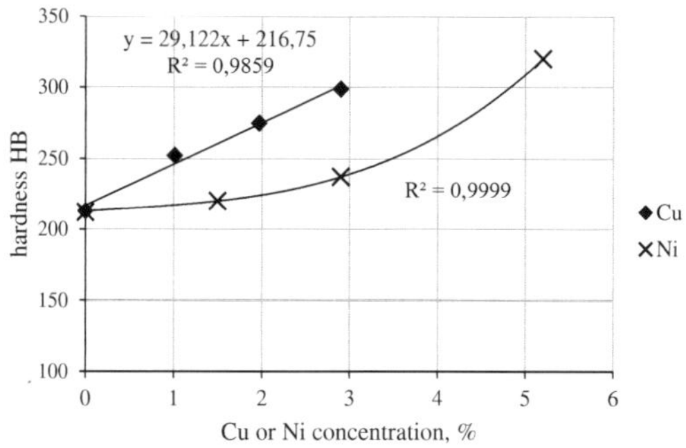

Fig. 4 Hardness of vermicular graphite cast iron vs copper and nickel concentration

Fig. 4 indicates that copper causes an increase in hardness of vermicular cast iron by approximately 29 HB per 1% Cu. The concentration of up to about 3% nickel causes an increase in hardness of the cast iron of about 8 units HB per 1% Ni. Further more intense increase in hardness is due to the precipitation of martensite.

Microhardness tests showed that Cu improves pearlite microhardness of about 14 HV0,1 per 1% Cu probably due to reduce the diffusion of carbon in austenite. The increase in concentrations of Ni did not result in a significant increase in microhardness of pearlite.

References

[1] S. Pietrowski: Solidification of Metals and Alloys. (2000) 279-292.

[2] E. Guzik: Archives of Foundry Engineering 10 (2010) 95-100.

[3] M. Górny, M. Kawalec, G. Sikora: Archives of Foundry Engineering. Special Issue 14, (2014) 139-142.

[4] E. Guzik, T. Kleingartner: Archives of Foundry Engineering 9, (2009) 55-60.

[5] M.S. Soiński, P. Mierzwa: Archives of Foundry Engineering 11, (2011) 133-138.

[6] Z. Andrsova, L. Volesky: (2012). COMAT 2012. 21.-22. 11. 2012. Plzeň, Czech Republic, EU. 2015.

[7] L. Choong-Hwan, G. Byeong-Choon: G.). Met. Mater. Int. 17, (2011) 199-205.

[8] P.I. Popov, I.G. Sizov: Metal Science and Heat Treatment 48, (2006) 272-275.

Proceedings of the 72nd World Foundry Congress
21-25th May 2016, Nagoya, Japan

High performance cast parts. Advanced feeding systems for metallurgical optimization

J.Prat[1], G.Artola[2], G.Trillo[2] and J.Izaga[2]
[1] ASK-Chemicals España, S.A.U., C/ Autonomía nº4-2º, 48940 Leioa (Bizkaia), Spain
[2] IK4-Azterlan, Aliendalde Auzunea 6, 48200 Durango (Bizkaia), Spain

Foundry sector, as a whole, has incorporated relevant technological advances into productive processes. Nevertheless, the specific field of feeding system design criteria hasn't improved in the same proportion. The use of simulation tools doesn't always come together with the optimization of key indicators. Part rework costs remain unchanged, yield values hardly reach 50% and the metallurgy of the processed alloys is kept under nearly unchangeable principles.
With this starting point, a deep review of the feeding systems that are employed for the manufacture of cast steel components is carried out, to lead in the end, to the identification and assessment of the metallurgical advantages that can be exploited in the most thermally affected zones and to the net/gross yield optimization.
A complex variable matrix has been defined and studied, taking into consideration high added value materials, their metallurgy both in liquid and solid state, novel feeding devices such as core-risers, riser-paddings and EXACTCAST™ (patented) mini-risers, and advanced simulation concepts.
The most important achieved results are exposed which, taking advantage of the new feeding technologies, allow reaching yields over 75%, with an optimized metallurgical quality in the final product.

Keywords: *Mini-risers, thermal modulus, metallurgical quality, net/gross yield.*

1. Introduction

Previous work from the authors [1] has been referred to advanced feeding systems in small cast steel parts. In this work, it has been considered necessary to apply the previously developed knowledge, in order to solve the feeding problems in big steel castings.

The couple of terms "advanced feeding systems/ big sized parts" in this paper and the results that are presented, are properly backed-up by industrial practice.

2. Experimental procedure

The previously developed calculation methodologies, mainly based in the concept of thermal modulus, are applied. Thus, it has been necessary to characterize, the from thermo-physical point of view, the mixtures employed in the new feeding devices.

2.1 Feeding system design

The conception of the feeding systems must answer the thermal modulus and volume requests, so that shrinkage related incidences are minimized. These studies have been performed for super-duplex alloys and carbon steels.

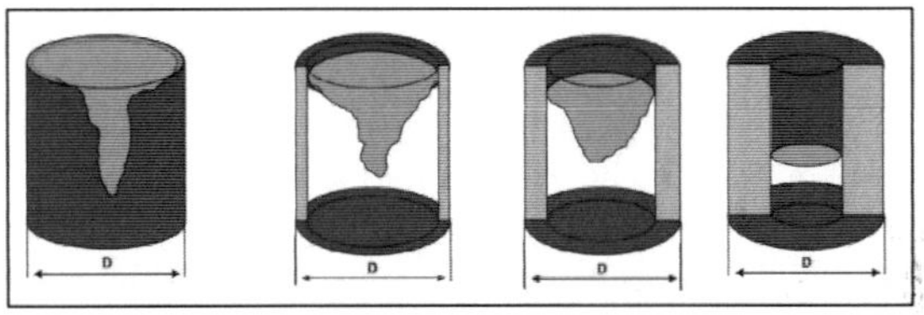

Fig. 1. Different mixtures sand, insulating, exothermic-insulating and mini-risers [1].

2.2 Molding systems

A variable of special relevance, not always taken into account, is the relationship between the molding system and the superheating needs. It has been verified that the molding system exerts a great influence on the metal's castability and, in consequence, on the superheating requirements, that can be very unfavorable in terms of shrinkage defects.

Fig. 2. Castability tests.

Regarding the feeding system design, the application of the exothermic paddings has been found remarkable since, combined with mini-risers, offer great benefits in terms of yield and postprocessing costs.

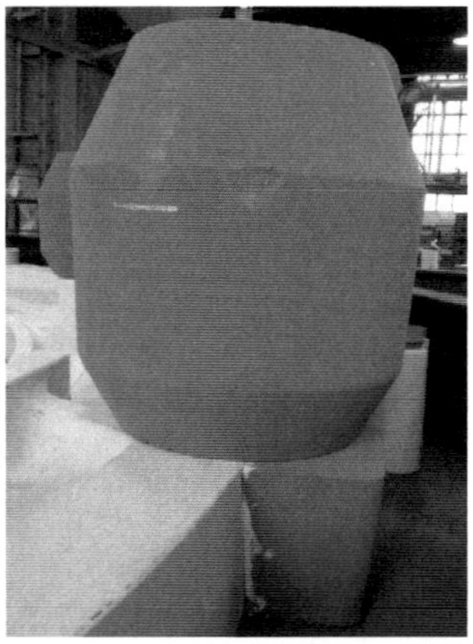

Fig. 3. Mini-riser with an 34.500cc exothermic padding.

2.3 Industrial trials

Once the feeding systems are designed, the corresponding manufacturing batches are performed, so that conventional and advanced feeding systems can be compared. As an average value, 77% yield has been achieved with the advanced feeding solutions.

3. Results and discussion

The presented results are related to big sized cast components that have been processed in several alloys.

3.1 Hot tearing

The severe thermal gradients that are generated in section changes, frequently lead to cracks of thermal origin. A fine tuning of the simulation models allows realistic predictions that can be used as hot tearing indicators.

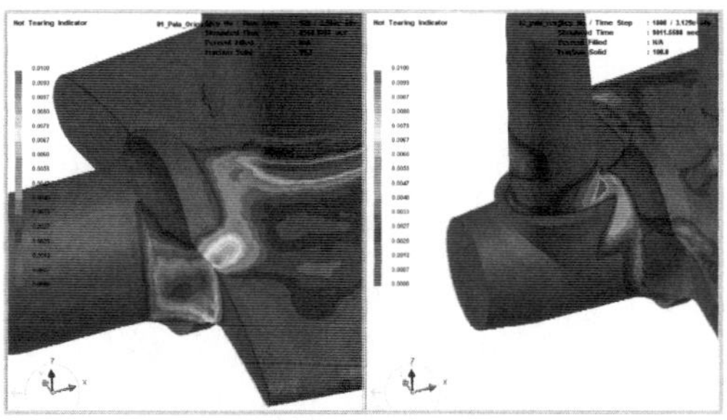

Fig. 4. Thermal stress analysis.

It has been verified that the thermal gradients and the sharp section changes in the set "part-feeding system" are critical and, if the thermal characteristics of the feeding system are properly designed, hot tearing can be strongly reduced and even disappears when using mini-riser.

3.2 Influence on microstructure

Super-duplex alloys are especially sensitive to the presence of micro-precipitates that are directly related to a reduction in mechanical properties and corrosion resistance. Employing mini-risers is very beneficial in order to reduce these negative effects, as they allow a more precise control of solidification sequence.

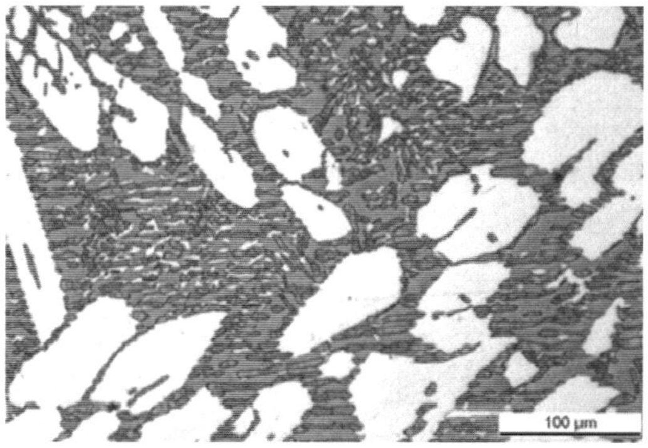

Fig. 5. Micro-precipitates with standard risers.

4. Conclusions

The development of advanced feeding systems (mini-risers and paddings) is a competitiveness key point, both in terms of profitability (net/gross yield) and in terms of technological advantages. The industrial application of mini-risers is an efficient and effective tool for the optimization of the metallurgical quality of cast parts.

Acknowledgements

Our most grateful acknowledgement to the foundries that have helped in developing the studies behind this paper, Fundiser, S.A., Fondesal, S.A., Ebroacero, S.A. and Pyrsa, S.A.

References

[1] A.Carrasco, D. Peña, G.Trillo, J.Prat, J.Izaga, M. Manzanares : 71st World Foundry Congress, Bilbao (2014).

[2] J. Prat: VIII International Foundry Technical Forum, Bilbao (2013).

Proceedings of the 72nd World Foundry Congress
21-25th May 2016, Nagoya, Japan

Effect of coating thickness on Melt Filling Rate of Cast Iron in Evaporative Pattern Casting Process

Toru Maruyama[1], Gou Nakamura[2], Mitsuyoshi Tamaki[2] and Keisuke Nakamura[2]
[1] Department of Chemistry and Materials Engineering, Kansai University, Japan
[2] Graduate School of Science and Engineering, Kansai University, Japan

The increment of a melt filling rate of cast iron on EPC process was investigated. The melt filling rate, the decomposition rate of a polystyrene formed pattern were measured during the casting experiment. A copper alloy was also cast in the same method for comparing with the case of cast iron. The melt filling rate of the cast iron was lower than that of the copper alloy. Liquefied resin was formed by thermal decomposition of the formed pattern. A volume of gas gap was increased with increasing melt temperature. The melt filling rate increased with increasing coating permeability, and more increased with increasing the coating thickness. Misrun tends to occur in the case of cast iron melt although the coating permeability increased. In the case of a thick coating, misrun was prevented.

Keywords: *Cast iron, Coatings, Evaporative pattern casting, Permeability, polystyrene.*

1. Introduction

Molten metal is filled into mold by exhausting thermal decomposition products through castings in evaporative pattern casting process. Coating permeability is important factor for melt filling [1, 2], and higher permeability is recommended because large amount of thermal decomposition gas forms during casting. Liquid products also form, and can penetrate into the coatings. The coating permeability may decrease by filling porous parts in the coatings with the liquid products. The coatings with large thickness are expected to absorb the liquid products for exhausting the liquid products from a gas gap [3], especially in aluminum alloy melt whose temperature is smaller than that of cast iron melt. However, there are few reports about influences of absorption of liquid products on filling of cast iron melt. Cast iron melt was poured into a polystyrene form coated with same permeability coatings whose thickness was changed even the coating permeability was same.

2. Experimental Procedures

The casting design of bottom pouring method was applied to avoid turbulence flow of molten metals. A density of the polystyrene foamed pattern was 15.4 kg m^{-3}. Dimension of a columnar pattern was 40 mm in outer diameter and 250 mm in height. Melt tough sensors and gas pressure ports were inserted into the pattern. The sensor points were set at 5mm, 65mm, 125mm, 185mm and 245mm form bottom of the pattern. For collecting the thermal decomposition products, the products were flowed from the gas gap to a sampling bin with PTFE tube inserted in the formed pattern at 245mm from bottom. The patterns were coated by coatings whose thickness was 0.25 mm and permeability was changed from 1.10 to 4.84. Coatings of 0.72 mm in thickness of 1.29 in permeability were also applied for understanding the influence of coating thickness on the melt filling. Cast iron melt (CE: 4.3) and Cu-25mass%Sn alloy melt were cast into the mold at 1473K and 1173K respectively.

3. Results and discussion

Figure 1 shows the relationship between elapsed time after pouring Cu-25%Sn alloy melt at 1173K and thermal decomposition distance and melt filling distance from bottom of pattern whose permeability is 3.71. The pattern decomposition and the melt filling progressed with elapsed time. The pattern decomposition and the melt filling were almost finished in 3 s and 9 s, respectively. Figure 2 shows the relationship between elapsed time after pouring cast iron melt at 1473K and thermal decomposition distance and melt filling distance from bottom of pattern whose permeability is 3.71. The progresses were similar to 1173K in melt temperature. However, rates of both the pattern decomposition and the melt filling were lower than those of 1173K even the melt temperature is 300 K higher than Cu-25%Sn alloy.

The cast iron melt was not filled to 250 mm and misrun occurred.

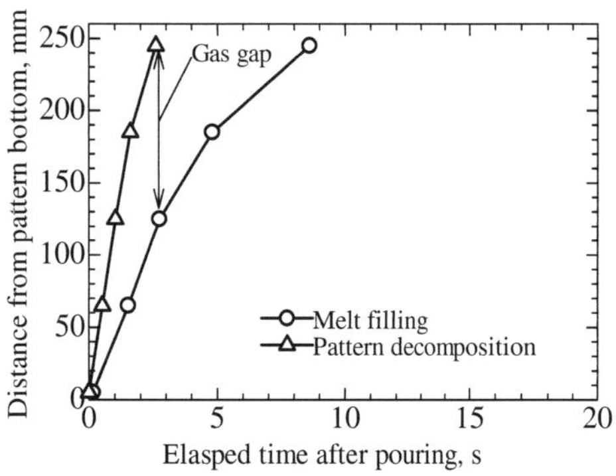

Fig. 1 Relationship between elapsed time after pouring Cu-25%Sn alloy melt at 1173K and thermal decomposition distance and melt filling distance from bottom of pattern whose permeability is 3.71.

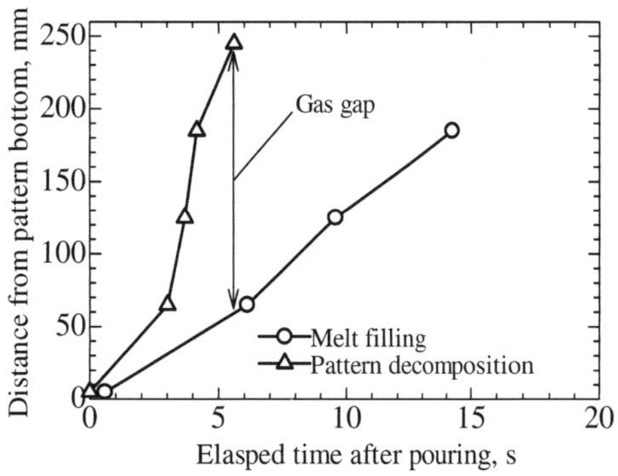

Fig. 2 Relationship between elapsed time after pouring cast iron melt at 1473K and thermal decomposition distance and melt filling distance from bottom of pattern whose permeability is 3.71.

The gas gap thickness in the case of 1473K was larger than that in the case of 1173 K. Large gas gap thickness means that rate of exhaust of thermal decomposition products is also large, and large melt filling rate is expected. However, the melt filling rate of the cast iron melt was considerably-slow. It is suggested that exhaust of thermal decomposition gas is inhibited from any cause. For understanding state in the gas gap, the thermal decomposition products in the gas gap during casting were collected. The collected products were gas, mist and liquefied resin. The amount of liquefied resin from cast iron melt was much larger than that from Cu-25%Sn alloy melt. In the case of the cast iron melt, it is considered that the liquefied resin penetrates into the coatings and clogging in the coatings occurs.

Figure 3 shows the influence of coating permeability and thickness on melt filling times of cast iron and Cu-25%Sn alloy. The melt filling times of both alloys decrease with increasing coating permeability and with increasing coating thickness. Also, misrun did not occur in the case of thicker coating thickness. Large permeability and thickness means large porosity volume in coatings. It is suggested that large porosity volume in coatings absorbs the liquefied resin formed by thermal decomposition and the clogging in the coatings is prevented.

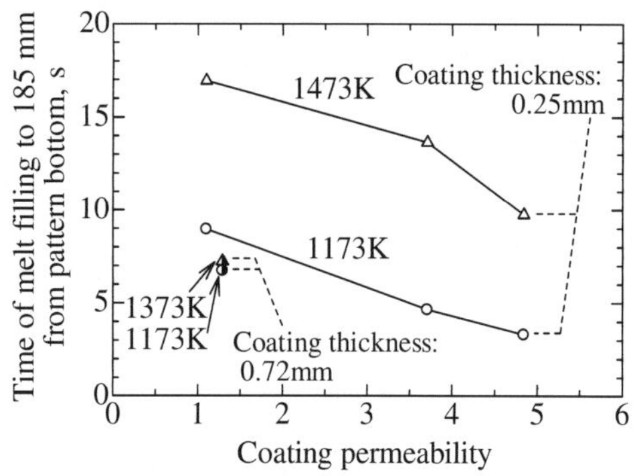

Fig. 3 Influence of coating permeability and thickness on melt filling.

4. Conclusions

The melt filling rates of cast iron and copper alloy were measured in changing coating permeability and thickness. The following results and conclusions were obtained:

(1) In cast iron melt at 1473 K, thermal decomposition rate and melt filling rate are smaller than those in Cu-25%Sn alloy melt at 1173 K.

(2) The larger permeability and thickness of coatings increases, the larger melt filling rate increases.

References

[1] C. A. Goria, G. Serramoglia, G. Caironi and G. Tosi: Trans. Am. Foundry Soc., 94 (1986) 589-600.

[2] J. J. Green, C. W. Ramsay and D. R. Askeland: Trans. Am. Foundry Soc., 106 (1998) 339-347.

[3] J. Yang, T. Huang and J. Fu: Trans. Am. Foundry Soc., 106 (1998) 21-26.

O-126

Proceedings of the 72nd World Foundry Congress
21-25th May 2016, Nagoya, Japan

Residual Stress Measurement in Large Grey and Ductile Iron Castings

Lars Noppenau
MAN Diesel & Turbo SE, Two-stroke Engineering, Production Support, Copenhagen, Denmark

Simulation of the casting process has become an increasingly important tool to improve the design of the various cast components in a modern large-bore two-stroke diesel engine and is therefore used extensively in the development and design phase.

In order to better understand the state of residual stresses in our cast components and to compare results from commercial stress simulation software with real life data, residual stress measurements have been performed using the ring-core method.

The results show that the ring-core method is viable and that predictions by the simulation software is relatively close to the measured stress levels and more important, the tendencies and distribution of the stresses correspond very well.

Keywords: Residual Stress Measurements, Ring-core Method, Cast Iron, Simulation.

1. Introduction

The residual stress measurements have been performed on cylinder frames, as these are a vital component in the two-stroke engine structure and have large variances in section thickness.

In total 32 measurements have been performed on three different cylinder frame castings, additionally, 10 measurements were performed in our laboratory in Copenhagen on a stress relieved grey iron bar subjected to a known load.

The measurements have all been compared to simulations with MAGMASOFT®, using the unified creep material model. It should be noted that for none of the castings on which the measurements have been performed, have we had access to detailed data on the cooling conditions, these are therefore based on qualified guesses.

2. Ring-core method

The method used for the residual stress measurements described in this paper is the ring-core method, in which a strain gauge is fixed to the surface of interest, and a drill (in this case mechanical) is used to remove the material surrounding the strain gauge, and by this relieving the stresses, so that the inverse stress-state is established by the strain gauge.

1. Dial for measuring drilling depth.
2. Wire from the strain gauge going through the drill and connecting to the terminals.
3. Drill with magnetic base.
4. Core drill.
5. Strain gauge.

Fig. 1: Setup for the measurements

3. Measurements

3.1 20 ton grey iron cylinder frame

The measurements were performed on a 6S50MC-C cylinder frame (EN-GJL-250 alloyed with copper) cast by Shanghai Hulin Heavy Industry Co., Ltd. The measurements were performed at the foundry, after cleaning of the casting.

Figure 2 show the results of the measurements performed in the transverse direction compared to the results of the simulation.

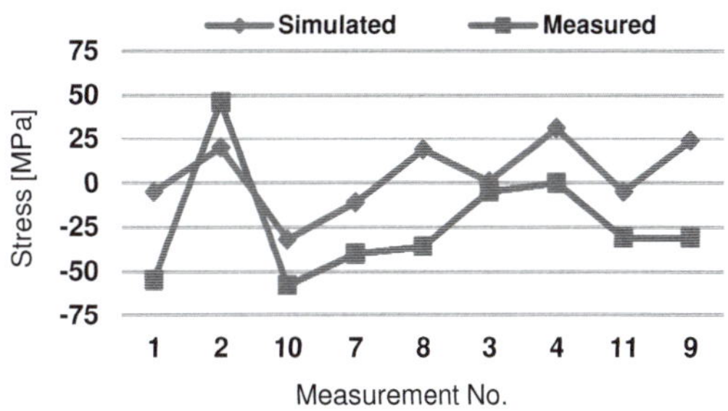

Fig. 2: Transverse results, simulated vs. measured

3.2 10 ton ductile iron cylinder frame

The measurements were performed at Mitsui Engineering & Shipbuilding Co., Ltd., on a ductile iron 3S50MC-C test casting (EN-GJS-500-7) cast by

Mitsui Meehanite Metal Co., Ltd. Sasayama Manufacturing Section.

In total 13 measurements were performed, there were however some problems resulting in unusable results, still 8 reasonably good and stable measurements were performed.

Figure 3 show the results of the remaining measurements, in the longitudinal direction, compared to the results of the simulation.

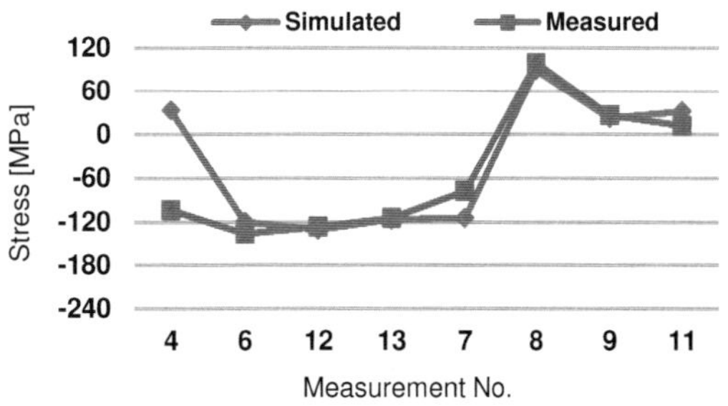

Fig. 3: Longitudinal results, simulated vs. measured

Except for measurement no. 4, the match between the measured and simulated results is exceptionally good.

3.3 Tests on loaded grey iron bar

After the problems experienced during the former measurements, 10 measurements were performed in our laboratory in Copenhagen, on a stress relieved grey iron (EN-GJL-250) bar subjected to a known load, to test different drills and the effects of grinding.

The force was adjusted to ensure a load of 75 MPa in the cross-section of the test bar.

The results from all 9 successful measurements is shown in figure 4. It can be seen that the reproducibility of the measurements is very good.

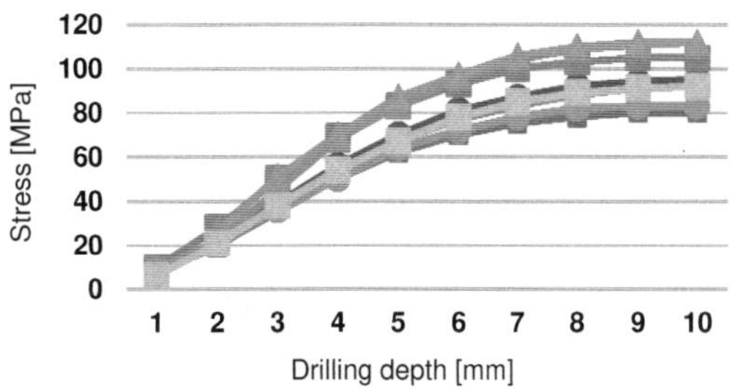

Fig. 4: All measurements

The measurements showing higher than the actual 75 MPa were the measurements on which the bar had been exposed to excessive grinding or on which an unsuitable drill was used.

3.4 25 ton ductile iron cylinder frame

The measurements were again performed at Mitsui Engineering & Shipbuilding Co., Ltd., this time on a ductile iron 6S50ME-B cylinder frame (EN-GJS-500-7) cast by Mitsui Meehanite Metal Co., Ltd. Sasayama Manufacturing Section.

10 measurements were performed, all in a relatively thin (30 mm) section.

The longitudinal results and the results from the simulation can be seen in figure 5, again the tendency of the distribution of stresses in the various points of the cylinder frame matches very well between the simulated and measured results.

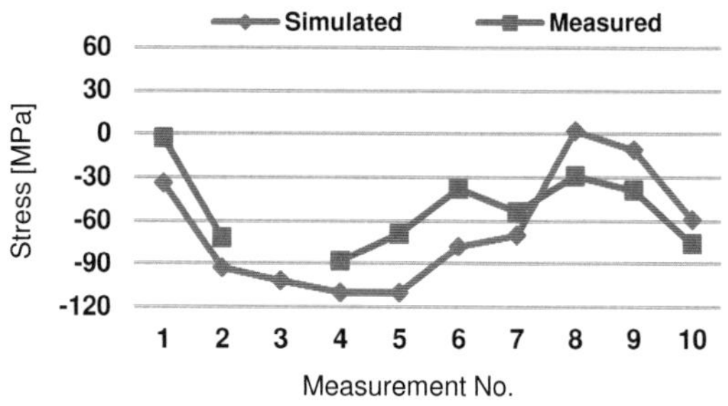

Fig. 5: Longitudinal results, simulated vs. measured

4. Conclusions

After initial problems in performing the measurements, we now belive that we are able to consistently perform good measurements using the ring-core method on grey and ductile iron castings where a relatively large plateau of even stresses is present.

The tests in our laboratory proved that the method gave the expected results and also showed that excessive grinding and/or the state of the drill have an influence on the measured strains.

Generally, and considering our limited knowledge of the shake out procedure and cooling conditions of the castings, the prediction of stresses by MAGMASOFT® has been very good, which has given us increased confidence in using simulation as part of our development and design process.

Proceedings of the 72nd World Foundry Congress
21-25th May 2016, Nagoya, Japan

Process improvement by optimal use of feeding systems

Stefan A. Fischerl[1], Calogero Vassallo[2]

[1]ASK Chemicals Deutschland GmbH, Reisholzstrasse 16-18, 40721 Hilden/ Germany
[2] ASK Chemicals Deutschland GmbH, Reisholzstrasse 16-18, 40721 Hilden/ Germany

1. Abstract

The development of feeding technology has improved very much in recent decades. This development started with very simple- shaped sand-molded natural risers, the shape was modified over the years to increase the feeding performance of these so-called natural risers. Another important big step was the introduction of exothermic material to increase the performance of the riser, the development continued with introduction of the mini-riser technology end of the 1970´s.

Modern foundries are facing demands for castings with great complexity and high requirements for process reliability. A few examples of such challenges affecting many foundries are reduction of emissions, elimination of casting defects, and supply of consistently high quality, very often combined with the pressure of cost savings. These requirements can be met by qualified, motivated personnel and outstanding technologies, driven by top-level research and development focusing on the elaboration of efficient and ecofriendly solutions, without loss of performance. A good example of this is the mini-riser technology that combines several patents and offers unique efficiency to foundries.

Reliable support for the foundry industry to maintain its strong position and competitiveness on the global markets by refining feeding systems on a continuous basis is the objective of the foundry suppliers.

***Keywords:** Feeding, miniriser, yield optimization, reduced fettling costs.*

2. Introduction

From time immemorial, the shrinkage of metal during the cooling-down from casting temperature to ambient temperature has had to be corrected to ensure the casting quality.

3. The basic of feeding

3.1 Solidification and feeding direction

The solidification of a casting starts at the coldest point of the casting geometry and runs towards the hotter zones. This is called "aligned solidification".

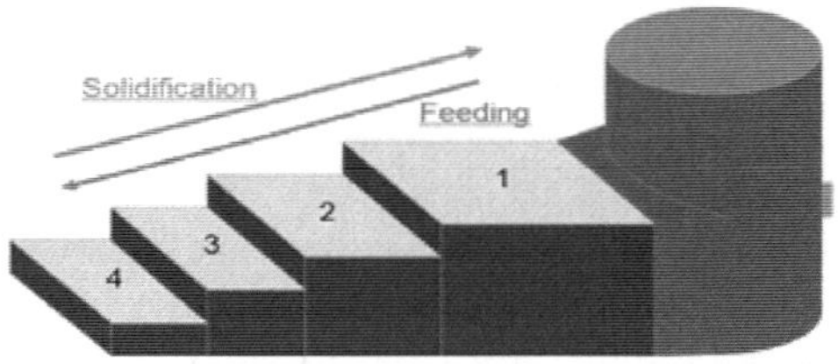

Fig. 1 Solidification and feeding directions

3.2 Modulus and feeding volume

To get an idea of the solidification times of the different sections and the risers applied, the calculation of the modulus was developed.

4. Development of feeding systems

4.1 Natural risers

The easiest way to feed a casting is to set a cylindrical riser on top of or next to the hottest zone of the casting.

4.2 Exothermic solutions and riser sleeves

By using the exothermic reaction of the thermite process, also called the Goldschmidt method, it has been possible to introduce additional heat to the riser.

4.3 Mini-risers – optimal use of feeding systems

The development of the mini-risers was a big step towards improving the performance of the riser and optimizing the yield of the castings. The basis of this development was the knowledge, that the wall thickness of an exothermic riser sleeve has a big influence on the behavior of the shrinkage in the riser.

Besides the main purpose of the riser to avoid shrinkage defects in the casting, the reduction of fettling costs acquired increasing significance over time.

5. Increased productivity for green sand molding lines

5.1 Accurate breaker edge through the implementation of metal necks

To make the application easier, a new kind of mini-riser with a loose metal neck was developed. The hole in the top was closed by a special plastic cap.

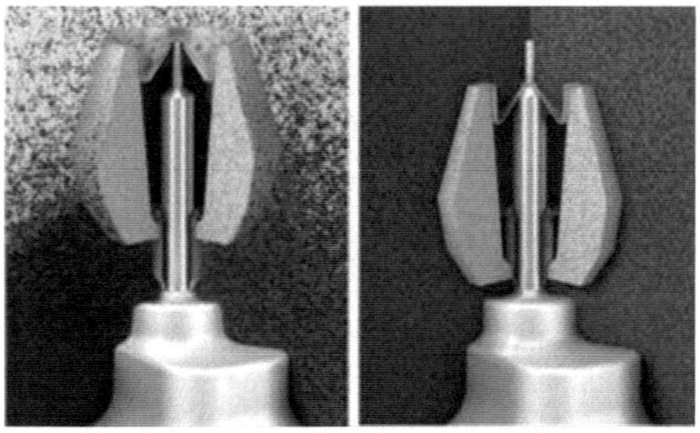

Fig.2 Riser with pin before and after compaction

5.2 Avoiding of the spring back effect

When using metal necks which are positioned outside the riser the compaction forces can cause elastic deformation of the metal neck, and thereby mold fissures can appear

This defect can be eliminated completely with the loose metal neck.

5.3 Influence of the metal neck on the riser performance

The installation of the metal neck into the inside of the riser has the positive effect that the metal neck is mainly not exposed to any forces during the compaction.

5.4 Elimination of disruptive riser particles in the mold

During the compaction process, rigid pins usually destroy the riser and the crumbling riser particles can fall into the mold. The introduction of risers which are fitted with a cover of non-friable material solves this problem.

6. Process improvement by using the precise riser composition

6.1 The riser composition

Improvement of the riser compositions is an essential part of achieving better production processes and finding modern and environmentally friendly feeding technologies.

6.2 Fiber-free composition

Conventional sleeves encountered mainly in steel casting contain fibers and rice husks which could cause casting defects and contaminate the casting with fibers through erosion.

Fig.3 rice husk and fibers (left), low density aluminum silicate ceramics (right)

6.3 Fluorine-free composition

A further step to expand the feeding technology was done by introduction of a new fluorine-free riser composition.

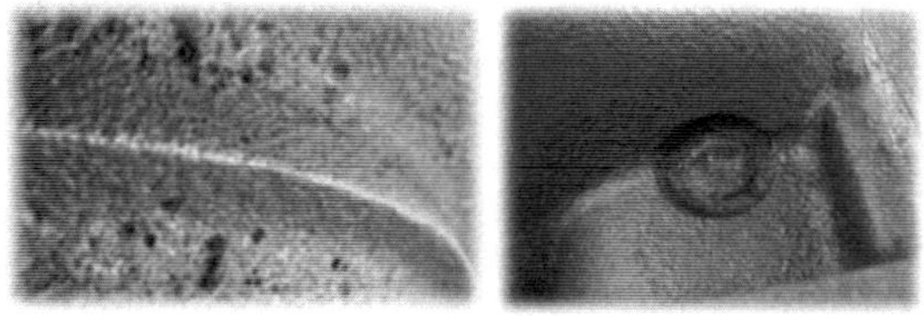

Fig.4 Fluorine-induced surface defects

7. Environment, health and safety (EHS)

The reduction of fluorine by using fluorine-free risers can have a remarkably positive effect on the ecological footprint of the foundry industry.

8. Conclusion

The challenges for foundries in today's highly competitive world markets are not only driven by economic aspects but also by current expectations of eco-friendly solutions.

Modern riser solutions involving new riser geometries and modern riser compositions are the outcome of the long-term cooperation between riser-producer and the foundries.

The suppliers of the foundry industry are not simply dealers any longer. The suppliers are also partners in development and improvement of the casting processes, partners who do not cease in their activities to provide the foundries with high-end solutions in order to ensure their prominent positions on the markets.

References

[1] E. Brunnhuber, *Gießerei Lexikon*

[2] N. Chvorinov "*Theory of the Solidification of Castings*", Giesserei, 1940, Vol 27

Proceedings of the 72nd World Foundry Congress
21-25th May 2016, Nagoya, Japan

Research on ion carbonitriding and wear resistance of cast iron

LI Jilin[1], ZHANG Fang[1], LI Yongjie[2]
[1]Institute of Materials Processing, Guangzhou General Research Institute for industrial technology, Guangzhou 510650, China; Guangzhou Yi Yin New Material Technology co., LTD,Guangzhou 510530, China;

In this study, the microstructure and phase, wear resistance of ion carbonitrided cast iron were investigated. Casting crankshaft consists of ductile cast iron and grey cast iron samples were subjected to quenching and tempering pretreatment. Ion carbonitriding were carried out at 570 ℃ for 6h, followed by cooling in the furnace. The result showed that there are bright layer which thickness is 120~210μm, appeared on the surface of samples. What's more, after ion carbonitriding, the hardness of samples were improved significantly which can enhance wear resistance greatly. In addition, in comparison with carbonitrided grey cast iron, wear resistance of carbonitrided ductile cast iron shown better performance.

Key words: *cast iron; carbonitriding; wear resistance;*

1. Introduction

In order to improve the hardness, abrasion resistance, fatigue resistance of the cast iron, chemical heat treatment on the surface are often used in industrial production. Carburizing, nitriding technique has been widely used in steel production. However, little information is available regarding the carbonnitriding technology applied to cast iron. In this paper, two different cast irons were both processed by the same carbonitriding method. Furthermore, compared these two kinds of cast irons which was processed by carbonitriding regarding their characteristics and wear resistance. The results of this paper strive to provide the theoretical basis for related theory and practical application.

2. Experimental

2.1 Experimental material

The chemical composition of the investigated are given in Table 1. Samples were divided into 5 groups, 1 #, 2 #, 3 # samples are different spheroidizing effect of centrifugal casting ductile iron samples, and, 4#, 5# are sand casting crankshaft gray cast iron. The 4# is HT250 (CuCr Low alloyed gray iron), 5# samples for HT300.

Table 1 Chemical composition of samples (wt. %)

	C	Si	Mn	S	P	Mg	Ce	Cu	Cr
1#	3.51	2.34	0.35	0.009	0.021	0.036	0.014	----	----
2#	3.54	2.48	0.35	0.015	0.022	0.036	0.014	----	----
3#	3.57	2.53	0.35	0.011	0.021	0.036	0.016	----	----
4#	3.30	1.86	0.92	0.020	0.050	----	----	0.52	0.22
5#	3.25	1.91	0.92	0.055	0.048	----	----	----	----

2.2 Experimental methods and procedure

Before ion carbonitriding, quenching and tempering treatment was necessary for all samples. All samples were quenching in the electric furnace with temperature 920 °C, holding 1.5 hours, followed by air cooling to room temperature. Then, tempering at 560 °C for 2h, followed by air cooling to room temperature.

Ion carbonitriding was carried out in the LDMC-50AQK plasma nitriding furnace at 560 °C for 6h; Wear resistance test was carried out in the UMT-3(Nano Tribometer) friction and wear pilot facility.

3. Result and analysis

3.1 Microstructure after ion carbonitriding

The microstructure after ion carbonitriding was shown in Fig1. After ion carbonitriding, there are birght layer and black diffusion layer appeared on the samples surface, and center for the matrix metal. It is similar to the microstructure of ion carburizing, nitriding structure. The matrix metal was tempered sorbite. In addition, compared with the samples without ion carbonitriding, Sorbite of the carbonitrided samples evenly distributed in the matrix metal and ferrite phase is almost disappeared. It illustrated that the carbonitriding process has benefit to restrain matrix metal grain growth .

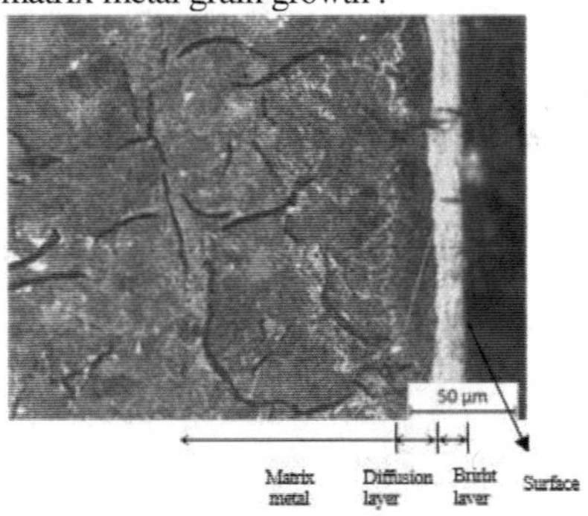

Figure 1 Microstructure of carbonitrided samples

3.2 The phase composition of carbonitrided surface layer

X-ray diffraction (XRD) spectrum of carbonitrided surface layer of 3#, 4# samples were shown in Fig 2 and 3. Both the ductile cast iron and grey cast iron have a multiphase structure, consisting of γ′ (Fe_4N), ε (Fe_3N), Fe_3C phase. These compounds affected the chemical composition of the surface, which has played an important role in improving the performance of samples. In a vacuum furnace, adding DC between workpiece which is cathode and furnace wall which is anode to make the rarefied gas ionization for forming H^+, N^+, $C4^+$. Then, the accelerated ions bombard cathode work piece. In addition, ion carbonitriding has strong capability providing C, N. The characteristics of cast iron which containing high carbon and silicon, played an obstacles role in infiltration of C, N atoms of carbonitriding. Then, a series of reactions will take place in the workpiece surface. Fe will combine with C, N atoms to form Fe-C, Fe-N compound. Since the back scattering effect, Fe-N compounds deposited at the cathode surface again. Under the effect of ion bombardment and thermal activation, in turn, it will decomposes into Fe-Fe_2N-Fe_3N-Fe_4N, and producing active atoms, the active N atoms seeping into the work piece with existing active C atoms. Furthermore, because Fe_3C and ξ（Fe_3N） have a similar structure, C atoms can replace the N atoms from ξ（Fe_3N）,then get Fe_3C [1].

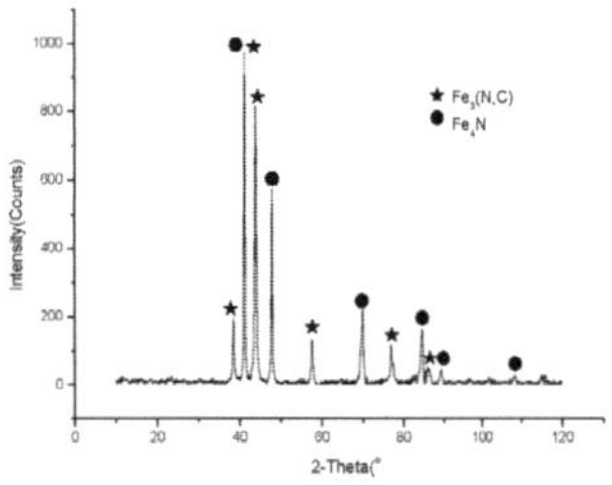

Figure 2 X-ray diffraction spectrum of carbonitrided surface layer of 3# samples

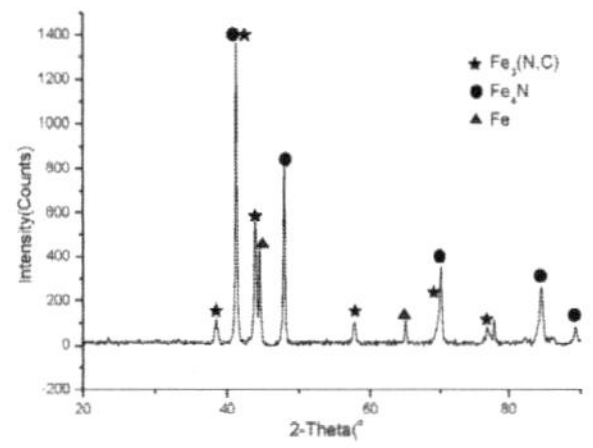

Figure 3 X-ray diffraction spectrum of carbonitrided surface layer of 4# samples

3.3 Test of hardness and thinkness of carbonitrided surface layer

Ladder hardness method is used to determine the thickness of carbonitrided layer. Setting the microhardness tester with loading 50g, holding for 15s. Hardness test will be processed every 30μm.

The carbonitriding layer thickness could be estimated by the hardness change value. The thickness of carburized layer, respectively as follows:1# : 210μm; 2#: 180μm; 3#: 210μm; 4#: 150μm; 5#: 120μm. showed that the thickness of carburized layer for ductile cast iron was thicker than the gray cast iron, so, it can concluded that the effect of carbonitriding on ductile iron is more better.

Compared with the matrix metal, the hardness of the white layer and carbonitriding layer were improved significantly according to the microhardness value change with the depth from the surface. The surface hardness of ductile iron can be as high as 700-900 HV. The surface hardness of gray cast iron samples is 350-450 HV.

3.4 Wear resisitance analysis

The wear mass loss was used to represent the abrasion performance of the samples. The comparison of the wear loss between the carbonitriding treatment samples and the samples without carbonitriding shown in Fig 4. Compared to the samples without carbonitriding, the wear loss of carbonitrided was smaller. That is to say, ion carbonitriding can improve the wear resistance. This is because of that there are a lot of γ′(Fe_4N), ε(Fe_3N), Fe_3C phase distributed on the surface of carbonitrided samples which can improve the hardness of the samples, so it can enhance the wear resistance of the samples. In addition, the matrix metal is tempered sorbite which has excellent mechanical properties, with the high strength and toughness[2]. Compared to the carbonitrided grey cast iron samples, the wear loss of carbonitrided ductile cast iron samples is smaller. It instructionsed that after ion carbonitriding, the wear resistance of ductile iron is better than grey cast iron. The carbonitrided layer has high hardness after ion carbonitriding, and the abrasion performance improved to some extent.

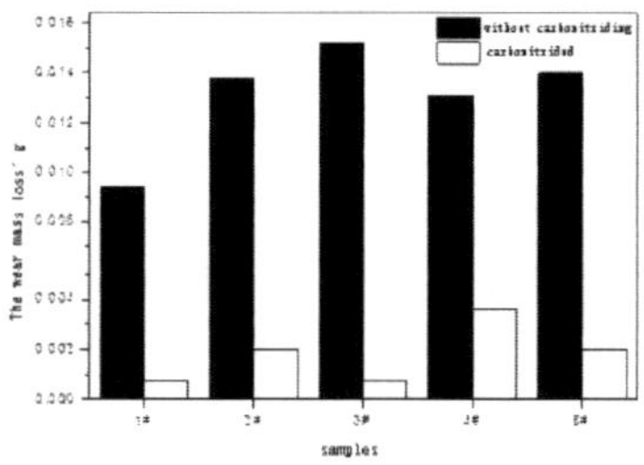

Figure 4 Wear lose of carbonitrided samples and the samples without carbonitriding

4.Conclusions

1) After ion carbonitriding, the hardness of matirix materials and surface improved greatly. The surface hardness of ductile cast iron layer can be as high as 700-900 HV. The surface hardness of grey cast iron is 350-450 HV.

2) After ion carbonitriding, there were bright layer and black diffusion layer appeared in the surface. The bright layer is the mixture of γ′(Fe_4N), ε(Fe_3N) and Fe_3C ,and it enhanced the surface properties of materials significant.

3)Compared with the samples without carbonitriding, the wear resistance of the ductile cast iron and grey cast iron can increased significant through the ion carbonitriding treatment. And it will enhance the fatigue and wear resistance, prolong its service life. Furthermore, the ductile cast iron shows more excellent wer resistance.

References

[1] Sun Dingguo Han Li, Zhao Cheng. Influence of Nitrogen on Compound Layer in Ion Nitrocaburizing [J]. China surface engineering: 2003 tobacco (6) : 24-26.

[2] Song Rui. High-temperature wear behavior of 3Cr2W8V hot-forging die steel [D]. Jiangsu, jiangsu university, 2009.

Proceedings of the 72nd World Foundry Congress
21-25th May 2016, Nagoya, Japan

Inoculation – Higher potency alternatives to Barium Calcium alloys

Gunnar Maltsev Hansen[1], Andy Plowman[2] and Yusman Rochamasi[3]
[1] Elkem Foundry R&D, Kristiansand, Norway
[2] Elkem Foundry Technical Customer Service, Singapore
[3]Elkem Sales Asia, Singapore

At this moment, around 60% of the new iron castings of the world are under production in Asia, and it is a market that is rapidly growing. The predominant Asian choice for inoculation, when not simply Ferro Silicon, has traditionally been a Barium Calcium inoculant. Aside from alloy cost, some reasons are ease of use, availability, ability to provide sufficient casting quality for most iron grades and casting types. However, the Asian market is evolving: Quality demands from end customers are increasing. New iron grades demand new process solution. The availability of raw materials forces changes in base iron composition. Cast iron must also meet the challenge from other alternative materials. For many foundries struggling to adapt to this new situation, higher potency inoculation can be instrumental in resolving these issues.

Keywords: *Ductile iron, Grey iron, inoculants.*

1. Introduction

CaBa inoculants almost has as long a history as ductile iron. In addition, while well-known, low cost, versatile, and readily available, they do not come without their own inherit problems.

Firstly, they are of low-medium potency. Typical addition rates are high, increasing cost and additional Si units; which is undesirable for some foundries.

They offer poor protection from shrinkage defects in the final castings. It is important for foundry men, as processes evolve, to weigh any advantage against total cost, and be aware of the potential reduction in final properties.

2. Nucleation, inoculation and active elements

Ductile irons are live, dynamic materials, which go through many different complex stages as the material solidifies, based on a multitude of factors from raw materials to process temperatures to alloy addition [1]. The process of nucleation is a general term for the mechanisms governing the growth sites of graphite. The higher the amount of nucleation sites available in the melt, the greater the graphite growth. Without available nucleation sites, graphite will not grow, and carbon end up as undesirable random inclusions or iron carbide.

Since the 1920s, the addition of inoculants to cast iron has been known to improve the properties and reproducibility of iron. Inoculation is a method that influences what happens to the graphite in the iron as it cools and solidifies, as graphite precipitation and growth during solidification has a strong effect on the final properties of the iron. By introduction of active elements, such as Ca, Ba or Ce, it is possible to minimize under-cooling and increase the number of nucleation sites during solidification. [1]

All inoculants will fade over time due to a coarsening of inclusions and result in increased undercooling, greater chill, reduced number of nodules, lower nodularity and a general degradation of properties. Thus, it is beneficial to add the inoculant as late in the process as possible [2]. Different inoculants and active elements have different characteristics and resistance to fading [3].

Calcium although of medium potency and fast fading is an essential element for cast iron inoculation for its role in graphite nucleation and chill control.

Barium is also a low/medium potency element that promotes graphite growth, but with a good fade resistance. It is often used in combination with Ca.

Other elements prevalent in ductile iron inoculants are Al, Zr, Mn, Ce, La and Bi. Some high potency inoculants are also coated with S and O to provide a late stage boost to nucleation potential [4,5].

3. Shrinkage and inoculation

One of the issues foundries struggle with for some types of casting is shrinkage defects. Shrinkage occurs when an internal shell forms that cannot be deformed by the difference in pressure with the atmosphere during the later stages of solidification [6], leaving voids that can severely reduce the properties of castings. While there are many valid strategies to avoid shrinkage [4, 6, 7], the choice of inoculant has been shown to have a strong effect [4, 5].

One recent internal study [8] has made similar observations, finding a CaBa inoculant far more shrinkage prone than a Ce-based inoculant at the same addition rate. The image below shows the shrinkage result from two different tests, with different inoculation schemes.

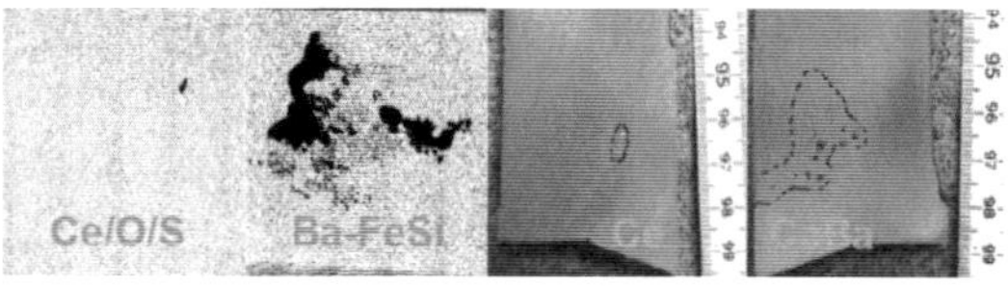

Fig. 1: Shrinkage samples; Study 1 [4] (l), Study 2 [8] (r)

Part of the reason for the difference in shrinkage characteristics can be found in the difference in nodule count and size distribution. Having a sufficient amount of small, late forming nodules is helpful in countering the contraction in last liquid phases before solidification [9]. Below are structure images from the shrinkage samples of study 2 in fig.1.

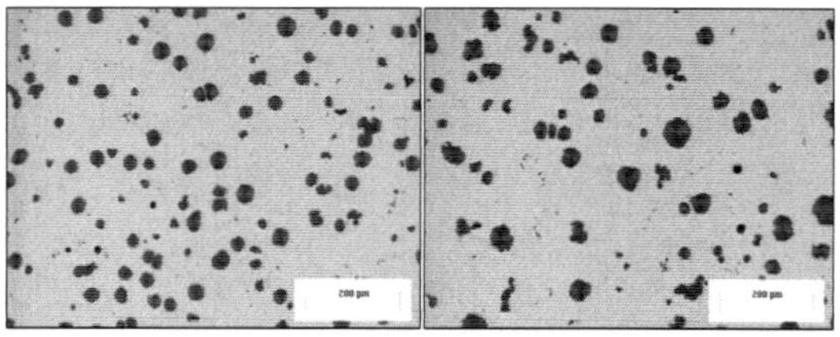

Fig. 2: Structure Ce(l) and CaBa(r)-inoculant., w. RE-free MgFeSi [9]

A powerful tool that has helped many foundries to identify and combat shrinkage as well as optimizing their inoculation solution in general is thermal analysis. It is possible to predict the kind of shrinkage behaviour shown here using thermal analysis curves [6,9].

4. Case study: Ductile iron turbocharger housing

An automotive foundry producing crankshafts in ductile iron grade 800/2 was having problems with shrinkage in a 30mm thick section. The MgFeSi (6.0%Mg/0.5%RE) with a 0.4% addition of BaCa inoculant in the transfer ladle and a 0.1% addition of Ba/Ca inoculant in the metal stream was replaced with a single 0.2% high potency Ce-based O&S-coated inoculant in the metal stream. As a result the shrinkage was reduced from 15% to 3% of the castings.

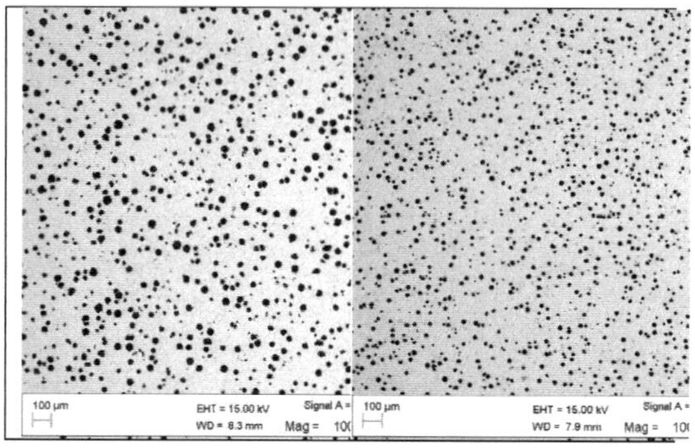

Fig. 3: Structure of old(l) and new(r) process

5. Conclusion

Calcium Barium inoculants still have a place in today's industry. If there are no complex, shrinkage prone casting geometries, if there are no difficult specifications to be achieved and if only a simple process using one inoculant for everything preferred. Then a CaBa inoculant might adequately cover those needs. However, foundries looking to solve issues of shrinkage, fading and increasing demands on iron properties should look to higher potency inoculation alternatives.

References

[1] T. Skaland, *Nucleation mechanisms in ductile iron*, AFS, (2005).

[2] T. Skaland, *Fading of inoculants in cast iron*, (1992)

[3] *Effect of minor and trace elements in cast iron*, Elkem Technical information sheet #12

[4] T. Skaland, *A new approach to ductile iron inoculation*, AFS (2001)

[5] I. Reposing, et.al., *Analysis of cooling and contraction curves to identify the influence on shrinkage behaviour of ductile iron*, Keith Millis, DIS, (2003)

[6] R. Sillén, *Shrinkages in iron castings*, DIS vol 2, (2002)

[7] D. White, *Avoiding shrinkage defects and maximizing yields in ductile iron*, AFS, (2012)

[8] Internal test report, Elkem (2015)

[9] D.S.White, *Studies of Nuclei in Ductile Iron*, presentation, DIS conference, Vancouver, (2010)

Proceedings of the 72nd World Foundry Congress
21-25th May 2016, Nagoya, Japan

Influence of Amount of Chunky Graphite on the Mechanical Properties of Spheroidal Graphite Cast Iron

Hideaki Nakayama[1], Bai-Rong Zhao[2], Noriaki Furusato[3], Satoru Yamada[4]
[1,2,3,4] Department of Research & Development, I Metal Technology Co.,Ltd, 4-2 Kitakandatsu-Machi, Tsuchiura-City, Ibaraki-Pref, Japan

Chunky graphite is generated in the heavy sectioned castings [1], and will be detrimental to the mechanical properties of cast irons, especially ductility and toughness [2]. However, there is little review showing change of mechanical properties by the amount of chunky graphite quantitatively.

In this study, the cylindrical casting which contained some chunky graphites was made. As base metals, highly pure pig iron and steel scrap with Fe-Si and RE-Si alloys were used. We took some specimens from the casting, and measured mechanical properties by the tensile strength test and the brinell hardness test.

Tensile strength was not affected until the chunky graphite amount of about 20%. However, when the amount exceeded this percentage, it started to get down gradually, and eventually decreased about 50MPa which is lower than the only spheroidal graphite region. Elongation was not decreased so much when the chunky graphite amount was up to about 20%, however when the amount exceeded 20%, it decreased rapidly to about 40% and remained the same. On the other hand, yield strength and hardness were not affected.

Keywords: *Cast Iron, Chunky graphite, Mechanical properties.*

1. Introduction

The cast iron parts used for construction machinery and industrial vehicles have always required reliability and durability. Consequently, they are often made of thick-walled spheroidal graphite cast iron for higher rigidity, high load, high repeated load, and resistance against shock, vibration, and heat. However, in the thick-walled spheroidal graphite castings, chunky graphite is often generated due to cooling rate and chemical composition. Chunky graphite is generally known as one of the abnormal graphite which is detrimental to mechanical properties. However, the relation between amount of chunky graphite and mechanical properties is not well known quantitatively. In this study, the influence of amount of chunky graphite on the mechanical properties of spheroidal graphite cast iron are investigated.

2. Experimental Method

As base metals, highly pure pig iron and steel scrap were melted in the 100kg high frequency induction furnace. Their fundamental chemical compositions are Fe- 3.7%C- 3.5%Si- 0.2%Mn- 0.02%P- 0.005%S. Spheroidal treatment was carried out using the sandwich method with 1.3wt% of graphite spheroidizer, 0.2wt% of Fe-Si alloy, and 0.1wt% of RE-Si alloy. Immediately after treatment, the samples were provided to chemical composition analysis, and the melt at 1603K with 0.1wt% of stream inoculant was poured into a cylindrical casting, its thickness of 70mm, its outside diameter of 200mm and its height of 100mm. To determine the mechanical properties, 20 tensile test pieces of JIS No.4 were extracted from cylindrical casting and used for the tensile test. Their brinell hardness were measured at their grip parts. The cross-sections, 5mm apart from fracture surface of a tensile test pieces, are observed for the chunky graphite. Then the chunky graphite area ratio (hereafter referred to as R_{CG}) were measured in theire microstructures by the image analysis quantitatively. The definition equation is shown in (1).

$$R_{CG} = (A_C / A_T)\times 100 \ (\%) \qquad (1)$$

where A_C is the chunky graphite area, and A_T is the all graphite area. R_{CG} was measured in five visual fields at a magnification 100. The average of these was set as the amount of chunky graphite of each test piece.

3. Results and Discussion

Figure 1 shows the typical microstructures. The matrix structures were ferrite. Chunky graphite was found to be distributed unevenly in the cylindrical casting so that R_{CG} ranged from 0 to 80%.

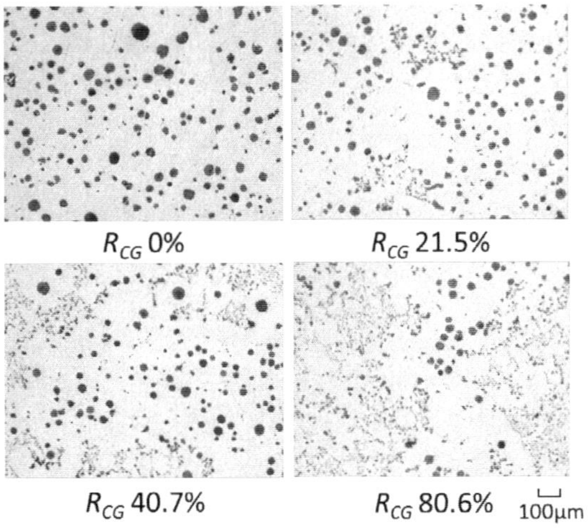

Fig.1 Typical microstructures of each chunky graphite area ratio of cylindrical casting.

Figure 2 shows the relation between R_{CG} and mechanical properties. Both yield strength and hardness remain uniform with increasing R_{CG}, and no influence of the chunky graphite were found. On the other hand, the tensile strength was found to form a moderate S-curve with increasing R_{CG}. It did not change until R_{CG} 20%, then from there to about R_{CG}40%, it started to decrease gradually and became uniform thereafter. In addition, the tensile strength of the spheroidal graphite region was 506MPa while that of the part with a large amount of chunky graphite was 457MPa, indicating a decrease of about 10%. Elongation was found to form a rapid S-curve with increasing R_{CG}. It did not decrease so much to R_{CG} of about 20%, then from there to about R_{CG}40%, it started to decrease rapidly and became uniform thereafter. The elongation of the spheroidal graphite region was 25%, while that of the part with a large amount chunky graphite was 5%, indicating about 1/5.

Tensile strength did not decrease when R_{CG} was less than 20% , when the amount of chunky graphite was very small, only the elongation was considered to be decreased due to fracture in the stage after the maximum stress of the stress-strain curve. As the amount of chunky graphite increases to a certain level, it affects before the maximum stress, resulting in the decrease of both the elongation and tensile strength.

Generally the elongation of cast iron is greatly affected by the shape of the graphite. While chunky graphite consists of refined graphites independent from each other two-dimensionally. Figure 3 shows the backscattered electron image of chunky graphite region on the fracture surface by EPMA. It can be seen that chunky graphite is three-dimensionally joined and distributed. Consequently, the fracture morphology of the chunky graphite regions is thought to resemble flake graphite and compacted vermicular graphite. We considered that when the amount of chunky graphite exceeds 20 %, break up the continuity of the matrix caused to decrease elongation suddenly.

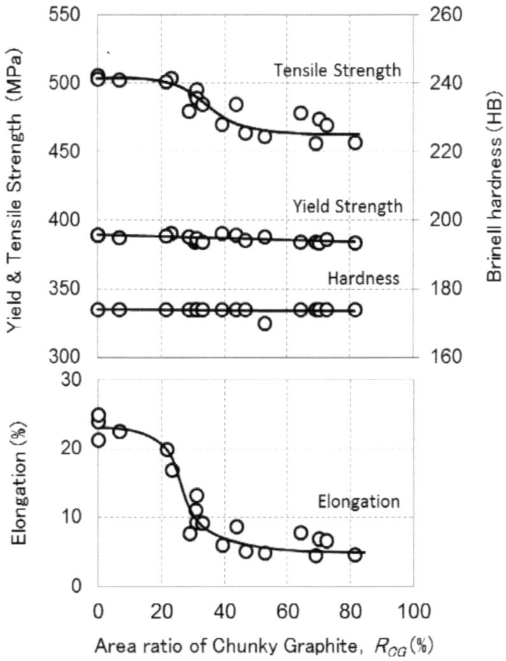

Fig.2 The relation between R_{CG} and mechanical properties.

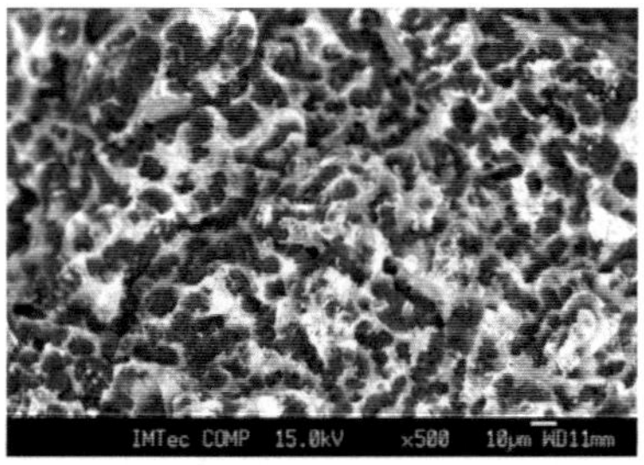

Fig.3 The backscattered electron image of chunky graphite region on the fracture surface by EPMA.

4. Conclusion

Spheroidal graphite cast iron with ferrite matrix, the amount of chunky graphite generated decreased tensile strength by about 10% most and elongation became about 1/5. Its effects on yield strength and hardness are very small. Chunky graphite had little influence of mechanical properties when the amount was less than 20%. (Omit the references.)

Proceedings of the 72nd World Foundry Congress
21-25th May 2016, Nagoya, Japan

Effect of Casting Skin Condition on Fatigue Strength of Gray Cast Iron

Shusaku Nasu[1], Satoshi Fujita[1], Noriaki Furusato[1], Satoru Yamada[1] and Sadato Hiratsuka[2]
[1] Department of Research & Development, I Metal Technology Co.,Ltd,
2-106-6 Goto, Waga-chou, Kitakami-City, Iwate-Pref, Japan
[2] Department of Material Science and Engineering, Faculty of Engineering Iwate University,
4-3-5 Ueda, Morioka-City, Iwate-Pref, Japan

The microstructure around casting skin of cast iron changes according to conditions such as the mold and cooling speed. When self-hardened mold is used, the all pearlite is obtained. When green sand mold is used, the microstructure of the pearlite and the ferrite is obtained. In this study, we conducted fatigue tests on two kinds of gray cast iron: one is the sample taken from casting skin consisting of fully pearlite and another one is the sample consisting of pearlite-ferrite mixture, and investigated the effects of fatigue strength. For comparison, tests were also conducted on the sample of perlite matrix whose surfaces have been machined, sampled from inside the material.

The fatigue strength of the casting skin compared with the thoroughly machined sample (from the internal part), in case of pearlite casting skin increased 20% and in case of pearlite-ferrite mixed decreased 12%. Because the tensile strength of ferrite is supposed to be lower than one of pearlite, fatigue strength became lower. Comparing thoroughly machined with casting skin both in pearlite matrix, the fatigue strength of casting skin was higher because the cooling rate was faster in the casting surface, graphite size became refiner and the number of eutectic cell was larger than the inside.

Keywords: *Cast Iron, Casting skin condition, Fatigue strength, matrix.*

1. Introduction

In some cases cast iron products are used with casting skin, there are need to investigate the effect of the casting skin on the mechanical properties. When the casting skin microstructure of each part of the gray cast iron cylinder head was investigated, most parts were found to be pearlite, and pearlite/ferrite mixtures and ferrite due to decarburization were observed in some parts. Otherwise, the actual tensile strength was measured inside the part which was wholly treated. The microstructure was found to be pearlite, and the strength near the casting skin is thought to be different from inside due to the influence of ferrite and the surface roughness. In this study, the fatigue tests were carried out using test pieces with two kinds of casting skin: pearlite and pearlite-ferrite mixed. For comparisons, fatigue tests were also conducted on pearlite samples wholly treated from the inside.

2. Experimental procedure

2-1. Sample preparation

The samples were prepared using 100kg high frequency induction furnace from 50% retuned scrap and 50% steel scrap. After melt, Fe-Si alloy which is equivalent to 0.35% of the molten metal was inoculated by the ladle inoculation, and the casting into the self-hardening mold and green sand mold. The self-hardened molds applied to get pearlite matrix, on the other hand green sand mold for pearlite and ferrite mixed matrix. The reasons why pearlite-ferrite mixed matrix structure appeared in the casting skin were caused by the chemical reactions between oxygen from the moisture in the green sand and carbon in cast iron melt to become CO or CO_2 gas and to be evaporated so that the carbon concentration of the casting skin became lower partly. Changes occur only near the casting skin and the inside is pearlite. Their target chemical compositions were 3.3%C -2.0%Si-0.7%Mn-0.2%Cr-0.25%Mo and 0.65%Cu.

2-2. Fatigue test

The fatigue test sample measured 4mm in thickness and 90 mm in length. It had flat plate with rounded notches in the center. Specimens had only one side casting skin and another side is machined. Figure 1 shows the test piece shape.

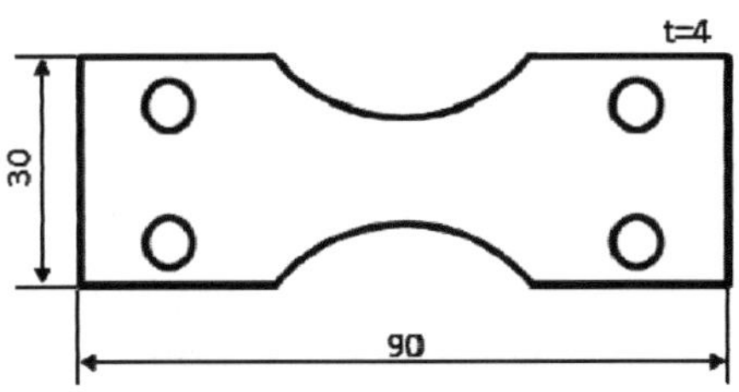

Fig.1 Test piece shape and dimention.

Plane bending fatigue tests were conducted at the following conditions : test temperature was set to room temperature, repeated rate is 1100cpm, complete pulsating condition of stress ratio R = 0. When the bending moment dropped 30% from the initial setting level, it was considered as crack. When there was no drop in stress, tests were conducted up to 1E+7 times. Figure 2 shows the state of the casting skin of the sample used in this test.

Fig.2 Microstructure of casting skin.

3. Experimental result

Figure 3 shows the tensile strength of each material. Tensile strength of pearlite casting skin (hereafter called S-P) was the highest and pearlite-ferrite mixed (hereafter called S-P/F) was the lowest. In addition, the inconsistency of the tensile strength of S-P/F was greater than other samples. It is considered that the reason for due to the difference in the ferrite formed on the casting skin according to the material. The machined sample (hereafter M-P) whose total surface had been machined and extracted from inside mold, had a consistent microstructure and stable tensile strength.

Figure 4 shows S-N curve of plane bending fatigue test. Compared with the fatigue strength of the M-P, in case of S-P increased 20% and in case of S-P/F decreased 12%. This trend was the same as the tensile strength.

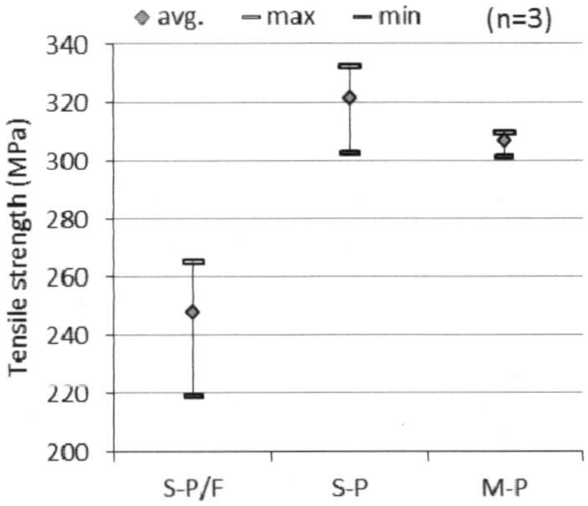

Fig.3 Tensile strength of each material.

Comparing thoroughly machined with casting skin both in pearlite matrix, the fatigue strength of casting skin was higher because the cooling rate was faster in the casting surface, graphite became finer and the number of eutectic cell was larger than the inside. In the case of gray cast iron, the cracks are thought to pass through the graphite/graphite material or graphite/matrix. It is considered that the number of eutectic cells was larger, the resistance to crack propagation increased. Figure 5 shows compared eutectic cell of the casting skin and the inside.

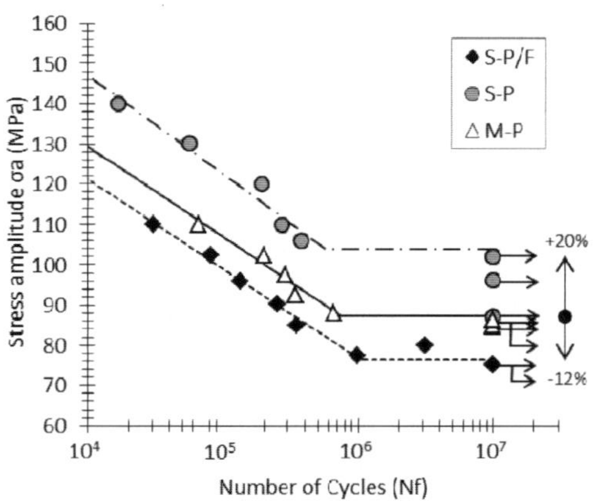

Fig.4 S-N curve of plane bending fatigue test.

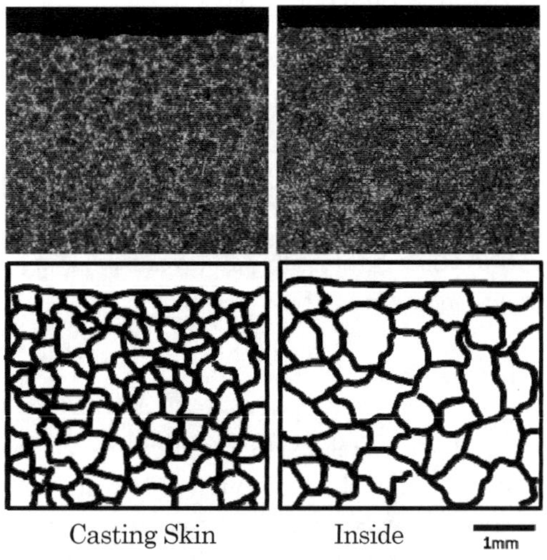

Fig.5 Eutectic cell of surface.

4. Conclusion

Casting skin microstructure becomes finer because cooling rate is faster than that of inside. It indicated the tendency of the fatigue strength was higher. However, the fatigue strength was reduced when the matrix is the perlite and the ferrite mixed.

Proceedings of the 72nd World Foundry Congress
21-25th May 2016, Nagoya, Japan

Cooling curves and graphite shapes of eutectic temperature region in gray cast iron

Taiki Tsuchiya[1], Katsutoshi Shigeno[1], Hirokazu Kawashima[1] and Kenji Matsuda[2]
[1] Matsubara co., Ltd.
[2] UNIVERSITY OF TOYAMA

Graphite shape is important to consider mechanical property in gray cast iron. Chemical composition and cooling rate affect graphite morphology.
In this work, the column-shaped castings having various volumes were used to measure cooling curves. Cooling curves were related to microstructure.

Keywords: *graphite shape. cooling curve, casting simulation*

1. Introduction

Graphite shape is important to consider mechanical property in gray cast iron. Chemical composition and cooling rate affect graphite morphology [1,2]. Graphite morphology in gray cast iron has been distinguished from A-type to E-type by ASTM [3].

In gray cast iron, inverse chill and D-type graphite caused by rapid cooling are observed at the final solidification part or thick part, both of which solidifies the most slowly.

We showed that inverse chill and D-type graphite form due to rapid cooling from eutectic temperature range (γ-Fe+G+L) with the experimental and simulated result of our previous work [4,5].

In this work, the column-shaped castings having two different volumes were used to measure cooling curves. We measured cooling curves at the center region and the other regions.

B-type graphite has been observed near the surface, but the mechanism of formation B-type graphite has not been clarified. We will also show simulated cooling curves of forming B-type graphite.

2. Experimental procedure

The melt tapped from cupola having ability to melt 8t/hr was held in forehearth for 10 minutes. Later the melt was transferred to a ladle with carbon inoculant. The melt corresponding to FC250 gray cast iron was poured to mold having two different shaped cavity.

Fig.1 shows the molds in this work. One is shell cup mold having φ30 column-shaped cavity (Fig.1 a), the other is CO_2 mold having φ50 column-shaped cavity (Fig.1 b). We called the castings from two molds φ30 shell cup cast and φ50 CO_2 cast, respectively.

The cooling curves were measured from the center to near the surface. In φ30 shell cup cast, they were measured at P1 and P2. In φ50 CO_2 cast, they were measured at P1, P2 and P3.

Casting simulation (ADSTEFAN) was used to study metal flow and solidification process.

Microstructures were observed with optical microscope. The matrixes were observed after etching with 3% nital.

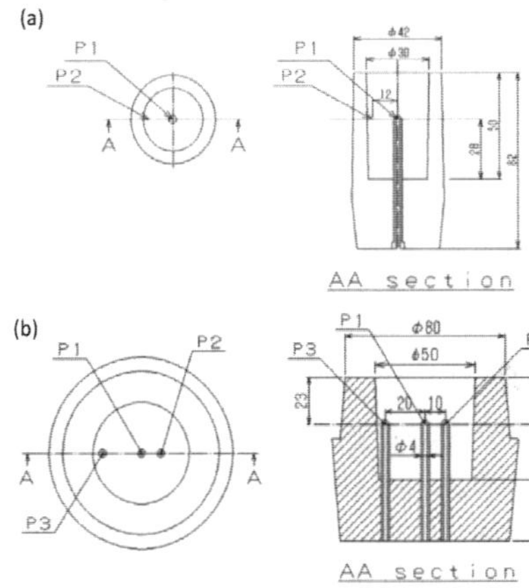

Fig.1 The molds in this work (a) φ30 shell cup cast, (b) φ50 CO_2 cast

3. Results and discussion

3.1 Cooling curves

Fig.2 shows the cooling curves in both of the castings. We could measure the cooling curves from center region to near the surface. The temperature at near the surface in both of castings were lower than center region.

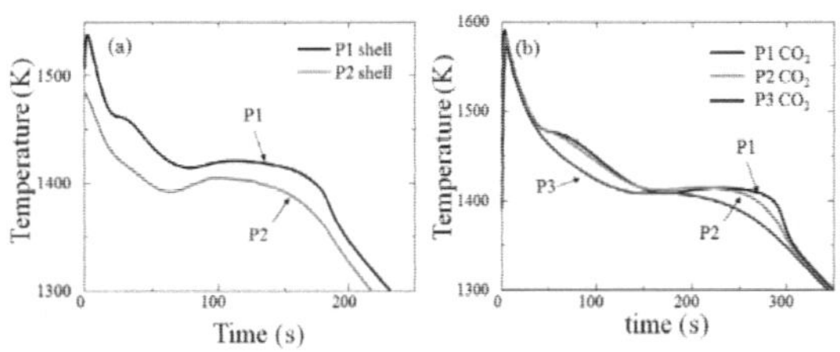

Fig.2 Cooling curves (a) φ30 shell cup, (b) φ50 CO_2 cast

3.2 Casting simulation

In casting simulation, the latent heat emmission patterns were determined by the relationship between temperature and solid fraction. In our previous work, we recomennded to use the model that has a certain temperature range in the solidification in accordanace with Fe-C-Si ternary equilibrium phase diagram [5]. In this work, the temperature of primary crystal and eutectic were set from experimental values. In this simulation, three differnt temperature ranges for eutectic reaction were designated as follows :

1) The case of 1K

According to Fe-C binary equlibrium phase diagram, the eutectic temperature is constant. Therefore, the temperature of eutectic reaction end was set on 1420K for φ30 shell cup cast and 1414K for φ50 CO_2 cast.

2) The case of 10K

According to Fe-C-Si ternary equlibrium phase diagram [4], Si content may establish the eutectic temperature rage. Therefore, the temperature of eutectic reaction end was set on 1411K and 1405K.

3) The case of 20K

We consider that the castings are multicomponent systems. Contents of P,Mn S, etc. may expand the temperature range. Therefore, the temperature of eutectic reaction end was set on 1401K and 1395K.

Fig.3 shows the results of casting simulation. The upper charts show the simulation results of φ30 shell cup cast in accordance with the three different temperature ranges described above. The bottom charts show the results of φ50 CO_2 cast. In the case of 1K, the simulated curves were differnt from experimetal result. In the case of 10K, The difference between experiment and simulation became smaller than the case of 1K. In the case of 20K, the simulation closed to experiment in both of castings.

At P1 in both of castings, D-type graphite caused by rapid cooling and A-type graphite were observed. According to the experimental and simulation results, rapid cooling was observed at final solidification part. Therefore, it influenced formation of D-type graphite at P1. As previous work suggests [5], we show that the temperature of eutectic has temperature range (γ-Fe+G+L) and rapid cooling occurs at final solidification part with the experimental and simulated result.

Fig.4 shows microstructures near the surface in both of castings. B-type graphite was observed. The matrix near the B-type graphite was ferrite. We consider that the formation of B-type graphite occured due to rapid cooling around 50s in cooling curve at P2.

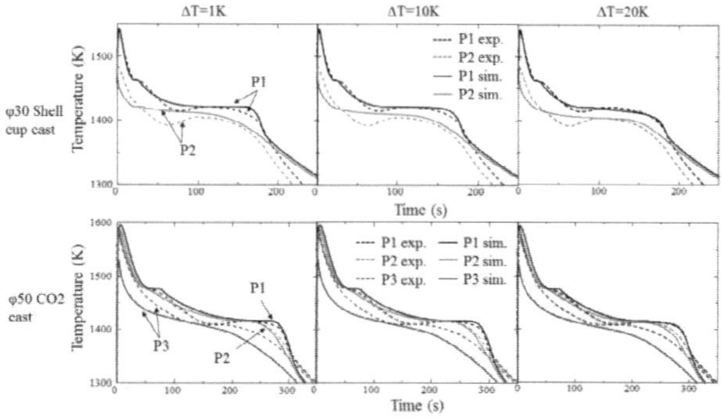

Fig.3 Simulation results of cooling curves with different eutectic temperature ranges

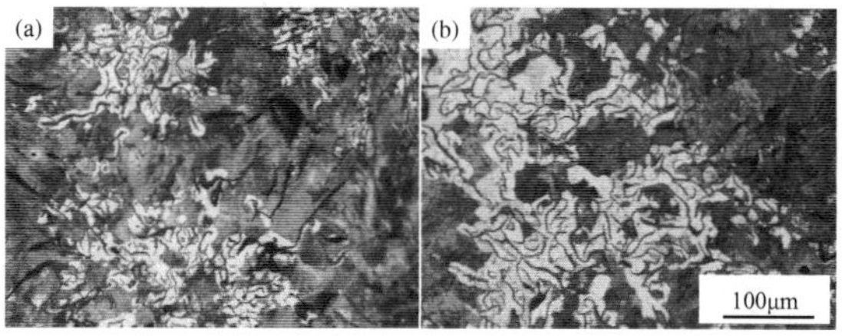

Fig.4 Microstructure near the surface in both of castings (a) φ30 shell cup, (b) φ50 CO_2 cast

4. Conclusions

We have done researches on the solidification in gray cast iron. The conclusions are as follows:

Throgh the measuring cooling curve and casting simulations, it was deduced the formation of D-type graphaite occurred due to the rapid cooling at final solidification part.

On the other hand, the formation of B-type graphite near the surface occurred due to the rapid cooling observed in early stage of solidification.

References

[1] H.Nakae, H.Shin, N.Hashihara: J.JFS 83 (1997) 562-569.

[2] M.Makimura, K.Sakai, Y.Nishiyama, M.Tanaka: J.JFS 56 (1984) 15-21.

[3] American Society for Metals: *Metals Handbook-Properties and Selection: Iron and steels-, 9th Ed. Vol.1*, (American Society of Metals, Metals Park, 1978) pp.13

[4] American Society for Metals: *Metals Handbook-Properties and Selection: Iron and steels-, 9th Ed. Vol.1*, (American Society of Metals, Metals Park, 1978) pp.4

[5] H.Kawashima, K.Shigeno, K.Tachibana : J.JFS 83 (2011) 3-6.

Proceedings of the 72nd World Foundry Congress
21-25th May 2016, Nagoya, Japan

Effect of Rare Earth Content on Fatigue Strength of Spheroidal Graphite Cast Iron with Different Thickness

Takafumi Funabiki[1], Kazumichi Shimizu[1], Shinsuke Nagaumi[1], Masato Shirai[2]
[1] Muroran Institute of Technology, 27-1, Mizumoto-cho, Muroran-shi, Hokkaido, 050-8585, Japan
[2] Iwamizawa Imono. Co., Ltd., 497-3, Yura, Kurisawa-cho, Iwamizawa-shi, Hokkaido, 068-0111, Japan

Spheroidal graphite cast iron is excellent in mechanical properties. The spheroidization ratio and nodule count of graphite is effect to mechanical properties. Therefore, rare earth used in order to improve the spheroidization rate and nodule count of graphite. However, the supply of rare earth is relies on imports from overseas in Japan. Need for spheroidal graphite cast iron with mechanical properties not inferior to conventional products without the use of rare earth is growing. Effect of rare earth is different by the thickness of the casting. Therefore, we investigated the effect on the fatigue strength by the rare earth content in the different thickness of spheroidal graphite cast iron.

Keywords: *Rare earth, Spheroidal graphite cast iron, Fatigue strength*

1. Introduction

Spheroidal graphite cast iron is widely used in various purposes due to its excellent in mechanical properties. In the case of spheroidal graphite cast iron, spheroidal graphite can be used to alleviate the stress concentration that may occur into the matrix. Consequently, spheroidal graphite cast iron exhibits an excellent mechanical so that it has been used as automobile parts. It has been reported that the fatigue other than tensile strength has bad effect on failure of machine component. So, consideration of fatigue strength is often required for the designing of machine component [1]. Moreover, it is clear from previous studies that the fatigue phenomenon has been confirmed for variety of materials under various conditions [2]. Additionally, previous studies have been reported that the starting point of failure is often at the variant part of shape.

Therefore, the purpose of this paper is to make clear the effect of RE contents on fatigue strength of spheroidal graphite cast iron by using the plane bending fatigue test.

2. Experimental condition

2.1 Materials

In order to investigate the influence of different RE content on its fatigue strength, spheroidal graphite cast iron are prepared by using different nodularizer which contain RE content 0% and 1.0%. The specimens were made from iron scraps, speroidizing in sandwich method. Table 1 shows the chemical composition of nodularizer. Shape of specimen is a plate sized with 120×40×50mm. The molten metal was poured into the mold at 1683K. Table 2 shows chemical composition of molten metal. Table 3 shows the spheroidization ratio is more than 80% in RE1.0% and RE0%. Fig. 1 shows microstructure of specimens. The microstructure is all ferrite in the specimens.

Table 1 Chemical composition of nodularizer.

mass%

	Si	Mg	Ca	Al	RE
RE1.0%	45.88	5.92	2.98	0.19	0.92
RE0%	45.90	5.80	2.90	0.20	-

Table 2 Chemical composition of molten metal.

mass%

RE	C	Si	Mn	P	S	Cu	Mg	Cr
1.0%	3.67	2.62	0.28	0.011	0.010	0.045	0.046	0.029
0%	3.68	2.66	0.28	0.012	0.010	0.045	0.039	0.029

Table 3 Spheroidization, nodule count and brinell hardness of each specimen.

	Spheroidization [%]	Nodule count [counts/mm^2]	Brinell hardness [HB]
RE1.0%	90	288	130
RE0%	90	281	128

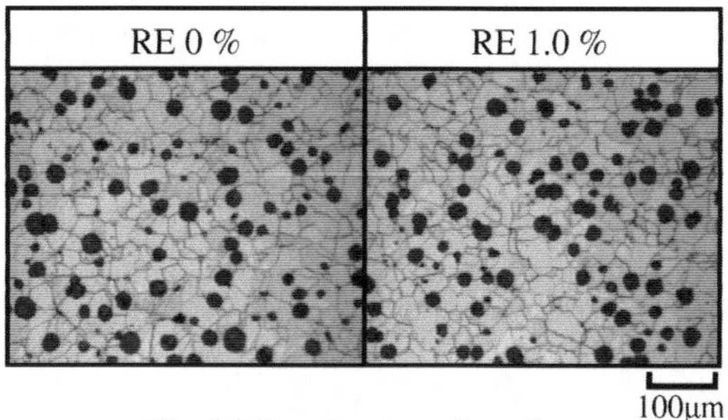

Fig. 1 Microstructure of specimens.

2.2 Experimental method

The fatigue tests were performed by using plane bending fatigue testing machine. The schematic diagram of the specimen is shown in fig. 2. The testing condition is shown in Table 4. The stop condition of fatigue test is the 10^7 time repetitions or fracture of specimen. Fatigue strength is the load stress if the specimens were not fractured. After fatigue test, the fracture surface was observed by using Scanning Electron Microscopy (SEM).

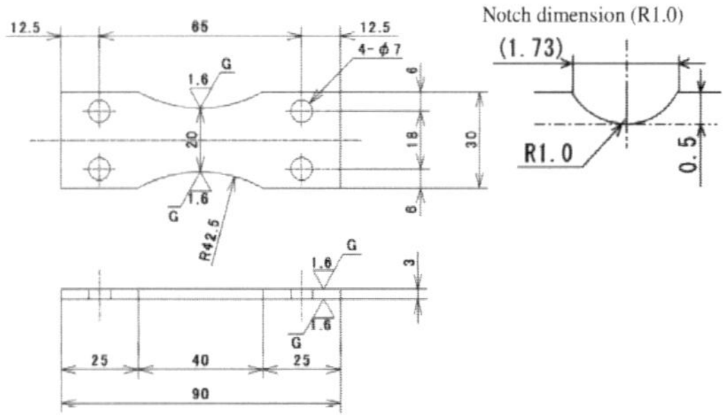

Fig. 2 Schematic diagram of test specimen.

Table 4 Condition of plane bending fatigue testing.

Loading cycle	1200 cycles per min
Stress ratio	-1
Maximum number of cycles	10^7
Test environment	Room temperature

3. Results and discussions

The S-N diagram and fatigue strength obtained from plate bending fatigue tests are shown in Fig. 3. RE1.0% became 234MPa in smooth specimens and 177MPa of notch specimens. RE0% became 190MPa of smooth specimens and 169MPa of notch specimens. The smooth specimens fatigue strength was reduced in RE0% but it became the same fatigue strength in the notch specimens.

And then, micro image of the fracture surface is observed by using SEM. Casting defect is confirmed in the specimens without RE content in Fig. 4. We evaluated each casting defect quantitatively by using $\sqrt{\ }$area parameter [3], stress intensity factor (ΔK) and thresh hold stress intensity factor (ΔK_{th}), respectively.

$$\Delta K = \alpha \Delta \sigma \sqrt{\pi \sqrt{area}} \qquad (1)$$

Where α denote a constant of place of defect and $\Delta\sigma$ is stress range. In addition, we can express the ΔK_{th} as the following equation.

$$\Delta K_{th} = 3.3 \times 10^{-3}(HV + 120)\left(\sqrt{area}\right)^{1/3} \qquad (2)$$

The values of ΔK / ΔK_{th} in the specimens without RE are about 1.1. Indicating that fatigue strength of specimens without RE is decreased to excluding RE.

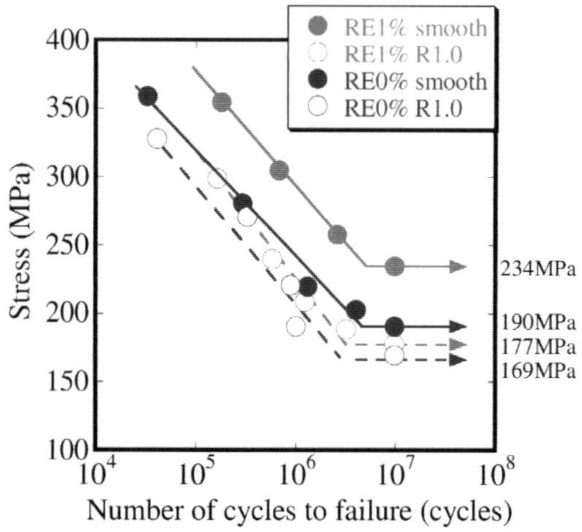

Fig. 3 S-N diagram

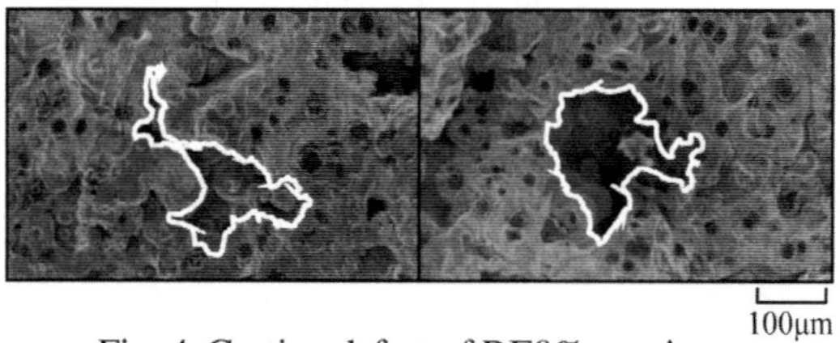

Fig. 4 Casting defect of RE0% specimens

4. Conclusions

The purpose of this paper was to make clear the influence of thickness and notch sensitive on the fatigue strength of spheroidal graphite cast iron. Results are shown below.

[1] As a result of fatigue test, the smooth specimens without RE have fatigue strength of 190MPa. In contrast, the specimens with RE have fatigue strength of 234MPa.

[2] The notch specimens without RE have fatigue strength of 169MPa. In contrast, the specimens with RE have fatigue strength of 177MPa.

[3] Values of $\Delta K/\Delta Kth$ in the specimens without RE are about 1.1, resulting in the fatigue strength decreasing of specimens without RE.

References

[1] Soc. of Materials Sci., Japan, Design fatigue hand book, (1995), 2, Yokendo

[2] H. Nishiya, Strength and fracture of fatigues, (1985), 13-16

[3] Y. Murakami, Effects of small defects and nonmetallic inclusions, (1973), 54, Yokendo

Proceedings of the 72nd World Foundry Congress
21-25th May 2016, Nagoya, Japan

Effect of Boron and Nitrogen Addition on the Solidification Microstructure and Hardness of High Speed Steel Type Mill Roll

Kengo Yoshimoto[1],Ryutaro Hara[1],Masahiro Yamamoto[1],Gen Ito[1],Kazunori Kamimiyada[2],
Ichihito Narita[1] and Hirofumi Miyahara[1]
[1] Department of Materials Science and Engineering, Kyushu University, Japan, Fukuoka 819-0379
[2] Nippon Steel & Sumikin Rolls Corporation, Japan, Kitakyushu 804-0002

Influence of boron and nitrogen addition as one of the alloying element on the formation of solidification structure and the improvement of hardness after thermal treatments was investigated for high-speed-steel alloys (Fe-2%C-5%Cr-5%Mo-5%V). The macrohardness of the quenched specimens gradually increases with increasing quenching temperature. Nitrogen addition helps to improve the hardness as carbon does. Boron addition is also contributed to the increasing of hardness. B-containing specimens maintained higher tempered-hardness than HSS and N-containing specimens quenched at the peak hardness. The specimen, which contains high volume fraction of residual austenite, shows the superior secondary hardening after the optimized tempering.

Keywords: *high-carbon high-speed steel, cast iron, carbide, heat treatment, martensite, solidification*

1. Introduction

For the last several decades, high-carbon high speed steel type cast alloys (HSS) which have hard carbides like MC, M_2C and M_7C_3 in the matrix have been used in cast mill rolls, because of the attractive thermal resistance and wear resistance imparted by the carbides [1,2]. However, further improvement of the durability of cast rolls under high pressure environment is still required, to decrease the replacement costs of the rolls as much as possible. It could be effective that generating hard intermetallic compounds in the matrix to improve wear resistance [3,4]. In this study, we focus on boron and nitrogen because both elements could create the hard compounds. Furthermore, boron is low cost alloying element, and some researchers have reported that boron addition can increase the hardness and improve the abrasion resistance of high chromium cast steel [5]. On the other hand, nitrogen tends to form hard compounds with Al, Cr, Nb, Ti, and V, and alloys including these elements exhibit great abrasion resistance by nitrogen addition [6].

In this study, influence of boron and nitrogen addition on the solidification microstructure, solute distribution and hardness after quenching and tempering treatments were evaluated for high-carbon high-speed steel type alloys.

2. Experimental Methods

Pure iron, carbon, Cr (99.99% purity) and ferroalloys (Fe-83%V, Fe-60%Mo) were melted in a high-frequency induction furnace at 1873K in an argon atmosphere, and cast into a permanent mold (ϕ15mm×200mm). N-containing specimen and B-containing specimens were prepared by the addition of Cr_2N and Fe-20%B, respectively. The chemical compositions of obtained as-cast specimens are listed in Table 1. The heat treatment of as-cast specimen was performed from 1123K to 1373K for 1 hour in an argon atmosphere and the subsequent quenching into an oil bath. Tempering treatment was also performed twice or three times from 673K to 873K for 1 hour, systematically. As-cast and heat-treated specimens were mirror-polished, etched using Murakami reagent and metallurgically analyzed by an optical microscope. Vickers macrohardness was measured at a load of 30kgf for 10s and evaluated with the amount of retained austenite, which has been measured by XRD (Rigaku, RINT2100) using Mo-Kα radiation.

Table 1 Chemical composition of specimens.

Specimen	Chemical compositions(mass%)						
	C	Cr	Mo	V	B	N	Fe
HSS	2.04	4.96	5.00	5.07	-	-	Bal.
0.1N	2.06	5.01	4.87	5.07	-	0.132	Bal.
0.1B	2.02	4.45	4.85	4.59	0.108	-	Bal.
0.5B	2.04	4.59	4.81	4.60	0.497	-	Bal.
1.0B	2.03	3.44	4.93	4.52	1.040	-	Bal.

3. Results and Discussion

Figure 1 shows the relation between Vickers hardness and quenching temperature. Vickers hardness increase toward the peak and then decrease with the elevation of quenching temperature in all specimens. Standard HSS specimens shows peak hardness of 900HV at 1223K. 0.1N specimen shows higher hardness than HSS from 1123K to 1223K. Microstructural analysis indicates that the formation of M_2CN carbonitride cause the increasing of carbon content and hardness of the martensite matrix in these temperatures. On the contrary, 0.1B specimen shows higher hardness from 1273K to 1373K, and that could be attributed to the lower amount of soft retained austenite.

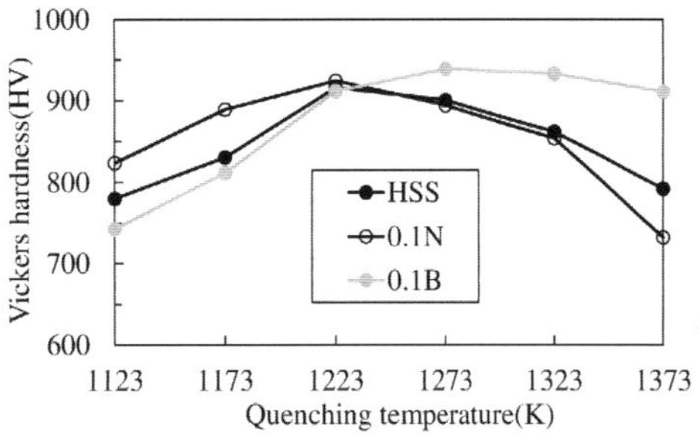

Fig. 1 Relation between Vickers hardness and quenching temperature.

In order to evaluate the influence of next tempering temperature on Vickers hardness, two quenching temperatures were selected; one was peak hardness temperature and the other was maximum temperature (1373K) in Fig. 1. The relation between hardness and tempering temperature were shown in Fig. 2. In Fig. 2(a), Hardness of HSS specimen decreases continuously. However, B-containing specimens maintained higher macrohardness than other specimens. In HSS and 0.1N specimens, the amount of solute alloying elements is not enough to produce the secondary hardening at this quenching temperature. Figure 2(b) shows relation between Vickers hardness and tempering temperature on specimens quenched at 1373K. HSS, 0.1N, and 0.1B specimens exhibit secondary hardening and 0.1B and 0.1N specimens increase 819HV at 773K and 853HV at 823K, respectively. The appropriate amount of retained austenite, carbonitride precipitation during tempering could cause the instable retained austenite and formation of martensite at quenching after tempering process.

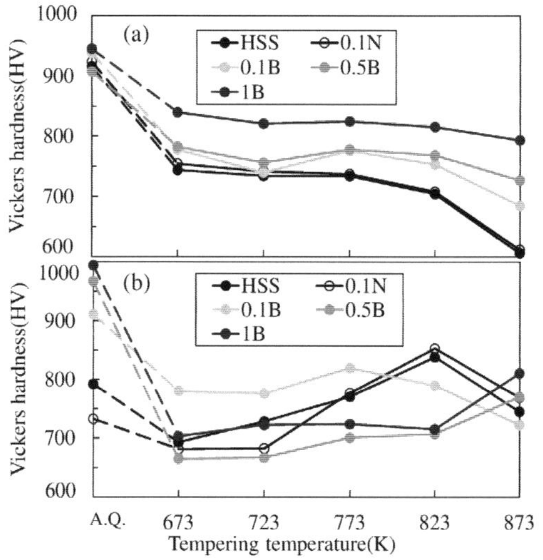

Fig. 2 Relation between Vickers hardness and tempering temperature on specimens (a) quenched at the peak hardness (b) quenched at 1373K.

4. Conclusion

The influence of boron and nitrogen addition on the solidification and hardness after heat treatments was investigated for high speed steel type cast alloys, and the following conclusions were obtained:

(1) B-containing specimens maintained higher tempered-hardness than HSS and N-containing specimens quenched at the peak hardness.

(2) Specimens tempered at 1373K exhibited secondary hardening, and N-containing specimen showed 15~30HV higher hardness than that of HSS standard specimen.

References

[1] K. Goto, Y. matuda, K. Sakamoto and Y. Sugimoto: ISIJ. Int. 32(1992)1184-1189.

[2] T. Kudo, S. Kawasima and R. Kurahashi: ISIJ. Int. 32(1992)1190-1193.

[3] K. Nakajima and Y. Yamamoto: Daido inst Technol. 31 (1995) 37-49.

[4] T. Fujita, S. Fujita, A. Sawamoto and K. Ogi: J.JFS .72 (2000) 662-667.

[5] C. Q. Guo, P. M. Kelly: Materials Science and Engineering A 352(2003) 40-45.

[6] M. B. Karamis, K. Yildizli and G. C. Aydin: Tribol. Int. 51 (2000) 18-24.

Proceedings of the 72nd World Foundry Congress
21-25th May 2016, Nagoya, Japan

Erosive wear characteristics of heat treated multi-component cast iron containing Cr, V, Mn and Ni at elevated temperature

Yao Zhang, Kazumichi Shimizu, Kenta Kusumoto
Muroran Institute of Technology
27-1 Mizumoto-cho, Muroran city, Hokkaido, 050-8585, Japan

High chromium cast iron, spherical vanadium carbide cast iron and multi-component white cast iron are used as wear resistant materials recently. However, they are extremely expensive because of the addition of Cr, Ni, Mo, V and W, etc. It's expected to reduce costs by suppressing the addition of these elements to develop an inexpensive wear resistant material which can be used at elevated temperature.

In this study, erosive wear characteristic at elevated temperature was investigated with 9 kinds of multi-component cast irons of varying V and Cr content after heat treatment. It has been found that erosive wear property increased with the increase of Cr addition and material with 9 mass % Cr showed the best erosive wear resistance at 873 K. There's no differences between erosive wear property of 5V-9Cr, 7.5V-9Cr and 10V-9Cr. So that, 5V-9Cr with excellent erosive wear property can reduce the amount of expensive vanadium which makes the production cost reduction possible by increasing the amount of relatively inexpensive chromium.

Keywords: *erosive wear, multi-component, cast iron, heat treatment, elevated temperature.*

1. Introduction

Wear is often divided into three types which are adhesive wear, corrosion wear and abrasive wear. Abrasive wear which is caused by solid particles can be further classified by the difference in wear type [1]. Among them, there is one type called erosion which is the phenomenon of damage or peeling of part of the material surface caused by the repeat collision of particulate material. Erosive wear happens not only at room temperature, the damage could be more serious at elevated temperature. Erosion that occurs in the inner surface of conduit cannot be judged from the appearance, so that periodic replacement and buildup welding in the thinning section are necessary ways to prevent accidents. From the viewpoint of reducing the manufacturing cost and maintenance fee, the development of wear resistant material is an urgent issue to be solved.

In previous research, high chromium cast iron, spheroidal vanadium carbide cast iron and multi-component white cast iron which contains carbides with high hardness showed small impact angle dependence and good erosive wear resistance. The reason for the wear resistance improvement is that the stress caused by the particle collision is dispersed and the plastic deformation of the material surface is also suppressed by the crystallized spheroidal vanadium carbides (VC). However, these wear resistant materials are extremely expensive because of the addition of Cr, Ni, Mo, V, W or other rare elements. It's expected to reduce manufacturing cost and maintenance fee by suppressing the addition of rear metals to develop an inexpensive wear resistant material[2-5].

In this study, erosive wear property of multi-component cast irons with varying V and Cr content was investigated at elevated temperature.

2. Experimental Method

2.1 Experimental Materials

9 kinds of multi-component cast iron with addition of about 3.0 mass % C, 4 mass % Mn, 1 mass % Si and varying V and Cr content (5, 7.5, 10 mass % V and 0, 4.5, 9 mass % Cr) were used as experimental materials. 50 kg raw materials were melted in high frequency induction furnace at 2023 K and then poured into a sand mold with a Y-shape of 53×113×125 mm after the process of spheroidizing by 95 mass % Ni - 5 mass % Mg. The specimens used in the erosion test were mechanically machined into flat plate with diameter of 50×50×10 mm. The condition of heat treatment was air-cooling after keeping the specimens for 7200 seconds at 1043 K for 10V-0Cr and 10V-4.5Cr, 1133 K for 7.5V-0Cr, 1193 K for 7.5V-4.5Cr, 1253 K for 10V-9Cr, 7.5V-9Cr, 5V-9Cr, 5V-4.5Cr and 5V-0Cr of which temperature is the best heat treatment temperature for each material [6].

2.2 Experimental Method

A high temperature erosive wear setup was used in this study. Alumina ball (Ø1.16 mm) which hardness is 1140 HV was used as impact particle. The velocity of the erodent particles was 30 m/s. The total amount of erodent particles was 8 kg. Erosive wear property of different multi-component cast irons was investigated at 873 K. The mass of the specimen was measured by electronic scale before and after each erosion test. Erosion rate was used to judge the erosive wear property[7]. Microstructure analysis, Vickers Hardness test, Energy Dispersive Spectrometer (EDS) and X-ray Diffraction (XRD) were undertaken to analysis the property of material.

3. Experimental results and discussion

According to microstructure analysis and X-ray diffraction results, the matrix of material is pearlite and austenite. According to EDS results, all the materials contain spheroidal vanadium carbides, and materials with Cr addition also contain lamella carbides which can be considered as M_7C_3. The hardness of the experimental materials is between 360~570 HV10 at room temperature.

Erosion rates of experimental materials at 873 K are as shown in Fig. 1. For materials with the same V additions (10, 7.5 and 5 mass %), erosion rates showed a smooth decrease with Cr content increased from 0 to 4.5 mass % (except for 10V-4.5Cr, which would be explained later), and followed by a sharp decline when Cr content continually increased to 9 mass %. Material with 9 mass % Cr addition revealed the best erosive wear performance, which has been improved approximately 44.7% (10V-9Cr), 41.1% (7.5V-9Cr) and 31.6% (5V-9Cr), comparing to 0 mass % Cr addition material separately.

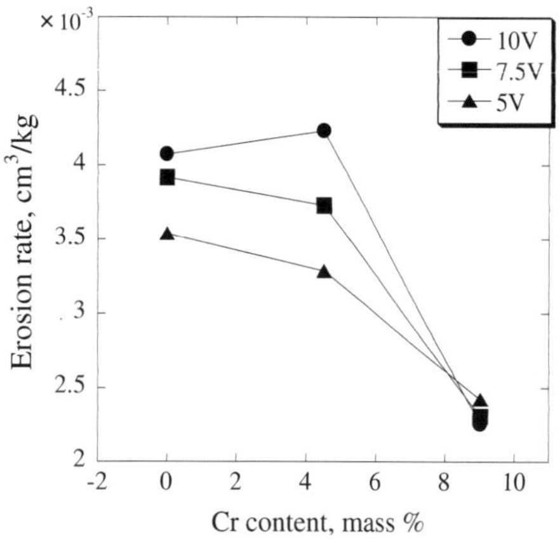

Fig. 1 Erosion rates of experimental materials.

For materials with 0 and 4.5 mass % Cr addition, erosion rates showed a slight decrease with the V content reduced which revealed a slight improvement in erosive wear resistance. While, for material with 9 mass % Cr, no differences were illustrated in erosion rate for materials with different V content. 10V-9Cr, 7.5V-9Cr and 5V-9Cr displayed the same high level in erosive wear resistance which indicated that the V addition can be reduced from 10 mass % to 5 mass % without any bad influence in erosive wear property.

On the other hand, according to the oxidation results, for materials with the same V content, the amount of oxidation was decreased with Cr addition increased, which can explain well that why erosive wear property increased with Cr addition increased. (Except for 10V-4.5Cr which had more defects than the other specimens to increase the area ratio of surface making the oxidation rate unusually high). Furthermore, for material with 0 and 4.5 mass % Cr, V promoted the formation of oxidation which reduced the erosive wear resistance. However, for material with 9 mass % Cr, the formation of oxidation was controlled and the oxidation amount of 5V-9Cr, 7.5V-9Cr and 10V-9Cr was extremely little at the same level.

4. Conclusions

In this study, erosive wear property of 9 kinds of multi-component cast iron at 873 K was investigated. It has been found that erosive wear property was improved with the increase of Cr addition and material with 9 mass % Cr showed the best erosive wear property. There's no significant differences between erosive wear property of 5V-9Cr, 7.5V-9Cr and 10V-9Cr. Therefore, 5V-9Cr with excellent erosive wear property makes the cost reduction possible that the amount of expensive vanadium can be declined from 10 mass % to 5 mass %.

References

[1] Japan society of corrosion engineering: *Erosion and Corrosion*, (1987) pp. 6.

[2] X. Yaer, K. Shimizu, et al. *Wear*. 264 (2008) pp. 947.

[3] H. Hara, K. Shimizu, et al. *J. JFS*. 86 (2014) pp. 43-48.

[4] K. Kusumoto, K. Shimizu, et al. *J. JFS*. 87 (2015) pp.16.

[5] V.G. Efremenko, K. Shimizu, et al. *Journal of Friction and Wear*. 34 (2013) pp. 466-474.

[6] Research report of PSTU (Priazovskyi State Technical University). (2013).

[7] I. Finnie. *Wear*. 3 (1960) pp. 87-103.

O-136

Effect of Joining Conditions on Joint Strength of Barium Titanate Brazed by Pure Molten Aluminum

Ikuzo Goto[1], Ken-ichi Ohguchi[1] and Setsuo Aso[1]
[1] Graduate School of Engineering Science, Akita University, Akita 010-8502, Japan

The effects of joining conditions on the joint strength of barium titanate brazed by pure molten aluminum were examined. Higher shear strength was obtained with the fracture of barium titanate for brazed specimens under longer holding times. This suggests that the joint strength is greater than the shear strength of barium titanate. On the other hand, lower shear strength with the interfacial delamination was obtained for the specimens under shorter holding times. Observations of the delaminated surface under longer holding times showed that fractured barium titanate pieces and grains were strongly bonded to the aluminum layer. Here, the cross-sectional oxygen distribution in the vicinity of the joint interface indicated the existence of a thicker alumina layer between barium titanate and pure aluminum under longer holding times. Therefore, interfacial bonding may be achieved through the progress of oxidation of molten aluminum during joining.

Keywords: *joining, pure metal, aluminum, ceramic, barium titanate*

1. Introduction

Positive temperature coefficient (PTC) ceramics such as barium titanate are used as electric heating elements in electric vehicles [1]. However, the use of electric heaters reduces the driving range of the vehicles. In conventional heaters, PTC ceramics contact heat diffusers through insulating layers and thermal grease. Direct bonding of the ceramics and pure aluminum diffusers by cast-in insertions would enable the manufacture of highly efficient heaters [2]. In this study, the effects of joining conditions on the joint strength of barium titanate brazed by pure molten aluminum were examined.

2. Experimental procedure

Brazed specimens were prepared by heating in a muffle furnace under several conditions and then cooling in air, as shown in Fig. 1 [3]. The joint strength of the specimens was evaluated using a shear test, as shown in Fig. 2 [4]. The fracture surface of the specimens was observed using an optical microscope and a scanning electron microscope (SEM). The cross-sectional distribution of the elements was also analyzed by an electron probe microanalyzer.

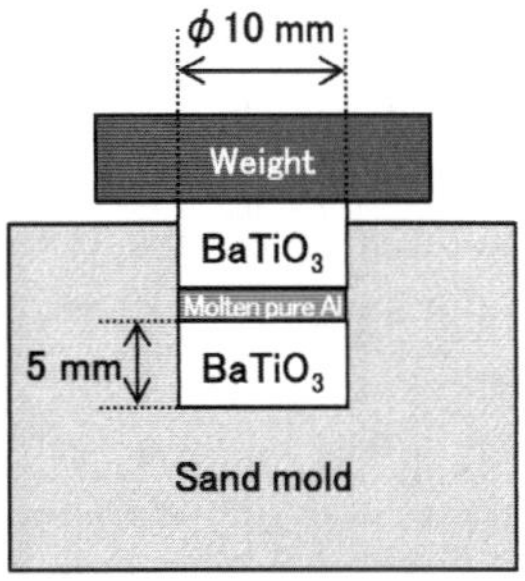

Fig. 1 Illustration of specimen preparation.

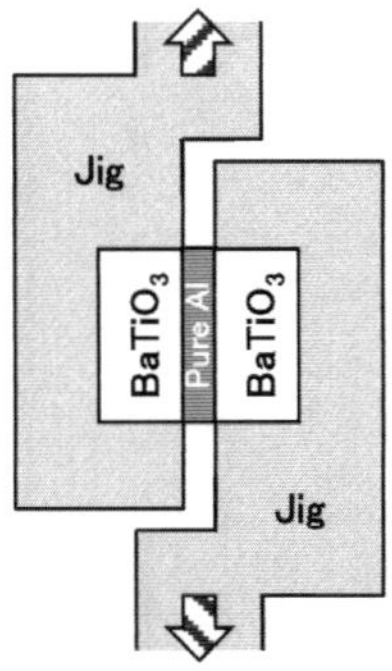

Fig. 2 Illustration of shear test.

3. Results and discussion

Figure 3 shows the relationship between the holding time and the shear strength at 780°C. For the specimens under longer holding times, higher shear strengths of more than 10 MPa were obtained with the fracture of barium titanate. This result suggests that the joint strength is greater than the shear strength of barium titanate. On the other hand, lower shear strength with interfacial delamination was obtained for the specimens under shorter holding times. This result suggests that the shear strength reflects the joint

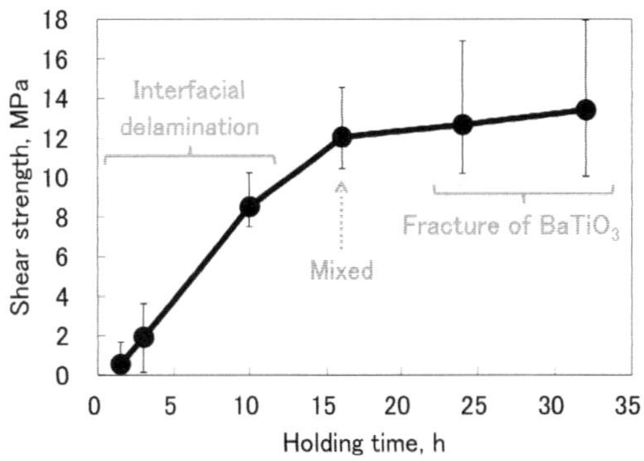

Fig. 3 Relationship between holding time and shear strength (Temp: 780°C, pressure: 12.7 kPa).

strength. Furthermore, the joint was not obtained at 720°C. Therefore, both joining conditions of a longer holding time and a higher temperature are necessary to obtain a higher joint strength.

Figures 4 and 5 show the comparisons of the macrostructure and microstructure of the delaminated surface of the specimens, respectively. Adhesion of fractured barium titanate pieces to the aluminum layer was observed for the delaminated surface near the periphery under the holding time of 16 h [Fig. 4(b)]. In addition, significant adhesion of the fractured micro grain was observed [Fig. 5(b)]. Here, Fig. 6 shows the comparison of the cross-sectional oxygen distribution in the vicinity of the joint interface near the periphery. An oxygen-rich layer was observed at the joint interface between barium titanate and pure aluminum under a holding time of 16 h [Fig. 6(b)]. This result indicated the existence of the thicker alumina layer due to a longer holding time and a higher temperature during joining [5,6]. Therefore, interfacial bonding may be achieved through the progressive oxidation of molten aluminum [7,8].

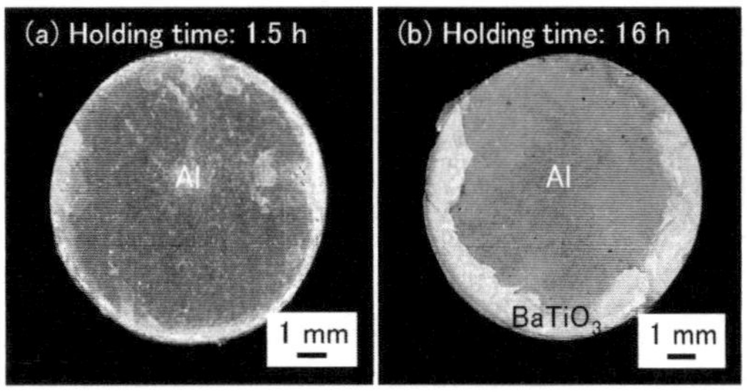

Fig. 4 Appearance of delaminated surface of the specimens (Temp: 780°C, pressure: 12.7 kPa).

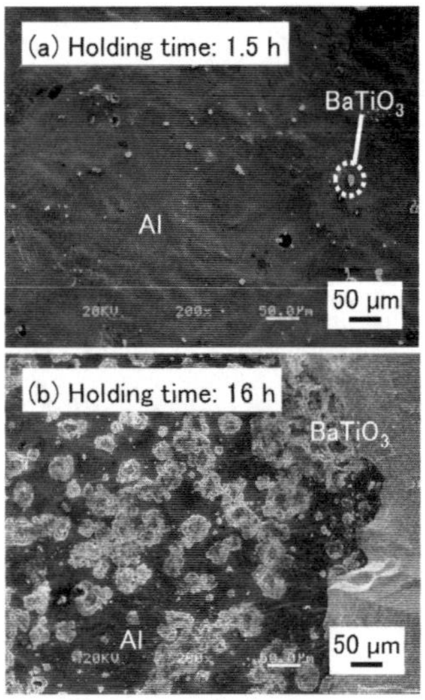

Fig. 5 SEM images of delaminated surface near the periphery (Temp: 780°C, pressure: 12.7 kPa).

Fig. 6 Cross-sectional O distribution in joint interface near the periphery (Temp: 780°C, pressure: 12.7 kPa).

4. Conclusions

(1) Higher shear strengths of more than 10 MPa was obtained with the fracture of barium titanate under longer holding times. Lower shear strength with interfacial delamination was also obtained under shorter holding times.

(2) Significant adhesion of fractured barium titanate pieces and grain to the aluminum layer was observed on the delaminated surface under a holding time of 16 h.

(3) The existence of a thicker alumina layer between barium titanate and pure aluminum in the joint interface was indicated under a holding time of 16 h.

Acknowledgements

This work was supported by the Japan Aluminium Association.

References

[1] Air-Conditioning & Refrigeration Systems Headquarters, Automotive Thermal Systems Department: Mitsubishi Heavy Industries Technical Review 46 (2010) 19-21.

[2] I. Goto, K. Anzai and S. Ideguchi: Q. J. Jpn. Weld. Soc. 30 (2012) 345-353.

[3] K. Suganuma: J. Mater. Sci. 26 (1991) 6144-6150.

[4] K. Otsuka, T. Michihiro and A. Mutou: J. JFS 70 (1998) 891-896.

[5] I. Haginoya and T. Fukusako: IMONO 54 (1982) 664-669.

[6] E. Isoyama: Aluminum Finishing Society of Kinki 117 (1986) 1-6.

[7] K. Suganuma, E. Saiz and A. P. Tomsia: J. Jpn. Inst. Met. 62 (1998) 92-97.

[8] S. Rattanachan, Y. Miyashita and Y Mutoh: J. Eur. Ceram. Soc. 23 (2003) 1269-1276.

Progressive Evaluation Method for Aluminium Alloys filtration process

Marek Brůna[1], Dana Bolibruchová[1], Augustín Sládek[1]
[1]University of Žilina, Department of Technological Engineering, Univerzitná 1, 01026 Žilina. Slovak Republic, E-mail: rapture.marko@gmail.com

Submitted work is focusing on the filtration of aluminum alloys with aim on filling time, flow turbulence and melt velocity in close vicinity to ceramic filters. These parameter are considered as the most important parameters during melt filtration and therefore needs to be analyzed in real conditions, but also by using modern simulation systems. Supported by numerical simulation software ProCAST, we were able to understand and observe the actions taking place in the close vicinity of filters, which cannot be analyzed by direct methods.

Keywords: *Filtration, simulation, aluminium alloys*

1. First stage of the experiment

In the first stage of the experiment, three different filters were chosen (with direct channels and with conical channels). Figure 1 shows the proposed gating system. Simple cuboid was chosen as the casting.

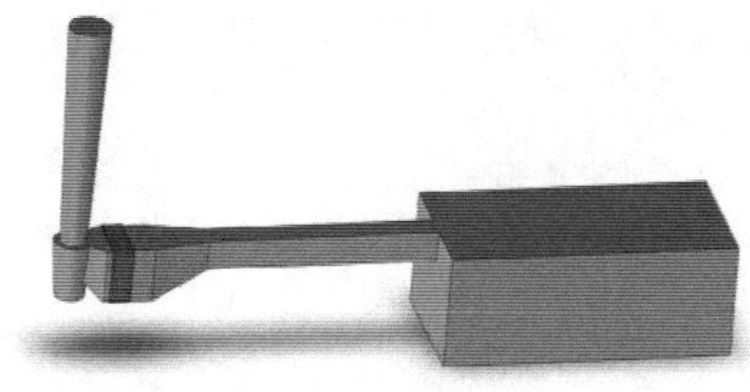

Fig. 1 Gating system with the filter location

We used an alternative (more complex) approach instead of the standard function to analyze the melt flow. Individual filters were imported to the simulation software including the precise channel geometry (based on the CAD files). This fact has a significant impact on the complexity of the entire process, especially on the creation of a 2D and 3D mesh. Input data were set as a model situation, constant for all three variants, which enabled their mutual comparison. To perform simulation was chosen AlSi7Mg0.3 alloy poured at 720 °C by flow rate 1 kg/s into green sand mold (room temperature 20 °C). Figure 2 shows the turbulence intensity at cross-section of the gating system for variant A (channels with equal cross-sections). The color scale expresses the melt turbulent energy level, where purple means zero kinetic turbulent energy and with graduated shades it passes to red, which represents 350 $cm^2.s^{-2}$. Larger turbulence occurs on the outflow side of the filter in the lower part than in the upper part. However, we will be able to draw specific conclusions in terms of turbulence intensity only upon comparison with the other variants.

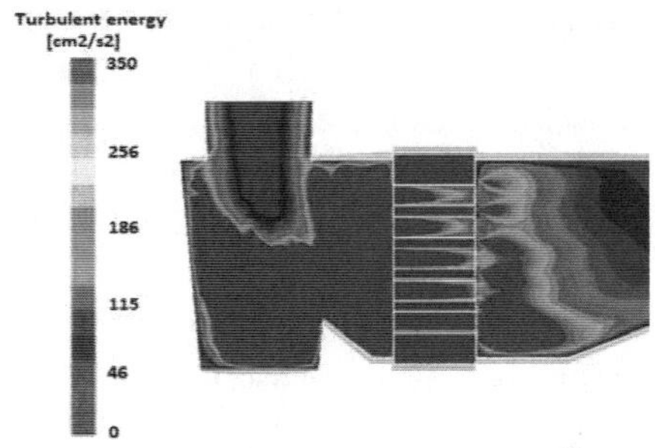

Fig. 2 Turbulence intensity – Variant A, time 0.6 sec.

The filling in Variant B is of a different nature compared to variant A (channels with conical cross-sections - fig. 3). Turbulence on the outflow side of the filter is significantly smaller. Results comparison will be described later on.

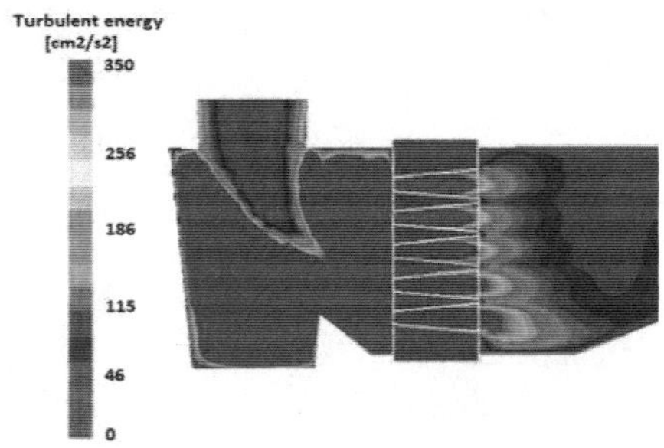

Fig. 3 Turbulence intensity – Variant B, time 0.6 sec.

Variant B1 shows the largest turbulence intensity on the outflow side compared to the other variants (fig. 4, the largest share of red). In terms of flow optimization and the associated turbulences, variant B can be considered as the optimal alternative based on the results criterions.

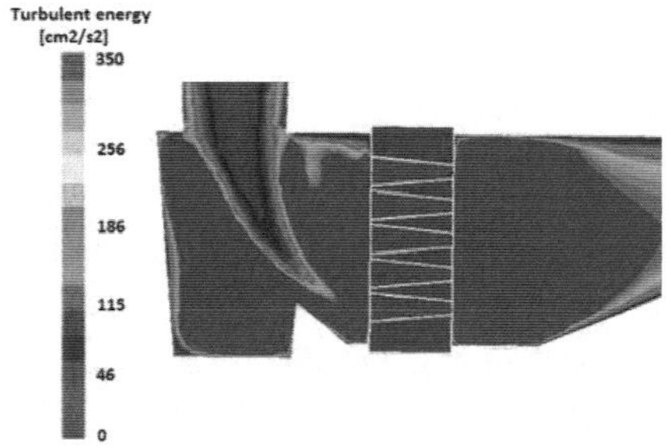

Fig. 4 Turbulence intensity – Variant B1, time 0.6 sec.

Variant B1 features the highest velocity on the outflow side, uneven flow and large turbulences. These turbulences are probably caused by nozzle-like shape of channels, which supports increase of velocity on the outflow side. This phenomena can be evaluated as negative effect leading to higher reoxidation and porosity nucleation during solidification [1].

2. Second stage of the experiment

During the second stage of the experiment were used 4 different ceramic filters (tab. 1). The objective of the second stage was to evaluate turbulence activity and velocity of the melt in the filter close vicinity (while flow area is changing due to amount of channels and their diameter).

Table 1 Parameters of the filters

	Filter 0810	Filter 0812	Filter 0735	Filter 0825
Channel diameter [mm]	Ø 3,2	Ø 2,3	Ø 2	Ø 2
Flow area [mm^2]	1351	1495	1348	1223

Filter impact on the melt velocity in the section plane leading through the middle of the gating system (Fig. 5) was investigated. The greatest deceleration of the melt occurs by introducing a 0810 filter. Comparison of the 0735 and 0825 filter simulations, clearly shows that melt velocity is affected also by the size of the flow area. Flow area affects the size of metallostatic pressure in the way that the larger the flow area, the more easily depressurization occurs. At the same time the melt velocity on the outflow side of the filter is also influenced to a great extent by the primary shock wave on the filter [1].

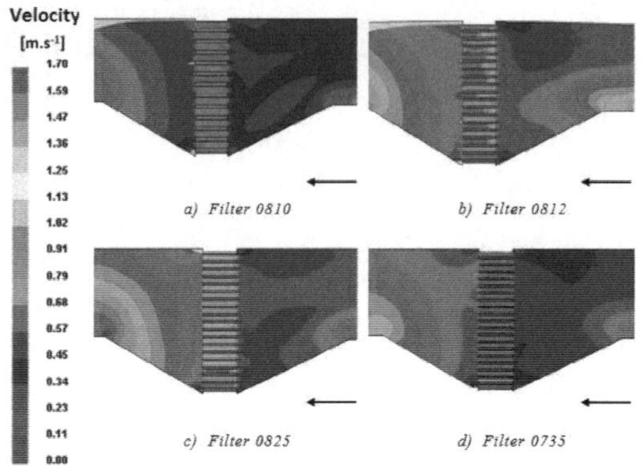

Fig. 5 Melt velocity – time 0.65 sec.

When evaluating turbulence, optimal results were achieved by a 0812 filter, which exhibits on the outflow side the lowest level of turbulent flow (Figure 6b). Although it has been proved that flowing through narrower filter channels accelerates the melt movement, the channel hydraulic diameter is so small that the Reynolds number reaches lower values than those for a 0810 filter having the largest channel diameter.

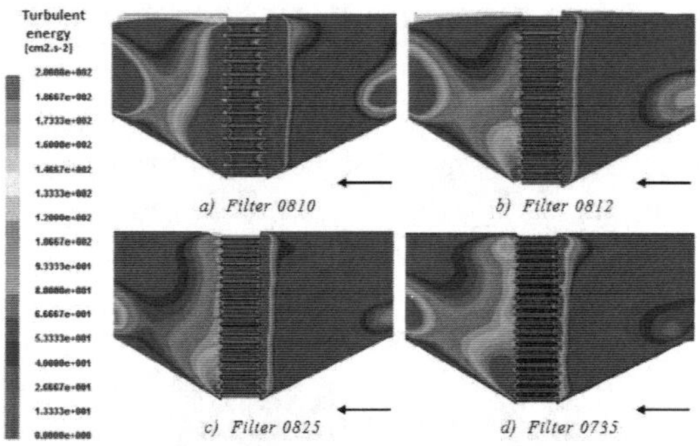

Fig. 6 Melt turbulence – time 0.65 sec.

3. Conclusion

The objective of this paper was to utilise numerical simulation for analysis of important parameters when casting aluminium alloys through ceramic filters. Partial objectives following the individual stages of experiments have confirmed that simulation is a useful tool even for this field of foundry industry.

References

[1] PASTIRČÁK, R. (2014). Effect of low pressure application during solidification on microstructure of AlSi alloys. In: Manufacturing technology. ISSN 1213-2489. Vol. 14, no. 3. pp. 397-402.

Proceedings of the 72nd World Foundry Congress
21-25th May 2016, Nagoya, Japan

Grain Refining Efficiency of Al Cast by Novel Al-$Al_{2.7}Ni_{0.3}Ti$ Refiners

Kazunori Maekawa[1], Yoshimi Watanabe[1] and Hisashi Sato[1]
[1] Graduate school of Engineering, Nagoya Institute of Technology
Gokiso-cho, Showa-ku, Nagoya, 466-8555, Japan

In this study, heterogeneous nucleation behavior of $L1_2$ cubic structure of $Al_{2.7}Ni_{0.3}Ti$ intermetallic compound particles having small misfit with fcc Al was studied. Al-$Al_{2.7}Ni_{0.3}Ti$ grain refiners were fabricated by spark plasma sintering (SPS) method using gas atomized $Al_{2.7}Ni_{0.3}Ti$ intermetallic compound particles. Grain refining efficiency of Al cast by the fabricated grain refiners with $L1_2$ structure nucleation sites was investigated.

Keywords: *aluminum, grain refiner, $L1_2$ structure, $Al_{2.7}Ni_{0.3}Ti$, spark plasma sintering (SPS)*

1. Introduction

Grain refinement is widely used in Al foundries to promote a fine equiaxed grain structure with improved mechanical properties. Generally, grain refinement of Al and Al alloys can be achieved by adding grain refiners, such as Al-Ti, Al-Ti-B and Al-Ti-C alloys with heterogeneous nucleation sites of Al_3Ti, TiB_2 and TiC particles, into the melt. There should be two approaches to enhance the refining performance of grain refiners. One is achieved by increasing the number of heterogeneous nucleation sites in the melt [1]. Another one is achieved by decreasing lattice registry between heterogeneous nucleation site and the Al [2]. Meanwhile, it is known that alloying with a certain amount of transition element such as V, Cr, Mn, Fe, Co, Ni, Cu and Zn causes the transformation from the $D0_{22}$ type tetragonal structure of Al_3Ti into high-symmetry $L1_2$ cubic structure [3-6]. Since the lattice constant of $Al_{2.7}Ni_{0.3}Ti$ intermetallic compound with $L1_2$ structure is a = 0.394 nm [7], disregistry value between $Al_{2.7}Ni_{0.3}Ti$ intermetallic compound and Al can be calculated to be 2.71 % for all planes, which is much smaller than that for platelet plane of Al_3Ti phase [2]. Therefore, it is expected that the $Al_{2.7}Ni_{0.3}Ti$ intermetallic compound particle with $L1_2$ structure become favorable heterogeneous nucleation sites for Al cast. In the previous study, grain refining performance of Al-$L1_2$ type $Al_{2.7}Fe_{0.3}Ti$ refiner has been studied [8]. In this study, $Al_{2.7}Ni_{0.3}Ti$ intermetallic compound with $L1_2$ cubic structure was chosen as the heterogeneous nucleation site, since the disregistry value between $Al_{2.7}Ni_{0.3}Ti$ and Al is smaller than that of $Al_{2.7}Fe_{0.3}Ti$ and Al. First, $Al_{2.7}Ni_{0.3}Ti$ intermetallic compound particles were prepared by gas atomization. Then, Al-$Al_{2.7}Ni_{0.3}Ti$ grain refiners were fabricated by SPS method. The grain refining performance of fabricated refiners for pure Al cast was investigated.

2. Experimental Procedure

$Al_{2.7}Ni_{0.3}Ti$ intermetallic compound particles with $L1_2$ structure were prepared by gas atomization method. The particles in the range of 75-150 μm were sintered with pure Al particles by SPS method. Volume fraction of $Al_{2.7}Ni_{0.3}Ti$ particles within the refiner was fixed to be 10 vol.%. Microstructural evolution for the $Al_{2.7}Ni_{0.3}Ti$ particles and the refiner was studied by scanning electron microscope (SEM) with energy dispersive X-ray spectrometer (EDS). Phase analysis was carried out by the X-ray diffraction (XRD) technique using Cu Kα radiation. The refiner was added into the pure Al melt at 750 °C. Subsequently, the melt was stirred for 30 s. After a certain holding time, the melt was poured into a steel mold. Microstructural obsevations were carried out with an optical microscope. The mean size of α-Al grains was calculated by using a mean linear intercept techniqe. The mechanical property of Al cast was evaluated by vickers hardness test.

3. Results and Discussion

SEM microphotograph of $Al_{2.7}Ni_{0.3}Ti$ particles prepared by gas atomization is shown in Fig. 1. As shown in Fig. 1, most of the particles are spherical. Dendrite structure is also observed on the surface. EDS and XRD analyses revealed that these particles have the stoichiometric chemical composition of $Al_{2.7}Ni_{0.3}Ti$ and $L1_2$ structure.

SEM microphotograph of the Al-10vol.% $Al_{2.7}Ni_{0.3}Ti$ refiner fabricated by SPS method is shown in Fig. 2. It is seen from the figure that the spherical shaped particles are successfully embedded within the Al matrix. Also, the clear interface is shown between

$Al_{2.7}Ni_{0.3}Ti$ particles and Al matrix. It was also confirmed that there is no reaction between $Al_{2.7}Ni_{0.3}Ti$ particles and Al matrix by analyses of EDS.

Macrographe of unrefined pure Al sample cast is shown in Fig. 3 (a). Fig. 3 (b) shows the macrostructure of Al cast refined by the Al-10vol.% $Al_{2.7}Ni_{0.3}Ti$ refiner, where holding time is 690 s. The Al cast without refiner has coarse and inhomogeneous grains, as shown in Fig. 3 (a). On the other hand, the grain size becomes smaller and more homogeneous in the Al cast with the Al-10vol.% $Al_{2.7}Ni_{0.3}Ti$ refiner, as shown in Fig. 3 (b).

Fig. 4 shows the effect of holding time on the grain size of α-Al. The grain size of casts added the refiners reaches to minimum at 690 s and then an ascending trend with the holding time. Since $Al_{2.7}Ni_{0.3}Ti$ intermetallic compound phase cannot exist within the Al matrix in equilibrium state, the decomposition of $Al_{2.7}Ni_{0.3}Ti$ intermetallic compound particles occurs during the casting, which may result in the fading behavior.

Fig. 1 Backscattered electron composition image of $Al_{2.7}Ni_{0.3}Ti$ powder.

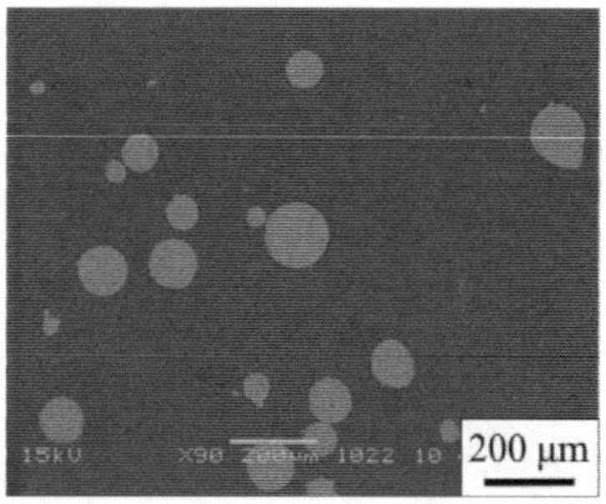

Fig. 2 Backscattered electron composition image of Al-10vol.% $Al_{2.7}Ni_{0.3}Ti$ refiner.

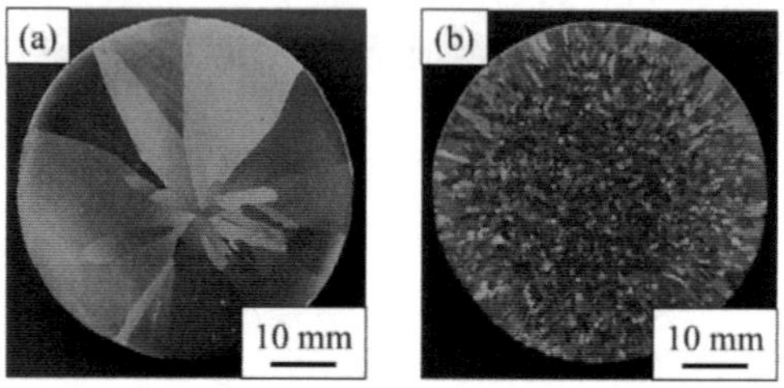

Fig. 3 Photographs of Al casts with and without Al-10vol.% $Al_{2.7}Ni_{0.3}Ti$ refiner. (a) without refiner, (b) with the refiner (holding time 690 s)

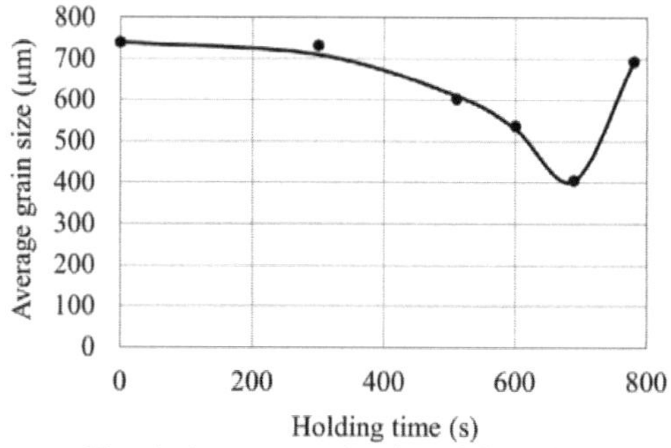

Fig. 4 Average grain size of the Al casts plotted against the holding time.

4. Conclusions

Al-10vol.% $Al_{2.7}Ni_{0.3}Ti$ refiners are able to be fabricated by SPS method. The grain size of Al cast becomes smaller and homogeneous microstructure could be achieved in the Al casts with the refiners. Therefore, usefulness as the heterogeneous nucleants of $Al_{2.7}Ni_{0.3}Ti$ was indicated.

References

[1] Z. Zhang, S. Hosoda, I. S. Kim and Y. Watanabe: Mater. Sci. Eng. A425 (2006) 55-63.

[2] Y. Watanabe and H. Sato: J. Jpn Inst, Light Met. 64 (2014) 157-163.

[3] J. Tarnacki and Y-W. Kim: Scr. Metall. 22 (1988) 329-334.

[4] Y.Nakayama and H. Mabuchi: Intermetallics. 41 (1993) 41-48.

[5] S. Liu, R. Hu and C. Wang: J. Appl. Phys. 74 (1993) 3204-3210.

[6] S. M. Kim, M. Kogachi, A. Kameyama and D. G. Morris: Acta Metall. Mater. 43 (1995) 3139-3149.

[7] P. Villars and L. D. Calvert: *Pearson's Handbook of Crystallographic Date for Intermetallic Phases* (ASM International, Materials Park, OH, 1997).

[8] Y. Watanabe, T. Hamada and H. Sato: Jpn J. Appl. Phys. 55 (2016).

Proceedings of the 72nd World Foundry Congress
21-25th May 2016, Nagoya, Japan

Development of elasto-plastic-creep constitutional equation for a die casting alloy and validation of thermal stress analysis

Hidetoshi Shiga[1], Hiroshi Kambe[1], Makoto Yoshida[2], Yuichi Motoyama[3]
[1] Powertrain Technology and Prototype Development Department Nissan Motor Co, Ltd., Japan
[2] Kagami Memorial Research Institute for Materials Science and Technology Waseda University, Japan
[3] Digital manufacturing process group National Institute of Advanced Industrial Science and Technology, Japan

In order to predict the residual stress in shape castings, elasto-plastic models have been used. However, these models are not able to express the recovery, which is a characteristic behavior of alloys at high temperature. This often causes over estimation of the residual stress of the castings. In this study, an elasto-plastic-creep model is developed to express the recovery of the casting during cooling. Under the conditions of some temperatures and strain rates, tensile tests were conducted to identify the creep properties. In order to validate the thermal stress analysis with the developed model, an original device, called "Equipped I-Beam Caster" was also developed to measure thermal stress and temperature of the JIS ADC12 (ISO Al Si 11 Cu3(Fe)) casting continuously during cooling. As a result, compared to the elasto-plastic model, the elasto-plastic-creep model predicts more accurately not only the thermal stress during cooling but also the residual stress.

Keywords: *Thermal stress analysis, Constitutive equation, Residual stress*

1. Introduction

In order to predict the residual stress in shape castings, elasto-plastic constitutional equation (E-P) models have been used. However, these models are not able to express the recovery, which is a characteristic behavior of alloys at high temperature. This often causes over estimation of the residual stress of the castings.

In this study, an elasto-plastic-creep constitutional equation (E-P-C) model is developed to express the recovery of the casting during cooling.

In order to validate the thermal stress analysis with the developed model, an original device was also developed to measure thermal stress and temperature of the ADC12 casting continuously during cooling.

2. Measurement data for validation

In order to validate the thermal stress analysis with the developed model, an original device, called "Equipped I-Beam caster" (Fig.1) was also developed to measure thermal stress and temperature of the casting continuously during cooling.

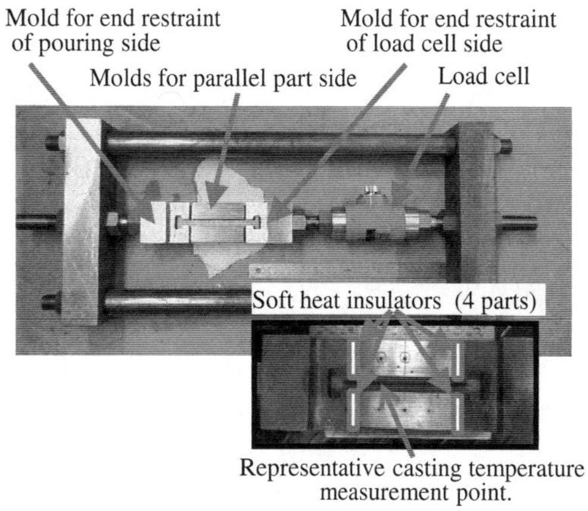

Fig. 1 Appearance of Equipped I-Beam Caster.

Cavity in this device is designed as I-shape. Both ends of the cavity, which are T-shaped, are constrained in metal molds for end restraint. To the frame, one metal mold is connected directly and the other is connected through 30kN load cell. Soft heat insulation with a thickness of 2mm is installed in a gap between end restraint metal mold and mold for parallel part side of cavity. This allows the metal molds on both ends to displace as the casting shrinks in the cavity. The size of parallel part of the cavity is 10mm x 10mm x 100mm. The end restraint metal molds are made with JIS SUS430 (ISO X6Cr17). The molds for parallel part side are made with JIS SUS304 (ISO X5CrNi18-10). Block-shaped insulations with a thickness of 15mm cover the upper and lower surfaces of the cavity.

With the device, the displacement of the metal molds on the ends and the load on the load cell were measured during cooling process of the ADC12 casting. The temperatures of the casting and the metal mold were also measured.

3. Identification of constitutional equation model

An E-P-C model was developed to express the recovery of the casting during cooling.

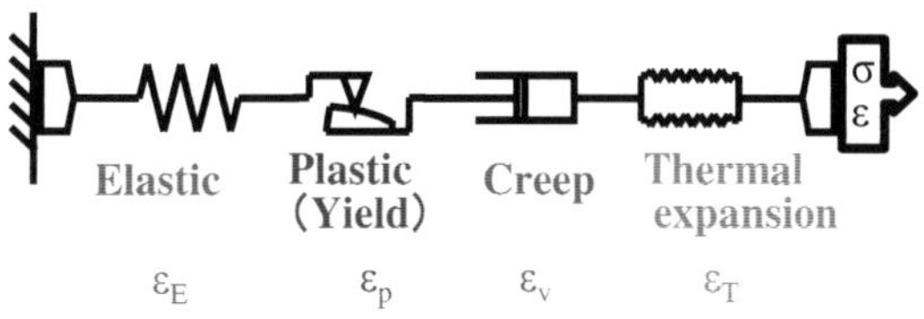

Fig.2 One dimensional schema drawing of E-P-C model.

Fig.2 shows one dimensional schema drawing of E-P-C model. σ is stress and ε is total strain which is sum of strains by elastic(ε_E), plastic(ε_p), creep(ε_v), and thermal expansion(ε_T).

Under the conditions of various temperatures and strain rates, tensile tests were conducted to measure the plastic and creep properties. The specimens of ADC12 were fabricated by the PF die casting process. After the solution treatment at 450°C for 3.6 ks, for removing the natural aging, the specimens were cooled to the aimed temperatures. In the identified E-P-C model, creep contributes to deformation above 250°C and plasticity does not work above 300°C.

For a comparison, two E-P models, dependent and nondependent on strain rate respectively, were identified from the results of the same test.(Each of them is called E-P<DSR> , E-P<NDSR>)

4. Thermal stress analysis

For validation of developed model, load history calculated with thermal stress analysis with boundary conditions based on actual temperature and actual displacement, was compared with actual load history.

Temperature history data necessary for the thermal stress analysis was created with filling analysis and solidification analysis under the actual casting conditions. To match temperature history from the analyses with the actual measured temperature history, interfacial heat transfer coefficient was adjusted in filling analysis and solidification analysis.

Then the thermal stress analysis was performed using the constitutional equation models. The temperature history data acquired in the solidification analysis was used for the constraint condition of the analysis. The measured displacement of the end restraint metal molds of equipped I-Beam caster was also used for same purpose.

5. Result of validation

Fig.3 shows a comparison between the calculated load history and the actual measured load history. Load histories were calculated in both analyses with considering and without considering the recovery. The case in which the recovery is taken into consideration by E-P-C model shows a smaller load error and its output is closer to the actual measured value. Compared to the E-P models, the E-P-C model predicts more accurately not only the thermal stress during cooling but also the residual stress.

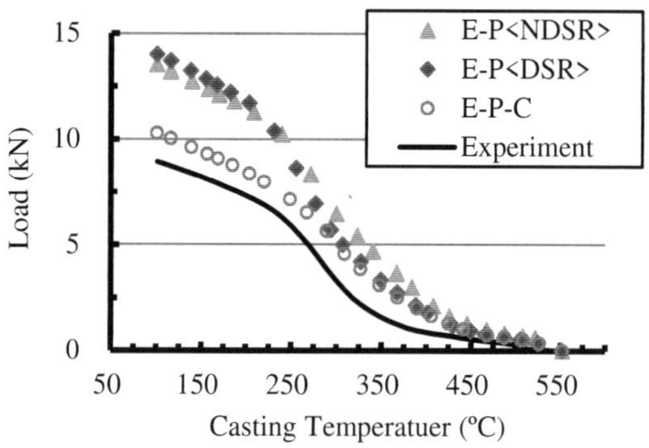

Fig.3 Comparison of the simulated and the measured thermal load during cooling.

6. Conclusion

In order to determine prediction accuracy in a thermal stress analysis of ADC12 casting, the outcomes of experiment with equipped I-Beam caster and the result of the thermal stress analysis employing the constitutional equation model in which the recovery was considered were continuously compared by the load of the casting.

It is found that the use of the constitutional equation model considering the recovery leads to a higher prediction accuracy.

References

[1] Yuichi Motoyama : Reports of the 161st Japan foundry Engineering Society meeting (2012) 21,22

[2] Hidetoshi Shiga, Takeshi Sato, Hiroshi Kambe, Yuichi Motoyama, Makoto Yoshida : Journal of Japan foundry Engineering Society 7 (2015) *pp*.453-459

Proceedings of the 72nd World Foundry Congress
21-25th May 2016, Nagoya, Japan

Numerical Simulation of Mechanical Wave Propagation in Casting Solidification Process

Wu Shiping[1], Wang Rujia[1], Chen Wei[1], Cheng Weixin[1], Chen chenglong[1]
[1] The material science and Engineering in Harbin Institute of Technology, Harbin, China, 150001

This paper is mainly devoted to the numerical simulation of mechanical wave propagation in casting solidification process. Firstly, this paper found the mathematical physics model of elastic wave propagation in the elasticity, viscosity and viscoelastic medium. Then, the above equation were solved by finite difference method. Finally, the boundary conditions of casting/mold interface, to realize the mechanical wave propagation in the alloy melt numerical simulation. To achieve the numerical simulation of mechanical waves propagating in the alloy melt. Studied mechanical wave propagation in the alloy melt solidification process. Research shows that mechanical wave propagation attenuation trend in a casting. The site of minimum vibration of the mechanical wave is not the center of the casting in the process of solidification, but the position between the center and the reflection surface.

Keywords: *mechanical wave, casting, solidification process, propagation.*

1. Introduction

An effective measure to improve the solidification structure, eliminate or reduce the casting defects has been recognized by applying an external field in the metal solidification process. It has been 100 years to improve the microstructure and mechanical properties by means of vibration solidification [1-4]. The scientific problem of the impact of vibration on metal solidification is the problem how the elastic wave propagate in viscoelastic two-phase media. In this paper, the author uses the numerical simulation technology to study the propagation of vibration wave in the alloy melt, and the influence of vibration wave in the solidification process.

2. Mathematical physics model

2.1 Constitutive relation

We used the liquid-solid two phase region rheometer by self-designed to measure rheological model of ZL205A in liquid-solid phase and solid alloy. It is showed that the rheological characteristics of ZL205A can be described by the five element model (H1- (N1|S1) - (H2|N2)) in the whole temperature ranges below the liquidus.

In this paper, we consider the force of the alloy does not exceed the yield stress during the metal solidification process. In this case, the constitutive relation of the alloy is H1- (H2|N2).

2.2 Wave equation

According to the "corresponding rule" of the elastic medium and the viscoelastic medium suggested by Auld in 1990, we can obtained 1D (one dimensional) wave equation.

$$\begin{cases} \dfrac{\partial v_x}{\partial t} = \dfrac{1}{\rho}\dfrac{\partial \tau_{xx}}{\partial x} \\ (\mu_1 + \mu_2)\dfrac{\partial \tau_{xx}}{\partial t} + \mu^{'}\dfrac{\partial^2 \tau_{xx}}{\partial t^2} = (\mu_1\lambda + \mu_2\lambda + 2\mu_1\mu_2)\dfrac{\partial v_x}{\partial x} \\ +(\mu_1\lambda^{'} + \mu_2\lambda^{'} + \mu^{'}\lambda + 2\mu_1\mu^{'})\dfrac{\partial^2 v_x}{\partial x\partial t} + \mu^{'}\lambda^{'}\dfrac{\partial^3 v_x}{\partial x\partial t^2} \end{cases} \quad (1)$$

2.3 Boundary conditions

The boundary of the wave field calculation in the casting solidification process can be divided into three categories:

(1) Mold / air interface

The interface only have speed without stress,

$$\sigma_x = \tau_{xy} = \tau_{xz} = 0 \quad (2)$$

(2) Mold / earth interface

The casting of the interface only transmitted stress without speed.

$$u_x = u_z = 0 \quad (3)$$

(3) Casting/ Mold interface

When the wave transmits to the interface, the velocity and stress at the two sides of the interface are distributed. It is decided by the relative size of the wave impedance (ρv) between the two sides. The interface does not need to do special treatment.

3. Propagation simulation and experiment

The propagation law of the mechanical wave in a 1D high temperature solid metal was obtained by the numerical simulation of the wave equation with MATLAB. We used Ricker wavelet as pulse source, frequency is 780Hz, and calculating time is 12.5ms. The length of the rod is 2.8m, and the measuring temperature is 490°C. The simulation result and the experiment result are showed in Fig.1.

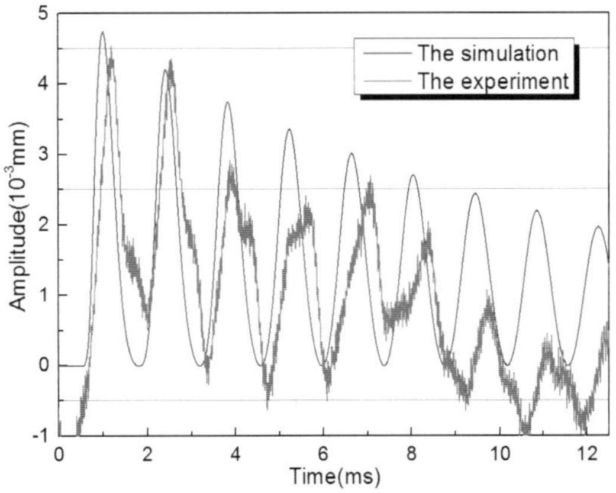

Fig. 1 Numerical simulation and experiment results of wave propagation in high temperature metal rod

We can see that the mechanical wave propagation in the high temperature metal rod is more serious attenuation and the amplitude is becoming smaller and smaller. The velocity of wave in the metal rod is about 4000 m/s when temperature is 490 °C. By comparing the pulse signal derived from the experimental to detected signal in the experiment, the velocity and the attenuation rate in the simulation conform to the experiment result well.

4. Numerical simulation in 1D casting

In this paper, the author has carried on the numerical simulation of the mechanical wave propagation in the metal solidification process of 1D metal casting. The temperature in the both ends of 1D casting is 550°C, and the center is 650°C. The gradient temperature is linear from the edge to center. The vibration source was set in the left end. The experimental structure and points distribution showed in Fig.2.

We can see from Fig.3, the initial wave amplitude is small, only 0.3mm, the amplitude of the wave gradually decays in propagation process. But the source continues to provide energy, leading to amplitude get bigger in the casting. The casting center is not the wave amplitude minimum position, although the central part of the decay is the most serious. With the supply of the source energy, the amplitude is not small. The site of the minimum amplitude is located away from the vibration 0.75m. Because of the energy supplement was inadequate from the source, the reflection of the wave was almost completely attenuated, and the site obtained the smallest amplitude in the last.

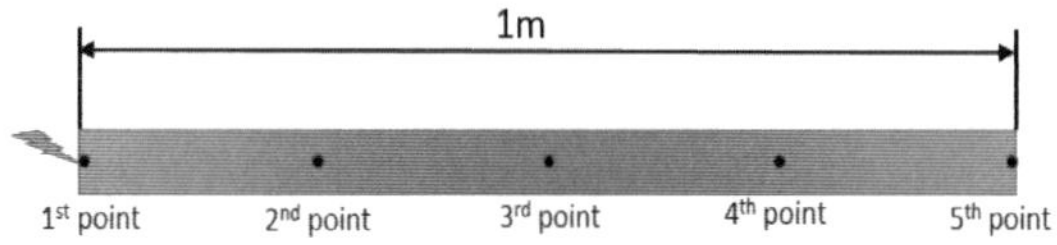

Fig.2 Experimental structure and points distribution

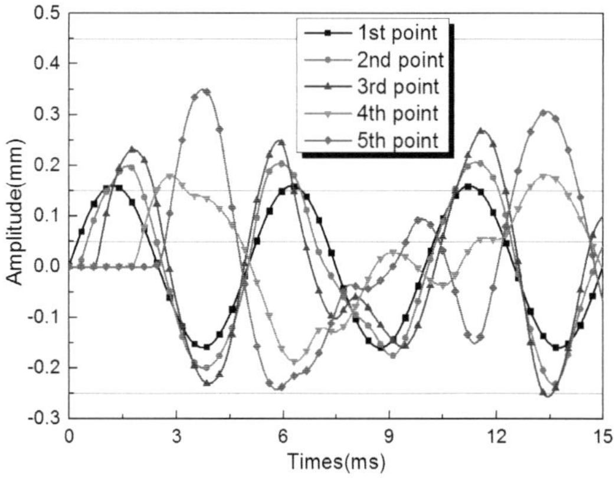

Fig. 3 The amplitude variation with time of each point in 1D metal solidification process

5. Conclusion

In this paper studied the propagation of mechanical waves in the casting solidification process, and obtained the following conclusions :

(1) The rheological characteristics of high temperature solid and solid-liquid phase of ZL205A alloy can be described with the mechanical model of five element (T=H1-(S1|N1) - (H2|N2));

(2) The mechanical wave decays rapidly in the high temperature alloy, and the propagation time is very short.

(3) The site of minimum vibration of the mechanical wave is not the center of the casting in the process of solidification, but the position between the center and the reflection surface.

Acknowledgement

This work was supported by the National Natural Science Foundation of China (No. 51475120)

References

[1] Fisher T P: Br. Foundryman, 66(1973) 71-84.

[2] Zhang Z Q, Le Q C and Cui J Z: Applied Mechanics and Materials. 105(2012) 1616-1619.

[3] Currey D A, Pickles C A and Dofasco: J. Mater. Sci. 23(1988)3756-3763.

[4] Eskin G I and Eskin D G: CRC Press, 2014.

Proceedings of the 72nd World Foundry Congress
21-25th May 2016, Nagoya, Japan

Predicting Thermal Fatigue Life of Aluminum Die Casting Molds - An Experimental and Computational Approach

Sebastian Müller[1], Helge Pries[1] and Klaus Dilger[1]
[1] Braunschweig University of Technology, Institute of Joining and Welding, Braunschweig, Germany

The cost effectiveness of the die casting process correlates to a large extent with the lifetime of die casting molds. Hence, ensuring a predictable and adequate lifetime of aluminum die casting molds is of high importance for foundries worldwide. Within the scope of this work, methods to quantify the thermal fatigue life in critical areas of an aluminum die casting mold were developed. The results can serve to support engineers in designing optimized die casting molds.

Keywords: *lifetime, thermal fatigue, die casting mold, hot work tool steel, simulation*

1. Approach of the developed lifetime assessment

The work is based on strain-life fatigue analysis. As depicted in Fig. 1, an extensive number of further aspects were taken into account in order to ensure the relevance of the approach for the die casting process. The identified correlations were considered by correction factors.

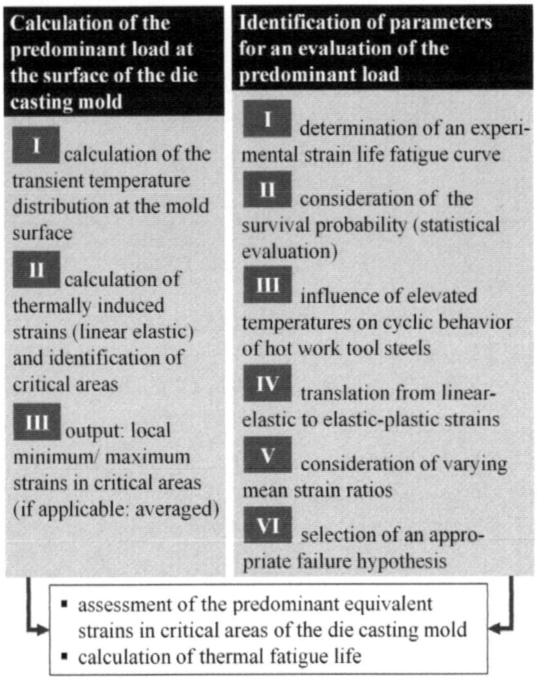

Fig. 1 Approach of the developed lifetime assessment

2. Experimental methods: cyclic material properties

In order to obtain strain-life curves for the hot work tool steel AISI H11, cyclic strain-controlled tests were conducted. Within the scope of this work, well established aviation-specific standards respectively code-of-practices for thermomechanical fatigue (TMF) testing were considered [1]. The experiments were carried out with a servo-hydraulic fatigue testing machine (maximum load: 50 kN). For the measurement of the specimen mechanical strain, a laser extensometer P50 (Fiedler Optoelektronik GmbH) was used. The assessment of the recorded hysteresises were performed by automated evaluation routines. In this context, Fig. 2 shows the strain-life curve and the corresponding parameters for the strain life equation according to Coffin, Manson and Morrow. This data constitutes the basis for an evaluation of the predominant load. Moreover, it should be taken into account that the aspects II to VI (cf. Fig. 1) have a significant influence on the calculation of the thermal fatigue life.

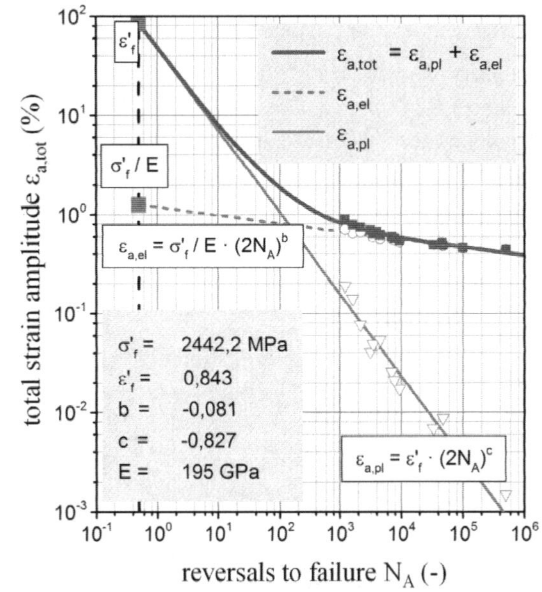

Fig. 2 Strain life fatigue curve of AISI H11

This shall apply in particular to the application of a statistical evaluation of the experimental results. For example, by adopting a survival probability of 90 %, the calculated lifetime for the strain amplitude $\varepsilon_{a,tot}$ = 0,5 % is reduced from approx. 45.000 cycles to approx. 20.000 cycles.

Numerous series of experiments were performed for the purpose of comparing the fatigue behavior of the conventional hot work tool steel AISI H11 with the fatigue behavior of a specially developed steel grade (DIEVAR). A statistical evaluation revealed that there is no significant difference between both hot work tool steels. This statement refers to the examined strain amplitudes and to the specific boundary conditions of the strain-controlled fatigue test series.

An essential part of this work also investigated the influence of the mean strain on the strain-life fatigue of the hot work tool steel. Based on the experimental results, it was suggested to establish a correlation between the relative strain-life and the dimensionless mean stress relation σ^*. However, it must be stated that the influence of this parameter is relatively low when compared to the other aspects, mentioned in Fig. 1.

3. Computational methods: lifetime assessment of a die casting mold

The plausibility of the gained experimental results and correlations were validated by the use of an existing die casting mold. For the calculation of the predominant load, the finite element software ANSYS was applied. Sound assumptions must be made on the boundary conditions since they are essential for a meaningful calculation of thermal fatigue life. Within the scope of this work, data from [2,3] was taken into consideration. Fig. 3 shows a detailed view of the predominant von Mises equivalent strains at the cavity surface of the ejector die.

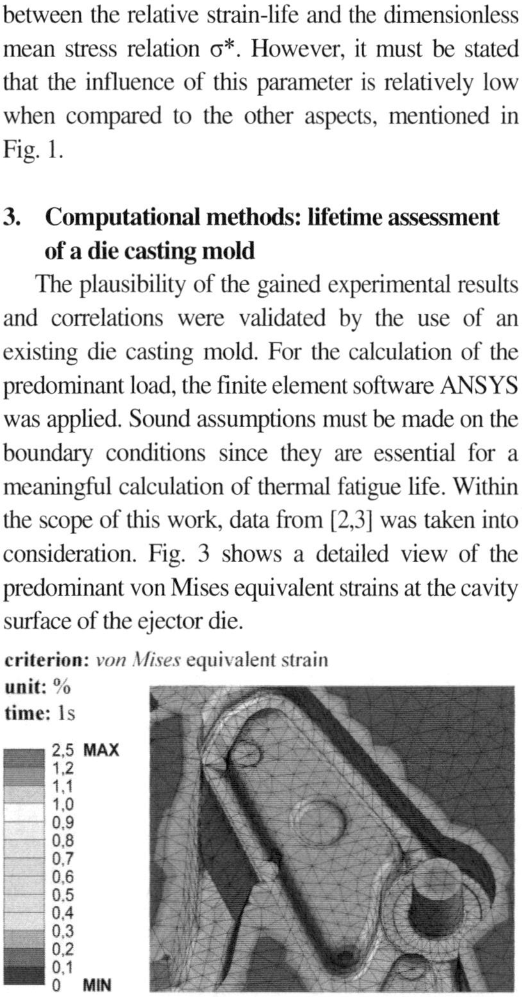

Fig. 3 Load at the cavity surface (ejector die)

Fig. 3 clearly illustrates that the critical areas correspond to the notched areas at the cavity surface. Calculated equivalent strains mostly exceed the linear-elastic range of the hot work tool steel (in critical local areas greater than 1,2 %). Hence, the depicted results emphasize the importance of a translation from linear-elastic to elastic-plastic strains, if the finite element analysis was based on a linear-elastic material model.

By comparing experimental and computational results, a good accordance between the predicted and the actual lifetimes was found.

4. Summary and outlook

Within the framework of this work, thermal fatigue life in critical areas of an aluminum die casting mold has been quantified. A large number of aspects were considered in order to ensure the relevance of the approach for the die casting process.

It has to be taken into account that static/cyclic strengths of hot work tool steels are inhomogeneous due to limitations in the production and heat treatment processes. The predominant load at the cavity surface varies. Hence, load assumptions are still fraught with considerable uncertainties. Consequently, no universal approach can easily predict accurate thermal fatigue life. However, in any case, the depicted method allows a good risk assessment. Furthermore, the presented approach can also contribute to the optimization of the production process since it helps to identify parameters that are sensible with regard to a long lifetime. The modular approach of the work allows future enhancements by correction, refinement or even by the consideration of further correlations that are of relevance for thermal fatigue life.

References

[1] *Test methods for metallic materials – Constant amplitude strain-controlled low cycle fatigue testing*. AECMA Standard prEN 3988, Edition P1 (1998).

[2] Z. Guo, S. Xiong, S. Cho, J. Choi: *Interfacial heat transfer coefficient between metal and die during high pressure die casting process of aluminum alloy*, (Frontiers of Mechanical Engineering in China, 2007) Volume 2, Issue 3, pp. 283-287.

[3] J. Röse, A. Gebauer-Teichmann, B. Scholtes: *The effect of conventional mould spraying and micro-spraying on heat transfer and separating layer formation for high pressure die casting processes*, (International Foundry Research, 2014) Volume 66, Issue 1, pp. 16-23.

Proceedings of the 72nd World Foundry Congress
21-25th May 2016, Nagoya, Japan

Optimization of geometry of ceramic gating systems for large-size ferrous castings

Michal Szucki[1], Jozef Szczepan Suchy[1], Tomasz Goraj[1], Jacek Sobczyk[2], Marek Pater[3], Janusz Lelito[1], Pawel Leszek Zak[1]
[1]AGH University of Science and Technology, Department of Foundry Processes Engineering, Reymonta 23, 30-059 Krakow, Poland
[2] Strata Mechanics Research Institute of the Polish Academy of Sciences, Reymonta 27, 30-059 Krakow, Poland
[3]GE Power, Stoczniowa 2, 82-300 Elblag, Poland

The purpose of the study was analysis of existing solutions and development of the new ones in the scope of designing ceramic gating systems for large-size castings. The current study covers the numerical simulations and experimental investigation of liquid - air flows. The authors focused on phenomena occurring during liquid metal motion in characteristic parts of ceramic gating systems. Analyses were aimed at determination of the optimal geometry of fireclay shapes allowing to obtain laminar flow and to decrease the amount of entrained air.

The results confirmed that the geometry of the currently used gating system components can promote turbulence formation. Thus, the authors proposed its modification, which allowed to limit the air entrainment phenomenon. Technological recommendations, allowing to reduce turbulences accompanying the liquid metal flow in the gating system, were also presented.

Keywords: *gating system, fireclay shapes, air entrainment, numerical simulations.*

1. Introduction

Foundry engineering is a branch of science and industry encompassing a wide range of technologies and production methods significantly differing from each other. An interesting example is production of specific large-size ferrous castings. Such castings are usually required to be of high quality, hence their production is costly and time consuming. Their characteristic feature is using ceramic gating systems at the mould preparation and filling stage. These systems are built of ready prefabricated ceramic products, the so-called fireclay shapes [1,2].

In the case of large-size castings special issues constitute turbulences accompanying the liquid metal flow in the gating system. These are caused by high stream velocity as well as by dimensions of the system [3]. These turbulences, in turn, contribute to air entrained in liquid, which leads to forming casting defects that are difficult to remove [4]. Utilization of ready ceramic systems simplifies development of technology and mould preparation, however, it can be increase these unfavourable effects. Thus, it seems justified to search for a more optimal geometry of ceramic gating systems.

2. Numerical simulations

The first part of the current study covers the numerical simulation of liquid metal (steel) - air flows in characteristic parts of ceramic gating system performed in the Flow-3D software (Fig. 1).

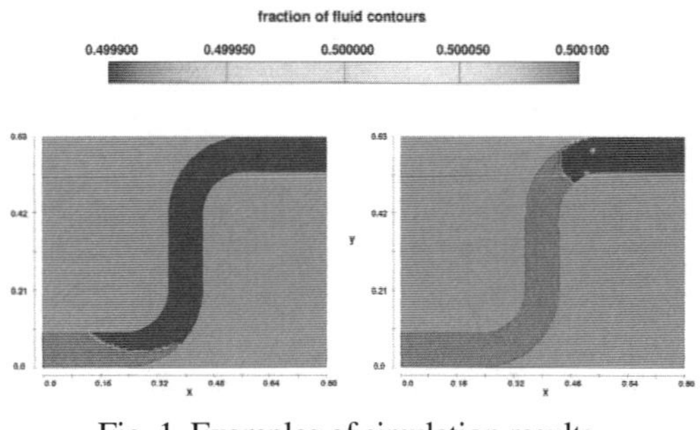

Fig. 1. Examples of simulation results

The results have shown that the highest amount of air is entrapped in liquid when stream is split or suddenly changes direction of the motion. These processes accompanied by flow turbulences occur in fireclay shapes such as tees and elbows. Hence the geometry and position of these gating system components have significant meaning for nature and kinetics of flows.

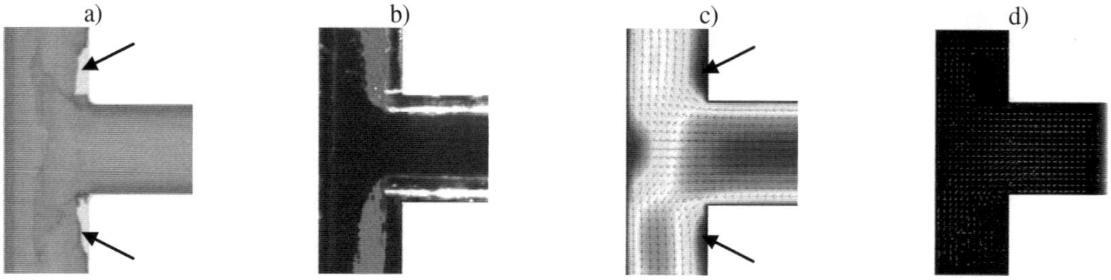

Fig. 2. Gas bubbles: a) - simulation, b) - experiment and dead zones: c) - simulation (2D), d) - PIV analysis

3. Experimental investigation

In the second part of this study the simulation results were compiled with the data obtained during experimental tests. Two-phase (model liquid - air) and single-phase (model liquid) flows in model systems composed of pipes, elbows and tees were analyzed. At this stage of the research authors used a test stand equipped with a ultra-fast digital camera [5].

3.1 Two-phase flows

Experimental evaluation has confirmed the simulation results (Fig. 2). Studies of two-phase (water - air) flows have also shown that tee type fireclay shapes have particularly negative impact on the flow behavior. It has been found that during pouring in these gating system components large gas areas are formed (Fig. 2b). The air entrapped in such regions is gradually released in the form of small bubbles, which is especially dangerous, as bubbles may not have enough time to leave the mould cavity (before solidification begins).

3.2 Single-phase flows

In order to clarify this phenomenon we also performed an analysis of single-phase flows in model systems by the use of PIV method (Particle Image Velocimetry). It showed that during the liquid flow in certain areas of gating system such as elbows or tees "dead zones" (where velocity is close to zero) are formed (Fig 2d). Hence such regions remain unfilled (filled with air) for a longer period (during pouring). This process depends strongly on the orientation (typically horizontal or vertical) of fireclay shapes in relation to the gravity vector.

4. Geometry optimization

Based on the numerical and experimental studies the authors suggested modification of gating system geometry. Among others, the tees ("T" shape) were replaced by a "Y" shapes. This allowed for a significant reduction in the volume of entrained air (Fig. 3).

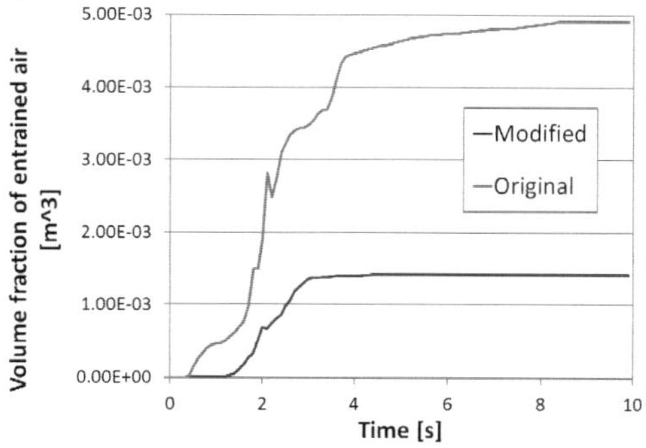

Fig. 3. Comparison of volume of entrained air for the gating system with the use of original ("T") and modified shapes ("Y")

5. Conclusion

Currently used ceramic gating system can promote the air entrainment effect. In components such as elbows and tees the turbulence and "dead zones" are created which, in consequence, lead to formation of gas defects in castings. Significant reduction of the entrained air volume is possible by changing the orientation and geometry of the fireclay shapes.

Acknowledgements

The authors acknowledge financial support from the statement contract no. 11.11.170.318 task 1.

References

[1] T. Goraj, M. Szucki, J. Lelito and J.S. Suchy: Arch. of Foundry Eng. 14 (2014) pp. 25-40.

[2] B. Grabowska, M. Szucki, J. S. Suchy, S. Eichhilz and K. Hodor: Polimery 58 (2013) pp. 39-44.

[3] M. Blair, R.W. Monroe and C. Beckermann: *The Effect of Pour Time and Head Height on Air Entrainment,* (1993).

[4] Y. Kuriyama, K. Yano and S. Nishido: Fluid Dyn., Comp. Model. and App. (2012) pp. 575-588.

[5] M. Szucki: *Application of the lattice Boltzmann method for modelling of liquid metal motion,* (PhD thesis, 2014).

O-143

Solidification of the casting in a sand mould with forced cooling

Paweł L. Żak[1], Józef Sz. Suchy[1], Jerzy Zych[1], Janusz Lelito[1],
Michał Szucki[1], Paweł Malinowski[1]
[1] AGH University of Science and Technology,
Faculty of Foundry Engineering,
Reymonta 23, 30-059 Kraków, Poland

The casting solidification is one of the most important stages of the manufacturing of casted details. During solidification a microstructure is formed. Final microstructure will affect the mechanical properties of the casting. One of the parameters influencing the casting microstructure, besides the chemical composition, is a cooling rate. By controlling the cooling rate one may affect the type of phases and their size. Therefore so it's important to choose adequate cooling conditions especially in the case of large-size casting that are cast in sand moulds. Due to the thickness of the casting wall the control of the cooling rate becomes difficult. In order to avoid unnecessary scrap computer systems of foundry processes modelling may come with aid [1, 2].
This paper presents an analysis of the solidification process of the ductile iron in sand mould with forced cooling, based on computer simulations done in the Magma Soft. At the stage of numerical examination different simulation parameters were considered. The results of numerical studies showed that under forced cooling the casting can be faster taken out of the mould. Differences in the casting rate at eutectoid reaction allows to control microstructure of the casting [3,4].

Keywords: *solidification, ductile iron, cooling condition, simulation.*

1. Introduction

Quality of casted details depends mostly on the cooling and solidification process run. Especially in the case of ductile cast iron several stages of microstructure forming can be distinguished. The analysis undertaken in this work was to analyse the cooling and solidification process of ingot mould (Fig. 1) for permanent ingot casting.

In presented work the influence of proposed cooling system (Fig. 1) on casting solidification time as well as the time to removal from the mould was analysed. The casting can be shake out from the mould as soon as the temperature in the ingot mould drops below 400 °C.

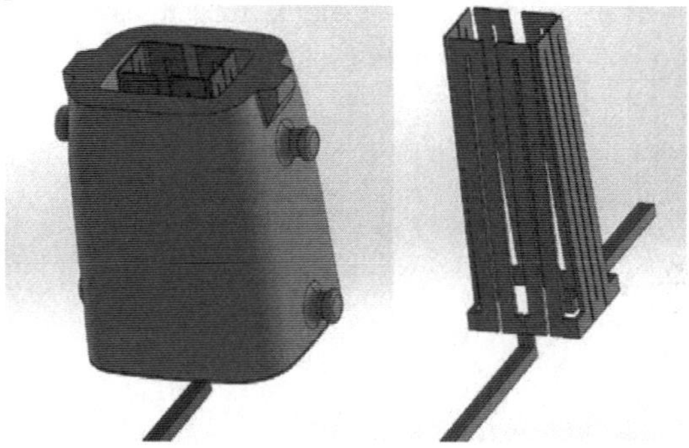

Fig. 1. Ingot mould with cooling system (left hand side) and cooling system (right hand side)

2. Numerical model

2.1 Boundary - initial conditions

The casting geometry as well as design of cooling system was provided by KrakOdlew foundry. Height of ingot mould was 2100 mm its maximal width 2088 mm. It was made of GJS-400 (Table 1) ductile cast iron. Pouring temperature was 1310 °C. Mould, core and pumped cooling liquid (mixture of air and water) temperatures at the beginning of the process were equal to 20 °C. Pumping pressure for the cooling liquid was different for new versions of simulations. Base value was 6 atm. Mould and core was made by furan. Cooling system cover was made by steel plate 50 mm thick.

Table 1. Composition of GJS-400 cast iron (wt.%) – Fe balance [5]

C	Si	Mn	Mg	Cu	P	Cr
3.75	1.8	0.36	0.035	0.013	0.04	0.01

2.2 Simulations

Simulations were performed for case without cooling system and for the cooling systems under different pumping pressures of the cooling liquid : under 4, 6 and 8 atmospheres.

3. Simulation results and discussion

During described work four versions of simulations were performed. Results were analysed under the following conditions: influence of forced cooling on the shake out time, eutectoid reaction rate, core cooling and crystallization process. Results are gathered in the graphs and tables.

In the figure 2 influence on crystallization curve is shown. The influence is not significant, however the system without forced cooling crystallization time is longest.

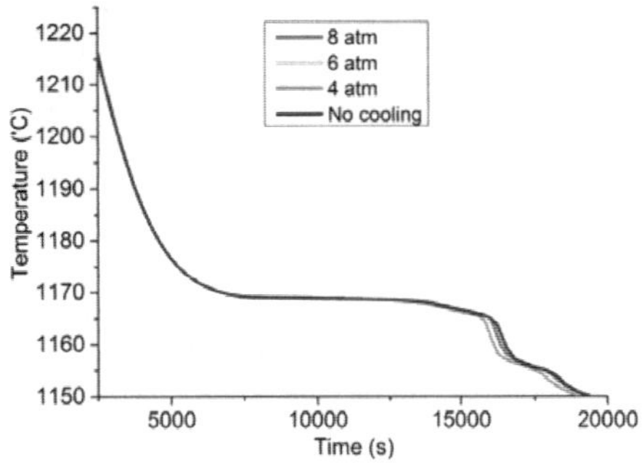

Fig. 2. Forced cooling influence on crystallization process

In the figure 3 influence on core temperature distribution after 4th day of cooling is shown. The cooling system changes behaviour of the core which is much warmer than the casting in the case without forced cooling. Forced cooling lowers significantly temperature near to casting wall.

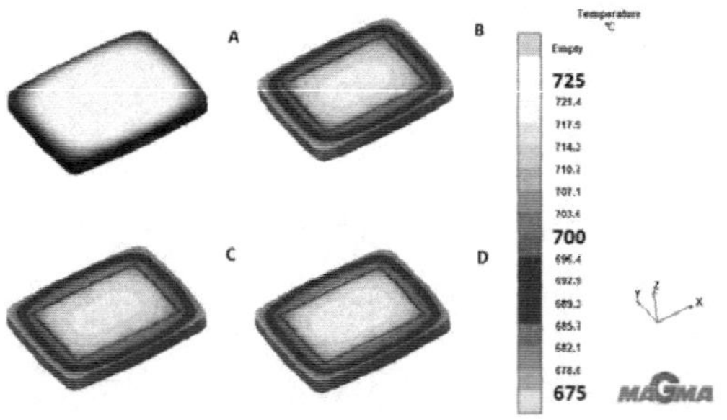

Fig. 3. Core temperature distribution after 4 days of cooling: A – without forced cooling; B – pumping pressure 8 atm.; C – pumping pressure 6 atm.; D – pumping pressure 4 atm.

In table 2 cooling rate at eutectoid reaction temperature is gathered. Differences in the cooling rates can lead to the conclusion that forced cooling may influe on more amount of pearlite

Table 2. Cooling rate at eutectoid reaction, (10^{-4}K/s)

No cooling	4 atm.	6 atm.	8 atm.
-6.87	-7.08	-7.1	-7.18

In table 3 casting shake out times are gathered. The forced cooling makes shake out waiting time about 7% shorter.

Table 3. Casting shake out times (10^4 s)

No cooling	4 atm.	6 atm.	8 atm.
88.28	84.84	84.32	82.15

4. Conclusions

Forced cooling has no significant influence on crystallization process. It has impact on cooling process of the casting. It is important as it reduces the core temperature and makes cooling rate at eutectoid reaction greater. In effect the microstructure of ductile cast iron will be different. Another beneficial impact is reduction of shake out time. This has influence on foundry production capacity and income.

Acknowledgements

Program Badań Stosowanych

Authors acknowledge National Centre for Research and Development in Poland for financial support under grant: PBS3/B5/47/2015.

References

[1] User Manual MAGMASOFT, Aachen, 2002.

[2] R. Hawranek, J. Lelito, J. S. Suchy, P. Żak: Archives of Metallurgy and Materials 54 (2009), 351-358.

[3] J. S. Suchy, J. Lelito, B. Gracz, P. L. Żak, China Foundry 9 (2012) 184-190.

[4] P. K. Krajewski, G. Piwowarski, P.L. Żak, W. K. Krajewski: Archives of Metallurgy and Materials 59 (2014), 1405-1408

[4] J. Fourie, J. Lelito, P. L. Żak at all: Archives of Metallurgy and Materials 60 (2015) 2437-2446.

[5] MAGMASOFT database

Proceedings of the 72nd World Foundry Congress
21-25th May 2016, Nagoya, Japan

Increasing the Capabilities of Computer Process Modeling With Applied Programming Interface

S. Ravi, J. Thiel
University of Northern Iowa, Cedar Falls, IA, USA

Abstract
Computer based process modeling has long been used for simulating the casting of metals. However, the majority of the analysis of the casting process has been limited to the metallic casting rather than the mold that contains the metal. High temperature characterization of molding materials has yielded an increased understanding of the condition of the mold and core during solidification. Properties such as thermal expansion, high temperature reactions, and changes in surface viscosity and decomposition gasses have yielded an increased understanding of the casting process. By using actual high temperature mechanical property data, the accuracy of the solidification modeling is improved and the range of capabilities is increased. The use of applied programming interface code can extend the software's capabilities and yield additional information on the condition of the mold before, during and after solidification. Although once limited in release, new versions of the process modeling software will include the capability to both optimize process variables and develop material specific datasets that increase both the accuracy and extend the capabilities of the analysis. Using applied programming interface code, the range of capabilities of the process modeling software can be extended to include information on veining defects, final casting dimensions, and decomposition gas generation.

Keywords: *Casting simulation, prediction analysis, veining, porosity, dimensional accuracy*

Introduction

Process simulation software is widely used in the foundry industry to predict and analyze the final casting integrity. Presently, the databases included in the software are generic and are limited with regards to the temperature dependent properties of sand. Improving the accuracy of the temperature dependent properties of molding materials will increase the overall accuracy of the process simulation software.

The material properties in this research were determined using a combination of direct measurement using analytical equipment and inverse optimization. Properties such as the coefficient of thermal expansion (CTE), specific heat capacity and surface viscosity were measured directly using advanced analytical techniques. Thermal conductivity, thermal diffusivity and heat transfer coefficients were determined by inverse optimization using automated software.

Veining defects are primarily seen in silica sand cores or molds, due to its thermal expansion profile. Previous research has determined that the volumetric contraction observed in silica sand after the alpha-beta phase transition at 573 C (1163 F) leads to a surface tensile stress that cause cracks in the mold or core surface [1]. The current research focuses on the use of application programming interface to predict vein strain and hence, the propensity of veining in castings. Thermal expansion data was used to calculate the differential volume in sand cores and an algorithm was developed to calculate the strain on the surface of the core at higher temperature.

Accounting for dimensional variation in castings has traditionally been performed using rules such as shrink rules, which fail to take into account the variation of the mold or core dimensions at high temperature. The current research uses a process simulation software to accurately determine casting dimensions from room temperature pattern dimensions. Although this method considers only one variable currently, it provides a valuable and for certain applications, accurate prediction of final casting dimensions.

Gas porosity in castings is often a result of gaseous combustion product from core or mold resin. A new testing methodology was developed at the University of Northern Iowa which measures the temperature dependent gas evolution based on weight loss. Although this methodology is in a developmental stage, it provides repeatable and quantifying data for gas generation due to organic sand binders.

Testing Methodology

Thermal expansion tests were run using the University's high temperature aggregate dilatometer. The samples were heated to 1650 C (3002 F) at a heating rate of 15 C per minute in a ceramic sample holder. Surface viscosity was calculated from the deformation recorded from the dilatometer. The specific heat capacity and sample weight loss was calculated using a high temperature Differential Scanning Calorimetry/Thermo gravimetric Analysis unit (DSC/TGA).

Surface strain was simulated in the MAGMASoft v4.4 casting simulation software using API functions, which uses the thermal expansion results. Dimensional accuracy of castings was calculated using thermal expansion and core displacement results from Magmasoft. Step-cone castings were poured to compare the predicted and actual results. A coordinate measuring machine was used to measure the dimensions of the cores and castings. Phenolic Urethane Cold-Box with a total resin content of 1.0%, 1.2% and 1.6% were tested for gas evolution. The sample weight loss results were used to calculate the gas evolution from the sand samples.

Discussion of Results

The vein-strain simulation and test steel casting results for baseline silica sand are shown in figures 1. The casting simulation indicates that the vein strain at the surface of the core is equal or greater to a value of 0.01 in/in. (cm/cm). If the sand at the surface fails to exhibit sufficient tensile strength, then the strain induced by the subsurface sand volume increase will raise stress above the tensile strength of the sand and cause cracking. A good correlation was observed between the predicted and actual casting results.

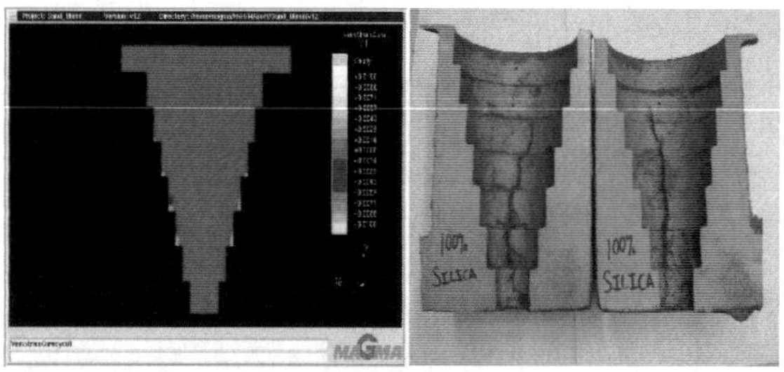

Fig. 1. Veinstrain comparison

Figure 2 shows the agreement in final casting dimensions using the modified thermal expansion data and the actual casting dimensions as determined by the CMM.

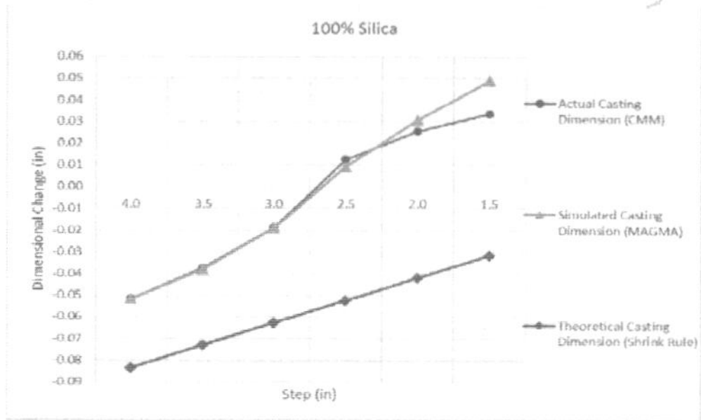

Fig. 2. Baseline silica dimensional accuracy

Figure 3 shows the comparison of gas evolution results between the traditional gas pressure tester and the new method for the 1.0% PUCB sand sample. Slight differences can be seen in between the two methods. The total gases evolved from the traditional method and the TGA method were 11.42 cc and 10.93 cc respectively.

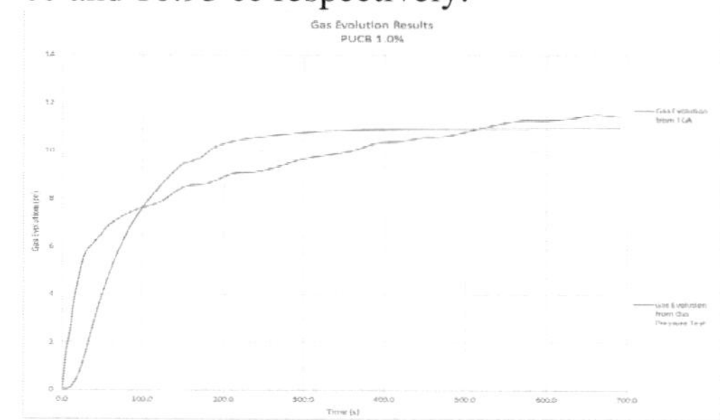

Fig. 26. Comparison of gas evolution results

Conclusion

The research demonstrated that veining defects can results from the thermal expansion of the molding aggregates. It was also shown that final casting dimensions can be estimated using accurate temperature dependent properties of molding aggregates. The measurement of the amount of gas emitted from organic material added to foundry sands based on the thermo- gravimetric analysis has the ability to directly compare various resins and organic sand additives. The results of the testing can also be used in process modeling software to serve as a predictor of gas porosity defects.

Acknowledgements

The authors wish to acknowledge the efforts of the student employees of the University of Northern Iowa Metal Casting Center for their assistance in the experiments required in the research.

References

1. J. Thiel, S. Ravi, Causes and Solutions to Veining Defects in Iron and Steel Castings, AFS Transactions 2014.

Proceedings of the 72nd World Foundry Congress
21-25th May 2016, Nagoya, Japan

Optimization of Gating System Design for High Pressure Die Casting to Reduce Air Entrapment Defects

Ken'ichi Kanazawa[1], Ken'ichi Yano[1], Ryusei Kawatani[2], Jun'ichi Ogura[2] and Yasunori Nemoto[3]
[1] Department of Mechanical Engineering, Mie University, Tsu, Japan
[2] Yamaha Motor Co., Ltd., Iwata, Japan
[3] Flow Science Japan, Inc., Tokyo, Japan

This study aimed to optimize a gating system design for high pressure die casting (HPDC) to minimize air entrapment defects using mold filling simulation. The optimized gating system design and another conventional design were applied to actual metal molds, and casting experiments using an HPDC machine with each of the gating system designs were conducted. Eventually, the results showed that the optimized gating system design can excellently decrease air entrapment defects in products.

Keywords: *HPDC, CFD, optimum design, sprue, runner.*

1. Introduction

High pressure die casting (HPDC) has a number of advantages and has been used to manufacture automobile parts and many other industrial products. However, in HPDC, when molten metal is plunged into a metal mold, the molten metal can entrap a large amount of air and this generates serious gas porosity defects in products. Since these air entrapments into molten metal can be occurred also in the sprue and runners, the gating systems should be designed to minimize such air entrapments [1, 2].

In the present study, we performed an optimum gating system design for HPDC to reduce air entrapment defects. The optimized design was applied as a metal mold, and its effectiveness was demonstrated through casting experiments using an actual HPDC machine.

2. Optimization of gating system design

2.1 Problem definition

Figure 1 shows the shapes of the products and gating system designs used in the paper. The shape of a product part is a thin plate 3 mm thick, which was designed to prevent air entrapment in itself. The internal structure of the mold has three cavities of product and the shape of the runner is divided into the three directions. The gating system design shown in Fig. 1 (a) is called as the standard shape, which is used to be compared with the optimum shape that is to be derived later.

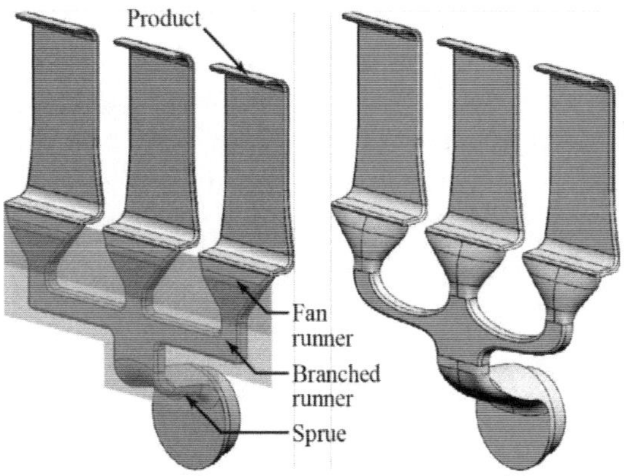

(a) Standard Shape (b) Optimized Shape

Fig. 1 Shapes of product and gating system design.

In the optimization, to simplify the design problems and reduce the computational time, we divided the gating system into three types of parts: the sprue, the branched runner and the fan runner, which were also designed separately. Each of the parts were evaluated using mold filling simulations based on computational fluid dynamics (CFD).

For evaluation, we defined three objective functions J_1, J_2 and J_3. J_1 was for evaluating air entrapment occurred in each part of the gating system. Thus J_1 was defined as the amount of remaining air in the cavity at the time when the fluid flow became almost a steady state. J_2 and J_3 were for evaluating the uniformity of the fluid arrival and the fluid velocity vectors at the outlet, respectively, which were for considering the continuity of flow when connecting the sprue and runners each other.

Finally, the optimization problem was defined as a multi-objective problem to minimize these objective functions J_1, J_2 and J_3. The problem was solved by using the nonparametric shape optimization algorithm based on a genetic algorithm that we had proposed [3].

2.2 Optimization result

Figure 1 (b) shows the shape derived in the optimization. The optimized gating system design had a round shape as a whole, and this can be expected to efficiently decrease air entrapment defects.

For confirmation, we conducted mold filling analyses in the region of the gating system design and product. As a result, the optimized shape was verified to have a high reduction effect on air entrapment in the product parts on the left and right sides except for the center part, because of the shape of the branched runner in particular.

3. Casting Experiments

3.1 Method

In the experiments, a cold-chamber type HPDC machine with the mold clamping force of 135 t was used. A metal sleeve with the inner radius of 60 mm and the length of 220 mm and metal molds with the standard and optimized gating system design were also used. As casting conditions, the four conditions A, B, C and D listed in Table 1 were selected and used. Here, $t_{pouring}$ is the time required for pouring with the ladle, t_{stl} is the time from the completion of pouring to the start of injection (i.e. shot-time-lag), and $v_{injection}$ is the injection velocity of the plunger. We acquired ten castings as samples for each mold shape and each casting condition. Thus a total of 80 castings were sampled in the experiments.

Table 1 Parameters of Conditions A, B, C and D for pouring and injection.

Condition	$t_{pouring}$ [s]	t_{stl} [s]	$v_{injection}$ [cm/s]
A	2.4	1.0	20
B	2.9	1.1	32
C	2.9	2.0	32
D	2.9	1.6	20

Each of the acquired castings was separated into the product parts and the others at the gates, and these product parts were heat treated to expand the air inside. Then the density ρ of each product part was measured based on Archimedes' principle. Moreover, the volume of internal air per the weight of each product, which is called the specific air volume, was calculated as the following equation:

$$v = \frac{1}{\rho} - \frac{1}{\rho_m}, \quad (1)$$

where ρ_m is the density of the aluminum alloy, which is 2.74 g/cm^3 in this paper. We finally used this specific air volume v as the evaluation value for the experiments.

3.2 Results

Figure 2 shows the mean values of specific air volume of the product parts on the left and right sides, which had a smaller amount of air entrapment when using the optimized shape in the mold filling analyses. The result shows that the castings molded using the optimized shape had averagely a high density, small amount of air entrapment and thus good quality in each of the casting conditions.

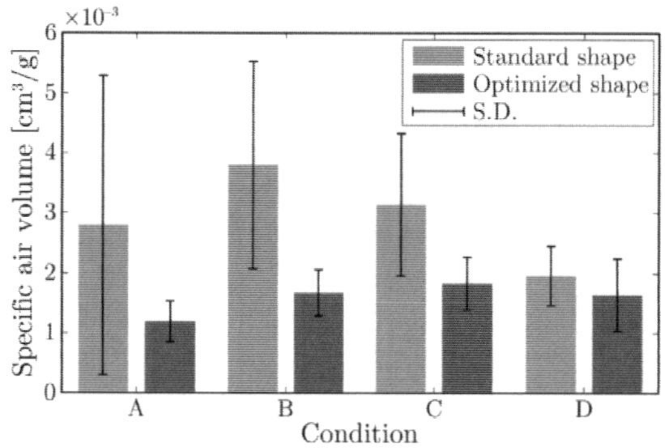

Fig. 2 Mean values of specific air volume of product parts for each condition and gating design system.

4. Conclusions

We optimized a gating system design including a sprue and runners for HPDC in order to decrease air entrapment defects. The performance of the optimized design was evaluated through experiments using an actual HPDC machine. The results showed that the optimized design decrease air entrapment defects in products compared to a conventional design.

References

[1] B. H. Hu, K. K. Tong, X. P. Niu and I. Pinwill: J. Mater. Process. Technol., 105 (2000) pp. 128-133

[2] B. D. Lee, U. H. Baek and J. W. Han: J. Mater. Eng. Perform., 21 (2012) pp. 1893-1903.

[3] K. Kanazawa, K. Yano, J. Ogura and Y. Nemoto: Proc. ASME Int. Mech. Eng. Congr. & Expo. (2014) paper No. IMECE2014-37419

Proceedings of the 72nd World Foundry Congress
21-25th May 2016, Nagoya, Japan

Core Shooting Simulation – to the Economic and Environmental Advantage of the Foundry

Christof Nowaczyk[1], Fumihiko Kimura[2]

[1]ASKChemicals GmbH, Reisholzstrasse 16-18, 40721 Hilden/ Germany
2 ASK Chemicals Japan Co. Ltd. 8th Floor, Bashamichi 450 Building, 50, Otamachi 4-Chome Naka-Ku, Yokohama 231-0011/ Japan

The global competition demands ever-improving quality with shorter development and production times at lower costs from the companies; the constant upgrade of an increasingly diverse product range is the rule nowadays and presents a challenge to the caster that is at least just as great. In this situation, computer programs, such as simulation software, can help lower costs, reduce development times and design optimized stable processes.

Keywords: *Simulation, Core shooting, Economic, Environmental Advantage*

While the development process from the idea to production led in the past from the drawing board via model construction, test casting and various adjustments to the finished product, computer-aided design (CAD), simulation, computer-aided manufacturing (CAM) and prototyping are used today. In brief, we speak of computer-aided engineering (CAE). With regard to model construction and also, specifically, the development and design of casting systems, this has certainly been the case for some years now. We are all familiar with the advantages and possibilities that simulation methods offer in this context.

However, the simulation of core production must still be considered to be relatively new. But do we need this type of simulation? Surely, nobody knows more about their core business, i.e. their core production, than experienced casters themselves. Nevertheless, we must ask ourselves – is this true? Do we really know what happens and whether we have designed the most optimum setup?
You could almost say that this sheds light on one of the last dark areas of our casting processes, and that this helps us to master our "core business" even better.

Two key simulation steps are distinguished in core shooting simulation. The first is the simulation of the filling process of the core box, i.e. the actual shooting of the core. The second is possible or necessary gassing, i.e. a through-flow of gas through a cavity of a core box with any type of filling.

The simulation software Arena-Flow® is the only software on the market that can depict the actual interaction between particles (sand grains are particles) and the flow medium (air) in a realistic manner. It can depict problem areas with insufficient compaction very clearly (Figure 1). The compaction problem illustrated here is caused by a venting situation that is not ideal.

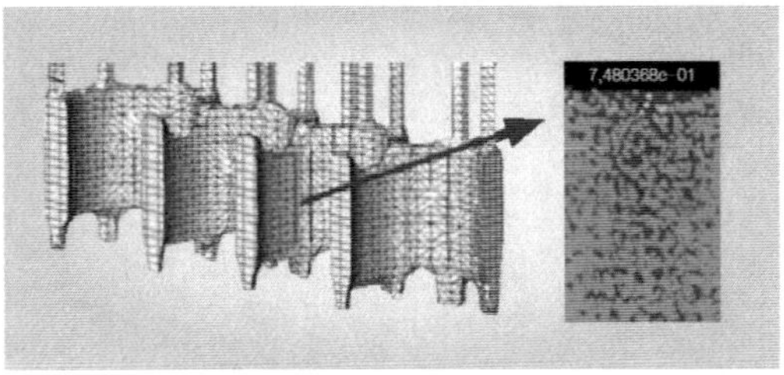

Fig. 1: Crankcase water jacket with poorly compacted area between cylinders 2 and 3

This example also clearly shows that the filling dynamics or the filling behavior depends primarily on the flow conditions of the air in a core box. This flow behavior can be illustrated by flow vectors and shows very clearly where insufficient compaction or problems with gassing can be expected.

In Figure 2, areas with insufficient airflow are shown in dark blue. With regard to gassing a core, this type of evaluation by means of simulation provides an initial insight into whether the process is homogeneous.

If the lower box area already displays poor flow conditions, it can be assumed with certainty that gassing problems will occur.

What are known as family core boxes are often designed to go with existing core shooting machines. Production is then frequently faced with the problem that the compaction of certain cores or areas of cores is insufficient, which often leads to increased cleaning effort or even considerable rework of the cast parts.

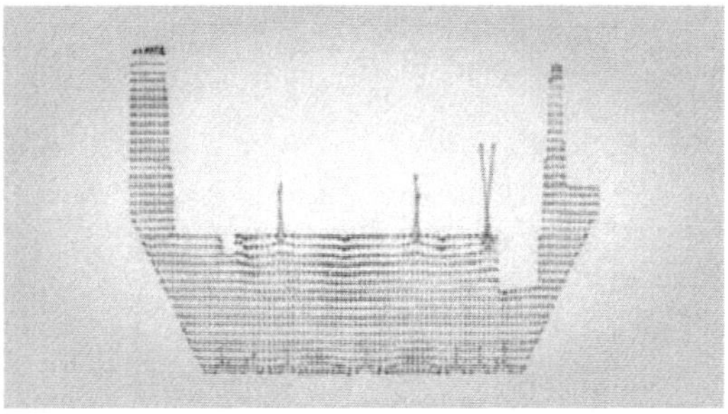

Fig. 2: Airflow in a core box

In most cases, this is caused by the interaction between the geometry of the shooting head, which is installed, accepted as given and not considered further, and the actual setup of the core box and the arrangement of the shooting nozzles.

Figure 3 shows clearly that the existing geometry prevents the back left area of the core from being filled completely, since the amount of mold material that can flow through the sand magazine in the given time is not sufficient.

Such a situation inevitably leads to significant additional costs, which could have been avoided by performing a corresponding simulation beforehand.

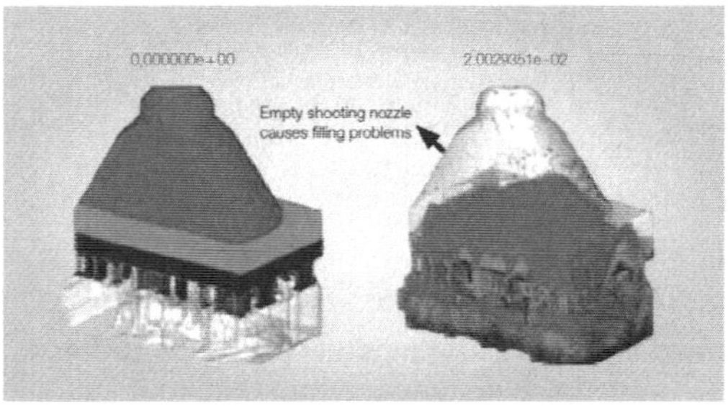

Fig. 3: Analysis of the interaction between the shooting head and the core box

A further practical example (Fig. 4 and 5), based on an oil duct core in a design by AUDI AG, shows the potential that core shooting simulation offers in terms of saving costs and resources. As part of a customer project, the task was to check an existing setup and to optimize it if necessary before constructing new core boxes. This project was supported and promoted not only by the foundry itself, but also by the client, AUDI AG. As an OEM that does not itself operate a foundry in which cores are used, AUDI AG consistently relies on simulation as a means of achieving stable processes, both in the foundry and in the purchasing companies later on.

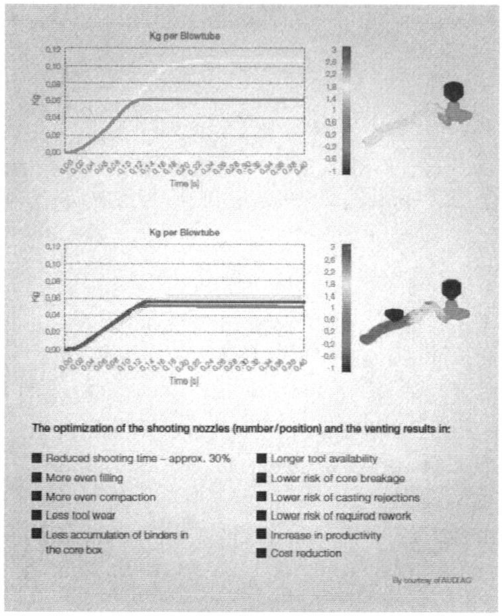

Fig. 4: Shooting nozzle dynamics

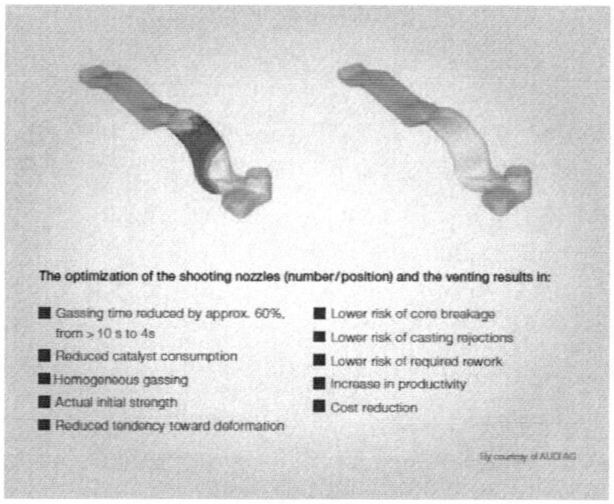

Fig. 5: Simulation of gassing

Summary

The simulation of core production offers enormous potential in terms of development and production – be it in the context of tool development, ensuring that this tool will perform in the most optimum way and as desired prior to expensive tool production, or be it as an aid for detecting error causes and savings potential.

The simulation helps to plan, implement and operate stable processes. However, it also helps to improve cycle times and reduce amine consumption by optimizing gassing cycles. This enables an increase in productivity and a reduction of resource consumption to occur.

In the highly competitive international foundry market, this represents a distinct contribution to increasing or maintaining competitiveness.

Proceedings of the 72nd World Foundry Congress
21-25th May 2016, Nagoya, Japan

Lightweight, high-strength aluminum AlZnMgCu alloy castings.

Stanisław Pysz[1], Robert Żuczek[1], Edward Czekaj[1], Jarosław Piekło[2], Maria Maj[2],
[1] Foundry Research Institute, 73 Zakopianska St., Krakow, Poland
[2] University of Science and Technology, Faculty of Foundry Engineering
Corresponding author: stanislaw.pysz@iod.krakow.pl,

The paper presents research work and application of AlZnMgCu alloys in the swing arm as cast element. The chemical composition was developed to obtain optimal mechanical properties (UTS >400MPa) and elongation ($A_5 \approx 1.5\%$) along with the proper heat treatment procedure. Investment casting was chosen as a manufacturing method. Test samples were obtained directly from the casting structure. The results are very similar to the tensile strength UTS in the cast, which is about 420 MPa, with the results obtained from the simulation analysis of the casting using the MAGMASOFT software. The fatigue strength simulation of the swing arm were performed in the ANSYS software. The research work shoed that the ration of the fatigue strength to the tensile strength is 0,4, which is higher compared to the generally accepted ration of 0.2 - 0.25 for the aluminum alloys.

Keywords: *AlZnMgCu alloy, computer simulation, fatigue tests*

1. Introduction

The increased focus on the reduction of the emission of the CO_2 gases in the atmosphere is forcing the vehicle designers to reduce the fuel consumption which is connected to the overall reduction of weight [1]. The solution, could be an application of high strength aluminum alloys, which were commonly used in plastic working and known as 7xxx series to be used as castings. The use of the alloys in the manufacturing process of high-strength castings is still at the initial research stage. The strength properties of the alloys generally increase along with the amount of the Zn and Mg. However, along with the increase of the amount of these elements, particularly for Zn content above 7%, the tendency of hot cracking during the solidification and rapid cooling process increases. This is the technological issue. The high strength of the AlZnMgCu alloys results from a phenomenon of the separation of the particles of the certain phase from the supersaturated solution. The phase which affects the strength properties is η – $MgZn_2$. The literature analysis allowed to explore the dynamics and kinetics of the formation of phases such as: T (Al_6CuMg_4), S (Al_2CuMg) η ($AlZn_2$), which have a significant effect on the final properties which are dependent on the content of Zn, Mg and Cu. The study was based primarily on the use of the JMatPro program [2-4]. Increased amount of Zn is preferred due to the influence on the strength which is based on the decreasing of the amount of T phase. The T phase inhibits the fine yield in the casting. Therefore, the content of Zn in the alloy was determined in the range of 5 to 6%. Also, Cu and Mg content have significant influence on the final properties of the alloy which must be well chosen, because of the increase of Mg increases the amount of unfavorable T (Al_6CuMg_4) phase. While increasing the Cu content increases the separation of the S (Al_2CuMg) phase occurs.

2. Research procedure

The alloy with chemical compositions presented in the table 1 was chosen in the research work:

Table 1 Chosen chemical compositions

No	Zn	Mg	Cu	Fe	Si
1	**5,53**	**2,37**	**1,56**	**0,2**	**0,1**
2	**4,70**	**1,50**	**0,71**	**0,10**	**0,06**

Based on the DSC curves the fallowing heat treatment procedure parameter was applied:

- Solutioning – 450 °C, 2h + 530 °C, 2h + 560 °C, 2 h, cooling in the water in 80 °C
- Aging – 120 °C, 4h + 150 °C , 8h,

Experimental castings were prepared by the investment casting process without and with additional cooling by dipping the ceramic mold after pouring in a specially prepared cooling medium (Fig. 1). The research work was performed in order to obtain parameters: mechanical properties, the size of microporosities, interphase precipitations, influence of solidification rate and fatigue strength curve. Conducted numerical analysis of the casting process allowed to determine the final casting properties including tensile strength and distribution of microporosity. The results were compared with the experimental results. The next step was the fatigue analysis of the swing arm for 10^8 cycles.

Tabel 2 Mean properties of the samples - composition 2

Additional Cooling	UTS MPa	YS MPa	A_5 %	Grain µm
No	**422**	**403**	**0,5**	**138**
Yes	**443**	**404**	**1,4**	**110**

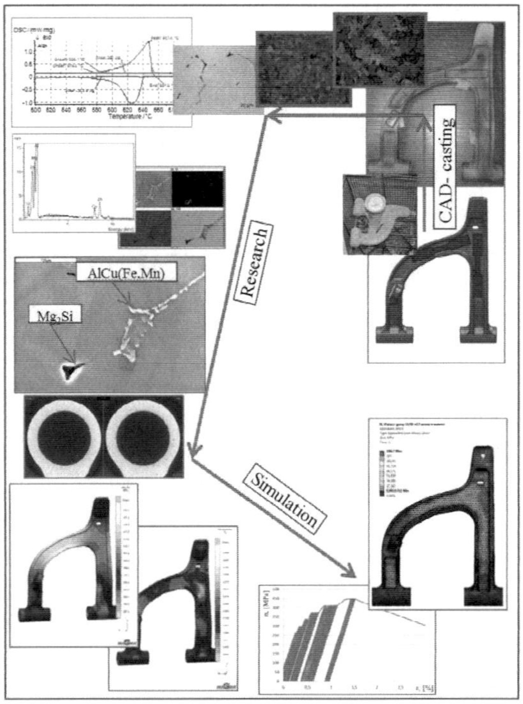

Fig. 1. Diagram of conducted research procedure.

3. Results

Expected tensile strength UTS (near 420 MPa) in the casting from AlZnMgCu alloys is possible to achieve (Table 2). Unfortunately the elongation achieved in the composition 1 is highly unsatisfactory. The chemical composition 2 is characterized with higher elongation with decreased tensile strength. It has been observed that the alloys have a high tendency to create the microporosity. This is a result of width crystallization range $\Delta T = T_{lq}$ (637 °C) – T_{sol} (475,5 °C) = 167,5 °C. High solidification rate influence on the grain refinement which results in the increased tensile strength. The calculated mean value of the grain after the heat treatment is:

- Without additional cooling - 138 µm,
- With additional cooling – 110 µm.

In the case of the precipitation hardening alloys like AlZnMgCu, very important is the heat treatment process. The procedure should be multistage in order to dissolve the intermetallic phases in the matrix and create dispersion precipitates which are strengthening the microstructure. The separation of brittle and hard intermetallic phases were observed at the grain boundaries. This is a factor lowering the elongation of the alloy. Comparison of the results of the computer simulation with the results of the test samples obtained from the castings indicates good compatibility. The strength of the cast element is varied depending on the solidification rate and ranges from 390 MPa to 435 MPa. A similar agreement is obtained in comparison of the micro-porosity determined in the computer simulation and computer tomography evaluation. The microporosity of the swing arm is approximately 1.5%. The maximum value of the deformation of the swing arm structure does not exceed the permissible range during the fatigue testing.

4. Conclusions

- AlZnCuMg alloys are suitable in the casting process as a material for enhanced operating condition elements,
- Further studies will be conducted for the increase of the elongation and corrosion resistance.

Acknowledgements

The work was funded by the National Center for Research and Development – Poland , No. PBS1/A5/9/2012.

References

[1] "Application of Lightweighting Technology to Military Aircraft, Vessels, and Vehicles", The National Academies Press, Washington, D.C, 2012.

[2] Z. Guoa, N. Saunders, A.P. Miodownik, J.-Ph. Schille: "Modelling of materials properties and behaviour critical to casting simulation", Materials Science and Engineering A 413–414 (2005) 465–469.

[3] John F. Chinella, Zhanli Guo: "Computational Thermodynamics Characterization of 7075 7039, and 7020 Aluminum Alloys Using JMatPro". Army Research Laboratory, ARL-TR-5660, September 2011.

[4] Jingjing Yu and Xiaomei Li: "Modelling of the Precipitated Phases and Properties of AlZnMgCu Alloys". Journal of Phase Equilibria and Diffusion Vol. 32 No. 4 2011.

O-149

Proceedings of the 72nd World Foundry Congress
21-25th May 2016, Nagoya, Japan

RAPID DEVELOPMENT OF EXPORT CASTINGS IN OUR FOUNDRY BY SIMULATION TECHNIQUES – SOME INDUSTRIAL CASE STUDIES

Tapan Roy - Prof MICME (U.K.)
Assistant .General. Manager – Methods,
Texmaco Rail & Engg. Ltd., Steel Fdy. Div. Belgharia, Kolkata – 700056 (India)
Cell No- +91-9836700315, Email – tapan.roy@texmaco.in

Introduction

In the Steel Foundry of Texmaco Rail & Engineering Limited castings for Ukraine Railways, American Railways and Mining Industries for Australia have been developed. The requirements of international customer are to get quality castings as per their stringent Specification within committed delivery time including development of tooling, samples etc. So, to cope up with this situation policy has been formulated in such a way that can deliver quality castings within committed delivery schedule..

1. Tooling development at our vendor end with close monitoring by methods departments.

First we made the relevant method drawings describing the parting line, pattern print design, core print design, number of cores involved and their locking arrangement to eliminate extra fins in a casting, number of loose pieces in core boxes, special instructions for a part of castings either it will be made from cores or self system adopted, contraction allowances, draft allowances specified on pattern and core boxes along with machining allowances and if the casting is a plate type casting requires camber allowances to be specified in the method sheet.
and pattern print to make the pattern and core box model.
Pl. refer the Fig.1 for draft analysis on a pattern green colour showing positive and red colour showing negative draft. and the model of the constructions of core boxes taking into consideration different loose pieces and their stripping arrangements.

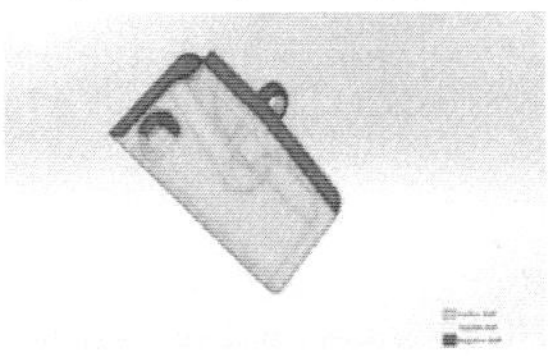

Fig.1 Draft analysis on pattern green colour showing positive and red colour showing negative draft

2 *Finalisation of method of gating and riser design by simulation & experience by prediction chances of defects and suitable corrective measure*

The second step of the development is the finalisation of method of gating and riser design by simulation & experience by prediction of chances of defects and suitable corrective measure to be taken on computer screen to eliminate these defects. refer the Fig.2 for the flow and solidification simulation of side frame casting.

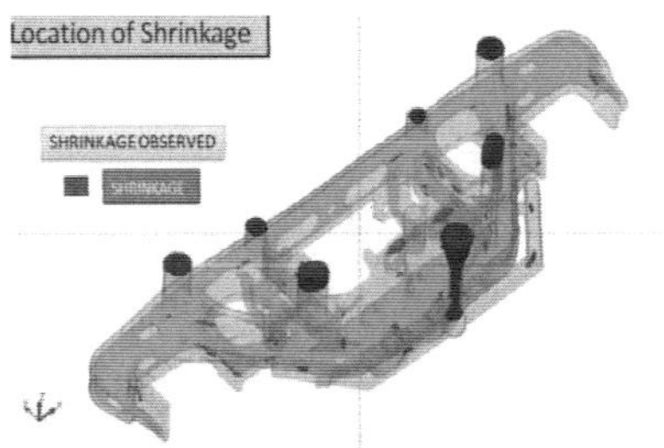

Fig. 2 Flow and solidification simulation view of side frame

3. ***Defect prediction like hot tear defects or other solidification and flow related defects on computer screen and to take corrective actions to eliminate these defects***

The third step is the Defect prediction like hot tear defects or other solidification and flow related defects on computer screen and to take corrective actions to eliminate these defects.

Case study -Hot tear defects in Bolster casting- Technical Discussions

Fig.3 shows the simulation result of Centre pivot area of Bolster castings. The view shows that the hot metal at this junction whose temperature is 1470 °C which is well within solidus range and is in mushy form and solidifies latter causing hot tearing due to contractional hindrances by mould or core materials. But after positioning of external chills Fig. 4 at this area the temperature at this area found 1380 °C as per Fig.3, which indicates faster cooling of this area and no hot tear found as per Fig 5.

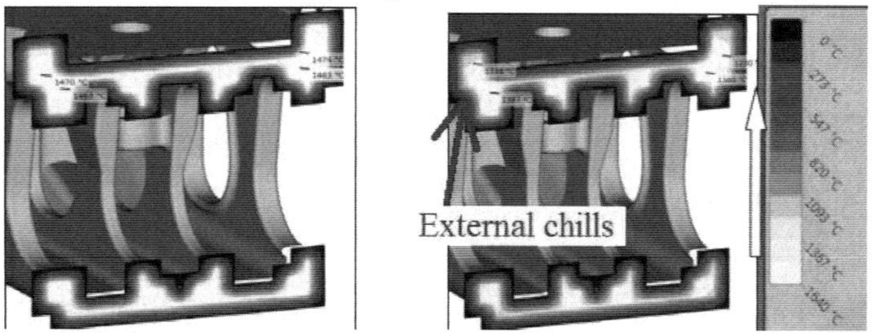

Fig.3 Showing hot zone (temp.1470°C) & No hot zone (temp.1380°C)

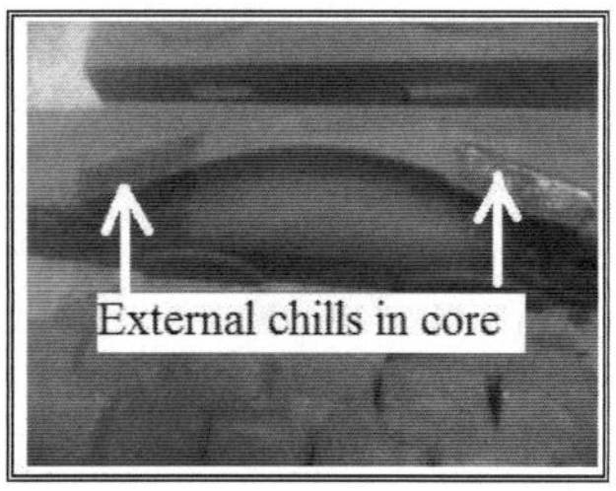

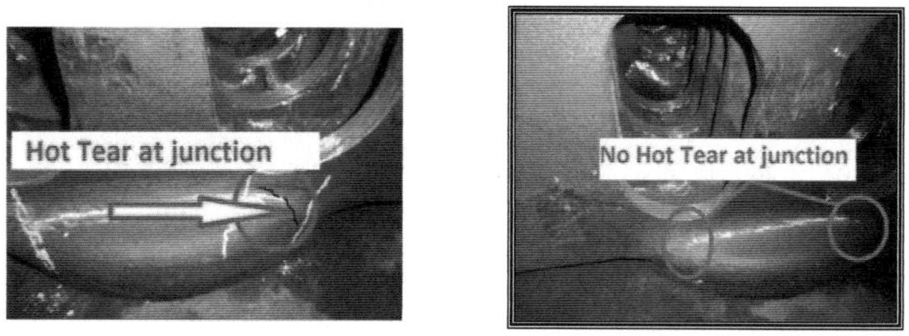

Fig4 External chills to arrest hot tears *Fig 5 Hot tear before and after in real casting*

Conclusion

Simulation results help to predict the defects during solidification and during flow of molten metal. In this way the number of trials for a newly developed item has been reduced considerably.

It has been seen that where the total development time in conventional system is 4 to 5 months but here the total development time by incorporating CNC make metallic tooling, methoding done by simulation results etc. Development time reduces to 1 to 1.5 months.

Acknowledgement

This work was supported by the practical study in the Steel Foundry Division of Texmaco Rail & Engineering Limited., Kolkata, India.

References

1. Directional Solidification of Steel Castings – By R.Wlodawer
2. Principles of Metal castings – By Heine, Loper and Rosenthal. Second edition- McGraw-Hill.
3. Simulation of the components is done by Auto Cast-X1-Flow+ CAD Simulation Techniques, developed by Indian Institute of Technology, Mumbai (India) & ADSTEFAN, Japanese casting simulation software developed by HITACH, JAPAN.

Proceedings of the 72nd World Foundry Congress
21-25th May 2016, Nagoya, Japan

Design of a remote operation control system for vibration suppression of a liquid container by an overhead traveling crane

Akihiro Kaneshige[1], Yu Yamauchi[1], Satoshi Ueki[1], Takanori Miyoshi[2], Kazuhiko Terashima[2]
[1] National Institute of Technology, Toyota College, JAPAN
[2] Toyohashi University of Technology, JAPAN

In this research, a remote-operation control system for a liquid (molten metal) container transport by an overhead crane system is proposed. The proposed control system in this research allows an operator to control a speed of an overhead crane freely and to feel a swing of the rod instinctively through the reaction force, and can be controlled from everywhere used via the internet. Also, it is able to suppress liquid vibration (sloshing) in a container. The usefulness of the proposed control system is demonstrated by an experiment.

Keywords: *overhead traveling crane, vibration suppression of the liquid container, remote operation*

1. Introduction

For a liquid container transfer of molten metal by an overhead crane system in the steel and the casting industries, the construct of transfer system that can be transfer to the destination in a short time and suppress sloshing (liquid vibration) in a container, has been desire from the view point of both safety and efficiency.

On the other hand, a remote-control can also be applied in a various fields, and research and development for it systems are carried out in many fields [1],[2],[3]. In particular, Miyoshil[2] introduced a wave filter in the scattering theory to constrain the norm and were able to apply scattering theory to the system including non-passive material by using the small gain theorem of norm. We have designed a remote control system of a liquid container to be transported by an overhead traveling crane, but it does not implement the sloshing suppression in a container [3].

A design method the remote-operation control system for a liquid container transported by an overhead crane system is suggested. The proposed control system allows an operator to control the speed of an overhead crane freely and to feel a swing of the rod instinctively through the reaction force. In remote control, the information of the distant place is very important. Particularly, feeling the reaction force for operation is important like watching the rod's movement with cameras. And the proposed control system can be used via general internet and can be controlled from everywhere. Also, it is able to suppress the sloshing in the container.

The operator in a good working environment can be perform work comfortably by using the proposed the remote-operation control system. Furthermore, the proposed control system is able to suppress the sloshing, unskilled operator is able to operate easily. The usefulness of the proposed remote-operation control system is demonstrated by an experiment.

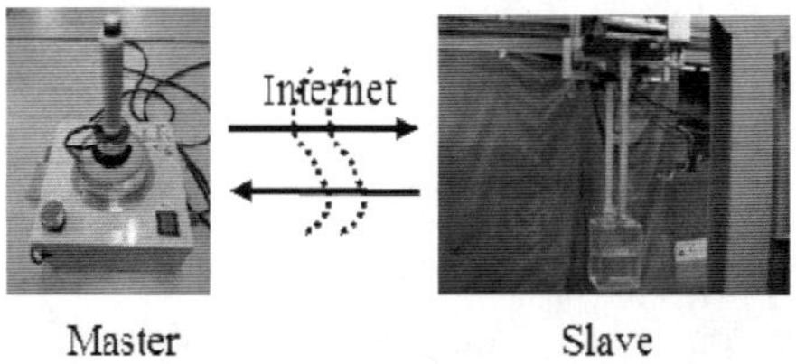

Fig. 1. Overview of the remote-control system

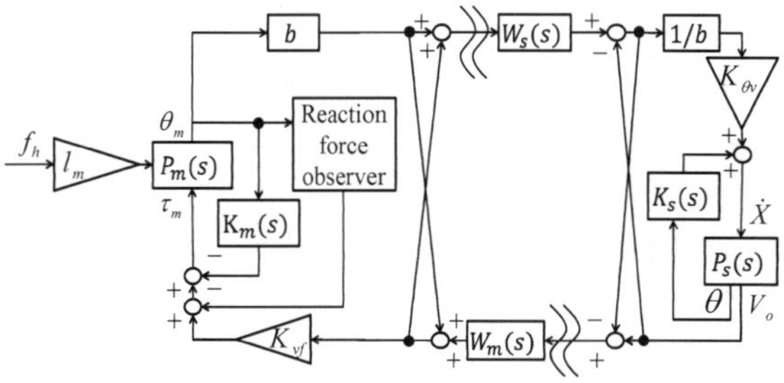

Fig. 2. Block diagram of the remote-control system.

2. Remote-operation control system

Figure 1 and 2 shows overview and block diagram of the remote-operation control system. The remote-operation control system is constructed based on a design method Miyoshi[2]. The proposed control system can be used via general internet and can be controlled from everywhere. In Fig.2, $P_m(s)$ and $P_s(s)$ are the model of master and slave respectivly, these model are expressed as Eq.1 and Eq.2.

$$P_m(s) = \frac{\theta_m(s)}{\tau_m(s)} = \frac{1}{J_m s^2 + D_m s + l_m K_h} \quad (1)$$

$$P_s(s) = \frac{\theta(s)}{\dot{X}(s)} = \frac{ms}{mls^2 + mg} \quad (2)$$

where J_m is a moment of inertia, D_m is a viscosity coefficient, l_m is a paddle length of the master, K_h is a input gain. m is a mass of container, l is a length of rod for hanging the container. Also, $K_m(s)$ is a controller for reaction force, $K_s(s)$ is a controller for sloshing supression. $W_m(s)$ and $W_s(s)$ are wave filter(phase lead compensator), $K_{\theta v}(s)$ and $K_{vf}(s)$ are propotional gain, b is conversion factor.

The operator provides input angle θ_m by operating the paddle of the master, and the cart velocity $\dot{X}$ of crane proportional to the angle θ_m is entered. The swing angle θ of rod for hanging the container is detected as an output value from the cart velocity, and the velocity of container V obtain same manner. The container velocity will be feed back to the operator as a reaction force of the master.

3. Experimental results

The proposed system is verified two aspects of a remote-operation control experiment between NIT,T.C(National Institute of Technology, Toyota College) and T.U.T.(Toyohashi University of Technology). The Master arm is installed in NIT,T.C and an overhead crane Slave is installed in TUT, with the internet acting as the communication line in these experiments. There is a straight distance of approximately 50 km between NIT,T.C and TUT. There are constant time delays of approximately 50ms.

An experimental result is shown in Fig.3. The operator tried to transport the liquid container a distance of 0.6m and 1.5m approximately from the start point. The operator was manipulated crane while looking at the web camera. From the experimental results, the proposed remote-operate control system was confirmed to be effective in suppressing the rod vibration and sloshing

4. Conclusion

In this research, the remote-operation control system for a liquid container transported by an overhead crane system was designed. This designed control system is able to suppress the sloshing, unskilled operator is able to operate easily.

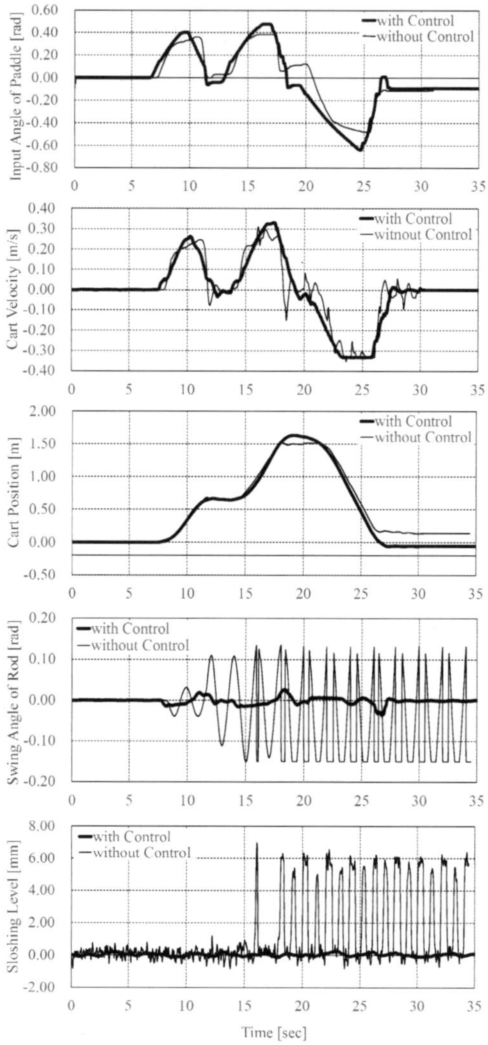

Fig. 3. Experimental results.

Acknowledgements

The authors would like to thank Prof. Dr. Martin Buss of Munich University of Technology, Germany, for its cooperation.

References

[1] S.Majima, Design method for a controller of a distance learning calligraphy system via public networks, TENCON 2004, IEEE Region 10 Conference, vol.4, (2004) 593-596.

[2] T. Miyoshi, K.Terashima and M. Buss, A Design Method of Wave Filter for Stabilizing Non-Passive Operation System, IEEE Int. Conf. on Control Applications, (2006) 1318-1324.

[3] A.Kaneshige, Y.Kawasaki, S.Ueki, T.Miyosi and K.Terashima, A Design of the Tele-operation Control System for Liquid Container Transfer by an Overhead Traveling Crane, Proc. of 11th Int.Conf. on Remote Eng.and Virtual Instrumentation, (2014) 29-30.

Proceedings of the 72nd World Foundry Congress
21-25th May 2016, Nagoya, Japan

High-precision Automatic Pouring Machine with Adaptive Control Parameters

Yoshiyuki Noda[1], Makio Suzuki[2] and Kazuhiro Ohta[2]
[1] Department of Mechanical Engineering, University of Yamanashi,
4-3-11, Takeda, Kofu, Yamanashi, 400-8511, Japan
[2] Sintokogio, Ltd.,
3-1, Honohara, Toyokawa, Aichi, 442-8505, Japan

This study contributes an advanced control system in tilting-ladle-type automatic pouring machine in casting industry. In order to reduce the preliminary experiments for tuning the pouring control system, and pour precisely the molten metal even if the pouring condition is varied, we propose the adaptive pouring control system which the control parameters are automatically tuned for pouring precisely in this study. In this proposed approach, the online model parameters identification and control parameters updating system are integrated with the pouring control system based on the mathematical model of the pouring process. The effectiveness of the proposed pouring control system is verified by the simulations with the tilting-ladle type automatic pouring machine. This paper provides valuable information to reduce casting defects caused by pouring process.

Keywords: *Automatic pouring machine, Model Based Controller Design, Environmental Adaptive Control*

1. Introduction

In recent years, an automatic pouring machine has been exploited to improve the working environment of the pouring process and to produce stable casting products. Furthermore, the automated pouring process has been required that molten metal is quickly and precisely poured into the mold. Many pouring control approaches for satisfying such requirements have been proposed in recent studies[1]. However, the conventional pouring control systems need the preliminary experiments for tuning the control parameters. These experiments take an immense amount of time and effort. Moreover in case that the pouring conditions are changed from the conditions of the preliminary experiments, it is difficult to pour precisely the molten metal by the pouring machine with the conventional control systems.

Therefore, in order to reduce the preliminary experiments for tuning the control parameters, and pour precisely the molten metal even if the pouring condition is varied, we propose the adaptive pouring control system which the control parameters are automatically tuned for pouring precisely in this study. In this proposed approach, the online model parameters identification and control parameters updating system are integrated with the pouring control system based on the mathematical model of the pouring process. These model parameters are identified automatically by minimizing the error of practical data and simulation data from the model at every pouring motion, and the control parameters according to the mathematical model of the pouring process are updated online to the controller. In this study, minimizing the error of practical data and the simulation data is performed by Gauss-Newton method for identifying the parameters in a short time. Therefore, the parameters identification and updating can be accomplished before next pouring. Even if the pouring conditions are changed, the molten metal can be poured precisely by the proposed adaptive pouring control system. The effectiveness of the proposed pouring control system is verified by the simulations with the tilting-ladle type automatic pouring machine.

2. Adaptive Pouring Control System

The proposed pouring control system is shown in Fig. 1. In order to control precisely the flow rate of outflow liquid from the ladle, the feedforward control with the inverse model of pouring process is constructed to the automatic pouring control system. The inverse models can be derived from the model of automatic pouring machine as

$$T_m \frac{d\omega(t)}{dt} + \omega(t) = K_m u(t) \quad , \qquad (1)$$

$$A(\theta(t))\frac{dh(t)}{dt} = -q(t) + \left\{ h(t)\frac{\partial A(\theta(t))}{\partial \theta(t)} + \frac{\partial V_s(\theta(t))}{\partial \theta(t)} \right\}\omega(t) \quad , \qquad (2)$$

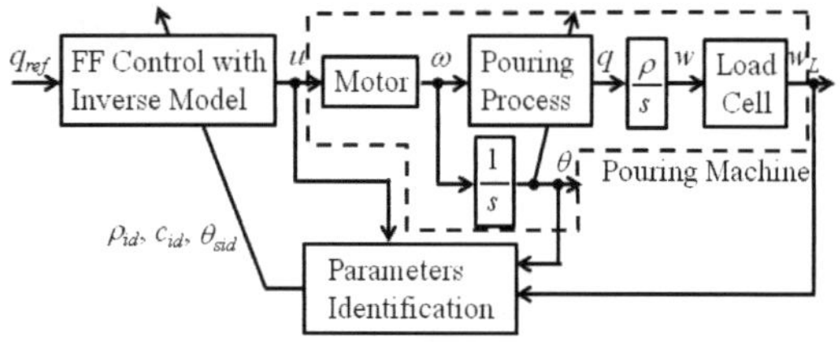

Fig. 1 Pouring Control with Adaptive Control Parameters

$$q(t) = \int_0^{h(t)} cL_f(h_b)\sqrt{2gh_b}\,dh_b\,, \qquad (3)$$

$$\frac{dw(t)}{dt} = \rho q(t)\,, \qquad (4)$$

$$T_L \frac{dw_L(t)}{dt} = -w_L(t) + w(t)\,, \qquad (5)$$

where ω[rad/s] and θ[rad] are the tilting angular velocity and angle, respectively. T_m[s] and K_m[rad/s/V] are the time constant and gain of the motor, respectively. h[m], q[m^3/s], A[m^2] and V_s[m^3] are the liquid height on the pouring mouth, the flow rate, the area of upper surface of the liquid in the ladle, and the liquid volume under the pouring mouth, respectively. L_f[m], c, and g[m/s^2] are the width of the pouring mouth, the flow rate coefficient, and the acceleration of gravity, respectively. w[kg] and ρ[kg/m^3] are the weight and the density of outflow liquid, respectively. w_L[kg] and T_L[s] are the weight of outflow liquid detected by the load cell and the time constant of the load cell, respectively. The most parameters of model can be obtained from geometry of the ladle and specifications of the motor and the load cell. However, the flow rate coefficient c and the density ρ must be identified by the preliminary experiments. And, the tilting angle θ_s starting liquid outflow also is identified by the experiments.

In this study, the model parameters c, ρ and θ_s are identified online for reducing the preliminary experiments and adapting the pouring control system to the pouring condition as shown in Fig. 1. In the parameters identification, the cost function is minimized as

$$\begin{aligned}&\begin{bmatrix} c_{id} & \rho_{id} & \theta_{sid} \end{bmatrix} \\ &= \arg\min\left\{\int (w_{L\exp}(t) - w_{Lsim}(t))^2\,dt\right\}\end{aligned} \qquad (6)$$

where w_{Lexp} and w_{Lsim} are the weights of outflow liquid detected load cell in the experiments and simulations, respectively. Gauss-Newton approach is used for minimizing the cost function.

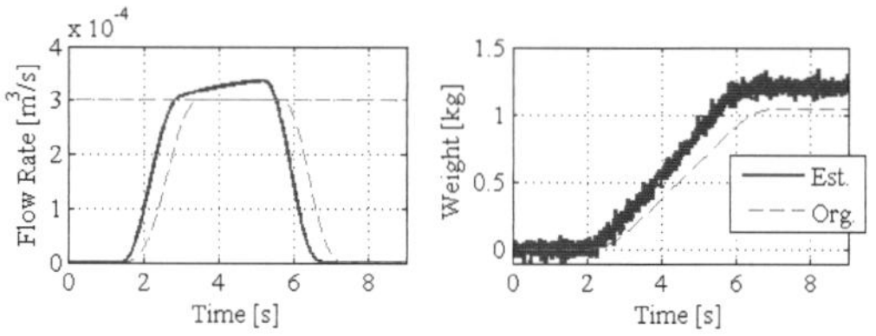

Fig. 2 Simulation results on first pouring

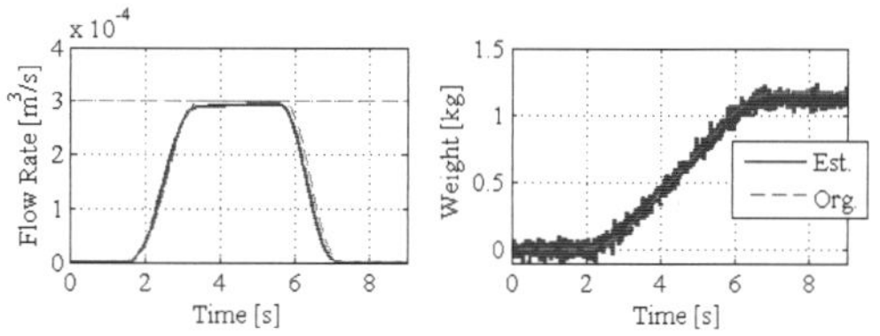

Fig. 3 Simulation results on 6th pouring

3. Simulation Results

The effective of the proposed adaptive control system is verified by the simulations with the model of the automatic pouring machine. In the simulations, the pouring conditions are given as c=0.97, ρ=1.0x10^{-3}[kg/m^3], and θ_s=33.1xπ/180[rad]. On the other hand, the control system based on the pouring model with the parameters c=0.8, ρ=0.9x10^{-3}[kg/m^3], and θ_s=32.1xπ/180[rad] is constructed in the first simulation. Fig. 2 shows the simulation results on first pouring. The left and right graphs show the flow rates and the weight of outflow liquid detected by load cell, respectively. And, the green broken lines are the simulation results with correct parameters. The blue solid lines are those with initial parameters. Since the model parameters in the control system are different to the correct model parameters, it is difficult to pour precisely. The automatic pouring process with the proposed control system is simulated repetitively. Fig. 3 shows the simulation results on 6th pouring. The arrangements of Fig. 3 are same as Fig. 2. The parameters are identified precisely by the proposed control system, even if the model parameters are varied.

4. Conclusions

The model based pouring control system with adaptive control parameters is proposed in this study. By performing repetitively, the liquid can be poured precisely.

References

[1] Y. Noda, K. Terashima, Modeling and Feedforward Flow Rate Control of Automatic Pouring System with Real Ladle, Journal of Robotics and Mechatronics , Vol.19, No.2, pp 205-211 , 2007.

Proceedings of the 72nd World Foundry Congress
21-25th May 2016, Nagoya, Japan

Optimum Feedforward and Model Predictive Control of Molten Metal Pressure in Greensand Mold Press Casting

*T. Takahito[1], R. Tasaki[1], K. Hashimoto[2] and K. Terashima[1]
[1] Department of Mechanical Engineering, Toyohasi University of Technology, Hibarigaoka 1-1, Tempaku, Toyohashi, 441-8580, Japan
[2] Sintokogio, Ltd., Honohara 3-1, Toyokawa 442-8505, Japan
*E-mail: takako@syscon.me.tut.ac.jp

This paper presents a novel technique for controlling the molten metal pressure in the casting production process, in order to realize high productivity and high-quality castings. We propose a pressure feedback control design using a model predictive control method (MPC), and also present a simulation study. The MPC method is a useful control technique that predicts a future pressure by using the nominal model. Several experiments have been done to demonstrate the effectiveness of the proposed approach comprising optimum feedforward control and model predictive feedback control. Good performance has been achieved with respect to both high-speed response and robustness for model uncertainty.

Keywords: *Press casting, Quality control, Pressure control, Model predictive control, Parameter identification*

1. Introduction

The greensand press casting method is a novel technique for controlling the molten metal pressure in the casting production process. Moreover, this is a high-speed, high-quality method capable of achieving a production yield exceeding 90%. In greensand mold press casting, for which our group invented this novel process [1], molten metal is first poured into a lower mold, then the upper mold is lowered toward the lower mold, and finally the two molds are matched. Since casting defects such as physical metal penetration, are often caused by the high pressure in a high-speed pressing process, pressure control is needed for defect-free production. However, direct measurement of liquid pressure, such as by a contact-type pressure sensor, is difficult in a high-temperature metal process.

Feedforward control cannot adapt to viscosity change dependent on temperature. Tasaki et al. proposed sequential pressure control by feedforward control [2]. But there are some problems concerning control time, disturbance, and robustness for model uncertainty. Hence, in this paper, we present a new method to estimate the in-mold pressure using the pressing reaction force from load-cell, and then construct a press feedback control system.

2. Press casting process

The system configuration and the experimental equipment used in this study are shown in Fig. 1. We set up a servo drive pressing machine, a transparent acrylic mold, and colored and viscous water with CMC: carboxymethylcellulose. The actual liquid pressure inside the mold can be measured by a contact-type sensor of a pressure gauge. Moreover, we measure the reaction force of the molten metal using a force sensor installed between the upper mold and the lifting device. Then we can estimate the pressure using a reaction force of the molten metal. A feedback control system using the estimated pressure is newly constructed for continuous high-quality production.

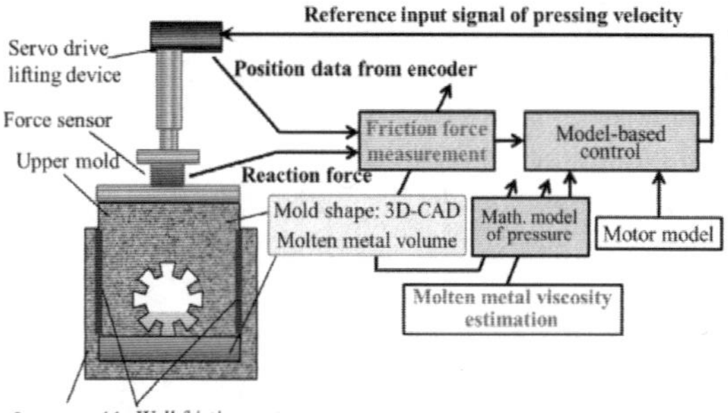

Fig. 1 Pressing casting process with pressure feedback control system

3. Pressure estimation

During pressing, wall frictional resistance generated between the upper mold and the lower mold, is an unpredictable disturbance for pressure estimation. Considering the force-pressure relation, the actual pressure can be estimated from Eq. (1) by subtracting frictional resistance from the force sensor, as follows:

$$P = (F - mg - m\ddot{z} - \mu(\dot{z})Lz)/A, \qquad (1)$$

$$\mu(\dot{z}) = \begin{cases} \mu_d & (|\dot{z}| > \dot{z}_{lim}) \\ \mu_s - \frac{\mu_s - \mu_d}{\dot{z}_{lim}} |\dot{z}| & (otherwise) \end{cases} \qquad (2)$$

where, Eq. (2) is the mathematical model of frictional resistance, molten metal pressure: P(Pa), force sensor value: F(N), the upper mold mass: m(kg), gravitational acceleration: g(m/s^2), the upper mold position: z(m), speed: $\dot{z}$, acceleration: $\ddot{z}$, the friction coefficient $\mu(\dot{z})$ (-), friction part length: L(m) and the cross-sectional area of upper mold: A(m^2).

An example of the pressure estimation result in pressing experiments with viscous water is shown in Fig. 2. The dashed-dotted line is actual pressure measured by the contact-type sensor, the dotted line is measured pressure from a force sensor, and the solid line is observed pressure by the proposed estimation method during pressing.

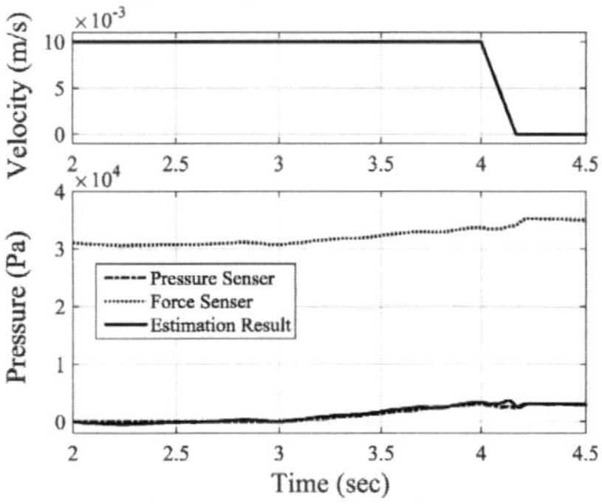

Fig. 2 Pressure estimation result

4. Viscosity estimation

In the pouring process before pressing, poured molten metal is often subject to volume error or temperature change. Therefore, the parameter of volume V_p (m^3) and viscosity λ related to temperature T (degrees C) must be identified for high-quality iron casting and construction of the MPC system in our study. The base model for MPC is derived by the extension of Bernoulli's theorem, represented by

$$P_B = \rho \frac{A_M^2(e_h)}{A_S^2(e_h)} \left(\ddot{z} z + \left(\frac{1}{2} + \lambda(T) \frac{L(e_h)}{D(e_h)} \right) \dot{z}^2 \right) + \rho g f(V_p, z) \quad (3)$$

As shown in Eq. (3), the pressure estimation requires information of z, $\dot{z}$, $\ddot{z}$, V_p and λ. These unknown parameters can be identified by fitting calculation using the method of least squares. Here, the observed data by Eq. (1) and the model output by Eq. (3) for pressure behavior within a specified short time at the beginning of pressing are used. Table 1 shows the parameter identification results for each condition of viscosity η = 0.001, 0.5, 1.0, 5.0 and for a setup liquid height h_0 = 0.05(m).

Table 1 Identified parameters

Actual viscosity η (Pa·s)	Viscosity: λ	Volume: V_P (m^3)	Initial liquid height: h_0 (m)
0.001 (water)	0.9	0.406	0.0518
0.5	2.0	0.408	0.0520
1.0	2.6	0.417	0.0532
5.0	6.9	0.422	0.0538

5. Model predictive control (MPC)

Pressure control must be satisfied under the pressure constraint condition for high-quality sand mold casting. The MPC method predicts the future output first, calculates the optimum input for keeping the constraint, and then realizes good performance of the real output.

In our MPC approach, we build linear predictive model control that is calculated at high speed, therefore is effective in real-time feedback control.

Eq. (4) is the response of the system. The response τ should be equal to the reference pressure trajectory τ_r,

$$\boldsymbol{\tau} = \boldsymbol{y}_f + \boldsymbol{S}\boldsymbol{u} \quad (4)$$

the optimal control input $\boldsymbol{u}$ is given by

$$\boldsymbol{u} = \left(\boldsymbol{S}^{\mathrm{T}}\boldsymbol{S}\right)^{-1} \boldsymbol{S}^{\mathrm{T}} \left(\boldsymbol{\tau}_r - \boldsymbol{y}_f\right) \quad (5)$$

where $\boldsymbol{y}_f$ is the free response in the future, $\boldsymbol{S}$ is the unit step response matrix. The MPC simulation result is shown in Fig. 3. This simulation result confirm that the MPC is very effective for the liquid pressure control during pressing in the press casting process.

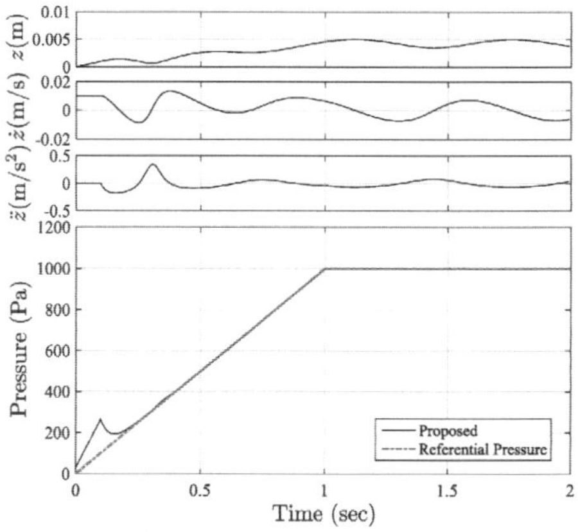

Fig. 3 MPC simulation result

6. Conclusion

We proposed a pressure feedback control system using MPC in this paper, and also showed its good performance, namely new process of continuous high-quality iron casting. We are now preparing comparative experiments for the optimum feedforward control and our proposed feedback MPC in our laboratory.

References

[1] K. Terashima, et al.: Novel Creation and Control of Sand Mold Press Casting "Post-Filled Formed Casting Process", Foundry Trade Journal International (2009), 183 (3670), pp. 314-318

[2] R. Tasaki, et al.: Modelling and Control of Pressurized Molten Metal in Press Casting, Journal of Mechanics Engineering and Automation (2012), Vol.2, No.2, pp. 77-84

Proceedings of the 72nd World Foundry Congress
21-25th May 2016, Nagoya, Japan

Robotic Applications of Foundry Industry in China

Yang Yang[1,2], Yongquan Zheng[1,2], Yan Xu[1,2], Mo Yang[1], Boyang Lu[1], Fenfen Liu[1]
[1] Chaint Robotics Co., Ltd. China, 410717
[2] Hunan Province Industry Robots Demonstration Base, China, 410717

Abstract: For decreasing the labor intensity and producing higher quality products ultimately, a proposal has been raised for an industrial robotic casting solutions. The industrial robotic casting production process plans to include molding, post processing and cold-working. With the help of robots, firstly the molding stage implements intelligent sand core processing. Then, the post processing stage implements intelligent rough casting finishing. Finally, the cold working stage realizes intelligent casting deburring, chamfering, cleaning and blowing by robots. The rich experience and successful applications have testified a promising future for foundry industry with a full combination of robotic, automation and information technology.

Keywords: *Industrial robots, Foundry, Intelligent equipment*

1. Introduction

Foundry industry, the cornerstone of automobile, oil, steel, electric, vessel and textile etc., is of critical importance. Since year 2000, the casting production in China has become the first place in the world [1]. Nevertheless, the domestic foundry industry still experiences the low level of relative technology and product quality which have been improved in developed countries to some extent [2]. Furthermore, the demographic dividend is near to being cashed out, which increases the labor risk in foundry corporations. Consequently, the traditional method "poor efficiency and dirty work environment" seems to be unable to meet the recent requirements. It is urgent indeed to construct a modern casting production line with high efficiency and environment friendly [3,4]. This paper will focus on the research about how to implement robotic technologies into full processes of casting production.

2. Robots utilized in casting production processes

The whole casting production process can be re-planned into molding, post processing and cold working stages. The rich robots application experience as the following description fully indicates the success on solving the aforementioned problems.

2.1 Molding stage

In the molding stage, the sand core intelligent processing (including sand core transport, setting, assembling, dip-coating, deburring and mold separation) can be implemented with the help of industrial robots. The key technologies include the following points:

1) Dynamic Finite Element Analysis. The optimized clamping position and force could be calculated and it fulfils some goals like more accurate clamping, sand core protection and higher precision of core-assembly.
2) 3D laser vision technology. Non-contact sand core recognition, measurement and positioning can be achieved, which increases the system flexibility and ultimately reduces the relative cost as well. Meanwhile, it also effectively combat the precision problem caused by mechanical wear.
3) High flexible clamp helps fulfil the requirement of multi-type production.

Figure 1 shows the sandcore dip-coating in rail industry (left) and the sandcore transportation in automobile industry (right) with the help of industrial robots.

Fig. 1 Robots used in rail and automobile industry

2.2 Post processing stage

The intelligent rough casting finishing (including casting finishing, polishing, transport and stacking) is implemented in post processing stage. The key technologies contain:

1) Error Detection and Robot Path Correcting

System (EDRC), which will help implement automatic finishing under casting and clamping deviation.

2) Feedback and alarming system, which will help robots implement automatic and safety work with uncertain and different casting characteristics, in order to prevent some serious accidents and protect devices.
3) Offline programming technology based on expertise experience database. This technic guarantees the fitting precision and decreases labor hours [5]. Meanwhile, the advance simulation and optimization avoid some serious risks, the engineering change can be designed off line, which reduces rework hours and decreases negative effects.

The post processing technologies have been utilized successfully in automobile, rail and aerospace area. For instance, one automobile plant had a 1.5 times increase in the production ability and the number of workers decreased from 62 to 20.

Fig. 2 Robots used for the post processing

Figure 2 represents the overall layout of the intelligent rough cylinder finishing unit (left) and specialized robots arranged inside this unit (right), these robots will carry proper tools to implement rough casting finishing.

Especially, these research results had been awarded in 2014"IERA AWARD" and collected as a part of "2014 Chinese High-tech results" in 2014 by the Chinese Academy of Engineering.

2.3 Cold-working stage

In cold-working stage, it is mainly to complete some intelligent procedures including the casting deburring, chamfering, cleaning, blowing and so on, which realizes the intelligent seamless interfaces. The key technologies include:

1) The robotic casting deburring and chamfering. The high flexibility of robot can adapt to complex multi-type casting, break the bottleneck while the traditional pattern is difficult to achieve.
2) The robotic casting cleaning and blowing effectively combat some key problems, e.g. Cleaning indicators are extremely difficult to reach the standard where usually occurs in the internal cavity of the workpiece, voids, corners and so on. Compared to traditional washing machine, the flexibility and efficiency of the system has been increased.

Figure 3 shows the industrial robots implemented for the cold-working stage like crankshaft deburring, chamfering (left) and ship motor cylinder washing (right).

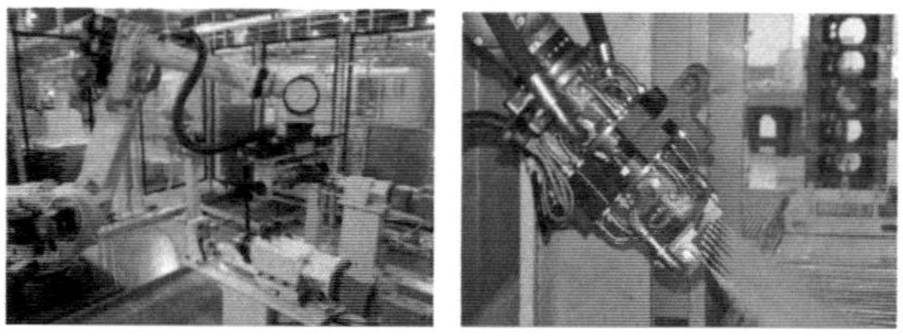

Fig. 3 Robots used for the cold-working stage

3. Conclusion

In traditional casting processing model, there are many problems like safety, health, harmful dust and low work accuracy. In order to guarantee the high product quality, the industrial robot is the best choice due to flexible procedures for multi-type castings, and meanwhile the industrial robot can perform jobs that might be hazardous to implement or be difficult to keep high accuracy. Recently, robotic and automation become a definite trend especially in automotive industry. In addition, the industrial robot can be much smarter by combining with internet of things, big data, cloud computing and mobile internet etc., which could promote the development of smart foundry industry in terms of intelligent casting and green foundry eventually.

References

[1] W. P. Chen, J. Luo, Y. Y. Li: *Net Shape Casting Technology and The Energy Conservation and Emissions Reduction*, (Week Proceedings of 2009 China casting activities, 2009).

[2] H. Liu, P. Xu: *Robots applied in casting production*,(24th Annual casting congress of Chongqing, China, Feb. 2014).

[3] X. L. Liu:*The Modern Casting Cleaning Method*, (Foundry Equipment and Technology,2011)pp. 6-7.

[4] X. X. Zhang:*Industrial robots application in foundry industry* ,(Machinist Metal Forming,2014)pp. 18-21.

[5] R. Zheng: *Robot Offline Programming System Design and Research*,(World Manufacturing Engineering & Market,2010)pp. 90-92.

Proceedings of the 72nd World Foundry Congress
21-25th May 2016, Nagoya, Japan

Generation of optimal trajectories for a pouring Robot considering Control system used in industry

Akos Csiszar[1], Markus Birkhold[2], Kazuhiko Terashima[3], Armin Lechler[2]

[1]Graduate School of Excellence Advanced Manufacturing Engineering University of Stuttgart, Germany, akos.csiszar@gsame.uni-stuttgart.de

[2]Institute for Control Engineering of Machine Tools and Manufacturing Units (ISW), University Stuttgart

[3]Department of Mechanical Engineering, Toyohashi University of Technology, Aichi, Japan

Metal casting is a highly productive manufacturing methodology. A typical application for metal casting is the production of motor blocks, with a yield of up to 25.000 units per day. There are various molding and casting methods, one such method is the gravitational casting. Gravitational casting can be either performed by a specialized casting machine or by an industrial robot with casting equipment attached, e.g. a ladle. In both cases programming the pouring motion is a tedious task usually performed via a teach-in process by an operator. In this paper we propose a general methodology to generate the casting motion via a process model.

Keywords: *Pouring Process, Robot Control, Optimal Trajectory Generation.*

1. Introduction

In the scientific literature control systems for pouring-robots and -mechanisms are intensely studied. Papers such as [1–4] contribute to solving sloshing suppression and prescribed flow rate assurance by redefining the closed loop control systems of the pouring-robot or -mechanism. This is a well suited approach, however it has the disadvantage that it is not easily implementable on currently used industrial robot controllers and industrial acceptance of any other control system than the classical cascaded torque-velocity-position faces difficulties.

This paper presents an alternative solution to the prescribed flow rate assurance problem. Instead of addressing "how to assure the prescribed flow rate for a pouring process" this paper addresses "how to assure the prescribed flow rate for a pouring process with a

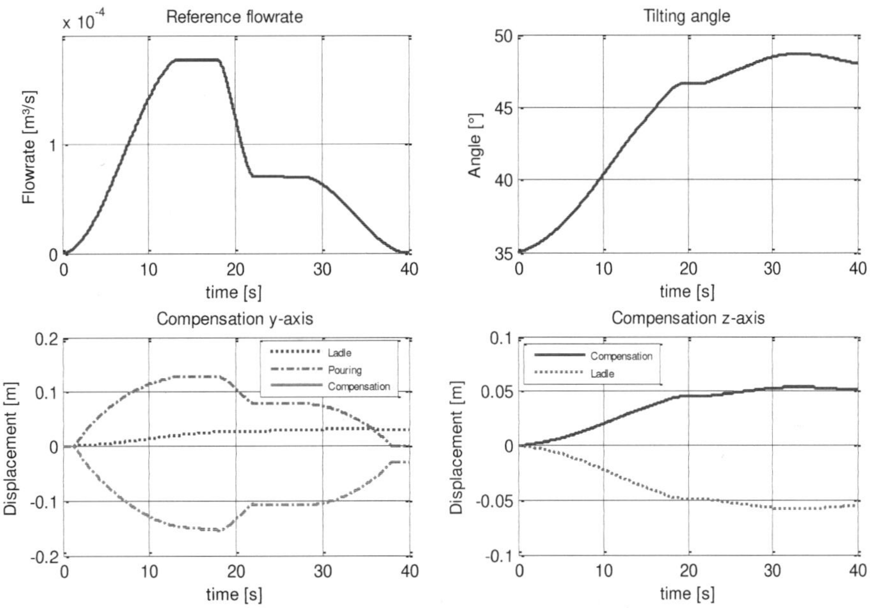

Figure 1: Reference flow rate and Cartesian curves

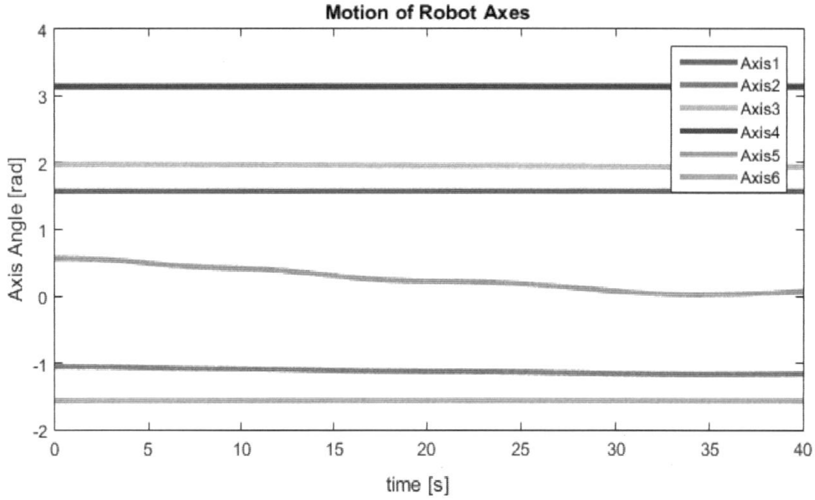

Figure 2: Robot axis movement

currently widely used industrial robot controller". The formulated question shows that a trade-off has been made with regard to performance in order to assure compatibility. The presented method does not perform better than dedicated control systems for sloshing suppression and prescribed flow assurance. However it makes existing robot controllers capable to suppress sloshing and assure flow rate.

2. Model based trajectory generation

To achieve this we have implemented the flow rate estimation equations of [2, 5] in a separate optimization module. This module computes the necessary motion to achieve the desired flow rate pattern. Figure 1 shows a sample pattern and the corresponding Cartesian motion to achieve the give flow rate and pour at a given position.

The Cartesian curve obtained with the above presented module is in function of time and therefore, a robot executing such a task must have the option to accept time based reference values. Major robot manufacturers offer real-time communication interfaces though which a robot accepts reference values both in joint space and in Cartesian space from an external real-time reference values generator. The RSI interface from Kuka and the LLI interface from Stäubli can serve as examples for such interfaces Figure 2 presents the motions of the axis of a Kuka KR-500 industrial robot arm while executing the above presented pouring task. The values are obtained from simulation. Since the reference values only change the coordinates X, Y and Phi, the angles of axis A1, A4 and A6 remain unchanged thought the pouring process.

3. Discussion and future works

As Figure 2 shows, the proposed method is capable of transforming the reference motion, obtained from model equations in Figure 1, directly into inputs for the robot motion controller. The presented method is therefore highly useful for pouring applications utilizing standard robots. It is no longer necessary to program the pouring motion by teach in, the motion can be directly computed from the model equations. Because the robot continuously receives new set points, the method can also be used in a feedback manner. Feedback control will be investigated in future studies.

References

[1] "K. Terashima and K. Yano: "Sloshing analysis and suppression control of tilting-type automatic pouring machine" Control Engineering Practice. Volume 9(6), 2001, 607-620," 2001.

[2] "Y. Noda and K. Terashima: "Estimation of Flow Rate in Automatic Pouring System with Real Ladle Used in Industry" In ICROS-SICE International Joint Conference. (2009) 354-359," 2009.

[3] "Y. Noda and K. Terashima: "Falling Position Control of Outflow Liquid for Automatic Pouring System with Tilting-Type Ladle" In IFAC MMM. (2007)53-58," 2007.

[4] K. Terashima and K. Yano, "Sloshing analysis and suppression control of tilting-type automatic pouring machine," Control Engineering Practice, vol. 9, no. 6, pp. 607–620, 2001.

[5] Yoshiyuki Noda, Yusuke Matsuo, Kazuhiko Terashima, Yufan Zheng, "A Novel Flow Rate Estimation Method Using Extended Kalman Filter and Sensor Dynamics Compensation with Automatic Casting Pouring Process,"

Proceedings of the 72nd World Foundry Congress
21-25th May 2016, Nagoya, Japan

Calciner-Free Green Sand Reclamation by Using Advanced Mechanical Reclamation System

Takafumi Oba[1], Junichi Iwasaki[1], Kazuya Abe[1], and Tatsuyuki Aoki[1]
[1] Sintokogio, LTD. Toyokawa Works, 3-1 Honohara, Toyokawa, Aichi 442-8505, Japan

In most green sand/core sand systems, new sand is usually added as cores, and waste sand is discharged as dust or removed at shotblasts. If their amounts are completely equal, they must be balanced. However, in most cases, amount of entered sand is much larger than lost after shakeouts. This unbalance makes overflowed sand. Unfortunately, overflowed sand has insufficient properties for replacing new sand. Consequently, overflowed sand must be recycled or wasted. Some large foundries reclaim thermally their waste sand as core sand, and other large foundries recycle it for steel making material or cement material. Both methods have various difficulties for small to medium foundries. Therefore, they must waste their overflowed sand for landfill. Moreover, landfill cost is unbelievably more expensive compared to purchase new sand. Authors developed the cutting-edge mechanical reclamation system that is effective for green sand reclamation with eliminating use of calciners. This presentation reports the principle of the system and results of some case studies. The results shows sand reclaimed by this system is usable as alternative to sand reclaimed by thermal system and unused sand. These advantages contribute not only waste reduction but also defect elimination caused by poor quality cores.

Keywords: *Greensand Reclamation, Mechanical, Core Sand, RCS, PUCB*

1. Principles and Mechanism

This reclamation system, *Roller Squeezing Type Mechanical Reclaimer*, consists of a pair of ceramic rollers and a rotating drum. Combination of attrition and squeeze works on removal of oolitics and fines effectively. At first, this system had developed for *tough chemically bonded sand processes* such as Phenol Nobake or Sodium Silicate processes [1-3]. However, after numerous experiments, authors established the mechanical green sand reclamation system without calciners.

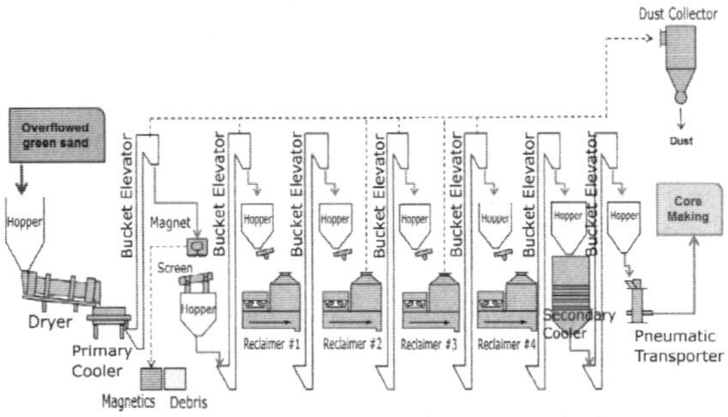

Fig. 1 Schematic Diagram of a Standard Plant

Fig. 1 shows the schematic diagram of standard reclamation system for overflowed green sand.

2. Case Studies

2.1 Reclamation for Resin Coated Sand (RCS)

RCS, used for shell cores, is one of the most popular applications of green sand reclamation in Japan. Authors conducted one comparison test between *Roller Squeezing Type Mechanical Reclaimer* and conventional reclaimer for a customer. Conventional method is to combust the sand by a calciner and then to scrub it by a numerous passes of attrition type reclaimers. On the other hand, developed system possesses four or six cell *Roller Squeezing Type Mechanical Reclaimers*. Test conditions are shown in Table 1, and test result for transverse strength is shown in Fig.2, and that for SiO_2 purity is shown in Fig. 3.

Table 1. Test Conditions from Green Sand to RCS

Throughput	Motor Power	Pass
4.0 MTPH	30kW/Pass (1st to 4th) 22kW/Pass (5th, 6th)	6 Passes

Fig. 2 and Fig. 3 clearly show that R*oller Squeezing Type Mechanical Reclaimer* achieves superior transverse strength and SiO_2 purity compared to conventional method using a calciner.

These results mean *Roller Squeezing Type Mechanical Reclaimer* enables to remove adhered Oolitics as well as fines more efficiently than conventional method. Oolitics is very hard to remove and enables to be burnt by combustion. Consequently, combination of attrition and squeezing shows much more effective performance compared to conventional method.

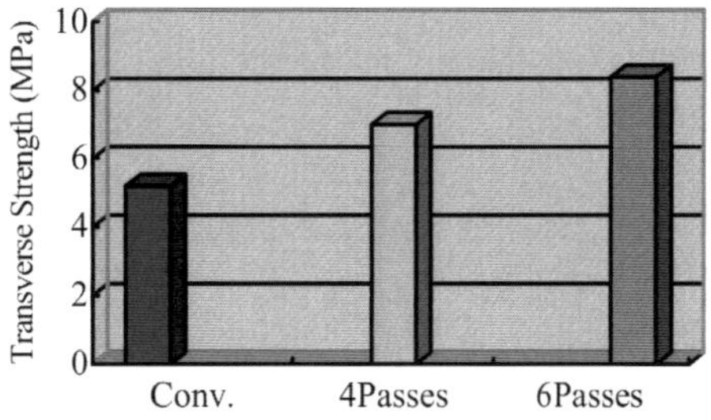

Fig. 2 Transverse Strength Comparison

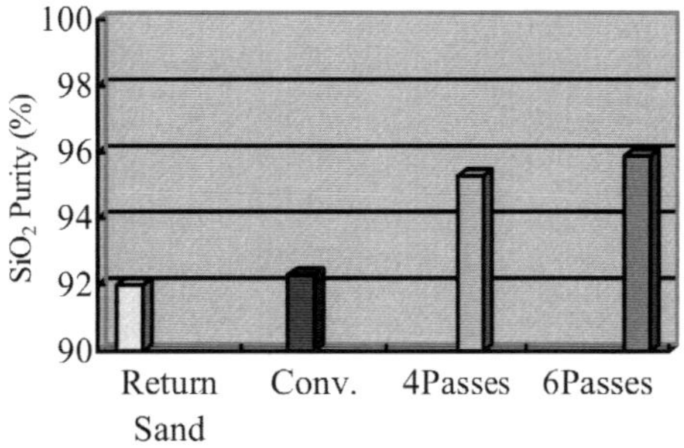

Fig. 3 SiO_2 Purity Comparison

2.2 Reclamation to Phenol Urethane Cold Box (PUCB) Sand

PUCB is popular core molding process for green sand mold all over the world. Authors conducted another reclamation test from waste green sand, core sand, and sand discharged from the shotblast mixture to PUCB cores for a customer. Test conditions are shown in Table 2, and Fig. 4 shows the transverse strength comparison between unused sand and reclaimed sand.

Table 2 Test conditions from Mold Sand Mixture to PUCB

Throughput	Motor Power	Pass
5.0 MTPH	30kW/Pass	4Passes
Sand Mixture		
Overflowed sand	Core Sand	Sand from Shotblast
50%	30%	20%

From the result shown in Fig. 4, *Roller Squeezing Type Mechanical Reclaimer* achieves superior values at 4 Passes in each condition compared to unused sand.

This result means that *Roller Squeezing Type Mechanical Reclaimer* does not removes oolitics and fines but also polish sand grainsurface and improves its shape to rounded.

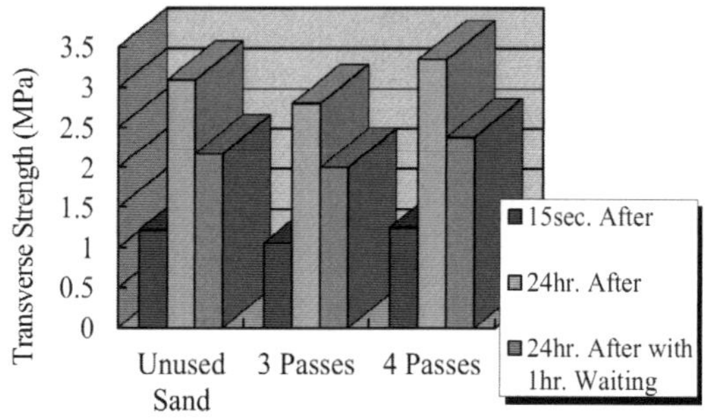

Fig. 4 Transverse Strength Comparison

3. Conclusions

Authors conducted numerous tests from green sand or sand mixture contains much green sand to core sand or mold sand. From the results mentioned above, *Roller Squeezing Type Mechanical Reclaimer* enables to eliminate calciners for reclaiming green sand. This system achieves to manufacture superior quality reclaimed sand only by mechanical reclamation.

Case study results clearly show the sand reclaimed by this system superior core strength and SiO_2 purity compared to the sand reclaimed by calciners and unused sand. These results mean sand reclaimed by this system enables to make more quality cores than that reclaimed by calciners. More quality cores must improve casting quality as well by eliminating casting defects caused by poor quality cores.

Authors expect this system to improve casting quality as well as to reduce waste sand.

References

[1] Oba, Iyoda, Iwasaki and Aoki: AFS Transactions 107 (1999) 89-93.

[2] Oba, Hagata and Nakayama: *Development of Mechanical Sand Reclamation Technology Using Roller Squeezing Method*, (159th Japan Foundry Society Meeting, Matsue Japan, 2011) pp12.

[3] Oba, Hagata and Nakayama: Journal of Japan Foundry Society 84 (2012) 533-538.

Proceedings of the 72nd World Foundry Congress
21-25th May 2016, Nagoya, Japan

New Melting System of Briquetted Aluminium Chip Achieves Energy-Saving and Environmental Improvement

Hirotoshi Yabumoto
2nd Engineering & Designing Dept., Engineering Division, Sanken Sangyo Co., Ltd., Hiroshima, Japan

Every cast product produces machining chip (hereinafter referred to as chip) in the machining process. Unlike foundry returns such as rejected products, it is not an easy task to melt chip, because the chip is wet with coolant and has thin sections, which leads to much formation of dross/oxide when melted in a reverberatory or a shaft melting furnace. Also, coolant in the chip generates fumes negatively affect the environment. So, traditionally chip just after being crushed has been dried in a drying oven such as rotary kiln system before melting. Our conventional chip drying/melting system also consists of a rotary kiln dryer system and our patented submergence vortex melting system (for more details, see our website: http://www.sanken-sangyo.co.jp) which can achieve the highest metal recovery of up to 98%. However, even our advanced rotary kiln system still has some drawbacks: a large installation space and a large amount of maintenance are required. Then, we focused on a briquetting machine which was primarily used to reduce the volume of chip for easy transportation. In 2013, we began to develop a new chip drying/melting system with less installation space, reduced energy consumption and the highest metal recovery rate, utilizing briquetted chip.

Keywords: *chip, briquette, drying, melting*

1. Introduction

Every rotary kiln system used for drying chip has the inherent risks of a dust explosion or a metallic combustion caused by dust/fine in the chip as well as a fire caused by condensed coolant in the exhaust system. The reason why we focused on a briquetting machine was to eliminate these risks and minimize the installation space required.

Of course, in the days of 2013 and before as well, wet briquetted chip was melted together with ingot and foundry returns, however, almost no tests were conducted to demonstrate what percentage of the briquetted chip being charged could be actually recovered as aluminium molten metal.

In any case, when trying to continuously and efficiently melt non-dried, wet briquetted chip as it is, we are faced with the following issues to be solved: (1) potential risk of a steam explosion when briquette is quickly and fully submerged in the metal bath, and (2) when briquette is charged into the metal bath (the briquette will float on the bath) or onto the dry hearth, in any event, the said briquetted chip will spring back over time prior to melting under such a hot circumstance, resulting in a worsened recovery rate due to excess oxidation.

2. Basic study

For the realization of a successful briquette melting, we set up three agendas: (1) energy-saving drying system utilizing sensitive heat of the bath metal, (2) detoxifying of fumes generating from the drying process, and (3) the dried briquette is to be melted in a submergence vortex with a higher recovery rate which is conventionally achieved.

First, we studied the properties of briquetted chip and conducted drying tests of them in/on aluminium molten metal. Also, we assumed that any steam explosion can be avoided so far as the briquette is not fully submerged in the molten metal, and we verified this assumption during testing. In this study/test, a briquette with 80 mm dia. x 80 mm high was used.

2.1 Briquette - density vs coolant content

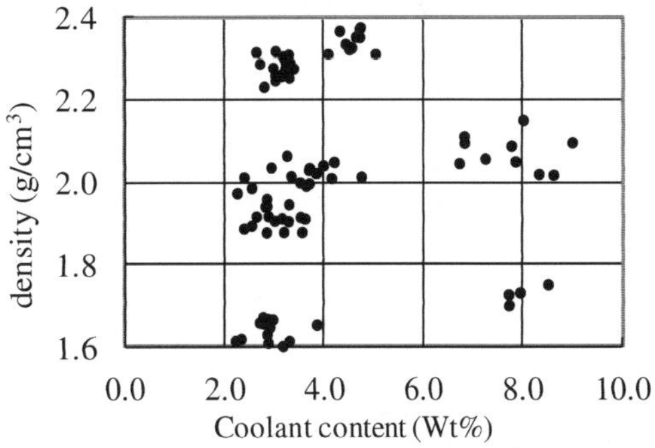

Fig1 Relationship density and coolant content

As shown in Fig. 1, we investigated the relationship between density and coolant content. The figure indicates that the coolant content drops to less than 5.5 wt. % and varies in a narrow range as the density grows, therefore, the briquetting machine requires an ability to constantly produce a dense briquette.

2.2 Briquette – drying time required

We had information already obtained by our past separate chip drying tests that moisture in a chunk of chip wet with coolant evaporates above 200℃ and residual oily ingredients evaporates above 400℃ at the center of the chunk. In our tests, the briquette was immersed in the molten metal by roughly 2/3 and the center temperature of the briquette reached 400℃ within 45 seconds. Fig. 2 shows the test results. During testing, no sign of steam explosions were seen.

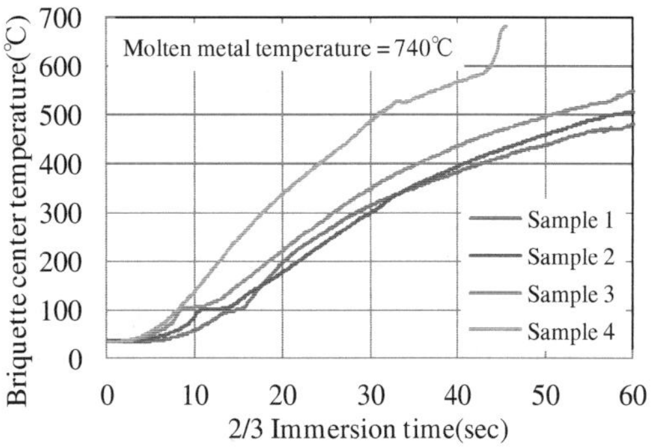

Fig.2 Immersion time vs coolant content

3. Development to actual drying/melting system

3.1 Concept of briquette drying/melting system

Based on our study and past experiences, we designed a new system which can meet the following criteria: (1) the briquette is to stay in the molten metal while not being fully submerged for more than 45 seconds, (2) then, it is to be quickly melted, and (3) the fumes are to be completely incinerated within the afterburner. Fig. 3 shows a schematic of the system.

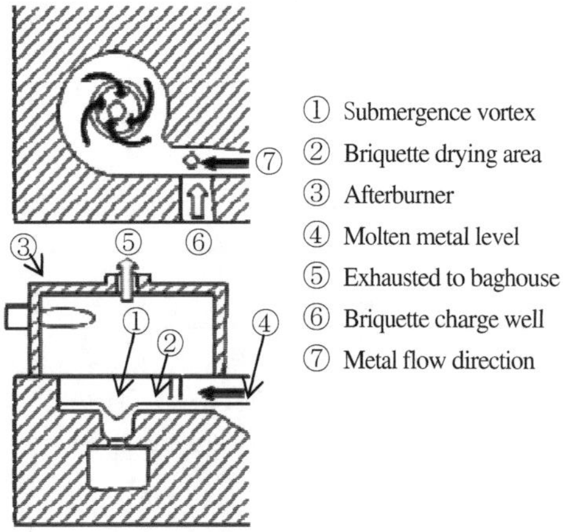

Fig. 3 Schematic of the new system

The briquette is continuously fed via a briquette charge well into metal flow generated by a mechanical pump and stays there for 45 seconds or more in a drying area which is located in a vortex periphery and has a shallow metal depth not allowing the briquette to be fully submerged, and the briquette is then quickly submerged in the vortex to be melted. The fumes including dioxins, generating from the drying process are completely incinerated in the afterburner which is mounted above the vortex chamber, and are then exhausted to the baghouse. To minimize the coolant content, the briquetting machine is expected to produce a briquette with a density of over 2.2 g/cm^3.

3.2 Commissioning results of the new system

We developed the concept to a new actual drying/melting system with a melting capacity of 700 kg/h and a metal temperature of 750℃. A briquette melting scene can be seen in Fig. 4 as we expected. Also, a performance comparision between the new system and our conventional system is shown in Table 1. With the new system, we have achieved energy saving by approx. 60% compared to our conventional one and obtained a 98% of metal recovery rate as high as the conventional one. Furthermore, istallation space was reduced by approx. 60%.

Both the new system and the conventional system are available, according to our customer's specific requirements.

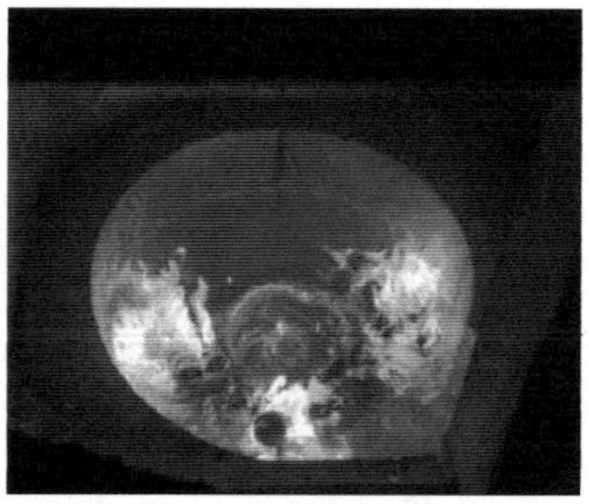

Fig. 4 Drying/melting scene of briquette

Table 1 Performance comparison (our product basis)

Description	New System	Our Conventional
Energy consumption *	116 kWh/t	465 kWh/t
Power consumption *	100 kWh/t **	70 kWh/t
Metal recovery;	Up to 98%	Up to 98%
Installation space (ratio)	0.4	1

* Except for melting process. ** Including the briquetting machine

Acknowledgements

In closing, we wish to acknowledge Central Motor Wheel Co., Ltd., Aichi, Japan, co-developer of the new system, and without whose cooperation this paper would not have been possible.

O-157

Sustainable Manufacturing in Foundry- A practical approach

Rajesh goel, Dy. General Manager

Sigma Electric Manufacturing Corporation pvt. Ltd, India

Abstract:

"What we are producing today should not be produced at the cost of future generation." Rapid industrialization & economic growth, continues to improve living standard but it may not continue for future generation. As foundries are recycling the metals, foundries thinks them as a part of the green manufacturing before the inception of word "green''. Foundry industry handles large quantities of processed waste materials and has the potential to emit a large quantity of carbon dioxide and other pollution in atmosphere. In reality foundries has just scratched the surface of green manufacturing and there is long way to move forward.

As foundries are mostly driven single owners and generally not considered manufacturing industries, so there is a need to shift paradigm in foundries sector from traditional manufacturing model, in which environmental activities were seen only as a additional cost; natural resources are viewed as freely available. For manufacturer; sustainability has been emerging as a new competitive requirement to achieve differentiation in market. This paper proposes a framework for development of practical model for foundries for sustainable manufacturing and improves overall performance. However, developing and maintaining a focus on sustainable manufacturing is new and difficult challenge for foundries.

Key words: -Sustainable Manufacturing, Foundries, Energy saving,

1. Introduction

After World War II, the manufacturers were struggling to survive with very limited resources. The pollution related to foundries is important due to their high potential risk to environment and human health. The main waste from casting is the spent foundry sand, which is generated at very large quantities during core and mold preparation.

Sustainable manufacturing highlights the road map of the industries for achieving performance improvement through sustainable development and its impact of organizational competitive outcomes. Some of the benefits of Sustainable manufacturing are listed below

- Controls and reduces material waste
- Preserves capital and saves money,
- Improves productivity and increases savings,
- Easy adaptability to changing rules in environmental regulations and legislation.

2. Literature Review:

Lovins (2003) suggests that from the 100% of natural capital extracted to make a product in the United States, commonly only 7% of materials become products that we end users see or use, meaning 93% becomes waste within industrial processes (this includes extractive and manufacturing waste). Out of the 7%, 1% becomes durable and 6% becomes waste from customer first use. Going further, of these 1% consumer durables, only 0.02% is recycled or remanufactured and the balance of 0.98% becomes persistent waste from disposal (typically landfill).

Beyond Greening – Strategies for Sustainable Manufacturing:-

30 years ago environmentalist Paul Ehrlich made a power full observation about SD. ***Environmental Burden (EB) = Population (P) X Affluence (A) X Technology (T)***

Stage one: Pollution prevention

This stage is governed by principal "Prevention is better than cure". Pollution prevention focuses on minimizing or eliminates the waste before it created. Pollution prevention of Foundry Industry Like other manufacturing industries, foundries are looking towards the trends of the future to understand the challenges which will be faced in coming year.

Stage two – Product Stewardship

The organization's need to look for Design for Environment (DFE), a tool for creating products that are easier to recover, reuse, or recycle is becoming increasingly important.

Stage three – Clean technology

Foundries with their eyes on future can begin to plan for and invest in tomorrow's technologies. The simple fact is that existing technology base in many industries is not environmentally sustainable. Cleaner technologies are desperately needed in foundries.

Process benefits of Sustainable manufacturing:-

- Material saving resulting from more substitution, reuse or recycling
- Better utilization of by-products, conversion of waste into valuable forms.
- Lower energy consumption during production process.
- Reduce material storage and handling cost.
- Elimination of cost of activities involved in discharge or waste handling, transportation and disposal.

Product benefits of Sustainable manufacturing:-

- Higher quality, more consistent products and lower products cost
- Lower packaging cost, more efficient use of resources by products.
- Lower net cost of products disposal to customer.
- Higher product resale and scrap value.

Implementation of Sustainable Manufacturing in Foundries

If we look individual continuous improvement projects, it will yield to 1% -2% saving, however it will create a great impact cumulative. Indian foundries have taken a big drive to save energy and there is great focus by Govt of India to reduce the energy consumption in foundries.To reduce energy consumption the following steps can be taken to realize the results of sustainable manufacturing.

There is some basic and important step for saving:

1. Reuse of waste heat from furnace flue gases
2. Improvement in Air exhaust system
3. Modern lighting system
4. Compresses air pressure control
5. Reduction in cost of poor quality
6. Reduction in loss of molten metal
7. Improvement in overall system efficiency

Conclusions:

Once we have achieved the focus to think green in all aspects of our foundry, we are open to the possibilities that were only waiting for us to seize them. This focus, along with the pragmatic application of existing technology, will guarantee that the foundry industry is recognized as a sustainable leader by other industries and the public. It the time to roll up the sleeve and be on our toes to make foundries greener and efficient to save the existence of foundries. Truth is that we have big opportunities ahead of us, wheatear we are a large / medium or small foundry.

Reference

[1] Che B. Jounga, John Carrell and Prabir Sarkara. Categorization of indicators for sustainable manufacturing. Elsevier Ecological Indicators 24 (2012) 148–157.

[2] Marc a. Rosen, hossam a. Kishawy, sustainable manufacturing and design: concepts, practices and needs, sustainability 2012, 4, 154- 174; doi:10.3390/su4020154

[3] Yiigit Cisem. Life cycle assessment in ferrous foundry industry, January 2013

Proceedings of the 72nd World Foundry Congress
21-25th May 2016, Nagoya, Japan

Wear Behavior of Carbon Fiber Reinforced Aluminum Alloy Composite

Muhammad Faiz Bin Zainuddin [1] and Kazunori Asano[2]
[1] Graduate School of Science and Engineering, Kindai University,
3-4-1 Kowakae, Higashiosaka City, Osaka 577-8502, Japan
[2] Department of Mechanical Engineering, Faculty of Science and Engineering, Kindai University,
3-4-1 Kowakae, Higashiosaka City, Osaka 577-8502, Japan

Effects of the reinforcement with Pitch-based carbon fiber on the wear characteristic of aluminum alloy have been investigated. The composites were fabricated by squeeze casting. Wear tests were carried out under dry sliding conditions using pin-on-disk method. Wear loss of the aluminum alloy was greatly reduced by the fiber reinforcement. The friction coefficient during the wear test of the composite was stable compare to that of the unreinforced alloy and the roughness values of the worn surface of the alloy became smaller by the reinforcement. These results indicate that the carbon fiber acted as a solid lubricant on the worn surface.

Keywords: *Aluminum, Carbon Fiber, Composite, Squeeze Casting, Wear.*

1. Introduction

Reinforcing with non-metallic fiber has been proposed to improve the properties of aluminum alloy. Carbon fiber was widely used as the reinforcement because it has a high strength, rigidity, low density and low thermal expansion. Pitch-based carbon fiber is one of the most attractive materials as the reinforcement because its thermal conductivity is higher than that of the PAN-based carbon fiber [1]. Although the properties of the carbon fiber reinforced aluminum alloy are widely studied [2-3], PAN-based carbon fibers were used as the reinforcement in these reports. No research has been found on the wear characteristics of the composites reinforced with the pitch-based carbon fibers. In the present study, aluminum alloy composites reinforced with pitch-based carbon fibers were fabricated by the squeeze casting, and then the effects of the fiber reinforcement on the wear characteristics of the aluminum alloy were discussed.

2. Experimental procedure

2.1 Materials

The JIS-AC8A aluminum alloy (Al-12Si-1Mg -1Cu-1Ni) was used as the matrix metal. Short pitch-based carbon fiber was used as a reinforcement. In order to fabricate the fibrous preform, the carbon fibers are mixed with water, an inorganic binder (silica sol) and an organic binder (PVA) and then pressed in the mold, followed by drying at 458 K for 2 h. The fiber volume fraction in the preform was 30 vol%. The composite was fabricated by the squeeze casting. The mold was firstly preheated with the preform at 673 K, the molten AC8A alloy (973 K) was then poured into the mold and quickly pressed by a plunger (40 MPa for 1 min) to be infiltrated into the preform. Figure 1 is the optical micrograph of the composite, showing that fibers are randomly oriented.

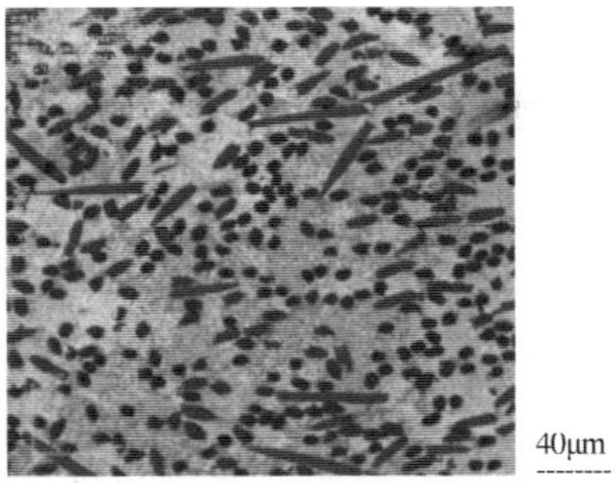

Fig. 1 Optical micrograph of the composite.

2.2 Wear test

The composites were subjected to wear test by the pin-on-disk method. AC8A alloy was used as a comparative material. Aluminum and composite were cut into the pin specimen (φ5×15 mm). High carbon chromium bearing steel material (JIS-SUJ2) (hardness: 188 Hv) is processed into a disk shape for the counterface. Wear tests were carried out under a contact load of 19.6 N and at the sliding velocity (υ) of 0.3 m/s and 3.0 m/s. The mass loss and the coefficient of friction (μ) during the wear test were also measured. Subsequently, the worn surfaces of the pin specimen and the counterface were observed using microscope and the roughness values (center line average roughness, R_a) of the worn surfaces of the specimens were measured.

3. Results and discussion

Figure 2 shows the relationship between the sliding distance and the amount of wear loss of the composite and unreinforced AC8A alloy (pin specimens). By reinforcing the alloy with the carbon fibers, wear loss of the AC8A alloy was greatly reduced.

The coefficient of friction of the specimens (μ) during the wear test is shown in Fig. 3. Error bars represent the range between the maximum and minimum values. The range was getting smaller by reinforcing with the carbon fibers, indicating the change in μ for the composite was small throughout the wear test.

Figure 4 shows the optical micrographs of the worn surfaces of the counterface after the wear test. When υ was 0.3 m/s, worn marks became unclear by the fiber reinforcement. When υ was 3.0 m/s, the adhesion of the aluminum on the surface seems to be slightly decreased by the fiber reinforcement.

Figure 5 shows the surface roughness values (R_a) of the disk specimens after the wear test. The average values and the range values was slightly decreased by the fiber reinforcement at every sliding velocity.

At a low υ, the wear loss, the change in the coefficient of friction during test and the surface roughness of the AC8A alloy and the counterface decreased by the reinforcement. These facts indicate that the carbon fibers act as a solid lubricant on the worn surfaces thus preventing the adhesion of the matrix on the counterface. At a high υ, although the change in the coefficient of friction of the AC8A alloy decreased by the reinforcement, the surface roughness of the AC8A alloy and the wear loss of the counterface did not significantly change by the reinforcement. This is probably due to the fact that the AC8A alloy melted by the heat that were produced by the friction on the surface thus welded and covered on the worn surface of the counterface.

4. Conclusions

The wear loss of the AC8A alloy was greatly reduced by the carbon fiber-reinforcement. Although the reinforcement had no effect on the average values of coefficient of friction, the range in the values during the wear test decreased by the fiber reinforcement. These results was probably due to the carbon fibers which act as a solid lubricant on the worn surface.

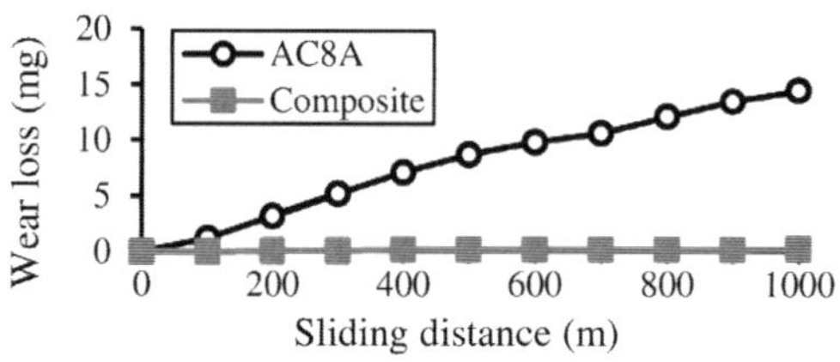

Fig. 2 Relationship between sliding distance and wear loss of AC8A alloy and composite.

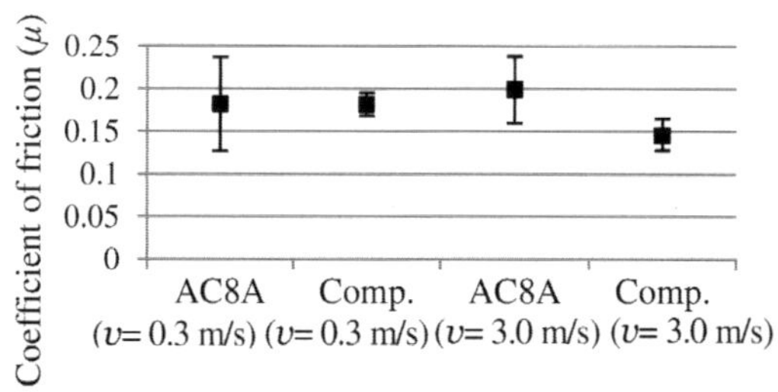

Fig. 3 Coefficient of friction of AC8A alloy and composite under wear test.

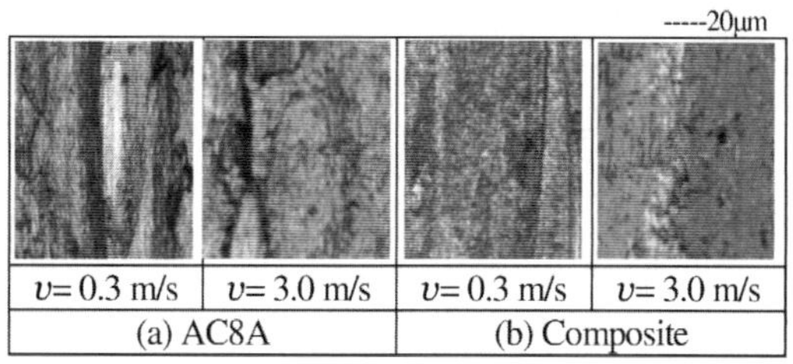

Fig. 4 Worn surfaces of the disk specimens after wear test.

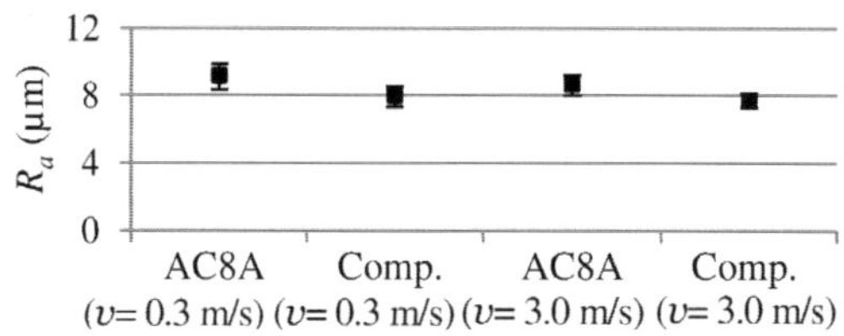

Fig. 5 Surface roughness of disk specimens after wear test.

References

[1] S. Li and C. Chaou: Metall. Mater. Trans. 35 (2004) 2153-2160.

[2] H. Dvir, J. Jopp and M. Gottlieb: *Estimation of polymer–surface interfacial* (J. Colloid. Interf. Sci, 1972) pp. 802-805.

[3] U. Zielke, K. J. Hüttinger and W. P. Hoffman: *Surface-oxidized carbon fibers*, (Surface Structure and Chemistry, Karlsruhe, 1996) pp. 983-998.

Fabricated of Metal-bonded Grinding Wheel with Cubic Boron Nitride Abrasive Grains by Centrifugal Casting

Takayasu Sugiura[1], Yuta Suzuki[2], Motoko Yamada[1], Hisashi Sato[1], Yoshimi Watanabe[1] and Hideaki Tsuge[3]
[1] Graduate School of Engineering, Nagoya Institute of Technology,
Gokiso-cho, Showa-ku, Nagoya, Aichi, 466-8555 Japan
[2] Department of Mechanical Engineering, Nagoya Institute of Technology,
Gokiso-cho, Showa-ku, Nagoya, Aichi, 466-8555 Japan
[3] Industrial Research Institute of Gifu Prefecture, 1288, Oze, Seki, Gifu, 501-3265 Japan

Al/cubic boron nitride (cBN) metal-bonded grinding wheel for drilling of carbon fiber reinforced plastic (CFRP) was fabricated by the centrifugal mixed-powder method (CMPM), which is application of centrifugal casting. Drilling performance of fabricated grinding wheel is tested using the gyro-driving grinding wheel system.

Keywords: *Centrifugal casting, composite, cBN, carbon fiber reinforced plastic (CFRP), drilling*

1. Introduction

Nowadays, carbon fiber reinforced plastic (CFRP) is increasingly used as a structural material in the aerospace industry due to its superior properties such as low weight, high strength, good toughness and so on. Drilling of CFRP is the most common post process of structural parts of aircraft fields. However, CFRP is known as a difficult to machining material, fiber pullout, delamination, burrs and splintering have become problems in machining CFRP by common drills.

Recently, gyro-driving grinding wheel system has been developed as a new CFRP drilling system [1,2]. This system provides holes with high precision on CFRP without delamination and burr formation. In the gyro-driving grinding wheel system, a grinding wheel was used, instead of drill bits for drilling.

In our previous study, Al-bonded grinding wheel with diamond abrasive grains was fabricated by CMPM [3]. In this study, cBN grinding wheel for drilling CFRP has been fabricated by the CMPM, since cBN is known as a diamond competing superhard material. Drilling performance of fabricated metal-bonded grinding wheel is also tested using the gyro-driving grinding wheel system.

2. Experimental procedure

Pure Al particles of 99.9% purity having 75 to 150 μm diameter and abrasive grains (cBN abrasive grains or diamond abrasive grains) having 125 to 150 μm diameter were mixed. The volume fraction of the abrasive grains in the mixed-powder was 25 vol.%. After that, the mixed-powder was put into a mold which was made by lost-wax method and the prepared mold was set in a compact vacuum centrifugal casting machine, show in Fig.1. After setting the mold, pure Al ingot of 99.9% purity was melted at 1300 °C by heating coil under vacuum condition. After melting, the motor was rotated to apply centrifugal force and molten Al was poured into the mold by its centrifugal force. After air cooling, the metal-bonded grinding wheel with 20 mm in outer diameter was obtained. The casting conditions are show in Table 1.

The microstructural observations of the fabricated samples were carried out with scanning electron microscope (SEM). CFRP drilling test was conducted at peripheral wheel speed of 7000 rpm, spindle speed of 2800 rpm and feed rate of 5 mm/min.

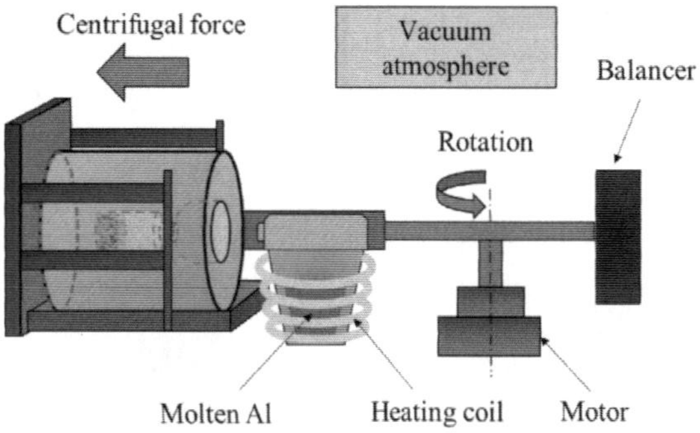

Fig. 1 Schematic illustration of centrifugal mixed-powder method.

Table 1 Casting conditions

Samples	Al/cBN	Al/diamond
Amount of Al particles [g]	2.73	
Amount of abrasive grains [g]	1.19	1.17
Amount of Al ingot [g]	30	
Casting temperature [°C]	1350	
Gravity multiples [G]	85	

3. Results and Discussion

Al/cBN and Al/diamond samples were successfully fabricated by CMPM. Figure 2 shows SEM images of abrasive grains in (a)Al/cBN and (b)Al/diamond samples. As seen in Fig. 2, it was observed that cBN abrasive grains were embedded in Al matrix without space between Al matrix and cBN abrasive grains. On the other hand, the space between Al matrix and diamond abrasive grains were observed. Therefore, better bonding force between the abrasive grains and the Al matrix is expected for the Al-cBN sample.

Figure 3 shows photos of the 30th hole on CFRP plate drilled by (a)Al/cBN and Al/diamond griding weels. As seen in Fig 3 (a), some burrs were observed for the hole drilled by the Al/cBN grinding wheel. On the other hand, as seen in Fig. 3 (b), there was no burr for the hole made by Al/diamond grinding wheel. From the results, Al/diamond grinding wheel was able to more high quality drilling than Al/cBN grinding wheel.

Figure 4 shows surface roughness of the machined CFRP as a function of number of machined holes. As seen in Fig. 4, it was found that the surface roughness values obtained by CFRP drilling test with Al/cBN and Al/diamond samples were lower than the requirement for certification.

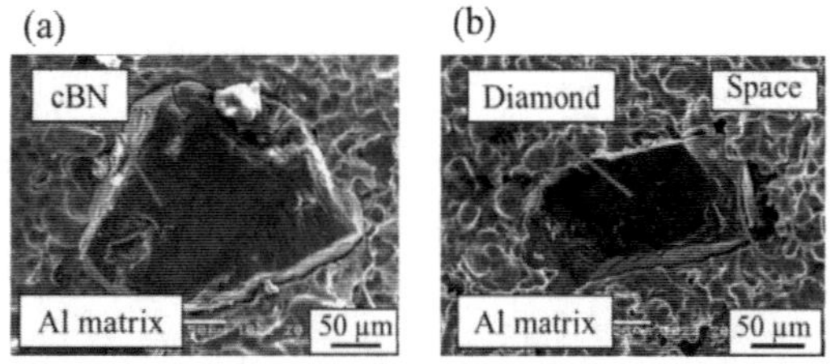

Fig. 2 SEM images of abrasive grains in (a)Al/cBN and (b)Al/diamond samples.

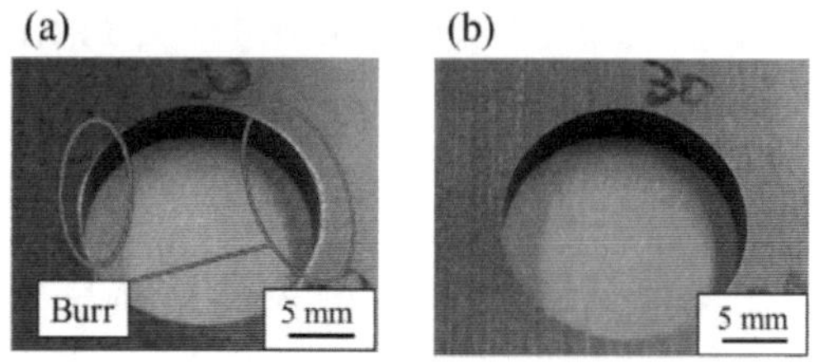

Fig. 3 Photos of the CFRP plate after making 30 holes with (a)Al/cBN and (b)Al/diamond samples.

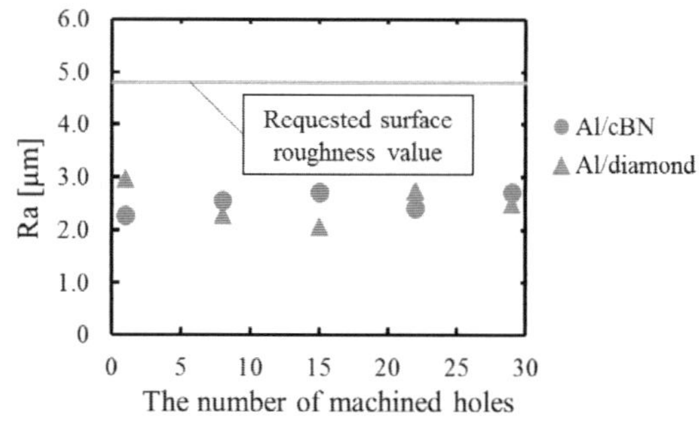

Fig. 4 Surface roughness of the machined CFRP

4. Conclusions

Al/cBN and Al/diamond grinding wheels have been successfully fabricated by CMPM. From obtained results, it was found that the surface roughness values obtained by CFRP drilling test with Al/cBN and Al/diamond grinding wheels were lower than the requirement for certification. It is also found that Al/diamond grinding wheel was able to more high quality drilling than Al/cBN grinding wheel.

References

[1] K. Takekoshi: J. Jpn. Soc. Mech. Eng. 114 (2011) 446-447.

[2] Y. Watanabe, H. Sato, H. Tsuge: J. Jpn. Soc. Mech. Eng. 115 (2012) 61.

[3] M. Shibuya, H. Sato, E. Miura and Y. Watanabe: J. FGMs. 25 (2011) 1-6.

Proceedings of the 72nd World Foundry Congress
21-25th May 2016, Nagoya, Japan

Microstructure And Mechanical Property Of Mg-Al-Ca Alloys Produced By High Pressure Casting.

Manabu Mizutani[1], Katsuhito Yoshida[1], Nozomu Kawabe[1], Yoshikazu Yamanishi[2], Hiroki Kako[3], Emi Yanagihara[4], Seiji Saikawa[4], Kenji Matsuda[4]

[1] 3rd Engineering Department, Magnesium Alloy Development Division, SUMITOMO ELECTRIC INDUSTRIES, LTD, 1-1-1, Koyakita, Itami-shi, Hyogo 664-0016, Japan
[2] Under Graduate School of Science and Engineering for Education, University of Toyama, 3190, Gofuku, Toyama, 930-8555, Japan
[3] Graduate School of Science and Engineering for Education, University of Toyama, 3190, Gofuku, Toyama, 930-8555, Japan
[4] Graduate School of Science and Engineering for Research, University of Toyama, 3190, Gofuku, Toyama, 930-8555, Japan

The microstructure and the mechanical property was investigated using by Mg-Al-Ca castings which is an automotive engine component with different Ca contents. With increasing of Ca content, 0.2%P.S. was increased and elongation was decreased. It was thought to be an cause that the amount of γ-(Mg, Al)$_2$Ca phase was increased. With such change of microstructure, heat resistance was also improved.

Keywords: *Mg-Al-Ca alloy, high pressure casting, microstructure, mechanical property.*

1. Introduction

In recent years, reducing the fuel consumption is getting a critical issue of the automobile industry. Reducing the weight of cars is one of the ways to improve the fuel efficiency, therefore the application of light materials such as Mg alloys to auto parts is highly expected. Mg alloys have the lowest density and the highest specific strength among practically used metals. Mg-6.0%Al-0.3%Mn (AM60) and Mg-9.0%Al-1.0%Zn (AZ91) are two typical alloys used in the auto industry. These alloys are applied for seat frames and door frames of some cars. The most effective parts in reducing the weight of cars are powertrain components [1].

However AZ91 and AM60 alloy are insufficient in heat resistance which is necessary for the powertrain components. Various heat resistant Mg alloys have been developed to improve the heat resistance. Mg-Al-Ca alloys are one of the promising alloy because it does not contain expensive rare earth elements and has high recyclability as well as high heat resistance [2, 3].

In this study, the effect of Ca contents on the microstructure and mechanical property of Mg-Al-Ca castings produced by high pressure casting was investigated.

2. Experimental procedure

The experimental alloys were Mg-7.0%Al-1.0% Ca (AX71) and Mg-7.0%Al-3.0%Ca (AX73). In addition, Mg-9.0%Al-1.0%Zn (AZ91) alloy was prepared as the comparative material. To achieve this composition, the following raw materials were weighed: pure Mg, pure Al, pure Ca, pure Zn, pure Sr, Mg-Mn and Al-Be master alloy. These alloys were melted in an electric furnace, while blowing protective gas to the surface of molten metal. The alloy codes and the chemical compositions of the castings are shown in Table 1. The samples were an automotive engine component of about 1 kg produced by cold chamber die-casting machine.

The microstructures of the castings were observed with optical microscope (OM) and scanning electron microscope with energy dispersed X-ray spectrum (SEM-EDS). The tensile test was carried out using by Instron Autograph tensile testing machine at R.T. with a cross-head speed of 1mm/min.

Table 1 The alloy codes and the chemical compositions of the castings (mass%).

Alloy code	Al	Ca	Zn	Sr	Mn	Fe	Mg
AX71	7.71	1.01	-	0.25	0.23	0.001	Bal.
AX73	6.90	2.78	-	0.25	0.20	0.006	Bal.
AZ91	8.97	-	0.65	-	0.19	0.001	Bal.

3. Results and Discussion

The optical micrographs of the castings in low and high magnification are shown in Fig.3 and 4, respectively. The microstructures mainly consisted of the primary crystallized α-Mg phase (white region) with dendritic shape and eutectic phase (dark gray region) crystallized at the last stage of solidification on the grain boundaries and the cell gaps in all the alloys. Al_6Mn phases (black region) were dispersed uniformly in the matrix of all alloys.

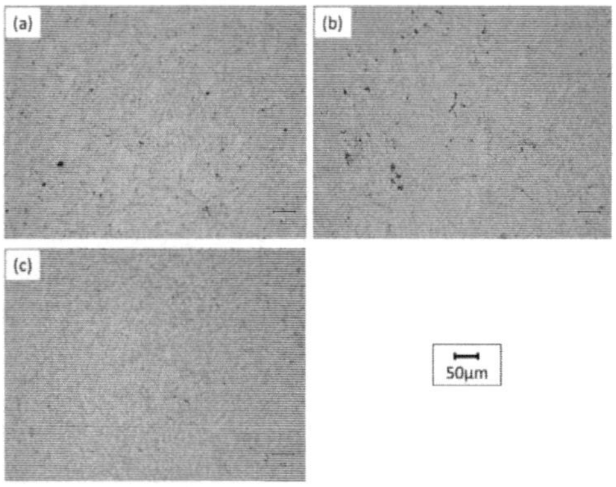

Fig. 1 Optical micrographs in low magnification of (a) AX71, (b) AX73 and (c) AZ91 castings.

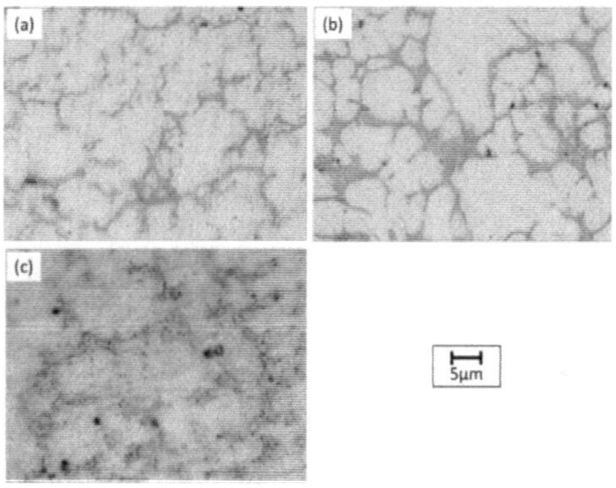

Fig. 2 Optical micrographs in high magnification of (a) AX71, (b) AX73 and (c) AZ91 castings.

Fig.3 shows the mechanical property of all the castings. Elongation was slightly low in all the castings. As the one of cause, it is thought that the castings included defect inside, because these castings were near-net shaped actual parts. With increasing of Ca contents from 1.0 to 3.0%, 0.2%P.S. was increased about 20MPa, on the other hand, elongation was decreased about 1.3%.

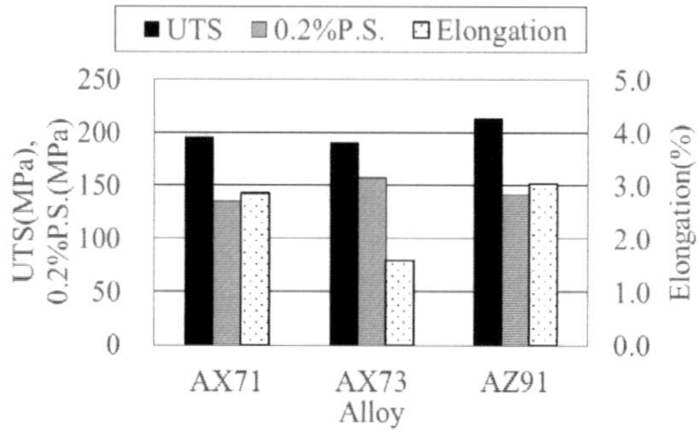

Fig. 3 Mechanical property of AX71, AX73 and AZ91 castings.

In order to clarify the cause of such changes of the mechanical property, crystallized phases were identified based on the observation results of microstructure. The amount and kind of crystallized phase of each castings are shown in Table 2. Because the amount of γ-(Mg, $Al)_2Ca$ phase was increased with increasing of Ca contents, the grain boundary and cell gaps were strengthened and heat resistance was improved.

Table 2 Amount and kind of crystallized phase in castings. (much) + + + + , + + + , + + , + , - (little).

Alloy	Crystallized phase			
	α-Mg	β-$Mg_{17}Al_{12}$	γ-$(Mg,Al)_2Ca$	Al_6Mn
AX71	+ + + +	+ +	+ +	+
AX73	+ + + +	-	+ + +	+
AZ91	+ + + +	+ +	-	+

4. Conclusion

Mg-Al alloy was strengthened and heat resistance of this alloy was improved by adding Ca, due to the generation of γ-(Mg, $Al)_2Ca$ phase

References

[1] S.Saikawa. Japan Institute of Light Metals. 60 (11) (2010) 571-577.

[2] W.Blum et al., Comparative study of creep of the die-cast Mg-alloys AZ91, AS21, AS41, AM60 and AE42, Materials Science and Engineering A319–321 (2001) 735–740.

[3]S.M. Zhu et al., Creep properties of a Mg–Al–Ca alloy produced by different casting technologies, Materials Science and Engineering A 483–484 (2008) 583–586.

Proceedings of the 72nd World Foundry Congress
21-25th May 2016, Nagoya, Japan

Prediction of hot tearing for a partially solidified bismuth bronze by using finite element method analysis

Akira Matsushita[1], Tomoaki Nakazawa[1,2], Toshimitsu Okane[3] and Makoto Yoshida[1,4]
[1] Department of Mechanical Engineering, Graduate School of Waseda University, 3-4-1 Shinjuku-ku Okubo, Tokyo 169-8555, Japan
[2] Toyota Motor Corporation, 1 Toyota-Cho, Toyota City, Aichi Prefecture 471-8571, Japan
[3] National Institute of Advanced Industrial Science and Technology, 1-1-1 Tsukuba Umezono, Ibaraki 305-8568, Japan
[4] Kagami Memorial Research Institute for Materials Science and Technology, Waseda University, 2-8-26 Shinjuku-ku Nishiwaseda, Tokyo 169-0051, Japan

The aim of this study is the prediction of solidification cracking, or hot tearing, for bismuth bronze by fluid flow, solidification, and thermal stress analysis. Constitutive equation required for the analysis was experimentally constructed as a viscoelastic equation by using tensile and relaxation test in the partially solidified state. For the purpose of predicting crack defects, some parameters were examined as cracking criteria, e.g. hot tearing indicator (HTI) which is the equivalent plastic strain during solidification [1]. The prediction results were compared with crack locations of the experimental castings by a copper mold in order to validate the analysis. As a result, there was a reasonable correlation between the distribution of HTI and cracking occurred during solidification.

Keywords: *semi-solid, solidification cracking, hot tearing, thermal stress analysis, viscoelastic*

1. Introduction

Recently, there is a strong need to predict cracking for increasing productivity of castings. Computer Aided Engineering (CAE) becomes an effective tool to predict these defects. In the crack prediction using CAE, the constitutive equation of solidifying alloy and the criteria for crack defects are required.

In this study, viscoelastic constitutive model was adopted for the crack prediction of partially solidified bismuth bronze. Firstly, Young's modulus and parameters of Norton's law in the partially solidified states were obtained as a function of fraction solid using tensile and relaxation tests during solidification. Secondly, casting tests of a water meter housing were executed using a copper mold. Thirdly, Analyses of fluid flow, solidification, and thermal-stress were performed. Lastly, crack locations of the obtained castings were compared with the analytical predictions by some criteria, such as HTI.

2. Experimental

The material used for this study is bismuth bronze. Its composition is given in Table 1.

Table 1 Composition of the test alloy (mass%).

Sn	Zn	Bi	Ni	Pb	P	S	Cu
6.3	4.3	1.63	0.12	0.08	0.03	0.002	Bal.

To validate analytical prediction of cracking, the copper mold casting was carried out [2]. The shape of the specimen is water meter housing as shown in Fig. 1. The pouring temperature was 1200 °C and the mold was held at 220 °C. The temperature of the casting was measured at the neck parts in Fig. 1.

3. Analysis

3.1 Constitutive equation

In this study, Maxwell constitutive model was used. The constitutive equations were as follows:

$$\varepsilon_{\mathrm{total}} = \varepsilon_{\mathrm{elastic}} + \varepsilon_{\mathrm{creep}} \quad (1)$$

$$\varepsilon_{\mathrm{elastic}} = \sigma / E \quad (2)$$

$$\dot{\varepsilon}_{\mathrm{creep}} = A\sigma^{n} \quad (3)$$

The total strain ($\varepsilon_{\mathrm{total}}$) was composed of the elastic ($\varepsilon_{\mathrm{elastic}}$) and the creep ($\varepsilon_{\mathrm{creep}}$) strain. In the partially solidified state, plastic strain which causes strain hardening was assumed to be negligible. For example, as to the Al-4.5%Cu alloy, Magnin et al. [3] showed that strain hardening coefficient in the partially solidified state had zero value. σ is stress, E is Young's modulus, n is stress exponent, and A is material constant. These mechanical properties of the bismuth bronze were obtained by unload, tensile, and stress relaxation tests during solidification as a function of temperature [4]. On copper alloys, there was no precedent for the experimental construction of the rate-dependent constitutive equation.

3.2 Finite element simulation

The experimental casting of water meter housing was simulated as shown in Fig. 1. A complete model was used in this work. The ProCAST 2013 software was used to perform fluid flow, solidification, and thermal stress analysis by finite element method (FEM). The simulated fields were computed with approximately 780,000 elements (mesh sizes range about 10 mm). Linear tetrahedral elements was chosen for the mesh.

The thermophysical properties of the cast and mold materials were calculated using JMatPro 8.0 software. Heat transfer coefficients were adjusted so that a temperature history of casting obtained by analysis should fit the experimental result.

4. Results and discussions

As shown in Fig. 1, the simulated distribution of HTI was found to be in reasonable agreement with the crack locations observed in the casting experiment. Consequently, in the copper alloy, HTI was capable of predicting hot tearing.

In the experiment, there was also external shrinkage on the upper surface part of the casting. Fig. 2 shows the relationships between HTI and upper surface deformation in z-direction, or external shrinkage displacement, in the analysis. The more external shrinkage occurred under the higher HTI condition.

Incidentally, the simulated distribution of maximum principal stress suggested only cracking at neck parts.

5. Conclusions

To predict the hot tearing for the bismuth bronze, fluid flow, solidification, and thermal stress analysis were performed by FEM. Viscoelastic constitutive equation was chosen for the analysis. The rate-dependent constitutive equation for the copper alloy was constructed by the experiment firstly in the world.

Comparing between simulated HTI and experimental result was suggested that HTI was able to predict the hot tearing and external shrinkage.

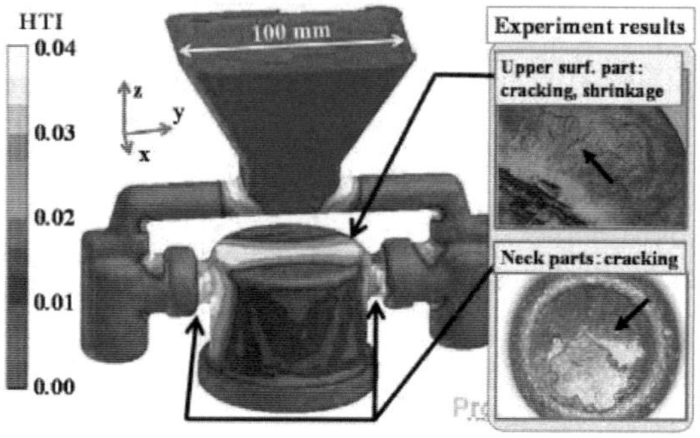

Fig. 1 Distribution of HTI and crack locations.

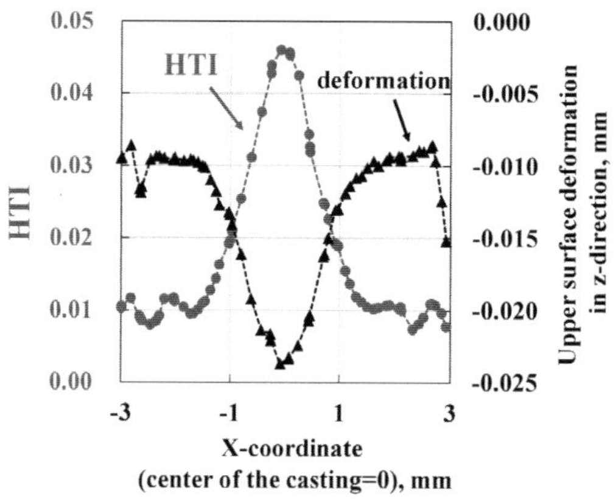

Fig. 2 Relationships between HTI and simulated external shrinkage displacement.

Acknowledgements

The authors gratefully acknowledge T. Maruyama and R. Yoshida (from Kansai University)I. Akashi and T. Akashi (from Akashi Gohdoh Inc.), S. Yamashita (from Joetsu Bronzl Corporation), H. Mori (from Azbil Kimmon Haramachi Co., Ltd.), and T. Kodate (from Rikokihara Corporation) for help with the experiments.

References

[1] J.Guo and J. Z. Zhu: Proceedings of the 5th Decennial International Conference on Solidification Processing, Sheffield, July 2007, pp.549-553

[2] T. Nakazawa et al.: Reports of 163th JFS Meeting, 24(2013), pp.25

[3] B. Magnin et al.: Materials Science Forum, 217–222 (1996), pp. 1209–1214.

[4] A.Matsushita et al.: Reports of 167th JFS Meeting, 85(2015), pp.86

Proceedings of the 72nd World Foundry Congress
21-25th May 2016, Nagoya, Japan

Tilting Motion Control of Automatic Pouring Ladle with Weir for Liquid Vibration Suppression

Atsushi Ito[1], Ryosuke Tasaki[1], Makio Suzuki[2], Kazuhiro Ohta[2] and Kazuhiko Terashima[1]

[1] Department of Mechanical Engineering, Toyohashi University of Technology, 1-1, Hibarigaoka, Tempaku, Toyohashi, Aichi, 441-8580, Japan
[2] Sintokogio, Ltd, 3-1, Honohara, Toyokawa, Aichi, 442-8505, Japan

This paper presents analysis results of molten metal vibration for the teapot ladle of an automatic pouring system and proposes suppression control during ladle tilting. First, by means of CFD simulations, we analyzed responses of molten metal behavior using the teapot ladle. In this analysis, trapezoidal shape input was designed as back tilting velocity and applied to simulations. As a result, molten metal in a teapot ladle showed not only sloshing but also flow oscillation beneath the weir. Secondly, it was assumed that two vibrations were equivalent to a second-order lag system. Characteristic parameters were identified from residual vibrations and control input was designed based on its process parameters using the Input Shaping control approach. The effectiveness of the proposed suppression control input was evaluated by comparison with the non-controlled case using CFD simulations. The results provide valuable information that can be used to reduce casting defects caused by the pouring process.

Keywords: *automatic pouring system, teapot ladle, filling accuracy, liquid vibration control, Input Shaping control*

1. Introduction

Tilting-type automatic pouring systems are automation equipment developed to enhance safety in factories and improve yields of cast products [1]. One solution is to drive ladle quickly during transferring or pouring in casting processes, e.g., transferring the ladle from a melting furnace to molds and tilting during pouring of molten metal. It is important to suppress sloshing, i.e., surface vibration in the liquid container.

The literature includes many achievements concerning analysis or sloshing suppression [2]-[4]. Of course, sloshing suppression control approaches are also applied for the pouring process of casting. Some achievements have been applied to back-tilting motion for cutting off pouring [5]. The cut off process has a deep relationship with the accuracy of filling volume into molds, and faster cutting is important to shorten a cycle time [6].

Previous studies, however, show suppression control approaches for a ladle with a simple shape. Therefore, those approaches have not yet been applied to a teapot ladle; there is a weir inside the ladle, which separates body and nozzle sides [7]. To our knowledge, there are no proposals of suppression control of liquid vibration using a liquid container with a weir in the literature. However, this problem is deeply related to filling accuracy of molten metal in a mold. Therefore, this paper presents vibration analysis of molten metal for a teapot ladle and design of the suppression control input for back-tilting motion.

2. Liquid vibration suppression for teapot ladle

The teapot ladle shown in Fig. 1 was assumed as the target of vibration suppression control. It is an example of the teapot ladles used in the tilting-type automatic pouring process. The ladle is constructed of two parts, namely, a cylindrical body part and a thin part extending from the body to the nozzle.

Behavior of molten metal in the ladle was simulated using computational fluid dynamics (CFD) analysis. In this work, the CFD application used was FLOW-3D ® (Flow Science Inc.) [8]. Casting iron was chosen as the target molten metal.

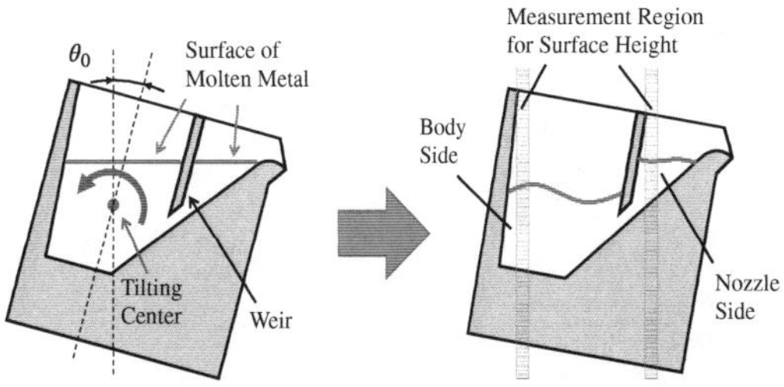

(a) Before tilting (b) After tilting

Fig. 1 Simulation condition for teapot ladle

2.1 Analysis result by trapezoidal input

At first, backward tilting motion for cut-off pouring was simulated using the reference input; the ladle was tilted to a target angle from an initial posture θ_0. In this paper, θ_0 was 15 deg. Before backward tilting, as shown in Fig. 1(a), the ladle was filled with molten metal without any flow outside the ladle. The reference input of backward tilting was assumed as a trapezoidal pattern. In this work, the reference input was assumed as follows: maximum velocity -4deg/s, maximum acceleration ±11.05 deg/s^2, and target tilting angle -5 deg.

Residual vibration of the liquid surface was measured at points close to the wall of the ladle, as shown in Fig. 1(b). FFT analysis of residual vibration revealed that vibration responses had two peak frequencies: 1.15 Hz and 0.34 Hz. 1.15 Hz vibration was first order-mode sloshing. It was mainly induced in the surface of the body side. 0.34 Hz vibration was flow oscillation of molten metal beneath the weir between body and nozzle sides, similar to a U-tube manometer. It was shown that backward tilting motion of the ladle induced not only sloshing but also volume moving vibration when teapot ladle was used.

2.2 Vibration Suppression by Input Shaping control

The Input Shaping approach is one of the vibration suppression controls; amplitudes and phases of multiple inputs are designed and superposed so that all residual vibrations eliminate one another [9]. Shaped input is dependent on natural frequency and damping ratio of vibration response. In this work, in which the Input Shaping approach was used, shaped input was applied to suppress the two vibration modes referred to in the previous subsection.

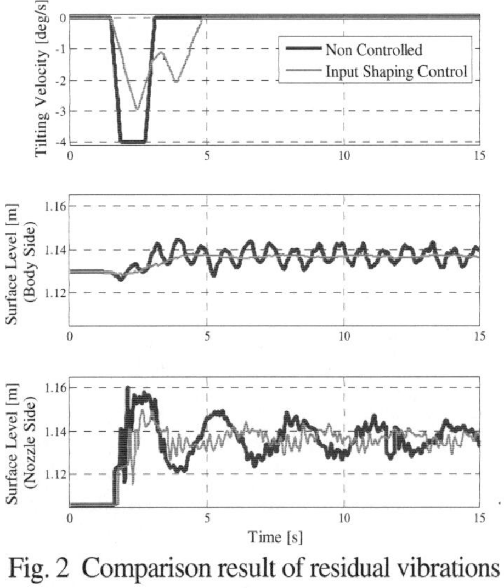

Fig. 2 Comparison result of residual vibrations

The controlled response and the non-controlled one are compared in Fig. 2. This figure compares inputs and their responses between the non-controlled response and the proposed one. The thick blue plot shows the result of the non-controlled response, and the thin magenta plot shows the result of the proposed approach. The proposed velocity input achieved good suppression control for the two vibration modes.

3. Conclusion

This paper presented CFD analysis results for backward tilting motion using a teapot ladle; cut-off pouring induced not only sloshing but also flow oscillation similar to a U-tube manometer. In addition, we applied Input Shaping control to suppress two vibration modes. The proposed control effectivity suppressed the amplitude of residual vibration.

In the next phase of this work, we will endeavor to completely suppress liquid vibrations and to clarify the relativities of a center position of a ladle tilting. Finally, we intend to verify the effectiveness of the proposed approach by experiments using actual pouring systems with molten metal.

References

[1] E. Neuman and D. Trauzeddel: *Foundry Trade Journal*, July (2002) pp. 23-24.

[2] M. Grundelius and B. Bernhardsson: *Proceedings of IEEE international conference on control applications* (1999) pp. 1654-1659.

[3] H. Takahara, S. Tamura, J. Fukuda, Y. Hikita and K. Kimura: *Journal of Environment and Engineering*, Vol. 6, No. 2 (2011) pp. 340-351.

[4] R. Shibuya, Y. Noda, Y. Maeda and K. Terashima: *Journal of Japan Foundry Engineering Society*, Vol. 86, No. 1, (2014) pp. 12-18.

[5] K. Terashima, K. Yano: *IFAC Journal of Control Engineering Practice*, Vol.9, No.6, (2001) pp. 607-620.

[6] Y. Noda, K. Terashima, M. Suzuki and H. Makino: *Preprints of IFAC Workshop on Automation in Mining, Mineral and Metal Processing*, (2009) pp.70-75.

[7] J. Thielke: *ASM Metals Handbook*, Vol. 15, (2002).

[8] C. W. Hirt: *Los Alamos Science Laboratory Report*, LA-5852, (1975).

[9] N. C. Singer and W. P. Seering: *Journal of Dynamic Systems, Measurement, and Control*, Vol. 112, No. 1, (1990) pp. 76-82.

Proceedings of the 72nd World Foundry Congress
21-25th May 2016, Nagoya, Japan

Multi-spiral flow benchmark for small thin-wall castings supported by computer simulation

Himanshu Khandelwal, PhD Research Scholar
K. H. Renukananda, PhD Research Scholar
Dr. B. Ravi, Institute Chair Professor
Mechanical Engineering Department,
Indian Institute of Technology Bombay, Mumbai

Fluidity of molten metal is an important measure of castability, and is useful for optimizing the gating and pouring parameters to ensure fully filled castings. In this work, an improved benchmark is proposed to measure metal fluidity in thin wall castings. It has three spirals of 1 mm, 2 mm and 3 mm thickness; all originating from a central sprue. The fluidity pattern was fabricated in stainless steel, and was used to create sand molds by alkyd urethane no-bake binder system, common in foundry industry today. To demonstrate the use of the proposed benchmark, Al-Si12 alloy was melted and poured at various superheat temperatures to obtain test castings with different extent of filling. The effect of thickness and superheat on metal fluidity are presented, which can be used to determine the optimal pouring temperature for a given range of casting thickness. The same geometry was also used to visualize the flow of molten metal inside the mold using casting simulation software; and the results were compared with experimental values.

Keywords: *Fluidity, thin-wall, simulation, benchmark*

1. Introduction

The increasing need for weight reduction is leading to research in thin wall castings. The main challenge is to ensure molten metal flows in thin sections, avoiding misrun and cold-shut defects [1,2]. The distance of flow of molten metal before it stops by solidification, is called fluidity [3]. Several tests have been proposed to measure fluidity, including: (i) Spiral (ii) Vacuum fluidity (iii) Strip fluidity, and (iv) Various other multi-channel fluidity tests [1,3–6]. Since all these tests are based on a single stream of metal flowing through a single section of constant thickness, they are not very useful to predict flow behavior in industrial parts with branched flow and varying section thickness. Thus there is a need to device a standard test for fluidity to overcome the above limitation, which has been taken up in the present work.

2. Multi-Spiral Fluidity Test

The proposed benchmark is shown in Fig. 1(a). It has three spiral channels of thickness 1 mm, 2 mm and 3 mm. A central common sprue (lower Φ=6 mm, upper Φ=7.5 mm, height=22 mm, pouring basin=15 mm) lets molten metal flow simultaneously through the three channels. Thus branching of flow and fluidity in sections of different thickness can be studied. This benchmark is especially suited to thin wall castings.

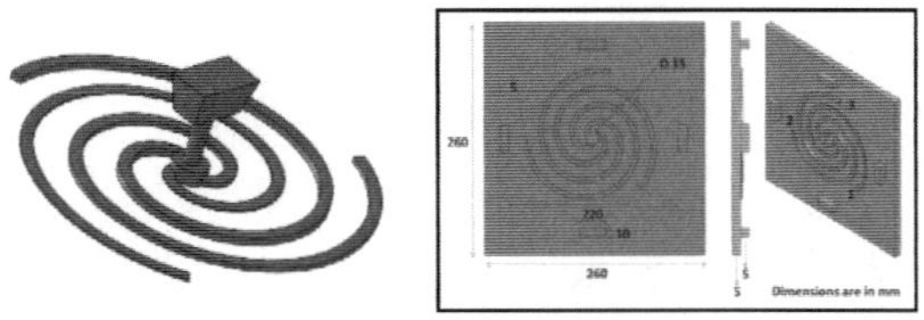

Fig. 1: CAD model of multi-spiral fluidity benchmark, and its match-plate pattern

3. Casting Experiments

To demonstrate the multi-spiral benchmark, a match-plate pattern was designed and fabricated in stainless steel by micro milling (Fig. 1). Silica sand molds (AFS 75 mesh size) were prepared using the pattern and three part no-bake alkyd urethane binder system. AlSi12 (LM6) alloy was melted and poured using a tabletop furnace with bottom opening, minimizing air contact and fall in temperature. Three castings were poured at temperature of 620℃, 660℃ and 700℃, shown in Fig. 2. The experiments were repeated and the average flow length measured for each section of each casting is given in Table 1.

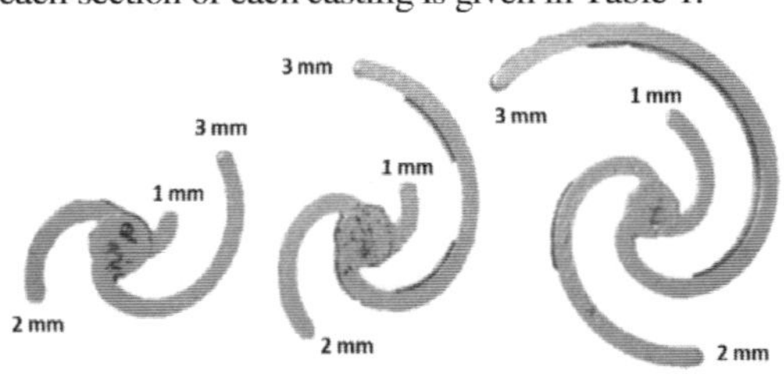

Fig. 2: Multi-spiral castings for fluidity study

Table 1: Results of fluidity experiments

Expt. No.	Pouring temperature (℃)	Flow length (mm) for Section thickness		
		1 mm	2 mm	3 mm
1	620	11	36	61
2	660	26	67	121
3	700	55	116	187

The results clearly show that flow length increases with thickness as well as the pouring temperature. Fig 3 represents the plot of flow length with respect to the pouring temperature for different values of thickness. Fig 4 represents the plot of flow length with respect to thickness for different values of pouring temperature. These plots are useful for two purposes: (i) casting design for manufacturability, in which the section thickness can be optimized for a given set of process parameters (mainly, gating layout and pouring temperature), and (ii) optimizing casting process parameters (gating and pouring) for a given section thickness of cast part.

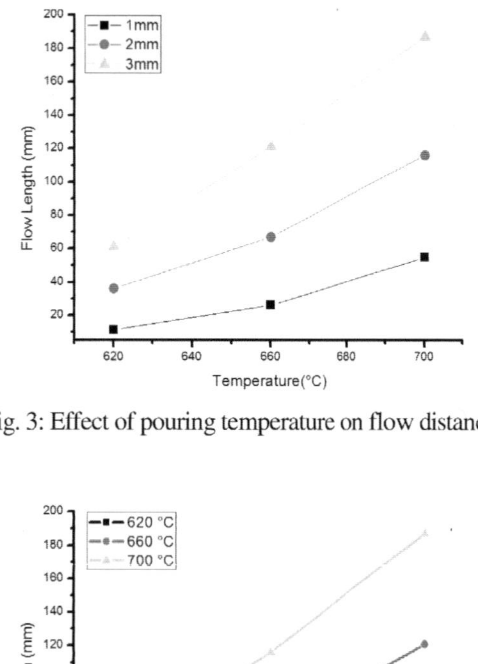

Fig. 3: Effect of pouring temperature on flow distance

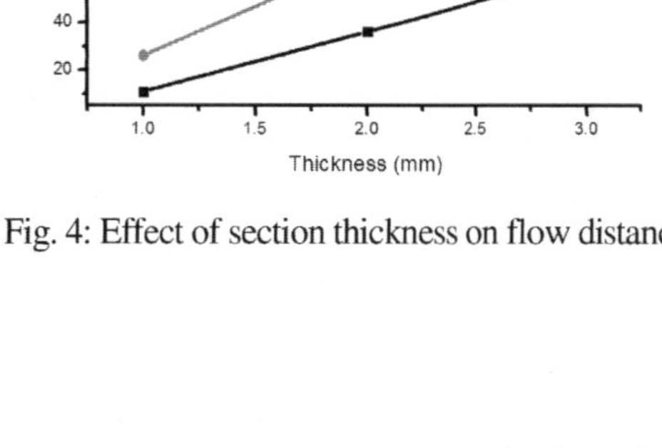

Fig. 4: Effect of section thickness on flow distance

4. Computer Simulation

Numerical simulation of mold filling and casting solidification was performed using FLOW+ module of AutoCAST-X1 software [7]. A mesh size of 0.54 mm was selected, giving 18.79 million cells for the total volume of casting and mold. The simulations were performed for all three pouring temperatures and the results are shown in Fig. 5. The sequence and extent of flow stoppage in simulation matches fairly well with experimental results, thus establishing the usefulness of multi-spiral as benchmark for simulation software.

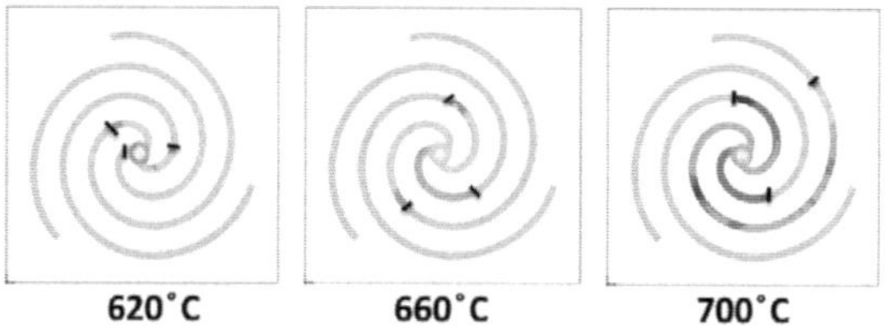

Fig. 5: Effect of section thickness on flow distance

5. Conclusion

The paper presented a multi-spiral benchmark for determining and understanding metal fluidity in thin wall castings, with branching of sections with varying thickness. The flow-length was found to increase with section thickness and pouring temperature; the values of fluidity for a given combination of thickness and temperature are useful for casting design (optimal wall thickness) as well as process optimization (gating and pouring parameters). Further, the benchmark results are also useful for verifying and comparing different software programs for metal casting simulation. The proposed fluidity benchmark can be used to produce castings with different combinations of casting alloy and process parameters, to develop a fluidity database that will be very useful in practice.

References

[1] A.H. Zadeh and J. Campbell: Trans. Foundry Soc. 022 (2003) 115–124.

[2] K.R. Ravi et al.: J. Alloys Compd. 456 (2008) 201–210.

[3] M. Di Sabatino et al.: Int. J. Cast Met. Res. 18 (2005) 59–62.

[4] A.H. Zadeh and J. Campbell: Int. J. Cast Met. Res. 17 (2004) 201–205.

[5] H. Khandelwal and B. Ravi, Indian Foundry J. 7 (2014) 23–29.

[6] M. Górny: J. Iron Steel Res. Int. 19 (2012) 52–59.

[7] AutoCAST, http://www.autocast.co.in, 2015.

Proceedings of the 72nd World Foundry Congress
21-25th May 2016, Nagoya, Japan

Numerical Simulation of Aeration Sand Filling-High Pressure Squeeze Molding Method

ZHANG Qingdong[1], NI Changjiang[1], JING Tao[1], WU Junjiao[1] and HIROYASU Makino[2]
[1] School of Materials Science and Engineering, Tsinghua University, Beijing, China
[2] Sintokogio Ltd, Nagoya, Japan

Aeration sand filling-high pressure squeeze molding method consists of two processes: aeration sand filling process with low pressure and the subsequent high pressure squeeze molding process. For numerical simulation, the aeration sand filling process has been simulated with a self-developed procedure based on Finite Difference Method(FDM), while a commercial software Abaqus based on Finite Element Method(FEM) has been used for the simulation of the high pressure squeeze molding process. Sand flowing state during aeration sand filling process can be seen from the simulation results. The simulation results can be used as the guideline for the production practice in order to produce a high quality casting. In this paper, some cases of 3-D numerical simulation and the correlative benchmark testings are reported in details.

Keywords: *green sand; aeration sand filling-high pressure squeeze molding; numerical simulation.*

1. Introduction

Aeration sand filling is a new sand filling method of green sand molding. For some unique advantages, such as better mold quality, less energy consumption and less pattern wear, aeration sand filling has been successfully applied in compound molding method - aeration sand filling-high pressure squeeze molding and also has been studied carefully [1,2].

The 3-D numerical simulation of aeration sand filling-high pressure squeeze molding process consists of two parts: aeration sand filling and high pressure squeeze molding. The aeration sand filling process is a typical gas-solid two phase flow problem. Different mechanical models, for example, discontinuous model [3, 4], continuous model and distinct cluster model [5], have been used to deal with this problem. The high pressure squeeze molding process is a static sand compaction problem and its simulation can be completed with some FEM software.

2. Numerical Simulation Model

The 3-D numerical simulation model of flaskless horizontal parting aeration sand filling-high pressure squeeze molding process could be divided into two parts: aeration sand filling and high pressure squeezing. The aeration sand filling part is a typical gas-solid two phase flow problem. The high pressure squeezing part is a static sand compaction problem, its simulation can be completed with FEM software.

2.1 The Aeration Sand Filling Part

Aeration sand filling is a typical two-phase flow phenomenon because sand flow and compressed air flow are mutually affected. As the sand phase can be treated as "fluid", the continuous mathematical model is more suitable for its higher accuracy and lower performance requirement for computer. Eulerian approach is used to describe mass, momentum, energy and the interactions.

Due to the limit of pages, the governing equations for aeration sand filling part are not listed in this paper. They were reported in details by Li Hua [1, 6].

2.2 The High Pressure Squeezing Part

As the nonlinear characteristics of molding sand and its large deformation during the high pressure squeezing part, a commercial FEM software Abaqus has be chosen for the simulation of the high pressure squeezing part. The revised Drucker-Prager/Cap model is used for the mechanical model, which has been successfully used in simulation of the squeezing process of soil [7].

The model parameters of the revised Drucker-Prager/Cap model are listed in Table 1. The experimental results reported by Li Hongliang [8] are referenced to get the isotropic compression stress-strain curve.

Table 1 Parameters of the revised Drucker-Prager /Cap model [9]

d	β	R	α	K	Yong's modulus	Poisson ratio
1.6KPa	20.8°	0.1	0.01	1	30MPa	0.35

3. Comparison of Simulation Results and Experimental Results

For the experiment, the step model is used (Figure 1(a)). The air pressure is set to be 0.1MPa during the aeration sand filling process and the squeeze pressure is set to be 1.0 MPa for the high pressure squeeze molding process.

For the simulation, the parameters are set to be the same as in the experiment. To combine the aeration sand filling process with the high pressure squeeze molding process into a complete molding process, the sand bulk density distribution after aeration sand filling has been used as the bridge between aeration sand filling and squeezing [10].

In the experiment, the mold strength of the sand mold is measured on the parting surface and the cut face (Figure 1(b)).

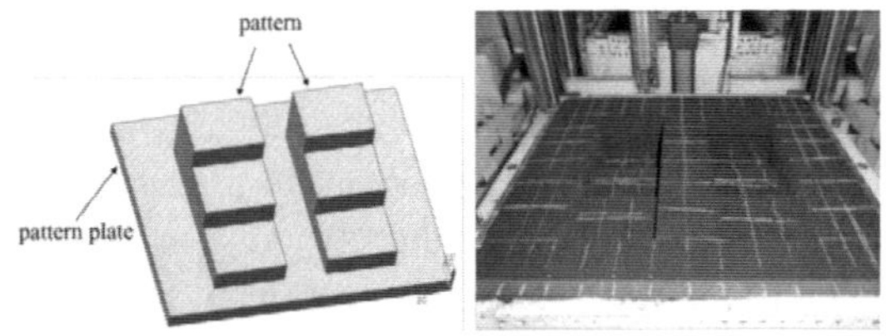

(a) (b)

Fig. 1 (a) The step model, (b) The corresponding drag mold.

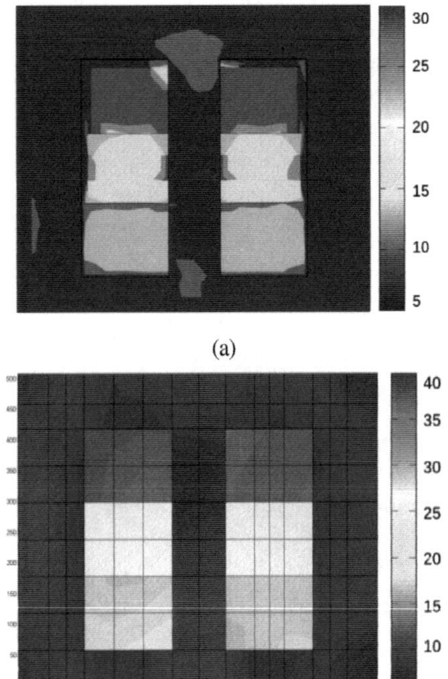

(a)

(b)

Fig. 2 Comparison between the simulation results (a) and experimental results (b) (unit: N/cm^2)

It can be seen from Fig. 2 that the experimental results and the simulation results have the same tendency, the strongest part and the weakest part are the same. When the height of the steps goes higher, the mold strength increases. But the simulation results are about 30% lower than the experimental results.

4. Conclusions

(1) A 3-D numerical simulation procedure for the aeration sand filling-high pressure squeeze molding process has been developed based on the two-phase flow theory and soil mechanics.

(2) The simulation results can give some guidelines for production practice and provide useful information to improve quality of casting. As for the differences between the simulation results and the corresponding experimental results, practical situation and complexity of aeration sand filling-high pressure squeeze molding method should be taken into account carefully.

Acknowledgements

This work was in part supported by Sintokogio Ltd. The experiment has been done by LI Zhonglin and LONG Jiangyou of Tsinghua University.

References

[1] LI Hua, WU Junjiao, HUANG Tianyou, MAKINO Hiroyasu. 3-D Numerical simulation for aeration sand filling process. *The 69th World Foundry Congress* (2010).

[2] M. Hirata, K. Sugita, "New sand filling method in flaskless molding and its controls," AFS Transactions (2005):319-326.

[3] H. Makino, Y. Maeda and H. Nomura, International Journal of Cast Metals Research (1997).

[4] Alexander Barth, "Computer simulation optimizes coremaking operation," Casting Plant & Technology (2008):46-52.

[5] Xiangjun Liu, Xuchang Xu, "Modeling of dense gas-particle flow in a circulation fluidized bed by distinct cluster method," Power Technology (2009):235-244.

[6] LI Hua, "The Experimental Study and 3-D Numerical Simulation on Shoot-Squeeze Molding Method," Doctor of Engineering Dissertation, Tsinghua University, Beijing (2011).

[7] Mizuno E, "Cap Models in Soil Mechanics," *Preprints-Asce Convention & Exposition* (1982).

[8] LI Hongliang, "The experiment study and numerical simulation on low pressure shoot-high pressure squeeze moulding," Doctor of Engineering Dissertation, Tsinghua University, Beijing (2005).

[9] Simulia Corporation, "Abaqus Help Document," Simulia Corporation, France (2006).

[10] ZHANG Yifei, "Experimental Study and Numerical Simulation on Shoot Squeeze Moulding and Air Impact Moulding," Doctor of Engineering Dissertation, Tsinghua University, Beijing (2003).

Proceedings of the 72nd World Foundry Congress
21-25th May 2016, Nagoya, Japan

Impact behaviors of SS400 by spherical and tetrahedral solid particle through Finite Element simulation

Lei Xiao[1], Kazumichi Shimizu[1] and Kenta Kusumoto[1]
[1] Muroran Institute of Technology, Muroran, Hokkaido, Japan, 050-0071

Erosion is a kind of material degradation, which is largely caused by solid particle impacts, so research on impact behaviors by solid particle is very important. In this study, the impact behaviors of SS400 when impact angles ranges from 10deg to 90deg was investigated by Finite Element (FE) simulation. Sphere and tetrahedron were selected as shapes of solid particles. The FE models were established in explicit dynamic software ANSYS/LS-DYNA. Erosion tests were also conducted to verify the FE results. It can be found that the Von Mises stress curve and erosion rate curve under different impact angles have same tendency, and the Von Mises stress of tetrahedral particle impact is much higher than that of spherical particle impact, with highest value occurring when impact angle is around 20~40deg.. The simulation results agrees with erosion tests results.

Keywords: *Erosion, Sphere, tetrahedron, Finite Element (FE) simulation, Erosion test*

1. Introduction

Erosion by solid particles impact has been a serious problem in many industrial fields, such as machinery, metallurgy, energy, chemistry and aviation etc.. Erosion is caused by large amount of solid particles impact on the surface of material, not caused by single particle impact, but single particle impact is fundamental for the research on erosion of material, and has been conducted widely in previous researches [1-3].

In this study, impact behaviors of ductile material SS400 by single spherical and tetrahedral particle impact was investigated by Finite Element (FE) simulation. The 3D FE model including solid particle and workpiece were established in explicit software ANSYS/LS-DYNA. Through analyzing the Von Mises stress after impacting under different impact angles, the impact behaviors of SS400 impacted by spherical and tetrahedral solid particle under different impact angles were studied. Erosion tests were also conducted to verify the FE simulation results.

2. Erosion tests

In erosion tests, spherical and angular steel grits were used as erodent particles, which were shown in Fig. 1. Test conditions were summarized in Table 1. The specimens were weighed using an electronic scale before and after tests to get the mass loss during erosion. Volumetric erosion rate was used to measure the erosion, its formula were given below:

$$E_V = (m_0 - m_1)/\rho M$$

E_V: Erosion rate, (cm^3/kg)
m_0: mass of specimen before test, (kg)
m_1: mass of specimen after test, (kg)
ρ: density of specimen, (kg/cm^3)
M: mass of erodent particles, (kg)

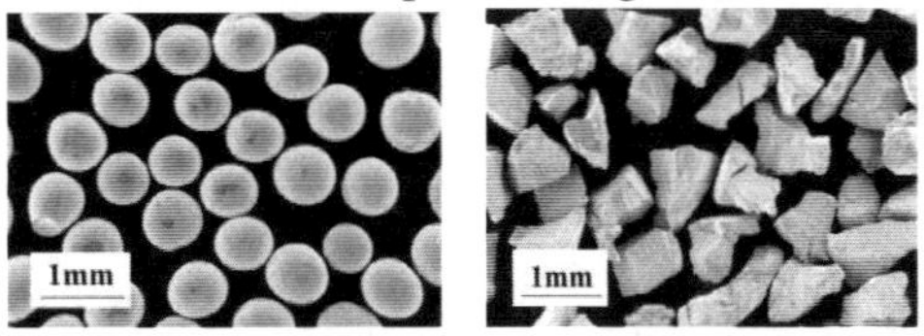

(a) Spherical grits (b) Angular grits
Fig. 1 Appearance of steel grits

Table 1 Erosion test conditions

Test variable	Test condition
Dimensions of specimen	10×10×12.5mm
Velocity of erodent particles	20m/s
Average diameter of steel grits	700μm
Hardness of steel grits	420 HV1
Erosion angle	10, 20, 30, ~90deg.
Test duration	3600s
Test temperature	Room temperature

3. FE simulation

The FE models and its dimension are shown in Fig.2. In simulation, the impact area on workpiece was refined to improve its accuracy of simulation results. Nodes on all the surfaces of workpiece except upper surface were fixed, solid particle impacted on the upper surface of workpiece.

Bilinear Kinematic Hardening was selected as the constitutive model of SS400. The solid particle is rather harder than workpiece, so it was set as rigid body to improve the computing speed. The material parameters are listed in Table 2. The friction coefficient between solid particle and workpiece was set as 0.1. The initial velocity of solid particle is 20m/s. The impact angle θ ranged from 10deg to 90deg in an interval of 10deg. The simulation time of sphere impact and tetrahedron impact is set as 0.01ms and 0.05ms respetively.

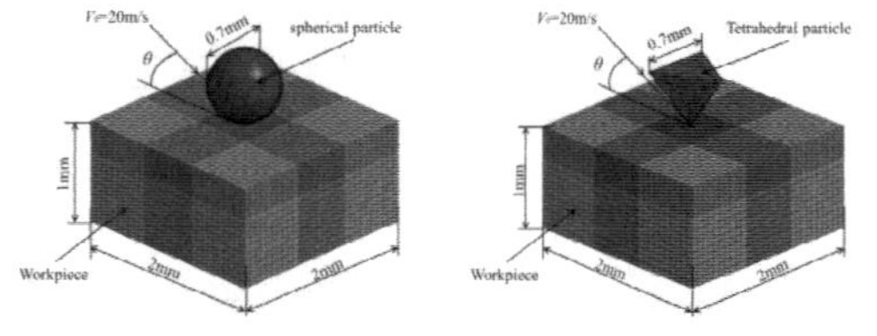

(a) Sphere impact (b) Tetrahedron impact

Fig. 2 FE models and dimension

Table 2 Material Parameters

Parameter	Workpiece (SS400)	Particle (Rigid body)
Density(kg/m^3)	8000	8000
Young's Modulus(GPa)	210	210
Possion ratio	0.3	0.3
Yield stress(GPa)	0.2	—
Tangent Modulus(GPa)	0.2	—
Hardening parameter	1	—

4. Results and discussion

Von Mises stress of FE simulation under different impact angles are shown in Fig. 3, it can be seen that the Von Mises stress of tetrahedral particle impact is much higher than that of spherical particle impact. That is because in tetrahedral particle impact, the solid particle impacts on the workpiece with its angular point, so kinetic energy of tetrahedral particle is concentrated on a very small impact area, however in spherical particle impact, the impact area is a part of spherical surface, which is much larger than impact area of tetrahedral particle impact.

Stress is a cause to make erosion occur, the higher stress, the more erosion, therefore erosion of angular particle impact is more than that of spherical particle impact. The erosion rate under different impact angles are shown in Fig. 4.

The curves tendencies of Fig. 3 and Fig. 4 are roughly same, the curves of angular/tetraheral particles impact show much more changeable than those of spherical particles impact. The highest value of all curves occur around 20~40deg.

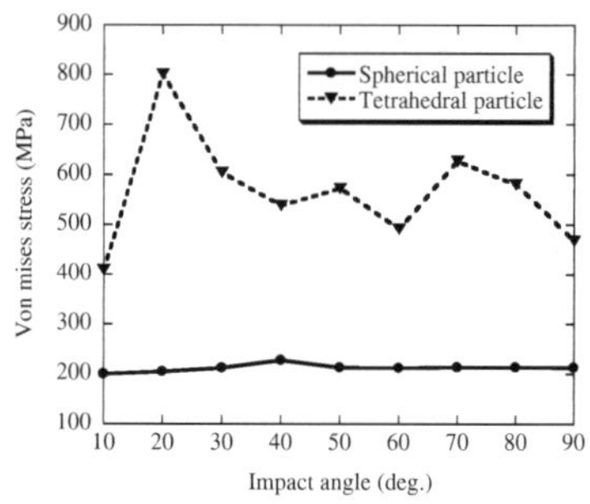

Fig. 3 Result of FE simulation

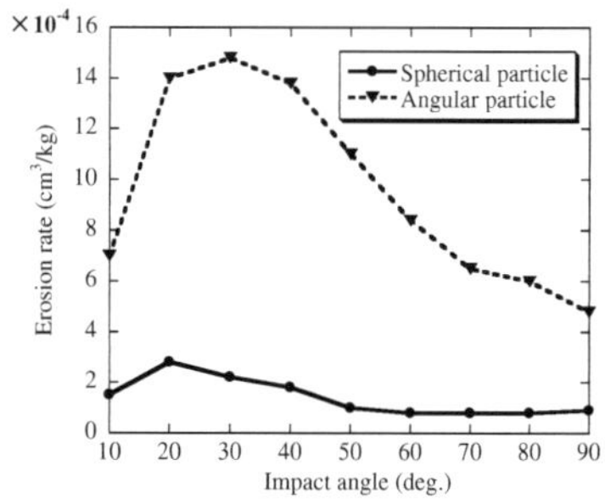

Fig. 4 Result of erosion tests

The difference between erosion tests and FE simulation comes from following reasons: (1)In erosion test, workpiece was impacted by large number of erodent particles, but in FE simulation, workpiece was impacted by single solid particle. (2)In erosion test, the shapes of angular particles was diverse, but in FE simulation, the shape of angular particle was simplified as regular tetrahedron.

5. Conclusions

The impact behaviors of SS400 impacted by spherical and tetrahedral particles were studied through FE simulation. The Von mises stress of tetraheral particles impact is higher than that of spherical particles impact, with highest Von Mises stress occurring when impact angle is around 20~40deg. The simulation results agrees with erosion tests results.

References

[1] K. Shimizu, T. Noguchi, H. Seitoh, E. Murakami: Wear. 233 (1999) 157-159.

[2] K. Shimizu, T. Noguchi, H. Seitoh, M. Okada, Y. Matsubara: Wear. 250 (2001) 779-784.

[3] K. Shimizu, T. Noguchi, M. Okada: Journal of Japan Foundry Society. 73 (2001) 362-366.

Proceedings of the 72nd World Foundry Congress
21-25th May 2016, Nagoya, Japan

FEM analysis of single impacts of spherical particles on SUS630 stainless steel at elevated temperature

Kenta Kusumoto[1], Kazumichi Shimizu[1]

[1]Muroran Institute of Technology, 27-1, Mizumoto-cho, Muroran-city, Hokkaido, Japan.

A surface damage that caused by impact of solid particles is called erosive wear. High temperature erosive wear is one of the main failures of industrial plants. This study investigated that impact angle dependently on high temperature erosive wear by Finite-element-method (FEM). LS-DYNA suitable for contact problems was used in this study. As a result of focused on equivalent plastic strain and Mises stress, the more increase of temperature, the higher equivalent plastic strain and the more increase of impact angle dependency. Therefore, it's possible to test the efficacy of FEM analysis of impact angle dependency for high temperature erosive wear.

Keywords: *erosive wear, impact angle dependency, equivalent plastic strain, mises stress.*

1. Introduction

Wear is divided into adhesive wear, corrosive wear and abrasive wear [1]. Abrasive wear refers to a solid particles causing damage, which can be subdivided by the type of wear. Among them, a surface damage that caused by impact of solid particles is called erosion or erosive wear. The phenomenon becomes a serious problem for bended section of pneumatic transportation pipe, valve, turbine blade and fun, and so on. The dominant factors that influence erosion have been reported by many researchers that there have features of erodent, impact angle, impact speed, and mechanical properties of target material, e. g. hardness, microstructure, rigidity, young's rate, etc.. Many researchers reported that erosive wear of materials is greatly influenced by erodent features[2]. Shimizu and his coworkers investigated the dependency of impact angle on erosive wear based on 2D Finite element model[3], the results uncovered the variation of plastic deformation with respect to different impact angles. However, because the Finite element model was established in 2D axial symmetry way, the particle was simplified as cylinder instead of sphere, so it was difficult to get the stress distribution.

This study investigated impact angle dependently on erosive wear by Finite-element-method (FEM). LS-DYNA suitable for contact problems was used in this study.

2. Experimental procedure

2.1 Analysis method

Analysis conditions are shown below of specimens. Impact velocity of the particles was 20m/s. The contact time of the collision is 0.01ms. The dimensions of the impact material are 10 × 10 × 10mm. Test specimen was SUS630 stainless steel. The erodent particle used in this study was a 1mm alumina ball. Impact angle was analyzed from 10deg. to 90deg..

The particle was assumed to be rigid. However, since a prescription of the elastic material properties was needed for the contact algorithm, the elastic constants of SUS stainless steel (according to the results of tensile test at each temperature) were used to define the particle material. Table 1 shows the analysis conditions of the material.

Table 1 Analysis conditions of the material.

Temperature, K	Mass density, kg/mm^3	Young's modulus, GPa	Poisson ratio	Yield stress, MPa
R.T.	7.74	300	0.272	950
573		250	0.272	820
873		210	0.300	510
1173		72.3	0.320	110

2.2 Experimental method

In the present research, to understand the erosive wear in the region of high temperature, a new high temperature erosion test machine was manufactured.

A high temperature erosion machine was used to test the single impact and erosive wear of target materials in the present study. Spherical shaped erodent particles (alumina ball) were used, they are spherical shaped steel grits with average diameter 1.16mm, Vickers hardness 1140HV. The examined air speed was 100m/s, and impact angle ranged 30, 60 and 90 deg. respectively. Test Pieces size were 10×10×10mm and 50×50×12.5mm. All the erosion tests were conducted at sevaral temperature. Before and after the test the amounts of specimens were weighed with an electronic scale and then calculate volumetric loss. Erosion rate refers to a value calculated from the volumetric loss divided by total feed of erodent. After the single impact test, the indentation of the material surface was observed by microscope._Impression of depth, mises stress and equivalent plastic strain were analyzed. For verification of the analysis result, a single impact experiments were conducted. The Correlation was obtained in result of experiment and analysis.

3. Results and discussion

The results of a single particle impact test and the finite element analysis were compared. As a result, they showed a good correlation (coefficient of correlation value was 0.96~0.99 elevated temperature). Therefore, it's determined that the analysis result is reasonable.

Then, shear strain energy and Mises stress of the material surface has been focused, impact angle dependency of erosive wear of SUS630 was verified.

Results of erosive wear tests were shown in Fig. 1. Mises stress and erosion rates were the largest at low angle side. Similar trends were observed in the ratio of Mises stress and impact angle dependence of the erosive wear test. Also, it showed the same tendency even at room temperature, 573K and 873K. Since a good correlation has been obtained, the calculation of the equivalent plastic strain of the material surface was performed. The result is shown in Fig. 2. From Fig. 1 and Fig. 2, erosion rate and the amount of equivalent plastic strain increased at a low angle side. Also, with increasing test temperatures, the amount of equivalent plastic strain increased. From these, result of experimental and analysis were similar trend. From these results, erosion rate and impact angle dependency were closely related to the plastic deformation of the material surface. It's possible to verify the impact angle dependence of erosive wear at high temperature by performing a single impact analysis considering mechanical properties.

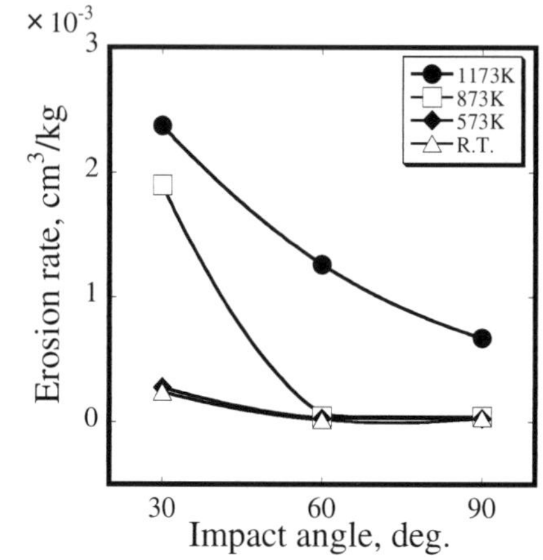

Fig. 1 Experimental result.

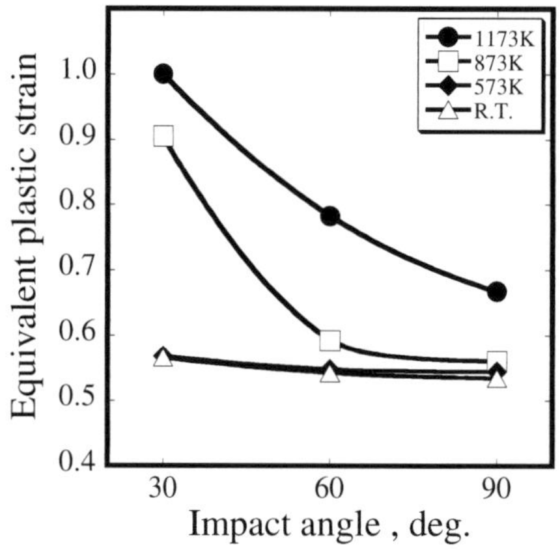

Fig. 2 3D FEM analysis result.

4. Conclusion

In this study, it was to clarify the relationship between the impact angle dependence of erosive wear and the mechanical properties at high temperature.

(1) An impact angle dependency at high temperature environment is mechanically verifiable by performing a single impact FEM analysis.

(2) With increasing temperature, the ratio of the Mises stress and amount of equivalent plastic strain wear increased. These are the factors that erosion rate increases at the low angle side at elevated temperature.

Reference

[1] Japan society of corrosion engineering: Erosion and Corrosion (1987) 6.

[2] K. Shimizu, T. Noguchi, T. Kamata, H. Takasaki: Transactions of AFS 104 (1996) 511.

[3] K. Shimizu, T. Noguchi, H. Seitoh, M. Okada, Y. Matsubara: Wear 250 (2001) 779.

Proceedings of the 72nd World Foundry Congress
21-25th May 2016, Nagoya, Japan

Production of thin walled compacted graphite iron castings using different moulding materials

Marcin Górny[1], Rafał Dańko[2], Janusz Lelito[3], Magdalena Kawalec[4] and Gabriela Sikora[5]
[1,2,3,4,5] AGH-University of Science and Technology, Reymonta Str. 23, 30-065 Krakow, Poland

This article addresses the effects of six mould materials in obtaining thin walled compacted graphite iron castings with a wall thickness of 3 mm. During this research, the following materials were analyzed: fine silica sand, coarse silica sand, cerabeads, molohite and also insulated materials in the shape of spheres; including LDASC sand (Low density alumina-silica ceramic). Their thermophysical characteristics were examined, including their thermal conductivity, diffusivity and specific heat as well as their granulometric parameters. In addition, mould materials were examined by a Scanning Electron Microscope. Microstructural changes of TWCGI were evaluated by analyzing quantitative data sets obtained by an image analyzer. Moreover, the temperature profile in mould materials was determined during mould filling and metal cooling. The cooling rate of metal in the mould cavity could be effectively reduced by applying insulating sands to obtain the desired compacted graphite fraction and microstructure of thin walled castings.

Keywords: *Compacted graphite cast iron, thermal conductivity, graphite morphology, mould materials.*

1. Introduction

Compacted graphite cast iron (CGI) is an alloy with attractive mechanical and thermophysical properties [1-5]. Moulding materials play an important role [6-9] in shaping the structure of thin walled castings. The main component of these moulding materials is the moulding sand matrix-mostly quartz and, in the case of the regeneration process, the reclaimed sand. In published literature, there is limited data [3,10] concerning production and the shaping of the structure of thin walled compacted graphite iron castings using different moulding sand matrixes.

2. Experimental

The chemical composition of the investigated iron was: C = 3.60%, Si = 2.65%, Mn = 0.02%, P = 0.02%, S = 0.01%, Mg = 0.02%. The pouring temperature was approximately 1400°C. Experimental castings applied in investigations were test blocks, according to the ASTM A 536-84 Standard, with plate section sizes of g=3 mm at the bottom end.

Six casting moulds were made of moulding sands marked with symbols: **S-1** through **S-6** (**S 1** – fine silica sand, **S 2** – coarse silica sand, **S 3**– microspheres, **S 4** – Cerabeads, **S 5** – LDSC sand and **S 6** – Molohite). All tested moulding sands were prepared with a water glass M145 binder with flodur hardener.

3. Results and discussion

3.1 Granulometry and SEM-EDS analysis

Granulometry and SEM-EDS analysis indicate that the smallest value of the average particle diameter, and the largest surface area were attained for S 5 sand. SEM studies showed that most of the sand grains from an S 5 and S 3 population have a regular - spherical shape. Moreover, from SEM observations follows that both sands, namely S 3 and S 5 represents spheres, which are empty inside.

3.2 Thermal analysis

On the basis of the actual temperature curves, it follows that the insulating molds (S 3, S 5) have favorable thermophysical parameters from the point of view of heat accumulation. This manifests itself in obtaining a much higher maximum temperature at the selected measuring points. The result of this would be a much lower cooling rate of metal in the mould cavity.

3.3 Thermal Conductivity

The behaviour of mould based on coarse silica sand (S 2) indicates high instability of its thermophysical properties as compared to other moulds. In the case of a mould–based on fine silica sand (S 1) a similar instability, though to a much lesser degree, is visible. High values of thermal conductivity of S 4 leads to low thermal resistance (defined as the ratio of the mould thickness to thermal conductivity), thereby favouring a higher cooling rate of the metal in this mould cavity. However, in the case of insulation moulds (S 3 and S 5), the highest values of thermal resistance (approximately 6 times higher compared to cerabeads, S 4) was attained, which significantly slows down the cooling rate of casting.

3.4 As-cast Microstructure

Figure 1 shows the exhibited microstructures found in thin walled castings obtained in different moulds.

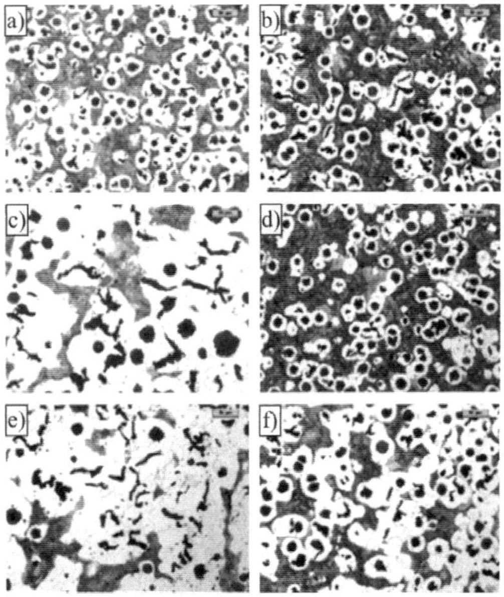

Fig. 1 The exhibited cast iron microstructures for castings with a wall thickness of 3 mm received in moulds: a) S 1, b) S 2, c) S 3, d) S 4, e) S 5 and e) S 6. Nital etchant used, mag. 200x

Thermophysical properties of moulding materials clearly show that the use of insulation sands (S 3, S 5) significantly affect values of thermal resistance, which slow down the cooling rate of metal in the mould cavity significantly. It has been reported [10] that when LDASC sand is used as a casting mould instead of silica sand, the cooling rates in thin walled castings with a 2 mm wall thickness decrease 5 fold. Compacted graphite as well as ferrite fractions are effectively increased, by employing insulating moulds (S 3, S 5).

It is recommended that this research be continued in terms of the heat transfer, reactions on the interface of the metal-mould and finally the surface quality. This data would assist in the effort made in the validation of the production of thin walled castings, where high dimensional accuracy of moulds is required. Moreover, as a result of large changes in the cooling rate in thin wall castings -may appear chill, misrun and cold laps.

4. Conclusions

Thermal data indicates high instability of thermophysical properties of moulds based on coarse silica sand. In turn, thermophysical properties of insulating sand (S 3, S 5) show high stability of thermophysical properties and low thermal conductivity (high thermal resistance) in the entire investigated temperature range. Studies have shown that moulds made with insulating sands possess thermal conductivity on average to be more than five times lower compared to those of silica sand. Metallographic examinations of thin-walled castings indicate a significant impact of various moulding materials in their microstructural formation. When using moulding materials S 3, S 5, ie. insulating moulds, it is much easier to get higher compacted graphite as well as ferrite fractions.

Acknowledgements

This work was supported by Polish NCN project no. 2013/09/B/ST8/00210.

References

[1] W. Guesser, T. Schroeder, S. Dawson: AFS Trans. 109 (2001) 1-11.

[2] D. Holmgren, R. Kallbom, I.L. Svensson: Metall. Mater. Trans. A 38(2) (2007) 268-275.

[3] M. Górny, J. Lelito, M. Kawalec, G. Sikora: ISIJ International 55(9) (2015) 1925-1931.

[4] E. Guzik: Archives of Foundry Engineering 10(3) (2010) 95-100.

[5] L. Sofroni, I. Riposan and I. Chria: Proceedings of the 2nd International Symposium on the Metallurgy of Cast Iron, Geneva (1974) 179-196.

[6] R.E. Showman, R.C. Aufderheide: AFS Trans. 112 (2004) 823-830.

[7] R. Danko, M. Holtzer: Archives of Metallurgy and Materials 55(3) (2010) 787-794.

[8] T. Midea, J.V. Shah: AFS Trans. 110 (2002) 121-136.

[9] H. Wolff, S. Engler, A. Schrey, G. Wolf: Advanced Engineering Materials 5(1-2) (2003) 55-58.

[10] M. Górny, M. Kawalec, G. Sikora and H. Lopez: ISIJ International 54(10) (2014) 2288-2293.

Proceedings of the 72nd World Foundry Congress
21-25th May 2016, Nagoya, Japan

Application of Rare Earth Less and Rare Earth Reduced Spherodizer

Yusuke Kiguchi[1], Takahiro Sato[2]
[1] Head Plant Engineering Dept., Aisin Takaoka Co., Ltd.
[2] Casting Production Engineering Dept., Aisin Takaoka Co., Ltd.
1 Tennoh, Takaokashin-machi, Toyota, Aichi, 473-8501 Japan

The reduction in rare earth (hereinafter referred to as "RE") exports has created a crisis situation where it is difficult to procure spherodizers (Fe-Si-Mg-RE). Since REs contribute to fading suppression and the increase in the number of graphite grains [1-3], we used RE-less and RE-reduced spherodizers containing Ba as the substitute for RE, because Ba has higher oxygen affinity than Mg, and combined inoculants in order to confirm material properties, chill, and internal defects. RE-less spherodizers had satisfactory material properties, but chill and internal defects worsened. RE-reduced spherodizers exhibited similar material properties to the mass-produced spherodizers. Since RE-less spherodizers demonstrated large supercooling and a short eutectic solidification time from recalescence, we deduced that the RE-less spherodizers had a low number of graphite nuclei.

***Keywords**: Spheroidal graphite cast iron, spherodizer, rare earth ,fading, number of graphite grains*

1. Introduction

In 2010, the imports of RE elements became less, making it difficult to secure spherodizer (Fe-Si-Mg-RE). Therefore, studies were conducted on the RE-less and RE-reduced spherodizer.

2. Experimental procedure

2.1 Verifying the effects of RE

As a preliminary experiment to confirm the fading suppression effect of RE, the following process was performed: The return materials, steel scraps, graphite electrodes, and Fe-75% Si alloy were melted in a 100 kg high-frequency melting furnace at 1550 °C in order to produce sample material with the final composition of 3.6% C-2.7% Si-0.2% Mn. Then, the sandwich method was applied to treat various spherodizers (RE = 0%, 0.5%, 1.0%, and 1.6%), and the sample material was poured into a 20-mm-diameter round bar at 1-minute intervals.

2.2 Studies on the application of RE-less and RE-reduced spherodizers

Ba was used as an RE replacement element of Mg because it had higher oxygen affinity than Mg. A spherodizer (Fe-45% Si-4.5% Mg-2.6% Ca-0.5% Ba) was used. Ba- and Zr-based primary inoculants and Zr-, S-, Fe-Si-, and Ba-based secondary inoculants were combined. The test specimens and products were prepared using the above materials. The materials were cast into 20-mm-diameter round bar, step bar, and Y-shaped test specimens, and products 15 minutes after performing the treatment to check the nodularity, the number of graphite grains, mechanical properties, chill, and internal defects. The same process was performed for the RE-reduced spherodizer.

3. Results

3.1 Results of confirming RE effects

Fig. 1 shows the relationship between the elapsed time after treatment and the nodularity of various spherodizers. As the amount of RE decreases, the nodularity degrades faster. If a spherodizer contains Ba, however, the spherodizer containing 0.6% RE achieves a nodularity that is equivalent to that of the mass-produced spherodizer containing 1.6% RE.

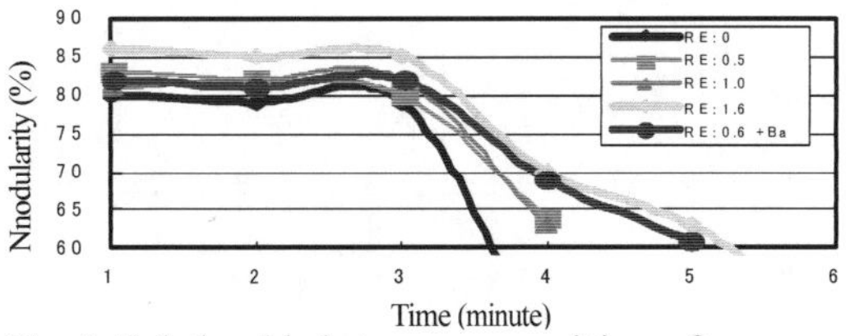

Fig. 1 Relationship between elapsed time after treatment and nodularity

3.2 Studies on the application of RE-less and RE-reduced spherodizers

1) Results of confirming the test specimens

Fig. 2 show the relationship between various inoculants and nodularity of 20-mm-diameter round bars. According to Fig. 2, the nodularity of spherodizers without secondary inoculation deteriorated to 70% or less while the nodularity of all spherodizers with secondary inoculation resulted in at least 80%. The number of graphite grains in the mass-produced spherodizers did not vary with wall thickness. In contrast, the number of graphite grains in the RE-less spherodizers containing the combined inoculants tested in this study decreased as the wall thickness increased. The mechanical properties of the RE-less spherodizers were not much different compared with ass-produced spherodizers. The material properties of the RE-less spherodizers were equivalent to those of conventional products due to the secondary inoculation. On checking chill using the 15 mm step bar test specimens, we found that chill was large in the absence of secondary inoculation. Even with secondary inoculation, chill worsened depending on certain combinations of the inoculants. Regardless of the conditions, the internal defects worsened considerably.

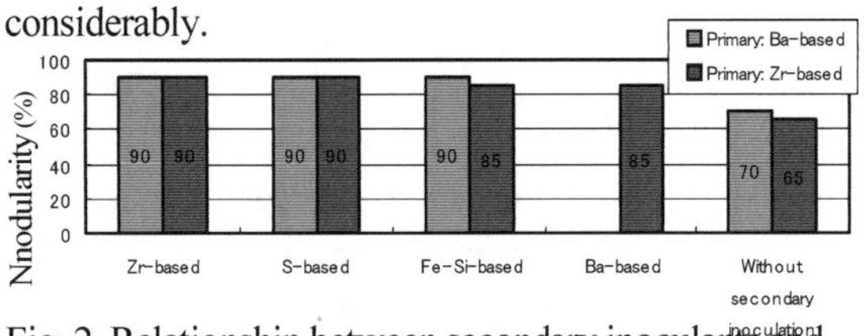

Fig. 2 Relationship between secondary inoculants and nodularity (20-mm-diameter round bar)

2) Results of confirming the products

We checked our company's main product, a differential case made of FCD600 material, which is characterized by large variations in wall thickness. Fig. 3 shows microstructure photos of this product.

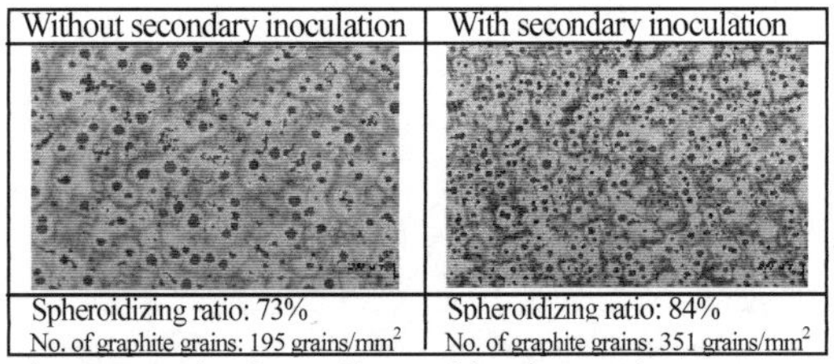

Fig. 3 Result of metallographic observation

Based on Fig. 3, we conclude that the RE-less spherodizer with secondary inoculation possesses satisfactory material properties. However, internal defects were increased. Chill was noted in the thin portion of this product regardless of the conditions. Based on the above, we determined that application of the RE-less spherodizer is difficult in terms of the casting quality. The spherodizer containing 0.8% RE and Ba-based secondary inoculant showed satisfactory casting quality.

4. Discussion

Internal defects and chill worsened when the RE-less spherodizers were used. Thus, it can be inferred that the solidified form of the spherodizers changes. Fig. 4 shows the results of a thermal analysis of mass-produced, RE-reduced and RE-less spherodizers.The spherodizers vary in their solidified form depending on whether RE is present or not. Supercooling caused rapid graphite nucleation in the spherodizers with 1.6% RE and 0.8% RE. It is inferred that a large number of graphite nuclei are present at the initial stage of eutectic solidification. On the other hand, recalescence of the RE-less spherodizers from supercooling is gentle. It is presumed that the casting quality deteriorates because of the low number of graphite nuclei at the initial stage of eutectic solidification.

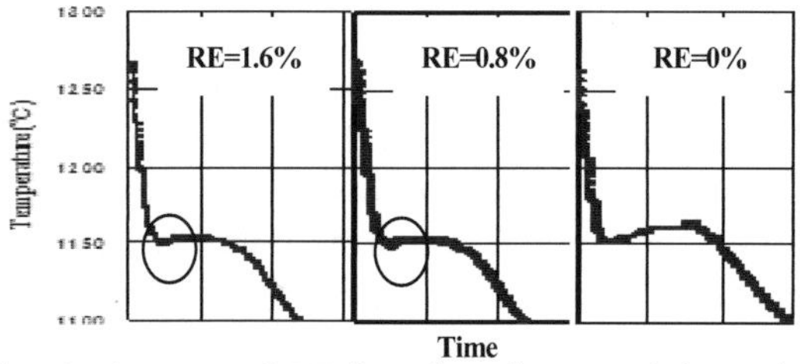

Fig. 4 Amount of RE in spherodizers and thermal analysis curves

5. Conclusions

We evaluated the microstructure, material properties, mechanical properties, internal defects, and chill of the RE-less and RE-reduced spherodizers using test specimens and products in an effort to reduce the amount of RE contained in spherodizers.

(1) The RE-less spherodizer suppressed fading with the addition of Ba and the secondary inoculation, showing satisfactory material properties. We confirmed, however, that spherodizers had differences in their solidified form, and internal defects and chill worsened.

(2) The solidified form of the RE-reduced spherodizer was similar to that of the mass-produced spherodizer. The internal defects and chill were significantly improved by secondary inoculation.

References

[1] T. Kusakawa, H. Lin, X. Xu: Report of the Castings Research Laboratory, Waseda University. (1985) 29-34

[2] W. A. Henning: Transactions of the American Foundry Society, 91 (1983) 71-76

[3] T. Kimura, C. R. Loper, Jr, H. H. Cornell: Transactions of the American Foundry Society, 89 (1981) 395-368

Proceedings of the 72nd World Foundry Congress
21-25th May 2016, Nagoya, Japan

Influence on the Fatigue Strength at Rare Earth content in the Spheroidizing agent

Shinsuke Nagaumi [1], Kazumichi Shimizu [1], Takafumi Funabiki [1], Kaito Todate [1] and Masato Shirai [2]
[1]Muroran Institute of Technology,
27-1, Mizumotocho, Muroran-city, Hokkaido, Japan, 050-8585
[2]Iwamizawa Imono. Co.,Ltd., 497-3, Yura, Kurisawa-cho, Iwamizawa-shi, Hokkaido, Japan, 068-0111

Mechanical properties of the spheroidal graphite cast iron are dependent on the spheroidized graphite. In the field of casting, rare earth (RE) is used in a spherical shape of the graphite. However, domestic production of RE has few, and rely on import. In this experiment, previous studied using a spheroidizing agent unified element other than RE as compared to the fabricated spherical graphite cast iron, RE test piece of spheroidizing agent in by performing a fatigue test the effect on me was investigated. Test piece the RE content of 0%, 0.3%, 0.5%, and with 1.0% of 4 kinds of smooth test pieces were subjected to fatigue tests by a plane bending fatigue tester. The results of fatigue test, fatigue strength was reduced by about 20-33% in the RE0% test pieces were compared with RE-containing test pieces. Result of cross-sectional observation of the fractured specimen by the test, the RE0% test piece inside a large casting defects are confirmed, the fatigue strength is significantly reduced was less affected by 0.3% to 1.0%.

Keywords: *Rare earth, Spheroidal graphite cast iron, Fatigue strength, casting defect, $\sqrt{}$area parameter.*

1. Introduction

Spheroidal graphite cast iron is widely used for automobile parts because it is excellent in mechanical properties, fatigue strength. Spheroidal graphite cast iron is known to graphitization nodularity exhibit good mechanical properties at 80%. Therefore, when producing spheroidal graphite cast iron by using a spheroidizing agent comprising a rare earth(RE) it has been attempted to improve the spheroidal graphite rate. However, RE is less production in Japan, and is dependent on imported from abroad, may become difficult to stably supplied. Therefore, development of a spheroidal graphite cast iron with a spheroidizing agent with reduced RE becomes a challenge.

A problem in the case of reduced RE content in spheroidizing agent is believed to affect the fatigue strength necessary for the strength member by a casting defects caused by thin part. In previous studies, RE content is reported to be a possible reduced to 0.3%. However, spheroidizing agents used are not uniform in composition, it believed that the influence of the elements might have occurred.

In this experiment, we produced a spheroidal graphite cast iron by using a spheroidizing agent unified element other than RE as compared to the previously used test pieces were subjected to fatigue tests. From the results, it was examined the effect of RE is on the test piece.

2. EXPERIMENTAL PROCEDURE

Test pieces shape, it was a flat plate shape of 120 × 40 × 5mm (Fig.1). We went a bending fatigue test. The test conditions were sinusoidal load repetition rate 20Hz. In fatigue test, the number of repetitions reaches 107 times, it was the load stress of unbroken of the test piece and fatigue strength σw. Test pieces fractured by the test, after sufficiently cooled at a temperature lower than the brittle-ductile transition temperature, was brittle fracture shocked. Specimens were separated by disruption was fracture observed using a scanning electron microscope (SEM).

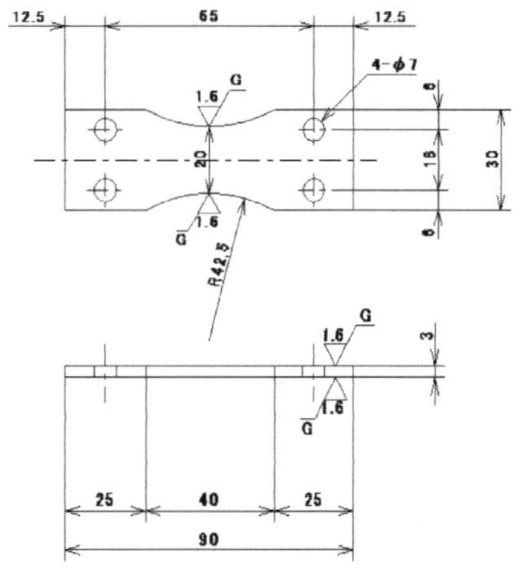

Fig.1 Test piece size

3. RESULTS AND DISCUSSIONS

Result of the tests were shown in Fig. 1. Test piece containing the RE0.3% was 321MPa, 0.5% was 287MPa, 1.0% was 306MPa, and 0% became 240MPa. RE content 0% test piece fatigue strength was reduced about 20 to 33% compared with the specimen containing the RE.

To investigate the cause of difference in the fatigue strength occurs for each test piece, the fracture surface observation was carried out using a scanning electron microscope (SEM). The SEM image of the entire fracture surface is shown in Fig. 3. Since a large casting defects were observed entirely in the fracture surfaces in the RE content of 0% of the test piece, it believed that lower significant fatigue strength occurs. Because the test piece containing the RE could not be confirmed is a large casting defects in the macro observation, we went the micro observation in the SEM. Result of the tests were shown in Fig. 4. Casting defects of about 100μm was confirmed in the test piece containing the RE. Investigate the effects of these casting defects have on the fatigue strength reduction, it was attempted to quantitatively evaluate that. Evaluation method is to organize the magnitude of casting defects of each test piece was evaluated in the stress intensity factor range that represents the strength of the stress field by using a √area parameter method (ΔK) and under critical stress intensity factor range (ΔKth).when ΔK its casting defects larger than ΔKth, it is a possibility to be the origin of the fracture. The results of measurement are shown in Table 1.The results are summarized in √area method, it showed no little difference in the value of ΔK / ΔKth.

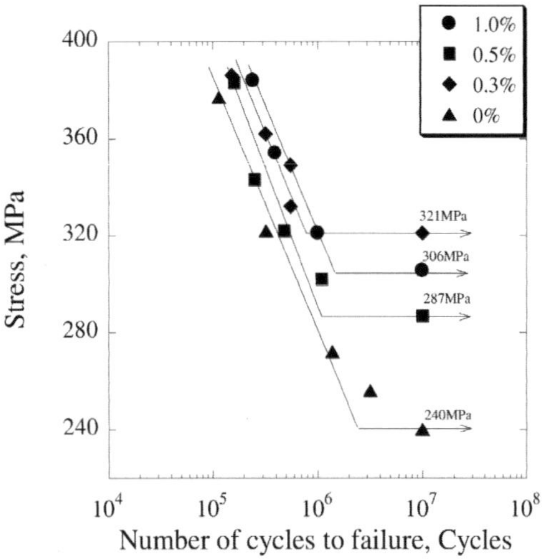

Fig. 2 S-N diagram.

Also, from the properties of the test piece, due to slightly higher nodularity-pearlite ratio and brinell hardness, as compared to other test pieces RE content 0.3% of the specimen is considered to fatigue strength is raised.

4. CONCLUSIONS

We investigated the cause reduced fatigue strength than the results obtained in the test.

1) RE content of 0% of the specimen as compared to the fatigue strength is reduced by about 20-33% and the test pieces containing the RE.
2) By adding the RE content of 0.3% or more of the spheroidizing agent, it can be assumed that it is possible to suppress a reduction in fatigue strength.
3) From the properties of the test piece, due to slightly higher nodularity and pearlite rate or brinell hardness, as compared to other test pieces RE content 0.3% of the specimen is considered to fatigue strength is raised.

References

(1) Nakae Hideo: new edition casting engineering 80, 2008, 149, Sangyo Tosho.

(2) Hideaki Nakayama, ZhaoBai-Rong, Kazuhiko Okada: casting engineering 84, 2012, 695.

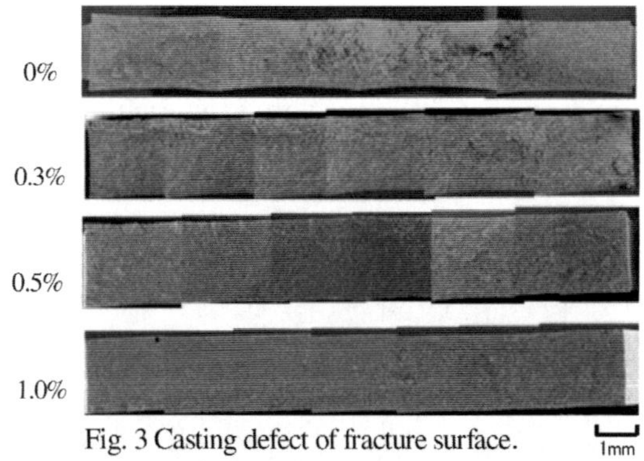

Fig. 3 Casting defect of fracture surface.

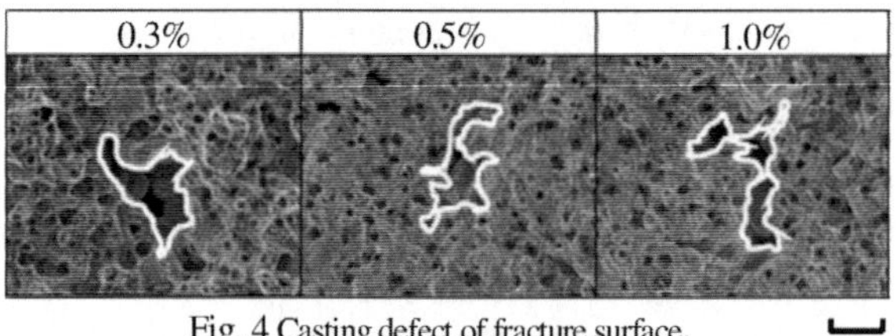

Fig. 4 Casting defect of fracture surface.

Table 1 Result of $\sqrt{area_{max}}$, ΔK and ΔK_{th}.

	Stress [MPa]	$\sqrt{area_{max}}$ [μm]	ΔK [MPa$\sqrt{m}$]	ΔK_{th} [MPa$\sqrt{m}$]	$\Delta K/\Delta K_{th}$
0.3%	332	206	8.45	6.76	1.25
0.5%	322	192	7.88	6.60	1.19
1.0%	321	218	8.39	7.21	1.17

Proceedings of the 72nd World Foundry Congress
21-25th May 2016, Nagoya, Japan

Erosive Wear Characteristics of WC Cast-in Insert Cast Irons on high temperature environment

Shunya Ishikawa[1], Kazumichi Shimizu[2] and Takafumi Funabiki[3]
[1-3]Muroran Institute of Technology,
27-1, Mizumoto-cho, Muroran-city, Hokkaido, Japan, 050-8585

It's revealed that erosion rate increases due to hardness of matrix decreases on high temperature environment. As the countermeasure, this study was focused WC cast-in insert cast irons. This study investigated that erosive wear properties of WC cast-in insert cast irons on high temperature environment by a high temperature erosion test machine. Erosion rate of WC cast-in insert cast irons decreased in relations to base metal. Square shaped WC transverse-mounted 5Nb-5Co cast iron was showed excellent wear resistance. It's considered that hardness of matrix was showed the highest value and the relative hardness was increased by WC.

1. Introduction

Material surface damage due to powder fluid collision called erosive wear. Intense erosive wear in turning shot liner has occurred on high temperature environment. There is a need for a development of materials that exhibit excellent wear resistance. It's revealed that erosion rate increases due to hardness of matrix decreases on high temperature environment. As the countermeasure, this study was focused WC cast-in insert cast irons. WC Cast-in Insert Cast Irons has been reported to decreases of erosive wear properties on high temperature environment. This study investigated that erosive wear properties of WC cast-in insert cast irons on high temperature environment by a high temperature erosion test machine.

2. EXPERIMENTAL PROCEDURE

The test specimens for erosion test were high chromium cast irons (Hi-Cr) and multi-component cast iron (5Nb-5Co). Size of WC are $\phi 5\times 50$, $\phi 8\times 50$, $\square 8\times 50$mm. Erosion test used high temperature erosion test machine in Fig.1. Spherical shaped alumina particles with average diameter 1.16mm, Vickers hardness 1140HV were used in this study. The examined air speed was 100 m/s, and the particle speed was 30 m/s with changing their impingement angle 30, 60 degree respectively. Erosion rate was used to the evaluation of wear resistant.

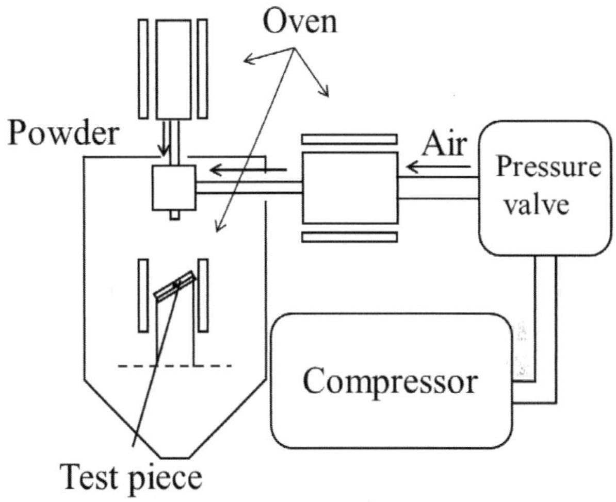

Fig. 1 high temperature erosion test machine

3. RESULTS AND DISCUSSIONS

Figure 2 and Figure 3 shows Erosion rate of WC cast-in insert cast irons decreased in relations to base metal. It's considered that hardness of matrix was showed the highest value and the relative hardness was increased by WC. Also, erosion rate of cylindrical shaped WC transverse-mounted cast iron and square shaped WC transverse-mounted cast iron decreased in relations to WC vertical-mounted cast iron and spherical shaped WC cast iron. It's considered area ratio of WC affected erosion rate. Square shaped WC transverse-mounted 5Nb-5Co cast iron was showed excellent wear resistance. Reaction layer base metal side harder than reaction layer WC side. From the result, It's significant for superior erosive wear resistance to formation reaction layer.

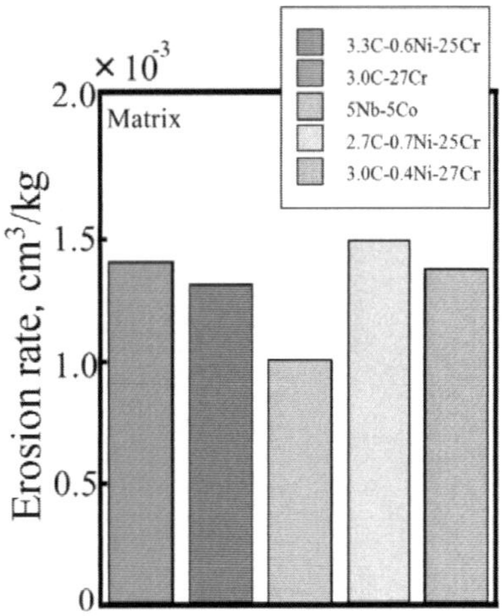

Fig. 2 Erosion rate of matrix.

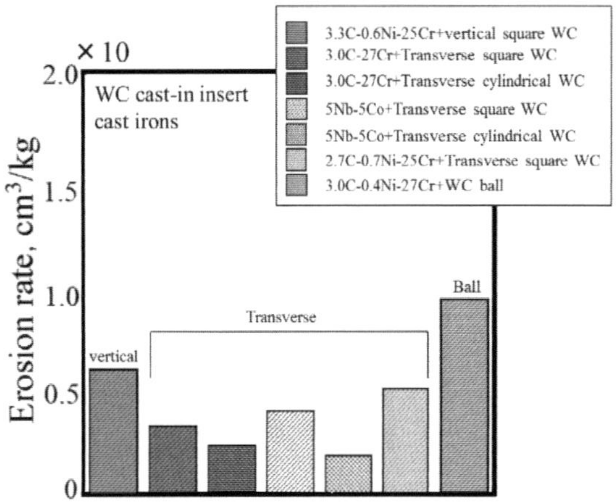

Fig.3 Erosion rate of WC cast-in insert cast irons.

Figure 4 shows Vickers hardness at 873K. Vickers hardness of 5Nb-5Co is higher than that of Hi-Cr. therefore, it's considered that Square shaped WC transverse-mounted 5Nb-5Co superior wear resistance. Also, between and WC base metal was observed. Wear depth of Square shaped WC transverse-mounted deeper than wear depth of cylindrical shaped WC transverse-mounted. Square shaped WC transverse-mounted 5Nb-5Co superior wear resistance. EDS surface analysis was performed detail in order to observe bonded surface of between WC and base metal in figure 5.

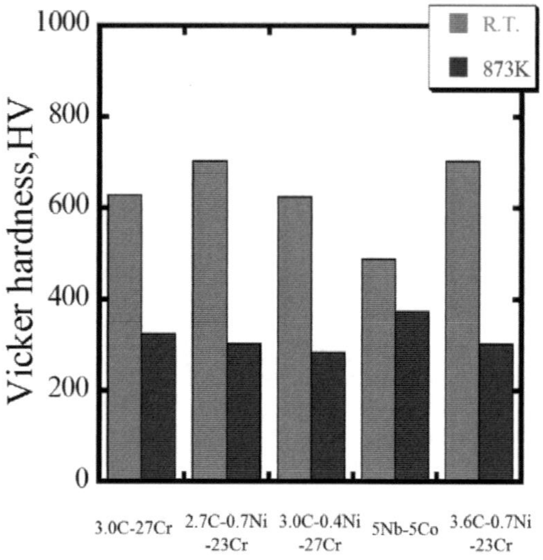

Fig. 4 Vickers hardness of Test material

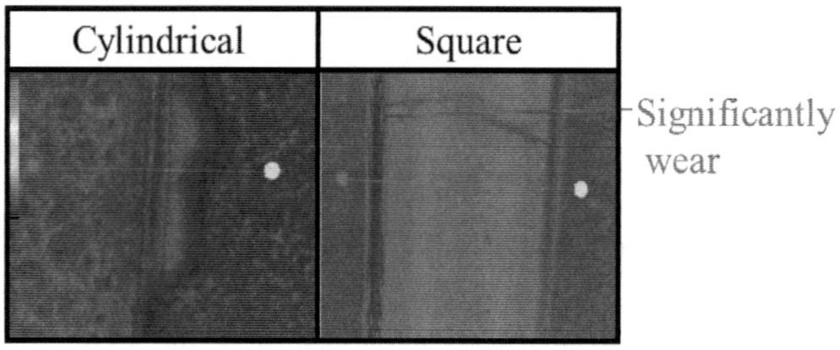

Fig.5 Profile result of microscope

4 CONCLUSIONS

- Erosion rate of WC cast-in insert cast irons decreased in relations to base metal. It's considered that hardness of matrix was showed the highest value and the relative hardness was increased by WC.
- Vickers hardness of 5Nb-5Co is higher than Vickers hardness of Hi-Cr cast iron. For that reason, erosive rate of Square shaped WC transverse-mounted 5Nb-5Co cast iron is highest.
- Then this study was tested on 873K, 1023K, 1173K. When the temperature is increase, erosion rate is increase and Vickers hardness is decrease. WC cast-in insert cast irons is a valid between 873K and 973K.

P-6

Development of Hybrid Bio-cokes for Cast Iron Melting and Establishment of Cupola Operation Method

Akihiro Oda[1], Shoichi Sato[2], Haruki Oikawa[3], Yuji Oyama[4], Hirotoshi Murata[5]
Isao Yonekura[6], Tamio Ida[7], Hiroshi Horie[8] and Sadato Hiratsuka[8]

[1] Nihon Kouken Co.,Ltd., Japan, [2] Negishi Kougyousyo Co.,Ltd., Japan, [3] Oiharu Fundry ,Inc, Japan,
[4]Nippon Piston Ring Co., Ltd.,Japan, [5] Naniwa Roki Co., Ltd. Japan,
[6] Oshu Casting Technology Exchange Center, Japan, [7] Kinki University, Japan, [8] Iwate University, Japan

To produce bio-coke from apple pomace as a raw material as fuel instead of purpose of coal coke, technology was studied. When the moisture content of the apple pomace decreased, the compression strength of the biocoke was increased. When compounding added limestone to apple pomace, bio-coke indicated more than 15 MPa of compression strength. It was to increase the formation temperature of the bio-coke to 423K from 403K and increased in the compression strength. The compression strength was more excellent than coal cokes. Hybrid bio-cokes for casts have been developed. After cast iron was melted by 0.7t/hr, 2t/hr and 6t/hr cupola using these bio-coke, operation method of the cupola by which temperature change and a chemical composition change become little was established.

Keywords: *Cast Iron, B*io-coke,, *Cupola*、 *Bentonite* 、

1. Introduction

Because the temperature in a furnace will be high temperature when coal cokes burn, coal cokes are used for cupola at present. The sulfur content of the coal cokes is becoming a lot recently. Coal cokes are becoming expensive.

CO_2 gas is discharged by fossil-fuel consumption of coal cokes, so we have to reduce greenhouse effect gas for a global environmental problem.

Bio-cokes is the solid fuel which can be formed out of a plant caused by photosynthesis. In case of photosynthesis, a plant absorbs carbon dioxide in the atmosphere. When burning this plant as fuel, the amount of released CO_2 will be the same amount as absorbed CO_2 gas by photosynthesis. In other words, it's possible to think as carbon neutral. Bio-cokes is watched as the substitution material of the coal cokes from such background.

Bio-cokes are primary made up of photosynthetic plant matter. Bio- cokes are the solid fuel of the domestic product with fixed calorific power, and the price becomes stable. Bio-coke does not produce harmful sulfides.

In this research, bio-cokes were made using bio-cokes production equipment of cokes made from the various biomass. The mechanical property of the made bio-cokes was investigated. Melting test was performed by a cupola as fuel of obtained bio-cokes.

2. Experimental procedure

2-1 Production of bio-cokes

Fig.1 shows a bio-cokes production equipment. The marc of apple, the chaff and the rice straw were made a raw material of the biomass. The drying time, crush time were changed every raw material. The formation pressure was changed from 20 in 30MPa.

The formation temperature was changed from 383K to 423K. Bentonite or slaked lime was added to marc of apple and hybrid biotechnology cokes were made too.

Chemical composition of bio-cokes was analyzed. Compression strength and specific gravity were measured.

Fig.1 Production epuipment

2-2 Melting of bio- cokes by cupola

Extraordinarily small cupola (0.05t/h), a small size cupola (0.7t/h), a medium size cupola (2t/h) and a large cupola (6t/h) were used for melting of cast iron.

Cokes used the bio-cokes made from marc of apple or hybrid bio-cokes. The cokes used for the other of coal cokes changed biotechnology cokes from 0 to 30%.

The temperature of the molten metal from a cupola was measured. A chemical composition of obtained molten metal was analyzed by a spectrometer.

3. Result

3-1 The quality of the bio-cokes

Fig.2 shows a relation between the specific gravity of the bio-cokes and the amount of bentonite. When increasing in the amount of bentonite addition, the specific gravity was increased. This is because the percentage of the bentonite with the large specific gravity becomes a lot more than bio-cokes.

More than 10 MPa of compression strength is needed for bio-cokes for cupolas to stand up to the weight of the metal in the furnace.

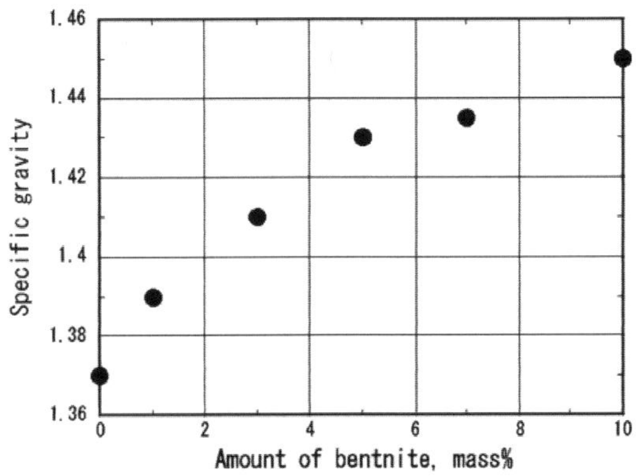

Fig.2 Relation between specific gravity and amount of bentnite

Fig.3 shows a relation between the compression strength of the bio-cokes and the amount of bentonite. Compression strength increases with increasing of amount of bentonite until 3%.

3-2 Melting of bio- cokes by cupola

When using an extraordinarily small cupola and a small cupola for melting test, it was possible to substitute for bio-cokes in 30%.

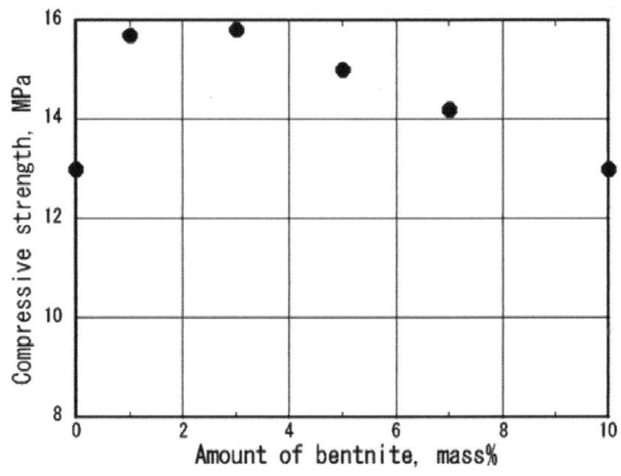

Fig.3 Relation between compressive strength and amount of bentonite

In case of a medium cupola, it could substitute for bio-cokes to 20%. In case of a large cupola, it could substitute for bio-cokes to 15%.

Fig.4 shows a change in chemical composition of molten metal when using a small cupola. Even if it increased in the substitution rate of the biotechnology cokes by operation of a small cupola, there were no big changes in C and Si amount in the molten metal.

When increasing in the substitution rate of the bio-cokes, the amount of sulfur decreased. Even if it increased in the substitution rate of the bio-cokes, the mechanical property did not change compared with 0% bio-cokes.

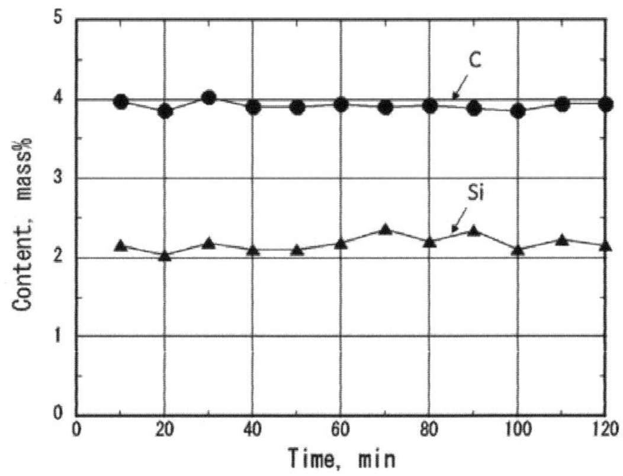

Fig.4 Change in chemical composition of molten metal

4. Conclusion

1) The specific gravity of the bio-cokes is increased by increase of the bentonite amount

2) The compressive strength is increased by increase of the bentonite amount to 3%.

3) It was possible to substitute for bio-cokes in 30%.

P-8

Proceedings of the 72nd World Foundry Congress
21-25th May 2016, Nagoya, Japan

Control of Remote Area Shrinkage in Ductile Iron Castings

Amjad Javaid[1] and Delin Li[1]
[1] CANMET Materials, Natural Resources Canada
183 Longwood Road South, Hamilton, Ontario, L8P 0A5, Canada

In this investigation, the most important process parameters to reduce or eliminate the formation of remote area (un-fed) shrinkage in ductile iron castings were determined. A Taguchi designed experiment was carried out that consisted of 7 factors at two different levels. Analysis of the results from the designed experiments showed that the only significant factors affecting the shrink and surface sink were final Si content and the tap temperature in the range studied. However, the large "error" factor in the Taguchi analysis indicates strong interactive effects, showing shrinkage to be a complex multi-variable problem.

Keywords: *Ductile iron, shrinkage, process variables*

1. Introduction

Prediction of shrinkage tendency in ductile iron is a complex multi-variable problem. There are conflicting reports in the literature [1-4] regarding the effect of various process variables on shrinkage. The major factors believed to influence the shrinkage behavior of ductile iron are: mold material/mold wall movement; chemical composition; tap/pour temperature and holding time; inoculation; Mg content of the melt; and casting geometry (surface area to volume ratio). The aim of this research work was to determine the optimum metallurgical conditions which will yield the minimum shrinkage or no shrinkage in ductile iron castings. Metallurgical conditions for this investigation included: pre-treatment, alloy treatment and post treatment of the iron melts. The effect of molding factors and casting geometry were beyond the scope of this work.

2. Experiment Procedure

A total of 16 melts were made in a Taguchi designed experiment that consisted of 7 factors at two different levels as shown in Table 1. The shrinkage test castings included a boss-plate, a step-plate, and a chill wedge, produced in CO_2 bonded silica sand molds. The test mold had two identical sets of castings on either side of the sprue, with two inoculation chambers in the runner to study the effect of different in-mold inoculants. The in-mold inoculants investigated were 75%FeSi (foundry grade), 75%FeSi with Bi, 75%FeSi with Zr, and an inoculation filter containing Zr inoculant. Most of the observations on shrinkage were made on an unfed boss-plate casting produced in a CO_2 bonded silica sand mold. Shrinkage and surface sink evaluation of the boss-plate castings was done by visual comparison, titration, grid, and x-ray (selected castings) methods. Carbon content was kept constant and Si varied at two different levels.

Table 1. Taguchi design of experiment summary.

Factors	Level #	
	1	2
Tap Temperature (°C)	1482	1538
Time after Treatment (min.)	1	10
Graphite Type	Amorphous	Crystalline
Mg Alloy Type	Standard	Special
Inoculation	0.3% ladle/ No Cover	0.1% Ladle/ 0.2% Cover
In Mold Inoculation	Yes	No
Final Silicon (%)	2.30	2.80

3. Results and Discussion

Analysis of the results from the designed experiments showed that the only significant factors affecting the shrink and surface sink were final Si content (Fig. 1) and the tap temperature (Fig. 2) in the

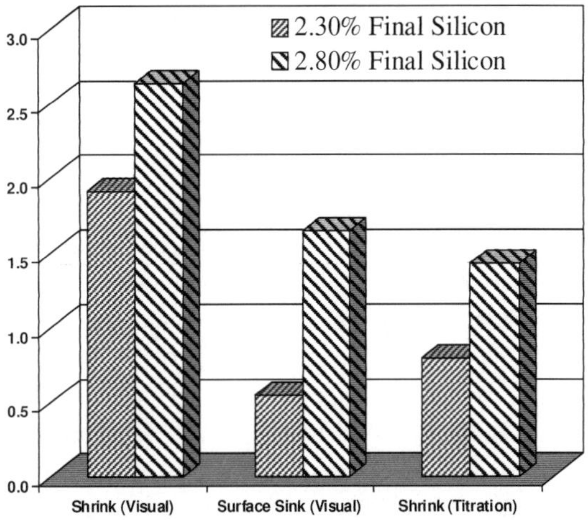

Fig. 1 Shrinkage increases with increase in Si content.

range studied. Inoculant type, holding time after Mg treatment, graphite additive type (crystalline or amorphous), Mg alloy type, and mode of ladle inoculation were all found to have little effect on shrinkage. However, the comparatively large error factor in analysis indicates extensive second order or higher order interactions, showing that the prediction of shrinkage tendency in ductile iron is a more complex multi-variable problem that cannot be so readily stated.

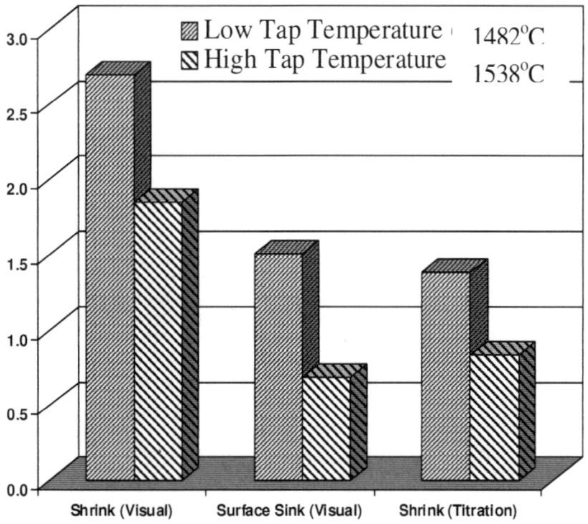

Fig. 2 Shrinkage decreases with increase in tap temp.

Of the 'non-Taguchi' variables, higher residual Mg and high nodule counts led to increased shrinkage. Shrinkage and surface sink increased with an increase in Si content from 2.3 to 2.8%. The wider mushy zone at high silicon content results in a thin or weak casting wall that cannot resist the internal pull and collapses to compensate for the shrinkage/porosity in the later stage of solidification resulting in the surface sink. The wider mushy zone at high silicon content also results in more intense intercellular segregation in the last to freeze areas, leading to increased shrinkage/porosity. The effect of high tap temperatures in reducing the shrinkage may be related to the fact that high tap temperatures lower the residual Mg level in the iron due to more rapid Mg fade after treatment and also lower the Mg recovery. The current finding that lower Mg content leads to less shrinkage is in accord with earlier work [3,4]. Also, superheating and holding the melt above 1538°C reduces its nucleation potential and so leads to lower nodule counts (low nodule counts were associated with decreased shrinkage tendency in this work).

In past work on nodule count effects, conflicting results have been presented [1-4]. The observed lower shrinkage at low nodule counts must be taken in context. The great majority of the nodule counts in this work would be considered satisfactory in normal practice, and many exceed what is normally obtained in commercial castings. The main recommendations coming from this work for reducing shrinkage in ductile iron castings are to keep the Si and Mg contents at the low end of their normal ranges, and use sufficient inoculation to avoid carbides and maintain a reasonable nodule count. If shrinkage problems appear, and nodule counts are high, it may be helpful to reduce the inoculation, *but only if other aspects of metal quality can be maintained.*

4. Conclusions

Of the variables investigated, the only ones shown by the Taguchi analysis to have a significant effect on shrinkage/surface sink were Si content and tap temperature. There was more shrinkage at the higher Si level, and less at the higher tap temperature. The large "error" factor in the Taguchi analysis indicates strong interactive effects, showing shrinkage to be a complex multi-variable phenomenon. Inoculant type, holding time after Mg treatment, graphite additive type (crystalline or amorphous), Mg alloy type, and mode of ladle inoculation were all found to have little effect on shrinkage. The increase in Si content from 2.30% to 2.80% had a particularly marked effect in increasing the shrinkage and surface sink. Of the 'non-Taguchi' variables, higher residual Mg and high nodule counts led to increased shrinkage.

Acknowledgements

The financial and in-kind support for this research project provided by the Ductile Iron Society is gratefully acknowledged.

References

[1] D. M. Stefanescu, et al: *AFS Transactions* 103 (1995) 189-197.

[2] J. F. Wallace et al: *AFS Transactions* 92 (1984) 765-784.

[3] *Ductile Iron Handbook* AFS, Des Plaines, IL, 1992.

[4] *Cast Iron Technology* by Roy Elliot, Butterworths and Co. (Publishers), London, 1988.

Proceedings of the 72nd World Foundry Congress
21-25th May 2016, Nagoya, Japan

Continuous Castings of Al-7mass%Si Alloy and Al-1.5mass%Mn Alloy by the Electromagnetic Vibration Technique

Takuya Tamura [1], Mingjun Li [1] and Koichi Takahashi [2]
[1] National Institute of Advanced Industrial Science and Technology (AIST), Nagoya 463-8560, Japan
[2] UACJ Corporation, Research & Development Division, Fukaya 366-8511, Japan

The electromagnetic vibration during solidification can refine the microstructures of cast aluminum alloys, cast magnesium alloys and gray cast iron etc. The purpose of this study is to develop the continuous casting using the electromagnetic vibration in aluminum alloys. It was found that the electromagnetic vibration can refine macrostructures of continuous casting rods.

Keywords: *aluminum alloy, electromagnetic vibration, continuous casting, macrostructure , refinement*

1. Introduction

The electromagnetic vibration casting technique is that the melt is vibrated by Lorentz force that results from static magnetic field and alternating electric current during solidification. In the AIST, it was found that the electromagnetic vibration during solidification can refine the microstructures of cast aluminum alloys [1], cast magnesium alloys [2] and gray cast iron [3] etc. On the other hand, in wrought aluminum alloys, it is important that the microstructures of continuous casting rods are refined before wrought operation. The microstructure refinement of continuous casting rods improves the mechanical properties of wrought products. Thus, the purpose of this study is to develop the continuous casting using the electromagnetic vibration in aluminum alloys in order to apply the electromagnetic vibration technique to the continuous casting of wrought aluminum alloys.

2. Experimental Procedures

The electromagnetic vibration casting technique has been developed by using a superconducting magnet [1-3]. However, in this study, a normal electromagnet is used because that is easy to apply to a continuous casting apparatus. The electromagnet is with a distance between the magnetic poles of 50 mm and able to produce a lateral magnetic field of 1.1 T.

Aluminum alloys were melted in an alumina crucible by a furnace which is placed over the magnetic poles. Molten metal was treated by Ar gas bubbling method. This molten metal was solidified in a stainless steel tube (with an outer diameter of 15 mm and an inner diameter of 13 mm) which was connected to the alumina crucible, and cast continuously by pulling-down device. Temperature of the stainless steel tube was controlled by heaters, and a continuous casting rod which came out of the stainless steel tube was cooled by sprayed water directly. The stainless steel tube was set between the magnetic poles and fixed as firmly as possible against mechanical vibrations induced by the electromagnetic vibrations. Electric current for the electromagnetic vibrations was supplied to semi-solid mushy zone which was in the stainless steel tube, by an electrode of the pulling-down device and a carbon electrode which was put in the molten metal. The electromagnetic vibrations were applied by 1.1T stationary magnetic field and 300A alternating electric current with a frequency of 150Hz.

3. Results

3.1 Al-7mass%Si Alloy

Al-7mass%Si rods with a diameter of 13 mm and a length of 280 mm were able to cast continuously cooled under the electromagnetic vibrations and without the electromagnetic vibrations. Figure 1 shows macrostructures of (a) non-vibrated rod and (b) vibrated rod. In the non-vibrated rod, coarse unidirectional solidified macrostructure was observed. However, in the vibrated rod, fine dispersed macrostructure was observed. It was found that the electromagnetic vibration during continuous casting can refine the macrostructures of Al-7mass%Si alloy.

Effects of the electromagnetic vibrations on mechanical properties of Al-7mass%Si continuous casting rods were investigated by tensile testing. The electromagnetic vibrations didn't affect the tensile strength. However, it was found that the electromagnetic vibrations improve scattering in the elongation and the reduction of area.

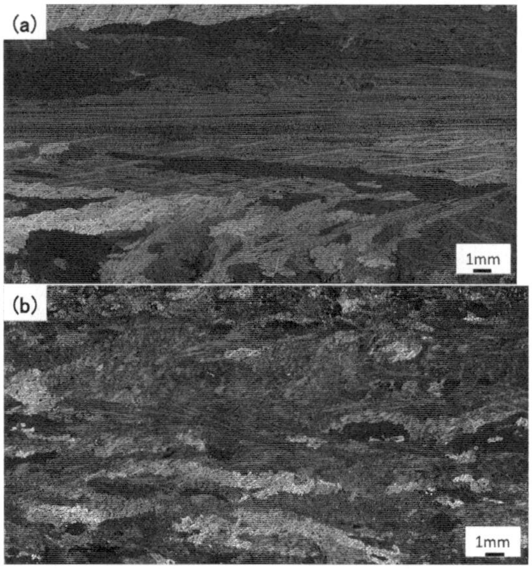

Fig.1 Macrostructures of (a) non-vibrated Al-7mass% Si rod and (b) vibrated Al-7mass%Si rod.

3.2 Al-1.5mass%Mn Alloy

Al-1.5mass%Mn rods with a diameter of 13 mm and a length of 280 mm were able to cast continuously without the electromagnetic vibrations. However, early in the continuous casting, surface wrinkle occurred, and defects on the surface occurred at the latter half. In contrast, when cooled under the electromagnetic vibrations, surface wrinkle decreased early in the continuous casting. However, at the latter half, defects on the surface became larger and the casting rod ruptured finally.

Figure 2 shows macrostructures of (a) non-vibrated rod and (b) vibrated rod. In the non-vibrated rod, coarse unidirectional solidified macrostructure was observed. However, in the vibrated rod, fine dispersed macrostructure was observed.

4. Conclusions

The continuous casting using the electromagnetic vibration in aluminum alloys has been developed. It was found that the electromagnetic vibration can refine macrostructures of continuous casting rods.

Acknowledgements

The authors thank Mrs. T. Yamaguchi, Mr. K. Chou and Mrs. C. Otomo for technical assistance.

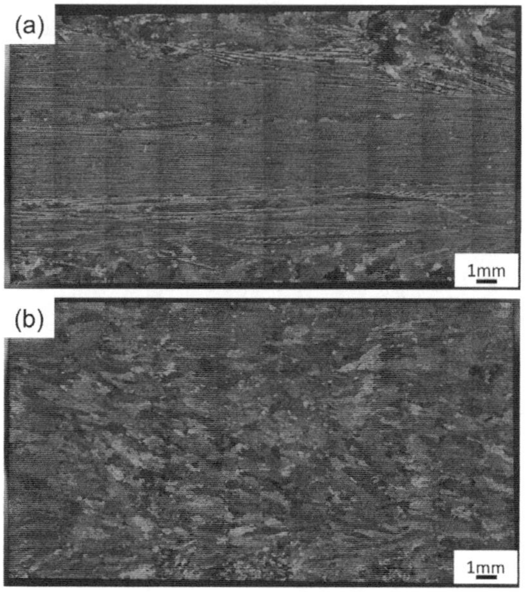

Fig.2 Macrostructures of (a) non-vibrated Al-1.5 mass%Mn rod and (b) vibrated Al-1.5mass% Mn rod.

References

[1] Y. Mizutani, Y. Ohura, K. Miwa, K. Yasue, T. Tamura and Y. Sakaguchi: Mater. Trans. 45 (2004) 1944-1948.
[2] M. Li, T. Tamura and K. Miwa: Acta Materialia 55 (2007) 4635-4643.
[3] A. Radjai and K. Miwa: Metall. Mater. Trans. A33 (2002) 3025–3030.

P-10

Proceedings of the 72nd World Foundry Congress
21-25th May 2016, Nagoya, Japan

Effect of reinforcement particles on the viscosity of the molten SiC_P/Al alloy composites

Masayuki Mizumoto[1], Yoshikazu Miyazaki[2] and Atsushi Yoshida[2]

[1] Department of Materials Science and Engineering, Iwate University
4-3-5, Ueda, Morioka, Iwate, JAPAN
[2] Graduate Student, Iwate University, JAPAN

To investigate the effect of reinforcement particles on the viscosity of the molten PRMMCs, the viscosity was evaluated by measuring the torque required to stir the molten PRMMCs. The torque increased with increasing the volume fraction and particle size of SiC particles. The torque changed with stirring time and the changing behavior of the torque was different in SiC particle size. However, the torque of molten PRMMCs with large SiC particles was high and decreased with stirring time, the torque of molten PRMMCs with small ones was low and almost constant. This difference is considered to be resulted from the difference of the settling behavior of SiC particles. The microstructures of the PRMMCs quenched after holding for various periods, showed that the larger the SiC particles size was, the faster the particles settled down. The result was in agreement with the settling behavior expected by the settling velocities of the SiC particles estimated using Stokes' law.

Keywords: *Metal matrix composite, Casting, Viscosity, Settling*

1. Introduction

Recently, the demands on the fabrication of particle reinforced metal matrix composite (PRMMC) components are increased due to their high mechanical and physical properties. The casting process is desirable to fabricate the PRMMC components because of its high productivity and formability. In the PRMMC casting process, it is known that the viscosity of the molten PRMMCs and the settling of the reinforcement particles will influence the castability and the characteristics of the PRMMC castings [1]. Some researchers have reported that the effect of reinforcement particles on the viscosity of molten PRMMCs in the stir casting process [2, 3], though there are few reports concerning the relationship between the viscosity of molten PRMMCs and the settling behavior of reinforcement particles.

In the present work, the viscosity of the molten PRMMCs was evaluated by measuring the torque required to stir the molten PRMMCs. In addition, the changing behavior of the torque was investigated associated with settling behavior of reinforcement particles in the molten PRMMCs.

2. Experimental procedure

The SiC_P/Al-Si alloy (AC3A) composites fabricated by Lanxide process were prepared for the experiments. The SiC particle size was varied from 15 to 75μm and the volume fraction of the SiC particles was varied from 10 to 30vol%. The experimental setup for the measurement of the torque was shown in Fig. 1. The torque of molten PRMMCs was measured

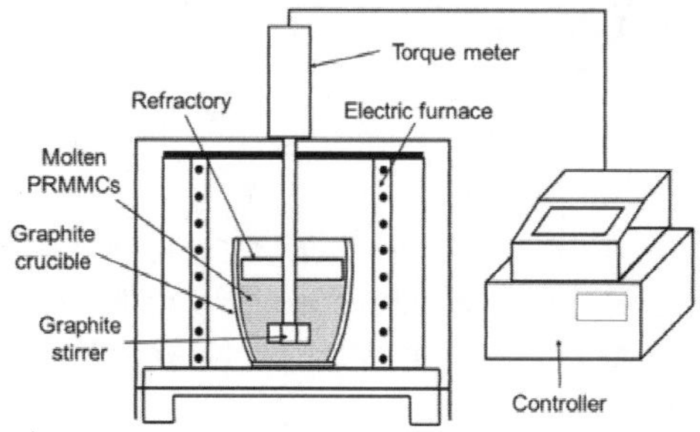

Fig. 1 Experimental setup for the measurement of the torque required to stir the molten PRMMCs.

by torque meter. The temperature of the molten PRMMCs was varied from 923 to 1023K. The rotational speed was fixed to 200rpm. To investigate the settling behavior of the SiC particles in the molten PRMMCs, the microstructure of the PRMMCs, which were quenched after holding at 973K for 1-5min, was observed. The volume fraction of SiC particles in the quenched specimens were evaluated by microstructure observation and image analysis.

3. Results and discussion

Fig. 2 shows the relationship between stirring time and stirring torque of molten PRMMCs containing 30vol% SiC particles with 15, 27 and 63μm diameter.

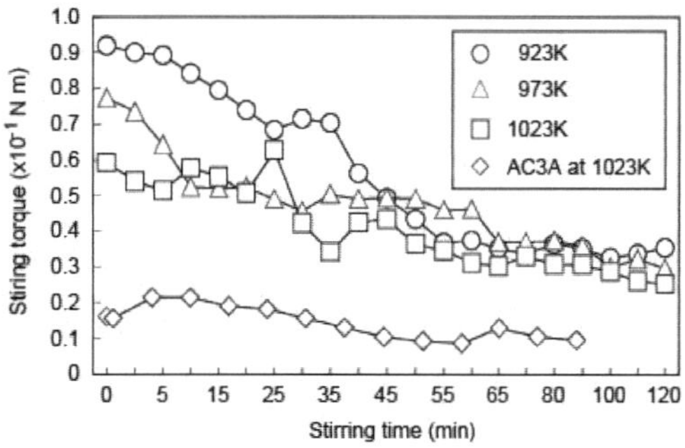

Fig. 2 Relationship between stirring time and stirring torque of 30vol% SiC_P/AC3A composites.

The stirring torque of all molten PRMMCs was higher than that of AC3A alloy melt and was increased with decreasing the temperature of molten PRMMCs. This increase in the stirring torque is considered to be caused by the decrease in the viscosity of AC3A alloy melt resulting from the temperature decrease. The effect of volume fraction and particle size of SiC particles on the stirring torque is shown in Fig.3. The

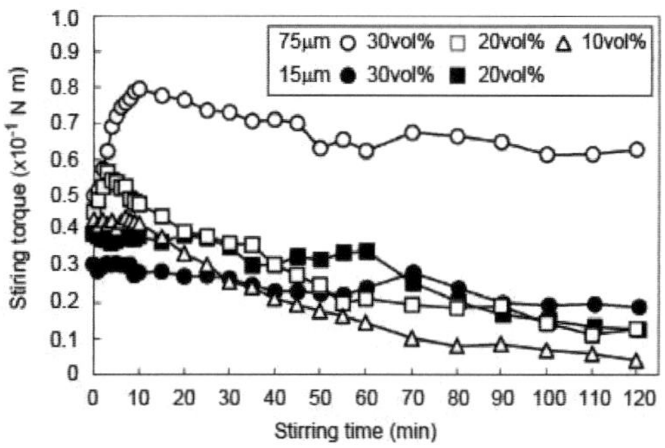

Fig. 3 Effect of volume fraction and particle size of SiC particles on stirring torque of SiC_P/AC3A composites at 973K.

stirring torque increased with increasing the volume fraction and particle size. The stirring torque changed with stirring time and the changing behavior of the stirring torque was different in particle size. However, the stirring torque of molten PRMMCs with large SiC particles was decreased with stirring time, the stirring torque of molten PRMMCs with small ones was almost constant. In addition, the peak of the stirring torque of molten PRMMCs with large SiC particles seems to shift to longer stirring time with increasing the volume fraction of SiC particles. These differences is considered to be resulted from the difference of the settling behavior of SiC particles. In order to investigate the dependence of the settling behavior of the SiC particles on particle size, the volume fraction of the SiC particles at the bottom and upper part of the quenched specimens was measured by image analysis. The results of image analysis are shown in Fig. 4. The

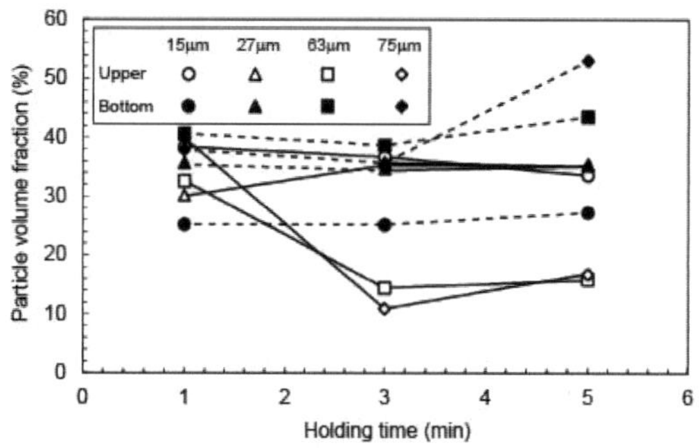

Fig. 4 Particle volume fraction in upper and bottom part of PRMMCs quenched after holding at 973K.

larger the reinforcement particles size was, the faster the particles settled down to the bottom. Then, the particle settling time was expected by the settling velocities of SiC particles using Stoke's law. The estimated particle settling time was 1.6, 2.3, 12.4 and 40min for 15, 27, 63 and 75μm diameter, respectively. These results were in roughly agreement with the settling behavior shown in Fig. 4. It is indicated that the changing behavior of the torque depends on the settling behavior of the SiC particles.

4. Conclusion

The stirring torque of molten PRMMCs is changed depending on the volume fraction and particle size of reinforcement particles. The changing behavior of the stirring torque of molten PRMMCs depends on the settling behavior of the reinforcement particles.

Acknowledgements

This work was supported by JSPS KAKENHI Grant-in-Aid for Scientific Research (C), 26420752.

References

[1] M. Emamy, R. Abbasi, S. Kaboli and J. Campbel: Int. J. Cast Metals Res. 22 (2009) 430-437.

[2] L. Drenchev, J. Sobczak and N. Sobczak: Colloids and Surfaces 197 (2002) 203-211.

[3] K.R. Ravi, V.M. Sreekumar, R.M. Pillai, C. Mahato, K.R. Amaranathan, R. Arul kumar and B.C. Pai: Materials and Design 28 (2007) 871–881.

Proceedings of the 72nd World Foundry Congress
21-25th May 2016, Nagoya, Japan

Changes in Fatigue Life Behaviour of Alloy A359 and A359-Based Composites Bulk-Reinforced with SiC Particles as a Function of the Reinforcement Content

Maria Maj[1], Adam Klasik[2], Krystyna Pietrzak[3], Andrzej Wojciechowski[3] and Jarosław Piekło[1]
[1] AGH University of Science and Technology, Faculty of Foundry Engineering 23 Reymonta str., 30-059 Krakow, Poland
[2] Motor Transport Institute, 80 Jagiellońska str., 03-301 Warsaw, Poland
[3]Institute of Precision Mechanics, 3 Duchnicka str., 01-796 Warsaw, Poland

The results of mechanical tests carried out by the proprietary modified low cycle fatigue test (MLCF) on alloy A359 and A359-based composites bulk-reinforced with SiC particles (20% and 30% volume fraction of the reinforcement) were disclosed. The described method was based on previous results of own studies, successfully validated for a wide range of materials such as steel, cast iron with spheroidal and flake graphite, ADI included, cast steel and non-ferrous metal alloys. From the results of mechanical tests it was concluded that MLCF method is particularly useful in all those cases when some heterogeneities are expected in the tested material. The effect of these inhomogeneities is practically totally eliminated, since fatigue life is estimated on one sample only and not, as in the case of the conventional LCF method, on at least several samples. Micro-structural research was also conducted to assess the observed effects. The results of microstructural examinations and fatigue tests enable drawing a conclusion that alloy reinforcement with SiC particles significantly improves the fatigue life behaviour of the resulting composite, but when certain content of the reinforcement is exceeded the fatigue life is reduced.

Keywords: *Fatigue, Composites, Microstructure.*

1. Introduction

Metal matrix composites (MMCs) are a very interesting group of materials, due to their light weight and attractive useful properties [1, 2]. Moreover, composites based on aluminium alloy matrix reinforced with SiC particles can be subjected to multiple remelting, if the Si content in alloy matrix equals minimum eight percent by weight. This composition prevents the formation of Al_4C_3 at the metal matrix/reinforcement interface, which is of crucial importance since Al_4C_3 precipitates in aluminium matrix of the composite adversely affect the mechanical properties of this material [1, 2]. The mechanical properties including fatigue life are also tested and determined before new materials are introduced to practical applications. Up to now, most commonly, the data on fatigue characteristics have been derived from the well-known fatigue tests carried out in a high- or low-cycle regime (HCF and LCF, respectively) [3-5]. As a result of these tests, complete Wöhler diagrams are obtained, but for both HCF and LCF, the accuracy of the measurements increases with the increasing number of samples available in the test. In this study it was decided to select as a test material the A359 (AlSi9Mg) aluminium alloy and respective composites based on this alloy, reinforced with 20 vol % and 30 vol % of SiC particles. Attention was focused on fatigue behaviour and microstructure effects.

2. Procedures and results

2.1 Chemical composition

The chemical composition of the tested materials is given in Table 1.

Table 1 Chemical composition of tested materials.

A359 (Al - residue)						
Si	Fe	Cu	Mn	Mg	Zn	Ti
8.58	0.12	0.03	-	0.46	-	0.11
A359 + 20 vol% SiC (Al - residue)						
9.20	0.12	<0.01	0.02	0.54	<0.01	0.10
A359 + 30 vol% SiC (Al - residue)						
9.30	0.18	0.01	0.02	0.56	<0.01	0.11

2.2 Fatigue life assessment and microstructure examination

The fatigue life was determined by an original modified low cycle fatigue method (MLCF) described in detail in [3, 5]. The fatigue tests were stress-controlled. The results are presented in diagrams (Fig. 1-3) and compared in Table 2.

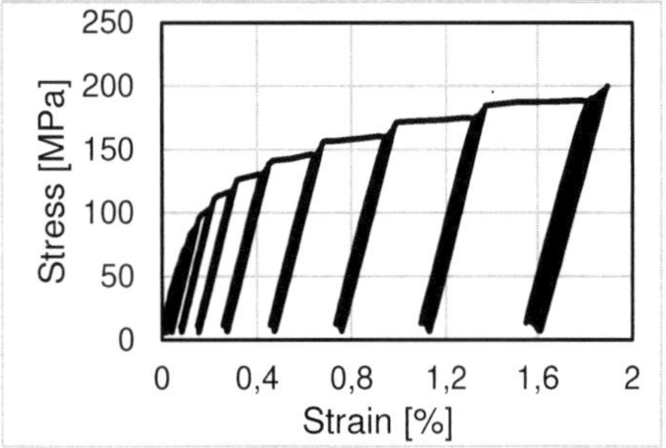

Fig. 1 The stress-strain diagram for A359 alloy.

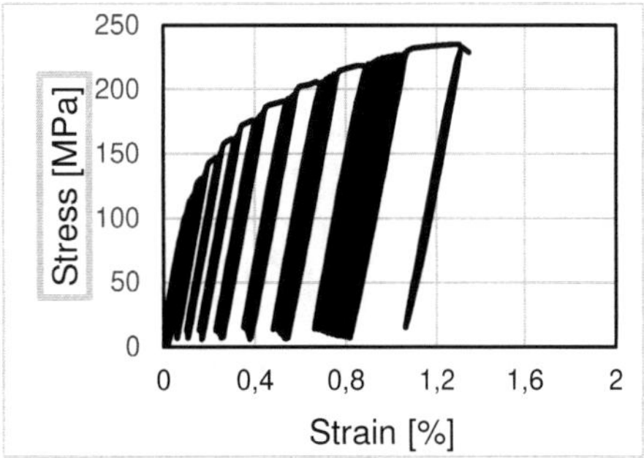

Fig. 2 The stress-strain diagram for A359 alloy reinforced with 20 vol % SiC.

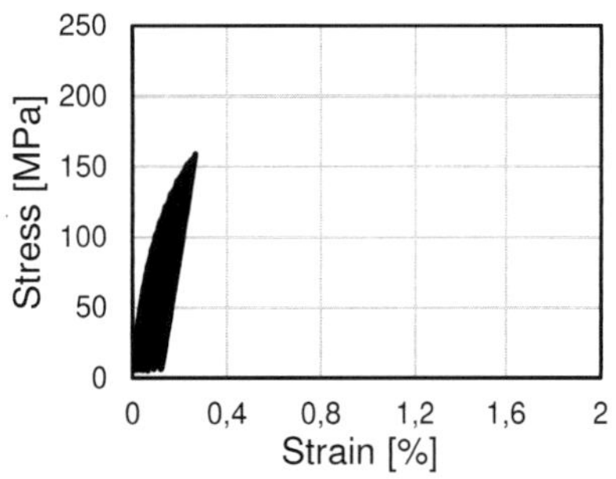

Fig. 3 The stress-strain diagram for A359 alloy reinforced with 30 vol % SiC.

Figures 1, 2 and 3 present the stress-strain relationship for A359 alloy, A359 alloy reinforced with 20 vol % SiC and A359 alloy reinforced with 30 vol% SiC, respectively. Attention deserves the fact that the highest fatigue strength was obtained for A359 alloy reinforced with 20 vol % SiC (Fig. 2, Table 2), and the lowest for A359 alloy reinforced with 30 vol % SiC (Fig. 3, Table 2). Microstructures of the investigated materials are shown in Fig. 4. It was characterized by the optimal content and distribution of SiC particles in alloy matrix. The most preferred type of microstructure was obtained in the composite containing 20 vol% SiC (Fig. 4b).

Table 2 Selected fatigue parameters.

Material	b	c	ε_{max}	K' MPa	n'
A359	-0.16994	-0.42018	0.01469	455.7	0.23257
A359 +20 vol% SiC	-0.11867	-0.28059	0.02242	636.8	0.21238
A359 +30 vol% SiC	-0.09231	-0.13129	0.04922	-	-

where: b – Basquin's exponent, c – fatigue ductility exponent, ε_{max} – maximum strain, K' – cyclic strength coefficient, n' – cyclic strain hardening exponent.

The particles were smaller than in the composite containing 30 vol% SiC. The composite with 30 vol% SiC was also characterized by a non-uniform distribution of the reinforcement particles in alloy matrix (Fig. 4c), which was the main cause of the reduced fatigue life.

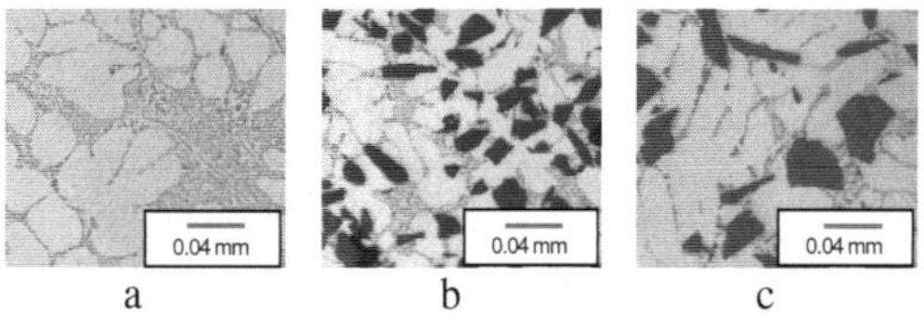

Fig. 4 Microstructure of A359 alloy (a), A359 + 20 vol % SiC (b), A359 + 30vol % SiC (c), 500x

3. Concluding remarks

Based on the obtained results it was concluded that the best microstructural and fatigue characteristics were obtained in the composite containing 20 vol % SiC.

References

[1] A. Klasik, J. Sobczak, K. Pietrzak, N. Sobczak, A. Wojciechowski: *Influence of multiple remelting of particulate reinforced cast aluminium composites on their properties and structure* Motor Transport Institute (Warsaw-Krakow 2011).

[2] D.J. Lloyd: Composite Science Technology 35 (1989) 159-179.

[3] M. Maj, A. Klasik, K. Pietrzak, D. Rudnik: Metalurgija = Metallurgy 54 (2015) no. 1, 207-210.

[4] M. Maj, K. Pietrzak: Archives of Metallurgy and Materials / Polish Academy of Sciences. Committee of Metallurgy. 58 (2013) iss. 3, 877-881.

[5] M. Maj, J. Piekło: Archives of Metallurgy and Materials / Polish Academy of Sciences. Committee of Metallurgy. 54 (2009) iss. 2, 393-397.

Casting Solidification Structure of Titanium and Titanium Alloys using Oxide Cements Mold

Yoshikazu Mantani[1] and Erina Okuda[1]
[1] National Institute of Technology, Suzuka College, Shiroko-cho 1, Suzuka, Mie, 510-0294, Japan

The purpose of this study is to evaluate the casting solidification structure of titanium and titanium alloys by using some oxide cements mold. The molten metals are commercial pure titanium, α+β type Ti-6Al-4V alloy, and β type Ti-22V-4Al alloy. These were melted by arc furnace. The mold was made from the lost wax process using commercial alumina, magnesia and zirconia cements.

As for manufactured castings, Vickers hardness distribution of the section and metallographic microstructure were investigated. From these results, it was discussed about the difference of solidification structure and the difference of reaction layer between oxide cements and titanium or titanium alloys.

Keywords: *titanium alloy, Vickers hardness, optical microstructure, ceramic cements mold*

1. Introduction

The working cost of titanium and titanium alloys were generally high, so that using casting process is one of the effective method to reduce the cost of titanium products. The study on casting of titanium and its alloys and manufacturing its products have already performed [1-3]. However, to deepen knowledge about casting solidification structure or reaction layer formed by each ceramics mold in titanium and its alloys more ever, it leads to improve the casting technology of titanium and its alloys.

The molten metals are commercial pure titanium (CPTi), α + β type Ti-6Al-4V alloy (Ti6Al4V), and β type Ti-22V-4Al alloy (Ti22V4Al). These titanium alloys ware ternary alloys indicate different phase constitution of the α and β phases with different content of Al and V. There is a metallurgical significance for discussing about the difference in the solidification structure by the difference between the additional element to titanium and the phase constitutions. Refractory cements mold using the casting were commercial alumina (Al. mold), magnesia (Mg. mold) and zirconia (Zr. mold). Although Zr. mold is considered to be suitable for the purpose from the Ellingham diagram, it is more expensive than other molds.

The purpose of this study is to evaluate the casting solidification structure of titanium and its alloys from the fundamental metallurgical scope.

2. Experimental Procedure

The used molten metals were CPTi, Ti6Al4V, and Ti22V4Al. The Al. mold, Mg. mold and Zr. mold were used in this casting process. The molds were made mix mud cements, and pattern made by wax was embedded in mix mud, and made them dry. The pattern was removed by heating. The molds were heated at 1473 K for Al. mold, 1573 K for Mg. mold, and 1673 K for Zr. mold, and hold at an hour for the purpose of strengthen the mold.

The molten metals were made by the vacuum ark furnace using a nonconsumable W-electrode. These were poured in the molds under the vacuum, and castings were made by removing the molds. The castings were embedded in epoxy and it was sliced for the purpose of consequent evaluation. **Fig.1** shows the sliced specimen of CPTi casting using Mg. mold.

The sliced specimens were polished until obtain the mirror surface using alumina abrasive powder. Vickers hardness test was performed with a load of 1kgf to obtain the differences between center of specimen and near the boundary of the molds. In addition, the change of Vickers hardness from the boundary of the molds to inner side were investigated with the step of 20μ m at a load of 0.05 or 0.1 kgf.

Cross section microstructure of the sliced specimens were observed by microscope after the etching. The etchant using this process is HF (5%vol.) + HNO_3 (5%vol.) + water (bal.).

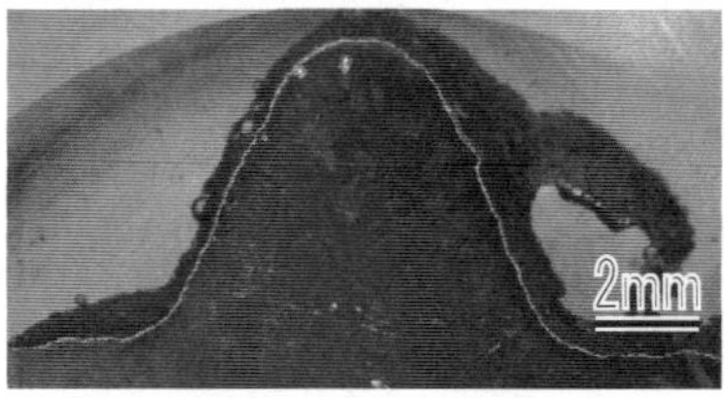

Fig.1 CPTi casting using Mg. mold.

3. Results and discussion

The change of Vickers hardness on center or edge of specimens with the kind of alloys and molds were shown in **Fig.2**. Although the hardness of CPTi using Al. mold indicated higher value at both edge and center, that of titanium alloys were not higher at the center of the specimens. In Ti6Al4V, the hardness of Mg. mold was comparatively higher. That of Zr. mold was lower at both edge and center. On the contrary in Ti22V4Al, that of Zr. mold was higher than that of Mg. mold. To evaluate the hardness distribution more detail, the hardness from the mold boundary was investigated. The change of Vickers hardness with the distance from the mold boundary were shown in **Fig.3**. In Ti6Al4V, the hardness of the area until 50 μm from the Mg. mold was especially high, and it was indicated over 500HV after that. On the other hand, substantial change of hardness from the Zr. mold was not occurred. In Ti22V4Al, substantial change of hardness was not occurred by using Mg. mold. On the contrary by Zr. mold, the area until 250 μm from the boundary indicated especially higher values, and it was indicated under 400HV after the inner side which was almost the same tendency of Mg. mold. These differences of the hardness from the mold boundary was considered to be caused by institutional depth of the oxygen and formation of a compound layer which was affected by the phase constitution or content of alloying elements of titanium alloys.

To compare and examine the difference of the hardness and the difference of the microstructure, observation by optical microscope was performed. **Fig.4** shows optical microstructure of the specimens from the mold boundary. Ti6Al4V using Mg. mold (a), boundary area of 50 μm form the mold was observed. The area is equivalent to the area of higher hardness. Typical acicular microstructure was formed at the inner side. On the other hand using Zr. mold (b), boundary area of 200 μm was observed, though the area did not affect the hardness considered from the results of hardness. Ti22V4Al using Mg. mold (c), any boundary area was not observed. Only of grain boundary of β phase was observed. On the contrary, that using Zr. mold (d), boundary area of 200 μm from the mold was observed. The area was considered to be the area that indicated especially higher hardness.

To discuss about the results more detail, there are a need to evaluate the elements distribution, oxide or compounds layer.

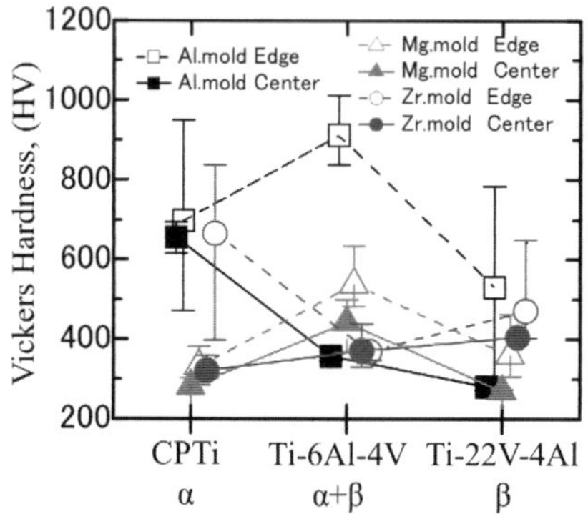

Fig.2 Change of Vickers hardness on center or edge of specimens with the kind of alloys and molds.

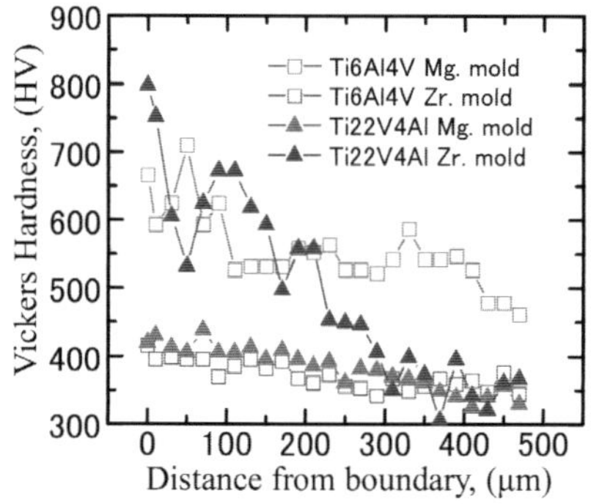

Fig.3 Change of Vickers hardness with the distance from the mold boundary.

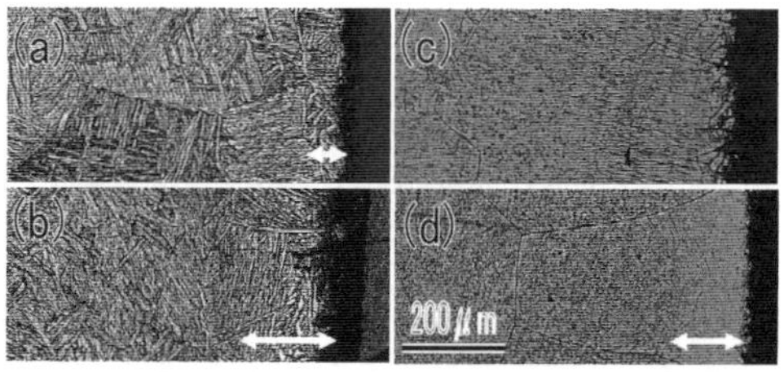

Fig.4 Optical microstructure of specimens from the mold boundary.

4. Conclusion

While the hardness of Ti6Al4V using Mg. mold was higher than Zr. Mold, reverse results were obtained by Ti22V4Al. Formation of reaction layer thickness or oxygen institutional depth are different from the alloying element content or phase constitution in a kind of titanium alloys. This metallurgical differences were considered to cause the hardness distribution.

References

[1] M. Niinomi: J. JFS 73 (2001) 784-790.

[2] Y. Tamaki and T. Miyazaki: J Jpn Prosthodont Soc. 42 (1998) 528-539.

[3] S. Hyunsun, M. Hattori and Y. Oda: *The Shikwa Gakuho* 103 (2003) 657-672.

Fatigue and hydrogen embrittlement properties of Al-Mg-Zn series casting alloy with Cu addition.

Yudai Nasu[1], Tadao Fukuta[2], Kiyotaka Obunai[2] and Koichi Ozaki[2]
[1] Graduate School of Okayama Prefectural University Department of Systems Engineering
[2] Okayama Prefectural University Department of Systems Engineering

The purpose of this study is to improve the fatigue and hydrogen embrittlement properties of Al-Mg-Zn series cast alloy by adding Cu. The weight fractions of Mg, Zn and Cu in the cast alloy were controlled to 3.6, 3.0 and 0.9 mass%, respectively. In order to obtain higher mechanical properties the aging heat treatment was conducted to casted alloy ingots. The fatigue strength of cast alloys was characterized under cyclic sinusoidal tensile load by using fatigue tester. The creep strength under corrosive environment of cast alloy was also measured to evaluate the hydrogen embrittlement characteristics. The surface of specimen dipped into NaOH solution was observed for measuring the amount of hydrogen generation by corrosion. Moreover, in order to discuss the effect of Cu addition on grain boundary strength, first principle calculation was conducted. Test results showed that the fatigue strength of proposed alloy was improved compared to that of conventional Al-Mg-Zn alloy. The creep strength of proposed alloy was also improved. Calculated results revealed that the grain boundary strength was improved when Cu was existed. These results revealed that the addition of Cu was effective to improve the fatigue characteristics of Al-Mg-Zn alloy by enhancing the grain boundary strength.

Keywords: *Fatigue property, Hydrogen embrittlement, Grain boundary strength, AC7A*

1. Introduction

Nowadays, aluminum casting wheel were often appeared in automotive to reduce the weight and inertia and also to put the grace for many user. Many of aluminum casting wheel were made from AC4CH and to obtain shiny surface, plating technique was required. From the viewpoint of environmentally friendly, Al-Mg-Zn series casting aluminum that can obtain shiny surface without plating have been developed by one of the authors[1]. However, the fatigue strength of Al-Mg-Zn series casting aluminum was considerably lower than that of AC4CH. In this study, in order to improve the fatigue strength of Al-Mg-Zn series casting aluminum, the Cu was used as new additive and its effect on fatigue strength was investigated. To applying the automotive wheel, various mechanical characteristics have to consider not only under lab environment but also under corrosive environment. Therefore, the creep strength of developed alloy was also investigated by controlled corrosive environment.

2. Experimental procedure

Table 1 show the chemical composition of alloys. The conventional Al-Mg-Zn alloy and developed alloy were casted to ingot. The solution heat treatment and age hardening treatment was conducted to casted alloy ingots. Table 2 shows the conditions for heat treatment of each test samples. The conditions for heat treatment were decided by investigating the strength and fracture strain of each treatment conditions. By using dumbbell type specimen, the tensile properties of cast alloys in which conditions for heat treatment were different were characterized under of 1.0 %/min of constant uniaxial strain. The fatigue strength of cast alloys was also characterized under cyclic sinusoidal tensile load by using hydraulic type fatigue tester. Stress ratio for fatigue test was set to 0.1. The creep strength under controlled corrosive environment of cast alloy was measured to evaluate the hydrogen embrittlement characteristics. In this study, the creep strength was defined as the applied stress when time until fracture reaches 240 hour. Constant tensile load was applied to small dumbbell type specimen with circumferential V notch by using lever type creep tester shown in Fig.1. The wet-dry cycle corrosive environment was simulated by dropping the NaOH solution to gauze wrapped around V notch. For measuring the amount of hydrogen generation under corrosive environment, coupon type specimen was also prepared. The tensile load was applied to polished specimen in NaOH solution, and its surface was observed.

Tabel 1 Chemical composition (mass%).

	Mg	Zn	Cu	Si	Al
AC7A+CuZn	3.6	3.0	0.9	-	Bal
AC7A+Zn	4.5	3.1	-	-	Bal
AC4CH	0.25-0.45	-	-	6.5-7.5	Bal

Tabel 2 Heat treatment conditions.

	Solution treatment	Age hardening treatment
AC7A+CuZn	753 K-6h	453 K-5h
AC7A+Zn	703 K-6h	453 K-5h
AC4CH	808 K-4h	433 K-4h

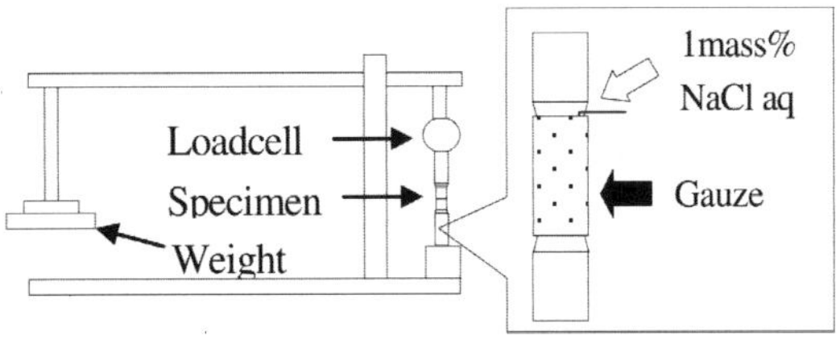

Fig. 1 Schematic diagram of apparatus.

3. Results and discussion

Tensile test results revealed that when applying the appropriate heat treatment, the tensile strength and fracture strain of developed alloy was almost same as that of conventional Al-Mg-Zn alloy. Figure 2 shows the S-N diagram of test samples. Test results showed that the fatigue limit of developed alloy was improved by adding the Cu and removing the Mg from conventional alloy. In order to reveal the morphology of Cu in alloy, FE-EPMA analysis was conducted. The FE-EPMA analysis of proposed alloy revealed that the added Cu was segregated to the grain boundary. One of the authors reported that the grain boundary cohesive energy evaluated by first principle calculation was improved when segregated Cu was existed at modeled grain boundary of Al-Mg-Zn series alloy[2]. These results suggested that the improvement in fatigue strength of developed alloy was explained by the improvement in grain boundary strength. Figure 3 also shows the results of the creep test under corrosive environment. The creep strength of developed alloy was improved compared with that of conventional alloy. The surface observation of specimen in corrosive environment revealed that when load was applied, the generated hydrogen by corrosion was absorbed into the specimen. The amount of hydrogen absorption of developed alloy was almost same as that of conventional alloy. These results suggested that the addition of Cu was effective to enhance the grain boundary strength even if the hydrogen was absorbed.

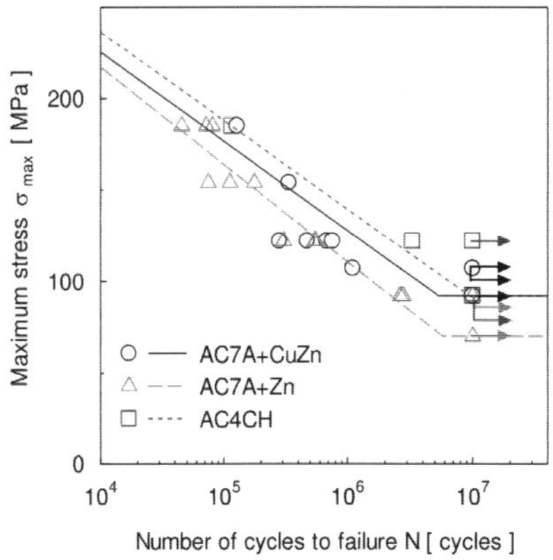

Fig. 2 S-N diagram.

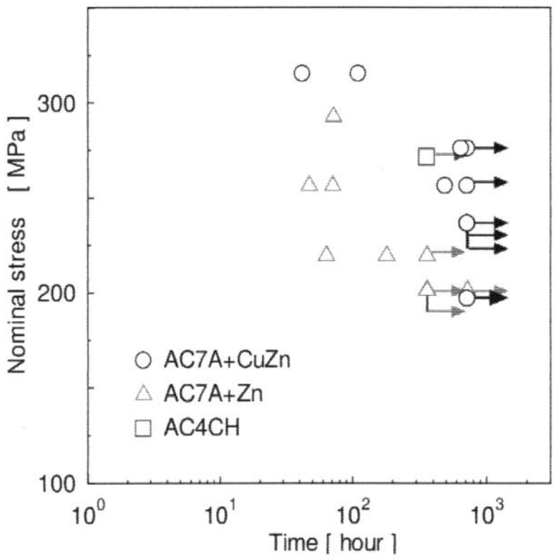

Fig. 3 Creep strength under corrosive environment

4. Conclusions

(1) The addition of Cu is effective in improving fatigue strength of Al-Mg-Zn alloy.

(2) The added Cu at grain boundary enhances the grain boundary strength, even if the hydrogen was absorbed.

Acknowledgements

The authors would like to thank hikari light metals for their donation of aluminum alloy.

References

[1] T.Fukuta et al: Transactions of the JSME, Series A,79-802 2013), 838-846

[2] T. Fukuta et al: *A Study on Grain Boundary of Al-Mg-Zn-Cu Series Alloy by First-Principles Calculation*, (25th Annual Meeting of MRS-Japan 2015, Yokohama, 2015), E2-P8-006

Shaping the Al-Si alloy microstructure and properties using the multipoint water mist cooling system

R. Władysiak[1], A. Kozuń[1], T. Pacyniak[1]
[1] Lodz University of Technology, Department of Materials Engineering and Production Systems, 1/15 Stefanowskiego Str., Lodz, Poland

The paper presents the test results of crystallization process, structure and mechanical properties of several near- and hyper-eutectic Al-Si alloys casted in metallic mold. In casting process was used the multi-point water mist cooling system of mold. The cooling system increases the intensity of heat transfer in a cast-casting die setting and causes the local occurrence of directional solidification effect. The paper presents the results of the crystallization process that was investigated by the TDA thermographic method with use of metallic probe that was cooled during the tests with water mist. It was shown that the use of mold cooling water mist enables the development of microstructure and mechanical properties of Al-Si alloys. In addition, a wide range of solidification temperature of hypereutectic Al-Si alloys particularly increases the possibilities to influence of cooling rate on the size and the phase morphology of microstructure and properties as well.

Keywords: *Casting Die, Cooling, Al-Si Alloy.*

1. Introduction

The ongoing work is a part of studies on the application of water mist system for multipoint sequential cooling of casting die to produce Al-Si alloys castings [1-6]. The essence of the research is the efficient cooling mist of water through evaporation of water droplets on a hot surface of the casting die. An analysis of knowledge subject and earlier studies indicate that the cooling of mold with the water mist stream enables the fast directional crystallization, shaping of microstructure and then achieving high quality casts made of neareutectic and hypereutectic Al-Si alloys [7]. Efficiency of heat transfer process is largely determined by the characteristics of generated stream, optimization amount of air and water in the mist stream and adequate spraying of water and also wall thickness of casting. The aim of the study was to investigate the effect of water mist cooling on microstructure and mechanical properties of Al-Si alloys: AlSi11 and AlSi20 (Table 1) that was casted in casting die cooling with use of multipoint opened system (Fig.1).

Table 1. Chemical composition of Al-Si alloys (mass%).

Alloy	Si	Mg	Cu	Fe	Mn	P	Al
AlSi20	21.5	0.01	0.2	0.22	0.01	0.03	Bal.
AlSi11	10.8	0.02	0.3	0.17	0.01	0.03	Bal.

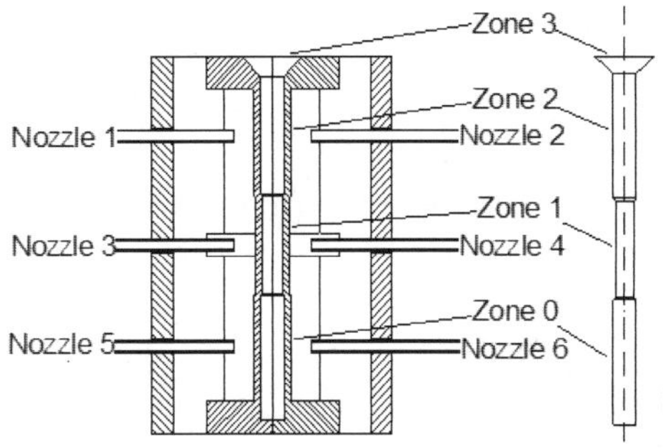

Fig. 1. Section of research casting die and casting, zones and nozzles of cooling system

2. Results

The crystallization of neareutectic alloy AlSi11 contets of 2 stages: of 1-preeutectic Al and 2 – Al+Si eutectic. The hypereutectic alloy AlSi20 in cooled casting die (Fig. 2a) starts from the initial crystallization of silicon walled crystals and also a dendritic silicon crystals (Fig. 2c). They are good nucleate pad for growing dendrites of Al phase. Farther lowering the temperature causes the alloy enter into a zone of eutectic coupled growth and in terms of irregular eutectic crystallized lamellar (Al+Si) that's the same like in the neareutectic alloy AlSi11 (Fig. 2b).

The results of hardness tests were presented in Fig. 2d. They proved that cooling of casting die with use of water mist significantly increases hardness of casting by refinement of structure and shaping additional dendritic phase of silicon that are characterized by high values of Vickers hardness. While tensile stress tests for AlSi11 alloy show that results are much better compared to non-cooled castings.

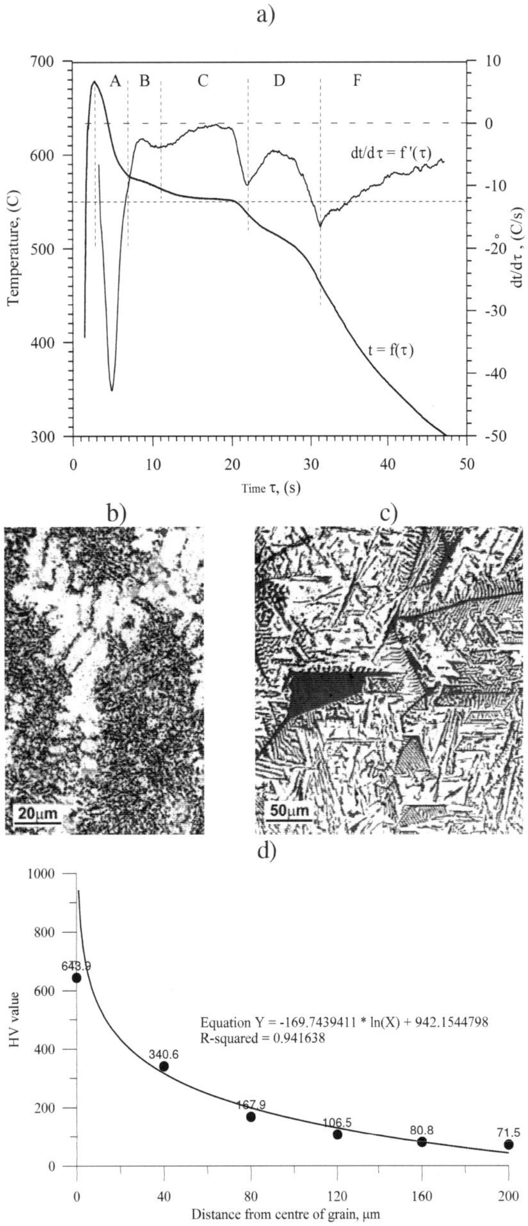

Fig. 1. Thermal and derivative curve (TDA) of AlSi20 alloy (a), microstructure of AlSi 11 alloy (b) and AlSi20 (c), Vickres hardness of dendritic grain of AlSi alloy (d)

The largest average values of Rm = 222 MPa were obtained for castings cooled using the sequence with priority for bottom part of the die and casting. The values of the other tested properties were as follows: $Rp_{0,2}$ = 117 MPa, A_5 = 2.7% and HB = 86. It is a result of decreasing of shrinkage defects and porosity in the casting and also of microstructure refinement due to the increased cooling of the liquid alloy during solidification of casting.

3. Conclusions

Al–Si alloys prepared by rapid cooling of melt were used to explain the effect of casting die cooling with water mist on the solidification process, the structure and mechanical properties of near- and hypereutectic Al-Si alloys. The study shows that the treatment is able to form preeutectic silicon dendrites and dendritic Si eutectic in the structure of hypereutectic Al-Si alloys. It was proved by thermal effects on the curves of solidification process.

The structure has a several times refining grain, compared to the casting obtained from uncooled permanent mold and from shell mold casting as well. The refinement and homogeneity of the structure were increased the mechanical properties of Al-Si alloys.

Acknowledgments

The article presents the findings of research included in the research project no. N N508 399237 financed by the science funding programme of the Ministry of Science and Higher Education of Poland.

References

[1] R. Władysiak: (2007). *Effective Intesification Method of Die Casting Process of Silumins.* Archives of Metallurgy and Materials, 52 (2007) 529-534.

[2] S. Pietrowski, R. Władysiak: *Result of cooling of dies with water mist.* Journal of Achievements in Materials and Manufacturing Engineering, 25, (2007) 27-32.

[3] R. Władysiak: *Water mist effect on cooling range and efficiency of casting die.* Archives of Foundry Engineering, 8(2004), 213-218.

[4] R. Władysiak: *Effect of water mist on cooling process of casting die and microstructure of AlSi11 alloy.* Archives of Metallurgy and Materials, vol. 55 (2010) 939-946.

[5] R. Władysiak: *Computer control the cooling process in permanent mold casting of Al-Si alloy.* Archives of Metallurgy and Materials, 58 (2013) 977-980.

[6] R. Władysiak: *Water mist effect on cooling process of casting die and microstructure of AlSi9 alloy.* Archives of Foundry Engineering, 10 (2010) 185–194.

[7] E. Fraś: Crystallization of alloys. PWN (2004).

P-15

Surface roughness of casting solidified in the frozen mold

Naoki Omura and Shuji Tada
National Institute of Advanced Industrial Science and Technology (AIST)
2266-98 Shimo-Shidami, Moriyama-ku, Nagoya 463-8560, Japan

Effects of frozen mold conditions such as the particle size of sand and the water content on the surface roughness of castings solidified in the frozen mold were demonstrated. The surface roughness of castings increases with increasing the particle size of sand. Furthermore, the surface roughness was also affected by the water content of mold. The fewer the water content, the smoother the surface of casting. When water content was 3 mass% (minimum value in this study), the arithmetic mean roughness of the casting was about 15 μm and was comparable with that fabricated using the green sand mold.

Keywords: *AC4CH aluminum alloy, water content, grain size*

1. Introduction

A frozen mold that is fabricated by freezing the mixture of only sand and water is a kind of sand mold. The frozen mold has many advantages such as reducing environmental loads and improving the casting environments because it doesn't include any organic compounds. Reduction of environmental load is desired in various fields nowadays, so the technique using the frozen mold in casting is received remarkable attention as a future casting process. Several studies on the properties of frozen molds such as strength, heat-transfer characteristic, cooling ability, fluidity and so on, have been reported [1-4]. However, there has been little information about the properties of casting fabricated using the frozen mold [5]. In this study, we examined effects of the mold conditions such as the grain size of silica sand and the water content of mold on the surface roughness of AC4CH aluminum alloy casting solidified in the frozen mold.

2. Experimental procedure

Silica sands having different mean particle sizes were used to fabricate frozen molds in this study. The grain size distributions of these silica sands are shown in Table 1. 5 mass% water was added to these silica sands, and they were blended for 180 seconds by a mixer to prepare the sand mixtures for sand mold.

Table 1 Grain size distribution and fineness number of sand used in this study

Grain size μm	850	600	425	300	212	150	106	75	53	<53	AFS. FN
R5	0.2	20.0	77.8	1.8	0.2						28.2
R6			0.2	8.6	52.4	34.6	3.8	0.4			58.3
R7					13.8	55.0	27.4	3.6	0.2		78.2
R8						10.0	56.8	23.8	7.4	2.0	117.9
GSM※			1.2	3.6	10.0	35.0	37.6	4.7	0.5	0.7	84.3

※Green sand mold

Furthermore, three kinds of sand mixtures consisted of R6 silica sand and different amount of water (3, 8 and 10 mass%) were prepared.

The sand mold was formed for each sand mixture using a plastic mold. The sand mold was frozen by keeping in the freezer held below 238K for more than 14.4 ks. Fig. 1 shows the photograph of frozen mold used in this study. Upper mold had a sprue and a degassing hole, and lower mold had a plate-like cavity. The cavity size of the plate is as follows; length: 135mm, width: 22mm, and height: 7mm.

The AC4CH aluminum alloy ingot was put in a graphite crucible and heated up to 993K by a high-frequency induction furnace. Then, the crucible was taken out from the furnace. When the melt temperature goes down at 983K, the molten aluminum alloy was cast into the frozen mold.

The surface roughness of the specimen solidified in the frozen mold was measured by a contact type surface roughness tester (SURFTEST SJ-310, Mitutoyo Corporation), and arithmetical mean roughness, Ra, was calculated. The measuring

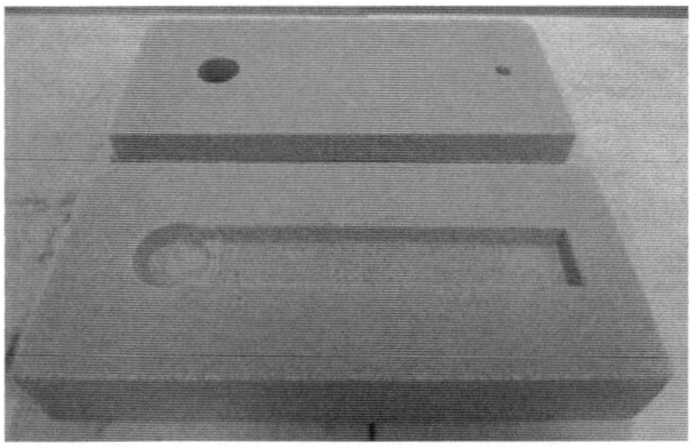

Fig. 1 Photograph of frozen mold.

conditions were as follows; the evaluated length was 16mm, the cut-off value was 8.0mm, and the measuring speed was 0.25mm/s. For comparison, the surface roughness of the casting fabricated in a green sand mold was also measured. The green sand mold consisted of silica sand, clay, additives, and 2.6 mass% water.

3. Results and discussions

The effects of grain size of silica sand and water content of mold on the surface roughness of castings solidified in frozen molds are shown in Fig. 2 and Fig. 3, respectively. For comparison, the data from the green sand mold were also included in these two figures. It is revealed that the surface roughness of casting fabricated by frozen mold decreases with a decrease in the particle size of the sand. Moreover, the scattering of the surface roughness increases with an increase of the particle size. The specimen fabricated using frozen mold prepared from the sand with large particle size (R5) showed a very rough surface. The height difference of the surface of this specimen exceeded the measuring range of the apparatus (360μm), making it impossible to acquire the surface roughness of this specimen. The frozen mold consists of only sand and water. So the particle size of sand directly influences on the surface roughness of the frozen mold. That is, the finer the sand is used, the smoother the surface of a frozen mold will be. This may be the reason that the surface roughness of castings solidified in the frozen mold varied with the particle size of sand.

In addition to the particle size of sand, water content plays the other important role in influences the surface roughness. It is clearly seen that the surface roughness of casting increases with increasing the water content of mold. When water content was 3 mass%, the casting showed relatively smooth surface. The arithmetic mean roughness value of this specimen was about 15 μm. This value was comparable with that fabricated using the green sand mold. In contrast, the specimen cast in the frozen mold which included water 10 mass% showed rough surface. The surface roughness of this specimen can't be measured because the height difference of surface goes beyond the measuring range. The more the frozen mold includes water, the more the vapor generates at casting process. It is considered that large amount of vaper produces high back pressure in the mold and has a negative impact on the surface roughness of castings.

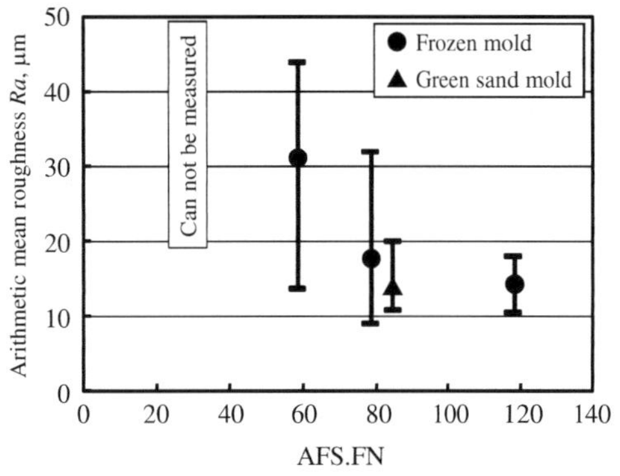

Fig. 2 Variation in the surface roughness of castings as a function of grain size of silica sand.

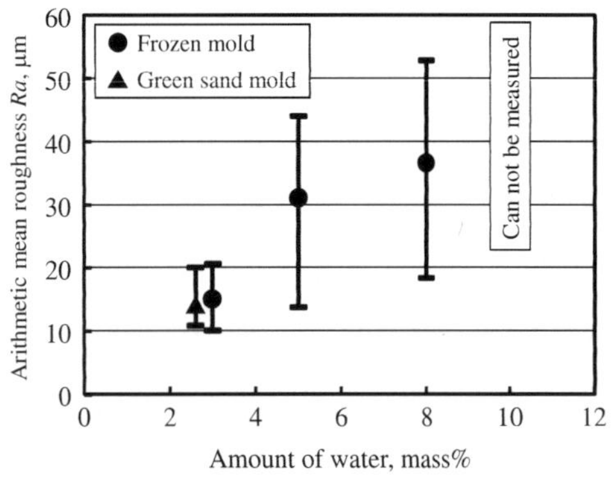

Fig. 3 Variation in the surface roughness of castings as a function of water conten in mold.

4. Conclusions

The surface roughness of castings fabricated using frozen mold was affected by the particle size of sand and the water content of mold. The frozen mold prepared from the fine sand and less amount of water can provide the casting with smooth surface.

References

[1] S. Tada, T. Nishio, and K. Kobayashi: Inter. J.Cast Metals Research 21 (2008) 260-264.

[2] T. Nishio, K. Kobayashi, and H. Nakayama: J. Jpn. Foundry Soc. 82 (2010) 350-355 (in Japanese).

[3] H. Nakayama, S. Tada, T. Nishio, and K. Kobayashi: J. Jpn. Foundry Soc. 81(2009) 371-376 (in Japanese).

[4] N. Omura, Y. Murakami, and S. Tada: J. Jpn. Foundry Soc. 85(2013) 659-664 (in Japanese).

[5] K. Shimizu, Y. Xinba, M. Tanaka, and H. Shudai: Mater. Trans. 50 (2009) 1128-1134.

Proceedings of the 72nd World Foundry Congress
21-25th May 2016, Nagoya, Japan

Influence on Corrosion Resistance of Additive Element for Environmental-Friendly Brass Alloy

Hiroshi Yamada[1], Masaaki Yamamoto[1]
[1] KURIMOTO,LTD, *Osaka 559-0021 Japan*

We have studied environmentally-friendly brass with good corrosion resistance required for water products. The aim of this study is to investigate the mechanical property and corrosion-resistance of the brass alloys, called KBS1 and KBS2. KBS1 and KBS2 can adapt to newly environmental regulation for lead in water supply works in all the countries of the world, and were based on 70%Cu-30%Zn added element such as Al, Ni, Sn and P. The mechanical property and corrosion resistance of KBS1 and KBS2 were compared with JIS copper alloys and were investigated the influence on the corrosion resistance of additive element, in this study. The results showed that the mechanical properties of KBS1 and KBS2 were equal to CAC203 alloy. In addition, the dezincification corrosion test specified by ISO 6509 and Jet-in-slit erosion-corrosion test indicated that both the KBS1 and KBS2 had superior corrosion resistance to CAC203 alloy. Therefore, these additive elements were suggested to be effective for corrosion resistance.

Keywords: *lead-free-brass, low-lead-brass, dezincification-corrosion, erosion-corrosion, corrosion resistance*

1. Introduction

Japan Industrial Standard (JIS) CAC203(CAC203) alloy containing 0.5 to 3 mass% of Pb has been used as water faucet materials for water supply because of its good castability, high mechanical properties and good machinability. However, in late years the application of CAC203 has become difficult because of Pb regulation for drinking water in all over the world. In addition, the use of CAC203 alloy in water service product is limited because it does not have good corrosion resistance.

We developed two kinds of 70%Cu-30%Zn base alloy contained Bi or low Pb as environmental-friendly brass, called KBS1 and KBS2, added Al, Ni, P and Sn. In this study, to evaluate influence on corrosion resistance of additive element, the mechanical properties and corrosion resistance of KBS1 and KBS2 were investigated by tensile test, dezincification corrosion test and erosion-corrosion test, respectively, and were compared with those of three copper alloys.

2. Experimental

2.1 Test Sample

As for the test samples, CAC203, CAC406, CAC804 and the developed brass alloys (KBS1 and KBS2) were casted in φ20mm of metal die and sand mold of JIS type A(JIS H5120). Chemical compositions of test materials are shown in Table1.

Table1 Chemical compositions of test materials

	Cu	Zn	Sn	Pb	Bi	Ni	Al	Si	P
CAC203	Bal.	37.8	0.2	2.1	-	-	0.02	-	-
CAC406	Bal.	5.4	4.4	4.8	-	0.2	-	-	0.020
CAC804	Bal.	20.2	0.1	-	-	-	-	3.25	0.086
KBS1	Bal.	30.6	-	-	0.7	0.4	1.3	-	0.045
KBS2	Bal.	29.1	1.2	0.18	-	-	0.8	-	0.035

2.2 Tensile test

Tensile tests of KBS1 and KBS2 casted in sand mold were performed according to Japanese Industrial Standard JIS Z2241-2011.

2.3 Dezincification corrosion test

Dezincification corrosion tests of CAC203, CAC406, CAC804, KBS1 and KBS2 alloys casted in metal die were conducted according to ISO 6509-1981.

2.4 Erosion-corrosion test (Jet-in-Slit test)

The erosion–corrosion behavior of CAC203, CAC406, CAC804, KBS1 and KBS2 alloys casted in metal die were examined using Jet-in-Slit test. The principle of Jet-in-Slit erosion-corrosion test is shown in Fig.1. As shown in Fig.1, Jet-in-Slit test can be used to evaluate the erosion-corrosion induced by the shear stress and fluid flow turbulence.

The test condition of Jet-in-Slit test is shown in Table 2. After the test, maximum corrosion depth and mass loss were measured.

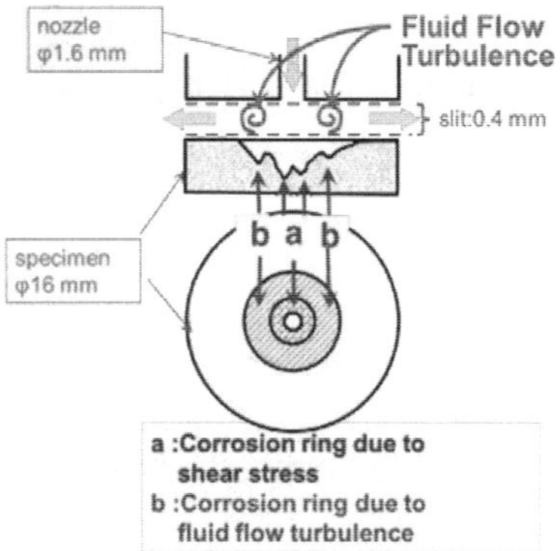

Fig.1 Principle of Jet-in-slit erosion-corrosion test

Table2 The test condition of Jet-in-Slit test

Test condition	
Test solution	1% $CuCl_2aq$
solution temp., (℃)	40
Flow rate (m/s)	3.3
Slit (μm)	400
time (hour)	2

3. Result

3.1 Tensile test

Tensile test results of KBS1 and KBS2 are sown in Table3, together with the tensile strength and elongation values of CAC203, CAC406 and CAC804 specified by JIS. As shown in Table 3, mechanical properties of KBS1 and KBS2 were inferior to those of CAC804; however, they were superior to those of CAC406. KBS1 and KBS2 meet the standard requirement of CAC203.

Table3 Tensile test results of KBS1 and KBS2 and standard value of JIS materials.

Alloy name	Value	Tensile Strength (MPa)	Elongation (%)
CAC203	Standard Value	more than 245	more than 20
CAC406	Standard Value	more than 195	more than 15
CAC804	Standard Value	more than 300	more than 15
KBS-1	**measured Value**	**301**	**40.5**
KBS-2	**measured Value**	**253**	**25.9**

3.2 Dezincification Corrosion test

The maximum dezincification corrosion depth of specimen is shown in Fig.2. CAC203 only exhibited a very large maximum dezincification depth of 426μm. The other four kinds of copper alloys, included KBS1 and KBS2, had the maximum dezincification depth of less than 30μm, showing good dezincification resistance.

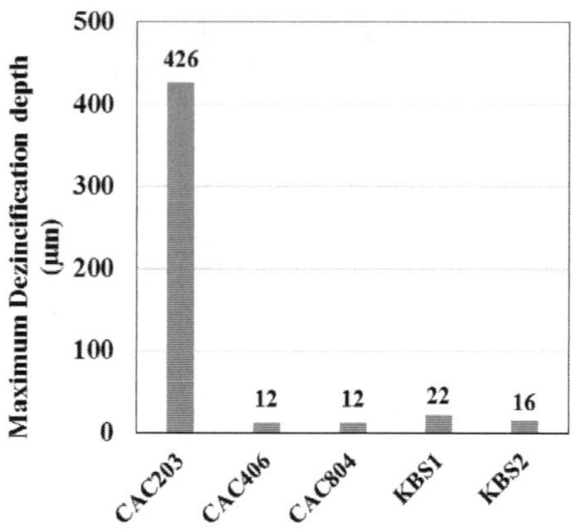

Fig.2 The maximum dezincification corrosion depth of specimen

3.3 Erosion-corrosion test (Jet-in-Slit test)

The results of Jet-in-Slit erosion-corrosion test are shown in Fig.3. In CAC203, CAC804 and KBS1, corrosion ring due to erosion-corrosion were observed. By comparing these three alloys, it could be found that KBS1 had a smaller maximum corrosion depth than the CAC203 and CAC804, suggesting that KBS1 had a better erosion-corrosion resistance than CAC203 and CAC804. In contrast, corrosion ring was not confirmed in CAC406 and KBS2, indicting that CAC406 and KBS2 had excellent erosion-corrosion resistance. Considering the two kinds of the corrosion test results, it could be suggested that KBS2 has the same corrosion resistance with CAC406 bronze castings.

Fig.3 Results of Jet-in-Slit erosion-corrosion test

Alloy	CAC406	CAC203	CAC804	KBS1	KBS2
Appearance	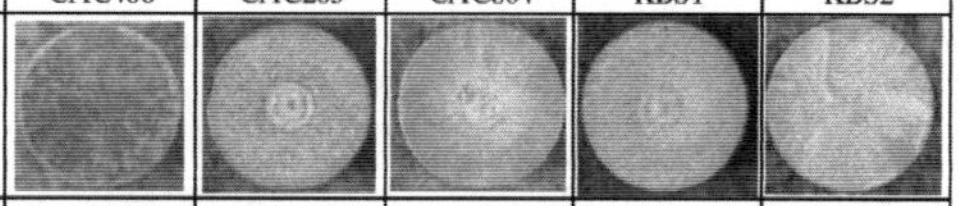				
mass loss (mg)	133	318	317	163	118
maximum corrosion depth (μm)	**25**	**498**	**502**	**287**	**33**

4. Conclusion

1) KBS1 and KBS2 have good mechanical property equal with JIS standard value of CAC203.

2) KBS1 and KBS2 show good corrosion resistance (including dezincification corrosion and erosion-corrosion resistances) as compared with CAC203 and CAC804

3) KBS2 has almost same erosion-corrosion resistance with CAC406 bronze castings.

4) Additive of Al, Ni, P and Sn in brass alloys were effective to corrosion resistance.

P-19

Impact of Speed of Cooling of Initial Phase (α) and of Eutectics (α + β) on Physical and Mechanical Properties of Al-Si-Mg Alloys

Jerzy Zych
AGH University of Science and Technology,
Faculty of Foundry Engineering,
Reymonta 23, 30-059 Kraków, Poland

ABSTRACT

The work shows the results of testing the speed of cooling cylindrical samples intensively cooled at one end. Designated cooling curves and their derivatives. The tests were subjected to alloy and A356.
Cooling speed varies in a wide range of both alpha phase crystallization and solidification eutectic. Increasing the cooling rate of temperature changes the location of T sol (start). At a higher speed of cooling of alloys achieves a higher density and hardness after heat treatment T6
Keywords: Al-Si-Mg alloys, speed cooling, cooling rate, mechanical properties

1.0. OWN RESEARCH

Assessing the impact of cooling speed on structure and properties of AlSi7Mg alloys on is very much the work of [1-7]. In the work of the cooling curves shall be recorded, calculates the speed of cooling and evaluates the structure of or properties of alloys. Similar studies were done by the author. The assessment of the speed of cooling of a ALSi7Mg alloy was conducted by preparing an alloy of the composition shown in table 1. Before pouring it into the mould, the alloy was modified by using TiB5 master alloys. In order to obtain high variability of the cooling speed, the casting technology shown in figure 1 was used in the tested sample. Samples in the form of a bar with a diameter of 25 mm touched the cooling plate with their lower ends, as shown on figure 1a. Thermocuples were placed every 5 mm, beginning from the cooling plate, so as to register the process of cooling of the metal.

Table 1. Composition of ALSi7Mg alloy (wt.%)

Chemical composition of the examined alloy [%]					
Si	Mg	Cu	Ti	Sr	Zn
7,67	0,48	0,06	0,206	0,0002	0,090

The registered processes of cooling are shown in figures 2 and 3. Analysis of the curves (of cooling) allows for the determination of thermocouples and average cooling speeds in the periods of solidifications in phase α as well as the coagulation of eutectics.

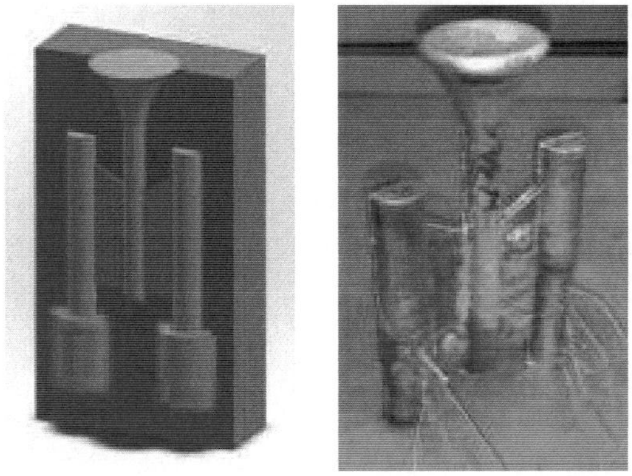

a/ b/

Fig.1 Test bar: a/ mold sands with chills, b/ casting whit thermocouples

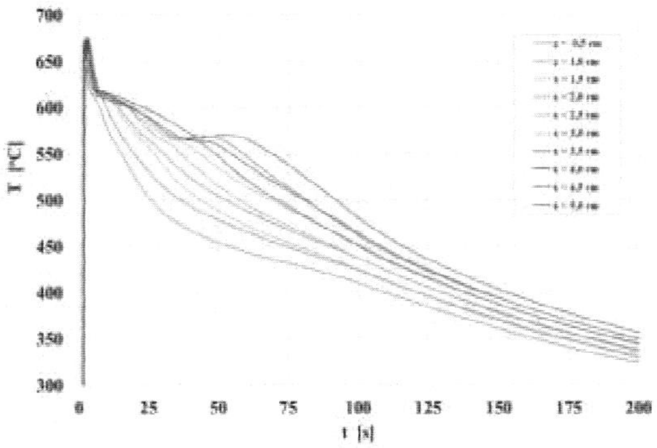

Fig. 2 Curves of cooling

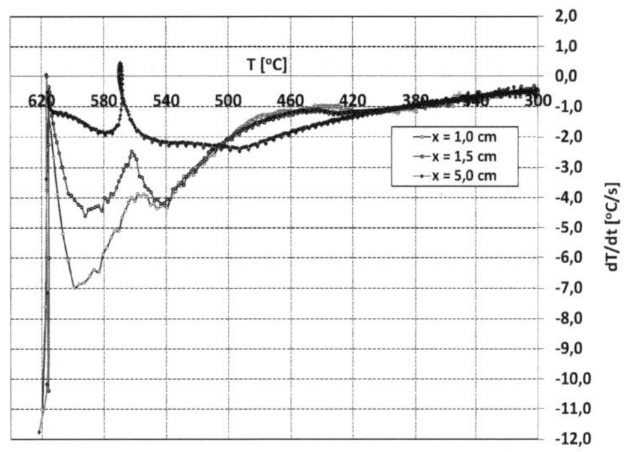

Fig. 3 Curves of the TDA

In figure. 3 cooling rate showing the custom which are determined in the following temperatures. In every field of cooling rate of change in broad ranges. Instantaneous speed determines microstructure phases: primary and eutectic. Influence of cooling rate on the position of the temperature T_{lik}, $T_{sol(s)}$, $T_{sol(f)}$ shows the figure 4. The samples were subjected to standard heat treating T6, after which the density of the alloy in function f' – distance from the cooling plate – and the hardness were determined. The results of this examination are shown in figure 5

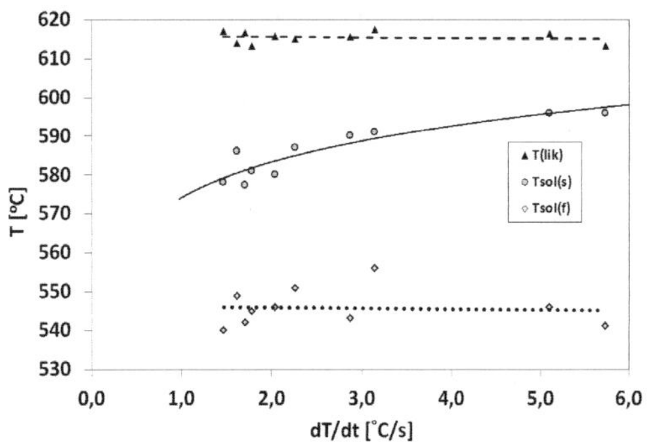

Fig.4. The impact of the cooling speed on the values of temperatures: T_{lik}, $T_{sol(s)}$, $T_{sol(f)}$ on the curve of cooling

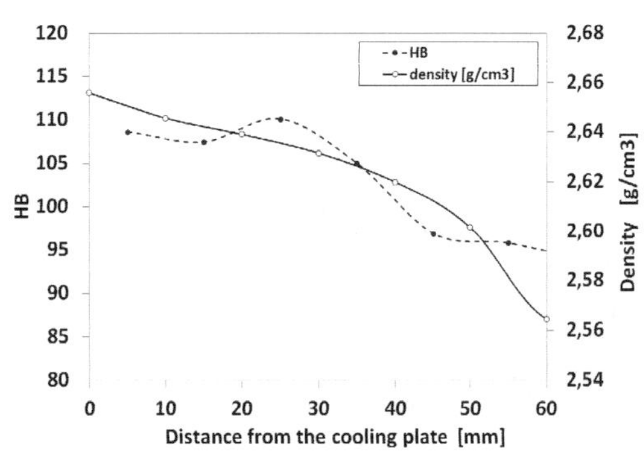

Fig 5. The impact of the cooling speed on the density and hardness alloys after heat treatment T6

2.0. CONCLUSIONS

From the results of the conducted examination, the following conclusions may be drawn:

- High cooling speeds (above 3° C/s) lead to the disappearance of the recalescens effect in the period of eutectic transformation
- Increase of cooling speed leads to the increase of the value of $T_{sol(s)}$ and has a small impact on the remaining characteristic temperatures (T_{Lik}, $T_{sol(f)}$).
- Higher cooling speeds cause the alloy to have higher density and hardness in the state after heat treatment T6.

REFERENS

[1] Djurdjevic M.B., Vicario I., Huber G.; Review of thermal analysis applications in aluminium casting plants, Revista de Metalurgia Vol 50, No 1, 2014.
[2] Seifeddine S., Sjölander E., Bogdanoff T.; On the Role of Copper and Cooling Rates on the Microstructure, Defect Formations and Mechanical Properties of Al-Si-Mg Alloys", Materials Sciences and Applications, 2013.
[3] Niklas A., Abaunza A., Fernández-Calvo A.I., Lacaze J. Suárez R.; Thermal analysis as a microstructure prediction tool for A356 aluminium parts solidified under various cooling conditions, China Foundry, Vol.8, 2011.
[4] Chen R., Shi Y., Xu Q., Liu B.; Effect of cooling rate on solidification parameters and microstructure of Al–7Si–0.3Mg–0.15Fe alloy, Transactionsof Nonferrous Metals Society of China, Vol. 24, No. 6, 2014.
[5] Pietrowski S., Władysiak R. Ocena krystalizacji siluminów podeutektycznych syntetycznych metodą ATD; Solidification of Metals and Alloy 1995.
[6] Thompson S.;Effect of cooling rate on solidification characteristics of aluminium alloys A356 and AA5182, Sheffield Hallam University, England, 1999.
[7] Anyalebechi P.N, Rouns T.N, Sanders R.S; Efect of cooling rate and grein refining on constituent phase particle in as-cast 3004 alloy. Light Metals. Vol 3. TMS. Wiley 2013.

Proceedings of the 72nd World Foundry Congress
21-25th May 2016, Nagoya, Japan

Mechanical properties with thixo-extrusion of the semi-solid 7075 Aluminum alloy

Jae-Chan Jo[1], Je-Hoon Lee[2], Dae-Hwan Kim[3], Su-gun Lim[4]

[1] Dept. of Materials Manufacturing Processes, Gyeongsang National University, 900 Gazwa, Jinju, Korea
[3] Dept. of Metallurgical and Materials Engineering, ReCAPT, Gyeongsang National University, 900 Gazwa, Jinju, Korea
[2, 4] Dept. of Metallurgical and Materials Engineering, Engineering Research Institute, Gyeongsang National University, 900 Gazwa, Jinju, Korea

*Corresponding author, e-mail: suglim@gnu.ac.kr

This study investigated on the mechanical properties with thixo-extrusion of semi-solid casted 7075 Aluminum alloy. The semi-solid alloys used for the experiments were prepared by cooling plate casting, and semi-solid cast alloy showed fine and globular microstructure more than permanent cast alloy. Difference of microstructures with extrusion direction showed that thixo-extrusion had equiaxed grain while hot extrusion had elongated grain along with extrusion direction. As a result, ΔHv value along with extrusion direction at thixo-extrusion is smaller than hot extrusion. Tensile test result of each extruded alloy showed that thixo-extrusion had low yield strength, low ultimate tensile strength and high elongation more than hot extrusion.

Keywords: 7075 Aluminum alloy, semi-solid casting, thixo-extrusion

1. Introduction

Al-Zn-Mg alloys have been used in the aerospace, the railroad cars and sporting goods which is required the specific strength. It is categorized for 2 groups according to Cu addition. A7075 that is typical alloy adding Cu have been used in structural materials requiring high strength, but when it was extruded or forged, due to high load, it has taken limitation on shape design and effectiveness reduction on life of mold [1-2].

Thixo-extrusion has been known for the alternative to solve the problem on the plastic forming of high strength Aluminum alloys. Because of rheological advantage of between cast and plastic forming taken by reheating process, the thixo-extrusion can decrease the extrusion pressure and reduce the problem on crystal grain anisotropy of the extruded alloys [3]. But some alloys that has solid-liquid coexistence state of small region is hard the application of thixo-extrusion, and it has difference of extrusion properties and mechanical properties of extruded alloys on reheating temperature at solid-liquid coexistence region [4-5].

In this study, to search reheating temperature of giving moderate liquid rate used for previous research result [6], and the microstructure and mechanical properties of the billets made by semi-solid casting method and permanent mold casting method was investigated.

2. Experiment

2.1 Semi-solid casting using slope cooling method

The composition of A7075 alloy used in this study is given in Table 1. The alloy melted at 740°C was treated by degassing process, then, when it was at 650°C, it was poured at mold of 570°C. And after it had a holding time during 60sec, it was quenched by water. When semi-solid casted alloys was poured at mold, slope cooling method was used.

2.2 Thixo-extrusion of reheated billet

The reheating of billet for thixo-extrusion was performed at 595°C, and The reheated billet was extruded for diameter 10mm at extrusion rate of 27:1, Ram speed of 1mm/s and heated extrusion container for 450°C.

The microstructure analysis of A7075 prepared at each conditions was observed by optical microscope, its mechanical properties measurement was performed by universal testing machine using tensile specimen made on ASTM E 8M.

Table 1. Chemical compositions of A7075 alloys (mass%).

Zn	Mg	Cu	Si	Fe	Mn	Zr	Al
6.07	2.50	0.41	0.03	0.12	0.28	0.09	Bal.

3. Result and Discussion

In Fig. 1(a) observed the microstructure of dendrite structure, but In Fig. 1(b) generally observed the microstructrue of spherical structure. Because the primary crystal nuclear made by slope cooling method provided heterogeneous nucleation site, the microstructure of semi-solid cast materials was more refined.

Fig. 2 showed the microstructure on reheating temperature of semi-solid cast materials. As reheating temperature was increased, the partially melted grain boundary was observed, because the melting point was decreased by alloying elemnet around grain boundary.

Table 2 showed maximum extrusion pressure and conditions of each extrusion. Partially melted result by reheating process showed significant decraese of extrusion pressure.

Table 3 showed tensile properties of hot-extruded materials and thixo-extruded materials. At both conditions, Tensile strength was not many difference, but fracture elongation showed significant difference.

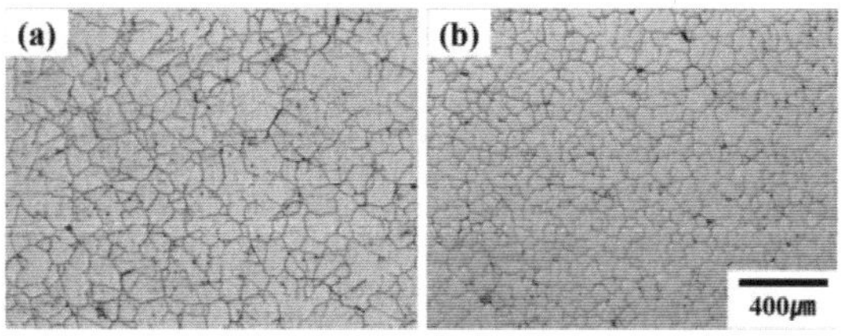

Fig. 1. The microstructures of (a) permanent mold casting, (b) semi-solid casting

Table 2. Extrusion conditions and maximum pressure

Extrusion Method	Rehearting Temp. (°C)	Extrusion Speed (mm/s)	$E.P._{max}$ (MPa)
Hot-extrusion	-	1	816
Thixo-extrusion	595	1	597

Table 3. Mechanical properties of each extrusded materials

Extusion Method	Yield Stress (MPa)	Tensile Stress (MPa)	Elongation (%)
Hot-extrusion	263	375	15.7
Thxio-extrusion	227	368	25.2

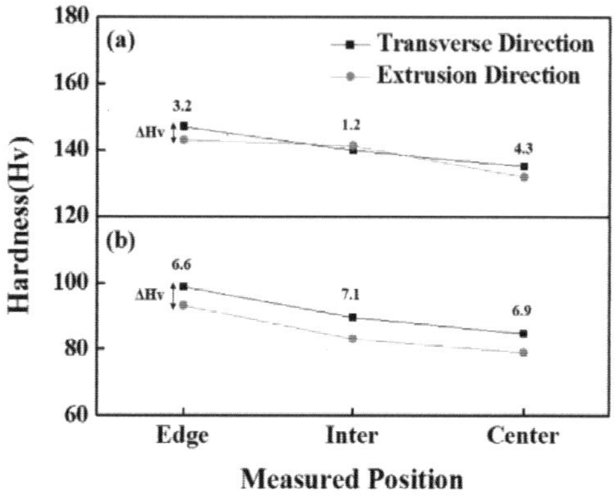

Fig. 2. Vicher`s hardness of (a) thixo-extrusded and (b) hot-extrusded materials; * **ΔHv = TD - ED**

Fig. 3 showed vicher`s hardness measurement results on extrusion direction and transverse direction of hot-extruded materials and thixo-extruded materials. The hardness of thixo-extruded materials was higher than hot-extruded materials. and ΔH_V value on hardness difference of between extrusion direction and transverse direction was smaller at thixo-extruded materials. As a result, the microstructure on extrusion direction at thixo-extrusion was spherical, and it means that microstructure had isotropy characteristic.

4. Conclusions

The following conclusions can be obtained from this investigation.

(1) The microstructure was mostly spherical by thixo-extrusion, the result showed decrease of hardness difference between center position and edge position of extrusion materials.

Acknowledgements

This work was supported by the Human Resource Training Program for Regional Innovation and Creativity through the Ministry of Education and National Research Foundation of Korea (2015-0633).

References

[1] Y. O. Yoon: RARE METALS. 25 (2006) 124-128.

[2] M. A. Bayoumi: Mater. Des. 30 (2009) 4469-4477.

[3] S. Y. Shim: Sur. Rev. Lett. 17 (2010) 185-188.

[4] J. P. Eom and D. E. Yoon: KFS. 17 (1997) 538-544.

[5] S. Y. Shim: Materials Transactions. 52 (2011) 862-867.

[6] J. H. Kim: Metals and material international. 16 (2010) 669-677.

Proceedings of the 72nd World Foundry Congress
21-25th May 2016, Nagoya, Japan

Effect of boron on the grain boundary morphology and creep property in nickel-based superalloy LESS 1

Jong-Cheon Yoon[1], Byung-IL Kang[1], Jeong-IL Youn[1], Young-Jig Kim[1]

[1]School of Advanced Materials Science and Engineering, Sungkyunkwan University, Korea

2066 Seobu-ro, Jangan-gu, Suwon, Gyeonggi-do 16419, Korea

To examine the correlation between microstructure and creep property by addition of boron, effect of boron on the grain boundary morphology and creep resistance in nickel-based superalloy LESS 1 is investigated. Scanning electron microscopy (SEM), transmission electron microscopy (TEM) and small-punch creep test are carried out. In case of boron-added LESS 1, so-called boro-carbides ($M_{23}(C, B)_6$) and γ` film ($Ni_3(Al, Ti)$) are precipitated at grain boundary (GB). Also creep resistance is considerably increased by addition of boron. The improvement of creep resistance can be attributed to the change of grain boundary morphology by addition of boron.

Keywords: *Nickel-based superalloy, creep, boron, grain boundary, boro-carbide*

1. Introduction

Nickel-based superalloys are under development as next generational materials for Hyper-Super Critical fossil power plant because of its outstanding mechanical properties at high temperature [1] and it has been proved that the additions of boron improves creep resistance of nickel-based superalloy by previous study. However, the mechanism about reinforcement have not been systemically studied by other research groups. Among them, the main explanations are divided. First, boron atoms are distributed at γ-$M_{23}C_6$ interface and this interface disrupts the movement of dislocation [2, 3]. Second, boron atoms are soluble in lattice of Ni_3Al and produce the strengthening effect [4, 5]. Finally, the reaction of carbide precipitation is influenced by boron [6]. As mentioned above, understanding rules of boron in nickel-based superalloy is not confirmed.

In this research, to understanding the role of boron, the existent position of boron and phase transition in nickel-based superalloy LESS 1 were investigated.

2. Experimental

LESS 1 alloy which designed by changing the composition of Cr, Al, Ti and W for control of TCP phases of INCONEL alloy 740 was used in this study. Its nominal chemical composition is Ni-22Cr-20Co-2Nb-2W-1.5Al-1.5Ti-0.3C (wt. %). The alloy was prepared by vacuum induction furnace. The experimental alloys LESS 1 and LESS 1 (B) are without and with B addition respectively. These specimens were homogenized at 1200 °C for 16 h and hot-forged at 1100 °C. The as-forged specimens were subjected to standard heat treatment: 1200 °C/0.5 h WQ + 800 °C/16 h AQ. Before starting investigation, actual chemical composition of LESS 1 and LESS 1 (B) was confirmed by EDS, OES and carbon/sulfur analyzer. Boron contents were 10 ppm and 110 ppm respectively.

Microstructure was observed by 7600F (JEOL) Field Emission Gun Scanning Electron Microscope (SEM) and JEM ARM 200F (JEOL) Transmission Electron Microscope (TEM).

Creep tests were carried out at 700 °C/50 kgf by a small-punch creep tester. The sample size was 10 x 10 x 0.5 mm.

3. Results and discussion

3.1. Microstructure characteristic

Grain size and grain interior between LESS 1 and LESS 1 (B) were similar. Fig. 1 shows the typical GB morphology of LESS 1 and LESS 1 (B). In case of LESS 1, Ti and Nb-enriched MC and Cr-enriched $M_{23}C_6$ carbides were partially distributed at the GB, whereas LESS 1 (B) revealed that Cr-enriched $M_{23}C_6$ carbides were continuously distributed at the GB. In general, the critical location of $M_{23}C_6$ carbides can improve the rupture strength. But rupture failure is usually initiated by desorption of γ- $M_{23}C_6$ interface [7].

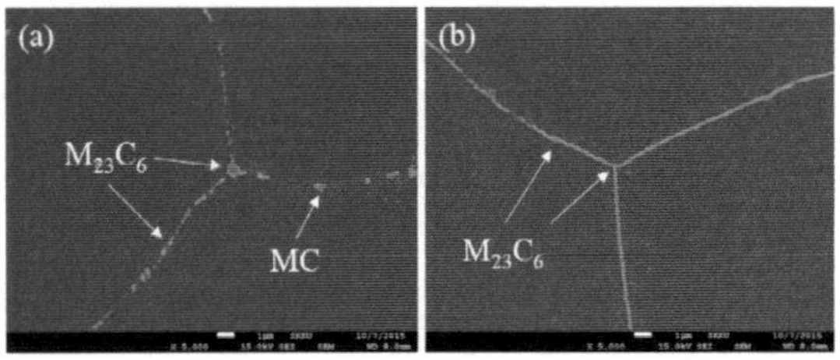

Fig. 1 SEM images showing the grain boundary morphology.

Results of TEM investigation in LESS 1 (B) are shown in Fig. 2. By the EDS analysis, coexisting titanium and aluminum which surrounding Cr-enriched carbides was observed (Fig. 2a). To identify the phase, SADP analysis was carried out and indicated that γ` film formed along the GB (Fig. 2b) which is believed to optimize the rupture properties [7].

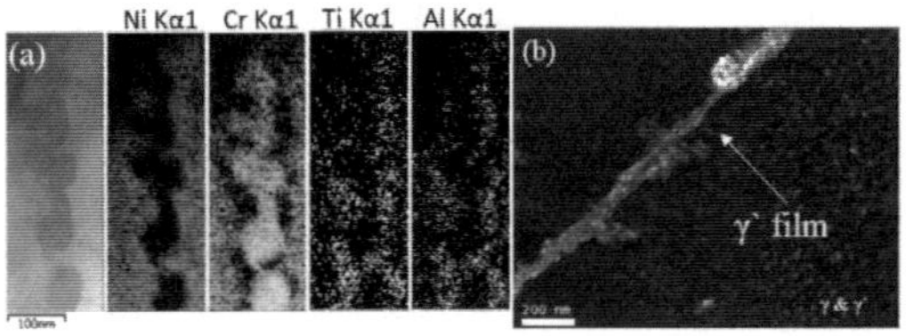

Fig. 2 TEM results of LESS 1 (B): (a) EDS, (b) dark field image of γ` film.

Fig. 3 shows distributions of carbon and boron by EELS. It confirmed that the $M_{23}C_6$-type carbides in the γ` film were not $Cr_{23}C_6$ but $Cr_{23}(C, B)_6$.

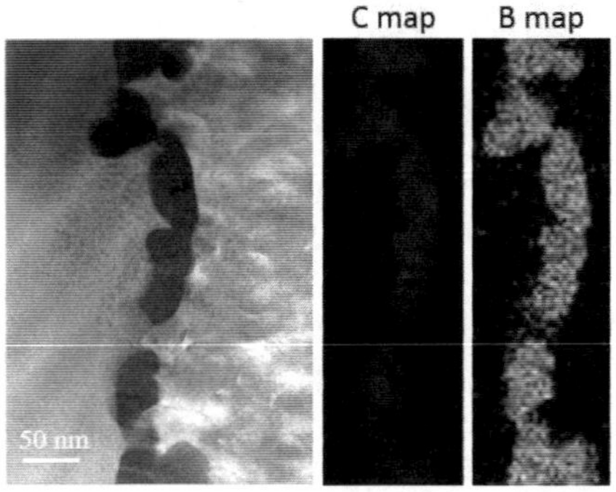

Fig. 3 Result of EELS showing the distribution of boron and carbon at the GB.

3.2 Creep behavior of LESS 1 and LESS 1 (B)

The result of creep tests is illustrated in Fig. 4. It is to be noted that in case of LESS 1 (B), test was aborted. Although test of LESS 1 (B) was aborted, it was confirmed that creep resistance of LESS 1 (B) was significantly improved.

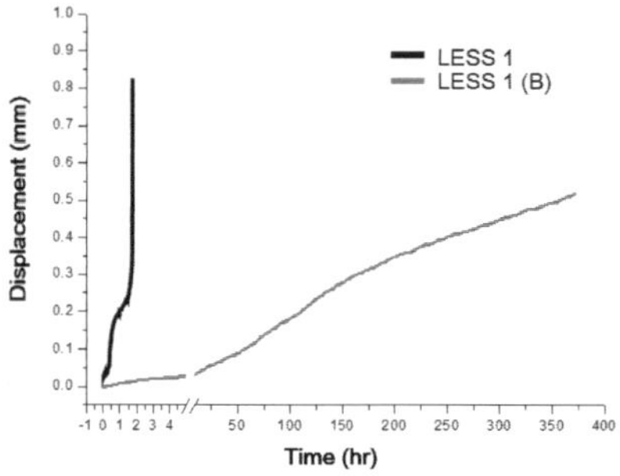

Fig. 4 Result of small-punch creep test at 700 °C/50 kgf.

4. Conclusion

Based on the results presented in this study, the following conclusions can be made:

1) After standard heat treatment, grain size and grain interior of LESS 1 and LESS 1 (B) are similar but reaction of carbide precipitation is changed by boron.
2) Source for Cr-enriched $M_{23}C_6$ is not only carbon but also boron, so that phase fraction of $M_{23}C_6$ is increased by addition of boron and it should be noted that the $M_{23}C_6$ mainly precipitates at grain boundary.
3) As a result of increased $M_{23}C_6$ carbides, concentrations of aluminum and titanium in γ matrix are simultaneously increased, so that it is supposed that γ` film is formed surrounding $M_{23}C_6$ carbides.
4) Creep resistance is considerably increased by addition of boron and it is assumed that the changes of grain boundary morphology make creep resistance improvable.

References

[1] Tresa M. Pollock: J. Propul. Power. Vol. 22, No. 2, (2006) 361-374.

[2] Changshuai Wang, YongAn Guo, Jianting Guo, Lanzhang Zhou: Mater. Sci. Eng. A 639 (2015) 380–388.

[3] L. Xiao, D.L. Chen, M.C. Chatuvedi: Mater. Sci. A 428 (2006) 1–11.

[4] C. S. LEE, G. W. HAN, R. E. SMALLMAN, D. FENG and J. K. L. LAI: Acta mater. Vol. 47, No. 6, (1999) 1823-1830.

[5] I. Baker, B. Huang, E. Schulson: Acta Metall. 36 (1988) 493–499.

[6] M. A. Burke, J. Greggi, Jr., and G. A. Whitlow: Scripta metallugica. Vol. 18, (1984) 91-94.

[7] C. T. Sims and W. C. Hagel: *The Superalloy*, (John Wiley & Sons, Inc, London, 1972) pp. 38 – 69.

Proceedings of the 72nd World Foundry Congress
21-25th May 2016, Nagoya, Japan

Damage Behavior of Micro Segregation in Al-Si Alloy by Use of Phase Contrast Imaging

Ryusuke Yamada[1], Hiroyuki Toda[1], Yasuko Besel [1], Kentaro Uesugi[2], Takeuchi Akihisa[3]
[1] Faculty of Engineering, Kyushu University, 744, motooka, nishi-ku, Fukuoka-city 819-0395, Japan
[2] Japan Synchrotron Radiation Research Institute (JASRI), 1-1-1, Kouto, Sayo-cho, Sayo-gun, Hyogo 679-5148 Japan

Some segregation in casting alloys are microstructural defects and degrade mechanical properties of the alloys. Therefore it is important to understand the influence of the distribution of segregation regions on fracture behavior. In this study, tensile test was carried out on an Al-Si cast alloy. Three-dimensional (3D) in-situ observation of its tensile fracture behavior was performed by means of the synchrotron X-ray computed tomography (CT) technique. Some segregation regions have similar density to that of the matrix. Phase retrieval processing, which is one of the phase contrast imaging techniques, enables us to distinguish such segregations based on X-ray CT tomographic images. Fracture path was identified by comparing unfractured and fractured 3D volumes. Grey values of each voxel in CT images correlate to the density of the material after the phase retrieval processing. Examining grey values of voxels on the extracted fracture path, it was found that fracture occurred preferably around the segregation part with high density.

Keywords: *Segregation, Phase contrast imaging, Die-cast alloy, Micro tomography, In-situ study.*

1. Introduction

Casting alloys unlike other forged or milled alloys being heat treated, has large scattering in material properties[1]. Segregation is one of the major causes of this variation[2]. It is the deviation of the alloying elements in the material and forms very complex three-dimensional network[3]. Some segregation region region could present as defects in the material and degrade its mechanical properties. The objective of this study is to evaluate correlation of three-dimensionally distributed segregation region and the fracture behavior of an Al-Si casting alloy. In-situ observation of tensile fracture behavior was carried out by means of the synchrotron X-ray CT technique. Since the some segregation regions in the Al-Si cast alloy have similar density to that of the matrix, it is difficult to distinguish them in absorption contrast imaging. In this study, phase retrieval processing based on X-ray CT images, was employed to visualize the segregations distinctively from the matrix.

2. Experiment

2.1 Material

Al-Si casting alloy with a composition of Al-11.82%Si-2.35%Cu-0.81%Fe-0.18%Mn-0.17%M g-0.06%Pb-0.04%Ni (mass %) was used in this study. A micro test piece for in-situ CT observation was cut by electrical discharge machining (EDM). Parallel section of the test piece had a prism shape with a cross section of 0.6×0.6mm2. Synchrotron X-ray CT imaging was performed at the undulator beam-line BL20XU of the SPring-8. In-situ observation during a tensile test was carried out at several strain levels as marked in Fig. 1. Phase contrast imaging was conducted at the third strain level just before the final fracture.

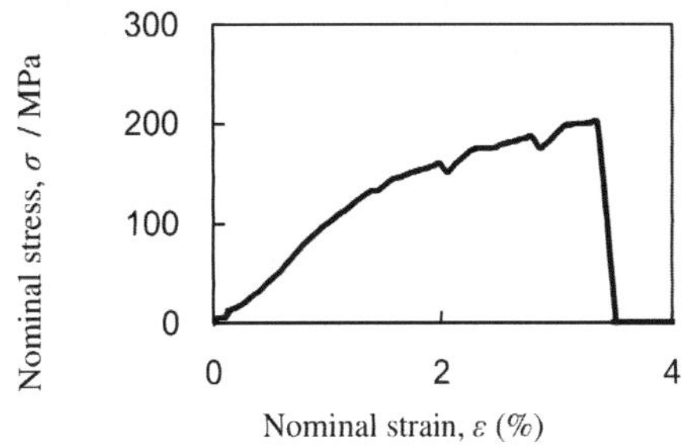

Fig. 1 Stress-strain curves in this tensile test

2.2 Phase retrieval processing

In order to visualize the segregation regions in the matrix, phase retrieval processing was applied in this study. It is based on Paganin's formula shown in Eq. (1). Details of this method is described somewhere else[5];

$$\varphi(x,\ y) = \frac{1}{2} ln \left[F^{-1}\left\{\frac{F(I(x,\ y))/I_0(x,\ y)}{\beta/\delta+[\lambda L/4\pi](v^2+u^2)}\right\}\right] \quad (1)$$

where I and I_0 are the intensity of transmitted and incident X-ray, respectively. λ is the wavelength and L is a camera length. δ and β are the real and the imaginary part of the complex refractive index of refraction of the X–ray, respectively. Δδ is the difference between the refractive indices of the two phases; in this study the matrix and the segregations. However, Δδ is related with the difference in densities of the two phases. Since there is scatter in the densities of the segregations in this material, Δδ was to be investigated.

3. Results and Disccusion

Figures 2(a) and (b) show the virtual cross sections of absorption and phase contrast images before no-loading. As seen in Fig. 2(b), difference in contrast was emphasized in the phase contrast image. The material with high density looks brighter in phase contrast images. In this material, the segregations contain heavier elements such as Cu or Fe than the matrix phase of Aluminum and look brighter in these images.

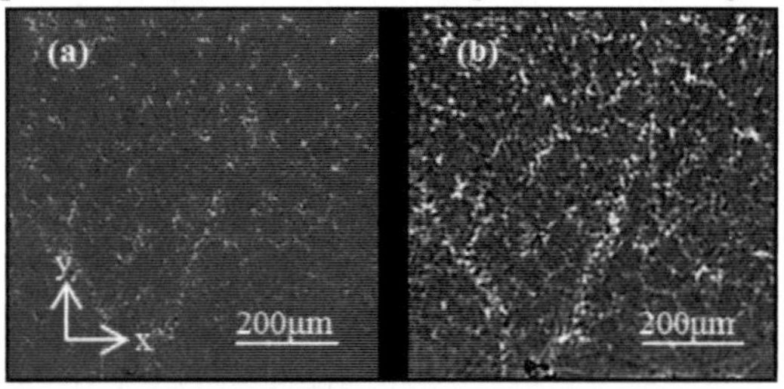

Fig. 2 (a) Absorption contrast image (b) Phase contrast image

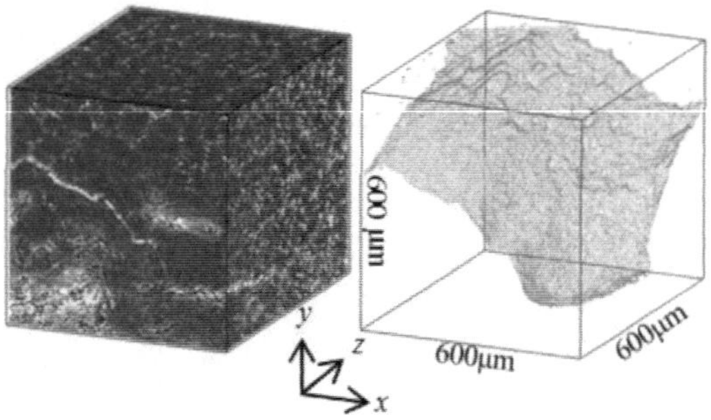

Fig. 3 (a) 3D volume and (b) extracted voxels around fracture path

Figure 3(a) shows the 3D volume observed just before the final fracture. Voxels surrounding fracture path were extracted from this volume as shown in Fig. 3(b). In Fig. 3(b), the voxels with grey values of higher part were colored in green and those with lower grey values in blue. It can be seen that most part of the fracture occurred around the high density part because of segregation. Moreover, the crack is developed complicatedly in 3D. This can be attributed to the segregation existed in this material. Figure 4 shows the grey value histogram of the voxels surrounding the fracture path (Fig. 3(b)) together with that of the full volume (Fig. 3(a)).

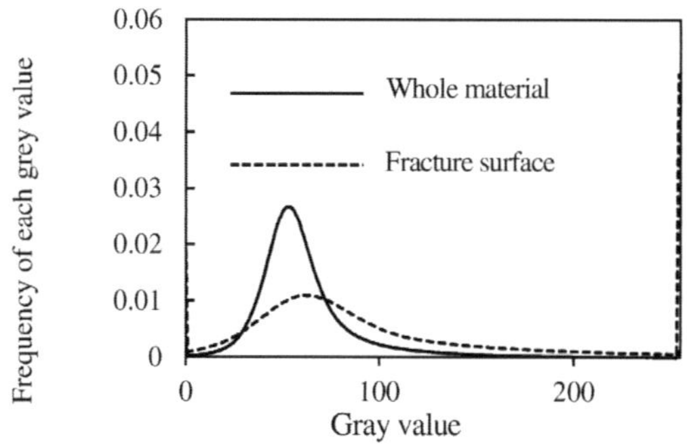

Fig. 4 histogram of grey value

It can be seen in Fig. 4 that the peak of the histogram of extracted voxels around the fracture surface was shifted to the right; i.e. they tend to have higher grey values than the whole material. From the observation results by means of phase and absorption contrast imaging technique together with X-ray CT images described above, it was found that the fractured area had higher grey values than that of the whole matrix. It suggests that fracture occurred preferably around the segregation region with high density in the Al-Si casting alloy.

References

[1] M. Okayasu, K. Kanazawa, N. Nishi, J. Jpn.: Foundry Eng. Soc:70 (1998), pp. 779-785

[2] X. S. Huang, L. J. He, G. B. Mi and P. J. Li: Materials Science and Technology 31 (2015), pp. 400-408

[3] H. I. Laukli, C. M. Gourlay, A. K. Dahle, O. Lohne: Materials Science and Engineering A 413–414 (2005), pp. 92-97

[4] M. Kobayashi, H. Toda, Y. Kawai, T. Ohgaki, K. Uesugi: Acta Mater 41 (2009), pp. 506-519

[5] D. Paganin, S.C. Mayo, T. E. Gureyev, P. R. Miller, S. W. Wilkins: Journal of microscopy 206(1) (2002), pp. 33-40

Proceedings of the 72nd World Foundry Congress
21-25th May 2016, Nagoya, Japan

Fabrication of Fine Diamond Dispersed Metal-Based Grinding Wheel by a Centrifugal Mixed- Powder Method

Eri Miura-Fujiwara[1], Hisashi Sato[2], Motoko Yamada[2] and Yoshimi Watanabe[2]
[1] Graduate School of Engineering, University of Hyogo, 2280 Shosha, Himeji, Hyogo, Japan
[2] Graduate School of Engineering, Nagoya Institute of Technology, Gokiso-cho, Showa-ku, Nagoya, Japan

Recent trend in aviation industry, weight reduction by using composite, such as a carbon fiber reinforced plastic (CFRP), has been positively promoted to save weight without sacrificing reliability. With increasing amount of usage of CFRP, improving quality and accuracy of CFRP drilling holes for a faster is one of the important requirements in order to ensure the safety of aircraft. To achieve the requirements, gyro driving grinding wheel system for CFRP drilling was proposed. For the grinding wheel, we have proposed functionally graded material (FGM) fabricated by a centrifugal mixed-power (CMP) method. The CMP method is one of the powerful and cost-effective processes to obtain composite materials by means of pressure impregnation. The mixed-powder method in addition to the centrifugal casting will make the fine particle dispersed FGM achievable. In this study, we attempted to fabricate Al/diamond and Cu/diamond composites by this method. Effect of casting temperature and metallic powder diameter on diamond abrasive powder dispersion was discussed.

Keywords: *functionally graded materials, metal-matrix-composite, Cu/diamond composite, Al/diamond composite, centrifugal casting method*

1. Introduction

Nowadays, carbon fiber reinforced plastic (CFRP) is starting to positively promoted for the next generation aircraft to save weight without sacrificing reliability since the composites are lightweight and have high specific strength. On the other hand, one of the major problems against application of CFRP to aircraft is machinability. Especially, improving quality of drilling hole is one of the urgent matters.

Takekoshi *et al.* proposed gyro-driving grinding wheel system for CFRP drilling as these requirements[(1)]. This system can provide high quality drilling hole by means of a two-axes rotating drive with a metal-matrix grinding wheel. For the grinding wheel, Watanabe *et al.* had proposed functionally graded materials (FGMs) fabricated by a centrifugal mixed-power (CMP) method [(2-4)]. The CMP method is a unique type of the centrifugal casting method, which can makes it possible to obtain FGMs on which surface fine reinforcement particles are dispersed.

In this study, we attempted to fabricate diamond dispersed Al or Cu matrix FGMs by the CMP method. Effect of casting temperature, centrifugal force and cast design on transverse particle distribution and effect of matrix powder diameter on particle distribution along depth direction were investigated in order to optimize casting condition for Al/nano-diamond and Cu/micron-diamond FGMs by the CMP method.

2. Experimental procedures

Mixed powder was prepared by mortar mixing with diamond powder and metallic powder. In this study, same element was used for the metallic powder and the matrix. Regarding Al/nano-diamond composites, Al power was employed for the matrix and nano-sized diamond powder was employed for dispersed particles. Mean primary diameter of nano-sized diamond powder ($d_{diamond}$) was approximately 50 nm. The used Al particle diameters (d_{Al}) were 425 - 850 μm, 250 - 245 μm, and 20 - 150 μm. After casting, matrix composition was 99.5 % Al. Cu powder and micron-diamond powder was prepared. Diamond powder with $d_{diamond}$ = 149 μm was used for this study, and particle size of Cu powder was less than 40 μm. From now on, nano-sized and micron-sized diamond will be abbreviated to nD and mD, respectively.

Two types of centrifugal casting machines were employed for the CMP method. Vertical-type and parallel-type centrifugal casting machines were used to Al/nD composites and Cu/mD composites, respectively. Details of the casting machines were shown in elsewhere[(2, 5, 6)].

3. Results and Discussions

3.1 Effect of casting temperature on powder distribution of the Al/nD composite

Fig. 1 shows side overview of the casts fabricated at the casting temperature at 984 K and 1038 K. The cast at 984 K, it seems that powder distribution was inhomogeneous in a longitudinal direction to the spin-mold. Powder was segregated and was mainly distributed within 50 mm from the runner side, where molten Al was infused, and less than 10 mm from motor side (the most far side from the inlet). On the other hand, at 1038 K, powder was successfully distributed uniformly along transverse direction. X-ray diffraction (XRD) profiles obtained from the cast surface indicated that nD kept diamond structure after casting, and no major reaction phase was observed on the sample surface. Therefore, it can be concluded that not only Al oxidation but also diamond-to-graphite transformation was sufficiently prohibited in the low pressure.

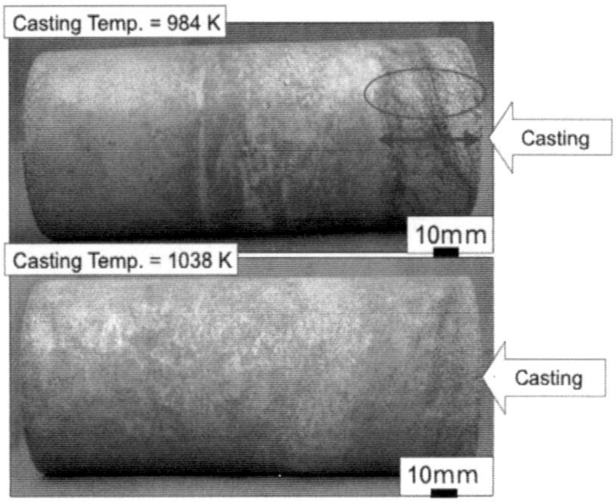

Fig. 1 Side overview of the Al/nD cast tubes fabricated at 984 K and 1038 K by the CMP method.

3.2 Fabrication of Cu/mD composites by CMP method

Initially, in case of poor wettability combination such as Cu and diamond, threshold impregnation pressure (P_c) can be roughly estimated in terms of wettability. According to Nishida[7], minimum spinning rate (N_p) in order to overcome P_c can be estimated by following equation;

$$N_p \geq \frac{1}{2\pi}\sqrt{\frac{2P_c}{\rho_m\left(r_1^2 - r_0^2\right)}}$$

ρ_m is density of molten metal; r_0 and r_1 are distance at medial surface of molten metal from rotation axis and that of mixed powder, respectively. Based on this equation, 34.7 G was applied for the centrifugal casting. In conclusion, Cu/mD FG composite was successfully obtained with 5 mm length of sprue at 1473 K of casting temperature. On the other hand, not only pre-form threshold stress also molten metal flow during casting played important role for the immobilization of diamond powder.

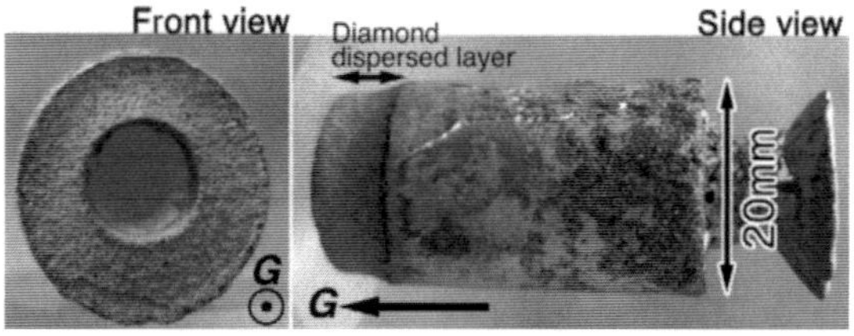

Fig. 2 Front and side overview of the Cu/mD cast grinding wheel fabricated by the CMP method.

4. Conclusions

Both diamond-particle-distributed Al and Cu FGM composites for a grinding wheel can be successfully obtained by CMP method.

Acknowledgements

This work was supported by Ministry of Education, Culture, Sports, Science and Technology, Regional Innovation Cluster Program (Global Type) "Tokai Region Nanotechnology Manufacturing Cluster".

References

1. Y. Watanabe, E. Miura-Fujiwara, H. Sato, K. Takekoshi, H. Tsuge, T. Kaga, N. Bando, S. Yamagami, K. Kurachi and H. Yokoyama: International Journal of Materials and Product Technology **42** (2011) 29-45.
2. Y. Watanabe, Y. Inaguma, H. Sato and E. Miura-Fujiwara: Materials **2** (2009) 2510-2525.
3. Y. Watanabe, S. Jamian, E. Miura-Fujiwara and H. Sato: Proc. of 11th Japan International SAMPE Symposium & Exhibition (JISSE-11) **SMS4-1** (2009) 1-5.
4. Y. Watanabe and H. Sato: Key Eng. Mater **434-435** (2010) 751.
5. T. Kunimine, M. Shibuya, H. Sato and Y. Watanabe: Journal of Materials Processing Technology **217** (2015) 294-301.
6. Y. Watanabe and H. Sato: *Review fabrication of functionally graded materials under a centrifugal force*, Ed., (InTech, 2011) pp.
7. Y. Nishida: *Introduction to Metal Matrix Composites*, 4th ed. Ed., (Corona Publishing Co. Ltd., Tokyo, Japan, 2008)pp.

Proceedings of the 72nd World Foundry Congress
21-25th May 2016, Nagoya, Japan

Interfacial Characteristics of Al-Cu Foam Cast Composites for High Conductivity Applications

Nam-Hoon Kim[1], Jeong-Min Kim[2*] and Se-Hyun Ko[3]
[1] Dept. Advanced Mater. Eng., Hanbat National Univ., Daejeon, Korea
[2] Dept. Advanced Mater. Eng., Hanbat National Univ., Daejeon, Korea
[3] Incheon R&D Center, Korea Institute of Industrial Technology, Incheon, Korea
*Corresponding author: jmk7475@hanbat.ac.kr

To optimize the conductivity and the weight lightning, Al-Cu composites were prepared through an suction casting procedure. Pure copper metal foam was infiltrated by aluminum melt by applying the vacuum, and then warm rolling was conducted to remove some remaing pores at the inferface between the Cu foam and the aluminum matrix. Even though the casting time was short, a significant dissolution of Cu into the melt was observed. And various AlCu intermetallic compounds were found to occur at the interface during the isothermal heating process after the casting. The average thickness of the AlCu intermetallic compound tended to increase in proportion to the heating time. The electrical and thermal conductivity of the cast composites were found to be comparatively low, mainly attributed to the incomplete interfacial contact.

Keywords: *aluminum composites, copper foam, conductivity, intermetallic compound*

1. Introduction

Metal matrix composites (MMCs) or hybrid structure metallic materials in which different metals with various properties are joined have recieved attention for many industrial applications. There are numerous methods to fabricate MMCs or the hybrid materials such as casting, accumulative roll bonding (ARB) and powder metallurgy processes [1-3]. Among the many available methods, compound casting is an econmic process in which solid state metal and liquid state metal are brought into contact [3,4].

Recently, copper foam materials are commericially available, so high conductivity Al/Cu composites may be easily fabricated by infiltrating aluminum melt into the copper foam. In this research a vacuum suction casting method was also employed to produce the Al/Cu foam composites, followed by isothermal heating processes to strengthening Al/Cu interfaces. This article aims to investigate the microstructural evolutions with respect to the heat treatments of cast Al/Cu composites and the resultant properties of them.

2. Experimental Procedure

A piece of copper foam (Metalfoam, Korea) was inserted in a quartz tube with the inner diameter of 8mm, and pure aluminum melt at the temperature of about 750°C was infiltrated into the prepositioned pure Cu foam by a vacuum suction apparatus. Then, the fabricated cast Al/Cu composite cylinder was warm-rolled (annealed at 300°C before rolling) and isothermally heated under different conditions.

The electrical and thermal conductivities of the specimens were evaluated by using a contact-type tester (Fischer, Sigmascope SMP10) and laser flash analyzer (Netzsch, LFA447), respectively. The microstructural analyses were performed using optical microscope, scanning electron microscope (SEM, JEOL, JSM-5610) equipped with energy dispersive X-ray spectrometer (EDS), and X-ray diffractometer (XRD, Rigaku, Smartlab).

3. Results and Discussion

3.1 Dissolution of Cu Insert

Significant dissolution of copper occurred during the suction casting process. Based on SEM-EDS analyses, Al-Al_2Cu eutectic phases were formed around the prepositioned copper foam due to the dissolution. This dissolution is known to be generally proportional to the square root of contact time, and can be increased by a dynamic agitation [5]. In this experiment, aluminum melt is suctioned and forced to flow through tortuous paths in the prepositioned copper metal foam. Therefore, it is assumed that the flow rate was high and the surface area of the Cu foam was also large, resulting in the significant dissolution.

3.2 Interfacial Reactions during Annealing

Apparently, diffusion and reactions took place at the Al/Cu interface during a isothermal heating process. Two distinct intermetallic compounds and one transient phase between them were observed in the heated specimens. The two compounds were identified as AlCu and Al_2Cu, respectively. The average thickness of the AlCu intermetallic compound was increased during the isothermal heating at 400°C.

Fig. 1 shows that the thickness of AlCu intermetallic compound heated at 400°C generally follows a parabolic relationship. This suggests that the reactive diffusion is controlled by volume diffusion because the square root of the compound thickness (l) is proportional to the heating time (t): $l^2 = kt$ where k is the parabolic coefficient [6-7].

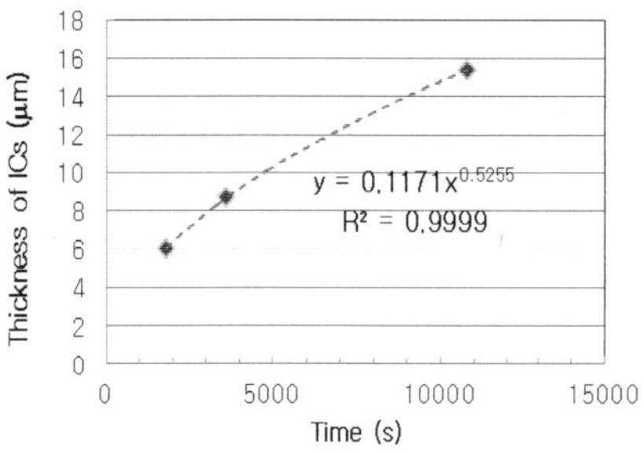

Fig. 1 Thickness variation of AlCu intermetallic compound formed at the Al/Cu interface of composites during isothermal heating at 400°C

3.3 Electrical and Thermal Conductivity

The electrical conductivity of composite specimens measured is compared with respect to the preparation method in Fig. 2. The conductivity is apparently higher in the case of higher rolling thickness reduction ratio, probably owing to the increased contact area with reduced porosity at the interface. Isothermal annealing could enhance the conductivity of both rolled specimens with different reduction ratios, howerver clearly greater improvement was made for the case of low reduction ratio. It is postulated that some pores diffused out and the interfacial contact became closer during the anealing. It has been also known that the conductivity is improved if the residual stress level is reduced.

The thermal conductivity of the cast composite specimens with 80% reduction ratio after annealing was also measured by a laser flash method, the obtained value was about 101.3(W/mK). Considering that the thermal conductivity of pure aluminum is 237(W/mK), the conductivity of cast composite seems to be considerably low.

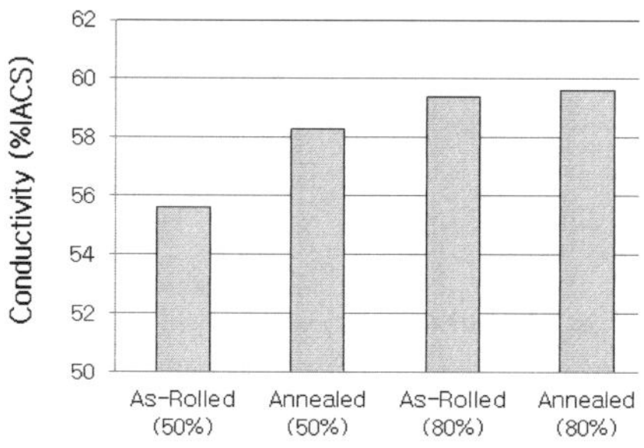

Fig. 2 Electrical conductivity of Al/Cu composites with different rolling thickness reduction ratios

4. Conclusions

A singnificant dissolution of copper preform into aluminum melt occurred during the suction casting, and some AlCu intermetallic compound formed at the insert/matrix interface. The average thickness of the intermetallic compound increased following a parabolic relationship.

The electrical and thermal conductivity of the hybrid spcimens were considerably low, implying an incomplete interficial bonding. Additional plastic forming was found to improve the conductivity a little, presumely attributed to an increased contact area.

References

[1] V.Y. Mehr, M.R. Toroghinejad and A. Rezaejan: Mater. Sci. Eng. A601 (2014) 40-47.

[2] I.K. Kim and S.I. Hong: Mater. Design. 47 (2013) 590-598.

[3] E. Hajjari, M. Divandari, S.H. Razavi, T. Homma and S. Kamado: Intermetallics. 23 (2012) 182-186.

[4] M. Divandari and A.R. Vahid Golpayegani: Mater. Design. 30 (2009) 3279-3285.

[5] M. Yan and Z. Fan: J. Mat. Sci. 36 (2001) 285-295.

[6] Y. Tanaka, M. Kajihara and Y. Watanabe: Mater. Sci. Eng. A445-446 (2007) 355-363.

[7] H.J. Kim, J.Y. Lee, K.W. Paik, K.W. Koh, J.H. Won, S.H. Choe. J. Lee, J.T. Moon and Y.J. Park: IEEE Trans. Comp. Packag. Technol. 26 (2003) 367-374.

P-26

Effect of Ultrasonic Frequency on Attenuation Coefficient for Effective Nucleation Enhanced Ultrasonic Melt Treatment

Jong Hwan LEE[1], Jeong IL YOUN[1], Young Ki LEE[1], Young Jig KIM[1*],
Jeong Wook PARK[2] and Woo Chun KIM[2]
[1] School of Advanced materials science and engineering, Sungkyunkwan University,
2066 Seobu-ro, Jangan-gu, Suwon, Gyeonggi-do 16419, Korea
[2] R&D Center, DR AXION Co., Ltd., 55 Sannam-gil, Ulju-gun, Ulsan 45010, Korea

Increasing of ultrasonic intensity is very essential to acquire higher efficiency of melt treatment for treat more melts in same time. Attenuation factor is contribution parameter of the intensity that depends on medium, and determining the efficient range of the ultrasonic. For A356 alloy, there is no research about ultrasonic attenuation, especially the effect of ultrasonic frequency on that. This study is focused on the evaluation of attenuation factor and efficient range of the ultrasonic at different frequency to improve efficiency of ultrasonic melt treatment. Firstly, ultrasonic with different frequencies were injected into Si oil to visualize attenuation and measure efficient range. Relationship between frequency and attenuation phenomena were analyzed from the experiments. The results were revised and confirmed through experiments on A356 with various ultrasonic frequency.

Keywords: *A356 alloy, Ultrasound, Attenuation coefficient, Ultrasonic intensity, Microstructure*

1. Introduction

Ultrasonic melt treatment can achieve grain refinement and degassing of the alloy at the same time, so the quality of castings can be improved. The dominant mechanism for the effect of ultrasonic melt treatment are cavitation effect and acoustic streaming due to the ultrasound. Cavitation effect and acoustic streaming are both related to intensity of ultrasonic that propagating through medium. The intensity of propagating ultrasonic, I can be expressed as follows [1-3]

$$I = I_0 \exp(-2\alpha x) \qquad (1)$$

where I_0 is initial intensity of ultrasonic, α is attenuation coefficient, x is propagated distance. Following this equation, as ultrasonic travels along the medium, the intensity of ultrasonic decreases. This phenomenon is called attenuation. Attenuation coefficient α is contribution parameter of the intensity that depends on medium, and determining the efficient range of the ultrasonic [4].

In this study, ultrasonic attenuation in different frequency and intensity is analyzed. Various conditions of ultrasonic induced to Si oil to measure penetration depth and analyze attenuation phenomena in medium for each condition. Also, ultrasonic induced to A356 melts to analyze attenuation by examine grain refinement efficiency.

2. Experimental

The ultrasonic inducing equipment is consists of an ultrasonic generator and control box, a transducer, a booster and a sonotrode. For each 15, 20 and 25 kHz frequency, boosters and sonotrodes have different length to match with resonant frequency. 350 cS standard viscosity Si oil was contained in transparent cylindrical acrylic container with 60 mm diameter. Ultrasonic sonotrode was dipped 20 mm in the Si oil. The penetration depth of ultrasonic was measured while inducing ultrasonic of each condition. Condition of each experiment is shown is Table 1. For following experiment, 2 kg of A356 alloy was melted in graphite crucible at 680℃. The prepared melt was withdrawn from the furnace and placed in air for ultrasonication. Ultrasonic was induced to melt at 660 °C for 180 s. Each ingot sample was cut perpendicularly. The grain size of each section was measured to analyze grain refinement effect of different ultrasonic inducing condition.

3. Result & discussion

Fig. 1 illustrates relationship of penetration depth and calculated intensity for each ultrasonic inducing condition. The result shows that, the penetration depth of ultrasonic is proportional to induced ultrasonic intensity. And at similar intensity, the penetration depth difference of each frequency is maximum 2 mm, so the difference is almost negligible. This means the effect of frequency to ultrasonic attenuation is relatively small than the effect of intensity. Fig. 2 shows the result of

Table 1 Conditions of induced ultrasonic in Si oil.

Frequency	Intensity (W/cm^2)	Amplitude (μm)
15 kHz	2567	13
	3668	19
	4918	23
20 kHz	2406	9
	3053	13
	4196	15
25 kHz	1806	6
	2286	8
	3415	9

injection of different frequency and intensity to A356 melts. The condition of induced ultrasonic, calculated intensity and grain refinement fraction is shown in Table 1. The result is similar to Si oil, as intensity increases, the grain refinement fraction also increases. But, in condition of 15 and 20 kHz 9μm amplitude, ultrasonic injection could not refine the grain. These ineffectiveness in low intensity is evidence of cavitation threshold. Generation of cavitation is essential for grain refinement of A356 alloy. But in 15 and 20 kHz 9 μm amplitude ultrasonic, the threshold intensity for cavitation bubble generation is unsatisfied, so refinement of grain did not occur. The cavitation threshold intensity is achieved over 1809 W/cm^2 intensity, and grain refining phenomenon expands by acoustic streaming. The acoustic streaming has tendency of proportional to intensity, so the ultrasonic intensity increases, grain refinement efficiency also increases.

4. Conclusion

In this study, an attempt has been made to analyze the effect of ultrasonic frequency on attenuation in A356 melts. The result of ultrasonic injection to Si oil revealed that ultrasonic frequency cannot affects attenuation. On the other hand, ultrasonic intensity can affect attenuation, as intensity increases, penetration depth also increases. The experiment for A356 melts supports this result. Grain refinement fraction of each ingot sample increases along the intensity. Additionally, intensity under 1157 W/cm^2, the grain did not refined. This is the evidence of existence for cavitation threshold in A356 melt.

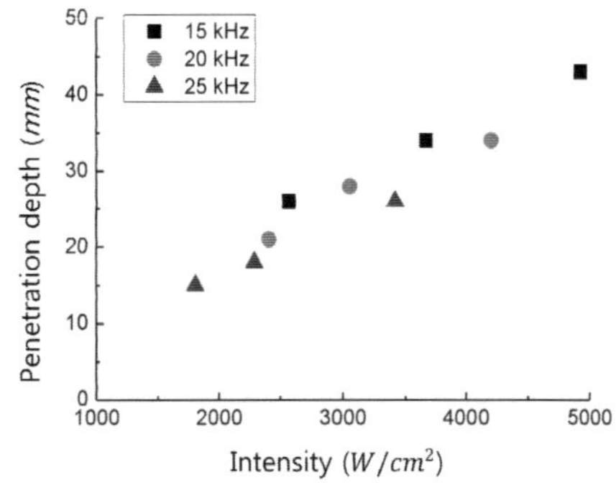

Fig. 1 Intensity vs. penetration depth in each frequency.

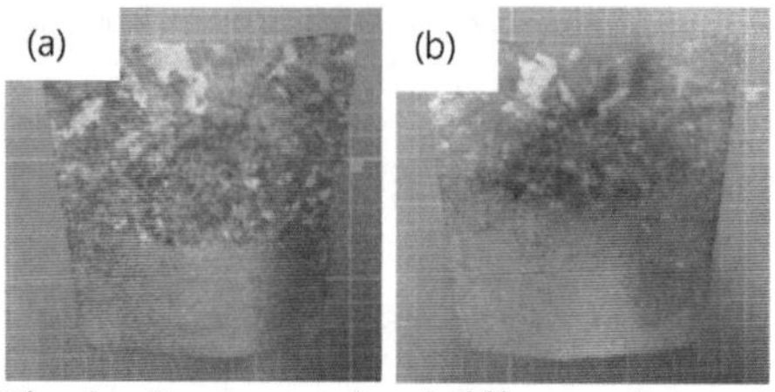

Fig. 2 Ingot sample of different ultrasonic condition: (a) 25 kHz 9 μm; (b) 20 kHz 15 μm.

Table 1 Ultrasonic inducing condition and calculated intensity and grain refinement fraction.

Inducing condition	Intensity (W/cm^2)	Grain refinement fraction (%)
15 kHz 9 μm	651	0
20 kHz 9 μm	1157	0
25 kHz 9 μm	1809	37
20 kHz 15 μm	4069	50
15 kHz 23 μm	5381	96

Reference

[1] G. I. Eskin, and D. G. Eskin: *Ultrasonic treatment of light alloy melts*, (CRC Press, Boca Raton, 2014) pp. 17-168

[2] V. Abramov, O. Abramov, V. Bulgakov and F. Sommer: Mater. Lett. 37 (1998) 27-34

[3] R. J. Farraro and R. B. McLellan: Metall. Trans. A 10.11 (1979) 1699-1702.

[4] M. Qian and A. Ramirez: J. Appl. Phys. 105 (2009) 013538

Proceedings of the 72nd World Foundry Congress
21-25th May 2016, Nagoya, Japan

USING VALUE STREAMING TO IMPROVE PRODUCTIVITY AND REDUCE LEAD TIME IN JOBBING FOUNDRY

K Mageza[1], C Banganayi [2]

[1] Metal Casting Technology Station, University of Johannesburg, kmageza@uj.ac.za, South Africa

[2] The Clever Idea Education & Engineering Services, cbanganayi@gmail.co.za, South Africa

Abstract:
Numerous South African foundries have very little or no exposure at all to the modern manufacturing principles of value streaming. The application of value streaming methods in the manufacturing environment is critical for the improvement of customer demands.
A value stream system includes all manufacturing steps and processes that occur to a given product from its inception through to its delivery to the customer. This paper focusses on the implementation of value streaming to reduce the delivery lead time in a local jobbing foundry. In this specific case study, seven deadly manufacturing waste were identified. A value stream mapping was implemented to improve productivity by reducing longer lead-time using value streaming approach. The current-state value stream mapping as-it is was drawn up to highlight the waste and eliminate them by implementation of a future-state value stream. The implementation of the future-state resulted in the evident improvement of the foundry melting process in terms of speed and reduced energy consumption. A proper production scheduling was also applied which reduced unnecessary work in progress inventory.

Keywords: *Value streaming, jobbing foundry, lead time.*

1. Introduction

Most South African foundries have little or no exposure to modern manufacturing principles in value streaming which is one of the key sources of competitive advantage. South Africa jobbing foundries are having challenges of delivering the products in the competitive short lead time due to wastes in the production system.

A Case study was carried out to (1) Identify waste which have tendency of increase cycle time (longer lead time) and (2) Reduce the identified wastes within the production system using value streaming mapping principles.

2. Literature Review

Value streaming is one of the effective ways to identify waste to create a value for the product. The purpose of the value stream is to find non-value-adding wastes, identify their root causes and prepare strategic plans for their elimination through planning and focusing on continuous improvement. Waste elimination is one of the most effective ways to increase the profitability of a company. The typical wastes found in manufacturing environments consist of: *Motion, Waiting, Transportation, and Correction/Rework, Over-processing, Inventory* and *Over-production [1]*.
A value stream is all the actions (both value-added and non-value added) currently required to bring product through its essential main flows: (1) the production flow from raw material into the arms of the customer and (2) the design flow from concept to launch [2].
VSM includes creating a map of the flow material through production and flow of information from the customer back to each production process. A current-state map of in-plant value streams the serves as the basis for developing future-state maps that omit non value adding steps while pulling resources through the system and smoothing flow. The difference between the current – state and potential future states provides a road map to start the implementation of performance improvements [3]. V. R. Murugananthan et.al [4], applied 5s technique in a foundry and this resulted in effective organization of the workplace, improvement of work's environment, elimination of losses connected with failures and breaks, improvement of the quality and Safety of work.

From VSM to remove the non-Valued activities during manufacturing and also to reduce manufacturing lead time.

3. Experimental Work

The value streaming mapping was mapped through the use of Micro Visio by direct observation of the process with aim of collecting, people, material and information flow and detailed operation steps within the system. Method was designed and used to identify the types of wastes associated with each process see table 1. This information was then weighted using numbers 1-3, 1 being severe waste and 3 less critical wastes. When waste was present at a specific process a **Yes** was filled in the table and when no waste was identified **No** was filled. A value streaming tools were used to reduce or eliminate wastes and then map the future value stream mapping.

4. Result and Discussion

Figure 1 shows the flow of the product life cycle from customer order to product shipment. The VSM shows that the total value adding time for making the pedestal products is *6 hours and 47 minutes* the non-value adding time is *20 hours and 53 minutes.* This account to only 24 % of time in the product life cycle being value adding. This means 76 % of time is spent on non-value adding activities.

Table 1 shows waste classification in each prodcution process.

Processes	Identified Wastes in the process						
	Transportation	Waiting	Motion	Inventory	Over-Production	Over-processing	Rework/Defects
Pre-moulding	Yes^2	Yes^1	Yes^2	Yes^3	no	no	Yes^1
Stripping	Yes^2	Yes^1	Yes^2	no	no	no	Yes^1
Core-Making	Yes^2	Yes^1	Yes^2	Yes^3	Yes^3	Yes^3	Yes^1
Mould assembly	Yes^2	Yes^1	no	no	no	no	Yes^1
Coating	Yes^2	Yes^1	no	no	no	Yes^3	Yes^1
Melting	Yes^2	Yes^1	no	no	no	Yes^3	Yes^1
Casting	Yes^2	Yes^1	no	no	no	no	Yes^1
Knock-Out	Yes^2	Yes^1	Yes^2	Yes^3	no	no	Yes^1
Fettling	no	Yes^1	Yes^2	Yes^3	no	no	Yes^1
Inspection	no	Yes^1	Yes^2	no	no	Yes^3	Yes^1

The results on table 1 show that waiting and rework/defects are the most critical/severe wastes because they occur in all processes, which is ranked 1. Figure 2 shows that by applying value streaming tools, such as *the single scheduling, pull system, line balancing* will promote first in first out (FIFO) and allow the upstream process to flow with less WIP (work in progress) inventory and reduce the lead time utilisation.

The figure 2 shows that by applying value streaming tool the process have potential of reducing the lead time.

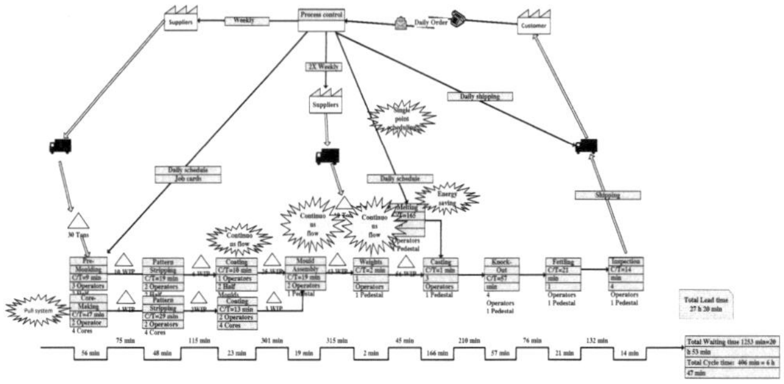

Fig 1: shows the current VSM and proposed changes of current stream map.

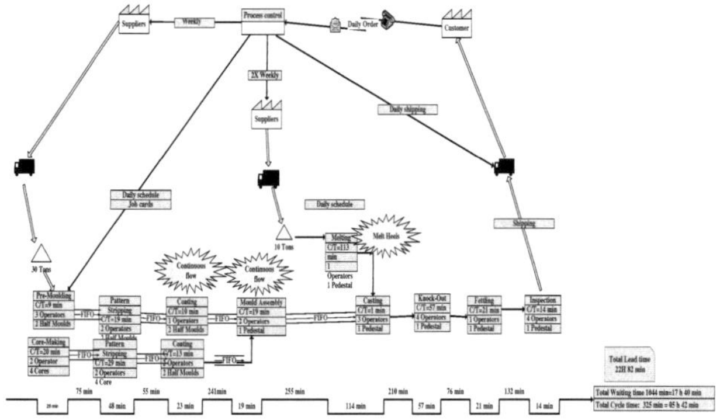

Figure: 2 shows the future value stream with reduce waste.

5. Conclusion and Recommendations

It can be concluded that the use of value streaming principle in the foundry has the capability to reduce lead time and allow local foundries to gain competitive advantage in the local markets. It can be recommended that South African foundries need to gain skills and competences in applying the value streaming principle in order to identify waste and eliminate within their processes.

6. References

[1] J Karilova (2010), Developing Standardised Work in Casting Department at Luvata Sweden AB Finspang, Master's Thesis, 1-88

[2] M Rother and J Shook 2003: Learning to see (value-stream mapping to create value and eliminate muda), The Lean Enterprise

[3] JP Womack and DT Jones 2003: Banish waste and create wealth in your corporation, pg. 57.

[4] V. R. Murugananthan, K. Govindaraj, and D.Sakthimurugan. Process Planning Through Value Stream Mapping In Foundry 201

The strength analysis in caulking fastening portion of automotive parts

Ryota Ueda[1], Kazumichi Shimizu[1] and Takafumi Funabiki[2]
[1]Muroran Institute of Technology,
27-1, Mizumoto-cho, Muroran-city, Hokkaido, Japan, 050-8585

In this study, we are focusing on the caulking for fixing the two metals. The method of fixing by plastically deforming the material is referred to as caulking fastening. This way, it is possible to low cost and shorten delivery time. However, since the weak rotational force at the junction, it has been reported that failure symptoms. Therefore, it is important to establish the strength evaluation method. So this time, we made a stress measurement of by the measurement and simulation of the slip torque in the two types of the miniature model and the full scale model. This model uses the shaft side and the ring side respectively FCD450, S45C tempering and quenching material. This analysis is intended to be to investigate the effect of the load applied in 1 second static structural analysis. Test in miniature model results showed a fixation for all interference. The results of shear stress, by tresca's theorem is shown that it's broken when fastening margin is 0.1 mm. Therefore, interference may be set to interference in the range of 0.1 ~ 0.05mm. This has been shown similar trend to full scale model.

Keywords: *caulking , fastening, fastening margin, slip torque, analysis*

1. Introduction

In this research, reported for caulking fastening method of the two parts. When joining the two parts, in the prior art is fixed by using bolts. However, there is a need for low-cost joining process from because it is costly. So, I will be paying attention to the caulking fastening. The method of fixing by plastically deforming the material is referred to as caulking fastening. This way, it is possible to low cost and shorten delivery time [1]. However, since the weak rotational force at the junction, it has been reported that failure symptoms. Therefore, it is important to establish the strength evaluation method. Therefore, in this study, analysis is performed for the purpose of practical use to the automotive products by swaging method of the differential case and ring gear. To the simulation analysis using the finite element method for the strength of the caulking fastening portion.

2. Analysis method and Experimental method

2.1 Analysis method

The torsional analysis using ANSYS a FEM analysis tool. It is intended to investigate the effect of load applied to one second. It shows the model to be used in the present analysis in Figure 1. Twist with adding a load to the key groove of ring side. In this analysis we use a complex model that was quenched FCD450 and S45C. However, it regard FCD450 as the shaft side and it treat S45C as ring side. Analysis object to investigate the stress analysis and contact of the crimped portion interference 0.1mm, 0.05mm, 0.025mm. To evaluate the strength of the caulked portion in the calculated shear stress. Also, friction coefficient of the contact surface is set to 0.25. Also, physical property values are as shown in Table 1.

2.2 Experimental method

In this study continuously performed to be twisted by the load on the ring of the key groove, and performs a torsional torque load test. It is shown in Fig. 2 the relationship between load and slip determination torque. Load will be 30~40% lower by value from the maximum slip torque. It is set as slip determination torque from the 2 percent lower value of torque at the start of the test. The test shall be conducted at ambient temperature. In addition, it will be when the number of repetitions is 10^6 times or slip and stop conditions. Also, physical property values takes a same value as the analysis.

Table 1 Each physical properties values

	Shaft	Ring
Material	FCD450	S45C
Longitudinal elastic modulus [GPa]	161	205
Linear expansion coefficient [1/℃]	1.05×10^{-5}	1.17×10^{-5}
Poisson's ratio [-]	0.3	0.25

3. Results and discussions

3.1 Analysis result

First, miniature model was set to the analysis object. It is observed after the analysis of the contact surface. There was no change in the state before and after the analysis in all of the fastening margin. Also, analysis results of the maximum shear stress that it is increased as the fastening margin increases. Next, full scale model was set to the analysis object. It is observed after the analysis of the contact surface. Analysis method is the same as a miniature model. Analysis result was a state of being fixed like the miniature model. In addition, the analysis result of the frictional stress and shear stress tended to analysis value increases with a large fastening margin. It was obtained the same tendency as miniature model. Analysis model is considered less likely to slip by a large fastening margin from the analysis result of the friction stress. Table 2 shows the Analysis results of shear stress in each model. The analysis result of shear stress is assessed by theorem of Tresca. Apply this to the analysis result as to whether or not it will be considered to break. In the case of 0.1mm, the shear stress should be set small fastening margin to exceed the allowable shear stress. Therefore, it is preferable that fastening margin is employed at between 0.1mm to 0.05mm.

Fig. 1 Analysis model

Table 2 Analysis results of shear stress

Model	Element	0.1mm (MPa)	0.05mm (MPa)	0.025mm (MPa)
Miniature model	Shaft	296.90	148.59	68.27
	Ring	855.19	417.57	239.29
Full scale model	Shaft	273.10	138.78	127.75
	Ring	585.00	326.79	159.43

3.2 Experimental result

The slip torque is subjected to torsion test of the two kinds of specimens. It shows the value of the slip torque of slip torque in Table 3. As a result, specimens of 210 N-m was slip determination at 5498 rotation. Meanwhile, the test piece of 631 N-m were obtained the results that do not break even when repeatedly load 10^6times.

Table 3 Measurements results of torque and slip determination value in the each specimen.

Fastening margin [mm]	4.27×10^{-3}	12.8×10^{-3}
Slip torque [N・m]	210	631
Torque after fastening [N・m]	95	426
Slip determination [N・m]	93	417

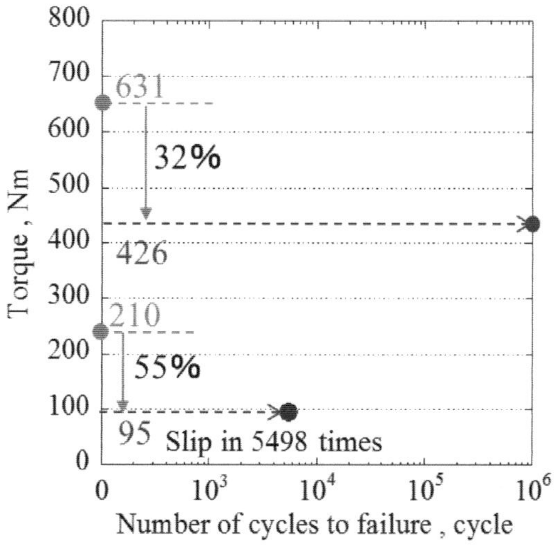

Fig. 3 The torsion fatigue test results of each specimen

4. Conclusion

This study was evaluated by simulation analysis the strength of the model that was produced in the interference of 0.1, 0.05, 0.025mm. Shear stress is evaluated by theorem Tresca, fastening margin is estimated to be broken at the time of 0.1mm. Therefor the interference it is preferable in the range from 0.1mm to 0.05mm.

References

[1] Kinoshita Hiroyuki, Nakanishi Tsutomu : Japan Society of Mechanical Engineers Proceedings 008-1(2000)73

Influence of the mulling condition to exert on the greensand nature

Kazunori Satou[1]
[1] Japan Foundry Engineering Society, Japan

Keywords: *greensand,mulling,compactability, tem-pering.*

The greensand adds water and regulates a compactability (CB). The adjustment of the CB is difficult and the greensand nature has a big looseness. From a flucuation factor to have an influence on the greensand nature, I chosed quantity of addition and addition time of the water. The mulling test which changed examined quantity of addition and addition time of the water. As a result, I knew that is important to tempering of greensand which the water adding performed a gross quantity early in mulling.

1. Introduction

The greensand process can mass-produce a cast. The process which can produce a high-quality cast. Quality stabilization of greensand is desired to pro -duce a highquality cast. Installation and automation are advancing the greensand process. A water controller is equipmentized as a representative of automation. Compactability (CB) adjustment is perfor med by a water controller and we're working on stabilization of the greensand nature. At the same time, an adjustment and management of greensand composition are being performed. But, looseness is big for the greensand nature and it's the real state of affairs to have trouble with stabilization of the gr eensand nature. The fluctuation factor which has an influence on the greensand nature is a-f.

a) Casting condition (pouring temperature, sand/metal weight ratio (S/M ratio), etc.)
b) Mulling condition (mulling time, addition amount, etc.)
c) Condition of mulling facility (Simpson type, agi-tator type, etc.)
d) Composition of return sand (Distribution of sand grains, Clay, Moisture, etc.)
e) Facility condition to treatment return sand (Cool ing of sand by water, Temperature of return sand, etc.)
f) Others (Tempering of greensand , Pre-mulling, etc.)

This report is here for investigating the size of the looseness of the greensand nature.

2. Experiment method

This experiment was made about "f" and a mulling condition of "b". A testing condition of"b" is indicated in the next.

2.1 Combination of sand

The material of greensand is composed a system sand 100%.

2.2 Test procedure of the mulling condition "b"

Mulling condition"b"changed addition time of the water and the addition amount and investigated the influence which gives it to the greensand. Mulling of greensand added water using a mulling machine and mulling for 4 minutes. Addition timing of the water and the addition amount were changed, and we inveti gated properties of greensand. The quantity of water was set to 2 standards of the 110cc and the 120cc. Lat er, the test procedure discuss an example of 120cc.

Mulling condition A did the whole quantity additi on of water with a systemsand and mulled for 4 minu tes, and we investigated properties of greensand. Mull ing condition B added 1/2 water of 120cc with a syst emsand and mulled for 2 minutes. After, added 1/2 water of 120cc and mulled for 2 minutes, and we inve stigated properties of greensand. Mulling condition C added 1/2 water of 120cc with a systemsand and mu lled for 2 minutes. After, added 1/3 water of 120cc and mulled for 1 minutes, and added 1/5 water of 120cc and mulled for 1 minutes, and we investigated proper ties of greensand.

2.3 Test procedure of the mulling condition "f"

We could preliminary mulling a systemsand and tempered greensand for 180 minutes per 60 minutes per 30 minutes after that. After, we could main mulling a systemsand and investigated properties of greensan d. Preliminary mulling added 3/12 water of 120cc, and mulled for 3 minutes. We put mulling sand in wellclo sed container and preserved it. This is the tempering of sand. Time of the tempering of sand was set to 180 mi nutes per 60 minutes per 30 minutes, and we mulled for the main mulling.

The tempering of sand per 30 minutes is Mulling condition D. The tempering of sand per 60 minutes is Mulling condition E. The tempering of sand per 180 minutes is Mulling condition F.

Preliminary mulling added 5/12 water of 120cc, and mulled for 3 minutes. We put mulling sand in wel lclosed container and preserved it. Time of the temper ing of sand was set to 180 minutes per 60 minutes per 30 minutes, and we mulled for the main mulling.

The tempering of sand per 30 minutes is Mulling condition G. The tempering of sand per 60 minutes is Mulling condition H. The tempering of sand per 180 minutes is Mulling condition I.

3. Experimental result

The test result is indicated on fig 1. The compression strength falls compared the B, C then the A which did the whole quantity addition of water. The permeability is also the similar tendency.

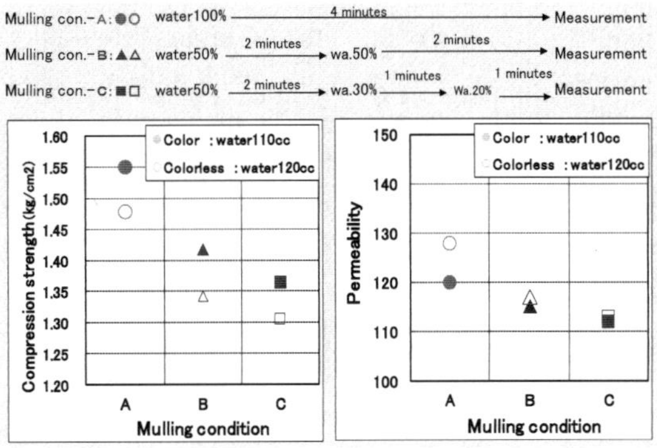

Fig. 1 Relation between compression strength and permeability in mulling condition A, B, C

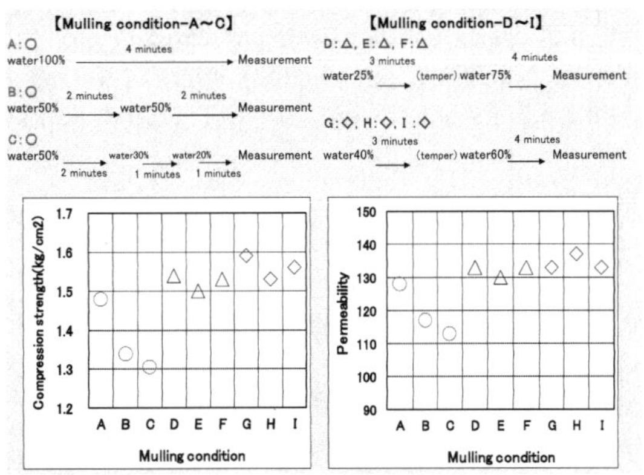

Fig. 2 Relation between compression strength and permeability in each mulling condition

The test result is indicated on fig 2. There are no big changes for compression strength of the D,E,F,G,H,I which went on the temper.

The permeability is also the similar tendency.

4. Summary

This report was investigated about the mulling con dition that I can think it influenced looseness of the gr eensand nature. When dividing and adding water in mulling, the compression strength and permeability indicate the decline tendency. When doing this mull ing after you could pre-mulling and we'll stay to keep for 180 minutes per 60 minutes per 30 minutes, the co mpression strength doesn't fall, and the looseness is also small.

Therefore that the mulling of a greensand is addi tion method of water and a temper is important.

References

[1] Y.Wakao, S.Katashima, T.Okakura : JFS *Study about control of system sand* Vo.45,July,1987

[2] Y.Wakao, S.Katashima, T.Okakura : JFS *The quality grade of the system sand and study about control* Vo.25, December,1980

[3] S.Katashima, Y.Wakao, T.Okakura : JFS *Management of greensand* Vo.38,October,1985

Proceedings of the 72nd World Foundry Congress
21-25th May 2016, Nagoya, Japan

Flow Rate Control of Automatic Pouring Machine by Iterative Learning Control

Minoru Yamada, Naoki Minemura
National Institute of Technology, Gifu college, Japan

This paper presents pouring flow rate control of the automatic pouring machine by iterative learning control. Pouring process has a feature of repeating motion of ladle tilting. So, iterative learning control is applied to pouring flow rate control. Iterative control systems can be viewed as 2D systems because systems have two kinds of dynamics: the one along time direction for each repetition index, and the one with respects to the repetition index. Therefore the iterative learning control system was designed by 2D control theorem. In this research, the motor attached to the ladle is controlled by output feedback based iterative learning control for flow rate control. The learning controller depends on the information of the input and the tracking error in both the present iteration and a finite number of previous iterations. A target angular velocity is calculated from a target flow rate by inverse model of pouring model. The output was able to be matched to the target angular velocity by a few times of repeating trial.

Keywords: *Automatic pouring machine, Flow rate control*

1. Introduction

In pouring process, molten metal is poured into the mold repeatedly. And the angle of ladle can be calculated from the desired flow rate. So we propose that the angle of the ladle is controlled by iterative learning control. Iterative learning control is a method to track the output of the dynamic system to the desired trajectory by modifying the input based on the trial error. The purpose of this paper is to apply iterative learning control to flow rate control of the automatic pouring machine.

2. Automatic pouring machine

Fig.1 shows the outline of the automatic pouring machine. The DC motor revolves the ladle, and its rotation angle is measured by the encoder. The weight of the liquid in the ladle is measured by the load cell. The microcontroller is used for signal sending and receiving with the PC, the DC motor and the load cell.

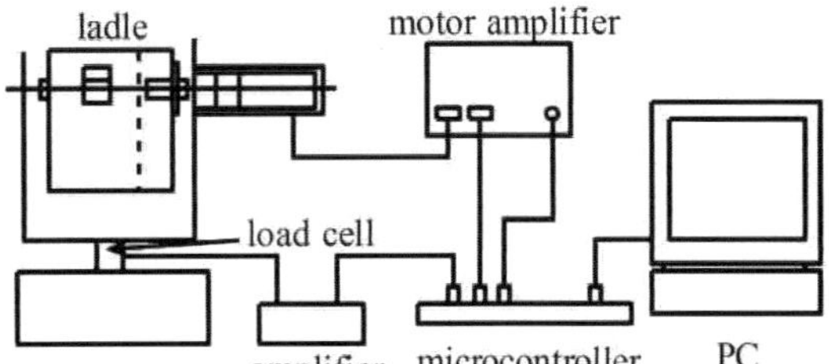

Fig. 1 Automatic pouring machine

3. Model following learning control

It's supposed that the automatic pouring machine is modeled by the discrete time system as follows.

$$\begin{aligned} x(t+1) &= Ax(t) + Bu(t) \\ y(t) &= Cx(t) + Du(t) \end{aligned} \tag{1}$$

where $\boldsymbol{x(t)}, \boldsymbol{u(t)}, \boldsymbol{y(t)}$ are the state, the control input and output respectively. And $\boldsymbol{t}$ is the time step. The following discrete time model system is considered as the model that outputs the target value that this controlled system should follow.

$$\begin{aligned} x_m(t+1) &= A_m x_m(t) + B_m r(t) \\ y_m(t) &= C_m x_m(t) + D_m r(t) \end{aligned} \tag{2}$$

where $x_m(t), r_m(t), y_m(t)$ are the state, the control input and output of the model system respectively.

The model following learning control that proposes by this research is considered as follows. We suppose that the system (1), the model system (2), initial conditions $x(0) = x_0$, $x_m(0) = x_{m0}$ and reference input $\{r(t), t = 0,1,\cdots,N\}$ are given. The objective is to design a controller such that the output $y(t)$ of (1) tracks the step response $y_m(t)$ of the model system (2). To achieve this objective the trial errors $\mathrm{y_m} - y(t)$ and the inputs the used in each trial are memorized. The inputs are modified based on the errors and the previous inputs to reduce the tracking error. After some repetitions the appropriate inputs are obtained and the outputs converge to the output of the model system.

When considering the model following learning control as mentioned above, the control system has two kinds of independent dynamics: the one along time direction for each repetition index, and the one with respects to the repetition index.Then the system (1) and the model system

(2) can be viewd as 2D discrete systems in the form

$$x(k,t+1) = Ax(k,t) + Bu(k,t)$$
$$y(k,t) = Cx(k,t) + Du(k,t) \quad (3)$$

$$x_m(k,t+1) = A_m x_m(k,t) + B_m r(t)$$
$$y_m(k,t) = C_m x_m(k,t) + D_m r(t) \quad (4)$$

where $x(k,t)$, $y(k,t)$, $u(k,t)$, $x_m(k,t)$, $y_m(k,t)$ are state, output and control input of the system, state and output of the model system espectively. And the tracking error of the k-th iteratin is defined as follow.

$$e(k,t) = y_m(k,t) - y(k,t) \quad (5)$$

Let $\Delta u(k,t)$ be the modification of control input of the k-th iteration, the general learning control law is expressed in the following.

$$u(k+1,t) = u(k,t) + \Delta u(k,t) \quad (6)$$

Now define

$$\eta(k+1,t) = x(k+1,t) - x(k,t)$$
$$\eta_m(k+1,t) = x_m(k+1,t) - x_m(k,t) \quad (7)$$

Then (3)-(7) can be rewriten as

$$\begin{bmatrix} \eta(k+1,t+1) \\ \eta_m(k+1,t+1) \end{bmatrix} = \begin{bmatrix} A & 0 \\ 0 & A_m \end{bmatrix} \begin{bmatrix} \eta(k+1,t) \\ \eta_m(k+1,t) \end{bmatrix} + \begin{bmatrix} B \\ 0 \end{bmatrix} \Delta u(t)$$

$$e(k+1,t) = \begin{bmatrix} -C & C_m \end{bmatrix} \begin{bmatrix} \eta(k+1,t) \\ \eta_m(k+1,t) \end{bmatrix} - D\Delta u(t) + e(k,t) \quad (8)$$

The controller that the traking error is converge to a zero after some repetitions is given by

$$\Delta u(k,t) = K_1 \begin{bmatrix} \eta(k+1,t) \\ \eta_m(k+1,t) \end{bmatrix} + K_2 e(k-1,t) \quad (9)$$

where K_1 and K_2 are given by the following equation [1].

$$[K_1 \quad K_2] = NY^{-1} \quad (10)$$

Symmetric matrices $Y > 0$ and $Z > 0$ can be determined by satisfying the LMI

$$\begin{bmatrix} Z - Y & 0 & Y\hat{A}_1^T + N^T\hat{B}_1^T \\ 0 & -Z & Y\hat{A}_2^T + N^T\hat{B}_2^T \\ \hat{A}_1 Y + \hat{B}_1 N & \hat{A}_2 Y + \hat{B}_2 N & -Y \end{bmatrix} < 0 \quad (11)$$

where

$$\hat{A}_1 = \begin{bmatrix} A & 0 & 0 \\ 0 & A_m & 0 \\ 0 & 0 & 0 \end{bmatrix}, \hat{A}_2 = \begin{bmatrix} 0 & 0 & 0 \\ 0 & 0 & 0 \\ -C & C_m & I \end{bmatrix}$$

$$\hat{B}_1 = \begin{bmatrix} B \\ 0 \\ 0 \end{bmatrix}, \hat{B}_2 = \begin{bmatrix} 0 \\ 0 \\ -D \end{bmatrix}$$

4. Simulation

We considered the angle of the ladle follows in the target angle which doesn't include the natural frequency of the liquid level to suppress shaking of molten metal in the ladle when a ladle revolves. So the model system includes the notch filter that suppresses shaking of the liquid level.

Simulation result is shown in Fig. 2 and Fig. 3. Rotation angle follows the output of the model system by an approximately 1st iteration.

5. Conclusion

Rotation angle of ladle can track to the output of the model system by using the model following learning control.

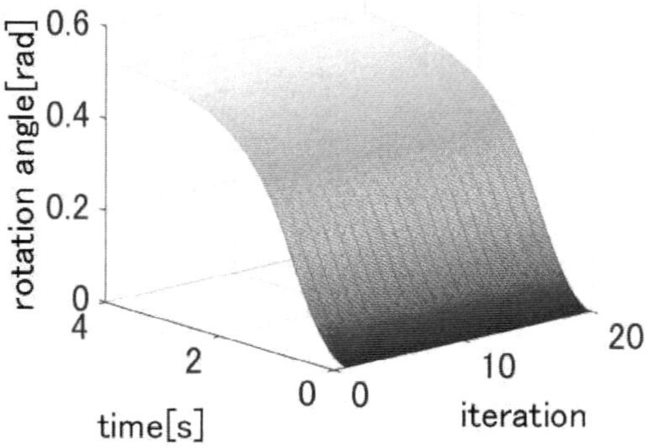

Fig. 2 Rotation angle of ladle

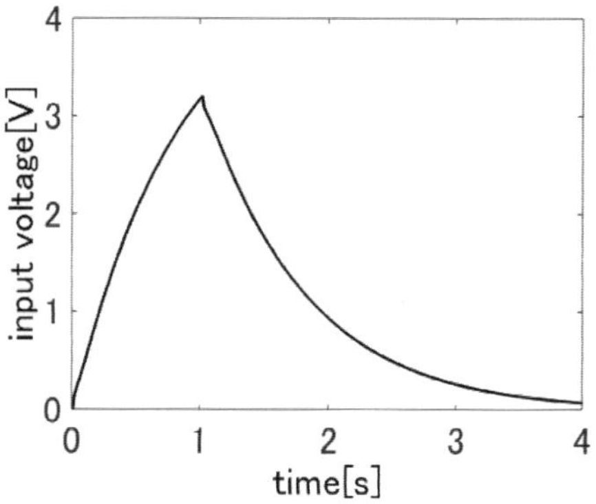

Fig. 3 Input voltage to the motor

References

[1] K. Gakowski, Eric Rogers, S. Xu, J. Lam, and D. H. Owens, "LMIs – a fundamental tool in analysis and controller design for discrete linear repetitive processes", IEEE Transactions on Circuits and Systems I: Fundamental Theory and Applications, No.49(6) pp.768-778 (2002).

P-32

Proceedings of the 72nd World Foundry Congress
21-25th May 2016, Nagoya, Japan

Effect of Nb and W addition on heat-resistance of austenitic cast steel

T.Higashisono[1], K.L.Ngo-huynh[2] and T. Okuyama[1]
[1] Kurume National College of Technology, 1-1-1 Komorino, Kurume, Fukuoka 830-8555, Japan
[2] Asahi Denki Seiko Steel Co., Ltd., 8 Nagadai, Nishio, Aichi 445-0031, Japan

This study is to clarify the combined effect of solid - solution strengthening by W and precipitation strengthening by Nb on thermal-shock resistance of austenitic cast steel at elevated temperature. It was found that simultaneous addition of Nb and W to cast steel retards the decomposition of crystallized $M_{23}C_6$. Also, precipitation of the fine secondary carbide ($M_{23}C_6$) and solid-solution by W lead to improvement of thermal-shock resistance of Nb-W steel, as compared with JIS SCH21 steel and Nb steel.

Keywords: *Austenitic cast steel, heat-resistance, carbide.*

1. Introduction

SCH21 steel, one of heat resistant austenitic cast steels, is widely used as heat treatment jigs, furnace components, etc. which are operated in range of 850 to 1000°C in oxidizing or carburizing environment. Exposure at the service temperature promotes the precipitation of sigma phase leading to high temperature brittleness. Increment of Ni content can improve heat-resistance of the steels, but it is not best choice due to high cost performance. Recently, it has been reported that Nb, Ti, Zr (precipitation strengthening elements) [1)] or W (solid solution strengthening element) [2)] provides higher creep-resistance for the steels. However, effect of simultaneous addition of Nb and W on heat-resistance of the steels has been little reported.

The aim of this study is to present the results of thermal-shock resistance and microstructure evolution of Nb-W steel during aging, as compared with Nb steel and SCH21 steel.

2. Experimental

All experimental steels used were alloy-designed by Thermo-Calc software (database: TCFE5) before casting. Carbon, which affects high temperature properties, was set as 0.32mass%, while the amount of other elements were adjusted to prevent high temperature embrittlement. The main component of the alloys are 1) Nb-W steel [Fe-19.9Ni-23.5Cr-0.5Nb-0.9W-0.32C], 2) Nb steel [Fe-18.9Ni-22.3Cr-1.3Nb-0.33C] and 3) SCH21 steel [Fe-20.2Ni- 23.0Cr-0.32C] .

Thermal-shock test was performed at 950°C up to 300cylces. Each cycle was heated for 0.6ks, and then water-quenched for 0.12ks. Disc-shape samples (diameter: 25mm and thickness:7mm) were cut from Y-block casting, as shown in Figure1. Coefficient of thermal expansion was measured by thermo-mechanical analyzer (TMA). Microstructural analysis of as-cast state and steels aged at 950°C for 50h, 100h and 200h was conducted in optical microscope, scanning electron microscope with energy dispersive spectrometry (SEM-EDS) and TEM, X-ray diffraction analyzer (XRD) and differential scanning calorimetry (DSC).

3. Results and Discussion

Figure1 shows results of deformation rate of samples after conducting thermal-shock test. Deformation rate was evaluated by equation, shown in Figure1 (a). The micro-cracks were found at around 120times in all samples. However, deformation rate of Nb-W steel is smallest as compared with Nb steel and SCH21 steel. Under condition of thermal-shock test (equivalent to steel aged for 50h), deformation rate of SCH21 steel is up to 6%, larger 6 times than that of Nb-W steel. As TMA result measured in the range of 150 to 950°C, the thermal expansion coefficient of Nb-W steel is smallest among steels. At the temperature of 950°C, the measured thermal expansion coefficient of Nb-W steel, Nb steel and SCH21 steel are 18.6, 18.9 and 20.5($\times 10^{-6}$/°C), respectively. It means that the simultaneous addition of Nb and W is effective to improvement of thermal shock-resistance of austenitic cast steel.

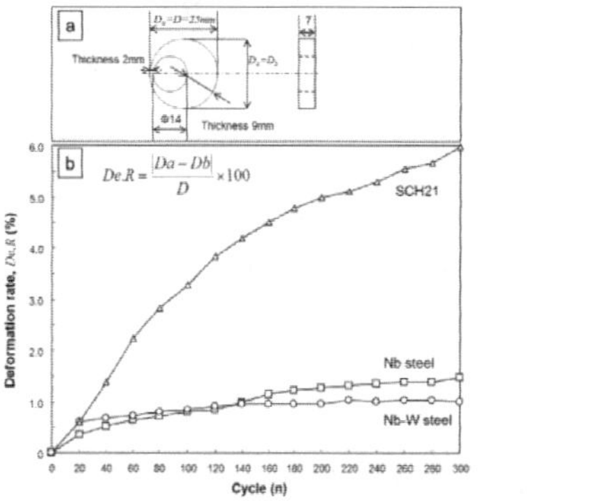

Figure1 Sample of thermal-shock test (a) and its deformation rate after thermal- shock test (b).

The microstructures of as-cast state and steels performed by 300cycles thermal-shock test are shown in Figure2. The microstructures of as-cast steels are composed of an austenitic matrix with network of crystallized carbides. It is found that secondary carbides precipitate in the vicinity of network, but appearance of void and crack is different in thermal-shocked steels. Cracks severely develop at not only dendrite boundary but also trans-granular boundaries in SCH21 steel, whereas micro voids appear along dendrite boundary and crystallized carbide. Juncture of void which develops crack is difficult to occur in Nb steel and Nb-W steel.

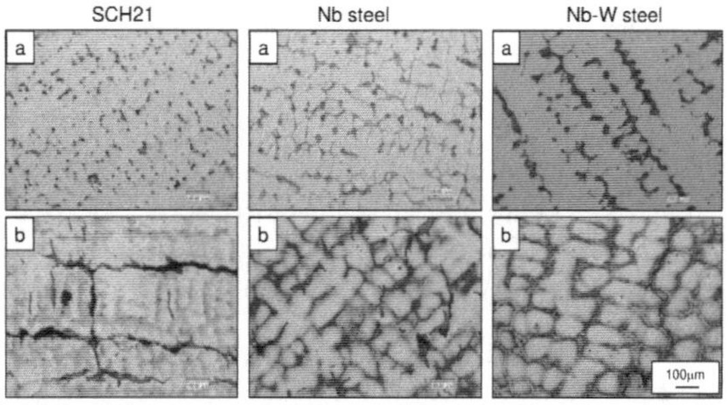

Figure2 Microstructures in as-cast steels (a) and steels thermal-shocked at 950°C for 300 cycles (b).

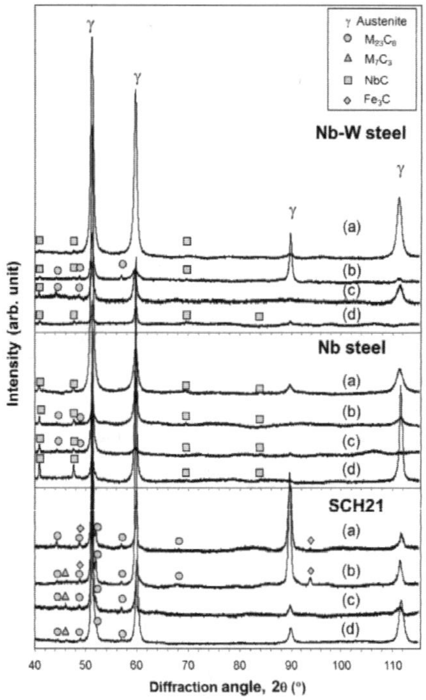

Figure3 Phase identification of as-cast steels (a) and steels aged at 950°C for 50h (b), 100h (c) and 200h (d) by XRD.

Microstructural evolution during aging is identified by XRD analysis, as shown in Figure3. Addition of Nb promotes rather crystallization of NbC than $M_{23}C_6$ in as-cast steel. As DSC results, two exothermic peaks appears at 500 and 600°C in Nb steel, which are reported as precipitation of secondary carbide NbC and $M_{23}C_6$, while only one peak appears at 600°C in Nb-W steel. Occurrence of secondary carbide ($M_{23}C_6$) could be detected in SEM-EDS microstructure of steels aging for 100h (Figure4). However, peaks of secondary carbide of $M_{23}C_6$ disappear and there is only peaks of NbC in Nb steel and Nb-W steel aged for 200h. Thermal stability of NbC plays role as increment of thermal shock resistance. On the other hand, W is completely dissolved in Nb-W steel; WC or W_2C was not found in matrix under the experimental conditions through XRD analysis, SEM-EDS and TEM. As a result, it is considered that the simultaneous addition of Nb and W to the steels restrains the development of crack during heating - cooling stage.

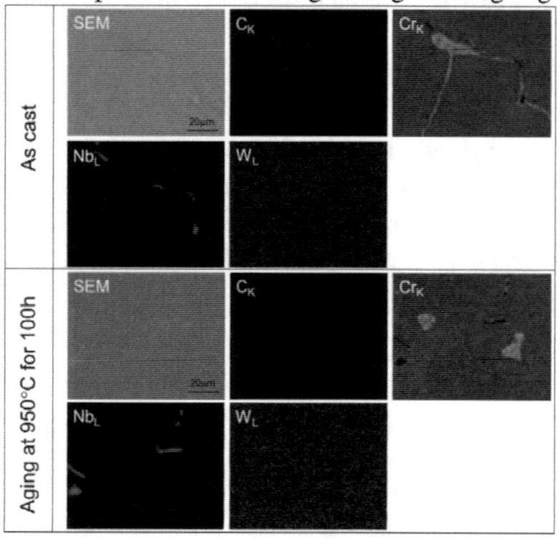

Figure4 Microstructures (SEM-EDS) of Nb-W steel.

4. Conclusion

The results obtained are as follows:

1) The simultaneous addition of Nb and W to austenitic cast steel restrains shape deformation and the development of crack during thermal shock as compared with the steel added Nb singly.

2) NbC crystallizes rather than $M_{23}C_6$ in as-cast steel. NbC is also thermally stabler than $M_{23}C_6$ at high temperature.

3) During aging or thermal-shock test, W distributes homogeneously in matrix as solid-solution. No occurrence of WC and W_2C was found in Nb-W steel.

References

[1] B.Piekarski, Materials Characterization, 47(2001)pp.181-186.

[2] J.S.Kim,Corrosion Engineering Section, 55(1999)pp.512-521.

Proceedings of the 72nd World Foundry Congress
21-25th May 2016, Nagoya, Japan

Influence of carbon content on erosive wear behavior of multi-component cast steel

Kimitoshi Nakamura, Kazumichi Shimizu, Kenta Kusumoto
Muroran Institute of Technology
27-1 Mizumoto-cho, Muroran city, Hokkaido, 050-8585, Japan

This study investigated the influence of carbon content on erosive wear behavior of multi-component cast steel. The multi-component cast steel with Cr, Mo, W and V of about 5 mass % with 1.0, 1.5 and 2.0 mass % C were prepared as the test material. The test specimens for erosive wear test were mechanically machined to be sized 50×50×10mm. Specimens were tested using a blast machine.

According to the results of erosion test, the erosion rate of specimen decreased with the increase of C content. The specimens of 2.0 mass % C showed excellent erosive wear resistance. The reasons of decreasing of erosion rate with increasing of C contents were determined by microstructure analysis and EDS analysis diffraction. According to the results, the matrix has changed from austenite to martensite with the increasing of carbon contents. Therefore, multi-component cast steel with 2.0 mass % C showed superior erosive wear resistance. It's considered that area ratio of carbide and matrix are important factors of erosive wear resistance.

Keywords: *Erosive wear, Erosion rate, Room temperature, Multi-component cast steel, Area ratio of carbide*

1. Introduction

The phenomenon becomes a serious problem for bended section of pneumatic transportation pipe, valve, turbine blade fan, and so on. Erosive wear in pipe line occurs in the inner wall of pipe and could not be judged from the exterior of pipe, so periodical exchange of parts and the inner wall of the pipe to be thinned down was built up through welding in order to avoid causing accidents. In order to reach an aim of high safety and low cost, it is important to develop wear resistant material and to estimate of life service during erosion. What's more, the influence of rodent is inevitable to life service estimation procedure [1-3]. Recently, many researchers have studied on the variety of alloys, and tempted to develop wear resistant materials [4-6].

This study investigated the influence of carbon content on erosive wear behavior of multi-component cast steel.

2. Experimental procedure

2.1 Materials

The chemical compositions of specimens are as shown in Table 1. The multi-component cast steel with Cr, Mo, W and V of about 5 mass % with 1.0, 1.5 and 2.0 mass % C were prepared as the test material. 50 kg raw materials were melted in high frequency induction furnace at 1908 K and then poured into a sand mold with a Y-shape of 53×113×125 mm after the process of spheroidzing by 95 mass % Ni - 5 mass % Mg. The specimens used in the erosion test were mechanically machined into flat plate with diameter of 50×50×10 mm. The heat treatment was performed for the purpose of secondary hardening, and the processing conditions are 4 hours keeping at 1223K for annealing, 2 hour at 1323K hardening, and 6 hours keeping at 798K for tempering.

Table 1 Chemical composition of specimens.

(mass%)

	C	Cr	V	Mo	W	Fe
1	1.0	5.0	5.0	5.0	5.0	Bal.
2	1.5	5.0	5.0	5.0	5.0	Bal.
3	2.0	5.0	5.0	5.0	5.0	Bal.

Si, Mn = 0.50 %, P, S = 0.020 %

2.2 Experimental method

A shot blasting machine was used to test the erosive wear property of target materials in the present study. Irregularly shaped erodent particles were in this experiment. They are irregularly shaped steel grits with average diameter 770μm and Vickers hardness of 810HV. Steel grits was used for very hard. The erodent particles were changed after each test because that the particles themselves also were eroded to be smaller during the experiment. The examined air speed was 100m/s, and the particle feed rate was measured with about 20g/s with changing their impingement angles 30, 60 and 90deg., respectively. All the erosion tests were conducted at room temperature in 3600sec.. Before and after the test, the mass amounts of specimens were weighed with an electronic scale and then calculated the volumetric loss. The erosion rate was introduced because it's more accurate to evaluate material removal by volumetric loss than by mass loss for specimens with different densities. The results of the wear test was discussed by Vickers hardness measurement and scanning electron microscope (SEM).

3. Experimental results and discussion

Fig. 1 showed the microstructure and area ratio of carbide measurement result of specimens. Matrix of material 1.0 mass % C was pearlite. Matrix of other specimens was martensite. Area ratio of carbide was calculated by the average of 5 photos in 400 times taking by SEM. Among all specimens, material with 2 mass % C showed the most area ratio of carbide.

Result of erosive wear test of 30deg. and Vickers hardness measurement are as shown in Fig. 2. According to the results, erosion rate decreased with C content increased, and material with 2 mass % C showed the best erosive wear property. The similar results can be get in other impact angles. Also hardness of the experimental materials increased with the C content increased.

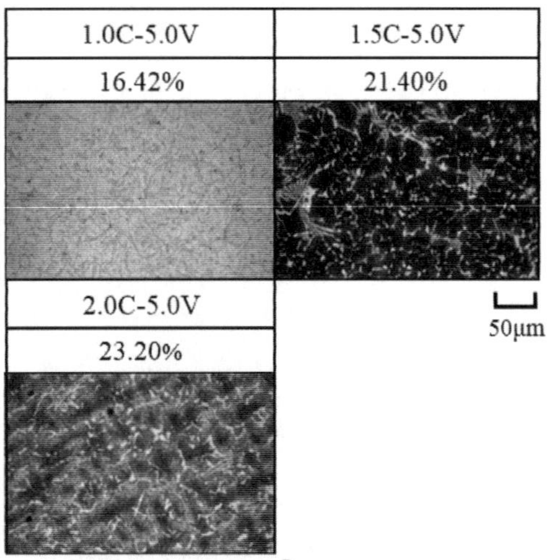

Fig. 1 Microstructure and area ratio of carbide measurement of specimens.

Therefore, multi-component cast steel with 2.0 % C showed superior erosive wear resistance. It can be considered that area ratio of carbide and matrix are important factors of erosive wear resistance.

4. Conclusions

This study investigated the influence of carbon content on erosive wear behavior of multi-component cast steel. And the following conclusions were obtained.

(1) Erosion rate decreased with C content increased. Especially, erosion rate of material with 2 mass % C showed the best erosive wear property. It is because that material with 2 mass% C has the highest hardness and the most area ratio of carbide compared with other specimens.

(2) It can be considered that area ratio of carbide and matrix are important factors of erosive wear resistance.

References

[1] K. Shimizu, T. Noguchi, et al. Trans. of AFS 104 (1996) pp. 511-515.

[2] K. Shimizu, T. Noguchi. Wear 176 (1994) pp. 255-260.

[3] M.A. Uusitalo, P.M.J. Vuoristo, et al. Wear 252 (2002) pp. 586-594.

[4] H. Wu, N.Sasaguri, M. Hashimoto, et al. J. of JFS 67 (1995) pp. 49-55.

[5] Y. Mastubara, N. Sasaguri, et al. J of JFS 71 (1999) pp. 183-189.

[6] X. Yaer, K. Shimizu, et al. Wear, 264 (2008) pp. 947-957.

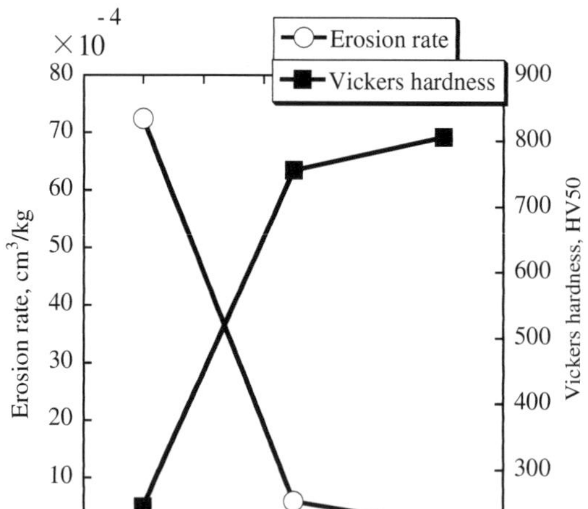

Fig. 2 Erosive wear test of 30deg. and hardness measurement.

Proceedings of the 72nd World Foundry Congress
21-25th May 2016, Nagoya, Japan

Effect of Misch-metal addition on Heat treatment of Semi-Solid Mg-7.5Al-0.3Mn alloy

Min-Su Jo, Tae-Jun Kim, Dae-Hwan Kim and Su-Gun Lim*

Dept. of Materials Manufacturing Processes Graduate School, ReCAPT, Gyeongsang National Univ., 501 Jinjudaero, Jinju-City, S.Korea

The misch-metal addition on the microstructure, aging behavior and hardness of semi-solid Mg-7.5Al-0.3Mn-based alloys were investigated. The semi-solid alloys used for the experiments were prepared by cooling plate casting. The microstructural analysis on heat treatment and semi solid casted alloys were carried out with the optical microscope, scanning electron microscope and energy dispersive spectroscopy. And, its mechanical properties were evaluated by Vickers hardness and tensile test. The microstructure of the as-cast alloys consists of α-Mg, $Mg_{17}Al_{12}$, phases formed by the addition of misch-metal, and the morphology of these intermetallic phases formed by the addition of misch-metal shows a shape of a needle.

Keywords: *Heat treatment, Misch-Metal, Semi-Solid, Mg alloy*

1. Introduction

In automotive industry, many studies for weight reduction of automotive parts have been conducted because it is possible to improve fuel efficiency of an automobile and reduce the emissions of carbon dioxide [1,2]. Magnesium alloys, as the lightest structural metal material, have excellent specific strength than that of other alloys such as aluminum and steel alloys. For this reason, it has a potential as a light-weight material for the application of the automobile parts [3,4]. The Mg-Al based alloys of Mg alloys is the most widely used commercial cast alloy due to their excellent castability and strength at room temperature. However, the alloys is limited to the use of the alloy because it has a low high temperature strength by the low melting point of β phases ($Mg_{17}Al_{12}$) formed by Al addition [5-7]. So, there are in progress many researches to improve the high temperature strength of the alloy by element addition (Si, RE, MM, etc.) and cast process [8-11]. Therefore, in this study, the effects of misch-metal addition on the heat treatment of semi-solid Mg alloys were investigated.

2. Design of Experiments

The used alloys for this study were prepared using Mg-7.5Al-0.3Mn alloys and misch-metal fabricated by a cooling plate method. The alloys were melted and stabilized at 720°C under a mixed CO_2 and SF_6 gas atmosphere. Heat treatment of the alloys was solution treated at 400°C for 8h. Then, aging treatment was performed at 200°C. And, the heated specimens were carried out the microstructure observation and Rockwell hardness test.

3. Results and Discussion

Fig. 1 shows the microstructure of Mg-7.5Al-0.3Mn-*x*MM alloys with the different casting method ((a,b) permanent mold casting and (c,d) semi-solid casting). As shown in Fig. 1, the alloys fabricated by permanent mold casting had coarse and dendritic grains by a relatively slow cooling rate but by semi-solid casting consisted of fine and global grains because of rapid cooling rate by inclined cooling plate method. And, addition of misch-metal to Mg-7.5Al-0.3Mn alloy (Fig 1(b), (d)) contributed to the formation of a rod or needle-like phases and reduction of β phases ($Mg_{17}Al_{12}$,) fraction. EDS analysis of the rod or needle-like phase, it shows that La, Ce and Al were the main constituents of the phase.

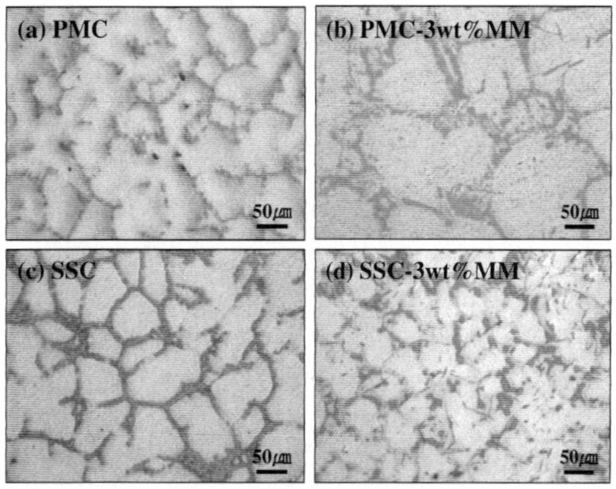

Fig. 1 Microstucture of Mg-7.5Al-0.3Mn alloys with different casting method

Fig. 2 shows the Hardness variation of Mg alloys after aging treatment at 200°C. With increasing aging time, the hardness of Mg alloys gradually increased and reached a maximum peak hardness of the alloys after 4hours.

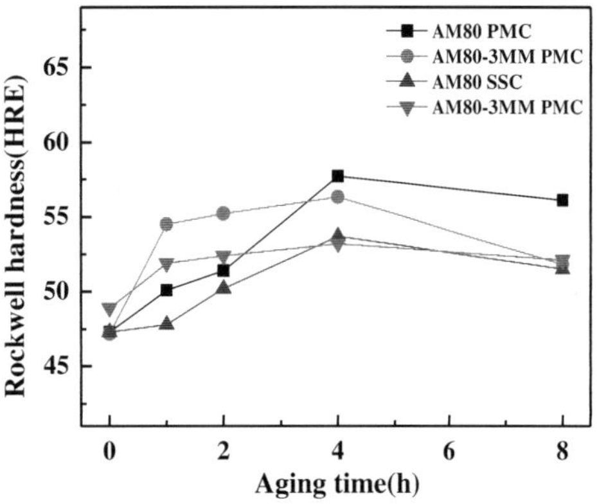

Fig. 2 Change in the Rockwell hardness of Mg-7.5Al-0.3Mn-xMM alloys after aging treatment

Fig. 3 shows the microstructure of semi-solid casted Mg-7.5Al-0.3Mn-xMM alloy aging heated at 200°C. In Fig. 3, the semi-solid cased Mg-7.5Al-0.3Mn alloy was gradually increased a discontinuous precipitate from the grain boundaries of Mg alloy with increasing aging time.

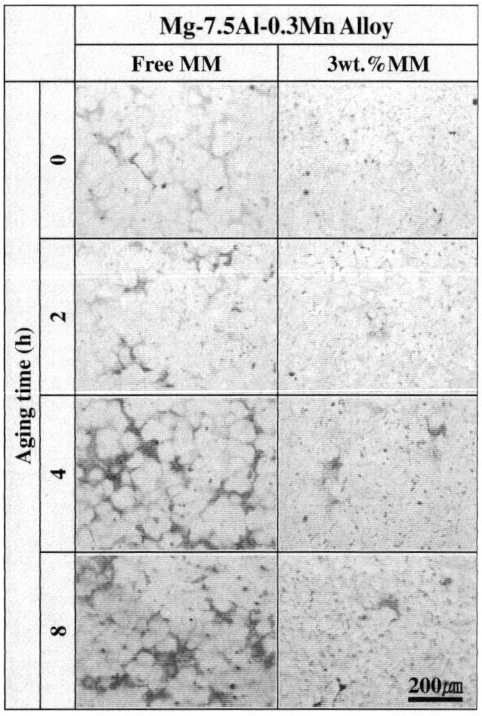

Fig. 3 Microstructure variation of semi solid casted Mg-7.5Al-0.3Mn-xMM alloys aging heated at 200°C

However, the semi-solid cased Mg-7.5Al-0.3Mn-3wt.%MM alloy was not observed a discontinuous precipitate from the grain boundaries in the early aging time. Also, it was confirmed that the discontinuous precipitation and growth of β phases was delayed by the addition of misch-metal during the heat treatment.

4. Summary

The addition of misch-metal to Mg alloy contributed to the formation of a rod or needle-like phases and reduction of β phase fraction.

The addition of misch-metal to Mg alloy was confirmed to delay the discontinuous precipitation of β phases and growth of the phases during the heat treatment of Mg alloy

Acknowledgements

This work was supported by the Human Resource Training Program for Regional Innovation and Creativity through the Ministry of Education and National Research Foundation of Korea (2015-0633).

References

[1] X. Li,P.Yang, L.-NWang, L.Meng, F.Cui, Mater. Sci. Eng. A 517 (2009) pp.160–169.

[2] Kwon-Sup Kim and Su-Gun Lim, *Korea J. Met. Mater.* 38(4) (2000) pp.529-533.

[3] A Yamashita, Z Horita, TG Langdon, Mater. Sci. Eng. A 300 (2001) p.142.

[4] Dae-Hwan Kim, Seung-Hwa Choi, Hee-Kyung Kim, Sung-Yong Shim, and Su-Gun Lim, *KFS*, 30(6) (2010) pp.235-240.

[5] Kainer Ku, Magnesium alloys and Technology. Weinheim: Wiley-VCH Verlag GmbH & Co. KGaA (2003) pp.2-12.

[6] A. Zafari, H.M. Ghasemi, R. Mahmudi, Materials and Design, 54, (2014) pp.544-552.

[7] B.H. Kim, S.W. Lee, Y.H. Park, I.M. Park, J. Alloys Compd. 493, (2010) pp.502-506.

[8] B. Kondori, R. Mahmudi, Mater. Sci. Eng. A 527, (2010) pp.2014–2021.

[9] Jinghuai Zhang, Shujuan Liu, Zhe Leng, Milin Zhang, Jian Meng, Ruizhi Wu, Mater. Sci. Eng. A, 528, (2011) pp.2670-2677.

[10] A.A. Luo, Int. Mater. Rev. 49 (2004) pp.13-30.

[11] M.O. Pekguleryuz, A.A. Kaya, Adv. Eng. Mater., 5, (2003) pp.866-878.

Proceedings of the 72nd World Foundry Congress
21-25th May 2016, Nagoya, Japan

Improvement for Strength of Cu-Ti Based Composites Containing Graphite Particles Fabricated by Centrifugal Mixed-Powder Method

Tomoaki Kayano[1], Hisashi Sato[2] and Yoshimi Watanabe[2]
[1] Graduate School of Engineering, Nagoya Institute of Technology
Gokiso-cho, Showa-ku, Nagoya, Aichi, 466-8555 Japan
[2] Nagoya Institute of Technology, Omohi College
Gokiso-cho, Showa-ku, Nagoya, Aichi, 466-8555 Japan

Centrifugal mixed-powder method (CMPM) is casting process combined of centrifugal casting and powder metallurgy. The CMPM has advantage that solid particles with low wettability for matrix melt can be composed into the matrix. Recently, parts of authors have successfully fabricated Cu-based composites containing graphite particles by using the CMPM. Also, it has been reported that Ti addition into Cu matrix can improve the wear resistance of the Cu-based composites containing graphite particles. However, those Cu/Ti-based composites containing graphite particles are not aged while Cu-Ti alloy is age-hardened alloy. Hence, it is expected that strength of the Cu/Ti-based composites containing graphite particles by the CMPM is further increased by aging treatment. In this study, Cu/Ti-based composites containing graphite particles have been fabricated by the CMPM, and then aging treatment is performed for this composite to improve its strength. The Cu/Ti-based composites containing graphite particles are successfully obtained by the CMPM. Hardness of the composites becomes maximum by aging treatment at 400 °C for 4 h. Moreover, TiC phase is observed in the aged Cu/Ti-based composites containing graphite particles. From the obtained results, it is found that aging treatment is effective to improve mechanical properties of the Cu/Ti-based composites containing graphite particles.

Keywords: Centrifugal mixed-powder method (CMPM), Centrifugal casting, Composite, Aging, Self-lubrication material, Wear

1. Introduction

Bearings are often used for rotary and reciprocating portions to prevent energy loss due to friction, and materials used for bearings are Fe and Cu alloys. As one of the improvement method for the lubrication between the bearing-parts, self-lubrication materials are proposed [1-3]. The self-lubrication materials are metal-based composites containing solid-lubricant particles such as graphite and molybdenum disulfide. The self-lubrication materials are usually fabricated by sintering [2]. However, since many of the bearing-parts are fabricated by centrifugal casting, processing method of the self-lubrication materials using centrifugal casting is demanded.

Parts of authors have recently developed Cu-based composites containing graphite particles by CMPM [3]. The CMPM is the casting process combined of centrifugal casting and powder metallurgy [4]. Furthermore, they have reported that Ti addition into Cu matrix can improve the wear resistance of the Cu-based composites containing graphite particles [5]. Although it is well known that Cu-Ti alloy is age-hardened alloy [6], aging treatment of the Cu/Ti-based composites containing graphite particles is not carried out [3].

In this study, the Cu/Ti-based composites containing graphite particles have been fabricated by the CMPM, and then aging treatment for this composite is performed. Hence, it is expected that strength of the Cu/Ti-based composites containing graphite particles by the CMPM can be improved by aging treatment.

2. Experimental Procedures

Cu/Ti-based composites containing graphite particles were fabricated by the CMPM. As first step, three kinds of mixed-powders of Cu powder (25 μm), Ti powder (32 μm or less) and graphite powder (50 μm) with different Ti concentration were prepared. Volume fraction of graphite in these mixed-powders was 25 vol.%. After that, the mixed-powder was inserted into mold, and subsequently centrifugal force was induced for the mold after melting Cu ingot. The centrifugal force was 35 *G*. Table 1 shows the casting conditions of the composites. Using the Cu/Ti-based composites containing graphite particles solution treatments were performed at 950 °C for 48 h and aging treatments were carried out at 400 °C for 1 ~ 4 h using vacuum furnace. Microstructural observations were made by a scanning electron microscope (SEM) for the obtained Cu/Ti-based composites. Moreover, micro-Vickers hardness tests were made for the composites.

Table 1 Casting conditions of sample fabrication.

Sample name	Heating temperature (℃)	Weight of mixed-powder (g)
Cu-2.5at.%Ti	1250	11.7
Cu-3.0at.%Ti	1250	11.6
Cu-5.0at.%Ti	1280	11.5

3. Results and discussion

Cu/Ti-based composites containing graphite particles are successfully fabricated by the CMPM. Hardness as a function of aging time is shown in Fig. 1. From this figure, the composites aged at 400 °C for 3 h has maximum hardness. Figure 2 is a backscatter electron compositional image of Cu-5.0at.%Ti sample aged at 400 °C for 3 h. As can be seen, some TiC phases are observed at interface between Cu/Ti matrix and graphite particle in the aged Cu/Ti-based composite containing graphite particles. Because of this, bonding strength between matrix and graphite would be improved by formation of TiC. Although Cu_4Ti phase is usually formed by aging treatment for Cu-Ti alloy, the Cu_4Ti phase can not be observed in Fig. 2. However, since correct age-hardened curves are obtained, very fine Cu_4Ti would be formed during aging treatment. Table 2 is results of wear tests for Cu-2.5at.%Ti alloy without graphite addition and Cu-2.5at.%Ti sample. It is seen that the Cu-2.5at.%Ti sample has lower frictional coefficient and higher wear resistance. Therefore, aging treatment for the Cu/Ti-based composites containing graphite particles is effective method to improve its mechanical properties.

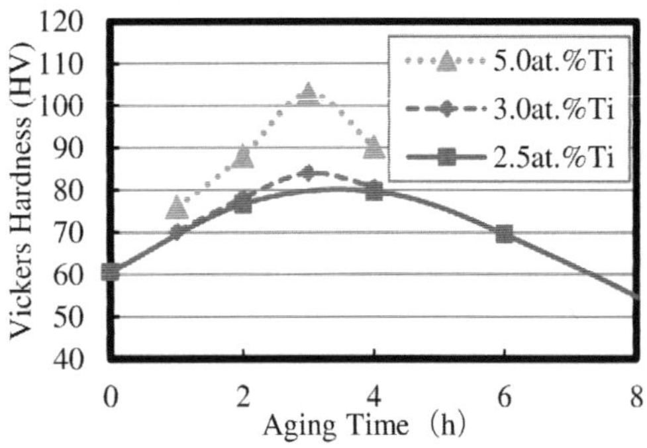

Fig. 1 Vickers hardness of aged Cu/Ti-based composites.

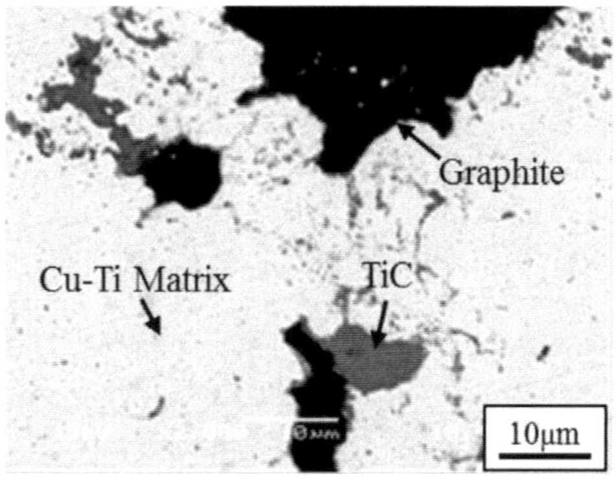

Fig. 2 Microstructures of Cu-5.0at.%Ti sample aged at 400 °C for 3 h fabricated by CMPM.

Table 2 Wear properties of Cu-2.5at.%Ti alloy and Cu-2.5at.%Ti sample

Sample name	Frictional coefficient	Cross-sectional area of wear groove [mm^2]
Cu-2.5at.%Ti alloy (without graphite addition)	1.06	0.047
Cu-2.5at.%Ti sample	0.26	0.019

Acknowledgement

This work was supported from Matching Planner Program by Japan Science and Technology Agency. This financial support was gratefully acknowledged.

References

[1] S. Maki, Y. Kyoso, Y. Harada and K. Mori: J. Mater. Process. Technol. 41 (2000) 1129-1133.

[2] K. Umeda, A. Tanaka and S. Takatsu: IEICE Technical Report 101 (2001) 33-38.

[3] H. Sato, W. Wei, K. Oguri, M. Yamada and Y. Watanabe: Mater. Sci. Forum, 783-786 (2014) 1579-1584.

[4] Y. Watanabe, Y. Inaguma , H. Sato and E. Miura-Fujiwara: Materials 2 (2009) 2510-2525.

[5]N.Shinozaki, Y.Katayama, Y.Umezawa and K.Wasai: J. JRICu 47 (2008) 187-191.

[6] S. Semboshi, T. Nishida and H. Numakura: Mater. Sci. Eng. A 517 (2009) 105-113.

P-37

Primary M_7C_3 Carbide Refinement of Hypereutectic High-Chromium White Cast Iron

Yuuki Kudo, Setsuo Aso, and Ikuzo Goto
Graduate School of Engineering Science, Akita University, 010-8502 Akita, JAPAN

Refining primary M_7C_3 hexagonal rod-like carbides is an effective method for improving the mechanical properties of hypereutectic high-chromium cast iron (HCCI). In this work, three hypereutectic HCCIs—4.0C–25Cr, 4.5C–28Cr, and 5.0C–31Cr—were poured into molds at 15°C, 100°C, and 200°C above the liquidus temperature of each alloy. The primary M_7C_3 carbides of each alloy poured at 15°C above the liquidus temperature were distributed randomly, without solidifying, along the heat flow. Therefore, the hexagonal cross section of the primary M_7C_3 carbides was primarily observed in the solidification structure. The average carbide size of the alloy poured at 15°C above the liquidus temperature was 25 μm, which was one-fourth the carbide size of the alloys poured at higher temperatures. Because the size of the primary carbides was refined when the alloys were poured at 15°C above their respective liquidus temperatures, the three aforementioned alloys were poured at this temperature to prepare four-step wedges. The carbide size decreased with step thickness. In particular, the primary carbide size of the 4.0C–25Cr alloy was most refined to approximately 20 μm. The primary M_7C_3 carbides were simultaneously solidified nondirectionally. These structures should improve the mechanical properties of the alloys.

Keywords: *abrasion, cooling rate, four-step wedge, hardfacing, M_7C_3 hexagonal rod-like carbide.*

1. Introduction

High-chromium cast irons (HCCIs) are generally hypoeutectic compositions because coarse primary M_7C_3 rod-like carbides formed at the hypereutectic composition often cause low brittleness [1]. By contrast, the hypereutectic hardfacing alloys are used in applications such as coal pulverizer liners and gravel pumps, which involve severe abrasion and little impact [2]. Generally, grain refining is an effective method for improving the mechanical properties of alloys; consequently, the hardfacing process, which features a faster solidification rate, is recommended for this application. Also, hypereutectic HCCIs improve the mechanical properties by inducing grain refining. Numerous researchers have reported that Ti is an effective element to reduce the size of primary M_7C_3 carbides because TiC serves as the heterogeneous nuclei for M_7C_3 carbides [3]. However, because the carbon content of a melt is decreased by the crystallization of TiC, the size and volume ratio of primary M_7C_3 carbides will inevitably be reduced [4]. The purpose of this research is to investigate the refining of the primary M_7C_3 carbides of hypereutectic HCCI by controlling the solidification conditions and clarify the limit of refinement by the relationship between the carbide size and the solidification conditions.

2. Experimental

2.1 Materials and processing

The alloy compositions are shown Table 1. The electrolytic iron, electrolytic chromium, electrolytic nickel, metallic silicon, and graphite were used as the raw materials. These materials were melted in an induction furnace and poured into two types of CO_2 process sand molds. The shape of the specimens obtained by these molds was a four-step wedge block with dimensions of 40 mm × 160 mm with a 10-mm step and a rectangular block with dimensions of 30 mm × 30 mm × 100 mm. The rectangular blocks of each alloy were poured at 15°C, 100°C, and 200°C above the liquidus temperature. The four-step wedge block of each alloy was poured at 15°C above the liquidus temperature.

Table 1 Chemical composition of the alloys used in this work (mass%).

C	Mn	Si	Ni	Cr	Mo	Cr/C
4.0	1.5	0.8	0.5	25	1.0	6.2
4.5	1.5	0.8	0.5	28	1.0	6.2
5.0	1.5	0.8	0.5	31	1.0	6.2

2.2 Observation of microstructure and carbides

Each block was cut in the longitudinal direction, and the resulting cross sections were polished and etched with a hydrochloric acid–picric acid mixture. The solidification microstructures near the center of the block or step were observed in the cross section of each block using an optical microscope. The carbide size and area fraction were determined using image processing software. The size of rod-like primary carbides was determined as their length in the short-axis direction.

3. Results and discussion

3.1 Effect of pouring temperature

Figures 1 and 2 show the macrostructures and microstructures, respectively, of the 4.0C–25Cr alloy specimens poured at 1355°C, 1440°C, and 1540°C. The macrostructure of the specimen poured at 1355°C is nondirectional, whereas the macrostructure of the 4.0C–25Cr alloy specimens poured at 1440°C and 1540°C is columnar. The primary M_7C_3 carbides with a large hexagonal rod-like structure are evident in Fig. 2 for the specimens poured at 1440°C and 1540°C. By contrast, the primary M_7C_3 carbides in the specimen poured at 1355°C solidified in a nondirectional fashion because of the lower thermal gradient in the molten metal. Therefore, the short-axis direction of the carbides was primarily observed. Figure 3 shows the effect of pouring temperature on the primary M_7C_3 carbide size; the primary M_7C_3 carbide size increases with pouring temperature.

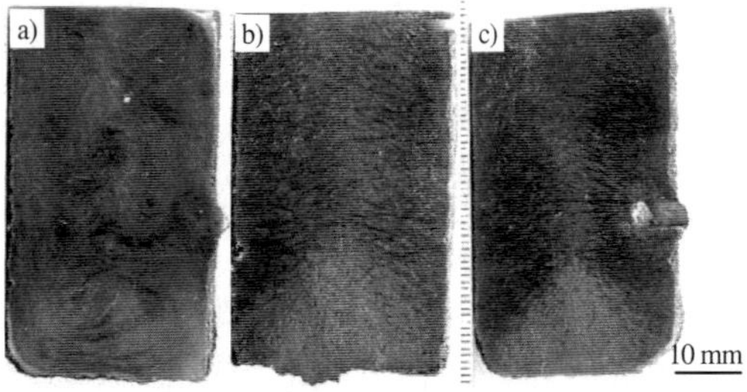

Fig. 1 Macrostructure of 4.0C–25Cr alloy specimens poured at a) 1355°C, b) 1440°C, and c) 1540°C.

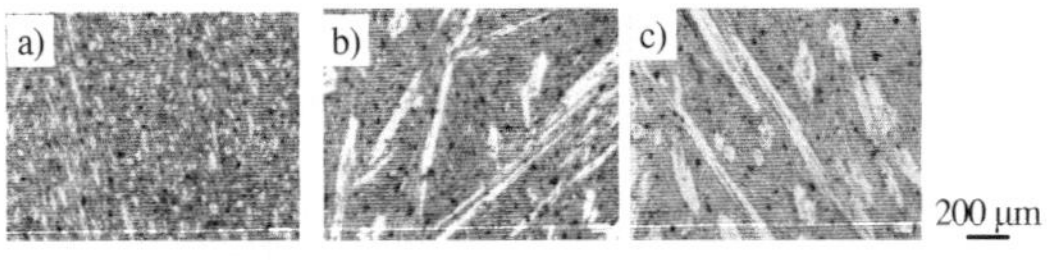

Fig. 2 Microstructure of 4.0C–25Cr alloy specimens poured at a) 1355°C, b) 1440°C, and c) 1540°C.

3.2 Effect of cooling rate

Figure 4 shows the effect of step thickness on the primary M_7C_3 carbide size of four-step wedge blocks of the three investigated alloy compositions poured at 15°C above the liquidus temperature of each alloy. The carbide size increases with step thickness and decreases with carbon content. The minimum carbide size is approximately 20 μm on the 10-mm-thickness step of the 4.0C–25Cr alloy. We measured the cooling rate of each alloy by constructing cooling curves for the temperature range between the pouring and eutectic temperatures. The size of primary carbides decreases inversely with the square root of the cooling rate. For the size of carbide to be 10 μm or less, a cooling rate of not less than 6°C/s is required.

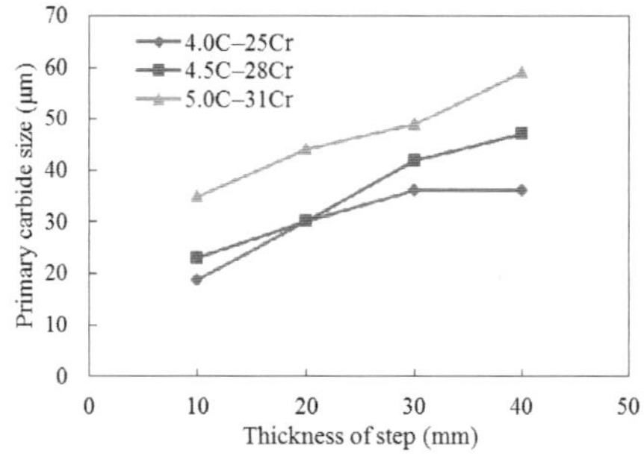

Fig. 3 Effect of pouring temperature on the primary M_7C_3 carbide size.

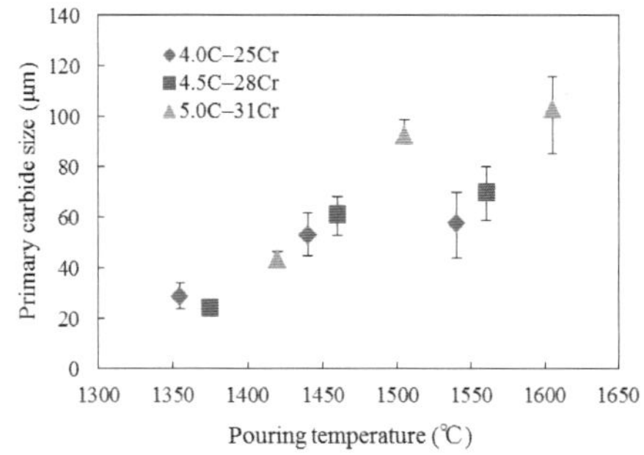

Fig. 4 Relationship between the step thickness of the four-step wedges and the primary M_7C_3 carbide size.

4. Conclusions

The size of primary carbides decreases with the pouring temperature and step thickness of the four-step wedge. Consequently, the size of the primary carbides decreases inversely with the square root of the cooling rate. The primary M_7C_3 hexagonal rod-like carbides solidified nondirectionally when each alloy was poured at 15°C above the liquidus temperature of the alloy.

References

[1] R.B. Gundlach: Friction, Casting, *ASM Handbook*, ASM International, 15 (2008) 896.

[2] P. Crook and H.N. Farmaer: Friction, Lubrication and Wear Technology, *ASM Handbook*, ASM International, 18 (1992) 758.

[3] H. Fu, X. Wu, J. Xing, Y. Lei and X. Zhi: J. Materials Engineering and Performance. 18 (2009) 1109-1115.

[4] Q. Liu, H. Zhang, Q. Wang, X. Zhou, P.G. Jönsson and K. Nakajima: ISIJ International. 52 (2012) 2210-2219.

Proceedings of the 72nd World Foundry Congress
21-25th May 2016, Nagoya, Japan

Age-hardening behavior of Al-7~10%Si-0.3%Mg alloy castings with water quenching and direct quenching

Shota Komura[1], Seiji Saikawa[2], Kenji Matsuda[2] and Susumu Ikeno[3]
[1] Department of Graduate School of Science and Engineering for Education, University of Toyama, Toyama 930-8555, Japan
[2] Department of Graduate School of Science and Engineering for Research, University of Toyama, Toyama 930-8555, Japan
[3]Hokuriku Polytechnic College, Uozu 937-0856, Japan

The solvus temperature of G.P. zones and/or clusters during the early stage of aging after solution treatment in Al-7%Si-0.3%Mg alloy was investigated by micro-vickers hardness measurement and transmission electron microscopy. In addition, the obtained result was compared the previous reported result of the Al-10%Si-0.3%Mg alloy. The alloys were cast by permanent mold process, homogenized at 813K for 36ks, cold rolled up to 68%, solution treated at 813K for 36ks, and then artificially aged at 348-523K after water-quenched (W.Q.) or direct-quenched into oil-bath at 348-523K for artificially aging (D.Q.). In the case of the aging temperature was highest; the peak hardness for the W.Q. condition was clearly higher than the D.Q. condition. This tendency is considered due to the difference of formation such as G.P. zones and/or clusters between W.Q. and D.Q. conditions. Also, these behaviors were similar to the previous reported result of the Al-10%Si-0.3%Mg alloy.

Keywords: *Al-Si-Mg alloy, casting, aging, direct quench, transmission electron microscopy*

1. Introduction

Al-7-10mass%Si alloys were added a small amount of Mg, such as AA356 and AA365 alloy which are widely applied to automotive components because of that has excellent castability and higher strength and ductility. Such actual components were often subjected to heat treatment for the purpose of improving the mechanical properties. However, there are few reports about precipitation behavior of the hypoeutectic Al-Si-Mg alloys.

From this reason, we previously investigated the solvus temperature of G.P. zones after solution treatment in Al-10%Si-0.3%Mg alloy [1]. As a result, its temperature in this alloy was estimated it is between from 373 to 423K.

In this study, we have investigated the solvus temperature of G.P. zones and/or clusters during the early stage of aging after solution treatment in Al-7%Si-0.3%Mg alloy. In addition, the obtained result was compared the previous reported result of the Al-10%-0.3%Mg alloy [1].

2. Experimental procedures

The target composition of the alloy was Al-7%Si-0.3%Mg-0.015Sr. The molten metal of this alloy was degassed by inserting Ar gas via a lance pipe and then poured into copper mold with a Y-block shaped cavity. Pouring and mold temperature were 973K and 433-443K, respectively. Immediately after casting, the castings was removed from the mold when the temperature of casting had dropped to 773K. And then, the castings quenched in ice water for 10s. Table 1 shows chemical composition of the alloy.

The bottom of obtained castings was homogenized at 813K for 36ks and then slow cooled in the electric furnace. Subsequently, it was cold rolled up to 68%, solution treated at 813K for 36ks, and then aged at R.T.(298K)-523K after quenched in ice water (hereafter referred to as the "W.Q. specimen"). On the other hands, it was quenched into oil-bath held at 348-523K or water of R.T. and then aged at its temperatures (hereafter referred to as the "D.Q. specimen"). The heat patterns of W.Q. and D.Q. specimens are shown schematically in Fig.1.

The hardness of specimens was measured by micro-vickers hardness tester (load: 2.94N, time: 10s). The transmission electron microscope (TEM) observation was performed by the accelerating voltage of 120kV.

3. Results and discussions

The age-hardening curves of the Al-7%Si-0.3%Mg alloy aged at 523K after water-quenched and direct-quenched are shown Fig.2. The peak-hardness of W.Q. and D.Q. specimens were measured the value of about 80HV and 57Hv, respectively, the both specimens were indicated that the clear difference of hardness was occurred.

Fig.3 shows the transmission electron micrographs of specimens peak-aged at 523K, it were observed along the [100] or [110] direction of the matrix. In the case of W.Q. specimens were observed, rod-shaped precipitates were aligned in the [100] direction of matrix in Fig.3 (a) and (b). The length of precipitates were approximately 200-300nm. In contrast, D.Q. specimens were occurred rod-shaped precipitates which were clearly coarser than W.Q. specimens. In addition, as shown by the arrow, the heterogeneous precipitates were also observed on dislocation lines in Fig.3 (d). From these result, it is estimated that the peak-hardness of D.Q. specimen was lower than W.Q. specimen. These age-hardening behaviors were similar to the previous reported result of the Al-10%Si-0.3%Mg alloy.

4. Conclusion

In the case of Al-7%Si-0.3%Mg alloy aged at 523K, the peak-hardness of W.Q. specimens were higher than D.Q. specimens. It is due to the difference of size and density of precipitation.

Reference

[1] S. Saikawa, R. Morioka, K. Matsuda, S. Ikeno, E. Yanagihara and S. Orii : J.JILM 64 (2015) 218-223.

Table 1 Chemical composition of the alloy

(mass %)

Alloy	Si	Mg	Fe	Ti	Sr	Al
Al-7%Si-0.3%Mg	6.90-7.30	0.29-0.32	0.11-0.13	<0.01	0.012-0.015	Bal.

Mn, Cu and Zn:<0.01

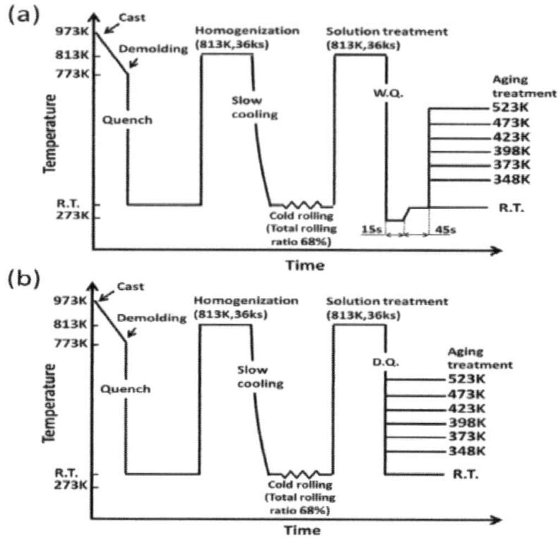

Fig.1 Heat-treatment procedure for the casting, homogenaization, cold rolling, solution treatment, quench and subsequent aging treatment. (a)water-quench and (b)direct-quench methods.

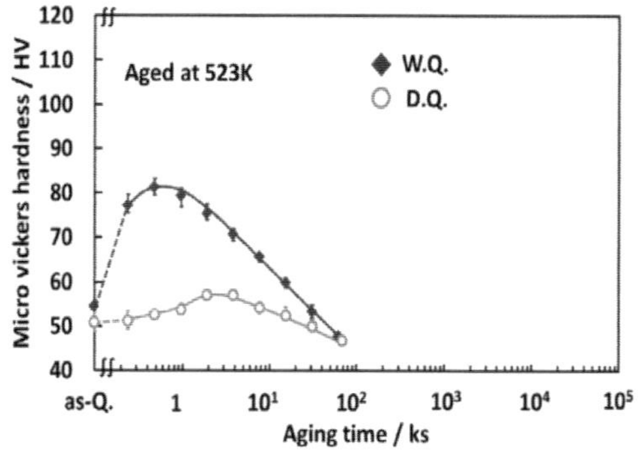

Fig.2 Age-hardening curves of the specimens aged at 523K after water-quenched and direct-quenched.

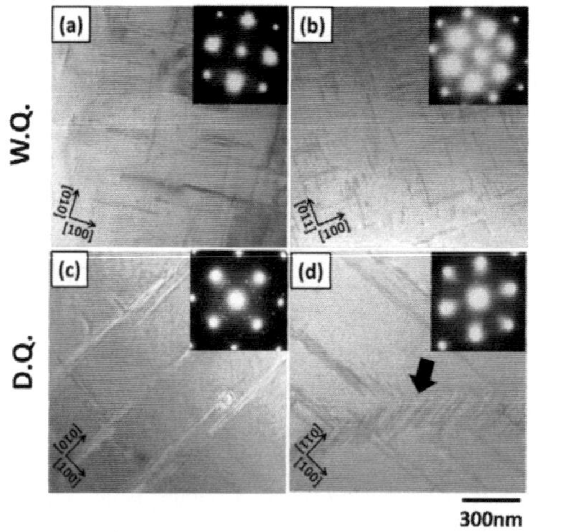

Fig.3 Transmission electron micrographs of the specimens peak-aged at 523K. (a) and (c) incidented beem direction parallels $[100]_m$, (b) and (d) incidented beem direction parallels $[110]_m$.

Proceedings of the 72nd World Foundry Congress
21-25th May 2016, Nagoya, Japan

Solidification Behavior of Mg-Al-Ca Alloys with Different Aluminum and Calcium Contents.

Hiroki Kako[1)], Seiji Saikawa[2)], Susumu Ikeno[3)], Kazunori Shimizu[4)] and Hiroaki Iwakawa[4)]
[1] Graduate School of Science and Engineering for Education, University of Toyama, 3190Gohuku, Toyama, Toyama930-8555, Japan
[2] Science and Engineering for Research, University of Toyama, 3190Gohuku, Toyama, Toyama930-8555, Japan
[3] Hokuriku Polytechnic Corporation, 1289-1Kawaberi, Uozu, Toyama937-0856, Japan
[4] Sankyo Tateyama Inc., Sankyo Material-Company, 8-3Nagonoe, Imizu, Toyama934-8515, Japan

Magnesium alloys have the characteristic with high specific tensile strength and lightweight properties. Therefore, it is used for auto mobile industry, aircraft industry, and electronic equipment parts, and a future use expansion attracts attention. In this study, behavior of solidification structure in Mg-0~12mass%Al-0~10%Ca alloys cast into sand mold were investigated by XRD measurement and OM, SEM-EDS observation. In the as-cast state of all alloys, solidification structures mainly consist of the primary crystallized α-Mg and secondary crystallized eutectic, such as Mg_2Ca, $Mg_{17}Al_{12}$ and Al_2Ca phases. The shape of primary crystallized α-Mg phase changed from cellular to dendritic, when Al and Ca content increased. On the other hands, shape of secondary crystallized Al_2Ca and Mg_2Ca phases were lamellar, while $Mg_{17}Al_{12}$ phase was crystallized at grain boundary and cell gaps with blocky shaped. Temperature stagnation region due to the latent heat release by crystallization of primary crystal with the increase of Al concentration and the Ca concentration is reduced.

Keywords: *Mg-Al-Ca alloys, microstructures, sand-mold, cooling curves*

1. Introduction

In recent years, lightweight of the vehicle body has been required in the automobile industry. Magnesium alloys have characteristic of high specific strength and lightweight properties. Therefore, the application of magnesium alloys have been expanded in the auto parts.

The conventional alloys Mg-Al-Zn(AZ) and Mg-Al-Mn(AM) series, have good strength at room temperature and ductility with sufficient corrosion resistance. However, the application of these alloys is limited to components operating at high- temperature below 393K because of their rapid degradation of creep properties.

The development of a variety of heat-resistant magnesium alloy has been carried out. Recently, rare metal-free heat-resistant Mg alloy has been attracting attention. Among them, Mg-Al-Ca-based alloy has the characteristics of high heat resistance and low cost.

We have examined such as the castability of Mg-Al-Ca alloys. As a result, hot tearing generated in the solid-liquid coexistence region was reduced by the increasing Al and Ca contents.

In this study, we investigated the influence of Al and Ca content on the solidification behavior and structure of Mg-Al-Ca alloy cast by sand mold.

2. Experimental procedures

The target composition is Mg-0-12mass%Al-0-10mass%Ca alloys. The Samples were melted using pure Mg, pure Al, pure Ca, Mg-Mn and Al-Be master alloys. These alloys were melted in an electric furnace while blowing protective gas (CO_2 + SF_6 gas) to the melt surface. Mold was molding a sand mold with a cavity of cylindrical shape as shown in Fig.1. To average cooling rate of casting, it was placed a thermocouple in the center of the cavity.

The observation specimen cut from the castings, and it was polished by grind emery paper of #80-2000. Then diamond paste with the diameter of 1μm was used. Finally, polished surfaces were etched in 1% nitric acid ethanol. These microstructures were examined using an optical microscope(OM), field emission scanning electron microscope(FE-SEM) equipped with an energy dispersive X-ray system(EDS) and X-ray diffraction method(XRD).

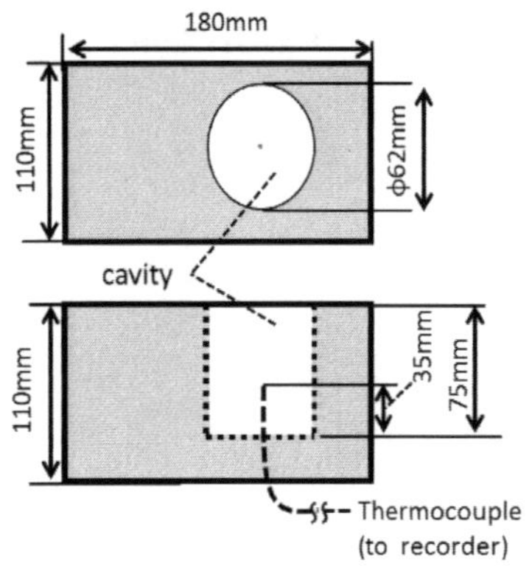

Fig.1 Schematic drawing of sand mold for measuring of cooling curves.

3. Results and discussion

The cooling curves of Mg-6%Al-0-10%Ca alloys cast by sand mold is shown in Fig.2. Fig.2 (a) occurs the crystallization of the primary crystallized α-Mg phase at about 893K.Then temperature stagnation by latent heat release has been confirmed to about 714K. In Fig.2 (b)-(d), crystallization temperature of the primary crystallized α-Mg phase with an increase in the Ca addition amount was confirmed that the decrease. In addition, the temperature stagnant due to the latent heat release was observed at about 789K to 807K by addition of Ca.

Fig.3 shows microstructure of Mg-6%Al-0-10%Ca alloys sand mold castings. Fig.3 (a) was observed matrix of α-Mg phase and non-equilibrium crystallized β-$Mg_{17}Al_{12}$ phase. Fig.3 (b)-(d) were observed in the grain boundaries and cell gap as fine or relatively coarse lamellar crystallized phase.

Fig.4 shows SEM-EDS analysis results of crystallized phase in the Mg-6%Al-6%Ca alloys cast by sand mold. These images show α-Mg phase with an interdendritic eutectic structure containing intermetallic phases that are identified as the Al_2Ca phase(1) and the Mg_2Ca phase(2).

4. Conclusions

Temperature stagnation region due to the latent heat release by crystallization of primary crystallized α-Mg phase with the increase of Al concentration and the Ca concentration is reduced.

References

[1] S.Saikawa. Japan Institute of Light Metals. 60(11)(2010)571-577.

[2] A.Suzuki, N.D. Saddock, J.W. Jones, T.M.Pollock. Acta Materialia. 53(2005)2823-2834.

[3] S.W.Xu, K. Oh-ishi, S.Kamado, H.Takahashi, T. Honma. Matrials Science and Engineering. A542(2012)71-78.

[4] H.Niwa et al., The Japan Light Metals Society 124, Spring meeting, (2013), 101-102

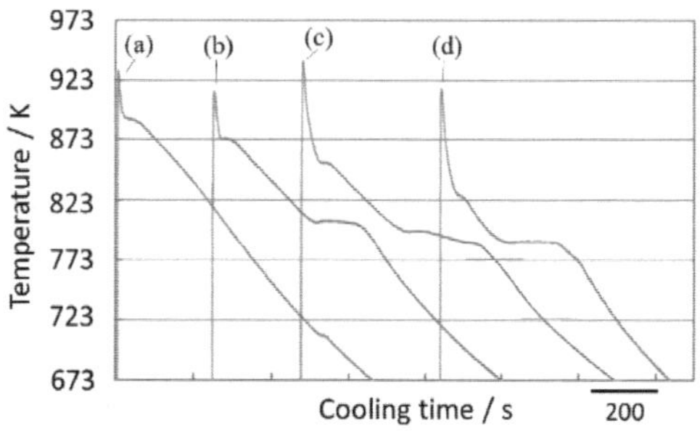

Fig.2 Cooling curves of Mg-6%Al-0-10%Ca alloys cast by sand mold: (a)6%Al-0%Ca, (b)6%Al-3%Ca, (c)6%Al-6%Ca, (d)6%Al-10%Ca.

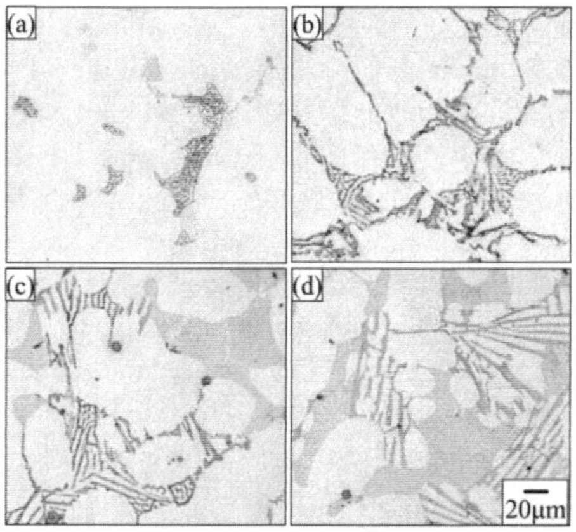

Fig.3 Optical micrographs of Mg-6%Al-0-10%Ca alloys sand mold castings: (a)6%Al-0%Ca, (b)6%Al-3%Ca, (c)6%Al-6%Ca, (d)6%Al-10%Ca.

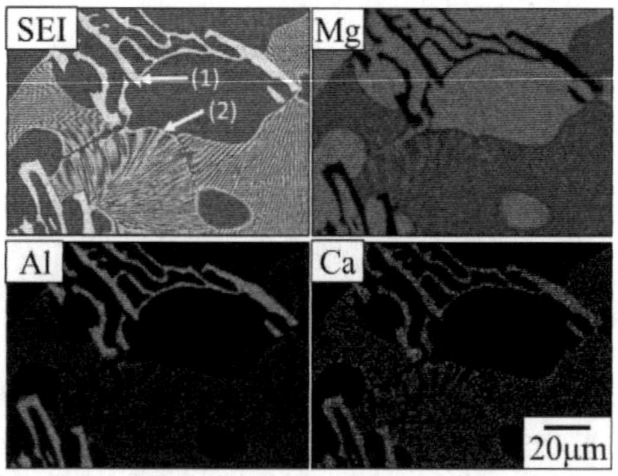

Fig.4 SEM-EDS analysis results of Mg-6%Al-6%Ca system alloys cast by sand mold.

Proceedings of the 72nd World Foundry Congress
21-25th May 2016, Nagoya, Japan

Two-Step Aging Behavior of Solution Treated Al-7-10mass% Si-0.3mass% Mg Alloys

Tomoya Ozasa[1], Shota Komura[1], Seiji Saikawa[2], Kenji Matsuda[2] and Susumu Ikeno[3]

[1] Department of Graduate School of Science and Engineering for Education, University of Toyama, Toyama 930-8555, Japan

[2] Department of Graduate School of Science and Engineering for Research, University of Toyama, Toyama 930-8555, Japan

[3]Hokuriku Polytechnic College, Uozu 937-0856, Japan

Two-step aging behavior of solution treated Al-7%Si-0.3%Mg and Al-10mass%Si-0.3mass%Mg alloys were investigated by micro-vickers hardness measurement and optical microscopy. The hardness of Al-10%Si-0.3%Mg alloy after the solution treatment was higher than Al-7%Si-0.3%Mg alloy one. In this reason, it was considered due to the increasing Si contents 7 to 10%. In the both alloys, the peak hardness of final aging stage after pre-aged at 423K in under aging region was higher than pre-aged at room temperature for 45s. The tendencies were estimated due to the influence of precipitate phases, such as clusters and G.P. zones during pre-aging region.

Keywords *Al-Si-Mg alloy, casting, solution treatment, pre-aging, two-step aging,*

1. Introduction

In recent years, lightweight material is required in the automobile industry. Especially, the hypo-eutectic Al-Si alloy which containing 7 to 10 mass% Si and a small amount of Mg has characteristics of higher ductility, mechanical properties and excellent castability. For the reasons, the alloy can be applied to die-casting components, such as rear frame of the automobile. Such actual die cast components were often applied to T5 treatment (artificial aging after casting) for the purpose of improving the mechanical properties. On the other hand, T6 treatment is expected to improve the higher than mechanical properties of the components with T5 treatment. These components are held at room temperature until heat treatment. In this process, the holding at room temperature is considered pre-aging treatment, heat treatment is considered final aging treatment. Therefore, the mechanical properties of the components in final-aging stage might change by pre-aging temperature and its time. However there are few reports about the effect of the pre-aging and final aging treatment on two step aging behavior of the hypo-eutectic Al-Si-Mg alloys.

In this study, we have investigated the two-step aging behavior of Al-7-10mass% Si-0.3mass%Mg alloys after solution treatment.

2. Experimental

The target composition of the alloys were Al-7mass%Si-0.3mass%Mg and Al-10 mass%Si-0.3mass%Mg. Small amount of Sr were added in order to modify of the Al-Si eutectic. The Ar gas was supplied to the molten metal for the degasification. Pouring was performed at 973K or 953K into cupper mold with a Y-block shaped cavity held at 433-443K. Immediately after casting, the castings was removed from the mold when the temperature of casting had dropped to 773K. And then, the castings quenched in ice water for 10s.

After the water-quenching, the specimens having dimensions of 10×10×2.5mm were cut out from the obtained casting. Solution treatment was carry out at 813K for 36.0ks and then the specimens quenced into ice water. The specimens pre-aged at 273K, room temperature(298K), 348K, 373K and 423K and then final-aged at 473K for various times. The hardness of specimens were measured by micro-vickers tester (load: 2.94N, holding time: 10s). The microstructure of specimens were observed by optical microscope after etched by 0.5% HF.

Table 1 Chemical compositions of the alloys (mass%)

alloy	Si	Mg	Fe	Ti	Sr	Al
Al-7%Si-0.3%Mg	6.91-6.98	0.31	0.12-0.13	<0.01	0.012-0.016	bal.
Al-10%Si-0.3%Mg	9.92-10.24	0.27-0.31	0.09-0.15	<0.01	0.009-0.0015	bal.

3. Results and discussion

The microstructures of Al-7%Si-0.3% Mg and

Al-10% Si-0.3% Mg alloy in the as-cast condition are shown in Fig.1. The microstructures of the both alloy mainly consists of the primary crystallized α-Al phase of white color and eutectic Al-Si phase of gray color on grain boundaries and cell gaps. With increasing Si contents from 7 to 10%, the volume of eutectic Al-Si phase were increased.

Fig.2. shows age hardening curves of the Al-7%Si-0.3%Mg and Al-10%Si-0.3%Mg alloy pre-aged at 423K after solutionized at 813K for 36.0ks. The as-quenched hardness of Al-10%Si-0.3%Mg alloy was slightly higher than Al-7%Si-0.3%Mg alloy one. In this reason, the defference of hardness was considered due to the volume of eutectic Al-Si phases in both alloys. The hardness of both alloys began to increase for about 0.5ks and then reach to peak for about 60.0ks. Moreover in the case of continue to aging process, the hardness was observed monotonically decreased.

Fig.3 shows Age hardening curves of the Al-7%Si-0.3%Mg and Al-10%Si-0.3%Mg alloy pre-aged at 423K for 0.48, 1.96 and 30.7ks and then final aged at 473K. As a comparison, the resalt of being pre-aged at 298K for 45s and then final-aging at 473K in those alloys. In the case of Al-7%Si-0.3%Mg alloy in Fig.3 (a).When the specimen pre-aged at 423K for 0.48 and 1.92ks, peak hardness in final aging stage was higher than the specimen pre-aged at 293K for 45s, the so-called positive effect was occured. In this reason, the two-step aging behavior was estimated due to the influence of precipitate phases, such as clusters and G.P. zones during pre-aging region. In case of pre-aged at 423K for 30.7ks, the hardness was constant until about 8ks and then decreased. Those behaviors were simmior to the result of the Al-10%Si-0.3Mg alloy.

4.Conclusions

In the both alloys, the occerrence of positive effect in the two-step aging behavior was revealed when the specimens pre-aged at 423K in the under-aging stage and then final-aged at 473K.

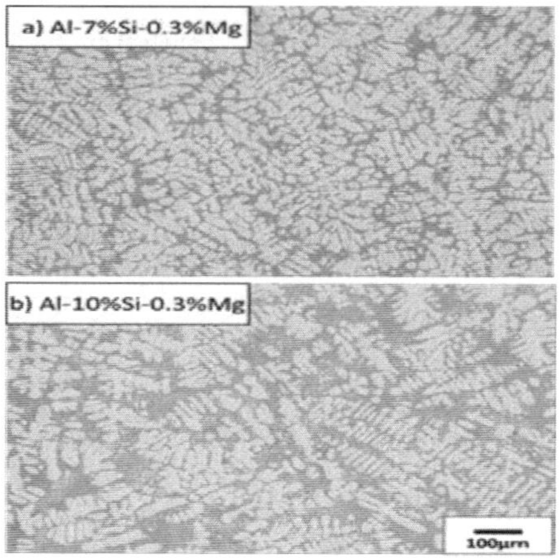

Fig1. The microstructures of Al-7% Si-0.3% Mg and Al-10% Si-0.3% Mg alloy in the as-cast condition.

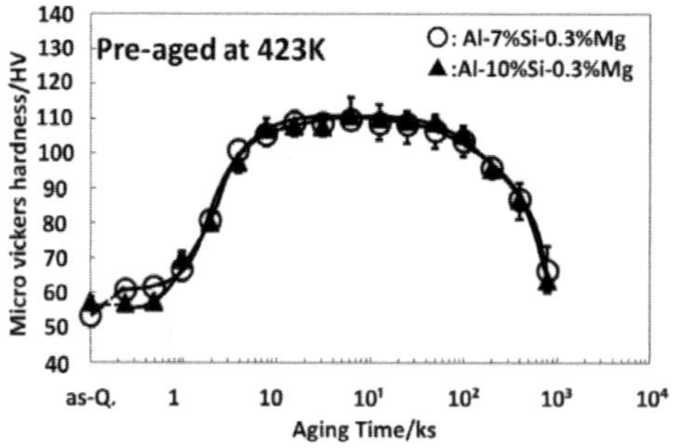

Fig.2 Age hardening curves of the specimens pre-aged at 423K solutionized at 813K for 36.0ks.

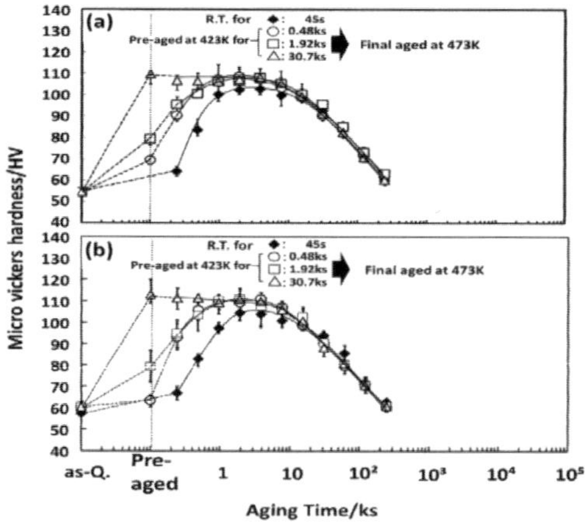

Fig.3 Age hardening curves of the specimens pre-aged at 423K for 0.48, 1.96and 30.7ks and then final aged at 473K.
(a)Al-7%Si-0.3%Mg alloy and (b) Al-10%Si-0.3%Mg allo

P-41

Microstructure and Heat-Resistant Property in Mg-3%Al-1%Si Alloy Castings

Yoshikazu YAMANISHI [1], Hiroki KAKO [2], Seiji SAIKAWA [3], Susumu IKENO [4]

[1] Under Graduate School of Science and Engineering for Education, University of Toyama3190Gohuku, Toyama, Toyama930-8555, Japan

[2] Graduate School of Science and Engineering for Education, University of Toyama, 3190Gohuku, Toyama, Toyama930-8555, Japan

[3] Science and Engineering for Research, University of Toyama, 3190Gohuku, Toyama, Toyama930-8555, Japan

[4] Hokuriku Polytechnic Corporation,1289-1Kawaberi, Uozu, Toyama937-0856, Japan

Magnesium alloys have the characteristic with high specific strength and lightweight property, it is widely used for auto mobile industry. Heat-resistant magnesium alloy is focused as a suitable material for weight reduction of the engine and power train parts in automotive field. In this study, microstructure and heat-resistant property in Mg-3mass%Al-1mass%Si alloy castings produced by permanent mold casting were investigated by optical microscope (OM), scanning electron microscopy (SEM) and measuring of bolt load retention at 423K. The specimens were prepared which simulated the bolt conclusion of true parts and investigated the change of the residual axis power when compressive stress was applied at a high temperature. As a result, bolt load retention of Mg-3mass % Al-1mass % Si (AS31) alloy was higher compared with conventional Mg alloy (e.g. AZ91 and AM60 alloys).

Keywords: *Mg-Al-Si alloy, microstructure, heat-resistant property, bolt road retention.*

1. Introduction

Magnesium alloys are lightest among practical alloys and have many good mechanical properties. Hence magnesium alloys draw attention as structural materials in automobile and railway industries. Mg-Al-Mn (AM) system alloys and Mg-Al-Zn (AZ) system alloys are used most frequently in magnesium alloys [1]. However, mechanical properties of these alloys are diminished at high temperature. Therefore, development of heat-resistant magnesium alloy was carried.

Table1 Chemical composition (mass%).

Coad	Target composition	Al	Si	Zn	Mn	Fe	Mg
AS31	Mg-3%Al-1%Si	3.60	0.90	–	0.09	0.018	Bal.
AM30	Mg-3%Al	3.05	–	–	0.07	0.001	Bal.
AM60	Mg-6%Al	5.31	–	–	0.06	0.003	Bal.
AZ91	Mg-9%Al-1%Zn	10.05	–	1.15	0.01	0.005	Bal.

In recent years, the rare metal-free heat-resistant Mg alloy attracts attention because it is highly recyclable. Among these, the most practical alloys are Mg-Al-Ca (AX) system alloys and Mg-Al-Si (AS) system alloys [2]. We have examined such as the casting of the AX alloys so far [3].

In this study, it is intended to focus on Mg-3mass % Al-1mass % Si (AS31) alloy, to clarify the microstructure and creep properties. We prepared the specimen which simulated the bolt conclusion of true parts and investigated the change of the residual axis power when compressive stress was applied at a high temperature. As a comparison, it was used Mg-3 mass % Al (AM30), Mg-6 mass % Al (AM60), Mg-9 mass % Al-1 mass % Zn (AZ91) and Al-10.5 mass % Si-2 mass % Cu (ADC12) alloys.

2. Experimental procedures

In this study, the specimens were cast as Mg-3 mass % Al-1 mass % Si alloy the target composition. Those were melted in an electric furnace under blowing protection gas (CO_2+SF_6). After that, these were gravity cast in a mold of iron having a cavity of the Y block shape. Table1 shows chemical composition of the samples.

The observation specimen cut from the castings. Then, those microstructure was observed by using OM and SEM. The phase of each alloys were identified using by X-ray diffraction (XRD).

Heat-resistant property was investigated by the high temperature bolt load retention test. The specimens were cut from the riser section of the Y-block shape casting. It was then processed into a ring-shaped outer diameter 20mm, inner diameter of 9mm, thickness 10mm. The test was measured between 10 ~ 12kN axial force, 64 ~ 256h holding time and 423K temperature. The length of the bolt was measured using a micrometer at room temperature.

3. Results and discussions

Fig.1 shows microstructure of the as-cast AS31 alloy, used by OM [a], [b] and SEM [c], [d]. AS31 alloy is compared matrix of alpha-Mg phase, the non-equilibrium crystallized beta-$Mg_{17}Al_{12}$ phases and Mg_2Si phases.

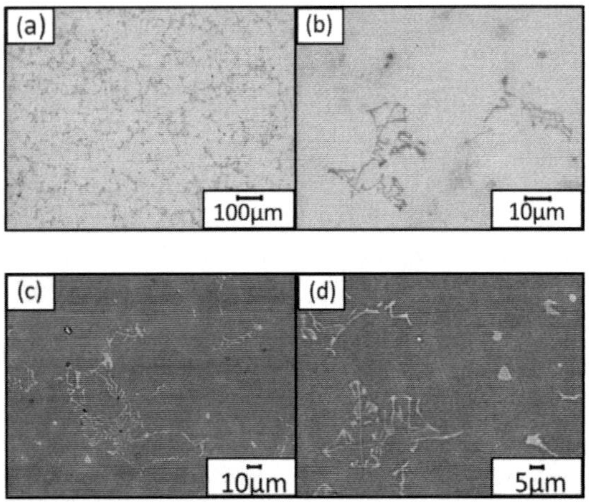

Fig.1 Optical microstructures [a] [b] and scanning electron micrographs [c] [d] of the as-cast AS31 alloy.

High temperature bolt load retention for 256h at 423K of AS31, AM system, AZ91 and ADC12 alloys are shown in Fig.2. Bolt load retention of Mg alloys were lower compared to Al alloy (Fig.2 [a]). In the Mg alloy, AS31 alloy showed the highest bolt road retention (Fig.2 [b]).

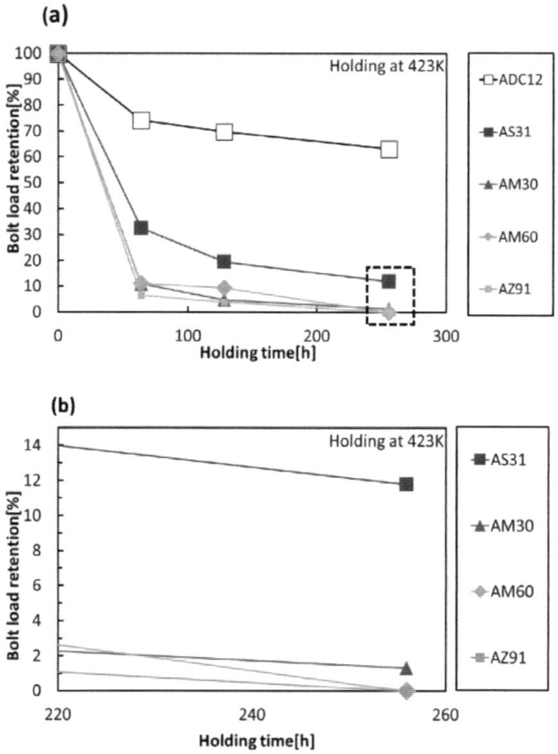

Fig.2 Bolt load retention of AS31, AM30, AM60, AZ91 and ADC12 alloys at 423K for 256h.
(b) shows an enlarged portion of square in (a).

Conclusion

Microstructure and heat-resistant property in Mg-3%Al-1%Si alloy castings were investigated heat-resistant property is AS31 superior with AZ91 alloy due to crystallized Mg2Si phases on the grain boundary cell gaps.

Acknowledgements

We thank Yasuhiro Shigenaga and Shota Komura for cooperation on performing this study.

References

[1] S Saikawa et al., Effects of microstructures on age-hardening of Mg–Al binary alloys, Journal of The Japan Institute of Light Metals, Vol. 62, No. 12(2012), 473–478

[2] W.Blum et al., Comparative study of creep of the die-cast Mg-alloys AZ91, AS21, AS41, AM60 and AE42, Materials Science and Engineering A319–321 (2001) 735–740.

[3] H.Niwa et al., Effect of solidification behavior on the hot tearing of Mg-Al-Ca system alloys, The Japan Light Metals Society 124, Spring meeting, (2013),101-102

P-42

Proceedings of the 72nd World Foundry Congress
21-25th May 2016, Nagoya, Japan

OPTIMIZING THE NICKEL AND COPPER ADDITIONS TO ADI CONSIDERING THE CASTING COOLING RATE

D. Wilk-Kołodziejczyk [1], A Kowalski, K. Jaśkowiec [1]
[1] Instytut Odlewnictwa, 30-418 Kraków, ul. Zakopiańska 73, Polska

The effect of Ni and Cu additions on the mechanical properties of ADI in castings with different wall thickness was discussed. A plan of experiment was developed using a combination of Ni additions in an amount of 0.5 to 1.5% and Cu additions in an amount of 0.5 to 1.0%. The content of other elements was as follows: C = 3.3-3.6%, Si = 2.2-2.6%, Mn = 0.10-0.13%, P up to 0.05%, S up to 0.020%. Four cast iron melts were made and keel blocks with 25, 50 and 75 mm wall thickness were cast. Spheroidization of cast iron was performed using a flexible wire (steel shell filled with magnesium). The casting molds were provided with a special system for filtration and inoculation. From each melt and from each keel block samples were taken for testing of the mechanical properties. After austempering Rm were measured. Relevant approximating polynomials Rm, = f (Ni, Cu) were derived.

Keywords: *nickel alloys, computer simulation, computed tomography, technological conversion*

1. Introduction

Whilst obtaining the ADI grades of 800-8 and 1000-5 in castings with the wall thickness of up to 30 mm poses no problems in the modern industry, obtaining these particular grades in heavy-walled castings (> 50 mm) has not been sufficiently well mastered yet [1,5,8, 11].

The overall objective of studies conducted on heavy-walled castings is to obtain the best cast iron ductility, which is influenced by many factors, among others, by the chemical composition (elimination of Mo and reducing to minimum the content of carbide-forming elements such as manganese, molybdenum, chromium and phosphorus), cast iron metallurgy (spheroidizing treatment and inoculation), and heat treatment parameters (properly selected time and temperature of the austenitizing and austempering treatment) [2-5, 9].

Recently, some improvement of the cast iron toughness, especially in massive castings, has been obtained by the secondary inoculation treatment carried out in a foundry mold to increase the nodule count and refinement degree. [4,8,9].

2. Preparing samples

The keel blocks of the following dimensions, molded in one box, were selected: 25x50x200 mm; 50x65x200 mm; 75x80x200 mm

Metal was fed from one ingate and one cross-gate. Each keel block was fed by two blind risers provided with atmospheric cores. Between the ingate and the cross-gate there was a chamber in which the inoculating and filtering block was placed. Figure 1 shows the principle of operation of the reaction chamber.

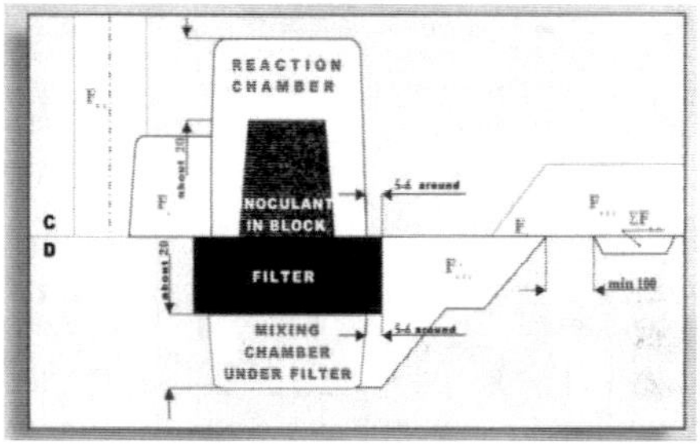

Fig. 1. The reaction chamber and the mixing chamber with inoculating and filtering block forming part of the gating system

A plan of experiment was applied in the study to enable development of mathematical relationships between Rm and a function of the Ni and Cu content[6]. Table 1 shows content of Mg, Ni and Cu.

Table 1 The chemical composition (Mg, Ni, Cu) of cast iron from different melts

Melt No.	Chemical composition, %		
	Mg	Ni	Cu
1	0.085	0.52	0.52
2	0.05	0.50	0.96
3	0.051	1.55	1.00
4	0.055	1.49	0.55

Altogether, four molds were poured from each melt, which gives the total of 16 keel blocks of the wall thickness of 25, 50 and 75 mm.

3. **Mathematical processing of test results**

Applying the developed plan of experiment, equations approximating the tensile strength Rm dependence on the content of Ni and Cu in keel blocks of different wall thickness were derived.

They were next used for plotting of the 3D diagrams illustrating the above mentioned relatioships (Figs. 2 – 3)

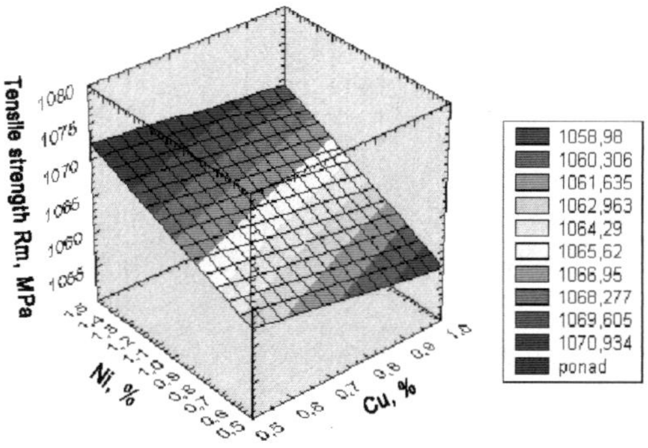

Fig. 2. Nickel and copper content vs ausferritic cast iron tensile strength. Keel block 50 mm

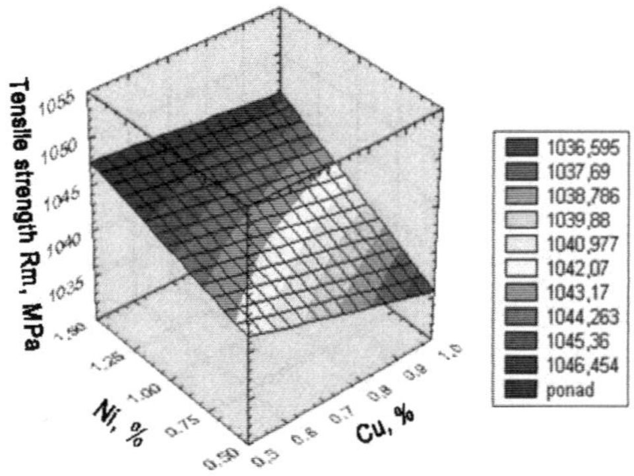

Fig. 3. Nickel and copper content vs ausferritic cast iron tensile strength. Keel block 75 mm

4. **Summary and conclusions**

- In-mold inoculation with the filtering and inoculating system gives very good results, as evidenced by the increase in nodule count and improvement of the cast iron ductility in keel blocks with 50 and 75 mm wall thickness.
- Contrary to previous studies, grades 800-8 and 1000-5 were successfully obtained in castings with 50 and 75 mm wall thickness.
- No significant increase in the mechanical properties of cast iron has been observed with the increasing addition of alloying elements. Studies have proved that the addition of 1% Ni and 0.5% Cu is sufficient to obtain grades 800-8 and 1005 in castings with the wall thickness of up to 75 mm. This conclusion requires, however, further confirmation in subsequent studies.

References

[1] Austempered ductile irons - their significance and present applications. Foundry Trade Journal, October 10, 1985, p. 277

[2] *Bowe K.H. et al.*: Temperaturbehandlungen und bruchmechanische Eigenschaften von bainitischem GGG-100 B. Konstruieren + Giessen 1987, vol.12, no. 4, p. 14

[3] *Dorazil E. et al.*: Influence of low and cryogenic temperatures on deformability and fracture behaviour of asustempered ductile cast irons under static and impact stresses. Giesserei - Praxis 1985, no. 8/9, p. 109

[4] *Fan Z., Smallman R.E.*: Carbide formation in austempered ductile iron alloyed with nickel and copper. 5th International Symposium on the Physical Metallurgy of Cast Iron, October 1994,

[5] *Faucher B. et al.*: Dynamic fracture toughness of austempered ductile irons. Transactions AFS, vol..95, 1987, p. 127

[6] *Górecka R.*: Teoria i technika eksperymentu. Politechnika Krakowska. Kraków 1995

[7] *Grech M., Young J.M.*: Impact properties of a Cu-Ni austempered ductile iron. Cast Metals 1988, no 1/2, p. 98

[8] *Hayrynen K.L. et al.*: Heavy section ADI: microsegregation, microstructure and tensile properties. Transactions AFS 1989, p.143

[9] *Hornung K.*: Four years experience in making gears of austempered iron (ADI). BCIRA International Conference: S.G. Iron - The Next 40 Years. Warwick, England 1987.

[10] *Hua-Qin Su, Xing-Li Guo, Schissler J.M.*: Influence of nickel on kinetics of bainitic transformation of the first stage of austempered ductile iron (ADI). 5th International Symposium on the Physical Metallurgy of Cast Iron, October 1994,

[11] *Janowak J.F., Morton P.A.*: A guide to mechanical properties possible by austempered 1,5% Ni-0,3% Mo ductile iron. Transactions AFS 1984, p. 489

Proceedings of the 72nd World Foundry Congress
21-25th May 2016, Nagoya, Japan

Characterization of duplex stainless steel casting with gadolinium as neutron absorbers for spent fuel storage applications

Byung-Moon Moon[1], Ji-Ho Ahn[1] and Hyun-Do Jung[1]
[1] Liquid Processing & Casting Technology R&D Group, Korea Institute of Industrial Technology, Incheon, 406-840, Korea

The aim of this study is to investigate the effects of gadolinium addition on the microstructure and mechanical properties of duplex stainless steel (DSS) fabricated using a normal casting method. The oxygen content in the cast DSS alloy with gadolinium decreased because of the high reactivity of gadolinium with oxygen. Notably, the ultimate tensile strength and strain at break of the cast alloy significantly increased with the addition of gadolinium from 919 ± 25 to 969 ± 8 MPa and from 24.8% ± 1.9% to 28.4% ± 1.1%, respectively.

Keywords: *Duplex stainless steel; gadolinium; casting; microanalysis; inclusions; mechanical characterization*

1. Introduction

Gadolinium (Gd) is one of more abundant rare earth metals and has special characteristics such as a high neutron cross-section [1]. The Gd has been used in neutron therapy applications for targeting tumor; and recently, it has been introduced as a neutron absorber material which can be used to shield nuclear reactors such as nuclear fuels [2]. The successful fabrication of DSS with Gd addition can take advantage of the rare earth metal characteristics to enhance corrosion resistance and mechanical properties of DSS for a variety of applications in the nuclear industry.

2. Method

The present study is focused on studying the Gd influence on the microstructure and mechanical properties of DSS. DSS with a pitting resistance equivalent number (PREN = wt.% Cr + 3.3 wt.% Mo + 30 wt.% N) 50 was used and 0.1 wt% of Gd was added and alloyed through casting [25]. The microstructure and inclusions of alloys were investigated, while the mechanical properties were examined by Charpy impact, hardness and tensile strength tests..

3. Results and Discussion

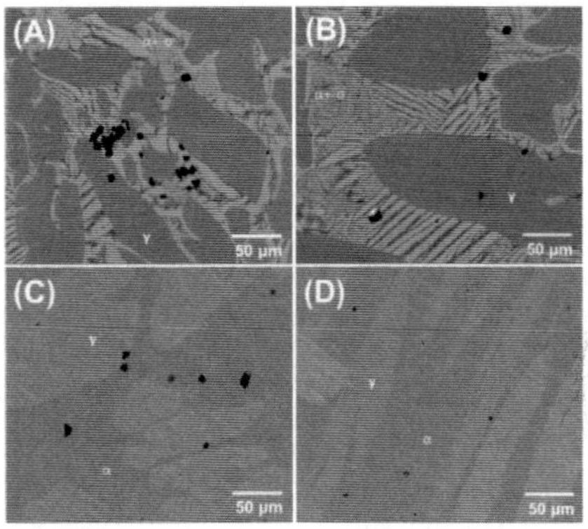

Fig. 1 Back scattered electron (BSE) images of (A), (C) Alloy I and (B), (D) Alloy II as casting (Left side is as cast and right side is after solution treatment

Figure 1 shows the BSE images of Alloy I and Alloy II before and after heat-treatment. The σ and χ phases in the ferrite matrix were observed before heat-treatment due to the slow cooling rate (Figure 1 (A) and (B)). The clusters of irregular-shaped inclusions (black particles) were found in Alloy I, whereas isolated globular inclusions (black and black-white particles) were detected in Alloy II, as shown in Figure 1 (A) and (B). Alloy I and Alloy II were heat-treated at 1130°C which is the predicted temperature for adjusting ferrite-austenite ratio (45:55) and dissolving the σ and χ phases. After water quenching, typical duplex structures were obtained from Alloy I and Alloy II as shown in Figure 1 (C) and (D). The austenite phase was evenly distributed as islands in the ferrite matrix, and no intermetallic phases were observed following double-solution annealing. The inclusions in Alloy I and Alloy II, however, were still detected which indicated that the remaining inclusions are in non-intermetallic forms such as oxides or nitrides which have high melting temperatures.

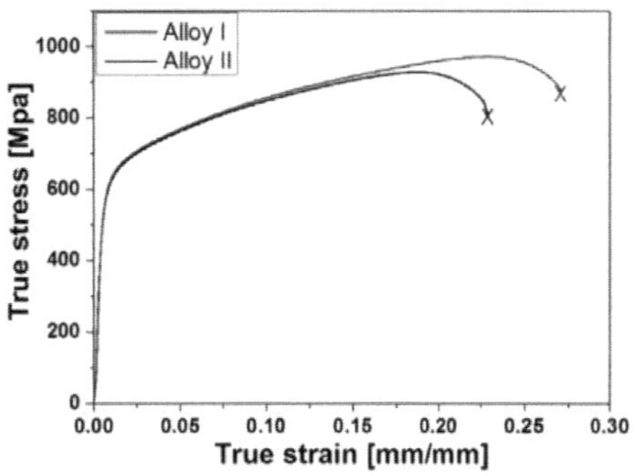

Fig. 2 Tensile stress versus strain curves of Alloy I and Alloy II

The typical stress versus strain curves of the fabricated duplex stainless steels by uniaxial tensile test is shown in Figure 2. For all cast alloys, the specimens possessed ductile fracture behaviors; the early stage elastic deformation region was followed by a plastic elongation prior to failure. The curves of Alloy I and Alloy II have no difference in elastic region with a yield strength of approximately 550 MPa and an elastic modulus of approximately 145 GPa which are reasonable values of cast duplex stainless steel alloys, whereas the mechanical properties of Alloy II in the plastic region seem to be improved greatly with higher ultimate tensile strength and strain at break [3].

The yield strength and elastic modulus of cast duplex stainless steel alloy slightly increased with the addition of gadolinium from 537±5 MPa to 557±12 MPa and from 143±16 GPa to 149±5 GPa, respectively. However, the ultimate tensile strength and strain at break of Alloy II (969±8 MPa and 28.4±1.1%) were significantly higher than those of Alloy I (919±25 MPa and 24.8±1.9%) by 5.4% and 14.5%, respectively ($p<0.05$).

4. Conclusions

Duplex stainless steel with stable microstructure and excellent mechanical properties was successfully fabricated by the addition of gadolinium. The oxygen content in cast duplex stainless steel alloy with gadolinium decreased by 16% due to the high reactivity of gadolinium with oxygen. The ultimate tensile strength and strain at break of cast alloy significantly increased with the addition of gadolinium from 919±25 MPa to 969±8 MPa and from 24.8±1.9% to 28.4±1.1%, respectively. This simple, but effective fabrication process of duplex stainless steel has great potential as a method to answer limitations of neutron absorbing structural materials as well as other various industry applications.

Acknowledgements

This work was supported by Ministry of Trade, Industry & Energy of the republic of Korea. (No. NS140028 and No. NS130039).

References

[1] Y. Choi, B.M. Moon, D.-S. Sohn, Nuclear Engineering and Technology, 45 (2013) 689-694.
[2] Z. Khan, Journal of the Southern African Institute of Mining and Metallurgy, 112 (2012) 309-321.
[3] D. Peckner, I.M. Bernstein, Handbook of stainless steels, McGraw-Hill New York, NY, 1977.

Proceedings of the 72nd World Foundry Congress
21-25th May 2016, Nagoya, Japan

Influence of graphite morphology, thermal history and S and Cu on ferrite/pearlite formation in cast iron

Hideo Nakae[1], Ying Zou[2] and Yuya Sato[3]
[1] Professor emeritus, Waseda Univ., Tokyo, Japan
[2] PhD. Student, Waseda Univ., Present address: Nanjing Tech University, Nanjing, China
[3] Student, Waseda Univ., Present address: Cyber Agent, Inc., Tokyo, Japan

Abstract: Ferrite and pearlite formation in cast iron is a more complicated phenomenon than that in steel. For example, we cannot observe any ferrite in hyper eutectoid steel, nevertheless, we often observe ferrite in cast iron in spite of its hyper eutectoid composition due to the silicon content. The eutectoid transformation in cast iron was then investigated using as-cast and heat-treated samples. The heating temperature of the cast iron determines the solubility of carbon in the austenite and the content significantly affects the eutectoid transformation. This is the main reason for the fully pearlite matrix in as-cast flake graphite cast iron samples. These results were analyzed and verified by a cooling curves analysis.

For discussing the difference in the matrix structure between the flake and spheroidal graphite cast iron, the influence of sulfur and copper was investigated using high purity flake graphite irons. The results show that sulfur is a very potent pearlite forming element; if the sulfur content is less than 0.0025%, ferrite is the major structure and for more than 0.01%S, the matrix is pearlite, but the copper did not affect the matrix structure of the flake graphite cast iron. This is the main reason why the bull's eye structure is so popular in spheroidal cast iron.

Keywords: eutectoid transformation, ferrite, pearlite, sulfur, copper, flake graphite, spheroidal graphite, diffusion of carbon

1. Introduction

The mechanical properties of cast iron can be controlled by the graphite morphology and matrix, especially, in spheroidal graphite cast iron, the matrix significantly affects them. A small number of papers have described the pearlite and ferrite formation mechanism in cast iron [1-3]. There is a significant difference in the ferrite formation between steel and cast iron. We cannot usually observe the ferrite in hyper eutectoid steel, nevertheless, we often observe ferrite in cast iron in spite of its hyper eutectoid composition due to its silicon content. The objective of the study described in this paper was the specific determination of the mechanism which controls the matrix structure in the flake and spheroidal graphite cast irons.

As is well known, the matrix of the as-cast flake graphite cast iron is mainly pearlite, nevertheless, that of the spheroidal graphite cast iron consists of ferrite and pearlite, called the bull's eye structure. After heat treatment, such as normalizing, the matrix of the flake graphite cast iron consists of a small amount of ferrite and also graphite; however, the spheroidal one is fully pearlite. Why? We called these phenomena due to the influence of the thermal history and graphite morphology.

If we consider the austenitization temperature during heat treatment for cast iron, the carbon content in the austenite is a function of the heating temperature due to the solubility of carbon in the austenite [4-6]. We then have to study the influence of the heating temperature on the matrix of the flake graphite cast iron.

We also considered that the significant difference in the effect of the alloying element between the flake and spheroidal graphite cast irons should be the activity of sulfur [6], then we discussed the influence of sulfur on the matrix of the flake graphite cast iron using high purity materials for making significantly low sulfur iron. Comparing the matrix structure with the spheroidal one, we discuss the effect of sulfur on the pearlite formation in the cast iron.

2. Experimental procedure

In this study, we carried out two experiments, i,e., experiment-1 and experiment-2. In experiment-1, the object was the specific investigation of the difference in the matrix formation between the flake and spheroidal graphite cast irons. We discuss the influence of the sulfur and copper on the matrix in the flake and spheroidal graphite cast iron using the samples, which were melted in a 20kW alumina -lined 7kg high frequency induction furnace. We prepared them using electrolytic iron, electrode graphite, semi-conductor grade pure Si and pure Cu for obtaining a significantly low sulfur cast iron for the base samples. Chemical reagent FeS was used as the sulfur additive to the flake graphite samples. The melt was inoculated with 0.3% 75%Si-3%Ca ferrosilicon, and then cast into a shell mold of 30mm inner diameter and 300mm length for observation of their microstructures.

The spheroidal graphite samples were prepared from the identical raw materials except for the FeS addition. The magnesium treatments were carried out with the addition of 2% ferrosilicon magnesium alloy,45%Si-3.86%Mg-1.75%Ca-1.37%RE-1.03% Al, and after the spheroidal treatments, the melt was inoculated with 0.3% of 75%Si-3%Ca ferrosilicon, then cast into shell molds of 30mm inner diameter and 300mm length the observation for their microstructures. The target chemical composition of the parent metal was 3.7 % C-2.0%Si-0.01%Mn -0.005%P- 0.0024%S.

The chemical compositions of the experiment-1 samples are listed in Table 1 in which the notations of FC and FCD mean the flake and spheroidal graphite cast iron, respectively. All of these samples were cut and polished in order to observe their microstructures.

Table 1 Chemical composition of experiment-1 samples. (mass%)

TP No.	C	Si	Mn	P	S	Cu	Mg
FC-S free, Cu free	3.70	2.02	<0.01	0.0048	0.0022	0.0016	<0.0002
S free-0.5Cu	3.64	1.95	〃	0.0043	0.0024	0.461	〃
S free-1.0Cu	3.55	1.97	〃	0.0045	0.0025	1.003	〃
-0.01S-0.5Cu	3.66	1.99	〃	0.0043	0.0083	0.529	〃
-0.02S-0.5Cu	3.61	2.01	〃	0.0047	0.0185	0.546	〃
FCD-Cu free	3.78	2.04	〃	0.0050	0.0026	0.0016	0.0424
-0.5Cu	3.93	2.00	〃	0.0054	0.0030	0.477	0.0486
-1.0Cu	3.78	2.09	〃	0.0050	0.0024	0.933	0.0463

In experiment-2, we studied the influence of the thermal history and copper on the ferrite and pearlite formation using flake graphite cast iron samples, comparing the as-cast samples to that of the heat treated ones. The cast iron melt was prepared as an commercial grade cast iron from a S25C bar, electrode graphite, metallic Si and pure Cu. The chemical composition of the S25C bar was as follows:
0.27%C-0.21%Si-0.46%Mn-0.014%P-0.017%S.
The target chemical composition of the parent metal was 3.5 % C-2.4%Si-0.4%Mn-0.013%P-0.013%S and their copper contents were changed with the copper addition from 0.25% to 1.0%. The melting process was nearly identical to that of experiment-1 except for S25C. The chemical compositions of experiment-2 samples are listed in Table 2.

These melts were cast into two 30mm inner diameter and 300mm length molds and four CE cup molds for each melting. The CE cup mold had a 30mm inner diameter and 50mm length and was equipped with a CA thermocouple at the center for obtaining the cooling curves; we labelled them the CE cup samples. The three CE cups were quenched in water during the eutectoid transformation and the fourth one was cooled in the mold to room temperature, these being the as-cast samples.

Table 2. Chemical composition of experiment-2 samples. (mass%)

TP No.	C	Si	Mn	P	S	Cu	Cr	Mo
0.25 Cu	3.59	2.42	0.41	0.013	0.013	0.25	0.039	0.015
0.5 Cu	3.37	2.41	0.40	0.013	0.012	0.49	0.039	0.015
1.0 Cu	3.48	2.42	0.41	0.013	0.013	0.95	0.039	0.015

These 300mm length bars were cut into 50mm length pieces and a 5mm diameter and 30mm depth hole drilled at the base center for the thermal analysis. These four samples, called HT samples, were heated to 850℃ or 950℃ for 1h and cooled from these temperatures in air, then the three samples were quenched, while the forth one was cooled to room temperature identical to that of experiment-1.

3. Results and discussion

3-1. Experiment-1

The influence of sulfur and copper on the matrix structure of the flake graphite cast iron is shown in **Fig. 1**. As can be clearly seen, if the sulfur content is more than 0.0083%, the matrix structure is fully pearlite, while, if the sulfur content is less than 0.0025%, the matrix structure is almost ferrite. Moreover, we did not recognize the copper effect on the pearlite formation in the flake graphite cast iron. As is well known, copper is a graphitizing element in cast iron. Nevertheless, we already reported that the effect of copper on the pearlite formation in spheroidal graphite cast iron can be explained by the formation of a copper thin film around the spheroidal graphite which prevents the diffusion of carbon to the graphite [7], then the matrix becomes fully pearlite.

If we consider the free sulfur content, which means the dissolved sulfur except for combined one such as MgS and MnS, the free sulfur content in the spheroidal graphite cast iron should be less than 0.0011% (or a_s = 0.008) [8], or a_s = 0.002 as Subramanian reported [9]. These values are much less than the 0.0025% in this experiment. These results show that the sulfur in the flake graphite iron is a very powerful pearlite stabilizing element, therefore, the bull's eye structure is very common in spheroidal graphite cast iron due to the ultra-low sulfur activity.

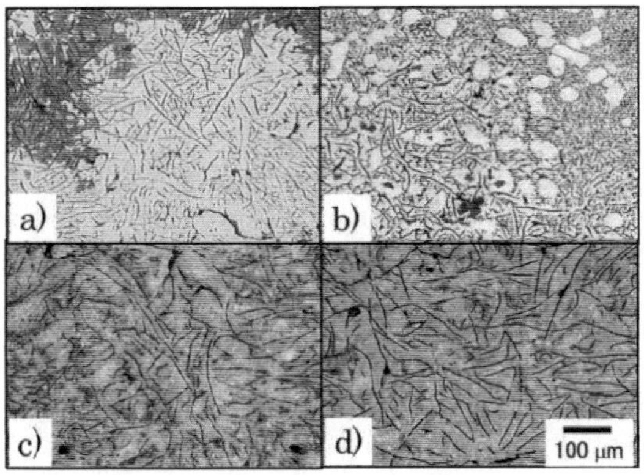

Fig. 1 Influence of S and Cu on matrix structure of as-cast flake graphite cast into 30mm samples. a) S & Cu free, b) S free-0.5Cu, c) 0.01S-0.5Cu, d) 0.02S-0.5Cu

Fig. 2 shows the influence of the graphite morphology and copper on the matrix structure for the S-free cast iron samples. As already mentioned, the copper in flake graphite cast iron is a ferrite stabilizing element, nevertheless, in the case of the spheroidal graphite cast iron, it is a very powerful pearlite stabilizing element as is well known. The copper in the flake graphite cast iron is a poor ferrite stabilizing element as shown in Figs. 1 and 2.

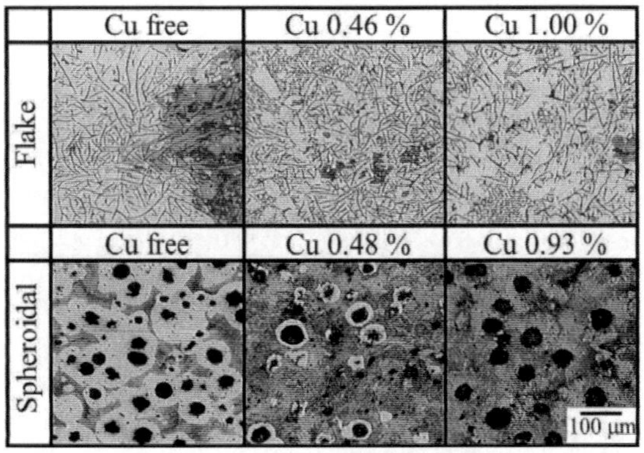

Fig. 2 Influence of graphite morphology and Cu on matrix structure for S-free 30mm samples.

3-2. Experiment-2

The graphite morphology and matrix structure of the as-cast flake graphite samples are shown in Fig. 3. These results are identical to that of Figs. 1 and 2, namely, the copper does not affect the graphite morphology and the matrix structure. The matrix of all the samples is pearlite due to the 0.013%S content as already mentioned. These results show that the pearlite stabilizing sulfur effect in the flake graphite cast iron is much stronger than the ferrite stabilizing effect of copper.

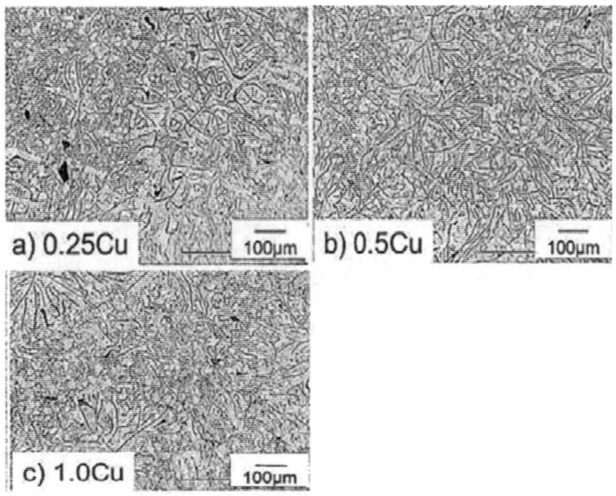

Fig. 3 Graphite structure of the as-cast 30mm samples. (0.013%S, full pearlite matrix).

Cooling curves, the differential curves, dT/dt, and the quenching temperature for the 0.25Cu CE cup samples are shown in Fig. 4. As can be clearly seen, the quenching interval was nearly periodic from ① -sample to ③-sample, and ④-sample was cooled in the CE cup to room temperature, thus, the cooling condition was identical to that of the 30mm bar as the cast sample and the matrix structure is fully pearlite as already described. Moreover, we can recognize the recalescence, in these figures, which should be the pearlite formation and these details will be discussed later.

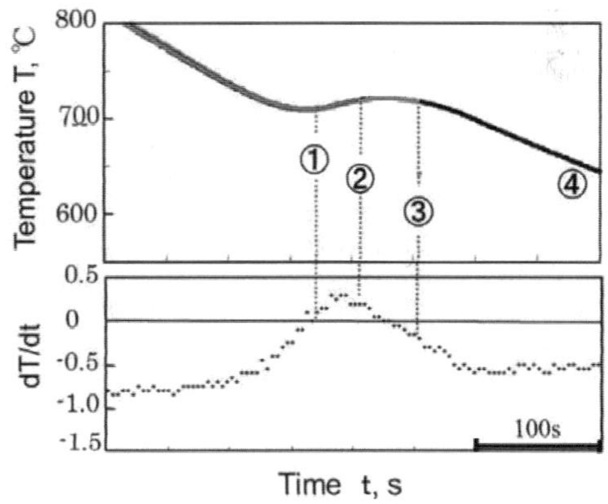

Fig. 4 Cooling curves, the differential curves during eutectoid transformation and quenching temperature of 0.25Cu CE cup samples, cooled from melts.

The microstructures of the ① to ④ samples are shown in Fig. 5. The microstructure of the ①-sample is the mixture of troostite and austenite, that of the ② is a small amount of pearlite in austenite, ③ is nearly a pearlite structure and a small amount of austenite, and that of ④ is fully pearlite as already mentioned.

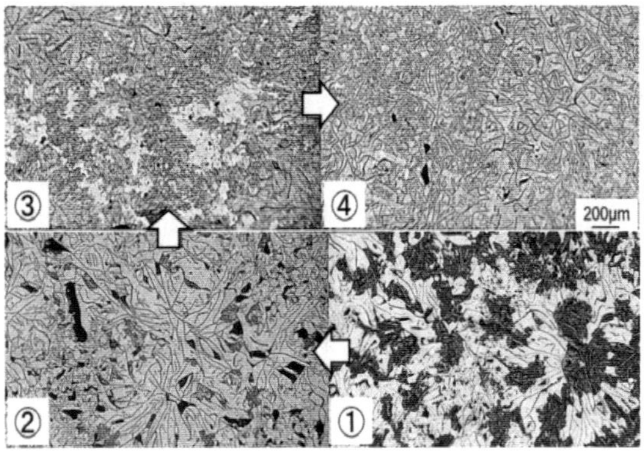

Fig. 5 Matrix structure of quenched 0.25Cu samples at different temperatures cooled from melts.

Four cooling curves of the 0.25Cu HT samples, cooled from 950℃, are shown in Fig. 6. Compared to Fig. 4, there is no clear recalescence in these cooling curves because the dT/dt values remain nearly zero at the onset of the eutectoid transformation. The influence of the thermal history on the cooling curves of the 0.25Cu HT samples, are shown in Fig. 7. The a) curve is the cooling curve of the CE cup, b) is that of the HT sample cooled from 950℃ and c) is that of the HT sample cooled from 850℃. The CE cup sample was cooled in the mold, but the HT samples were cooled without a mold, then the cooling rate of the HT samples was higher than that of the CE cup sample. The distinct recalescence in these cooling curves is recognized only in the CE cup sample, nevertheless, in the HT sample's ones, there is no recalescence.

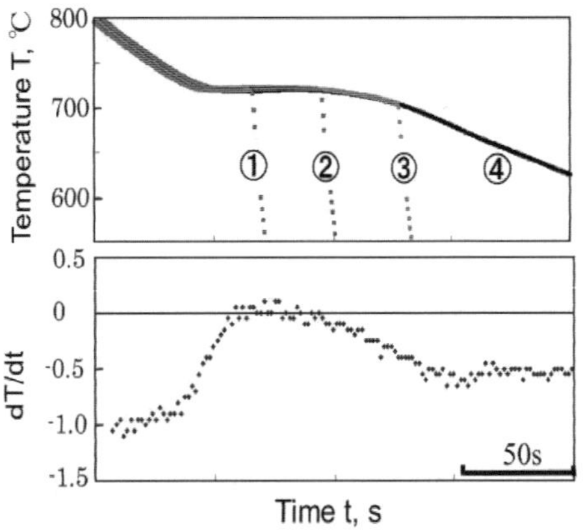

Fig. 6 Cooling curves, the differential curves and quenching temperature of 0.25Cu HT samples cooled from 950℃.

Fig. 8 shows the microstructure of these ④ samples. As can be seen clearly, that of the a) sample is fully pearlite, nevertheless, in the b) and c) samples, there are small amounts of ferrite around the A-type flake graphite. To confirm the influence of the thermal history on the pearlite fractions, we summarize all the data in Table 3. As clearly seen, we cannot find any pearlite formation effect by copper, however, the thermal history significantly affects the pearlite fraction. The matrix of these as-cast ones is fully pearlite, however, the pearlite fractions of the samples cooled from 950℃ and 850℃ are 60 to 70% and 39 to 47%, respectively. Why the thermal history is so significantly affected?

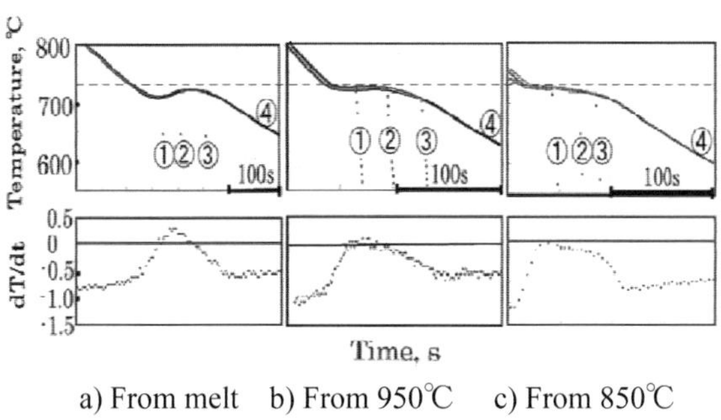

Fig. 7 Influence of thermal history on cooling curves of 0.25Cu CE and HT samples.

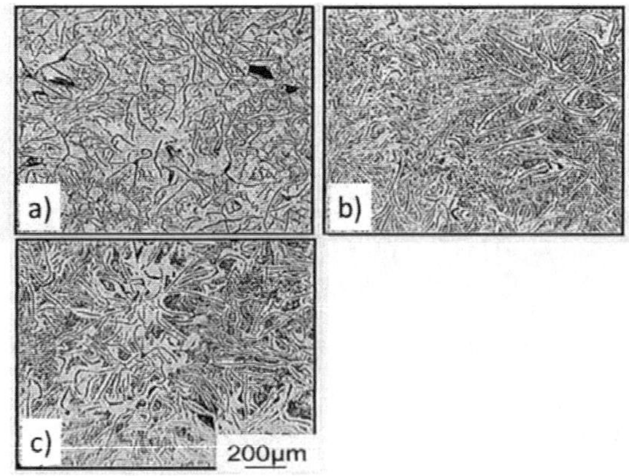

Fig. 8 Influence of holding temperature on matrix structure for air cooled 0.25Cu ④ samples in Fig. 7.

Table 3 Influence of thermal history and Cu content on pearlite fractions in flake graphite cast iron.

TP No.	As cast	Cooled from 950℃	Cooled from 850℃
0.25Cu	100%	65%	39%
0.5 Cu	100%	60%	41%
1.0 Cu	100%	70%	47%

Suppose that the solubility of carbon in the austenite can be estimated from the Fe-C-2.4%Si phase diagram with the function of the heating temperature as shown in Fig. 9. In this figure, we assumed that the heating temperature of the CE cup sample is 1150℃, thus the solubility becomes 1.6%, while that of the HT sample heated at 950℃ is 1.0% and that of the 850℃ sample is 0.7% as shown in this figure. We can then confirm that the solubility limit is the main factor which affects the ferrite and pearlite transformation in the cast iron.

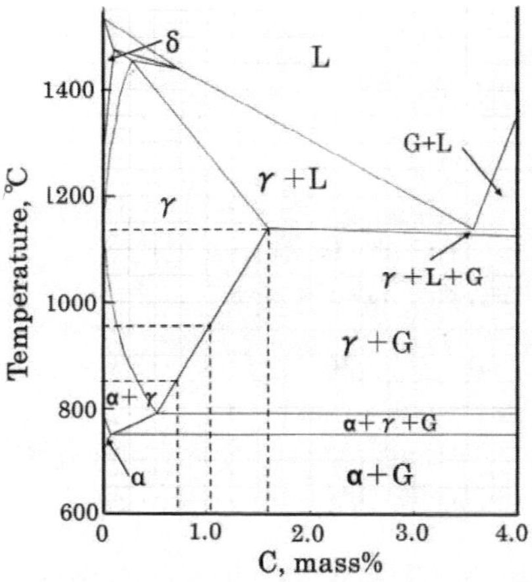

Fig. 9 Influence of heating temperature on solubility of carbon in austenite by the Fe-C-2.4%Si phase diagram

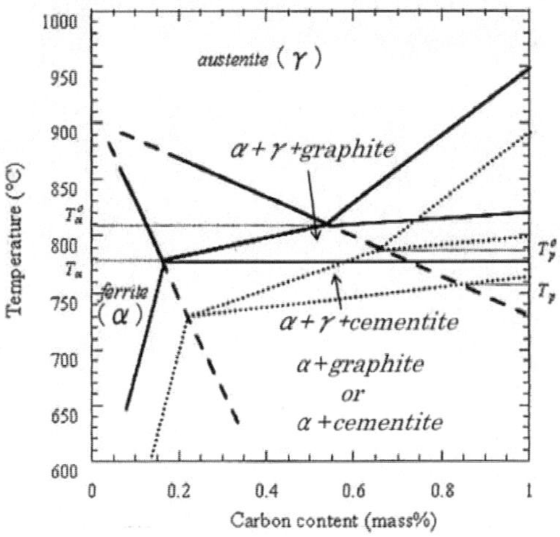

Fig. 10 Isopleth section of Fe-C-2.5%Si phase diagram reported by Gerval

For the discussion of the relationship between the cooling curves and the matrix structure, we have to use the stable and unstable eutectoid transformation diagram of cast iron [10,11], then we quoted Gerval's diagram, shown in Fig. 10 [10]. There are two eutectoid transformation regions, namely, a ferrite transformation area located above the pearlite transformation area. The transformations in Figs. 7 b) and c) commence at the ferrite formation followed by the pearlite formation. Nevertheless, we could not find any clear recalescence in these figures, but there is a clear recalescence in Fig. 7 a), namely the pearlite transformation as already shown in Fig. 8.

3-3. Influence of graphite morphology on matrix structure

The ferrite formation frequently occurs in the D-type graphite area and in dendrites as is well known shown in Fig. 11. These phenomena can be mainly explained by the diffusion distance of carbon from the austenite to the graphite. Namely, if the diffusion distance is short, such as in the D-type graphite area, the carbon in the austenite can easily diffuse to form the ferrite as show in Fig. 12. Under this condition, if the carbon content in the austenite is low, it easily transforms into ferrite, thus this is the main reason for the thermal hysteresis.

However, the pearlite occues at the eutectic cell boundry due to the micro-segrigation of manganise and silicon in the matrix [5,7].

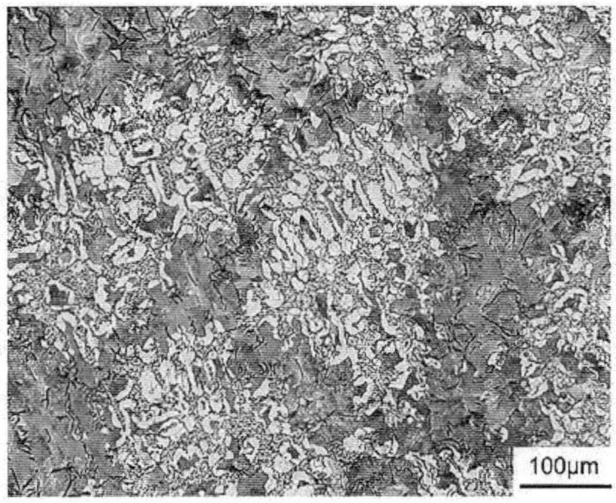

Fig. 11 Ferrite formation in D-type graphite area and dendrite area.

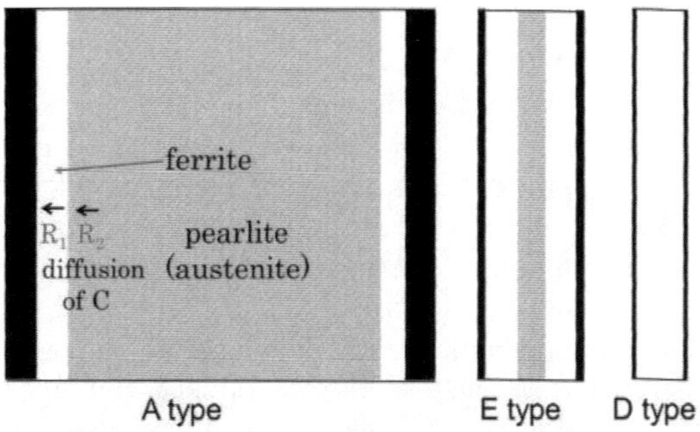

Fig. 12 Influence of graphite morphology on ferrite formation in flake graphite cast iron.

There is a significant difference in the ferrite formation mechanism between the flake and spheroidal graphite cast irons in the diffusion field, namely, rectangular coordinate diffusion in the flake graphite iron and spherical coordinate diffusion in the spheroidal graphite iron. This mechanism can be illustrate in Fig. 13. Therefore, if the cooling rate is high, such as a normalizing treatment, the pearlite structure can easily occur due to the spherical coordinate diffusion.

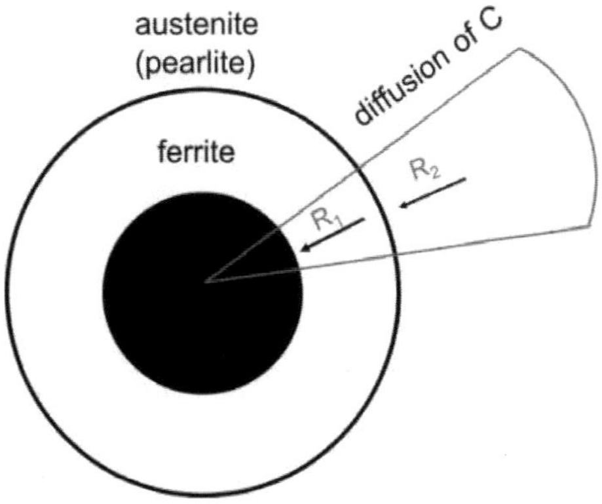

Fig. 13 Ferrite formation mechanism in spheroidal graphite cast iron.

The pearlite formation effect of Sn and Sb is due to the adsorption of these elements at graphite/ matirix interface, as already rerported by Johnson et al [12]. Nevertrhless, the effect of Cu is remtted only sheroidal graphite cast iron as alreay mentioned [7].

4. Conclusions

We discussed the influence of the graphite morphology, thermal hysteresis, and sulfur and copper contents on the ferrite/pearlite transformation in cast iron based on microstructure observations and their cooling curves. In this study, a number of experiments were conducted using the flake and spheroidal graphite cast irons. The conclusions are as follows:

1. The sulfur in the flake graphite cast iron is a powerful alloying element for the pearlite formation.
2. The copper is a pearlite stabilizing element in the spheroidal graphite cast iron, however, it is a ferritestabilizing element in the flake graphite cast iron.
3. The heating temperature during the heat treatment affects the carbon solubility in the austenite, and the temperature affects the ferrite/pearlite transformation. A low heating temperature produces more ferrite structure than that at a high temperature, then the fully pearlite structure can be formed in the as-cast samples.
4. A recalescence is observed during the pearlite transformation, but no significant recalescence can be observed in the ferrite/pearlite samples.
5. The ferrite and pearlite transformations are controlled by the sulfur content, the carbon diffusion distance and the field.

References

[1] G. Ohira and K. Ikawa: AFS Trans. 66(1958)526
[2] J. Lacaze, C. Wilson and C. Bak: Scandinavian J. Metallurgy 23(1994)151
[3] T. Skaland, Φ. Grong and T. Grong: Metallurgical Trans. A: 24A(1993)2347
[4] T. Szykowny and M. Trepzyńska: Archives of Foundry Eng. 10(2010,3)101
[5] R.-B. Zao, S. Ueno, S. Yamada, K. Sugita and H. Nakae: Intern's J. Cast Metals Res. 21(2008)62
[6] H. Nakae, K. Sigita and B.-R. Zhao: J. JFS 78(2006)457
[7] Y. Zou, K. Komada and H. Nakae: J. JFS 83(2011)378
[8] S. Jung, T. Ishikawa and H. Nakae: Materials Sci. and Eng. A 476(2008)350
[9] S. V. Subramanian, D. A. R. Kayn and G. R. Purdy: Mater. Res. Soc. Symp. Proc., 34(1985)47
[10] V. Gerval and J.Lacaze: ISIJ International 40(2000)386
[11] P. Mrvar, M. Petrič and J. Medved: Key Eng. Mater. 457(2011)163
[12] W.C.Johnson and B.V.Kovacs: Metallurgical Transaction A 9A(1978)219

AUTHOR INDEX

AUTHOR INDEX

AUTHOR INDEX

AUTHOR INDEX

AUTHOR INDEX

AUTHOR INDEX